G. C. Onwubolu, B. V. Babu

New Optimization Techniques in Engineering

Dear MyCopy Customer,

This Springer book is a monochrome print version of the eBook to which your library gives you access via SpringerLink. It is available to you at a subsidized price since your library subscribes to at least one Springer eBook subject collection.

Please note that MyCopy books are only offered to library patrons with access to at least one Springer eBook subject collection. MyCopy books are strictly for individual use only.

You may cite this book by referencing the bibliographic data and/or the DOI (Digital Object Identifier) found in the front matter. This book is an exact but monochrome copy of the print version of the eBook on SpringerLink.

Springer-Verlag Berlin Heidelberg GmbH

Studies in Fuzziness and Soft Computing, Volume 141

Editor-in-chief
Prof. Janusz Kacprzyk
Systems Research Institute
Polish Academy of Sciences
ul. Newelska 6
01-447 Warsaw
Poland
E-mail: kacprzyk@ibspan.waw.pl

Further volumes of this series can be found on our homepage: springeronline.com

Vol. 122. M. Nachtegael, D. Van der Weken, D. Van de Ville and E.E. Kerre (Eds.)
Fuzzy Filters for Image Processing, 2003
ISBN 3-540-00465-3

Vol. 123. V. Torra (Ed.)
Information Fusion in Data Mining, 2003
ISBN 3-540-00676-1

Vol. 124. X. Yu, J. Kacprzyk (Eds.)
Applied Decision Support with Soft Computing, 2003
ISBN 3-540-02491-3

Vol. 125. M. Inuiguchi, S. Hirano and S. Tsumoto (Eds.)
Rough Set Theory and Granular Computing, 2003
ISBN 3-540-00574-9

Vol. 126. J.-L. Verdegay (Ed.)
Fuzzy Sets Based Heuristics for Optimization, 2003
ISBN 3-540-00551-X

Vol 127. L. Reznik, V. Kreinovich (Eds.)
Soft Computing in Measurement and Information Acquisition, 2003
ISBN 3-540-00246-4

Vol 128. J. Casillas, O. Cord6n, F. Herrera, L. Magdalena (Eds.)
Interpretability Issues in Fuzzy Modeling, 2003
ISBN 3-540-02932-X

Vol 129. J. Casillas, O. Cord6n, F. Herrera, L. Magdalena (Eds.)
Accuracy Improvements in Linguistic Fuzzy Modeling, 2003
ISBN 3-540-02933-8

Vol 130. P.S. Nair
Uncertainty in Multi-Source Databases, 2003
ISBN 3-540-03242-8

Vol 131. J.N. Mordeson, D.S. Malik, N. Kuroki
Fuzzy Semigroups, 2003
ISBN 3-540-03243-6

Vol 132. Y. Xu, D. Ruan, K. Qin, J. Liu
Lattice-Valued Logic, 2003
ISBN 3-540-40175-X

Vol. 133. Z.-Q. Liu, J. Cai, R. Buse
Handwriting Recognition, 2003
ISBN 3-540-40177-6

Vol 134. V.A. Niskanen
Soft Computing Methods in Human Sciences, 2004
ISBN 3-540-00466-1

Vol. 135. J.J. Buckley
Fuzzy Probabilities and Fuzzy Sets for Web Planning, 2004
ISBN 3-540-00473-4

Vol. 136. L. Wang (Ed.)
Soft Computing in Communications, 2004
ISBN 3-540-40575-5

Vol. 137. V. Loia, M. Nikravesh, L.A. Zadeh (Eds.)
Fuzzy Logic and the Internet, 2004
ISBN 3-540-20180-7

Vol. 138. S. Sirmakessis (Ed.)
Text Mining and its Applications, 2004
ISBN 3-540-20238-2

Vol. 139. M. Nikravesh, B. Azvine, I. Yager, L.A. Zadeh (Eds.)
Enhancing the Power of the Internet, 2004
ISBN 3-540-20237-4

Vol. 140. A. Abraham, L.C. Jain, B.J. van der Zwaag (Eds.)
Innovations in Intelligent Systems, 2004
ISBN 3-540-20265-X

Godfrey C. Onwubolu
B. V. Babu

New Optimization Techniques in Engineering

Springer

Professor Godfrey C. Onwubolu
Chair of Department of Engineering
The University of the South Pacific
P.O. Box 1168
Suva
Fiji Islands
E-mail: onwubolu_g@usp.ac.fj

Professor B. V. Babu
Assistant Dean - Engineering Services Division (ESD) &
Head - Chemical Engineering & Engineering Technology Departments
Birla Institute of Technology & Science (BITS)
Pilani - 333 031 (Rajasthan)
India
E-mail: bvbabu@bits-pilani.ac.in;
bv_babu@hotmail. com

ISSN 1434-9922
DOI 10.1007/978-3-540-39930-8

Library of Congress Cataloging-in-Publication-Data

Onwubolu, Godfrey C.
New optimization techniques in engineering / Godfrey C. Onwubolu, B. V. Babu.
p. cm. -- (Studies in fuzziness and soft computing ; v. 141)
Includes bibliographical references and index.
1. Engineering--Mathemaical models. 2. Mathematica optimization. I. Babu, B. V. II. Title. III. Series.
TA342.059 2004
620'0015'196--dc22

Originally published by Springer-Verlag Berlin Heidelberg in 2004
MyCopy version of the originaledition 2004

Cover design: E. Kirchner, Springer-Verlag, Heidelberg
Printed on acid free paper 62/3020/M - 5 4 3 2 1 0
www.springer.com/mycopy

PREFACE

Presently, general-purpose optimization techniques such as Simulated Annealing, and Genetic Algorithms, have become standard optimization techniques. These optimization techniques commence with a single solution and then find the best from several moves made, and generally, past history is not carried forward into the present. Many researchers agree that firstly, having a population of initial solutions increases the possibility of converging to an optimum solution, and secondly, updating the current information of the search strategy from previous history is a natural tendency. Accordingly, attempts have been made by researchers to restructure these standard optimization techniques in order to achieve the two goals mentioned.

To achieve these two goals, researchers have made concerted efforts in the last one-decade in order to invent novel optimization techniques for solving real life problems, which have the attributes of memory update and population-based search solutions. This book describes these novel optimization techniques, which in most cases outperform their counterpart standard optimization techniques in many application areas. Despite these already promising results, these novel optimization techniques are still in their infancy and can most probably be improved. To date, researchers are still carrying out studies on sound theoretical basis and analysis to determine why some of these novel optimization techniques converge so well compared to their counterpart standard optimization techniques.

Interestingly, most books that have reported the applications and results of these novel optimization techniques have done so without sufficiently considering practical problems in the different engineering disciplines. This book, *New Optimization Techniques in Engineering* has three main objectives: (i) to discuss in the clearest way possible, these novel optimization techniques, (ii) to apply these novel optimization techniques in the conventional engineering disciplines, and (iii) to suggest and incorporate the improvements in these novel optimization techniques that are feasible as and when it is possible in the application areas chosen.

To achieve the first objective, Part I containing seven chapters have been written by the inventors of these novel optimization techniques or experts who have done considerable work in the areas (Genetic Algorithm, Memetic Algorithm, Scatter Search, Ant Colony Optimization, Differential Evolution, Self-Organizing Migrating Algorithm, Particle Swarm Optimization). Genetic Algorithm has been included for completeness since it is the progenitor of Memetic Algorithm. The contributors for Genetic Algorithm and Particle Swarm Optimization have been chosen, not as the inventors, but due to their expertise and contributions in these areas of optimization. To achieve the second objective, Part II contains several chapters in which researchers have applied these novel optimization techniques to different Engineering disciplines such as Chemical/Metallurgical Engineering, Civil/Environmental Engineering/Interdisciplinary, Electrical/Electronics Engineering, Manufacturing/Industrial Engineering, and Mechanical/Aeronautical Engineering. Firstly, the Engineering background is sufficiently given concerning the problem-domain, and then a novel optimization technique is applied. Consequently, Part II makes it easy for engineers and scientists to understand the link between theory and application of a particular novel optimization technique. To achieve the third objective, the possible improvements in these novel optimization techniques are identified, suggested and applied to some of the engineering problems successfully. Part III discusses newer areas, which are considered as extended frontiers.

The text serves as an instructional material for upper division undergraduates, entry-level graduate student, and a resource material for practicing engineers, research scientists, operations researchers, computer scientists, applied mathematicians, and management scientists. Those to purchase the book include upper division undergraduates or entry-level graduate students, academics, professionals and researchers of disciplines listed above, and libraries.

Godfrey C Onwubolu and B V Babu
June 2003

CONTRIBUTORS

B. V. Babu, Assistant Dean-Engineering Services Division & Head-Chemical Engineering & Engineering Technology Departments, Birla Institute of Technology and Science (BITS), Pilani-333 031 (Rajasthan), India. E-mail: bvbabu@bits-pilani.ac.in (CHAP. 1, 9, 10, 11 & 12)

Laxmidhar Behera, Assistant Professor, Department of Electrical Engineering, Indian Institute of Technology, Kanpur-208 016, India. E-mail: lbehera@iitk.ac.in (CHAP. 19)

Sergiy Butenko, Department of Industrial and Systems Engineering, 373 Weil Hall, University of Florida, Gainesville, FL 32611. E-mail: butenko@ufl.edu (CHAP. 29)

Antonella Carbonaro, Department of Computer Science, University of Bologna via Sacchi, 3, 47023 Cesena, Italy. Email: carbonar@csr.unibo.it (CHAP. 15)

Kai-Ying Chen, Assistant Professor, Department of Industrial Engineering, National Taipei University of Technology. E-mail: kychen@ntut.edu.tw (CHAP. 25 & 28)

Mu-Chen Chen, Professor, Department of Business Management, National Taipei University of Technology, No. 1, Section 3, Chung-Hsiao E. Road, Taipei 106, Taiwan, ROC. Email: bmcchen@ntut.edu.tw / iemcchen@yahoo.com.tw (CHAP. 25 & 28)

Maurice Clerc, France Télécom Recherche & Développement, 90000, Belfort, France. Email: Maurice.Clerc@ WriteMe.com (CHAP. 8)

Carlos Cotta, Associate Professor, University of Málaga, Departamento de Lenguajes y Ciencias de la Computación Complejo Tecnológico (Despacho 3.2.49), Campus de Teatinos 29071-Málaga. SPAIN. Email: ccottap@lcc.uma.es (CHAP. 3 & 27)

Kalyanmoy Deb, Professor, Department of Mechanical Engineering, Indian Institute of Technology, Kanpur, Pin 208 016, INDIA. E-mail: deb@iitk.ac.in (CHAP. 2)

Luca Maria Gambardella, Director, IDSIA, Istituto Dalle Molle di Studi sull'Intelligenza Artificiale Galleria 2 6928 Manno-Lugano, Switzerland. Email: luca@idsia.ch (CHAP. 5)

Fred Glover, MediaOne Chaired Professor of Systems Science, University of Colorado; Visiting Hearin Professor, University of Mississippi; Research Director of the Hearin Center for Enterprise Science; School of Business Administration University of Mississippi University, MS 38677. Email: fglover@bus.olemiss.edu (CHAP. 4 &13)

Lee Tong Heng, Department of Electrical & Computer Engineering, National University of Singapore, 4 Engineering Drive 3, Singapore-117 576, E-mail: eleleeth@nus.edu.sg (CHAP. 20)

Hanno Hildmann, University of Amsterdam, Department of Sciences, The Netherlands. Email: hanno@dfki.de (CHAP. 15)

Nguyen Van Hop, Sirindhorn International Institute of Technology, Thammasat University, P.O.Box 22, Klong Luang, Pathumthani 12121, Thailand. E-mail: vanhop@siit.tu.ac.th (CHAP. 14)

D. Nagesh Kumar, Associate Professor, Civil Engineering Department, Indian Institute of Science, Bangalore-560 012, India. E-mail: nagesh@civil.iisc.ernet.in (CHAP. 16)

Manuel Laguna, Leeds School of Business, University of Colorado,Boulder, CO 80309-0419, USA. Email: Manuel.Laguna@Colorado.edu (CHAP. 4 & 13)

Fabio de Luigi, Dept. Computer Science, University of Ferrara, Italy. Email: f.deluigi@unife.it (CHAP. 5)

Jouni Lampinen, Department of Information Technology, Laboratory of Information Processing, Lappeenranta University of Technology, P.O.Box 20, FIN-53851 Lappeenranta, Finland. E-mail: Jouni.Lampinen@lut.fi (CHAP. 6 & 26)

Alexandre Linhares, Adjunct Professor, Brazilian School of Business and Public Administration, FGV, Praia de Botafogo 190/426, Rio de Janeiro 22257-970. Email: linhares@fgv.br (CHAP. 18)

Vittorio Maniezzo, Professor, Department of Computer Science, University of Bologna via Sacchi, 3, 47023 Cesena, Italy. Email: maniezzo@csr.unibo.it (CHAP. 5 & 15)

Rafael Marti, Departamento de Estadistica e Investigación Operativa, Facultad de Matemáticas, Universidad de Valencia, C/ Dr. Moliner 50, 46100 Burjassot, Valencia, Spain. E-mail: Rafael.Marti@uv.es (CHAP. 4 & 13)

Alexandre de Sousa Mendes, Universidade Estadual de Campinas – UNICAMP Faculdade de Engenharia Elétrica e de Computação – FEEC Departamento de Engenharia de Sistemas - DENSIS C.P. 6101 - CEP 13083-970, Campinas - SP - Brazil E-mail: asmendes@yahool.com (CHAP.3, 18 & 27)

Pablo Moscato, Senior Lecturer and Postgraduate Director for Computer Science, School of Electrical Engineering and Computer Science, Faculty of Engineering and Built Environment, The University of Newcastle, Callaghan, 2308 New South Wales, Australia. Email: moscato@cs.newcastle.edu.au (CHAP. 3, 18, & 27)

Godfrey C. Onwubolu, Professor and Chair of Engineering, Department of Engineering, The University of the South Pacific, PO Box 1168, Suva, FIJI. Email: onwubolu_g@usp.ac.fj (CHAP. 1, 21, 22, 23, & 24)

Panos M. Pardalos, Professor and Co-Director, Center for Applied Optimization, Industrial and Systems Engineering Department, 303 Weil Hall, University of Florida, PO Box 116595, Gainesville, FL 32611-6595, Email: pardalos@ufl.edu / pardalos@cao.ise.ufl.edu (CHAP. 29)

Xiao Peng, Department of Electrical & Computer Engineering, National University of Singapore, 4 Engineering Drive 3, Singapore-117 576, E-mail: engp0525@nus.edu.sg (CHAP. 20)

K. Srinivasa Raju, Assistant Professor, Civil Engineering Group, Birla Institute of Technology and Science (BITS), Pilani-333 031, Rajasthan, India. Email: ksraju@bits-pilani.ac.in (CHAP. 16)

Anuraganand Sharma, Computer Section, Colonial, Suva, Fiji. Email: ANDS@Colonial.com.au (CHAP. 23)

Rainer Storn, Infineon AG, TR PRM AL, Balanstr. 73, D-81541 Muenchen, Germany / International Computer Science Institute 1947 Center Street, Suite 600, Berkeley, CA 94704-1198. E-mail: rainer.storn@infineon.com / storn@icsi.berkeley.edu (CHAP. 6)

Mario T. Tabucanon, Asian Institute of Technology, P.O. Box 4, Klong Kluang, Pathumthani 1210, Thailand. E-mail: mtt@ait.ac.th (CHAP. 14)

Prahlad Vadakkepat, Assistant Professor, Department of Electrical & Computer Engineering, National University of Singapore & General Secretary, Federation of International Robot-soccer Association, Singapore, E-mail: prahlad@nus.edu.sg (CHAP. 20).

Ivan Zelinka, Tomas Bata University in Zlin, Faculty of Technology, Institut of Information Technologies, Mostni 5139, Zlin 760 01, Czech Republic. Email: zelinka@ft.utb.cz (CHAP. 7, 17 & 26)

Acknowledgements

The process of producing a book is the outcome of the inputs of many people. We are grateful to all the contributors for their ideas and cooperation. Working together with such a number of contributors from remotely located distances is both challenging and interesting; there was such a speedy response from contributors, that they deserve to be acknowledged. On the publishers' side, we want to thank Thomas Ditzinger, Heather King and other colleagues at Springer-Verlag, Heidelberg, Germany for their enthusiasm and editorial hard work to get this book out to the readers.

BVB acknowledges the Director, Prof. S. Venkateswaran; Deputy Director (Administration), Prof. K.E.Raman; Deputy Director (Academic), Prof. L.K.Maheshwari; and Deputy Director (Off-Campus Programmes), Prof. V.S.Rao of BITS-Pilani for continuous encouragement, support, and providing him with the required infrastructure facilities in preparing this book. He thanks his colleagues and Ph.D. Students Mr. Rakesh Angira and Mr. Ashish Chaurasia for helping him in making some of the figures and proof reading.

Producing a book is not without the support and patience of some close people to the authors. In this regard, GCO thanks his wife, Ngozi and their children Chioma, Chineye, Chukujindu, Chinwe and Chinedu for their supporting-role and forbearance. In addition, the undergraduate and graduate students who worked under my supervision are acknowledged for their hardwork. BVB thanks his wife Shailaja for unconditional support and understanding, and his children Shruti and Abhinav for understanding the importance and seriousness of this project contribution by not complaining on reaching home late nights. BVB also thanks his parents Shri Venkata Ramana and Mrs. Annapurna for making him what he is today.

CONTENTS

Part II.3: Electrical/Electronics Engineering

Part II.5: Mechanical/Aeronautical Engineering

1 Introduction

Godfrey C Onwubolu and B V Babu

1.1 Optimization

Engineers, analysts, and managers are often faced with the challenge of making tradeoffs between different factors in order to achieve desirable outcomes. Optimization is the process of choosing these tradeoffs in the "best" way. The notion of 'different factors' means that there are different possible solutions, and the notion of 'achieving desirable outcomes' means that there is an objective of seeking improvement on how to find the best solution. Therefore, in an optimization problem, different candidate solutions are compared and contrasted, which means that solution quality is fundamental.

Many engineering problems can be defined as optimization problems, e.g. process design, process synthesis & analysis, transportation, logistics, production scheduling, telecommunications network, complex reactor networks, separation sequence synthesis, finding of an optimal trajectory for a robot arm, the optimal thickness of steel in pressure vessels, the optimal set of parameters for controllers, or optimal relations or fuzzy sets in fuzzy models, optimal network in energy integration, automated design of heat exchangers, mass exchanger networks, etc. The solutions to such problems are usually not easy to obtain because they have large search spaces. In practice, these real-life optimization problems are very 'hard' to solve since it is impossible to find the optimal solution by sampling every point in the search space within an acceptable computation time. For small size problems if all the points in a search space are examined, it is possible to find an optimum solution (which is not the case, anyway, in practice). However, for the class of 'hard' problems, which we have listed above, it is often difficult if not impossible to guarantee convergence or find the best solution in an acceptable amount of time. Another characteristic of many real-life engineering optimization problems is that they are 'discontinuous' (discrete) and 'noisy'.

The days when researchers emphasized using *deterministic search techniques* (also known as traditional methods) to find optimal solutions are gone. Deterministic search techniques will ensure that the global optimum solution of an optimization problem is found, but these techniques will not work when a complex, real-life scenario is considered. In practice, an engineer, an analyst, or a manager wants to make decision as soon as possible in order to achieve desirable outcomes. Moreover, when an optimal design problem contains multiple global are near global optimal solutions, designers are not only interested in finding just one global optimal solution, but as many as possible for various reasons. Firstly, a design suitable in one situation may not be valid in another situation. Secondly, designers may not be interested in finding the absolute global solution. Instead they are interested in a solution, which corresponds to a marginally inferior objective function value but is more amenable to fabrication. Thus, it is always prudent to know about other equally good solutions for later use. For example, an engineer may be given the task to find the most reliable way to design a telecommunications network where millions of people are subscribing to the telecommunication service, such as Internet. What is the point using a *deterministic search technique* to find the optimal or global solution, if it will take thousands of hours of computer time? Take another example: an engineer is given a task to drill several thousands of holes on a printed circuit board (PCB), within a short time in order to meet the customer requirement of minimum production time. The problem of drilling holes on PCBs with the objective of minimum production time could be formulated as the popular traveling salesman problem (TSP). Imagine using a *deterministic search technique* to find the optimal or global solution (the best route to take in drilling the holes from one point to the other), how much of computer time would be needed? There are numerous combinations of parameters, the values of which are available in discrete manner, are possible in the design of a shell-and-tube heat exchanger, which makes this design problem a large-scale discrete combinatorial optimization problem. Similarly, there are number of parameters and combinations of heat exchanger networks possible in energy integration problems at various stages of network proposal, starting from choosing minimum approach temperature, stream splitting, loop breaking, restoring minimum approach temperature, etc. The number of combinations in choosing the heat transfer parameters (effective radial thermal conductivity of the bed and wall-to-bed heat transfer coefficient) is innumerable in the design of most widely used industrial trickle bed reactors. Unless and until these heat transfer parameters are correctly estimated, there is a danger of hot spot formation that leads to reactor runaway. In the design of auto-thermal ammonia synthesis reactor, the heat released due to exothermic nature of the reaction has to be removed to obtain a reasonable conversion as well as to protect the catalyst life. At the same time, the released heat energy is utilized to heat the incoming feed-gas to proper reaction temperature. Here again, the choice of reactor length and the reactor top temperature play a crucial role for which the combinations are many. There are countless ways to make product-mix decisions in a factory that manufactures thousands of product types using common resources: which product-mix decision maximizes the total throughput? There are innumerable ways to organize a large-scale factory production schedule: which gives the best throughput? There are countless routes a missile could take in order to hit its target: which is the best route to take? What is the point for an optimization technique to take several hours to find the 'best' route if split of a second is what is required because the target is an enemy? There are countless trajectories for a robot arm in a pick-and-place problem in a three-dimensional space: which is the optimal trajectory to take? There are numerous options of thickness of steel in a pressure vessel when optimizing the cost of the pressure vessel: which is the optimal thickness? To find the optimal set of parameters for controllers, or optimal relations or fuzzy sets in fuzzy models is a colossal task: which are the optimal set of parameters and relations respectively? There are many other engineering problems with many equality and inequality constraints, which come under the above category. The following is only a representative, but not exhaustive, list of such problems: optimization of alkylation reaction, digital filter design, neural network learning, fuzzy decision making problems of fuel ethanol production, design of fuzzy logic controllers, batch fermentation process, multi sensor fusion, fault diagnosis and process control, optimization of thermal cracker operation, scenario integrated optimization of dynamic systems, optimization of complex reactor networks, computer aided molecular design, synthesis and optimization of non-ideal distillation systems, dynamic optimization of continuous polymer reactor, online optimization of culture temperature for ethanol fermentation, molecular scale catalyst design, scheduling of serial multi-component batch processes, preliminary design of multi-product non continuous plants, separation sequence synthesis, mass exchanger networks, generating initial parameter estimations for kinetic models of catalytic processes, synthesis of utility systems, global optimization of MINLP problems, etc. This list is unending. In practical scenario, where search space for solution is extremely large, it is inconceivable that today's computer power can cope with finding these solutions within acceptable computation time if *deterministic search techniques* are used.

Consequently, researchers in the area of optimization have recently de-emphasized using *deterministic search techniques* to achieve convergence to the optimum solution for 'hard' optimization problems encountered in practical settings. What is now emphasized is getting a *good enough* solution, within a reasonable time in order to achieve desirable outcomes of an enterprise. Modern researchers in optimization have now resorted to using good *approximate search techniques* for solving optimization problems because they often find near-optimal solutions within reasonable computation time. An engineer may be satisfied finding a solution to a 'hard' optimization problem if the solution is about 98% of the optimal (best) solution, and if the time taken is very 'reasonable' such that customers are none the less satisfied. We briefly discuss optimization techniques in general, and the framework for the well-established optimization techniques in the following subsections of this introductory chapter. Then new and novel optimization techniques are discussed, which form the building block for subsequent chapters that are devoted to applications of these techniques to hard optimization problems in different areas of engineering.

1.2 Stochastic Optimization Technique

There are two classes of approximation search techniques: tailored and general-purpose. Tailored and customized approximation search technique for a particular optimization problem may often find an optimal (best) solution very quickly. However, they fall short when compared with general-purpose types, in that they cannot be flexibly applied, even, to slightly similar problems. Stochastic optimization falls within the spectrum of the general-purpose type of approximation search techniques. Stochastic optimization involves optimizing a system where the functional relationship between the independent input variables and output (objective function) of the system is not known *a priori*. Hard and discrete function optimization, in which the optimization surface or fitness landscape is "rugged", possessing many locally optimal solutions is well suited for stochastic optimization techniques.

Stochastic optimization techniques play an increasingly important role in the analysis and control of modern systems as a way of coping with inherent system noise and with providing algorithms that are relatively insensitive to modeling uncertainty. Their general form follows the algorithm in Figure 1.1.

Developing solutions with the stochastic optimization technique has the following desirable characteristics:

- It is versatile, in that it can be applied with only minimal changes to other 'hard' optimization problems;
- It is very robust, being relatively insensitive to noisy and/or missing data.
- It tends to be rather simple to implement;
- It has proved very successful in finding good, fast solutions to hard optimization problems in a great variety of practical, real-life applications; and
- Development time is much shorter than for those techniques using more traditional approaches.

Steps involved:

1. Begin: Generate and evaluate an initial collection of candidate solutions S.
2. Operate: Produce and evaluate a collection of potential candidate solutions S' by 'operating' on (or making randomized changes to) selected members of S.
3. Replace: Replace some of the members of S with some of the members of S', and then return to 2 (unless some termination criterion has been reached).

Fig. 1.1. Generalized stochastic optimization techniques.

The existing, successful methods in stochastic iterative optimization fall into two broad classes: local search, and population-based search.

1.2.1 Local Search

In local search, a special *current* solution is maintained, and its neighbors are explored to find better quality solutions. The neighbors are new candidates that are only slightly different from the special *current* solution. Occasionally, one of these neighbors becomes the new *current* solution, and then its neighbors are explored, and so forth. The simplest example of a local search method is called stochastic hill climbing. Figure 1.2 illustrates it.

Steps involved:
1. Begin: Generate and evaluate an initial current solution S.
2. Operate: Make randomized changes to S, producing S', and evaluate S'
3. Replace: Replace S with S', if S' is better than S
4. Continue: Return to 2 (unless some termination criterion has been reached).

Fig. 1.2. The simple stochastic hill-climbing algorithm.

More sophisticated local search methods, in different ways, improve on stochastic hill climbing by being more careful or clever in the way that new candidates are generated. In simulated annealing, for example, the difference is basically in step 3: sometimes, we will accept *S'* as the new *current solution*, even if it is worse then *S*. Without this facility, hill-climbing is prone to getting stuck at local peaks. These are solutions whose neighbors are all worse than the current solution or plateau areas of the solution space where there is considerable scope for movement between solutions of equal goodness, but where very few or none of these solutions has a better neighbor. With this feature, simulated annealing and other local search methods can sometimes (but certainly not always) avoid such difficulties.

1.2.2 Population-based Search

In population-based search, the notion of a single current solution is replaced by the maintenance of a population or collection of (usually) different current solutions. Members of this population are first selected to be *current candidates solutions*, and then changes are made to these *current candidates solutions* to produce *new candidate solutions*. Since there is now a collection of current solutions, rather than just one, we can exploit this fact by using two or more from this collection at once as the basis for generating new candidates. Moreover, the introduction of a population brings with it further opportunities and issues such as using some strategy or other to select which solutions will be current candidates solutions. Also, when one or more new candidate solutions have been produced, we need a strategy for maintaining the population. That is, assuming that we wish to maintain the population at a constant size (which is almost always the case), some of the population must be discarded in order to make way for some or all of the new candidates produced via mutation or recombination operations. Whereas the selection strategy encompasses the details of how to choose candidates from the population itself, the population maintenance strategy affects what is present in the population, and what therefore is selectable. Figure 1.3 illustrates the population-based search method.

Steps involved:
1. Begin: Initialize and evaluate solution S(0).
2. Select: Select S(t) from S(t-1)
3. Generate: Generate a new solution space S'(t) through genetic operators
4. Evaluate: Evaluate solution S'(t) and replace S(t) with S'(t)
5. Continue: Return to 2 (unless some termination criterion has been reached).

Fig. 1.3. The population-based optimization algorithm

There are almost as many population-based optimization algorithms and variants as there are ways to handle the above issues. Commonly known evolutionary algorithms (evolution strategy, evolutionary programming, and genetic algorithm, for example), have the key characteristic of population-based algorithms.

1.3 Framework for Well-Established Optimization Techniques

In random search, information from previous solutions is not used at all. Consequently, performance of random search is very poor. However, in current, well-established optimization methods information gleaned from previously seen candidate solutions are used to guide the generation of new ones. In other words, intelligence gathered from previously seen candidate solutions, are used to guide the generation of new ones. Intelligence gathering of information related to previous events is a powerful tool in devising strategies for seeking solutions to related current events. This concept of utilization of intelligent information gathered from the past event in guiding solutions to a current event is now applied in economics, military, etc., and also in new and novel optimization methods. Most of the ideas discussed so far can be summarized in three categories (see Figure 1.4), which are used in well-established optimization methods (Corne *et al.*, 1999):

Classification:
Category 1: New candidate solutions are slight variations of previously generated candidates.
Category 2: New candidate solutions are generated when aspects of two or more existing candidates are recombined.
Category 3: The current candidates solutions from which new candidates are produced (in Category 1 or 2) are selected from previously generated candidates via a stochastic and competitive strategy, which favors better-performing candidates.

Fig. 1.4. Categories used in well-established optimization methods.

In genetic algorithm, for example, the 'slight variations' mentioned in Category 1 is achieved by so-called neighborhood or mutation operators. For example, suppose we are trying to solve a flow-shop schedule problem and we encode a solution as a string of numbers such that the jth number represents the order of job j in the sequence. We can simply use a mutation operator, which changes a randomly chosen number in the list to a new, randomly chosen value. The 'mutant' flow-shop schedule is thus a slight variation on *current candidate solutions* flow-shop schedule. In tabu search, for example, the 'slight variation' mentioned in Category I, is achieved by so-called neighborhood also called move operators. For example, suppose we were addressing an instance of the traveling salesman problem, then we would typically encode a candidate solution as a permutation of the cities (or points) to be visited. Our neighborhood operator in this case might simply be to swap the positions of a randomly chosen pair of cities.

The 'recombination' operators mentioned in Category 2, which produce new candidates from two or more existing ones, are achieved via crossover, recombination, or merge operators. Some operators involve more sophisticated strategies, in which, perhaps, two or more current candidate solutions are the inputs to a self-contained algorithmic search process, which outputs one or more new *candidate solutions*.

The randomized but 'competitive strategy' mentioned in Category 3 is used to decide which solution be a *current candidate solution*. The strategy is usually that the fitter a candidate is, the more likely it is to be selected to be a *current candidate solution*, and thus be the (at least partial) basis of a new candidate solution.

Several stochastic optimization tools have been enhanced in the last decade, which facilitate solving optimization problems that were previously difficult or impossible to solve. These stochastic optimization tools include Evolutionary Computation, Simulated Annealing (Laarhoven and Aarts, 1987), Tabu Search (Glover, 1995; 1999), Genetic Algorithms (Goldberg, 1989), and so on, all use at least one of these key ideas. Reports of applications of each of these tools have been widely published. Recently, these new heuristic tools have been combined among themselves and with knowledge elements, as well as with more traditional approaches such as statistical analysis (hybridization), to solve extremely challenging problems. Excellent resources for an introduction to a broad collection of these techniques are Reeves (1995) and Onwubolu (2002).

1.4 New & Novel Optimization Techniques

Continued investigation into local search and population based optimization techniques, and a consequent growing regard for their ability to address realistic problems well, has inspired much investigation of new and novel optimization methods. Often we can express these as different or novel ways of implementing Categories 1, 2 and 3 of section 1.3, but in other cases this would be quite difficult or inappropriate.

Chapter Two, introduced by Deb concerns genetic algorithm (Goldberg, 1989). Genetic Algorithm has been included for completeness since it is the progenitor of Memetic Algorithm.

Chapter Three, introduced by Moscato & Mendes, concerns Memetic Algorithms invented by Moscato (1999). This represents one of the more successful emerging ideas in the ongoing research effort to understand population based and local search algorithms. The idea itself is simply to hybridize the two types of method, and Memetic Algorithms represent a particular way of achieving the hybridization, and one, which has chalked up a considerable number of successes in recent years.

Chapter Four, introduced by Glover & Marti, is about Scatter Search invented by Glover (1995, 1999). Scatter Search places a particular focus on Category 2, that is, the use of previously generated candidate solutions in the generation of new ones. Similar to Differential Evolution, Scatter Search emphasizes the production of new solutions by a linear combination of the 'reference' solutions, although guided instead by combining intensification and diversification themes, and introducing strategies for rounding integer variables.

Chapter Five, introduced by Maniezzo and Gambardella concerns ant colony optimization developed by Dorigo (1992), which has drawn inspiration from the workings of natural ant colonies to derive an optimization method based on ant colony behavior. This has lately been shown to be remarkably successful on some problems, and there is now a thriving international research community worldwide further investigating the technique. Essentially, the method works by equating the notion of a candidate solution with the route taken by an ant between two (possibly the same) places. Ants leave a trail as they travel, and routes, which correspond to good solutions, will leave a stronger trail than routes, which lead to poor solutions. The trail strength affects the routes taken by future ants. We could view this as a very novel way to implement a combination of Categories 1, 2, and 3. Essentially, previously generated solutions (routes taken by past ants) affect (via trail strengths) the solutions generated by future ants. However, it makes more sense to view it for what it is a new and novel idea for optimization inspired by a process, which occurs in nature.

Chapter Six introduced by Lampinen & Storn concerns an algorithm called Differential Evolution (Storn and Price, 1995). Differential Evolution can be easily viewed as a new way to implement Category 2; that is, it is a novel and highly successful method for recombination of two or more current candidate solutions. Rather than new candidates solutions combining parts of their current candidates solutions in Differential Evolution a child is a linear combination of itself and its current candidates solutions in which the 'difference' between its current candidates solutions plays a key role. As such, Differential Evolution is necessarily restricted to optimization problems in which solutions are represented by real numbers. Recently, transformation scheme has been introduced so that Differential Evolution is generalized to deal with discrete optimization problems in which the solutions are represented by integers (Onwubolu, 2001); however, the power of even quite simple Differential Evolution algorithms to solve complex real-world problems provides great compensation for this restriction. Apart from its simplicity, it also has the flexibility to incorporate and improve upon the existing ten different strategies. It has the capability of finding the global optimum for complex and complicated problems of varied nature, ranging from optimization of continuous functions (Babu and Sastry, 1999) to large-scale discrete combinatorial optimization (Babu and Munawar, 2001). Very recently, a new concept of "nested DE" has been successfully implemented for the optimal design of an auto-thermal ammonia synthesis reactor (Babu *et al.*, 2002). This concept uses DE within DE wherein outer loop takes care of optimization of key parameters (NP – population size, CR – crossover constant, and F – scaling factor) with the objective of minimizing the number of generations required, while inner loop taking care of optimizing the problem variables. Yet complex objective is the one that takes care of minimizing the number of generations/function evaluations & the standard deviation in the set of solutions at the last generation/function evaluation, and try to maximize the robustness of the algorithm.

Chapter Seven concerns Self-Organizing Migrating Algorithm (Zelinka and Lampinen, 2000; Zelinka, 2001), originally proposed by Zelinka has drawn inspiration from existing algorithms like, evolutionary algorithms and genetic algorithms. However, in general, the design of Self-Organizing Migrating Algorithm is unique and original. Its principle is both interesting and novel. It offers a novel approach to optimization and combines global search (the attraction of individuals towards the fitter ones) and local search (the multiple jumps of each individual during one migration loop); hence it covers both exploration and exploitation. The temporary replacement of the cost function with a virtual cost function seems to be a unique approach to overcome the difficulties when local optima are populating search space. The idea of the search algorithm is certainly a contribution containing new scientific knowledge.

Chapter Eight concerns Particle Swarm Optimization (Kennedy and Eberhart, 1995; Clerc and Kennedy, 2002). An analogy with the way birds flock has suggested the definition of a new computational paradigm, which is known as particle swarm optimization. The main characteristics of this model are cognitive influence, social influence, and the use of constriction parameters. Cognitive influence accounts for a particle moving towards its best previous position, social influence accounts for a particle moving towards the best neighbor, and constriction parameters control the explosion of the systems' velocities and positions. Particle Swarm Optimization is a powerful method to find the minimum of a numerical function, on a continuous definition domain. As some binary versions have already successfully been used, it seems quite natural to try to define a framework for a discrete PSO. In order to better understand both the power and the limits of this approach, Clerc examines in detail how it can be used to solve the well-known Traveling Salesman Problem, which is in principle very "bad" for this kind of optimization heuristic.

1.5 The Structure of the Book

Part I introduces the new and novel optimization techniques. Part I containing six chapters have been written by the inventors of these novel optimization techniques or experts who have done considerable work in the areas.

Then, the applications of these new and novel optimization techniques in the different engineering-disciplines are discussed in Part Two.

Part II.1 puts together a collection of chemical/metallurgical engineering applications

Part II.2 puts together a collection of civil/environmental engineering applications

Part II.3 puts together a collection of electrical/electronic engineering applications

Part II.4 puts together a collection of manufacturing/industrial engineering applications

Part II.5 puts together a collection of mechanical/aeronautical engineering applications

Finally, Part III discusses newer or 'green' areas, which are considered as extended frontiers, not discussed in the book.

1.6 Summary

There are many more new optimization techniques we could have included in the succeeding chapters. Our selection of what to incorporate in this book is based partly on successful results so far in different areas of engineering. And partly on what we perceive as the 'generality' of the optimization technique. Several more new ideas and methods fit these criteria quite well.

What is left out is a result of constraints of time and space, but what remains within these pages is, we hope, a collection of some of the more novel and inspired new methods in optimization. We have assembled this material with three purposes in mind: first, to provide in one volume a survey of recent successful ideas in stochastic optimization technique, which can support upper-undergraduate and graduate teaching of relevant courses; second, to provide intellectual stimulation and a practical resource for the very many researchers and practitioners in optimization who wish to learn about and experiment with these latest ideas in the field; third, to provide in one volume a survey of recent successful applications of stochastic optimization techniques different areas of engineering practices.

References

Babu, B.V. and Sastry, K.K.N., 1999, Estimation of Heat Transfer Parameters in a Trickle Bed Reactor using Differential Evolution and Orthogonal Collocation, *Computers and Chemical Engineering*, 23, 327-339 *(Also Available via internet as .pdf file at http://bvbabu.50megs.com/about.html).*

Babu, B.V. and Munawar, S.A., 2001, Optimal Design of Shell-and-Tube Heat Exchangers using Different Strategies of Differential Evolution, *PreJournal.com - The Faculty Lounge, Article No. 003873*, posted on March 03 at website Journal *http://www.prejournal.com (Also Available via internet as .pdf files in two parts at http://bvbabu.50megs.com/about.html).*

Babu, B.V., Rakesh Angira, and Anand Nilekar, 2002, Differential Evolution for Optimal Design of an Auto-Thermal Ammonia Synthesis Reactor, Communicated to *Computers and Chemical Engineering*.

Clerc, M. and Kennedy, J., 2002, The Particle Swarm-Explosion, Stability, and Convergence in a Multidimensional Complex Space, *IEEE Transactions on Evolutionary Computation*, (6), 58-73.

Dorigo, M., 1992, *Optimization, Learning and Natural Algorithms*, Ph.D. Thesis, Dipartimento di Electronica, Politecnico di Milano, Italy.

Goldberg, D. E., 1989, *Genetic Algorithm in Search, Optimization & Machine Learning*, Addison Wesley, Workingham, England.

Glover, F., 1995, Scatter Search and Star-Paths: Beyond the Genetic Metaphor, Operational Research Spektrum, 17,125-137.

Glover, F., 1999, Scatter Search and Paths Re-linking, In *New Ideas in Optimization*, Corne, D., Dorigo, M., and Glover, F., (Eds.) Chapter 19, McGraw-Hill: London

Laarhoven, P. J. M., and Aarts, E. H. L., 1987, *Simulated Annealing: Theory and Applications*, Kluwer Academic Publishers: The Netherlands.

Kennedy, J., and Eberhart, R. C., 1995, Particle swarm optimization, *IEEE Proceedings of the International Conference on Neural Networks IV (Perth, Australia), IEEE Service Center, Piscataway, NJ*, 1942-1948.

Moscato, P., 1999, Memetic algorithms: a short introduction, In *New Ideas in Optimization*, Corne, D., Dorigo, M., and Glover, F., (Eds.) Chapter 14, McGraw-Hill: London

Onwubolu, G. C., 2001, Optimization using differential evolution, *Institute of Applied Science Technical Report, TR-2001/05.*

Onwubolu, G. C., 2002, *Emerging Optimization Techniques in Production Planning & Control*, Imperial College Press: London

Reeves, C. R. 1995, *Modern Heuristic Techniques for Combinatorial Problems*, (Ed.) McGraw-Hill (transfer from Blackwell Scientific, 1993)

Storn, R. and Price, K., 1995, Differential evolution - a simple and efficient adaptive scheme for global optimization over continuous spaces, *Technical Report TR-95-012, ICSI, March 1999*

(Available via ftp from ftp.icsi.berkeley.edu/pub/techreports/1995/tr-95-012.ps.Z).

Zelinka, I., and Lampinen, J., 2000, SOMA: Self-Organizing Migrating Algorithm, *3rd International Conference on Prediction and Nonlinear Dynamic*, Zlin, Czech Republic: Nostradamus.

Zelinka I., 2001, *Prediction and Analysis of Behavior of Dynamical Systems by means of Artificial Intelligence and Synergetic*, Ph.D. Thesis, Department of Information Processing, Lappeenranta University of Technology, Finland.

2 Introduction to Genetic Algorithms for Engineering Optimization

Kalyanmoy Deb

2.1 Introduction

A genetic algorithm (GA) is a search and optimization method which works by mimicking the evolutionary principles and chromosomal processing in natural genetics. A GA begins its search with a random set of solutions usually coded in binary string structures. Every solution is assigned a fitness which is directly related to the objective function of the search and optimization problem. Thereafter, the population of solutions is modified to a new population by applying three operators similar to natural genetic operators--reproduction, crossover, and mutation. A GA works iteratively by successively applying these three operators in each generation till a termination criterion is satisfied. Over the past couple of decades and more, GAs have been successfully applied to a wide variety of engineering problems, because of their simplicity, global perspective, and inherent parallel processing.

Classical search and optimization methods demonstrate a number of difficulties when faced with complex engineering problems. The major difficulty arises when one algorithm is applied to solve a number of different problems. This is because each classical method is designed to solve only a particular class of problems efficiently. Thus, these methods do not have the breadth to solve different types of problems often faced by designers and practitioners. Moreover, most classical methods do not have the global perspective and often get converged to a locally optimal solution. Another difficulty is their inability to be used in parallel computing environment efficiently. Since most classical algorithms are serial in nature, not much advantage (or speed-up) can be achieved with them.

The GA technique was first conceived by Professor John Holland of University of Michigan, Ann Arbor in 1965. His first book appeared in 1975 (Holland, 1975) and till 1985, GAs have been practiced mainly by Holland and his students. Exponentially more number of researchers and practitioners became interested in GAs soon after the first International conference on GAs held in 1985. Now, there exist a number of books (Gen and Cheng, 1997; Goldberg, 1989; Michalewicz, 1992; Mitchell, 1996) and a few journals dedicated to publishing research papers on the topic (including one from MIT Press and one from IEEE). Every year, there are at least 15-20 conferences and workshops being held on the topic at various parts of the globe. The major reason for GA's popularity in various search and optimization problems is its global perspective, wide spread applicability, and inherent parallelism.

In the remainder of the paper, we highlight the shortcomings of the classical optimization methods and then discuss the working principle of a GA by showing an illustration on a simple can design problem and a hand simulation. Thereafter, we argue intuitive reasons for the working of a GA. Later, we present a number of engineering applications, in which a combined binary-coded and real-parameter GAs are applied for solving mixed-integer programming problems. The description of GAs and their broad-based applicabilities demonstrated in this paper should encourage academician and practitioners to pay attention to this growing field of interests.

2.2 Classical Search and Optimization Techniques

Traditional search and optimization methods can be classified into two distinct groups: Direct and gradient-based methods (Deb, 1995; Reklaitis at el., 1983). In direct methods, only objective function and constraints are used to guide the search strategy, whereas gradient-based methods use the first and/or second-order derivatives of the objective function and/or constraints to guide the search process. Since derivative information is not used, the direct search methods are usually slow, requiring many function evaluations for convergence. For the same reason, they can be applied to many problems without a major change of the algorithm. On the other hand, gradient-based methods quickly converge to an optimal solution, but are not efficient in non-differentiable or discontinuous problems. In addition, there are some common difficulties with most of the traditional direct and gradient-based techniques:

- Convergence to an optimal solution depends on the chosen initial solution.
- Most algorithms tend to get stuck to a suboptimal solution.
- An algorithm efficient in solving one search and optimization problem may not be efficient in solving a different problem.
- Algorithms are not efficient in handling problems having discrete variables.
- Algorithms cannot be efficiently used on a parallel machine.

Because of the nonlinearities and complex interactions among problem variables often exist in complex search and optimization problems, the search space may have many optimal solutions, of which most are locally optimal solutions having inferior objective function values. When solving these problems, if traditional methods get attracted to any of these locally optimal solutions, there is no escape from it.

Many traditional methods are designed to solve a specific type of search and optimization problems. For example, geometric programming (GP) method is designed to solve only polynomial-type objective function and constraints (Duffin et al., 1967). GP is efficient in solving such problems but can not be applied suitably to solve other types of functions. Conjugate direction method has a convergence proof for solving quadratic functions, but they are not expected to work well in problems having multiple optimal solutions. Frank-Wolfe method (Reklaitis at el., 1983) works efficiently on linear-like function and constraints, but the performance largely depends on the chosen initial conditions. Thus, one algorithm may be best suited for one problem and may not be even applicable to a different problem. This requires designers to know a number of optimization algorithms.

In many search and optimization problems, problem variables are often restricted to take discrete values only. To solve such problems, an usual practice is to assume that the problem variables are real-valued. A classical method can then be applied to find a real-valued solution. To make this solution feasible, the nearest allowable discrete solution is chosen. But, there are a number of difficulties with this approach. Firstly, since many infeasible values of problem variables are allowed in the optimization process, the optimization algorithm is likely to take many function evaluations before converging, thereby making the search effort inefficient. Secondly, for each infeasible discrete variable, two values (the nearest lower and upper available sizes) are to be checked. For N discrete variables, a total of 2^N such additional solutions need to be evaluated. Thirdly, two options checked for each variable may not guarantee the optimal combination of all variables. All these difficulties can be eliminated if *only* feasible values of the variables are allowed during the optimization process.

Many search and optimization problems require use of a simulation software, involving finite element technique, computational fluid mechanics approach, solution of nonlinear equations, and others, to compute the objective function and constraints. The use of such softwares is time-consuming and may require several minutes to hours to evaluate one solution. Because of the availability of parallel computing machines, it becomes now convenient to use parallel machines in solving complex search and optimization problems. Since most traditional methods use point-by-point approach, where one solution gets updated to a new solution in one iteration, the advantage of parallel machines cannot be exploited.

The above discussion suggests that a traditional method is not good candidate for an efficient search and optimization algorithm. In the following section, we describe the genetic algorithm which works according to principles of natural genetics and evolution, and which has been demonstrated to solve various search and optimization problems in engineering.

2.3 Motivation from Nature

Most biologists believe that the main driving force behind the natural evolution is the Darwin's survival-of-the-fittest principle (Dawkins, 1976; Eldredge, 1989). In most situations, the nature ruthlessly follows two simple principles:

1. If by genetic processing an above-average offspring is created, it usually survives longer than an average individual and thus have more opportunities to produce children having some of its traits than an average individual.
2. If, on the other hand, a below-average offspring is created, it usually does not survive longer and thus gets eliminated from the population.

The renowned biologist Richard Dawkins explains many evolutionary facts with the help of Darwin's survival-of-the-fittest principle in his seminal works (Dawkins, 1976; 1986). He argues that the tall trees that exist in the mountains were only a feet tall during early ages of evolution. By genetic processing if one tree had produced an offspring an inch taller than all other trees, that offspring enjoyed more sunlight and rain and attracted more insects for pollination than all other trees. With extra benefits, that lucky offspring had an increased life and more importantly had produced more offspring like it (with tall feature) than others. Soon enough, it occupies most of the mountain with trees having its genes and the competition for survival now continues with other trees, since the available resource (land) is limited. On the other hand, if a tree had produced an offspring with an inch smaller than others, it was less fortunate to enjoy all the facilities other neighboring trees had enjoyed. Thus, that offspring could not survive longer. In a genetic algorithm, this feature of natural evolution is introduced somewhat directly through its operators.

The principle of emphasizing good solutions and eliminating bad solutions seems a good feature an optimization algorithm may have. But one may wonder about the real connection between an optimization procedure and natural evolution! Has the natural evolutionary process tried to maximize a utility function of some sort? Truly speaking, one can imagine a number of such functions which the nature may be thriving to maximize: Life span of a species, quality of life of a species, physical growth, and others. However, any of these functions is nonstationary in nature and largely depends on the evolution of other related species. Thus, in essence, the nature has been really optimizing much more complicated objective functions by means of natural genetics and natural selection than the search and optimization problems we are interested in solving. Thus, it is not surprising that a genetic algorithm is not as complex as the natural genetics and selection, rather it is an abstraction of the complex natural evolutionary process. The simple version of a GA described in the following section aims to solve stationary search and optimization problems. Although a GA is a simple abstraction, it is robust and has been found to solve various search and optimization problems of science, engineering, and commerce.

2.4 Genetic Algorithms

In this section, we first describe the working principle of a genetic algorithm. Thereafter, we shall show a simulation of a genetic algorithm for one iteration on a simple optimization problem. Later, we shall give intuitive reasoning of why a GA is a useful search and optimization procedure.

2.4.1 Working Principle

Genetic algorithm (GA) is an iterative optimization procedure. Instead of working with a single solution in each iteration, a GA works with a number of solutions (collectively known as a population) in each iteration. A flowchart of the working principle of a simple GA is shown in Figure 2.1. In the absence of any knowledge of the problem domain, a GA begins its search from a random population of solutions. As shown in the figure, a solution in a GA is represented using a string coding of fixed length. We shall discuss about the details of the coding procedure a little later. But for now notice how a GA processes these strings in a iteration. If a termination criterion is not satisfied, three different operators--reproduction, crossover, and mutation--are applied to update the population of strings. One iteration of these three operators is known as a generation in the parlance of GAs. Since the representation of a solution in a GA is similar to a natural chromosome and GA operators are similar to genetic operators, the above procedure is named as genetic algorithm. We now discuss the details of the coding representation of a solution and GA operators in details in the following subsections.

2.4.1.1 Representation

In a binary-coded GA, every variable is first coded in a fixed-length binary string. For example, the following is a string, representing *N* problem variables:

$$\underbrace{11010}_{x_1}\ \underbrace{1001001}_{x_2}\ \underbrace{010}_{x_3}\ \ldots\ \underbrace{0010}_{x_N}$$

The i-th problem variable is coded in a binary substring of length l_i, so that the total number of alternatives allowed in that variable is 2^{l_i}. The lower bound solution $x_i^{\min}$ is represented by the solution (00...0) and the upper bound solution $x_i^{\max}$ is represented by the solution (11...1). Any other substring s_i decodes to a solution x_i as follows:

$$x_i = x_i^{\min} + \frac{x_i^{\max} - x_i^{\min}}{2^{l_i} - 1} DV(s_i), \tag{2.1}$$

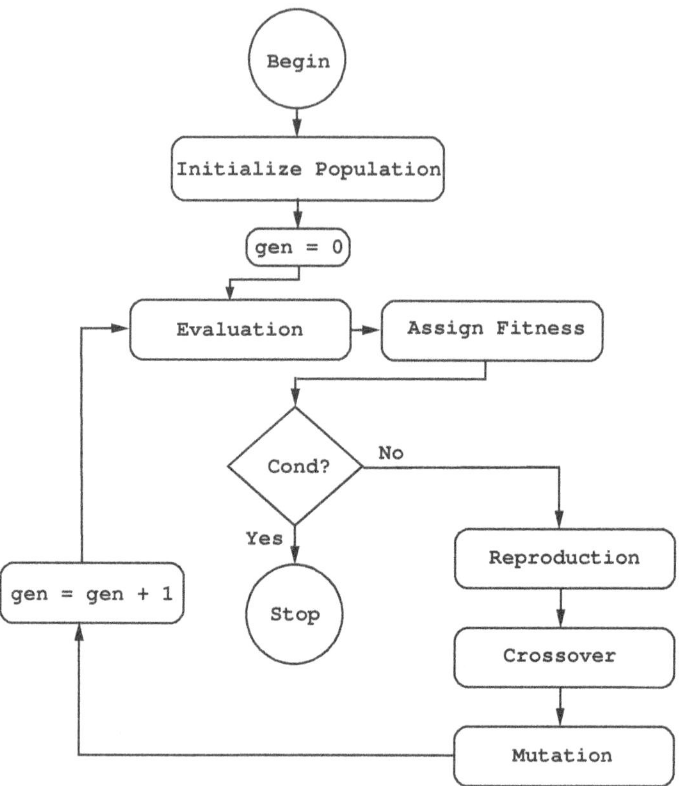

Fig. 2.1. A flowchart of working principle of a genetic algorithm

where $DV(s_i)$ is the decoded value of the substring s_i. The length of a substring is usually decided by the precision needed in a variable. For example, if three decimal places of accuracy is needed in the i-th variable, the total number of alternatives in the variable must be $(x_i^{\max} - x_i^{\min})/0.001$, which can be set equal to 2^{l_i} and l_i can be computed as follows:

$$l_i = \log_2 \left| \left(\frac{x_i^{\max} - x_i^{\min}}{\varepsilon_i} \right) \right|. \tag{2.2}$$

Here, the parameter ε_i is the desired precision in i-th variable. The total string length of a N-variable solution is then $l = \sum_{i=1}^{N} l_i$. Representing a solution in a string of bits (0 or 1) resembles a natural chromosome which is a collection of genes having particular allele values.

In the initial population, l-bit strings are created at random (at each of l positions, there is a equal probability of creating a 0 or a 1). Once such a string is created, the first l_l bits can be extracted from the complete string and corresponding value of the variable x_l can be calculated using Equation 2.1 and using the chosen lower and upper limits of the variable x_l. Thereafter, the next l_2 bits can be extracted from the original string and the variable x_2 can be calculated. This process can be continued until all N variables are obtained from the complete string. Thus, an l-bit string represents a complete solution specifying all N variables uniquely. Once these values are known, the objective function $f(x_l \ldots x_N)$ can be computed.

2.4.1.2 Fitness Assignment

In a GA, each string created either in the initial population or in the subsequent generations must be assigned a *fitness* value which is related to the objective function value. For maximization problems, a string's fitness can be equal to the string's objective function value. However, for minimization problems, the goal is to find a solution having the minimum objective function value. Thus, the fitness can be calculated as the reciprocal of the objective function value so that solutions with smaller objective function value get larger fitness. Usually, the following transformation function is used for minimization problems:

$$Fitness = \frac{1}{1 + f(x_1, \ldots, x_N)}. \tag{2.3}$$

There are a number of advantages of using a string representation to code variables. First, this allows a shielding between the working of GA and the actual problem. What a GA processes is l-bit strings, which may represent any number of variables, depending on the problem at hand. Thus, the same GA code can be used for different problems by only changing the definition of coding a string. This allows a GA to have a wide spread applicability. Second, a GA can exploit the similarities in a string coding to make its search faster, a matter which is important in the working of a GA and is discussed in Subsection 2.4.3.

2.4.1.3 Reproduction

Reproduction (or selection) is usually the first operator applied on a population. Reproduction selects good strings in a population and forms a mating pool. There exist a number of reproduction operators in the GA literature (Goldberg and Deb, 1991), but the essential idea is that above-average strings are picked from the current population and duplicates of them are inserted in the mating pool. The commonly-used reproduction operator is the *proportionate* selection operator, where a string in the current population is selected with a probability proportional to the string's fitness. Thus, the i-th string in the population is selected with a probability proportional to f_I. Since the population size is usually kept fixed in a simple GA, the cumulative probability for all strings in the population must be one. Therefore, the probability for selecting i-th string is $f_i / \sum_{j=1}^{N} f_j$, where N is the population size. One way to achieve this proportionate selection is to use a roulette-wheel with the circumference marked for each string proportionate to the string's fitness. The roulette-wheel is spun N times, each time keeping an instance of the string, selected by the roulette-wheel pointer, in the mating pool. Since the circumference of the wheel is marked according to a string's fitness, this roulette-wheel mechanism is expected to make $f_i / \bar{f}$ copies of the i-th string, where $\bar{f}$ is the average fitness of the population. This version of roulette-wheel selection is somewhat noisy; other more stable versions exist in the literature (Goldberg, 1989). As will be discussed later, the proportionate selection scheme is inherently slow. One fix-up is to use a *ranking* selection scheme. All N strings in a population is first ranked according to ascending order of string's fitness. Each string is then assigned a rank from 1 (worst) to N (best) and an linear fitness function is assigned for all the strings so that the best string gets two copies and the worst string gets no copies after reproduction. Thereafter, the proportionate selection is used with these fitness values. This ranking reproduction scheme eliminates the function-dependency which exists in the proportionate reproduction scheme.

The *tournament* selection scheme is getting increasingly popular because of its simplicity and controlled takeover property (Goldberg and Deb, 1991). In its simplest form (binary tournament selection as described in the next subsection), two strings are chosen at random for a tournament and the better of the two is selected according to the string's fitness value. If done systematically, the best string in a population gets exactly two copies in the mating pool. It is important to note that this reproduction operator does not require a transformation of the objective function to calculate fitness of a string as suggested in Equation 2.3 for minimization problems. The better of two strings can be judged by choosing the string with the smaller objective function value.

2.4.1.4 Crossover

Crossover operator is applied next to the strings of the mating pool. Like reproduction operator, there exists a number of crossover operators in the GA literature (Spears and De Jong, 1991; Syswerda, 1989), but in almost all crossover operators, two strings are picked from the mating pool at random and some portion of the strings are exchanged between the strings. In a single-point crossover operator, both strings are cut at an arbitrary place and the right-side portions of both strings are swapped among themselves to create two new strings, as illustrated in the following:

```
Parent1    0 0 | 0 0 0        0 0 | 1 1 1    Child1
                         =>
Parent2    1 1 | 1 1 1        1 1 | 0 0 0    Child2
```

It is interesting to note from the construction that good substrings from either parent string can be combined to form a better child string if an appropriate site is chosen. Since the knowledge of an appropriate site is usually not known, a random site is usually chosen. However, it is important to realize that the choice of a random site does not make this search operation random. With a single-point crossover on two l-bit parent strings, the search can only find at most $2(l-1)$ different strings in the search space, whereas there are a total of 2^l strings in the search space. With a random site, the children strings produced may or may not have a combination of good substrings from parent strings depending on whether the crossing site falls in the appropriate place or not. But we do not worry about this aspect too much, because if good strings are created by crossover, there will be more copies of them in the next mating pool generated by the reproduction operator. But if good strings are not created by crossover, they will not survive beyond next generation, because reproduction will not select bad strings for the next mating pool.

In a *two-point* crossover operator, two random sites are chosen and the contents bracketed by these sites are exchanged between two parents. This idea can be extended to create a multi-point crossover operator and the extreme of this extension is what is known as a *uniform* crossover operator (Syswerda, 1989). In a uniform crossover for binary strings, each bit from either parent is selected with a probability of 0.5.

It is worthwhile to note that the purpose of the crossover operator is two-fold. The main purpose of the crossover operator is to search the parameter space. Other aspect is that the search needs to be performed in a way to preserve the information stored in the parent strings maximally, because these parent strings are instances of good strings selected using the reproduction operator. In the single-point crossover operator, the search is not extensive, but the maximum information is preserved from parent to children. On the other hand, in the uniform crossover, the search is very extensive but minimum information is preserved between parent and children strings. However, in order to preserve some of the previously-found good strings, not all strings in the population are participated in the crossover operation. If a crossover probability of p_c is used then $100p_c\%$ strings in the population are used in the crossover operation and $100(1-p_c)\%$ of the population are simply copied to the new population. Even though best $100(1-p_c)\%$ of the current population can be copied deterministically to the new population, this is usually performed stochastically.

2.4.1.5 Mutation

Crossover operator is mainly responsible for the search aspect of genetic algorithms, even though mutation operator is also used for this purpose sparingly. Mutation operator changes a 1 to a 0 and vice versa with a small mutation probability, p_m:

$$00000 \implies 00010$$

In the above example, the fourth gene has changed its value, thereby creating a new solution. The need for mutation is to maintain diversity in the population. For example, if in a particular position along the string length all strings in the population have a value 0, and a 1 is needed in that position to obtain the optimum or a near-optimum solution, then the crossover operator described above will be able to create a 1 in that position. The inclusion of mutation introduces some probability of turning that 0 into a 1. Furthermore, for local improvement of a solution, mutation is useful.

After reproduction, crossover, and mutation are applied to the whole population, one generation of a GA is completed. These three operators are simple and straightforward. Reproduction operator selects good strings and crossover operator recombines good substrings from two good strings together to hopefully form a better substring. Mutation operator alters a string locally to hopefully create a better string. Even though none of these claims are guaranteed and/or tested while creating a new population of strings, it is expected that if bad strings are created they will be eliminated by the reproduction operator in the next generation and if good strings are created, they will be emphasized. Later, we shall discuss some intuitive reasoning as to why a GA with these simple operators may constitute a potential search algorithm.

2.4.2 An Illustration

We describe the working of the above GA further by illustrating a simple can design problem. A cylindrical can is considered to have only two parameters -- the diameter d and height h. Let us consider that the can needs to have a volume of at least 300 ml and the objective of the design is to minimize the cost of the can material. With this constraint and the objective, we first write the corresponding *nonlinear programming problem* (NLP problem) (Deb, 1999):

$$\left.\begin{array}{ll} \text{Minimize} & f(d,h) = c\left(\dfrac{\pi d^2}{2} + \pi d h\right), \\ \text{Subject} & g(d,h) = \dfrac{\pi d^2 h}{4} \geq 300, \\ \text{Variable bounds} & \begin{array}{l} d_{\min} \leq d \leq d_{\max}, \\ h_{\min} \leq h \leq h_{\max}. \end{array} \end{array}\right\} \quad (2.4)$$

The parameter c is the cost of the can material per square cm, and the decision variables d and h are allowed to vary in $[d_{min}, d_{max}]$ and $[h_{min}, h_{max}]$ cm, respectively.

2.4.2.1 Representing a Solution

Let us assume that we shall use five bits to code each of the two decision variables, thereby making the overall string length equal to 10. The following string represents a can of diameter 8 cm and height 10 cm:

$$\underbrace{01000}_{d}\ \underbrace{01010}_{h}$$

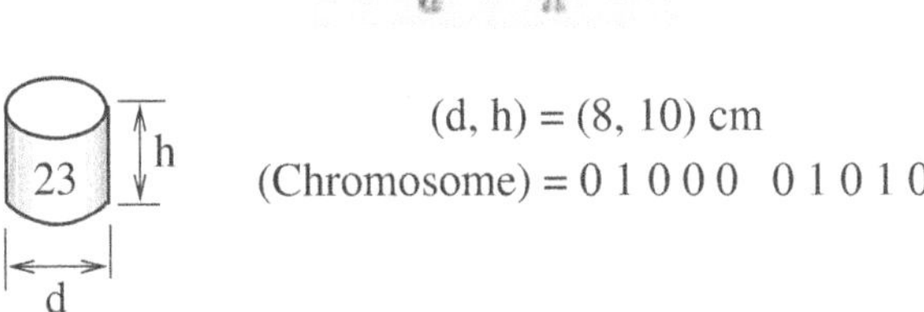

Fig. 2.2. A typical can and its chromosomal representation are shown. The cost of the can is marked as 23 units.

This string and corresponding decision variables are shown in Figure 2.2. Coding the decision variables in a binary string is primarily used to achieve a pseudo-chromosomal representation of a solution. For example, the 10-bit string illustrated above can be explained to exhibit a biological representation of a can having a diameter of 8 cm and a height of 10 cm. Natural chromosomes are made of many genes, each of which can take one of many different {allele} values (such as, the gene responsible for the eye color in a person's chromosomes may be expressed as black, whereas it could have been blue or some other color). When we see the person, we see the person's phenotypic representation, but each feature of the person is precisely written in his/her chromosomes -- the genotypic representation of the person. In the can design problem, the can itself is the phenotypic representation of an artificial chromosome of 10 genes. To see how these 10 genes control the phenotype (the shape) of the can, let us investigate the leftmost bit (gene) of the diameter (d) parameter. A value of 0 at this bit (the most significant bit) allows the can to have diameter values in the range [0, 15] cm, whereas the other value 1 allows the can to have diameter values in the range [16, 31] cm. Clearly, this bit (or gene) is responsible for dictating the 'slimness' of the can. If the allele value 0 is expressed, the can is slim, while if the value 1 is expressed, the can is 'fat'. Each other bit position or a combination of two or more bit positions can also be explained to support the can's phenotypic appearance, but some of these explanations are interesting and important, and some are not.

After choosing a string representation scheme and creating a population of strings at random, we are ready to apply genetic operations to such strings to hopefully find better populations of solutions.

2.4.2.2 Assigning Fitness to a Solution

In the absence of constraints, the fitness of a string is assigned a value which is a function of the solution's objective function value. In most cases, however, the fitness is made equal to the objective function value. For example, the fitness of the above can represented by the 10-bit string s is:

$$F(s) = 0.065\left[\frac{\pi(8)^2}{2} + \pi(8)(10)\right],$$
$$= 23,$$

assuming c=0.065. Since the objective of the optimization here is to minimize the objective function, it is to be noted that a solution with a smaller fitness value compared to another solution is better. The fitness of an infeasible solution is described later, but here we simply add a penalty in proportion to its constraint violation.

Figure 2.3 shows the phenotypes of a random population of six cans. The fitness of each can is also marked. It is interesting to that two infeasible solutions do not have internal volume of 300 ml and thus are penalized by adding an extra artificial *cost*. The extra penalized cost is large enough to cause all infeasible solutions to have a worse fitness value than that of any feasible solution. We are now ready to discuss three genetic operators.

2.4.2.3 Reproduction or Selection Operator

Here, we use the tournament selection, in which tournaments are played between two solutions and the better solution is chosen and placed in the mating pool. Figure 2.4 shows six different tournaments played

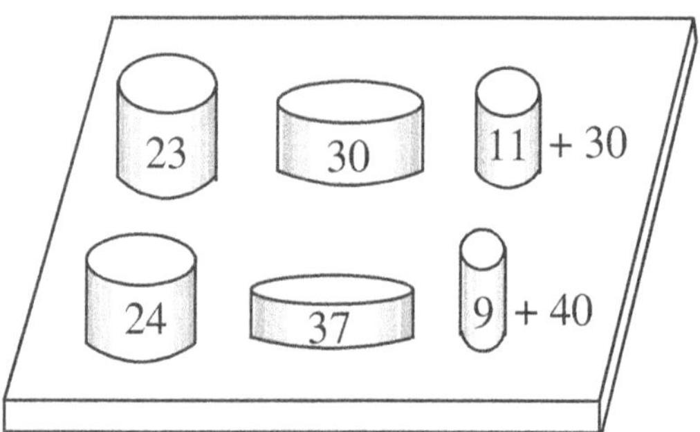

Fig. 2.3. A random population of six cans.

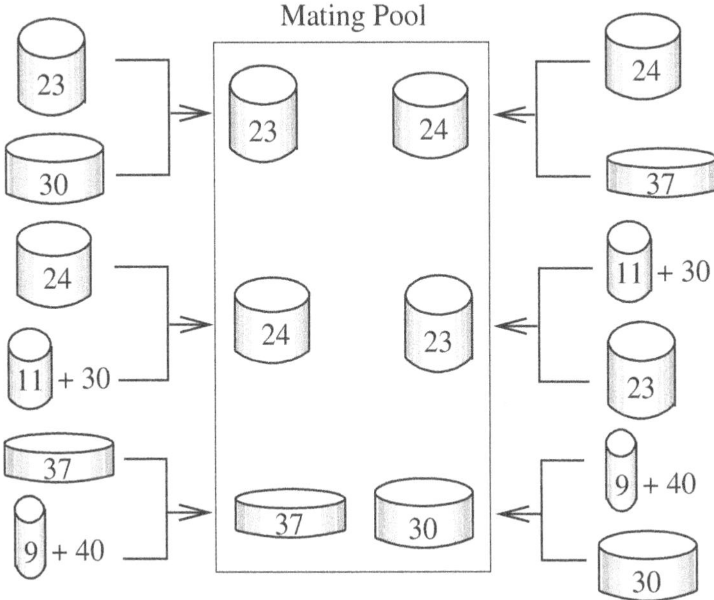

Fig. 2.4. Tournaments are played between the six population members of Figure 2.3. The population enclosed by the dashed box forms the mating pool.

between old population members (each gets exactly two turns). When cans with a cost of 23 units and 30 units are chosen at random for the first tournament, the can costing 23 units wins and a copy of it is placed in the mating pool. The next two cans are chosen for the second tournament and a copy of the better can is then placed in the mating pool. This is how the mating pool (Figure 2.5) is formed. It is interesting to note how better solutions (having less costs) have made themselves to have multiple copies in the mating pool and worse solutions have been discarded. This is precisely the purpose of a reproduction or a selection operator. An interesting aspect of the tournament selection operator is that just by changing the comparison operator, the minimization and maximization problems can be handled easily.

2.4.2.4 Crossover Operator

A crossover operator is applied next to the strings of the mating pool. Let us illustrate the crossover operator by picking two solutions (called parent solutions) from the new population created after the reproduction operator. The cans and their genotype (strings) are shown in Figure 2.6. The third site along the string length is chosen at random and the contents of the right side of this cross site are exchanged between the two strings. The process creates two new strings (called offspring). Their phenotypes (the cans) are also shown in this figure.

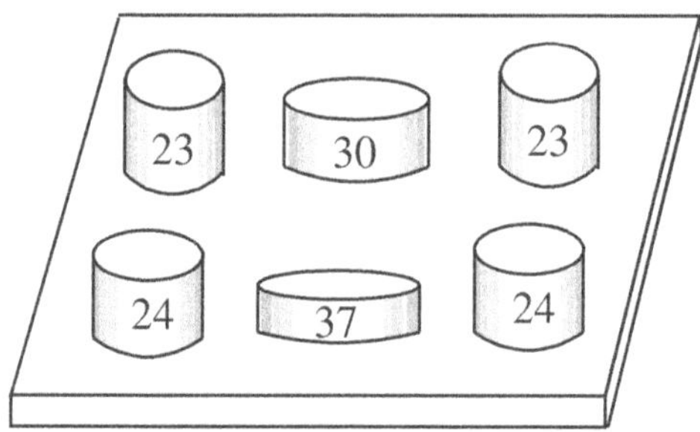

Fig. 2.5. The population after reproduction operation.

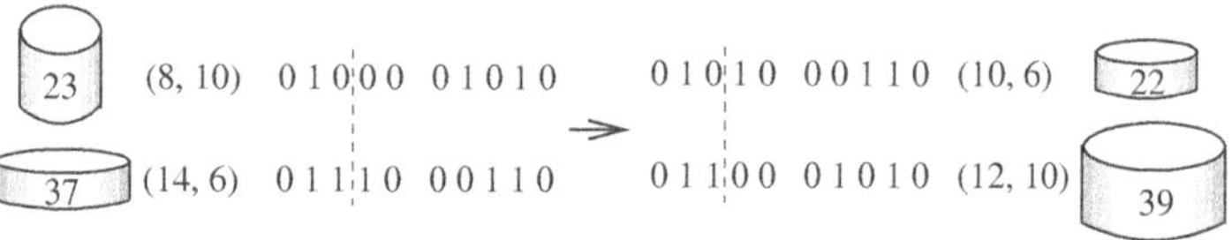

Fig. 2.6. The single point crossover operation.

It is important to note that the above crossover operator created a solution (having a cost of 22 units) which is better in cost than both of the parent solutions. One may wonder that if different cross sites were chosen or two other strings were chosen for crossover, whether we would have found a better offspring every time. It is true that every crossover between any two solutions from the new population is not likely to find offspring better than both parent solutions, but the chance of creating better solutions is far better than random. This is because the parent strings being crossed are not any two arbitrary random strings. These strings have survived tournaments played with other solutions during the earlier reproduction phase. Thus, they are expected to have some good bit combinations in their string representations. Since, a single-point crossover on a pair of parent strings can only create l different string pairs (instead of all 2^{l-1} possible string-pairs) with bit combinations from either strings, the created offspring are also *likely* to be good strings.

2.4.2.5 Mutation Operator

Figure 2.7 shows how a string obtained after the use of reproduction and crossover operators has been mutated to another string, thus representing a slightly different can. Once again, the solution obtained in the illustration is better than the original solution. Although it may not happen in all the instances of a mutation operation, mutating a string with a small probability is not a random operation since the process has a bias for creating only a few solutions in the search space.

22 (10, 6) 0 1 0 1 0 0 0 1 1 0 → 0 1 0 0 0 0 0 1 1 0 (8, 6) 16

Fig. 2.7. The bit-wise mutation operator. The fourth bit is mutated to create a new string.

2.4.3 A Hand Calculation

The working principle described above is simple, with GA operators involving string copying and substring exchange, plus the occasional alteration of bits. Indeed, it is surprising that with such simple operators and mechanisms, a potential search is possible. We will try to give an intuitive answer to such doubts and also remind the reader that a number of studies have attempted to find a rigorous mathematical {convergence proof} for GAs (Rudolph, 1994; Vose, 1999; Whitley, 1992). Even though the operations are simple, GAs are highly nonlinear, massively multi-faceted, stochastic and complex. There exist studies using Markov chain analysis which involves deriving transition probabilities from one state to another and manipulating them to find the convergence time and solution. Since the number of possible states for a reasonable string length and population size becomes unmanageable even with the high-speed computers available today, other analytical techniques (statistical mechanics approaches and dynamical systems models) have also been used to analyze the convergence properties of GAs.

In order to investigate why GAs work, let us apply the GA (with the proportionate selection operator) for only one-cycle to a numerical maximization problem (Deb, 2001):

$$\left.\begin{array}{ll} \text{Maximize} & \sin(x), \\ \text{Variable bounds} & 0 \le x \le \pi. \end{array}\right\} \quad (2.5)$$

We will use five-bit strings to represent the variable x in the range $[0, \pi]$, so that the string (00000) represents the $x = 0$ solution and the string (11111) represents the $x = \pi$ solution. The other 30 strings are mapped in the range $[0, \pi]$ uniformly. Let us also assume that we use a population of size four, the proportionate selection, the single-point crossover operator with $p_c = 1$, and no mutation (or, $p_m = 0$). To start the GA simulation, we create a random initial population, evaluate each string, and then use three GA operators, as shown in Table 2.1. The first string has a decoded value equal to 9 and this string corresponds to a solution $x = 0.912$, which has a function value equal to $sin(0.912) = 0.791$. Similarly, the other three stings are also evaluated. Since the proportionate reproduction scheme assigns a number of copies according to a string's fitness, the expected number of copies for each string is calculated in column 5. When the proportionate selection operator is actually implemented, the number of copies allocated to the strings is shown in column 6. Column 7 shows the mating pool. It is noteworthy that the third string in the initial population has a fitness which is very small compared to the average fitness of the population and is eliminated by the selection operator. On the other hand, the second string, being a good string, made two copies in the mating pool. The crossover sites are chosen at random and the four new strings created after crossover is shown in column 3 of the bottom table. Since no mutation is used, none of the bits are altered. Thus, column 3 of the bottom table represents the population at the end of one cycle of a GA. Thereafter, each of these stings is then decoded, mapped and evaluated. This completes one generation of a GA simulation. The average fitness of the new population is found to be 0.710, i.e. an improvement from that in the initial population. It is interesting to note that even though all operators used random numbers, a GA with all three operators produces a directed search, which usually results in an increase in the average quality of solutions from one generation to the next.

2.4.3.1 Understanding How GAs work

The string copying and substring exchange are all interesting and seem to improve the average performance of a population, but let us now ask the question: 'What has been processed in one cycle of a GA?' If we investigate carefully, we observe that among the strings of the two populations there are some similarities in the string positions among the strings. By the application of three GA operators, the number of strings with similarities at certain string positions has been increased from the initial population to the new population. These similarities are called *schema* in the GA literature. More specifically, a schema represents a set of strings with certain similarities at certain string positions. To represent a schema for binary codings, a triplet (1, 0 and *) is used; a * represents either 1 or 0. It is interesting to note that a string is also a schema representing only one string -- the string itself.

Table 2.1. One generation of a GA hand-simulation on the function sin(x).

Initial population						
String	DV[a]	x	$f(x)$	f_i/f_{avg}	AC[b]	Mating pool
01001	9	0.912	0.791	1.39	1	01001
10100	20	2.027	0.898	1.58	2	10100
00001	1	0.101	0.101	0.18	0	10100
11010	26	2.635	0.485	0.85	1	11010
	Average, f_{avg}		0.569			

Mating Pool	CS[c]	New population			
		String	DV[a]	x	$f(x)$
01001	3	01000	8	0.811	0.725
10100	3	10101	21	2.128	0.849
10100	2	10010	18	1.824	0.968
11010	2	11100	28	2.838	0.299
			Average, f_{avg}		0.710

[a] DV, decoded value of the string.
[b] AC, actual count of strings in the population.
[c] CS, cross site.

Two definitions are associated with a schema. The *order* of a schema H is defined as the number of defined positions in the schema and is represented as $o(H)$. A schema with full order $o(H) = l$ represents a string. The *defining length* of a schema H is defined as the distance between the outermost defined positions. For example, the schema H = (* 1 0 * * 0 * * *}) has an order $o(H)$ = 3 (there are three defined positions: 1 at the second gene, 0 at the third gene, and 0 at the sixth gene) and a defining length $\delta(H) = 6 - 2 = 4$.

A schema H_1 = (1 0 * * *) represents eight strings with a 1 in the first position and a 0 in the second position. From Table 2.1, we observe that there is only one string representing this schema H_1 in the initial population and that there are two strings representing this schema in the new population. On the other hand, even though there was one representative string of the schema H_2 = (0 0 * * *) in the initial population, there is not one in the new population. There are a number of other schemata that we may investigate and conclude whether the number of strings they represent is increased from the initial population to the new population or not.

The so-called schema theorem provides an estimate of the growth of a schema H under the action of one cycle of the above tripartite GA. Holland (1975) and later Goldberg (1989) calculated the growth of the schema under a selection operator and then calculated the survival probability of the schema under crossover and mutation operators, but did not calculate the probability of constructing a schema from recombination and mutation operations in a generic sense. For a single-point crossover operator with a probability p_c, a mutation operator with a probability p_m, and the proportionate selection operator, Goldberg (1989) calculated the following lower bound on the schema growth in one iteration of a GA:

$$m(H,t+1) = m(H,t)\frac{f(H)}{f_{avg}}\left[1 - p_c\frac{\delta(H)}{l-1} - p_m o(H)\right], \tag{2.6}$$

where $m(H, t)$ is the number of copies of the schema H in the population at generation t, $f(H)$ is the fitness of the schema (defined as the average fitness of all strings representing the schema in the population), and f_{avg} is the average fitness of the population. The above inequality leads to the schema theorem (Holland, 1975), as follows.

> Short, low-order, and above-average schemata receive exponentially increasing number
>
> of trials in subsequent generations.

A schema represents a number of similar strings. Thus, a schema can be thought of as representing a certain region in the search space. For the above function, the schema H_1 = (1 0 * * *) represents strings with x values varying from 1.621 to 2.330 with function values varying from 0.999 to 0.725. On the other hand, the schema H_2 = (0 0 * * *) represents strings with x values varying from 0.0 to 0.709 with function values varying from 0.0 to 0.651. Since our objective is to maximize the function, we would like to have more copies of strings representing schema H_1 than H_2. This is what we have accomplished in Table 2.1 without having to count all of these competing schema and without the knowledge of the complete search space, but by manipulating only a few instances of the search space. Let us use the inequality shown in equation (2.6) to estimate the growth of H_1 and H_2. We observe that there is only one string (the second string) representing this schema, or $m(H_1, 0) = 1$. Since all strings are used in the crossover operation and no mutation is used, $p_c = 1.0$ and $p_m = 0$. For the schema H_1, the fitness $f(H_1) = 0.898$, the order $o(H_1) = 2$, and the defining length $\delta(H_1) = 1$. In addition, the average fitness of the population is $f_{avg} = 0.569$. Thus, we obtain from equation (2.6):

$$\begin{aligned} m(H_1,1) &= (1) - \frac{0.898}{0.569}\left[1-(1.0)\frac{1}{5-1} - (0.0)(2)\right], \\ &= 1.184. \end{aligned}$$

The above calculation suggests that the number of strings representing the schema H_1 must increase. We have two representations (the second and third strings) of this schema in the next generation. For the schema H_2, the estimated number of copies using equation (2.6) is $m(H_2, 1) \geq 0.133$. Table 2.1 shows that no representative string of this schema exists in the new population.

The schema H_1 for the above example has only two defined positions (the first two bits) and both defined bits are tightly spaced (very close to each other) and contain the possible near-optimal solution (the string (1 0 0 0 0) is the optimal string in this problem). The schemata that are short, low-order, and above-average are known as the *building blocks*. While GA operators are applied on a population of strings, a number of such building blocks in various parts along the string get emphasized, such as H_1 (which has the first two bits in common with the true optimal string) in the above example. Note that although H_2 is short and low-order, it is not an above-average schema. Thus, H_2 is not a building block. This is how GAs can emphasize different short, low-order and above-average schemata in the population. Once adequate number of such building blocks are present in a GA population, they get combined together due to the action of the GA operators to form bigger and better building blocks. This process finally leads a GA to find the optimal solution. This hypothesis is known as the *Building Block Hypothesis* (Goldberg, 1989). More rigorous convergence analyses of GAs exist (Prugel-Bennett and Rogers, 2001; Rudolph, 1994; Shapiro, 2001; Vose, 1999).

2.4.4 Constraint Handling

As outlined elsewhere (Michalewicz and Schoenauer, 1996), most constraint handling methods which exist in the GA literature can be classified into five categories, as follows:

1. Methods based on preserving feasibility of solutions.
2. Methods based on penalty functions.
3. Methods biasing feasible over infeasible solutions.
4. Methods based on decoders.
5. Hybrid methods.

However, in most applications, the penalty function method has been used with GAs. Usually, an exterior penalty term (Deb, 1995, Reklaitis et al., 1983), which penalizes infeasible solutions, is preferred. Based on the constraint violation $g_j(\mathbf{x})$ or $h_k(\mathbf{x})$, a bracket-operator penalty term is added to the objective function and a penalized function is formed:

$$F(x) = f(x) + \sum_{j=1}^{J} R_j \langle g_j(x) \rangle + \sum_{k=1}^{K} r_k \left| h_k(x) \right|, \tag{2.7}$$

where R_j and r_k are user-defined penalty parameters. The bracket-operator ‹ › denotes the absolute value of the operand, if the operand is negative. Otherwise, if the operand is non-negative, it returns a value of zero. Since different constraints may take different orders of magnitude, it is essential to normalize all constraints before using the above equation. A constraint $g_j(\mathbf{x}) \geq b_j$ can be normalized by using the following transformation:

$$g_j(x) \equiv \underline{g_j}(x)/b_j - 1 \geq 0.$$

Equality constraints can also be normalized similarly. Normalizing constraints in the above manner has an additional advantage. Since all normalized constraint violations take more or less the same order of magnitude, they all can be simply added as the overall constraint violation and thus only one penalty parameter R will be needed to make the overall constraint violation of the same order as the objective function:

$$F(x) = f(x) + R\left[\sum_{j=1}^{J} \langle g_j(x)\rangle + \sum_{k=1}^{K} |h_k(x)|\right]. \tag{2.8}$$

When an EA uses a fixed value of R in the entire run, the method is called the *static* penalty method. There are two difficulties associated with this static penalty function approach:

1. The optimal solution of $F(\mathbf{x})$ depends on penalty parameters R. Users usually have to try different values of R to find which value would steer the search towards the feasible region. This requires extensive experimentation to find any reasonable solution. This problem is so severe that some researchers have used different values of R depending on the level of constraint violation (Homaifar et al., 1994), while some have used a sophisticated temperature-based evolution of penalty parameters through generations (Michalewicz and Attia, 1994) involving a few parameters describing the rate of evolution. We will discuss these *dynamically* changing penalty methods a little later.
2. The inclusion of the penalty term *distorts* the objective function (Deb, 1995). For small values of R, the distortion is small, but the optimum of $F(\mathbf{x})$ may not be near the true constrained optimum. On the other hand, if a large R is used, the optimum of $F(\mathbf{x})$ is closer to the true constrained optimum, but the distortion may be so severe that $F(\mathbf{x})$ may have artificial locally optimal solutions. This primarily happens due to interactions among multiple constraints. EAs are not free from the distortion effect caused due to the addition of the penalty term in the objective function. However, EAs are comparatively less sensitive to distorted function landscapes due to the stochasticity in their operators.

A recent study (Deb, 2000) suggested a modification, which eliminates both the above difficulties by *not* requiring any penalty parameter:

$$F(x) = \begin{cases} f(x), & \text{if } \mathbf{x} \text{ is feasible;} \\ f_{\max} + \sum_{j=1}^{J}\langle g_j(x)\rangle + \sum_{k=1}^{K}|h_k(x)|, & \end{cases} \tag{2.9}$$

Here, f_{max} is the objective function value of the worst feasible solution in the population. Figure 2.8 shows the construction procedure of $F(\mathbf{x})$ from $f(\mathbf{x})$ and $g(\mathbf{x})$ for a single-variable objective function. One fundamental difference between this approach and the previous approach is that the objective function value is *not* computed for any infeasible solution. Since all feasible solutions have zero constraint violation and all infeasible solutions are evaluated according to their constraint violations only, both the objective function value and constraint violation are not combined in any solution in the population. Thus, there is no

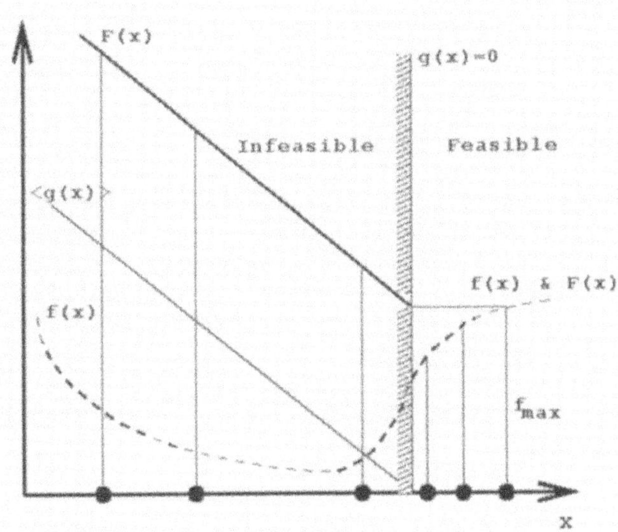

Fig. 2.8. A constraint handling strategy without any penalty parameter.

need to have any penalty parameter R for this approach. Moreover, the approach is also quite pragmatic. Since infeasible solutions are not to be recommended for use, there is no real reason for one to find the objective function value for an infeasible solution. The method uses a binary tournament selection operator, where two solutions are compared at a time, and the following scenarios are always assured:

1. Any feasible solution is preferred to any infeasible solution.
2. Among two feasible solutions, the one having a better objective function value is preferred.
3. Among two infeasible solutions, the one having a smaller constraint violation is preferred.

To steer the search towards the feasible region and then towards the optimal solution, the method recommends the use of a niched tournament selection (where two solutions are compared in a tournament only if their Euclidean distance is within a pre-specified limit). This ensures that even if a few isolated solutions are found in the feasible space, they will be propagated from one generation to another for maintaining diversity among the feasible solutions.

In the following section, we present an engineering case study.

2.5 Car Suspension Design Using Genetic Algorithms

Suspension systems are primarily used in a car to isolate the road excitations from being transmitted directly to the passengers. Although different types of suspension systems exist, we consider here the independent suspension system. We first consider the two-dimensional model of the suspension system and later show optimal solutions for the three-dimensional model.

2.5.1 Two-dimensional model

In a two-dimensional model of a car suspension system, only two wheels (one each at rear and front) are considered. Thus, the sprung mass is considered to have vertical and pitching motions only. The dynamic model of the suspension system is shown in Figure 2.9. For more information, refer to the detailed study (Deb and Saxena, 1997). The following nomenclature is used in the design formulation.

Sprung mass m_s, Front coil stiffness
k_{fs},
Front unsprung mass m_{fu}, Rear coil stiffness
k_{rs},
Rear unsprung mass m_{ru}, Front tyre stiffness
k_{ft},
Rear damper coefficient $\square_r$, Rear tyre stiffness
k_{rt},
Front damper coefficient $\square_f$, Axle-to-axle dis-
tance L,
Polar moment of inertia of the car J.

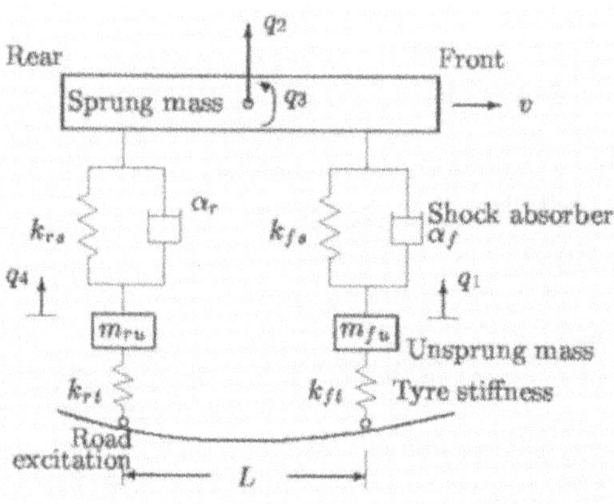

Fig. 2.9. The dynamic model of the car suspension system.

Since a suspension designer is interested in choosing the optimal dampers and suspension coils, we consider only four of the above parameters--front coil stiffness k_{fs}, rear coil stiffness k_{rs}, front damper coefficient α_f, and rear damper coefficient α_r --- as design variables. Considering the forces acting on the sprung mass and on the front and rear unsprung mass, we write the differential equations governing the vertical motion of the unsprung mass at the front axle q_1, the sprung mass q_2, and the unsprung mass at the rear axle q_4, and the angular motion of the sprung mass q_3 as follows (Deb, 1995):

$$\ddot{q}_1 = (F_2 + F_3 - F_1)/m_{fu}, \tag{2.10}$$

$$\ddot{q}_2 = -(F_2 + F_3 + F_4 + F_5)/m_s, \tag{2.11}$$

$$\ddot{q}_3 = [(F_4 + F_5)l_2 - (F_2 + F_3)l_1]/J, \tag{2.12}$$

$$\ddot{q}_4 = (F_4 + F_5 - F_6)/m_{ru}, \tag{2.13}$$

where the parameters l_1 and l_2 are the horizontal distance of the front and rear axle from the center of gravity of the sprung mass. Forces F_1 to F_6 are calculated as follows:

$$F_1 = k_{ft} d_1, \qquad F_2 = k_{fs} d_2, \qquad F_3 = \alpha_f \dot{d}_2,$$

$$F_4 = k_{rs} d_4, \qquad F_5 = \alpha_r \dot{d}_4, \qquad F_6 = k_{rt} d_3.$$

The parameters d_1, d_2, d_3, and d_4 are the relative deformations in the front tyre, the front spring, the rear tyre, and the rear spring, respectively. Figure 2.9 shows all the four degrees-of-freedom of the above system (q_1 to q_4). The relative deformations in springs and tyres can be written as follows:

$$d_1 = q_1 - f_1(t), \qquad d_2 = q_2 + l_1 q_3 - q_1,$$

$$d_3 = q_4 - f_2(t), \qquad d_4 = q_2 - l_2 q_3 - q_4.$$

The functions $f_1(t)$ and $f_2(t)$ are road excitations as functions of time in the front and rear tyre, respectively. For example, a bump can be modeled as $f_1(t) = A \sin(\pi t/T)$, where A is the amplitude of the bump and T the time required to cross the bump. When a car is moving forward, the front wheel experiences the bump first, while the rear wheel experiences the same bump a little later, depending on the axle-to-axle distance, L, and the speed of the car, v. Thus, the function $f_2(t)$ can be written as $f_2(t) = f_1(t-L/v)$.

The coupled differential equations specified in Equations (2.10) to (2.13) can be solved using a numerical integration technique to obtain the pitching and bouncing dynamics of the sprung mass m_s. At first, we use the bouncing transmissibility--the ratio of the bouncing amplitude $|q_2(t)|$ of the sprung mass to the maximum road excitation amplitude, A---as the objective of the design; later we shall use a more practical objective function. A practical guideline, often used in automobile industries, is to limit the maximum allowable vertical jerk experienced by the passengers. Another practical consideration is the piece-wise linear variation of spring and damper characteristics with displacement and velocity (Figure 2.10). For simplicity, we consider k_{rs}^a (when the deflection is between zero and a value δ) as the variable and assume other two spring rates, k_{rs}^b (when the deflection is more than δ) and k_{rs}^c (when the deflection is negative), as to vary in a fixed proportion with respect to k_{rs}^a. Dampers at front and rear are also considered to vary accordingly. However, the front suspension spring is assumed to vary linearly, as commonly followed in automobile industries. The following parameters of the car suspension system are used in all simulations:

$m_s = 730$ kg,	$m_{fu} = 50$ kg,	$m_{ru} = 115$ kg,
$k_{ft} = 15$ kg/mm,	$k_{rt} = 17$ kg/mm,	$L = 2.85$ m,
$l_1 = 1.50$ m,	$l_2 = 1.35$ m,	
$v = 5$ Kmph,	$J = 2.89(10^4)$ kg-m^2.	

The car motion is simulated over a sinusoidal bump having 500 mm width and 70 mm height. In all solutions, the spring rates are expressed in Kg/mm and damping coefficients are in Kg-s/mm. The following GA parameters are used:

Overall string length	: 40 (10 for each variable)
Population size	: 30
Crossover probability	: 0.8
Mutation probability	: 0.01

A fixed penalty parameter of $R = 100$ is used. With above parameters, the binary-coded GA finds the following solution:

$$k_{fs} = 4.53, \quad \alpha_f^a = 1.72, \quad k_{rs}^a = 2.86, \quad \alpha_r^a = 1.01.$$

The bouncing transmissibility for this solution is 0.315. The existing design for a car having identical data (as used in a renowned Indian automobile industry) is as follows:

$$k_{fs} = 1.56, \quad \alpha_f^a = 3.30, \quad k_{rs}^a = 1.45, \quad \alpha_r^a = 1.00.$$

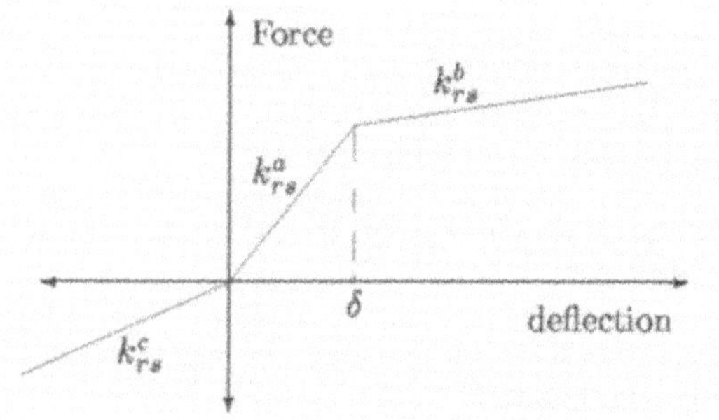

Fig. 2.10. Variation of rear spring rate with deflection.

The bouncing transmissibility for this existing design is 0.82. Comparing these two solutions, we notice that GA-optimized design has 64% lesser transmissibility than that in the existing design. The maximum jerk for this suspension is found to be 5.094 m/s^3, whereas the allowable limit is 18.0 m/s^3. To better appreciate the GA-optimized design, we plot the bouncing amplitude of the sprung mass, as the car moves over the bump in Figure 2.11.

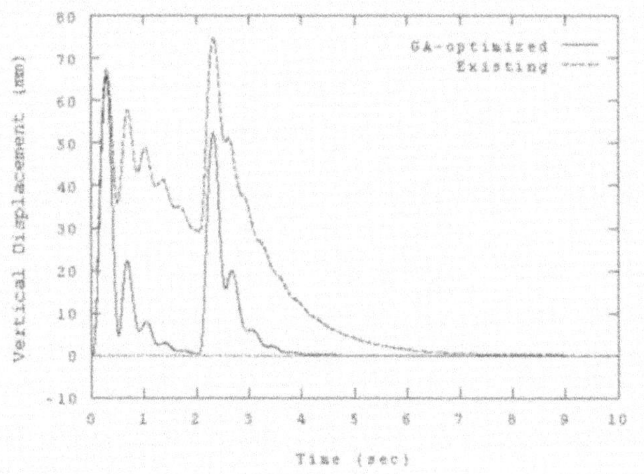

Fig. 2.11. Bouncing amplitude of the sprung mass as the car moves over a bump for existing and GA-optimized suspension.

2.5.2 Three-dimensional model

We now consider the three-dimensional model and introduce jerk and a set of frequency constraints to make the design procedure more realistic. Automobile industries design cars having front natural frequencies smaller than the rear natural frequencies. This is achieved to make the pitching oscillations die down faster. In this model, all four wheels are considered. Thus, the rolling motion of the sprung mass can also be studied. The sprung mass can have three motions--vertical bouncing, pitching, and rolling. Besides, each of the four unsprung masses will have a vertical motion. Thus, there are a total of seven second-order differential equations governing the motion of the sprung and unsprung masses, which can be derived using similar methods adopted in two-dimensional model. Since both left and right rear (or front) wheels have the same suspension system, the number of variables in three-dimensional optimal design model is also four. However, the dynamics of the car will be different than that in the two-dimensional case. Here, there are 14 nonlinear differential equations which are solved using a numerical integration procedure and all motions can be computed for a specified road profile.

In this model, m_s = 1,460 Kg and the wheel-to-wheel distance is 1.462 m are used. To make the problem more realistic, a polyharmonic road profile is used:

$$f(t) = \sum_{\lambda=\lambda_1}^{\lambda_n} A(\lambda)\sin(2\pi vt/\lambda - \phi). \tag{2.14}$$

The parameters $A(\lambda)$ is the amplitude of the road roughness which varies linearly with the wavelength (Demic, 1989):

$$A(\lambda) = A_0 + A_1\lambda_1, \tag{2.15}$$

where A_0 and A_1 are the coefficients which depend on the type of the road. For asphalt road, they are $A_0 = 6.44(10^{-4})$ and $A_1 = 3.14(10^{-5})$. The phase angle for the sine wave of a particular wavelength is randomly chosen between $-\pi$ to π, as follows:

$$\phi = 2\pi(RND - 0.5), \tag{2.16}$$

where *RND* is a random number between 0 and 1. In this study, we choose wavelengths to vary between 100 mm to 5000 mm. The velocity of the car is assumed to be 50 Kmph. ISO 2631 limits the extent of vibrations on different frequency levels which are allowed for different levels of comfort. The vertical motion (q_2) of the sprung mass is simulated by solving the governing equations of motion for the above mentioned realistic road. Thereafter, the vertical acceleration ($\ddot{q}_2$) is calculated by numerically differentiating the vertical motion of the sprung mass. The time-acceleration data is then Fourier-transformed to calculate the vertical acceleration as a function of the forcing frequency. The total area under the acceleration-frequency plot is used as the objective to be minimized.

The same GA parameters are used and the following suspension has been obtained:

$$k_{fs} = 1.45, \qquad \alpha_f^a = 0.14, \qquad k_{rs}^a = 1.28, \qquad \alpha_r^a = 0.11.$$

The front and rear natural frequencies of the suspension system are found to be 1.35 and 1.37 Hz. It is clear that the front natural frequency is smaller than that of the rear. This solution is close to the optimal solution since this design makes the front natural frequency almost equal to rear natural frequency, thereby making the constraint active. Moreover, the GA-optimized solution having smaller damping coefficients is also justified from the vibration theory--for a forcing frequency larger than the natural frequency, a suspension having a lower damping coefficient experiences a smaller transmissibility (Tse et al., 1983). The GA-optimized design is compared with that of the existing design by plotting the acceleration-frequency diagram in Figures 2.12 and 2.13. The figures show that the acceleration amplitude for the GA-optimized suspension is about an order of magnitude smaller. When the peak acceleration at significant frequencies are plotted on the ISO 2631 chart (Figure 2.14), it is observed that the comfortable exposure time for the existing design is only about 25 minutes, whereas that for the GA-optimized design is about 4 hours, thus giving a much longer comfortable ride.

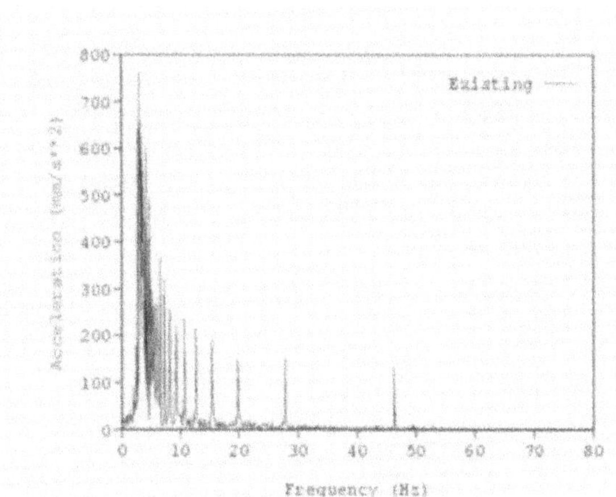

Fig. 2.12. Acceleration as a function of frequency for the existing design.

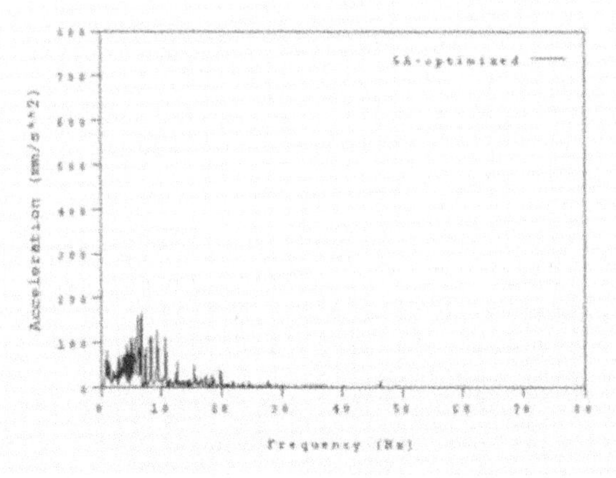

Fig. 2.13. Acceleration as a function of frequency for the GA-optimal design.

2.6 Real-Parameter Genetic Algorithms

In the real-coded GA, the problem variables are represented directly. Thus, a typical solution is a vector of real values, as follows:

$$(x_1 \ x_2 \ x_3 .... \ x_n)$$

In solving continuous search space problems using the binary-coded GA, there may exist a number of difficulties. First of all, since binary-coded GA discretizes the search space by using a finite-length binary string, getting arbitrary precision is a problem. Secondly, binary string coding introduces the so-called Hamming cliffs (nonlinearities) into the problem. Thirdly, the artificial string coding to problem variables causes design engineers and practitioners to have a hindrance in the understanding of GA's working

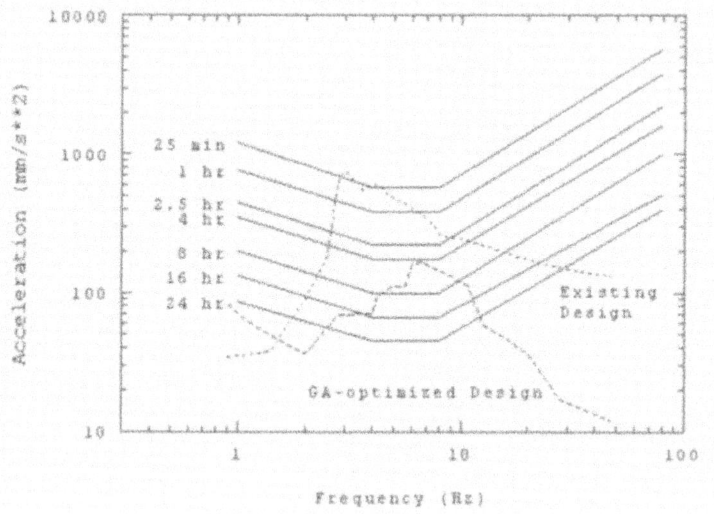

Fig. 2.14. Exposure times for existing and GA-optimized designs using ISO 2631 chart.

principles. The direct use of variables in the real-coded GA eliminates all difficulties stated above and makes the GA approach more user-friendly.

Since the problem variables are initialized directly, they can be used to calculate the fitness value. However, the crossover and mutation operators are different than that used in the binary-coded GA. Without going into the details, we first describe the crossover operator and then outline the mutation operator.

The crossover operator is applied variable-by-variable with a probability 0.5. If the i-th variable is to be crossed between two parents $x^{(t)}$ and $x^{(t+1)}$, we use the following simulated binary crossover (SBX) procedure to calculate two new children solutions $y^{(t)}$ and $y^{(t+1)}$ (Deb and Agrawal, 1995):

1. Create a random number u between 0 and 1.
2. Find a non-dimensionalized spread factor $\overline{\beta}$ for which

$$\int_0^{\bar{\beta}} p(\beta)d\beta = u. \tag{2.17}$$

The probability distribution $p(\beta)$ is given as follows:

$$p(\beta) = \begin{cases} 0.5(\eta_c + 1)\beta^{\eta_c}, & if\ \beta \le 1; \\ 0.5(\eta_c + 1)\dfrac{1}{\beta^{\eta_c+2}}. & if\ \beta > 1. \end{cases} \tag{2.18}$$

The non-negative factor β is defined as follows:

$$\beta = \left| \frac{y^{(t+1)} - y^{(t)}}{x^{(t+1)} - x^{(t)}} \right|.$$

The probability distribution $p(\beta)$ allows a large probability for creating children solutions near the parent solutions and less probability for solutions away from parent solutions. Another nice aspect of this crossover is that the same probability distribution can be used for any two parent solutions. The non-dimensionalized factor β takes care of this aspect.

3. The children solutions are calculated as follows:

$$y^{(t)} = 0.5\left[(x^{(t)} + x^{(t+1)}) - \bar{\beta}\left|x^{(t+!)} - x^{(t)}\right|\right]$$
$$y^{(t+1)} = 0.5\left[(x^{(t)} + x^{(t+1)}) + \bar{\beta}\left|x^{(t+1)} - x^{(t)}\right|\right]$$

It is clear that the distribution depends on the exponent η_c. For small values of η_c, solutions far away from the parents are likely to be chosen, whereas for large values of η_c, only solutions closer to the parents are likely to be chosen. Ideally, a good search algorithm must have a broad search (with a small η_c) in early generations and as the generations proceed the search must be focused on to a narrow region (with a large η_c) to obtain better precision in the solution. However, for brevity, we use a fixed compromised η_c = 2 for an entire simulation in this study. Figure 2.15 shows the probability distribution of creating offspring solutions in the case of variables with and without variable bounds. If a variable can take any value in the real space, the distribution shown in solid lines would be used and for variables with specific bounds, the distribution shown in dashed line would be used. The latter distribution is calculated by distributing the cumulative probability from (-∞, $x_i^{(L)}$]$ and [$x_i^{(U)}$, ∞) in the respective range.

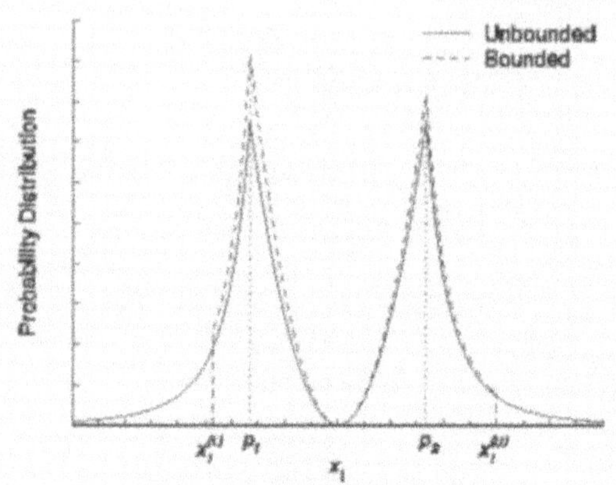

Fig. 2.15. Probability distribution for creating children solutions in continuous variables with and without variable bounds.

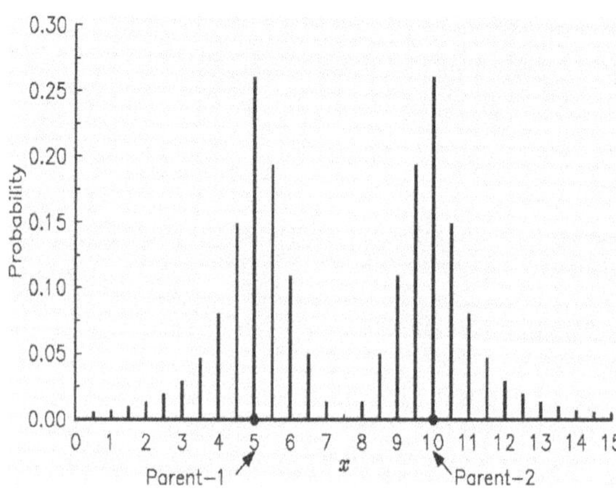

Fig. 2.16. Probability distribution for creating children solutions of discrete variables.

In order to handle discrete variables having arbitrary number of permissible values, a discrete version of the above probability distribution (equation 2.18) can be used. Figure 2.16 shows such a probability distribution, which creates only permissible values of the design variable.

In order to implement the mutation operator, the current value of a problem variable is changed to a neighboring value using a polynomial probability distribution having its mean at the current value and its variance as a function of the distribution index η_m. To perform mutation, a perturbance factor δ is defined as follows:

$$\delta = \frac{c-p}{\Delta_{\max}}, \tag{2.19}$$

where $\Delta_{\max}$ is a fixed quantity, representing the maximum permissible perturbance in the parent value p and c is the mutated value. Similar to the crossover operator, the mutated value is calculated with a probability distribution that depends on the perturbance factor δ:

$$p(\delta) = 0.5(\eta_m + 1)(1 - |\delta|)^{\eta_m}. \tag{2.20}$$

The above probability distribution is valid in the range $\delta \in (-1, 1)$. To create a mutated value, a random number u is created in the range (0, 1). Thereafter, the following equation can be used to calculate the perturbance factor $\bar{\delta}$ corresponding to u using the above probability distribution:

$$\bar{\delta} = \begin{cases} (2u)^{\frac{1}{\eta_m+1}} - 1, & if \quad u < 0.5; \\ 1 - [2(1-u)]^{\frac{1}{\eta_m+1}}, & if \quad u \geq 0.5. \end{cases} \tag{2.21}$$

Thereafter, the mutated value is calculated as follows:

$$c = p + \bar{\delta}\Delta_{\max}. \tag{2.22}$$

2.7 A Combined Genetic Algorithm

Based on the above two GA implementations, a combined GA can be developed to solve mixed-integer programming problems where some variables are discrete and some are continuous. They can be represented by using a mixed coding discrete variables can be coded using a binary substring (alternatively, discrete values can be directly used with a discrete version of SBX operator) and continuous variables can be used directly. Thereafter, mixed GA operators can be used to create new solutions. The variables represented using a binary substring, the single-point crossover can be used and the variables used directly, the real-coded crossover can be used. Similarly, appropriate mutation operator can also be used. This way, the best of both GAs can be combined and a flexible yet efficient GA can be developed. To illustrate the concept of a mixed coding, we consider the design of a cantilever beam having four variables:

((1) 14 23.457 (101))

The first variable (zero-one) can take only one of two values (either 1 or 0). The value 1 represents a circular cross-section and the value 0 represents a square cross-section of the cantilever beam. The second variable represents the diameter of the circular section if the first variable is a 1 or the side of the square if the first variable is a 0. This variable is discrete, since arbitrary section sizes are not usually available for fabrication. The third variable represents the length of the cantilever beam. Thus, it could be a continuous variable. The fourth variable is a discrete variable representing the material, which can take one of 8 pre-specified materials. For simplicity, a set of three binary digits (totaling 2^3 or 8 values) may be used to code this variable. Thus, the above string represents a cantilever beam made of the 5-th material from a prescribed list of 8 materials having a circular cross-section with a diameter 14 mm and having a length of 23.457 mm. With the above flexibility in design representation, any combination of cross sectional shape and size, material specifications, and length of the cantilever beam can be represented. This flexibility in the representation of a design solution is not possible with traditional optimization methods. Since each design variable is allowed to take only permissible values (no hexagonal shape will be tried or no unavailable diameter of a circular cross-section will be chosen), the computational time for searching of the optimal problem is also expected to be substantially less.

With such a representation scheme, a standard reproduction operator can still be used. However, for crossover and mutation operations, the respective genetic operator described above can be suitably used. For real variables, the SBX and the polynomial mutation operator and for integer variables represented by binary substrings, standard binary crossover and mutation operators can be used. In the following, we show a couple of engineering case studies.

2.7.1 Gear Train Design

A compound gear train is to be designed to achieve a specific gear ratio between the driver and driven shafts (Figure 2.17). The objective of the gear train design is to find the number of teeth in each of the four gears so as to minimize the error between the obtained gear ratio and a required gear ratio of 1/6.931 (Kannan and Kramer, 1993). All four variables are strictly integers. By denoting the variable vector $\boldsymbol{x} = (x_1, x_2, x_3, x_4) = (T_d, T_b, T_a, T_f)$, we write the NLP problem:

$$\text{Maximize} \quad f(x) = \left[\frac{1}{6.931} - \frac{x_1 x_2}{x_3 x_4} \right]^2,$$

$$\text{Variable bounds} \quad 12 \le x_1, x_2, x_3, x_4 \le 60,$$

$$\text{All } x_i\text{'s are integers.}$$

The variables are coded in the range (12, 60). We have used two different coding schemes. In Case I, the variables are coded directly allowing only integer values. In Case II, each variable can be coded in six-bit binary strings (having 2^6 or 64 values), so that variables take values between 12 and 75 (integer values) and four constraints ($60 - x_i \ge 0$, for $i = 1,2,3,4$) are added to penalize infeasible solutions. Table 2.2 shows the best solutions found in each case with a population of 50 solutions.

It is clear from the table that both runs with GeneAS have found much better solutions than the previously known solutions. By computing all possible gear teeth combinations (49^4 or about 5.76 million), it has been observed that the solution obtained by Case I is, in fact, the globally best solution.

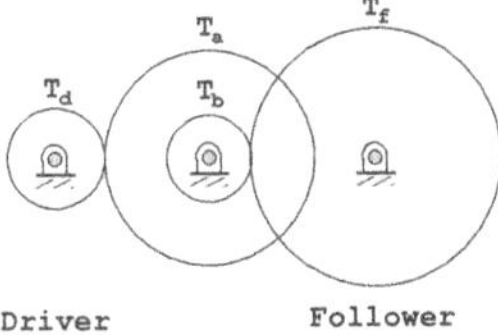

Fig. 2.17. A compound gear train.

Table 2.2. Optimal gear design solutions.

Design Variables	Obtained Solution			
	Case-I	Case-II	AL	BB
T_d	19	17	13	18
T_b	16	14	15	22
T_a	49	33	33	45
T_f	43	50	41	60
f	$2.7\times(10^{-12})$	$1.4\times(10^{-09})$	$2.1\times(10^{-08})$	$5.7\times(10^{-06})$
Error	0.001%	0.026%	0.11%	1.65%

2.8 A Spring Design

A helical compression spring needs to be designed for the minimum volume. Three variables are used: The number of coils N, the wire diameter d, and the mean coil diameter D. Of these variables, N is an integer variable, d is a discrete variable having 42 non-equispaced values and D is a continuous variable. A typical representation of a design solution is shown below:

$$\underbrace{(01101)}_{N}\ \underbrace{0.1480}_{d}\ \underbrace{1.563}_{D}$$

Denoting the variable vector $x = (x_1, x_2, x_3) = (N, d, D)$, we write the NLP problem:

$$\text{Min} \quad f(x) = 0.25\pi^2 x_2^2 x_3 (x_1 + 2)$$

$$\text{With} \quad g_1(x) = S - \frac{8KP_{\max} x_3}{\pi x_2^3} \geq 0,$$

$$g_2(x) = l_{max} - \frac{P_{max}}{k} - 1.05(x_1 + 2)x_2 \geq 0,$$

$$g_3(x) = x_2 - d_{min} \geq 0,$$

$$g_4(x) = D_{max} - (x_2 + x_3) \geq 0,$$

$$g_5(x) = D/d - 3 \geq 0,$$

$$g_6(x) = \delta_{pm} - \delta_p \geq 0,$$

$$g_7(x) = l_f - \frac{P_{max}}{k} - 1.05(N + 2)d \geq 0,$$

$$g_8(x) = \frac{P_{max} - P}{k} - \delta_w \geq 0,$$

x_1 is integer, x_2 is discrete, x_3 is continuous.

The parameters used above are as follows:

$K = \frac{4c-1}{4c-4} + \frac{0.615x_2}{x_3}$	$P = 300\text{lb}$	$D_{max} = 3\text{in}$
$k = \frac{Gx_2^4}{8x_1x_3^3}$	$P_{max} = 1000\text{lb}$	$\delta_w = 1.25\text{in}$
$\delta_p = \frac{P}{k}$	$\ell_{max} = 14\text{in}$	$\delta_{pm} = 6\text{in}$
$S = 0.189\text{Mpsi}$	$d_{min} = 0.2\text{in}$	$G = 11.5\text{Mpsi}$

The free length l_f is so chosen that the constraint g_7 becomes an equality constraint. The variable N is coded in five-bit strings so that the feasible values of N are integers between 1 to 32. The variable d is coded as a discrete variables coded directly. The variable D takes any real value. The flexibility of GeneAS allows these three different types of variables to be coded naturally. An initial population of 60 random solutions are created with a 5-bit binary string for N, a feasible discrete value for d, and a value of D in the range (1.0, 30.0) inch. However, any real value for D is allowed in subsequent iterations. Table 2.3 once again shows that the solution obtained by GeneAS in this problem has outperformed previously reported optimal solutions. The solution obtained by GeneAS is about 5% better than that reported by Sandgren (1988). When the discrete (integer) variable N is coded directly, a marginally better solution has emerged: $N = 9$, $d = 0.283$, and $D = 1.224$ with $f(x) = 2.661$.

Table 2.3. Optimal spring design solutions.

Design Variables	Obtained Solution		
	GeneAS	AL	BB
N	9	7	10
d	0.283	0.283	0.283
D	1.226	1.329	1.180
g_1	713.510	$-10,172.356$	5500.080
g_2	8.933	9.543	8.652
g_3	0.083	0.083	0.083
g_4	1.491	1.390	1.537
g_5	1.337	1.700	1.170
g_6	5.461	5.400	5.465
g_7	0.000	0.000	0.000
g_8	0.009	0.003	0.003
f	2.665	(*) 2.364	2.798

(*) The solution in Kannan and Kramer (1993) is not feasible, as it violates constraint g_1.

2.9 Advanced Genetic Algorithms

There exist a number of extensions to the simple GA described above. Interested readers may refer to the GA literature for details:

Micro GA: A small population size (of the order of 4 or 5) is used (Krishnakumar, 1989). This GA solely depends on the mutation operator, since such a small population cannot take advantage of the discovery of good partial solutions by a selecto-recombination GA. However, for unimodal and simple problems, micro-GAs are good candidates. For problems where the function evaluations are expensive, many researchers have used micro-GAs with a small population size in the expectation of finding a reasonable solution.

Knowledge-augmented GA: GA operators and/or the initial population is assisted with problem knowledge, if available. In most problems, some problem information is available and generic GA operators mentioned in this paper can be modified to make the search process faster (Davidor, 1991; Deb, Reddy and Singh, 2002).

Hybrid GA: A classical greedy search operator is used starting from a solution obtained by a GA. Since a GA can find good regions in the search space quickly, using a greedy approach from a solution in the global basin may make the overall search effort efficient (Powell and Skolnick, 1989).

Multimodal GA: Due to the population approach, GAs can be used to find multiple optimal solutions in one simulation of a GA run. In such a multi-modal GA, only the reproduction operator needs to be modified. In one implementation, the raw fitness of a solution is degraded with its ***niche count***, an estimate of the number of neighboring solutions. It has been shown that if the reproduction operator is performed with the degraded fitness values, stable subpopulations can be maintained at various optima of the objective function (Deb and Goldberg, 1989; Goldberg and Richardson, 1987). This allows GAs to find multiple optimal solutions simultaneously in one single simulation run.

Multi-objective GA: Most real-world search and optimization problems involve multiple conflicting objectives, of which the user is unable to establish a relative preference. Such considerations give rise to a set of multiple optimal solutions, largely known as the Pareto-optimal solutions or inferior solutions (Deb, 2001). Multiple Pareto-optimal solutions are found simultaneously in a population. A GA is an unique optimization algorithm in solving multi-objective optimization problems in this respect. In one implementation, non-domination concept is used with all objective functions to determine a fitness measure for each solution. Thereafter, the GA operators described here are used as usual. On a number of multi-objective optimization problems, this non-dominated sorting GA has been able to find multiple Pareto-optimal solutions in one single run (Deb, 2001; Fonseca and Fleming, 1993; Horn et al., 1994).

Non-stationary GA: The concept of diploidy and dominance can be implemented in a GA to solve non-stationary optimization problems. Information about earlier good solutions can be stored in recessive alleles and when needed can be expressed by suitable genetic operators (Goldberg and Smith, 1987).

Scheduling GA: Job-shop scheduling, time tabling, traveling salesman problems are solved using GAs. A solution in these problems is a permutation of N objects (name of machines or cities). Although reproduction operator similar to one described here can be used, the crossover and mutation operators must be different. These operators are designed in order to produce offspring's which are valid and yet have certain properties of both parents (Goldberg, 1989; Starkweather, 1991).

2.10 Conclusions

In this paper, we have described a potential search and optimization algorithm originally conceived by John Holland of University of Michigan, Ann Arbor, about four decades ago, but now gained a lot of popularity. A genetic algorithm (GA) is different from other classical search and optimization methods in a number of ways: it does not use gradient information; it works with a set of solutions instead of one solution in each iteration; it works on a coding of solutions instead of solutions themselves; it is a stochastic search and optimization procedure; and it is highly parallelizable. GAs are finding increasing popularity primarily because of their wide spread applicability, global perspective, and inherent parallelism.

The efficacy of genetic algorithms is shown by illustrating a number of engineering case studies. The standard binary-coded genetic algorithm can be combined with a real-parameter implementation to develop a combined GA which is ideal for solving mixed-integer programming problems often encountered in engineering design activities. Besides, the flexibility of GAs in solving a number of other types of search and optimization problems, such as scheduling and multi-objective optimization problems, is the main attraction of their increasing popularity in various engineering problem solving.

References

Davidor, Y. (1989). Analogous crossover. *Proceedings of the Third International Conference on Genetic Algorithms*, 98--103.

Dawkins, R. (1986). *The Blind Watchmaker*. New York: Penguin Books.

Dawkins, R. (1976). *The Selfish Gene*. New York: Oxford University Press.

Deb, K, Reddy, A. R., and Singh, G. (2002). Optimal scheduling of casting sequence using genetic algorithms. *KanGAL Report No. 2002002*. Kanpur, India: Kanpur Genetic Algorithms Laboratory, India.

Deb, K. (2001). *Multi-objective optimization using evolutionary algorithms*. Chichester, UK: Wiley.

Deb, K. (2000). An efficient constraint handling method for genetic algorithms. *Computer Methods in Applied Mechanics and Engineering, 186*(2--4), 311--338.

Deb, K. (1999). An introduction to genetic algorithms. *Sadhana, 24*(4), 293--315.

Deb, K. and Saxena, V. (1997). Car suspension design for comfort using genetic algorithms. In Thomas Back (Ed.) *Proceedings of the Seventh International Conference on Genetic Algorithms*, (East Lansing, USA), 553--560.

Deb, K. (1995). *Optimization for engineering design: Algorithms and examples*. Delhi: Prentice-Hall.

Deb, K. (1993). Genetic algorithms in optimal optical filter design. In E. Balagurusamy and B. Sushila (Eds.), *Proceedings of the International Conference on Computing Congress* (pp. 29--36).

Deb, K. and Agrawal, R. B. (1995) Simulated binary crossover for continuous search space. *Complex Systems, 9* 115--148.

Deb, K. and Goldberg, D. E. (1989). An investigation of niche and species formation in genetic function optimization, *Proceedings of the Third International Conference on Genetic Algorithms*, pp. 42-50.

Demic, M. (1989). Optimization of characteristics of the elasto-damping elements of a passenger car by means of a modified Nelder-Mead method. *International Journal of Vehicle Design, 10*(2), 136--152.

Duffin, R. J., Peterson, E. L., and Zener, C. (1967). *Geometric Programming*. New York: Wiley.

Eldredge, N. (1989). *Macro-evolutionary Dynamics: Species, niches, and adaptive peaks*. New York: McGraw-Hill.

Fonseca, C. M. and Fleming P. J. (1993). Genetic algorithms for multi-objective optimization: Formulation, discussion and generalization. In S. Forrest (Ed.), *Proceedings of the Fifth International Conference on Genetic Algorithms* (pp. 416--423).

Gen, M. and Cheng, R. (1997). *Genetic Algorithms and Engineering Design*. New York: Wiley.

Goldberg, D. E. (1989). *Genetic algorithms in search, optimization, and machine learning*. New York: Addison-Wesley.

Goldberg, D. E. and Deb, K. (1991). A comparison of selection schemes used in genetic algorithms, Foundations *of Genetic Algorithms*, edited by G. J. E. Rawlins, pp. 69-93.

Goldberg, D. E. and Richardson, J. (1987). Genetic algorithms with sharing for multimodal function optimization. *Proceedings of the Second International Conference on Genetic Algorithms*, 41--49.

Goldberg, D. E. and Smith, R. (1987). Non-stationary function optimization using genetic algorithms with dominance and diploidy. In J. J. Grefenstette (Ed.) *Proceedings of the Second International Conference on Genetic Algorithms*. New Jersey: Lawrence Erlbaum Associates. (pp. 59--68).

Homaifar, A., Lai, S. H.-V. and Qi, X. (1994). Constrained optimization via genetic algorithms. *Simulation 62*(4), 242--254.

Holland, J. H. (1975). *Adaptation in natural and artificial systems*. Ann Arbor: University of Michigan Press.

Horn, J., Nafploitis, N. and Goldberg, D. (1994). A niched Pareto genetic algorithm for multi-objective optimization. In *Proceedings of the First IEEE Conference on Evolutionary Computation*, pp. 82--87.

Kannan, B. K. and Kramer, S. N. (1995). An augmented Lagrange multiplier based method for mixed integer discrete continuous optimization and its applications to mechanical design. *Journal of Mechanical Design, 116*, 405--411.

Krishnakumar, K. (1989). Microgenetic algorithms for stationary and non-stationary function optimization. *SPIE Proceedings on Intelligent Control and Adaptive Systems, 1196*, 289--296.

Michalewicz, Z. (1992). *Genetic Algorithms + Data Structures = Evolution Programs*. Berlin: Springer-Verlag.

Michalewicz, Z. and Attia, N. (1994). Evolutionary optimization of constrained problems. In *Proceedings of the Third Annual Conference on Evolutionary Programming*, pp. 98--108.

Michalewicz, Z. and Schoenauer, M. (1996). Evolutionary algorithms for constrained parameter optimization problems. *Evolutionary Computation, 4*(1), 1--32.

Mitchell, M. (1996). *Introduction to Genetic Algorithms*. Ann Arbor: MIT Press.

Powell, D. and Skolnick, M. M. (1993). Using genetic algorithms in engineering design optimization with nonlinear constraints. In S. Forrest (Ed.) Proceedings of the Fifth International Conference on Genetic Algorithms}, San Mateo, CA: Morgan Kaufmann (pp. 424--430).

Prugel-Bennett, A. and Rogers, A. (2001). Modelling genetic algorithm dynamics. In L. Kallel, B. Naudts, and A. Rogers (Eds.) *Theoretical Aspects of Evolutionary Computing*. Berlin, Germany: Springer, 59--85. Reklaitis, G. V., Ravindran, A., and Ragsdell, K. M. (1983). *Engineering optimization methods and applications*. New York: John Wiley and Sons.

Rudolph, G. (1994). Convergence analysis of canonical genetic algorithms. *IEEE Transactions on Neural Network*. (pp 96--101).

Sandgren, E. (1988). Nonlinear integer and discrete programming in mechanical design. *Proceedings of the ASME Design Technology Conference*, Kissimee, FL, 95--105.

Shapiro, J. (2001). Statistical mechanics theory of genetic algorithms. In L. Kallel, B. Naudts, and A. Rogers (Eds.) *Theoretical Aspects of Evolutionary Computing*. Berlin, Germany: Springer, 87--108.

Spears, W. M. and De Jong, K. A. (1991). An analysis of multi-point crossover. In G. J. E. Rawlins (Eds.), *Foundations of Genetic Algorithms* (pp. 310--315). (Also AIC Report No. AIC-90-014).

Starkweather, T., McDaniel, S., Mathias, K., Whitley, D., and Whitley, C. (1991). A comparison of genetic scheduling operators. In R. Belew and L. B. Booker (Eds.) *Proceedings of the Fourth International Conference on Genetic Algorithms*. San Mateo, CA: Morgan Kaufmann. (pp. 69--76).

Syswerda, G. (1989). Uniform crossover in genetic algorithms. In J. D. Schaffer (Ed.), *Proceedings of the Third International Conference on Genetic Algorithms* (pp. 2--9).

Tse, F. A., Morse, I. E. and Hinkle, R. T. (1983). *Mechanical vibrations: Theory and applications*. New Delhi: CBS Publications.

Vose, M. D. (1999). *Simple Genetic Algorithm: Foundation and Theory*. Ann Arbor, MI: MIT Press.

3 Memetic Algorithms

Pablo Moscato, Carlos Cotta and Alexandre Mendes

3.1 Introduction

Back in the late 60s and early 70s, several researchers laid the foundations of what we now know as *Evolutionary Algorithms* (EAs) (Fogel et al. 1966; Holland 1975; Rechenberg 1973; Schwefel 1965). In these almost four decades, and despite some hard beginnings, most researchers interested in search or optimization – both from the applied and the theoretical standpoints – have grown to know and accept the existence – and indeed the usefulness – of these techniques. This has been also the case for other related techniques, such as *Simulated Annealing* (SA) (Kirkpatrick et al. 1983), *Tabu Search* (TS) (Glover and Laguna 1997), etc. The name *metaheuristics* is used to collectively term these techniques.

It was in late 80s that the term '*Memetic Algorithms*' (MAs) (Moscato 1989) was given birth to denote a family of metaheuristics that tried to blend several concepts from tightly separated – at that time – families of metaheuristics such as EAs and SA. The adjective 'memetic' comes from the term 'meme', coined by R. Dawkins (Dawkins 1976) to denote an analogous to the *gene* in the context of cultural evolution. Quoting Dawkins:

"Examples of memes are tunes, ideas, catch-phrases, clothes fashions, ways of making pots or of building arches. Just as genes propagate themselves in the gene pool by leaping from body to body via sperms or eggs, so memes propagate themselves in the meme pool by leaping from brain to brain via a process which, in the broad sense, can be called imitation."

The above quote illustrates the central philosophy of MAs: individual improvement plus population cooperation and competition, as they are present in many social/cultural systems. As it was the case for classical EAs, MAs had to suffer tough initial times, but they are now warmly received in community, as the reader may check by taking a quick look at the number of recent articles on MAs we review at the end of this chapter. It is often the case that MAs are used under a different name ('hybrid EAs' and 'Lamarckian EAs' are two popular choices for this). Not surprisingly in a rapidly expanding field as this is, one can also find the term MA used in the context of particular algorithmic subclasses, arguably different from those grasped in the initial definition of MAs. This point will be tackled in next section; anticipating further definitions, we can say that a MA is a search strategy in which *a population of optimizing agents synergistically cooperate and compete* (Norman and Moscato 1989). A more detailed description of the algorithm, as well as a functional template will be given in Sect. 3.2.

As mentioned before, MAs are a hot topic nowadays, mainly due to their success in solving many hard optimization problems, like the Traveling Salesman Problem (Merz and Freisleben, 2002b). A particular feature of MAs is greatly responsible for this: unlike traditional Evolutionary Computation (EC) methods, MAs are intrinsically concerned with exploiting *all available knowledge* about the problem under study; this is something that was neglected in EAs for a long time, despite some contrary voices such as (Hart and Belew 1991), and most notably (Davis 1991). The formulation of the so-called *No-Free-Lunch Theorem* (NFL) by (Wolpert and Macready 1997) made it definitely clear that a search algorithm strictly performs in accordance with the amount and quality of the problem knowledge they incorporate, thus backing up one of the *leiv motivs* of MAs (Norman and Moscato 1989).

The exploitation of problem-knowledge can be accomplished in MAs by incorporating heuristics, approximation algorithms, local search techniques, specialized recombination operators, truncated exact methods, etc. Also, an important factor is the use of adequate representations of the problem being tackled. These issues are of the foremost interest from an applied viewpoint, and will be dealt in Sect. 3.3.

As important as the basic algorithmic considerations about MAs that will be presented below, a more applied perspective of MAs is also provided in Sect. 3.4. The reader may be convinced of the wide applicability of these techniques by inspecting the numerous research papers published with regard to the deployment of MAs on the most diverse domains. For the purposes of this book, we will pay special attention to the application of MAs in Engineering-related endeavors. This chapter will end with a brief summary of the current research trends in MAs, with special mention to those emerging application fields in which MAs are to play a major role in the near future.

3.2 The MA Search Template

As mentioned in the previous section, MAs try to blend together concepts from (one or more) different metaheuristics, such as EAs and SA for instance. Let us start by those ideas gleaned from the former.

MAs are – like EAs – population-based metaheuristics. This means that the algorithm maintain a *population* of solutions for the problem at hand, i.e., a pool comprising several solutions simultaneously. Each of these solutions is termed *individual* in the EA jargon, following the nature-inspired metaphor upon which these techniques are based. In the context of MAs, the denomination *agent* seems more appropriate for reasons that will be evident later in this section. When clear from the context, both terms will be used interchangeably.

Each individual – or agent – represents a tentative solution for the problem under consideration. These solutions are subject to processes of competition and mutual cooperation in a way that resembles the behavioral patterns of living beings from a same species. To make clearer this point, it is firstly necessary to consider the high-level template of the basic populational event: a *generation*. This is shown next in Fig. 3.1.

Process Do-Generation (↑↓ *pop* : *Individual*[ ])
variables
 breeders, *newpop* : *Individual*[ ];
begin
 breeders ← Select-From-Population(*pop*);
 newpop ← Generate-New-Population(*breeders*);
 pop ← Update-Population(*pop*, *newpop*);
end

Fig. 3.1. The basic generational step

As it can be seen, each generation consists of the updating of a population of individuals, hopefully leading to better and better solutions for the problem being tackled. There are three main components in this generational step: *selection*, *reproduction*, and *replacement*. The first component (selection) is responsible (jointly with the replacement stage) for the competition aspects of individuals in the population. Using the information provided by an *ad hoc* guiding function (*fitness* function in the EA terminology), the goodness of individuals in *pop* is evaluated; subsequently, a sample of individuals is selected for reproduction according to this goodness measure. This selection can be done in a variety of ways. The most popular techniques are *fitness-proportionate* methods (the probability of selecting an individual for breeding is proportional to its fitness[1]), *rank-based* methods (the probability of selecting an individual depends on its position after ranking the whole population), and *tournament-based* methods (individuals are selected on the basis of a direct competition within small sub-groups of individuals).

Replacement is very related to this competition aspect, as mentioned above. This component takes care of maintaining the population at a constant size. To do so, individuals in the older population are substituted by the newly-created ones (obtained from the reproduction stage) using some specific criterion. Typically, this can be done by taking the best (according to the guiding function) individuals both from *pop* and *newpop* (the so-called "plus" replacement strategy), or by simply taking the best individuals from *newpop* and inserting them in *pop* substituting the worst ones (the "comma" strategy). In the former case, if $|pop| = |newpop|$ then

[1] Maximization is assumed here. In case we were dealing with a minimization problem, fitness should be transformed so as to obtain an appropriate value for this purpose, e.g., subtracting it from the highest possible value of the guiding function.

the replacement is termed *generational*; if $|newpop|$ is small (say $|newpop| = 1$), then we have a *steady-state* replacement.

The most interesting aspect in this generation process probably is the intermediate phase of reproduction. At this stage, we have to create new individuals (or agents) by using the existing ones. This is done using a number of reproductive *operators*. Many different such operators can be used in a MA, as illustrated in the general pseudocode shown in Fig. 3.2. Nevertheless, the most typical situation involves utilizing just two operators: recombination and mutation.

Process Generate-New-Population
($\downarrow$ *pop* : *Individual*[ ] , $\downarrow$ *op* : *Operator*[ ]) $\rightarrow$ *Individual*[ ]
variables
  buffer : *Individual*[ ][ ];
  j : [1..|*op*|];
begin
  buffer[0] $\leftarrow$ *pop*;
  for *j* $\leftarrow$ 1: |*op*| **do**
    buffer[*j*] $\leftarrow$ Apply-Operator(*op*[*j*], *buffer*[*j*-1]);
  endfor;
  return *buffer*[n_{op}];
end

Fig. 3.2. Generating the new population

Recombination is a process that encapsulates the *mutual cooperation* among several individuals where we see cooperation as exchange of information adquired during the individual search steps. Recombination algorithms typically employ at least two feasible solutions, but a higher number is possible (Eiben et al. 1994). The algorithm then uses the information contained in a number of selected solutions named *parents* in order to create new solutions. If it is the case that the resulting individuals (the *offspring*) are entirely composed of information taken from the parents, then the recombination is said to be *transmitting* (Radcliffe 1994). This is the case of classical recombination operators for bitstrings such as *single-point crossover*, or *uniform crossover* (Syswerda 1989). This property captures the *a priori* role of recombination as previously enunciated, but it can be difficult to achieve for certain problem domains (the TRAVELING SALESMAN PROBLEM – TSP – is a typical example). In those situations, it is possible to consider other properties of interest such as *respect* or *assortment*. The former refers to the fact that the recombination operator generates descendants carrying all *features* (i.e., basic properties of solutions with relevance for the problem attacked) common to all parents; thus, this property can be seen as a part of the *exploitative* side of the search. On the other hand, *assortment* represents the exploratory side of recombination. A recombination operator is said to be *properly assorting* if, and only if, it can generate descendants carrying any combination of compatible features taken from the parents. The assortment is said to be *weak* if it is necessary to perform several recombinations within the offspring to achieve this effect.

Several interesting concepts have been introduced in this description of recombination, namely, *relevant features* and *cooperation*. We will return to these points in the next section. Before that, let us consider the other operator mentioned above: *mutation*. From a classical point of view (at least in the genetic-algorithm arena (Goldberg 1989), this is a secondary operator whose mission is to "*keep the pot boiling*", continuously injecting new material in the population, but at a low rate (otherwise the search would degrade to a random walk in the solution space). Evolutionary-programming practitioners (Fogel et al. 1966) would disagree with this characterization, claiming a central role for mutation. Actually, it is considered the crucial part of the search engine in this context. This possibly reflects similar discussions on the roles of *natural selection* and *random genetic drift* in biological evolution.

In essence, a mutation operator must generate a new solution by partly modifying an existing solution. This modification can be random – as it is typically the case – or can be endowed with problem-dependent information so as to bias the search to probably-good regions of the search space. It is precisely in the light of this latter possibility that one of the most distinctive components of MAs is introduced: *local-improvers*. To understand their philosophy, let us consider the following abstract formulation: first of all, assume a mutation operator that performs a random minimal modification in a solution; now consider the graph whose vertices are solutions, and whose edges connect pairs of vertices such that the corresponding solutions can be obtained via the application of the mutation operator on one of them[2]. A local-improver is a process that starts at a certain vertex, and moves to an adjacent vertex, provided that the neighboring solution is better that the current solution. This is illustrated in Fig. 3.3.

Process Local-Improver (↑↓ *current* : *Individual*, ↓ *op* : *Operator*[ ])
variables
    new : *Individual*;
begin
    repeat
        new ← Apply-Operator(*op*, *current*);
        if ($F_g(new) <_F F_g(current)$) **then**
            current ← *new*;
        endif;
    until Local-Improver-Termination-Criterion();
    return *current*;
end

Fig. 3.3. Pseudocode of a Local-Improver

As it can be seen, the local-improver tries to find an "uphill" (in terms of improving the value provided by the guiding function F_g path in the graph whose

[2] Typically this graph is symmetrical, but in principle there is no problem in assuming it to be asymmetrical.

definition was sketched before. The formal name for this graph is *fitness landscape* (Jones 1995). Notice that the length of the path found by the local-improver is determined by means of a Local-Improver-Termination-Criterion function. A usual example is terminating the path when no more uphill movements are possible (i.e., when the current solution is a local optimum with respect to *op*). However, this is not necessarily the case always. For instance, the path can be given a maximum allowed length, or it can be terminated as soon as the improvement in the value of the guiding function is considered good enough. For this reason, MAs cannot be characterized as "*EAs working in the space of local-optima (with respect to a certain fitness landscape)*"; that would be an unnecessarily restricted definition. Indeed, the earlier MAs by Moscato and Norman in 1988-89, where not restricted to search with only local-optima as basis for recombination.

The local-improver algorithm can be used in different parts of the generation process, for it is nothing else than just another operator. For example, it can be inserted after the utilization of any other recombination or mutation operator; alternatively, it could be just used at the end of the reproductive stage. As said before, the utilization of this local-improver[3] is one of the most characteristic features of MAs. It is precisely because of the use of this mechanism for improving individuals on a local (and even autonomous) basis that the term 'agent' is deserved. Thus, the MA can be viewed as a collection of agents performing an autonomous exploration of the search space, cooperating via recombination, and competing for computational resources due to the use of selection/replacement mechanisms.

After having presented the innards of the generation process, we can now have access to the larger picture. The functioning of a MA consists of the iteration of this basic generational step, as shown in Fig. 3.4.

```
Process MA () → Individual[ ]
variables
    pop : Individual[ ];
begin
    pop ← Generate-Initial-Population();
    repeat
        pop ← Do-Generation(pop);
        if Converged(pop) then
            pop ← Restart-Population(pop);
        endif;
    until MA-Termination-Criterion();
end
```

Fig. 3.4. The general template of a MA

Several comments must be made with respect to this general template. First of all, the Generate-Initial-Population process is responsible for creating the initial

[3] We use the term in singular, but notice that several different local-improvers could be used in different points of the algorithm, or in different agents.

set of *pop* configurations. This can be done by simply generating *pop* random configurations or by using a more sophisticated seeding mechanism (for instance, some constructive heuristic), by means of which high-quality configurations are injected in the initial population (Surry and Radcliffe 1996; Louis et al. 1999). Another possibility, the Local-Improver presented before could be used as shown in Fig. 3.5:

```
Process Generate-Initial-Population ( ↓ μ : N ) → Individual[ ]
variables
    pop : Individual[ ];
    ind : Individual;
    j : [1..μ];
begin
    for j ← 1:μ do
        ind ← Generate-Random-Solution();
        pop[j] ← Local-Improver(ind);
    endfor;
    return pop;
end
```

Fig. 3.5. Injecting high-quality solutions in the initial population

There is another interesting element in the pseudocode shown in Fig. 3.4: the Restart-Population process. This process is very important in order to make an appropriate use of the computational resources. Consider that the population may reach a state in which the generation of new improved solution is very unlikely. This could be the case when all agents in the population are very similar to each other. In this situation, the algorithm will probably expend most of the time resampling points in a very limited region of the search space (Cotta 1997), with the subsequent waste of computational efforts. This phenomenon is known as *convergence*, and it can be identified using measures such as Shannon's entropy (Davidor and Ben-Kiki 1992). If this measure falls below a predefined threshold, the population is considered at a degenerate state. This threshold depends upon the representation of the problem being used (number of values per variable, constraints, etc.) and hence must be determined in an *ad-hoc* fashion. A different possibility is the use of a probabilistic approach to determine with a desired confidence that the population has converged. For example, in (Hulin 1997) a Bayesian approach is presented for this purpose.

Once the population is considered to be at a degenerate state, the restart process is invoked. Again, this can be implemented in a number of ways. A very typical strategy is keeping a fraction of the current population (this fraction can be as small as one solution, the current best), and substituting the remaining configurations with newly generated (from scratch) solutions, as shown in Fig. 3.6:

```
Process Restart-Population ( ↓ pop : Individual[ ] ) → Individual[ ]
variables
    newpop : Individual[ ];
    j, #preserved : [1..|pop|];
begin
    #preserved ← |pop| . %PRESERV;
    for j ← 1: #preserved do
        newpop[j] ← i^th Best(pop, j);
    endfor;
    for j ← (#preserved+1):|pop| do
        newpop[j] ← Generate-Random-Configuration();
        newpop[j] ← Local-Improver(newpop[j]);
    endfor;
    return newpop;
end
```

Fig. 3.6. A possible re-starting procedure for the population

The above process completes the functional description of MAs. Obviously, it is possible to conceive some *ad-hoc* modifications of this basic template that still could be catalogued as MA. The reader can nevertheless be ensured that any such algorithm will follow the general philosophy depicted in this section, and could be possibly rewritten so as to match the spirit of this template.

3.3 Design of Effective MAs

The general template of MAs we have depicted in the previous section must be instantiated with precise components in order to be used for as a heuristic for an specific problem. This instantiation has to be carefully done so as to obtain an effective optimization tool. We will address some design issues in this section.

A first obvious remark must be done: there is no single general approach for the design of effective MAs. This fact admits different proofs depending on the precise definition of *effective* in the previous statement. Such proofs may involve classical complexity results and conjectures if 'effective' is understood as 'polynomial-time', the NFL Theorem if we consider a more general set of performance measures, and even Computability Theory if we relax the definition to arbitrary decision problems. For these reasons, we can only define several *design heuristics* that will likely result in "well-performing" MAs, but without explicit guarantees for this.

Having introduced this point of caution, the first element that one has to decide is the *representation* of solutions. At this point it is necessary to introduce a subtle but important distinction here: representation and *codification* are different things.

The latter refers to the way solutions are internally stored, and it can be chosen according to memory limitations, manipulation complexity, and other resource-based considerations. On the contrary, the representation refers to an abstract formulation of solutions, relevant from the point of view of the functioning of reproductive operators. This duality was present in discussions contemporary to the early debate on MAs (e.g., Radcliffe 1992), and can be very well exemplified in the context of permutational problems. For instance, consider the TSP; solutions can be internally encoded as permutations, but if a edge-recombination operator is used (e.g., Mathias and Whitley 1992) then solutions are *de facto* represented as edge lists.

The above example about the TSP also serves for illustrating one of the properties of representations that must be sought. Consider that a permutation can be expressed using different *information units*; for instance, it can be determined on the basis of the specific values of each position. This is the *position-based* representation of permutations (Goldberg 1989). On the other hand, it can be determined on the basis of adjacency relationships between the elements of the permutation. Since the TSP is defined by a matrix of inter-city distances, it seems that edges are more relevant for this problem than absolute positions in the permutation. In effect, it turns out that operators manipulating this latter representation perform better than operators that manipulate positions such as *partially-mapped crossover* (PMX) (Goldberg and Lingle 1985) or *cycle crossover* (CX) (Oliver et al. 1987).

There have been several attempts for quantifying how good a certain set of information units is for representing solutions for a specific problem. We can cite a few of them:

- *Minimizing epistasis*: epistasis can be defined as the non-additive influence on the guiding function of combining several information units (see (Davidor 1991) for example). Clearly, the higher this non-additive influence, the lower the absolute relevance of individual information units. Since the algorithm will be processing such individual units (or small groups of them), the guiding function turns out to be low informative, and prone to misguide the search.
- *Minimizing fitness variance* (Radcliffe and Surry 1994a): This criterion is strongly related to the previous one. The fitness variance for a certain information unit is the variance of the values returned by the guiding function, measured across a representative subset of solutions carrying this information unit (Hofmann 1993). By minimizing this fitness variance, the information provided by the guiding function is less *noisy*, with the subsequent advantages for the guidance of the algorithm.
- *Maximizing fitness correlation*: In this case a certain reproductive operator is assumed, and the correlation in the values of the guiding function for parents and offspring is measured (Moscato 1989). If the fitness correlation is high, good solutions are likely to produce good solutions, and thus the search will gradually shift toward the most promising regions of the search space (Moscato 1993; Merz and Freisleben 2000). Again, there is a clear relationship with the previous approaches; for instance, if epistasis (or fitness variance) is

low, then solutions carrying specific features will have similar values for the guiding function; since the reproductive operators will create new solutions by manipulating these features, the offspring is likely to have a similar guiding value as well.

Obviously, the description of these approaches may appear somewhat idealized, but the underlying philosophy is well illustrated. It must be noted that selecting a representation is not an isolated process, but it has a strong liaison with the task of choosing appropriate reproductive operators for the MA as already recognized in (Moscato 89). Actually, according to the operator-based view of representations described above, the existence of multiple operators may imply the consideration of different representations of the problem at different stages of the reproductive phase. We will come back to this issue later in this section.

In order to tackle the operator-selection problem, we can resort to existing operators, or design new *ad hoc* operators. In the former case, a suggested line of action could be the following (Cotta 1998):

1. We start from a set of existing operators $\Omega = \{\omega_1, \omega_2, ..., \omega_k\}$. The first step is identifying the representation of the problem manipulated by each of these operators.
2. Use any of the criterions presented for measuring the goodness of the representation.
3. Select ω_i from Ω, such that the representation manipulated by ω_i is the more trustable.

This is called *inverse analysis of operators* since some kind of inverse engineering is done in order to evaluate the potential usefulness of each operator. The alternative would be a *direct analysis* in which new operators would be designed. This could be do as follows:

1. Identify different potential representation for the problem at hand (e.g., recall the previous example on the TSP).
2. Use any of the criterions presented for measuring the goodness of these representation.
3. Create new operators $\Omega' = \{\omega'_1, \omega'_2, ..., \omega'_k\}$. via the manipulation of the most trustable information units.

In order to accomplish the last step of the *direct analysis*, there exist a number of templates for the manipulation of abstract information units. For example, the templates known as *random respectful recombination* ($\mathbf{R}^3$), *Random Assorting Recombination* (RAR), and *Random Transmitting Recombination* (RTR) have been defined in (Radcliffe 1994). An example of the successful instantiation of some of these templates using the direct analysis in the context of flowshop scheduling can be found in (Cotta and Troya 1998).

The generic templates mentioned above are essentially *blind*. This means that they do not use problem-dependent information at any stage of their functioning. This use of blind recombination operators is traditionally justified on the grounds of not introducing excessive bias in the search algorithm, thus preventing extremely fast convergence to suboptimal solutions. However, this is a highly arguable point since the behavior of the algorithm is in fact biased by the choice of representation. Even if we neglect this fact, it can be reasonable to pose the possibility of quickly obtaining a suboptimal solution and restarting the algorithm, rather than using blind operators for a long time in pursuit of an asymptotically optimal behavior (not even guaranteed in most cases).

Reproductive operators that use problem knowledge are commonly termed *heuristic* or *hybrid*. In these operators, problem information is used to guide the process of producing the offspring. There are numerous ways to achieve this inclusion of problem knowledge; in essence, we can identify two major aspects into which problem knowledge can be injected: the selection of the parental features that will be transmitted to the descendant, and the selection of non-parental features that will be added to it[4].

With respect to the selection of parental features to be injected in the offspring, there exists evidence that *respect* (transmission of common features, as mentioned in the previous section) is beneficial for some problems (e.g., see (Cotta and Muruzábal 2002; Mathias and Whitley 1992)). After this initial transmission, the offspring can be completed in several ways. For example, (Radcliffe and Surry 1994a) have proposed the use of local-improvers or implicit enumeration schemas[5]. This is done by firstly generating a partial solution by means of a non-heuristic procedure; subsequently, two approaches can be used:

- *locally-optimal completion*: the child is completed at random, and a local-improver is used restricted to those information units added for completion.
- *globally-optimal completion*: an implicit enumeration schema is used in order to find the globally best combination of information units that can be used to complete the child.

Related to the latter approach, the implicit enumeration schema can be used to find the best combination of the information units present in the parents. The resulting recombination would thus be *transmitting*, but not necessarily *respectful* since it can be proved that these two properties are incompatible in general. However, we can enforce *respect by restricting the search to non-common* features. Notice that this would not be globally-optimal completion since the whole search is restricted to information comprised in the parents. The set of solutions that can be constructed using this parental information is termed *dynastic potential*, and for

[4] Notice that the use of the term 'parental information' does not imply the existence of more than one parent. In other words, the discussion is not restricted to recombination operators, but may also include mutation operators.

[5] Actually, these approaches can be used even when no initial transmission of common features is performed.

this reason this approach is termed *dynastically optimal recombination* (DOR) (Cotta and Troya 2002). This operator is monotonic in the sense that any child generated is at least as good as the best parent.

Problem-knowledge need not be necessarily only included via iterative algorithms. On the contrary, the use of constructive heuristics is a popular choice. A distinguished example is the *Edge Assembly Crossover* (EAX) (Nagata and Kobayashi 1997). EAX is a specialized operator for the TSP (both for symmetric and asymmetric instances) in which the construction of the child comprises two-phases: the first one involves the generation of an incomplete child via the so-called E-sets (subtours composed of alternating edges from each parent); subsequently, these subtours are merged in a single feasible subtour using a greedy repair algorithm. The authors of this operator reported impressive results in terms of accuracy and speed. It has some strong similarities with the recombination operator proposed in (Moscato and Norman 1992; Moscato 1993).

To some extent, the above discussion is also applicable to mutation operators, although these exhibit a clearly different role: they must introduce new information. This means that purely transmitting mechanisms would not be acceptable for this purpose. Nevertheless, it is still possible to use the ideas described in the previous paragraphs by noting that the 'partial solution' mentioned in several situations can be obtained by simply removing some information units from a single solution. A completion procedure as described before can then be used in order to obtain the mutated solution.

Once we have one or more knowledge-augmented reproductive operators, it is necessary to make them work in a synergistic fashion. This is a feature of MAs that is also exhibited by other metaheuristics such as *Variable Neighborhood Search* (VNS) (Hansen and Mladenovic 2001), although it must be emphasized that it was already included in the early discussions of MAs, before the VNS metaheuristic was formulated. We can quote from (Moscato 1989) discussing the implementation of MAs in parallel and distributed heterogeneous computer systems:

"*Another advantage that can be exploited is that the most powerful computers in the network can be doing the most time-consuming heuristics, while others are using a different heuristics. The program to do local search in each individual can be different. This enriches the whole, since what is a local minima for one of the computers is not a local minima for another in the network. Different heuristics may be working fine due to different reasons. The collective use of them would improve the final output. In a distributed implementation we can think in a division of jobs, dividing the kind of moves performed in each computing individual. It leads to an interesting concept, where instead of dividing the physical problem (assignment of cities/cells to processors) we divide the set of possible moves. This set is selected among the most efficient moves for the problem.*"

This idea of synergistically combining different operators (and indeed different search techniques) was exemplified at its best by the extraordinary achievements of Applegate, Bixby, Cook, and Chvatal in 1998. They established new breakthrough results for the MIN TSP that support our view that MAs will have a central role as a problem solving methodology. This team solved to optimality an instance of the TSP of 13,509 cities corresponding to all U.S. cities with populations of more than 500 people[6]. The approach, according to Bixby: "*...involves ideas from polyhedral combinatorics and combinatorial optimization, integer and linear programming, computer science data structures and algorithms, parallel computing, software engineering, numerical analysis, graph theory, and more*". Their approach can possibly be classified as the most complex MA ever built for a given combinatorial optimization problem.

These ideas have been further developed in a recent unpublished manuscript, "*Finding Tours in the TSP*" by the same authors (Bixby et al.), available from their web site. They present results on running an optimal algorithm for solving the MIN WEIGHTED HAMILTONIAN CYCLE PROBLEM in a subgraph formed by the union of 25 Chained Lin-Kernighan tours. The approach consistently finds the optimal solution to the original MIN TSP instances with up to 4,461 cities. They also attempted to apply this idea to an instance with 85,900 cities (the largest instance in TSPLIB) and from that experience they convinced themselves that it also works well for such large instances.

The approach of running a local search algorithm (Chained Lin Kernighan) to produce a collection of tours, following by the dynastical-optimal recombination method the authors named *tour merging* gave a non-optimal tour of only 0.0002% excess above the proved optimal tour for the 13,509 cities instance. We take this as a clear proof of the benefits of the MA approach and that more work is needed in developing good strategies for *Complete Memetic Algorithms*, i.e., those that systematically and synergistically use randomized and deterministic methods and can prove optimality.

We would like to close this section by emphasizing once again the heuristic nature of the design principles described in this section. The most interesting thing to note here is not the fact that they are just probably-good principles, but the fact that there is still much room for research in methodological aspects of MAs (e.g., see (Krasnogor 2002)). The open-philosophy of MAs make them suitable for incorporating mechanisms from other optimization techniques. In this sense, the reader may find a plethora of new possibilities for MA design by studying other metaheuristics such as TS, for example.

3.4 Applications of MAs

This section will provide an overview of the numerous applications of MAs. This overview is far from exhaustive since new applications are being developed continuously. However, it is intended to be illustrative of the practical impact of these optimization techniques.

[6] See: http://www.crpc.rice.edu/CRPC/newsArchive/tsp.html

3.4.1 NP-hard Combinatorial Optimization problems

Traditional *NP* Optimization problems constitute one of the most typical application areas of MAs. A remarkable history of successes has been reported with respect to the application of MAs to *NP*-hard problems such as the following: GRAPH PARTITIONING (Bui and Moon 1996, 1998; Merz and Freisleben 1998a, 1999b, 2000) MIN NUMBER PARTITIONING (Berretta et al. 2001; Berretta and Moscato 1999), MAX INDEPENDENT SET (Aggarwal et al. 1997; Hifi 1997; Sakamoto et al. 1997), BIN-PACKING (Reeves 1996), MIN GRAPH COLORING (Coll et al. 1999; Costa et al. 1995; Dorne and Hao 1998; Fleurent and Ferland 1997), SET COVERING (Beasley and Chu 1996), MIN GENERALISED ASSIGNMENT (Chu and Beasley 1997), MULTIDIMENSIONAL KNAPSACK (Beasley and Chu 1998; Cotta and Troya 1998; Gottlieb 2000), NONLINEAR INTEGER PROGRAMMING (Taguchi et al. 1998), QUADRATIC ASSIGNMENT (Brown et al. 1989; Carrizo et al. 1992; Merz and Freisleben 1997a, 1999a, 1999b), QUADRATIC PROGRAMMING (Merz and Freisleben 2002a; Merz and Katayama 2002), SET PARTITIONING (Levine 1996), and particularly on the MIN TRAVELLING SALESMAN PROBLEM and its variants (Freisleben and Merz 1996a, 1996b; Gorges-Schleuter 1989, 1991, 1997; Holstein and Moscato 1999; Katayama et al. 1998; Krasnogor and Smith 2000; Merz 2002; Merz and Freisleben 1997b, 1999b, 2002b; Moscato and Norman 1992; Radcliffe and Surry 1994b; Rodrigues and Ferreira 2001).

The problems cited above are viewed as "classical" NP-hard problems as the decisions versions appeared in Karp's notorious paper (Karp 1972) on the reducibility of combinatorial problems. In many of them the authors claim that their MAs are probably the best-known heuristic for the problem at hand. This is very important since these problems have been addressed with several different combinatorial optimization methods.

The MA paradigm is not limited to the above-mentioned problems. There exist additional "*non-classical*" combinatorial optimization problems that are difficult to solve in practice. In their resolution with heuristic methods MAs have revealed themselves as outstanding techniques. As an example of these problems, one can cite *partial shape matching* (Ozcan and Mohan 1998), *Kauffman NK Landscapes* (Merz and Freisleben 1998b), *spacecraft trajectory design* (Crain et al. 1999), *minimum weighted k-cardinality tree subgraph problem* (Blesa et al. 2001), *minimum k-cut problem* (Yeh 2000), *uncapacitated hub location* (Abdinnour 1998), *placement problems* (Hopper and Turton 1999; Krzanowski and Raper 1999; Schnecke and Vornberger 1997), *vehicle routing* (Berger et al. 1998; Jih and Hsu 1999), *transportation problems* (Gen et al. 1998; Novaes et al. 2000), and *task allocation* (Hadj-Alouane et al. 1999).

3.4.2 Telecomunications and networking

Another important class of combinatorial optimization problems are those that directly or indirectly correspond to telecommunication network problems. For example, we can cite: *frequency allocation* (Cotta and Troya 2001; Kassotakis et al. 2000), *network design* (Garcia et al. 1998; Runggeratigul 2001), *degree-constrained minimum spanning tree problem* (Raidl and Julstron 2000), *vertex-biconnectivity augmentation* (Kersting et al. 2002), *assignment of cells to switches in cellular mobile networks* (Quintero and Pierre 2003), and *OSPF routing* (Buriol et al. 2002).

Obviously, this list is by no means complete since its purpose is simply to document the wide applicability of the approach for combinatorial optimization.

3.4.3 Scheduling and Timetabling Problems

Undoubtedly, scheduling problems are one of the most important optimization domains due their importance in Production Planning. They thus deserve a section of their own, despite they could be included in the *NP*-hard class surveyed in the previous subsection.

MAs have been used to tackle a large variety of scheduling problems. We can cite the following: *maintenance scheduling* (Burke and Smith 1997, 1999a, 1999b), *open shop scheduling* (Cheng et al. 1999; Fang and Xi 1997; Liaw 2000), *flowshop scheduling* (Basseur et al. 2002; Cavalieri and Gaiardelli 1998; Murata and Ishibuchi 1994; Murata et al. 1996), *total tardiness single machine scheduling* (Mendes et al. 2002a), *single machine scheduling with setup-times and due-dates* (França et al. 1999, 2001; Lee 1994; Miller et al. 1999), *parallel machine scheduling* (Cheng and Gen M 1996, 1997; Mendes et al. 2002b; Min and Cheng 1998), *project scheduling* (Nordstrom and Tufekci 1994; Ozdamar 1999; Ramat et al. 1997), *warehouse scheduling* (Watson et al. 1999), *production planning* (Dellaert and Jeunet 2000; Ming and Mak 2000), *timetabling* (Burke et al. 1995, 1996, 1997, 1998; Burke and Newall 1997; Gonçalves 2001; Ling 1992; Monfroglio 1996b, 1996c; Paechter et al. 1996, 1998; Rankin 1996), *rostering* (De Causmaecker et al. 1999; Monfroglio 1996a), and *sport games scheduling* (Costa 1995).

3.4.4 Machine Learning and Robotics

Machine learning and robotics are two closely related fields since the different tasks involved in the control of robots are commonly approached using artificial neural networks and/or classifier systems. MAs, generally cited in these fields under the denomination of "genetic hybrids" have been used, i.e., in general optimization problems related to machine learning (for example, the training of artificial neural networks), and in robotic applications. With respect to the former, MAs have been applied to *neural network training* (Abbass 2001; Ichimura and Kuriyama 1998; Moscato 1993; Topchy et al. 1996; Yao 1993), *pattern recognition* (Aguilar and Colmenares 1998), *pattern classification* (Krishna and Narasimha-Murty 1999; Mignotte et al. 2000), and *analysis of time series* (Dos Santos Coelho et al. 2001; Ostermark 1999a).

In robotics, work has been done in *reactive rulebase learning in mobile agents* (Cotta and Troya 2000), *path planning* (Osmera 1995; Pratihar et al. 1999; Xiao and Zhang 1997), *manipulator motion planning* (Ridao et al. 1998), *time optimal control* (Chaiyaratana and Zalzala 1999), etc.

3.4.5 Engineering, Electronics and Electromagnetics

Electronics and engineering are also two fields in which these methods have been actively used. For example, with regard to engineering problems, work has been done in the following areas: *structure optimization* (Yeh 1999), *system modeling* (Wang and Yen 1999), *fracture mechanics* (Pacey et al. 2001), *aeronautic design* (Bos 1998; Quagliarella and Vicini 1998), *trim loss minimization* (Ostermark 1999b), *traffic control* (Srinivasan et al. 2000), *power planning* (Urdaneta et al. 1999; Valenzuela and Smith 2002), *calibration of combustion engines* (Knödler et al. 2002; Poland et al. 2001), and *process control* (Conradie et al. 2002; Zelinka et al. 2001).

As to practical applications in the field of electronics and electromagnetics (Ciuprina et al. 2002), the following list can illustrate the numerous areas in which these techniques have been employed: *semiconductor manufacturing* (Kim and May 1999), *circuit design* (Areibi 2001, 2002; Guotian and Changhong 1999; Harris and Ifeachor 1998; Weile and Michielssen 1999), *circuit partitioning* (Areibi 2000), *computer aided design* (Becker and Drechsler 1994), *multilayered periodic strip grating* (Aygun et al. 1997), *analogue network synthesis* (Grimbleby 1999), *service restoration* (Augugliaro et al. 1998), *optical coating design* (Hodgson 2001), and *microwave imaging* (Caorsi et al. 2002; Pastorino et al. 2002).

3.4.6 Problems involving optimization in molecules

This is a class of problems that involve nonlinear optimization techniques and global optimization challenges. We hope the reader can easily identify a common trend in the literature. Some authors continue referring to their technique as 'genetic', although they are closer in spirit to MAs (Hodgson 2000). This makes difficult to identify the trend from some abstracts available from database searches.

The Caltech report that gave the name to the field of MAs (Moscato 1989) discussed a metaheuristic that can be viewed as a hybrid of GAs and annealing based methods developed with M. G. Norman in 1988. In recent years, several papers applied hybrids of GAs with SA or other methods to a variety of molecular optimization problems (Bayley et al. 1998; Dandekar and Argos 1996; De Souza et al. 1998; Doll and VanHove 1996; Fu et al 1997; Gunn 1997; Jones et al. 1997; Kariuki et al. 1997; Landree et al. 1997; Li et al. 1997; Lorber and Shoichet 1998; Mackay 1995; Merkle et al. 1996; Miller et al. 1996; Pacey et al. 1999; Shankland et al. 1997, 1998; Tam and Compton 1995; White et al. 1998; Zacharias et al. 1998; Zwick et al. 1996). These are hybrid population approaches that cannot be catalogued as being 'genetic', but this denomination has appeared in previous work by Deaven and Ho (Deaven and Ho 1995) and then cited by J. Maddox in *Nature* (Maddox 1995) and is being used by researchers not aware of MAs.

Other fields of application include *cluster physics* (Niesse and Mayne 1996). Additional work has been done in (Deaven et al. 1996; Ho et al. 1998; Hobday and Smith 1997; Pucello et al. 1997; Pullan 1997; White et al. 1998). Other evolutionary approaches to a variety of *molecular problems* can be found in (Doll and VanHove 1996; Hartke 1993; Hirsch and Mullergoymann 1995; May and Johnson 1994; Meza et al. 1996; Raymer et al. 1997; VanKampen et al. 1996). Their use for design problems is particularly appealing (Clark and Westhead 1996; Kariuki et al. 1997; Willett 1995). They have also been applied in *protein design* (Desjarlais and Handel 1995; Lazar et al. 1997), *structure prediction* (Krasnogor et al. 2002; Krasnogor and Smith 2002), and *alignment* (Carr et al. 2002) (see also the discussion in (Moscato 1993) and the literature review in (Hodgson 2000).

This field is enormously active, and new application domains for MAs are continuously emerging. Among these, we must mention applications related to genomic analysis, such as *clustering gene-expression profiles* (Merz and Zell 2002), or *inferring phylogenetic trees* (Cotta and Moscato 2002).

3.4.7 Other Applications

In addition to the application areas described above, MAs have been also utilized in other fields such as, for example, *medicine* (Haas et al. 1996, 1998; Wehrens et al. 1993), *economics* (Li et al. 1996; Ostermark, 1999c), *oceanography* (Musil et al. 1999), *mathematics* (Reich 2000; Wei and Cheng 1999; Wei and Kangling 2000), *imaging science and speech processing* (Cadieux et al. 1997; Krishna et al. 1997; Li 1997; Mathias and Whitley 1994; Ruff et al. 1999; Yoneyama et al. 1999), etc.

For further information about MA applications we suggest querying bibliographical databases or web browsers for the keywords '*memetic algorithms*' (or its singular form) as well as '*hybrid genetic algorithm*'. We have selected some references for well-known application areas. This means that, with high probability, many important contributions, some of them new or some of them unknown to us, may have been inadvertently left out and we hope the authors would forgive us for important unvoluntary omissions.

3.5 Conclusions and Future Directions

We believe that MAs have very favorable perspectives for their development and widespread application. We have several good reasons as hypothesis. First of all, MAs are showing a great record of efficient implementations, providing very good results in practical problems as the reader may have already checked the references cited in the previous section. We also have reasons to believe that we are near some major leaps forward in our theoretical understanding of these techniques, including for example the worst-case and average-case computational complexity of recombination procedures. On the other hand, the ubiquitous nature of distributed systems, like networks of workstations for example, plus the inherent asynchronous parallelism of MAs and the existence of web-conscious languages like Java are all together an excellent combination to develop highly portable and extendable object-oriented frameworks allowing algorithmic reuse (Mendes et al. 2001).

The systematic development of other particular optimization strategies is a healty sign of the evolution of the field, particularly if we have a reductionist approach to metaheuristic development. If any of the simpler metaheuristics (SA, TS, VNS, GRASP, etc.) performs the same as a more complex method (GAs, MAs, Ant Colonies, etc.), an "elegance design" principle should prevail and we must either resort to the simpler method, or to the one that has less free parameters, or to the one that is easier to implement. This will be the major forces in heuristic creation and development. The existence of a simple method that works well should defy us to adapt the complex methodology to beat a simpler heuristic, or to check if that is possible at all. An unhealthy sign of current research, however, are the attempts to encapsulate metaheuristics on stretched confinements.

We think that there are several "learned lessons" from work in other metaheuristics. For instance, a Basic Tabu Search scheme (Glover and Laguna 1997) decides to accept another new configuration without restriction to the relative objective function value of the two solutions. This has lead to good performance in some configuration spaces where evolutionary methods and Simulated Annealing perform poorly. A classical example of this situation is the MIN NUMBER PARTITIONING problem (Berretta and Moscato 1999).

There are many open lines of research in areas such as co-evolution. In (Moscato 1999) we can find the following quotation:

"*It may be possible that a future generation of MAs will work in at least two levels and two time scales. In the short-time scale, a set of agents would be searching in the search space associated to the problem while the long-time scale adapts the heuristics associated with the agents. Our work with D. Holstein which will be presented in this book might be classified as a first step in this promising direction. However, it is reasonable to think that more complex schemes evolving solutions, agents, as well as representations, will soon be implemented.*"

At that time, we were referring to the use of a metaheuristic called *Guided Local Search* used in (Holstein and Moscato, 1999) as well as the possibility of co-evolving the neighborhood techniques by other means. Unfortunately, this was not studied in depth in Holstein's thesis. However, a number of more recent articles are paving the way to more robust MAs (Carr et. al 2002; Krasnogor 1999; Krasnogor and Smith 2001). Krasnogor has recently introduced the term *multimeme* algorithms to identify those MAs that also adaptively change the neighborhood definition (Krasnogor and Smith 2002). In collaboration with other colleagues he is applying the method for the difficult problem of *protein structure prediction*

(Krasnogor et al. 2002). Smith also presents a recent study on these issues in (Smith 2002).

More work is necessary, and indeed the protein folding models they are using are a good test-bed for the approach. However, we also hope that the researchers should again concentrate MAs for large-scale challenging instances of the TSP, possibly following the approaches of using population structures (Moscato and Tinetti 1992; França et al. 2001), self-adapting local search (Krasnogor and Smith 2000) as well as the powerful recombination operators that have been devised for TSP instances (Holstein and Moscato 1999; Merz 2002, Merz and Freisleben 2002b; Moscato 1993). We have also identified some problems with evolutionary search methods in instances of the TSP in which the entries of the distance matrix have a large number of decimal digits. This means that there is an inherent problem to be solved, for evolutionary methods to deal with fitness functions that have so many decimal digits. Traditional rank-based or fitness-based selection schemes to keep new solutions in the current population fail. It would be then reasonable to investigate whether some ideas from basic TS mechanisms could be adapted to allow less stringent selection approaches.

Multiparent recombination is also an exciting area to which research efforts can be directed too. From (Papadimitriou and Steiglitz 1982) we can read:

"The strategy developed by (Lin 1965) for the TSP is to obtain several local optima and then identify edges that are common to all of them. These are then fixed, thus reducing the time to find more local optima. This idea is developed further in (Lin and Kernighan 1973) and (Goldstein and Lesk 1975). It is intriguing that such an strategy, which has been around for more than three decades, is still not accepted by some researchers."

We think that the use of *multiparent* recombination with proven good properties is one of the most challenging issues for future developement in MAs, as well as for the whole EC paradigm.

References

Abbass H (2001) A memetic Pareto evolutionary approach to artificial neural networks. Lecture Notes in Computer Science 2256:1–12

Aggarwal C, Orlin J, Tai R (1997) Optimized crossover for the independent set problem. Operations Research 45:226–234

Aguilar J, Colmenares A (1998) Resolution of pattern recognition problems using a hybrid genetic/random neural network learning algorithm. Pattern Analysis and Applications 1:52–61

Abdinnour H (1998) A hybrid heuristic for the uncapacitated hub location problem. European Journal of Operational Research 106:489–499

Areibi S (2000) An integrated genetic algorithm with dynamic hill climbing for VLSI circuit partitioning. Proceedings of Data Mining with Evolutionary Algorithms, pp 97–102

Areibi S (2001) Memetic algorithms for VLSI physical design: Implementation issues. Proceedings of the 2nd WOMA – Workshop on Memetic Algorithms, pp 140–145

Areibi S (2002) The performance of memetic algorithms on physical design. Submitted to the Journal of Applied Systems Studies.

Augugliaro A, Dusonchet L, Riva-Sanseverino E (1998) Service restoration in compensated distribution networks using a hybrid genetic algorithm. Electric Power Systems Research 46:59–66

Aygun K, Weile D, Michielssen E (1997) Design of multi-layered periodic strip gratings by genetic algorithms. Microwave and Optical Technology Letters 14:81–85

Basseur M, Seynhaeve F, Talbi E (2002) Design of multi-objective evolutionary algorithms: Application to the flow-shop scheduling problem. Proceedings of the CEC'02 – Congress on Evolutionary Computation, pp 1151–1156

Bayley M, Jones G, Willett P, Williamson M (1998) Genfold: A genetic algorithm for folding protein structures using NMR restraints. Protein Science 7:491–499

Beasley J, Chu P (1996) A genetic algorithm for the set covering problem. European Journal of Operational Research 94:393–404

Beasley J, Chu P (1998) A genetic algorithm for the multidimensional knapsack problem. Journal of Heuristics 4:63–86

Becker B, Drechsler R (1994) Ofdd based minimization of fixed polarity Reed-Muller expressions using hybrid genetic algorithms. Proceedings of the IEEE International Conference on Computer Design: VLSI in Computers and Processor, pp 106–110

Berger J, Salois M, Begin R (1998) A hybrid genetic algorithm for the vehicle routing problem with time windows. Proceedings of the 12th Biennial Conference of the Canadian Society for Computational Studies of Intelligence, pp 114–127

Berretta R, Cotta C, Moscato P (2001) Forma analysis and new heuristic ideas for the number partitioning problem. Proceedings of the 4th MIC – Metaheuristic International Conference, pp 337–341

Berretta R, Moscato P (1999) The number partitioning problem: An open challenge for evolutionary computation? In: New Ideas in Optimization. McGraw-Hill, pp 261–278

Blesa M, Moscato P, Xhafa F (2001) A memetic algorithm for the minimum weighted k-cardinality tree subgraph problem. Proceedings of the 4th MIC – Metaheuristic International Conference, pp 85–90

Bos A (1998) Aircraft conceptual design by genetic/gradient-guided optimization. Engineering Applications of Artificial Intelligence 11:377–382

Brown D, Huntley C, Spillane A (1989) A Parallel Genetic Heuristic for the Quadratic Assignment Problem. Proceedings of the 3rd ICGA – International Conference on Genetic Algorithms, pp 406–415

Bui T, Moon B (1996) Genetic algorithm and graph partitioning. IEEE Transactions on Computers 45:841–855

Bui T, Moon B (1998) GRCA: A hybrid genetic algorithm for circuit ratio-cut partitioning. IEEE Transactions on Computer-Aided Design of Integrated Circuits and Systems 17:193–204

Buriol L, Resende M, Ribeiro C, Thorup M (2002) A memetic algorithm for OSPF routing. Proceedings of the 6th INFORMS Telecommunications Conference, pp 187–188

Burke E, Jackson K, Kingston J, Weare R (1997) Automated timetabling: The state of the art. The Computer Journal 40:565–571

Burke E, Newall J (1997) A phased evolutionary approach for the timetable problem: An initial study. Proceedings of the ICONIP/ANZIIS/ANNES'97 Conference, pp 1038–1041

Burke E, Newall J, Weare R (1996) A memetic algorithm for university exam timetabling. Lecture Notes in Computer Science 1153:241–250

Burke E, Newall J, Weare R (1998) Initialisation strategies and diversity in evolutionary timetabling. Evolutionary Computation 6:81–103

Burke E, Smith A (1997) A memetic algorithm for the maintenance scheduling problem. Proceedings of the ICONIP/ANZIIS/ANNES'97 Conference, pp 469–472

Burke E, Smith A (1999a) A memetic algorithm to schedule grid maintenance. Proceedings of the CIMCA'99 – International Conference on Computational Intelligence for Modelling Control and Automation, pp 122–127

Burke E, Smith A (1999b) A multi-stage approach for the thermal generator maintenance scheduling problem. Proceedings of the CEC'99 – Congress on Evolutionary Computation, pp 1085–1092

Burke E, Elliman DG, Weare RF (1995) A hybrid genetic algorithm for highly constrained timetabling problems. Proceedings of the 6th ICGA – International Conference on Genetic Algorithms, pp 605–610

Cadieux S, Tanizaki N, Okamura T (1997) Time efficient and robust 3-D brain image centering and realignment using hybrid genetic algorithm. Proceedings of the 36th SICE Annual Conference, pp 1279–1284

Caorsi S, Massa A, Pastorino M, Rafetto M, Randazzo A (2002) A new approach to microwave imaging based on a memetic algorithm. Proceedings of the PIERS'02 – Progress in Electromagnetics Research Symposium. Invited

Carr R, Hart W, Krasnogor N, Hirst J, Burke E, Smith J (2002) Alignment of protein structures with a memetic evolutionary algorithm. Proceedings of the GECCO'02 – Genetic and Evolutionary Computation Conference, pp 1027–1034

Carrizo J, Tinetti F, Moscato P (1992) A computational ecology for the quadratic assignment problem. Proceedings of the 21st Meeting on Informatics and Operations Research.

Cavalieri S, Gaiardelli P (1998) Hybrid genetic algorithms for a multiple-objective scheduling problem. Journal of Intelligent Manufacturing 9:361–367

Chaiyaratana N, Zalzala A (1999) Hybridisation of neural networks and genetic algorithms for time-optimal control. Proceedings of the CEC'99 – Congress on Evolutionary Computation, pp 389–396

Cheng R, Gen M (1996) Parallel machine scheduling problems using memetic algorithms. Proceedings of the IEEE SMC'96 – International Conference on Systems, Man and Cybernetics. Information Intelligence and Systems, pp 2665–2670

Cheng R, Gen M (1997) Parallel machine scheduling problems using memetic algorithms. Computers & Industrial Engineering 33:761–764

Cheng R, Gen M, Tsujimura Y (1999) A tutorial survey of job-shop scheduling problems using genetic algorithms. II. Hybrid genetic search strategies. Computers & Industrial Engineering 37:51–55

Chu P, Beasley J (1997) A genetic algorithm for the generalised assignment problem. Computers & Operations Research 24:17–23

Ciuprina G, Ioan D, Munteanu I (2002) Use of intelligent-particle swarm optimization in electromagnetics. IEEE Transactions on Magnetics 38:1037–1040

Clark D, Westhead D (1996) Evolutionary algorithms in computer-aided molecular design. Journal of Computer-aided Molecular Design 10:337–358

Coll P, Durán G, Moscato P (1999) On worst-case and comparative analysis as design principles for efficient recombination operators: A graph coloring case study. In: New Ideas in Optimization. McGraw-Hill, pp 279–294

Conradie A, Mikkulainen R, Aldrich C (2002) Intelligent process control utilising symbiotic memetic neuro-evolution. Proceedings of the CEC'02 – Congress on Evolutionary Computation, pp 623–628

Costa D (1995) An evolutionary tabu search algorithm and the NHL scheduling problem. INFOR 33:161–178

Costa D, Dubuis N, Hertz A (1995) Embedding of a sequential procedure within an evolutionary algorithm for coloring problems in graphs. Journal of Heuristics 1:105–128

Cotta C (1997) On resampling in nature-inspired heuristics (In Spanish). Proceedings of the 7th Conference of the Spanish Association for Artificial Intelligence, pp 145–154

Cotta C (1998) A study of hybridisation techniques and their application to the design of evolutionary algorithms. AI Communications 11:223–224

Cotta C, Moscato P (2002) Inferring phylogenetic trees using evolutionary algorithms. Lecture Notes in Computer Science 2439:720–729

Cotta C, Muruzábal J (2002) Towards a more efficient evolutionary induction of bayesian networks. Lecture Notes in Computer Science 2439:730–739

Cotta C, Troya J (1998) Genetic forma recombination in permutation flowshop problems. Evolutionary Computation 6:25–44

Cotta C, Troya J (1998) A hybrid genetic algorithm for the 0-1 multiple knapsack problem. In: Artificial Neural Nets and Genetic Algorithms 3. Springer-Verlag, pp 251–255

Cotta C, Troya J (2000) Using a hybrid evolutionary-A* approach for learning reactive behaviors. Lecture Notes in Computer Science 1803:347–356

Cotta C, Troya J (2001) A comparison of several evolutionary heuristics for the frequency assignment problem. Lecture Notes in Computer Science 2084:709–716

Cotta C, Troya J (2002) Embedding branch and bound within evolutionary algorithms. Applied Intelligence. To be published

Crain T, Bishop R, Fowler W, Rock K (1999) Optimal interplanetary trajectory design via hybrid genetic algorithm/recursive quadratic program search. Proceedings of the 9th AAS/AIAA Space Flight Mechanics Meeting, pp 99–133

Dandekar T, Argos P (1996) Identifying the tertiary fold of small proteins with different topologies from sequence and secondary structure using the genetic algorithm and extended criteria specific for strand regions. Journal of Molecular Biology 256:645–660

Davidor Y (1991) Epistasis variance: A viewpoint on GA-hardness. In: Foundations of Genetic Algorithms. Morgan Kaufmann, pp 23–35

Davidor Y, Ben-Kiki O (1992) The interplay among the genetic algorithm operators: Information theory tools used in a holistic way. Proceedings of 2nd PPSN – Parallel Problem Solving From Nature, pp 75–84

Davis L (1991) Handbook of Genetic Algorithms. Van Nostrand Reinhold Computer Library, New York

Dawkins R (1976) The selfish gene. Oxford University Press, Oxford

De Causmaecker P, Van Den Berghe G, Burke E (1999) Using tabu search as a local heuristic in a memetic algorithm for the nurse rostering problem. Proceedings of the 13th Conference on Quantitative Methods for Decision Making, abstract only, poster presentation

De Souza P, Garg R, Garg V (1998) Automation of the analysis of Mossbauer spectra. Hyperfine Interactions 112:275–278

Deaven D, Ho K (1995) Molecular-geometry optimization with a genetic algorithm. Physical Review Letters 75:288–291

Deaven D, Tit N, Morris J, Ho K (1996) Structural optimization of Lennard-Jones clusters by a genetic algorithm. Chemical Physics Letters 256:195–200

Dellaert N, Jeunet J (2000) Solving large unconstrained multilevel lot-sizing problems using a hybrid genetic algorithm. International Journal of Production Research 38:1083–1099

Desjarlais J, Handel T (1995) New strategies in protein design. Current Opinion in Biotechnology 6:460–466

Doll R, VanHove M (1996) Global optimization in LEED structure determination using genetic algorithms. Surface Science 355:L393–L398

Dorne R, Hao J (1998) A new genetic local search algorithm for graph coloring. Lecture Notes in Computer Science 1498:745–754

Dos Santos Coelho L, Rudek M, Junior OC (2001) Fuzzy-memetic approach for prediction of chaotic time series and nonlinear identification. Proceedings of the 6th On-line World Conference on Soft Computing in Industrial Applications

Eiben A, Raue P-E, Ruttkay Z (1994) Genetic algorithms with multi-parent recombination. Lecture Notes in Computer Science 866:78–87

Fang J, Xi Y (1997) A rolling horizon job shop rescheduling strategy in the dynamic environment. International Journal of Advanced Manufacturing Technology 13:227–232

Fleurent C, Ferland J (1997) Genetic and hybrid algorithms for graph coloring. Annals of Operations Research 63:437–461

Fogel LJ, Owens AJ, Walsh MJ (1966) Artificial Intelligence through Simulated Evolution. John Wiley & Sons, New York

França P, Mendes A, Moscato P (1999) Memetic algorithms to minimize tardiness on a single machine with sequence-dependent setup times. Proceedings of the DSI'99 – 5th International Conference of the Decision Sciences Institute, pp 1708–1710

França P, Mendes A, Moscato P (2001) A memetic algorithm for the total tardiness single machine scheduling problem. European Journal of Operational Research 132:224–242

Freisleben B, Merz P (1996a) A Genetic Local Search Algorithm for Solving Symmetric and Asymmetric Traveling Salesman Problems. Proceedings of the ICEC'96 – International Conference on Evolutionary Computation, pp 616–621

Freisleben B, Merz P (1996b) New Genetic Local Search Operators for the Traveling Salesman Problem. Lecture Notes in Computer Science 1141:890–900

Fu R, Esfarjani K, Hashi Y, Wu J, Sun X, Kawazoe Y (1997) Surface reconstruction of Si (001) by genetic algorithm and simulated annealing method. Science Reports of The Research Institutes Tohoku University Series A-Physics Chemistry And Metallurgy 44:77–81

Garcia B, Mahey P, LeBlanc L (1998) Iterative improvement methods for a multiperiod network design problem. European Journal of Operational Research 110:150–165

Gen M, Ida K, Yinzhen L (1998) Bicriteria transportation problem by hybrid genetic algorithm. Computers & Industrial Engineering 35:363–366

Glover F, Laguna M (1997) Tabu Search. Kluwer Academic Publishers, Boston

Goldberg DE (1989) Genetic Algorithms in Search, Optimization, and Machine Learning. Addison-Wesley

Goldberg D, Lingle Jr R (1985) Alleles, loci and the traveling salesman problem. Proceedings of 1st ICGA – International Conference on Genetic Algorithms, pp 154–159

Goldstein A, Lesk A (1975) Common feature techniques for discrete optimization. Computer Science Tech. Report 27, Bell Tel. Labs.

Gottlieb J (2000) Permutation-based evolutionary algorithms for multidimensional knapsack problems. Proceedings of the ACM Symposium on Applied Computing, pp 408–414

Glover F, Laguna M (1997) Tabu Search. Kluwer Academic Publishers, Boston, MA

Gonçalves JF (2001) A memetic algorithm for the examination timetabling problem. Proceedings of Optimization 2001 Conference, pp 23-25

Gorges-Schleuter M (1989) ASPARAGOS: An asynchronous parallel genetic optimization strategy. Proceedings of the 3rd ICGA – International Conference on Genetic Algorithms, pp 422–427

Gorges-Schleuter M (1991) Genetic Algorithms and Population Structures - A Massively Parallel Algorithm. PhD thesis, University of Dortmund, Germany

Gorges-Schleuter M (1997) Asparagos96 and the traveling salesman problem. Proceedings of the ICEC'97 – International Conference on Evolutionary Computation, pp 171–174

Grimbleby J (1999) Hybrid genetic algorithms for analogue network synthesis. Proceedings of the CEC'99 – Congress on Evolutionary Computation, pp 1781–1787

Gunn J (1997) Sampling protein conformations using segment libraries and a genetic algorithm. Journal of Chemical Physics 106:4270–4281

Guotian M, Changhong L (1999) Optimal design of the broad-band stepped impedance transformer based on the hybrid genetic algorithm. Journal of Xidian University 26:8–12

Haas O, Burnham K, Mills J (1998) Optimization of beam orientation in radiotherapy using planar geometry. Physics in Medicine and Biology 43:2179–2193

Haas O, Burnham K, Mills J, Reeves C, Fisher M (1996) Hybrid genetic algorithms applied to beam orientation in radiotherapy. Proceedings of the 4th European Congress on Intelligent Techniques and Soft Computing Proceedings, pp 2050–2055

Hadj-Alouane A, Bean J, Murty K (1999) A hybrid genetic/optimization algorithm for a task allocation problem. Journal of Scheduling 2:189–201

Hansen P, Mladenovic N (2001) Variable neighborhood search: Principles and applications. European Journal of Operational Research 130:449–467

Harris S, Ifeachor E (1998) Automatic design of frequency sampling filters by hybrid genetic algorithm techniques. IEEE Transactions on Signal Processing 46:3304–3314

Hart W, Belew R (1991) Optimizing an arbitrary function is hard for the genetic algorithm. Proceedings of the 4th ICGA – International Conference on Genetic Algorithms, pp 190–195

Hartke B (1993) Global geometry optimization of clusters using genetic algorithms. Journal of Physical Chemistry 97:9973–9976

Hifi M (1997) A genetic algorithm-based heuristic for solving the weighted maximum independent set and some equivalent problems. Journal of the Operational Research Society 48:612–622

Hirsch R, Mullergoymann C (1995) Fitting of diffusion-coefficients in a 3-compartment sustained-release drug formulation using a genetic algorithm. International Journal of Pharmaceutics 120:229–234

Ho K, Shvartsburg A, Pan B, Lu Z, Wang C, Wacker J, Fye J, Jarrold M (1998) Structures of medium-sized silicon clusters. Nature 392, 6676:582–585

Hobday S, Smith R (1997) Optimisation of carbon cluster geometry using a genetic algorithm. Journal of The Chemical Society-Faraday Transactions 93:3919–3926

Hodgson R (2000) Memetic algorithms and the molecular geometry optimization problem. Proceedings of the CEC'00 – Congress on Evolutionary Computation, pp 625–632

Hodgson R (2001) Memetic algorithm approach to thin-film optical coating design. Proceedings of the 2nd WOMA – Workshop on Memetic Algorithms, pp 152–157

Hofmann R (1993) Examinations on the algebra of genetic algorithms. Master Thesis, Technische Universitat Munchen, Institut fur Informatik

Holland J (1975) Adaptation in Natural and Artificial Systems. The University of Michigan Press

Holstein D, Moscato P (1999) Memetic algorithms using guided local search: A case study. In: New Ideas in Optimization. McGraw-Hill, pp 235–244

Hopper E, Turton B (1999) A genetic algorithm for a 2D industrial packing problem. Computers & Industrial Engineering 37:375–378

Hulin M (1997) An optimal stop criterion for genetic algorithms: A bayesian approach. Proceedings of the 7th ICGA – International Conference on Genetic Algorithms, pp 135–143

Ichimura T, Kuriyama Y (1998) Learning of neural networks with parallel hybrid GA using a Royal Road function. Proceedings of the IJCNN'98 – International Joint Conference on Neural Networks, pp 1131–1136

Jih W, Hsu Y (1999) Dynamic vehicle routing using hybrid genetic algorithms. Proceedings of the CEC'99 – Congress on Evolutionary Computation, pp 453–458

Jones G, Willett P, Glen R, Leach A, Taylor R (1997) Development and validation of a genetic algorithm for flexible docking. Journal of Molecular Biology 267:727–748

Jones T (1995) Evolutionary Algorithms, Fitness Landscapes and Search. PhD thesis, University of New Mexico, USA

Karp R (1972) Reducibility among combinatorial problems. In: Complexity of Computer Computations. Plenum, New York, pp 85–103

Kariuki B, Serrano-Gonzalez H, Johnston R, Harris K (1997) The application of a genetic algorithm for solving crystal structures from powder diffraction data. Chemical Physics Letters 280:189–195

Kassotakis I, Markaki M, Vasilakos A (2000) A hybrid genetic approach for channel reuse in multiple access telecommunication networks. IEEE Journal on Selected Areas in Communications 18:234–243

Katayama K, Hirabayashi H, Narihisa H (1998) Performance analysis for crossover operators of genetic algorithm. Transactions of the Institute of Electronics, Information and Communication Engineers J81D-I, 6:639–650

Kersting S, Raidl G, Ljubic I (2002) A memetic algorithm for vertex-biconnectivity augmentation. Lecture Notes in Computer Science 2279:101–110

Kim T, May G (1999) Intelligent control of via formation by photosensitive BCB for MCM-L/D applications. IEEE Transactions on Semiconductor Manufacturing 12:503–515

Kirkpatrick S, Gellat DC, Vecchi M (1983) Optimization by simulated annealing. Science 220: 671–680

Knödler K, Poland J, Zell A, Mitterer A (2002) Memetic algorithms for combinatorial optimization problems in the calibration of modern combustion engines. Proceedings of the GECCO'99 – Genetic and Evolutionary Computation Conference, pp 687

Krasnogor N (1999) Coevolution of genes and memes in memetic algorithms. Proceedings of the Graduate Student Workshop, Orlando, Florida, USA, July, pp 371

Krasnogor N (2002) Studies on the Theory and Design Space of Memetic Algorithms. Ph.D. Thesis, Faculty of Engineering, Computer Science and Mathematics. University of the West of England, United Kingdom

Krasnogor N, Blackburne B, Burke EK, Hirst JD (2002) Multimeme algorithms for protein structure prediction. Lecture Notes in Computer Science 2439:769–778

Krasnogor N, Smith J (2000) A memetic algorithm with self-adaptive local search: TSP as a case study. Proceedings of the GECCO'00 – Genetic and Evolutionary Computation Conference, pp 987–994

Krasnogor N, Smith J (2001) Emergence of profitable search strategies based on a simple inheritance mechanism. Proceedings of the GECCO'01 – Genetic and Evolutionary Computation Conference, pp 432–439

Krasnogor N, Smith J (2002) Multimeme algorithms for the structure prediction and structure comparison of proteins. Proceedings of the GECCO'02 – Genetic and Evolutionary Computation Conference, pp 42–44

Krishna K, Narasimha-Murty M (1999) Genetic k-means algorithm. IEEE Transactions on Systems, Man and Cybernetics, Part B 29:433–439

Krishna K, Ramakrishnan K, Thathachar M (1997) Vector quantization using genetic k-means algorithm for image compression. Proceedings of the 1997 International Conference on Information, Communications and Signal Processing, pp 1585–1587

Krzanowski R, Raper J (1999) Hybrid genetic algorithm for transmitter location in wireless networks. Computers, Environment and Urban Systems 23:359–382

Landree E, Collazo-Davila C, Marks L (1997) Multi-solution genetic algorithm approach to surface structure determination using direct methods. Acta Crystallographica Section B - Structural Science 53:916–922

Lazar G, Desjarlais J, Handel T (1997) De novo design of the hydrophobic core of ubiquitin. Protein Science 6:1167–1178

Lee C (1994) Genetic algorithms for single machine job scheduling with common due date and symmetric penalties. Journal of the Operations Research Society of Japan 37:83–95

Levine D (1996) A parallel genetic algorithm for the set partitioning problem. In: Meta-Heuristics: Theory & Applications. Kluwer Academic Publishers, pp 23–35

Li F, Morgan R, Williams D (1996) Economic environmental dispatch made easy with hybrid genetic algorithms. Proceedings of the International Conference on Electrical Engineering, pp 965–969

Li L, Darden T, Freedman S, Furie B, Baleja J, Smith H, Hiskey R, Pedersen L (1997) Refinement of the NMR solution structure of the gamma-carboxyglutamic acid domain of coagulation factor IX using molecular dynamics simulation with initial Ca2+ positions determined by a genetic algorithm. Biochemistry 36:2132–2138

Li S (1997) Toward global solution to map image estimation: Using common structure of local solutions. Lecture Notes in Computer Science 1223:361–374

Liaw C (2000) A hybrid genetic algorithm for the open shop scheduling problem. European Journal of Operational Research 124:28–42

Lin S (1965) Computer solutions of the traveling salesman problem. Bell System Technical Journal 10:2245–2269

Lin S, Kernighan B (1973) An Effective Heuristic Algorithm for the Traveling Salesman Problem. Operations Research 21:498–516

Ling S (1992) Integrating genetic algorithms with a prolog assignment program as a hybrid solution for a polytechnic timetable problem. Proceedings of 2nd PPSN – Parallel Problem Solving from Nature, pp 321–329

Lorber D, Shoichet B (1998) Flexible ligand docking using conformational ensembles. Protein Science 7:938–950

Louis S, Yin X, Yuan Z (1999) Multiple vehicle routing with time windows using genetic algorithms. Proceedings of the CEC'99 – Congress on Evolutionary Computation, pp 1804–1808

MacKay A (1995) Generalized crystallography. THEOCHEM – Journal of Molecular Structure 336:293–303

Maddox J (1995) Genetics helping molecular-dynamics. Nature 376, 6537:209–209

Mathias K, Whitley D (1992) Genetic operators, the fitness landscape and the traveling salesman problem. Proceedings of the 2nd PPSN – Parallel Problem Solving From Nature, pp 221–230

Mathias K, Whitley L (1994) Noisy function evaluation and the delta coding algorithm. Proceedings of the SPIE'94 – The International Society for Optical Engineering, pp 53–64

May A, Johnson M (1994) Protein-structure comparisons using a combination of a genetic algorithm, dynamic-programming and least-squares minimization. Protein Engineering 7:475–485

Mendes A, França P, Moscato P (2001) NP-Opt: An optimization framework for NP problems. Proceedings of the POM'01 – International Conference of the Production and Operations Management Society, pp 82–89

Mendes A, França P, Moscato P (2002a) Fitness landscapes for the total tardiness single machine scheduling problem. Neural Network World – an International Journal on Neural and Mass-Parallel Computing and Information Systems 2:165–180

Mendes A, Muller F, França P, Moscato P (2002b) Comparing meta-heuristic approaches for parallel machine scheduling problems. Production Planning & Control 13:1–6

Merkle L, Lamont G, Gates GJ, Pachter R (1996) Hybrid genetic algorithms for minimization of a polypeptide specific energy model. Proceedings of the ICEC'96 – International Conference on Evolutionary Computation, pp 396–400

Merz P (2002) A comparison of memetic recombination operators for the traveling salesman problem. Proceedings of the GECCO'02 – Genetic and Evolutionary Computation Conference, pp 472–479

Merz P, Freisleben B (1997a) A Genetic Local Search Approach to the Quadratic Assignment Problem. Proceedings of the 7th ICGA – International Conference on Genetic Algorithms, pp 465–472

Merz P, Freisleben B (1997b) Genetic Local Search for the TSP: New Results. Proceedings of the ICEC'97 – International Conference on Evolutionary Computation, pp 159–164

Merz P, Freisleben B (1998a) Memetic Algorithms and the Fitness Landscape of the Graph Bi-Partitioning Problem. Lecture Notes in Computer Science 1498:765–774

Merz P, Freisleben B (1998b) On the Effectiveness of Evolutionary Search in High-Dimensional NK-Landscapes. Proceedings of the ICEC'98 – International Conference on Evolutionary Computation, pp 741–745

Merz P, Freisleben B (1999a). A Comparison of Memetic Algorithms, Tabu Search, and Ant Colonies for the Quadratic Assignment Problem. Proceedings of the CEC'99 – Congress on Evolutionary Computation, pp 2063–2070

Merz P, Freisleben B (1999b) Fitness landscapes and memetic algorithm design. In: New Ideas in Optimization. McGraw-Hill, pp 245–260

Merz P, Freisleben B (2000) Fitness Landscapes, Memetic Algorithms and Greedy Operators for Graph Bi-Partitioning. Evolutionary Computation 8:61–91

Merz P, Freisleben B (2002a) Greedy and local search heuristics for the unconstrained binary quadratic programming problem. Journal of Heuristics 8:197–213

Merz P, Freisleben B (2002b) Memetic algorithms for the traveling salesman problem. Complex Systems. To be published

Merz P, Katayama K (2002) Memetic algorithms for the unconstrained binary quadratic programming problem. Bio Systems. To be published

Merz P, Zell A (2002) Clustering gene expression profiles with memetic algorithms. Lecture Notes in Computer Science 2439:811–820

Meza J, Judson R, Faulkner T, Treasurywala A (1996) A comparison of a direct search method and a genetic algorithm for conformational searching. Journal of Computational Chemistry 17:1142–1151

Mignotte M, Collet C, Pérez P, Bouthemy P (2000) Hybrid genetic optimization and statistical model based approach for the classification of shadow shapes in sonar imagery. IEEE Transactions on Pattern Analysis and Machine Intelligence 22:129–141

Miller D, Chen H, Matson J, Liu Q (1999) A hybrid genetic algorithm for the single machine scheduling problem. Journal of Heuristics 5:437–454

Miller S, Hogle J, Filman D (1996) A genetic algorithm for the *ab initio* phasing of icosahedral viruses. Acta Crystallographica Section D – Biological Crystallography 52:235–251

Min L, Cheng W (1998) Identical parallel machine scheduling problem for minimizing the makespan using genetic algorithm combined with simulated annealing. Chinese Journal of Electronics 7:317–321

Ming X, Mak K (2000) A hybrid hopfield network-genetic algorithm approach to optimal process plan selection. International Journal of Production Research 38:1823–1839

Monfroglio A (1996a) Hybrid genetic algorithms for a rostering problem. Software – Practice and Experience 26:851–862

Monfroglio A (1996b) Hybrid genetic algorithms for timetabling. International Journal of Intelligent Systems 11:477–523

Monfroglio A (1996c) Timetabling through constrained heuristic search and genetic algorithms. Software – Practice and Experience 26:251–279

Moscato P (1989) On evolution, search, optimization, genetic algorithms and martial arts: towards memetic algorithms. Technical Report C3P 826, Caltech Concurrent Computation Program, California Institute of Technology, Pasadena, USA

Moscato P (1993) An Introduction to Population Approaches for Optimization and Hierarchical Objective Functions: The Role of Tabu Search. Annals of Operations Research 41: 85–121

Moscato P (1999) Memetic algorithms: A short introduction. In: New Ideas in Optimization, McGraw-Hill, pp 219–234

Moscato P, Norman M (1992) A Memetic Approach for the Traveling Salesman Problem Implementation of a Computational Ecology for Combinatorial Optimization on Message-Passing Systems. In: Parallel Computing and Transputer Applications. IOS Press, pp 177–186

Moscato P, Tinetti F (1992) Blending heuristics with a population-based approach: A memetic algorithm for the traveling salesman problem. Technical Report 92-12, Universidad Nacional de La Plata, C.C. 75, 1900 La Plata, Argentina

Murata T, Ishibuchi H (1994) Performance evaluation of genetic algorithms for flowshop scheduling problems. Proceedings of the CEC'94 – Conference on Evolutionary Computation, pp 812–817

Murata T, Ishibuchi H, Tanaka H (1996) Genetic algorithms for flowshop scheduling problems. Computers & Industrial Engineering 30:1061–1071

Musil M, Wilmut M, Chapman N (1999) A hybrid simplex genetic algorithm for estimating geoacoustic parameters using matched-field inversion. IEEE Journal of Oceanic Engineering 24:358–369

Nagata Y, Kobayashi S (1997) Edge assembly crossover: A high-power genetic algorithm for the traveling salesman problem. Proceedings of the 7th ICGA – International Conference on Genetic Algorithms, pp 450–457

Niesse J, Mayne H (1996) Global geometry optimization of atomic clusters using a modified genetic algorithm in space-fixed coordinates. Journal of Chemical Physics 105:4700–4706

Nordstrom A, Tufekci S (1994) A genetic algorithm for the talent scheduling problem. Computers & Operations Research 21:927–940

Norman M, Moscato P (1989). A competitive and cooperative approach to complex combinatorial search. Technical Report 790, Caltech Concurrent Computation Program, California Institute of Technology, Pasadena, California, USA

Novaes A, De-Cursi J, Graciolli O (2000) A continuous approach to the design of physical distribution systems. Computers & Operations Research 27:877–893

Oliver I, Smith D, Holland J (1987) A study of permutation crossover operators on the traveling salesperson problem. Proceedings of the 2nd International Conference on Genetic Algorithms and their Applications, pp 224–230

Osmera P (1995) Hybrid and distributed genetic algorithms for motion control. Proceedings of the 4th International Symposium on Measurement and Control in Robotics, pp 297–300

Ostermark R (1999a) A neuro-genetic algorithm for heteroskedastic time-series processes: empirical tests on global asset returns. Soft Computing 3:206–220

Ostermark R (1999b) Solving a nonlinear non-convex trim loss problem with a genetic hybrid algorithm. Computers & Operations Research 26:623–635

Ostermark R (1999c) Solving irregular econometric and mathematical optimization problems with a genetic hybrid algorithm. Computational Economics 13:103–115

Ozcan E, Mohan C (1998) Steady state memetic algorithm for partial shape matching. Lecture Notes in Computer Science 1447:527–536

Ozdamar L (1999) A genetic algorithm approach to a general category project scheduling problem. IEEE Transactions on Systems, Man and Cybernetics, Part C (Applications and Reviews) 29:44–59

Pacey M, Patterson E, James M (2001) A photoelastic technique for characterising fatigue crack closure and the effective stress intensity factor. Zeszyty Naukowe Politechniki Opolskiej, Seria: Mechanika z.67, kol. 269/2001

Pacey M, Wang X, Haake S, Patterson E (1999) The application of evolutionary and maximum entropy algorithms to photoelastic spectral analysis. Experimental Mechanics 39:265–273

Paechter B, Cumming A, Norman M, Luchian H (1996) Extensions to a Memetic timetabling system. Lecture Notes in Computer Science 1153:251–265

Paechter B, Rankin R, Cumming A (1998) Improving a lecture timetabling system for university wide use. Lecture Notes in Computer Science 1408:156–165

Papadimitriou C, Steiglitz K (1982) Combinatorial Optimization: Algorithms and Complexity. Prentice-Hall, New Jersey

Pastorino M, Caorsi S, Massa A, Randazzo A (2002) Reconstruction algorithms for electromagnetic imaging. Proceedings of IEEE Instrumentation and Measurement Technology Conference, pp 1695–1700

Poland J, Knödler K, Mitterer A, Fleischhauer T, Zuber-Goos F, Zell A (2001). Evolutionary search for smooth maps in motor control unit calibration. Lecture Notes in Computer Science 2264:107–116

Pratihar D, Deb K, Ghosh A (1999) Fuzzy-genetic algorithms and mobile robot navigation among static obstacles. Proceedings of the CEC'99 – Congress on Evolutionary Computation, pp 327–334

Pucello N, Rosati M, D'Agostino G, Pisacane F, Rosato V, Celino M (1997) Search of molecular ground state via genetic algorithm: Implementation on a hybrid SIMD-MIMD platform. International Journal of Modern Physics C 8:239–252

Pullan, W (1997) Structure prediction of benzene clusters using a genetic algorithm. Journal of Chemical Information and Computer Sciences 37:1189–1193

Quagliarella D, Vicini A (1998) Hybrid genetic algorithms as tools for complex optimisation problems. Proceedings of the Second Italian Workshop on Fuzzy Logic, pp 300–307

Quintero A, Pierre S (2003) A multi-population memetic algorithm to optimize the assignment of cells to switches in cellular mobile networks. Submitted for publication

Radcliffe N (1992) Non-linear genetic representations. Proceedings of the 2nd PPSN – Parallel Problem Solving From Nature, pp 259–268

Radcliffe N (1994) The algebra of genetic algorithms. Annals of Mathematics and Artificial Intelligence 10:339–384

Radcliffe N, Surry P (1994a) Fitness Variance of Formae and Performance Prediction. Proceedings of the 3rd FOGA – Workshop on Foundations of Genetic Algorithms, pp 51–72

Radcliffe N, Surry P (1994b) Formal Memetic Algorithms. Lecture Notes in Computer Science 865:1–16

Raidl G, Julstron B (2000) A weighted coding in a genetic algorithm for the degree-constrained minimum spanning tree problem. Proceedings of the ACM Symposium on Applied Computing 2000, pp 440–445

Ramat E, Venturini G, Lente C, Slimane M (1997) Solving the multiple resource constrained project scheduling problem with a hybrid genetic algorithm. Proceedings of the 7th ICGA – International Conference on Genetic Algorithms, pp 489–496

Rankin R (1996) Automatic timetabling in practice. In: Practice and Theory of Automated Timetabling. First International Conference. Springer-Verlag, pp 266–279

Raymer M, Sanschagrin P, Punch W, Venkataraman S, Goodman E, Kuhn L (1997) Predicting conserved water-mediated and polar ligand interactions in proteins using a k-nearest-neighbors genetic algorithm. Journal of Molecular Biology 265:445–464

Rechenberg I (1973) Evolutionsstrategie: Optimierung technischer Systeme nach Prinzipien der biologischen Evolution. Frommann-Holzboog, Stuttgart, Germany

Reeves C (1996) Hybrid genetic algorithms for bin-packing and related problems. Annals of Operations Research 63:371–396

Reich C (2000) Simulation of imprecise ordinary differential equations using evolutionary algorithms. Proceedings of the ACM Symposium on Applied Computing 2000, pp 428–432

Ridao M, Riquelme J, Camacho E, Toro M (1998). An evolutionary and local search algorithm for planning two manipulators motion. Lecture Notes in Computer Science 1416:105–114

Rodrigues A, Ferreira JS (2001) Solving the rural postman problem by memetic algorithms. Proceedings of the 4th MIC – Metaheuristic International Conference, pp 679–684

Ruff C, Hughes S, Hawkes D (1999) Volume estimation from sparse planar images using deformable models. Image and Vision Computing 17:559–565

Runggeratigul S (2001) A memetic algorithm to communication network design taking into consideration an existing network. Proceedings of the 4th MIC – Metaheuristic International Conference, pp 91–96

Sakamoto A, Liu X, Shimamoto T (1997) A genetic approach for maximum independent set problems. IEICE Transactions on Fundamentals of Electronics Communications and Computer Sciences E80A, 3:551–556

Schnecke V, Vornberger O (1997) Hybrid genetic algorithms for constrained placement problems. IEEE Transactions on Evolutionary Computation 1:266–277

Schwefel HP (1965) Kybernetische Evolution als Strategie der experimentellen Forschung in der Stromungstechnik. Diplomarbeit, Technische Universitat Berlin, Germany

Shankland K, David W, Csoka T (1997) Crystal structure determination from powder diffraction data by the application of a genetic algorithm. Zeitschrift Fur Kristallographie 212:550–552

Shankland K, David W, Csoka T, McBride L (1998) Structure solution of ibuprofen from powder diffraction data by the application of a genetic algorithm combined with prior conformational analysis. International Journal of Pharmaceutics 165:117–126

Smith J (2002) Co-evolving memetic algorithms: Initial investigations. Lecture Notes in Computer Science 2439:537–548

Srinivasan D, Cheu R, Poh Y, Ng A (2000) Development of an intelligent technique for traffic network incident detection. Engineering Applications of Artificial Intelligence 13:311–322

Surry P, Radcliffe N (1996) Inoculation to initialise evolutionary search. Lecture Notes in Computer Science 1143:269–285

Syswerda G (1989) Uniform crossover in genetic algorithms. Proceedings of the 3rd ICGA – International Conference on Genetic Algorithms, pp 2–9

Taguchi T, Yokota T, Gen M (1998) Reliability optimal design problem with interval coefficients using hybrid genetic algorithms. Computers & Industrial Engineering 35:373–376

Tam K, Compton R (1995) GAMATCH - a genetic algorithm-based program for indexing crystal faces. Journal of Applied Crystallography 28:640–645

Topchy A, Lebedko O, Miagkikh V (1996). Fast learning in multilayered networks by means of hybrid evolutionary and gradient algorithms. Proceedings of International Conference on Evolutionary Computation and its Applications, pp 390–398

Urdaneta A, Gómez J, Sorrentino E, Flores L, Díaz R (1999) A hybrid genetic algorithm for optimal reactive power planning based upon successive linear programming. IEEE Transactions on Power Systems 14:1292–1298

Valenzuela J, Smith A (2002) A seeded memetic algorithm for large unit commitment problems. Journal of Heuristics 8:173–195

VanKampen A, Strom C, Buydens L (1996) Lethalization, penalty and repair functions for constraint handling in the genetic algorithm methodology. Chemometrics And Intelligent Laboratory Systems 34:55–68

Wang L, Yen, J (1999) Extracting fuzzy rules for system modeling using a hybrid of genetic algorithms and kalman filter. Fuzzy Sets and Systems 101:353–362

Watson J, Rana S, Whitley L, Howe A (1999) The impact of approximate evaluation on the performance of search algorithms for warehouse scheduling. Journal of Scheduling 2:79–98

Wehrens R, Lucasius C, Buydens L, Kateman G (1993) HIPS, A hybrid self-adapting expert system for nuclear magnetic resonance spectrum interpretation using genetic algorithms. Analytica Chimica ACTA 277:313–324

Wei P, Cheng L (1999) A hybrid genetic algorithm for function optimization. Journal of Software 10:819–823

Wei X, Kangling F (2000) A hybrid genetic algorithm for global solution of nondifferentiable nonlinear function. Control Theory & Applications 17:180–183

Weile D, Michielssen E (1999) Design of doubly periodic filter and polarizer structures using a hybridized genetic algorithm. Radio Science 34:51–63

White R, Niesse J, Mayne H (1998) A study of genetic algorithm approaches to global geometry optimization of aromatic hydrocarbon microclusters. Journal of Chemical Physics 108:2208–2218

Willett P (1995) Genetic algorithms in molecular recognition and design. Trends in Biotechnology 13:516–521

Wolpert D, Macready W (1997) No free lunch theorems for optimization. IEEE Transactions on Evolutionary Computation 1:67–82

Xiao J, Zhang L (1997) Adaptive evolutionary planner/navigator for mobile robots. IEEE Transactions on Evolutionary Computation 1:18–28

Yao X (1993) Evolutionary artificial neural networks. International Journal of Neural Systems 4:203–222

Yeh I (1999) Hybrid genetic algorithms for optimization of truss structures. Computer Aided Civil and Infrastructure Engineering 14:199–206

Yeh WC (2000) A memetic algorithm for the min k-cut problem. Control and Intelligent Systems 28:47–55

Yoneyama M, Komori H, Nakamura S (1999) Estimation of impulse response of vocal tract using hybrid genetic algorithm - a case of only glottal source. Journal of the Acoustical Society of Japan 55:821–830

Zacharias C, Lemes M, Pino A (1998) Combining genetic algorithm and simulated annealing: a molecular geometry optimization study. THEOCHEM – Journal of Molecular Structure 430:29–39

Zelinka I, Vasek V, Kolomaznik K, Dostal P, Lampinen J (2001) Memetic algorithm and global optimization of chemical reactor. Proceedings of the 13th International Conference on Process Control.

Zwick M, Lovell B, Marsh J (1996) Global optimization studies on the 1-d phase problem. International Journal of General Systems 25:47–59

4 Scatter Search and Path Relinking: Foundations and Advanced Designs

Fred Glover, Manuel Laguna and Rafael Martí

Scatter Search and its generalized form Path Relinking, are evolutionary methods that have been successfully applied to hard optimization problems. Unlike genetic algorithms, they operate on a small set of solutions and employ diversification strategies of the form proposed in Tabu Search, which give precedence to strategic learning based on adaptive memory, with limited recourse to randomization. The fundamental concepts and principles were first proposed in the 1970s as an extension of formulations, dating back to the 1960s, for combining decision rules and problem constraints. (The constraint combination approaches, known as surrogate constraint methods, now independently provide an important class of relaxation strategies for global optimization.) The Scatter Search framework is flexible, allowing the development of alternative implementations with varying degrees of sophistication. Path Relinking, on the other hand, was first proposed in the context of the Tabu Search metaheuristics, but it has been also applied with a variety of other methods. This chapter's goal is to provide a grounding in the essential ideas of Scatter Search and Path Relinking, together with pseudo-codes of simple versions of these methods, that will enable readers to create successful applications of their own.

4.1 Introduction

Scatter Search (SS) and Path Relinking (PR) are evolutionary methods that construct solutions by combining others by means of strategic designs that exploit the knowledge on the problem at hand. The goal of these procedures is to enable a solution procedure based on the combined elements to yield better solutions than one based on the original elements. SS and PR are rapidly becoming methods of choice for designing solution procedures for hard combinatorial optimization problems. A comprehensive examination of these methodologies can be found in

Glover (1997, 1999), Glover, Laguna and Martí (2000, 2002) and Laguna and Martí (2003).

In common with other evolutionary methods, SS and PR operate with a population of solutions, rather than with a single solution at a time, and employ procedures for combining these solutions to create new ones. The meaning of "combining," and the motivation for carrying it out, has a rather special origin and character in the SS/PR setting, however. One of the distinguishing features of this approaches is its intimate association with the Tabu Search (TS) metaheuristic, and hence its adoption of the principle that search can benefit by incorporating special forms of adaptive memory, along with procedures particularly designed for exploiting that memory. In fact, Scatter Search and Tabu Search share common origins, and initially SS was simply considered one of the component processes available within the TS framework. However, most of the TS literature and the preponderance of TS implementations have disregarded this component, with the result that the merits of Scatter Search did not come to be recognized until quite recently, when a few studies began to look at it as an independent method in the context of evolutionary procedures.

Unlike a "population" in genetic algorithms, the reference set of solutions in SS and PR is relatively small. A typical population size in a genetic algorithm consists of 100 elements, which are sampled by various schemes with a significant reliance on randomization to create combinations. In contrast, SS and PR chooses two or more elements of the reference set in a systematic way with the purpose of systematically creating new solutions. Randomization, where employed, is given a highly controlled and strategic character. (Some of the more recent GA proposals are beginning to incorporate some of the features of early SS and PR designs, including an adoption of some of the SS and PR terminology. However, a number of key elements continue to distinguish the SS and PR methods from GA approaches, with important consequences for performance.)

The process of generating combinations of a set of reference solutions in Scatter Search consists of forming linear combinations that are structured to accommodate discrete requirements (such as those for integer-valued variables). The resulting combinations may be characterized as generating paths between and beyond these solutions, where solutions on such paths also serve as sources for generating additional paths. Path Relinking is based on this conception of the meaning of creating *combinations* of solutions, but making reference to paths between and beyond selected solutions in neighborhood space, rather than in Euclidean space as in the case of Scatter Search. (More precisely, SS operates within a subset of Euclidean space determined by imposing certain constraints according to the problem setting.). The following principles summarize the foundations of the SS and PR methodologies:

- Useful information about the form (or location) of optimal solutions is typically contained in a suitably diverse collection of elite solutions.
- When solutions are combined as a strategy for exploiting such information, it is important to provide mechanisms capable of constructing combinations that extrapolate beyond the regions spanned by the solutions consid-

ered. Similarly, it is also important to incorporate heuristic processes to map combined solutions into new solutions. The purpose of these combination mechanisms is to incorporate both diversity and quality.

- Taking account of multiple solutions simultaneously, as a foundation for creating combinations, enhances the opportunity to exploit information contained in the union of elite solutions.

In this paper we describe the foundations of these closely related methods and discuss some extensions and advanced designs that have been proved effective on solving hard combinatorial optimization problems. Section 2 is devoted to the foundations of both methods, where we provide pseudo-code of simple versions of each. Extensions and advanced strategies are described in Section 3, to conclude the paper.

4.2 Foundations

4.2.1 Scatter Search

Scatter Search consists of five component processes:

1. A *Diversification Generation Method* to generate a collection of diverse trial solutions, using one or more arbitrary trial solutions (or seed solutions) as an input.
2. An *Improvement Method* to transform a trial solution into one or more enhanced trial solutions. (Neither the input nor the output solutions are required to be feasible, though the output solutions are typically feasible. If the input trial solution is not improved as a result of the application of this method, the "enhanced" solution is considered to be the same as the input solution.)
3. A *Reference Set Update Method* to build and maintain a *reference set* consisting of the b "best" solutions found (where the value of b is typically small, e.g., no more than 20), organized to provide efficient accessing by other parts of the solution procedure. Several alternative criteria may be used to add solutions to the reference set and delete solutions from the reference set.
4. A *Subset Generation Method* to operate on the reference set, to produce a subset of its solutions as a basis for creating combined solutions. The most common subset generation method is to generate all pairs of reference solutions (i.e., all subsets of size 2).
5. A *Solution Combination Method* to transform a given subset of solutions produced by the Subset Generation Method into one or more combined solutions. The combination method is analogous to the crossover operator in genetic algorithms but it must be capable of combining two or more solutions.

The structured combinations produced by Scatter Search are designed with the goal of creating weighted centers of selected sub-regions. These include non-convex combinations that project new centers into regions that are external to the original reference solutions. The dispersion patterns created by such centers and their external projections have been found useful in a variety of application areas.

Another important feature relates to the strategies for selecting particular subsets of solutions to combine. These strategies are typically designed to make use of a type of clustering to allow new solutions to be constructed "within clusters" and "across clusters". The method is generally organized to use improving mechanisms that are able to operate on infeasible solutions, removing the restriction that solutions must be feasible in order to be included in the reference set.

The basic procedure for a minimization problem given in Figure 1.4 starts with the creation of an initial reference set of solutions (*RefSet*). The Diversification Generation Method is used to build a large set of diverse solutions *P*. The size of *P* (*PSize*) is typically 10 times the size of *RefSet*. An *Improvement Method* is applied to each generated solution to obtain a better solution, which is added to *P* (often, though not invariably, as a replacement for the solution it was derived from). Initially, the reference set *RefSet* consists of *b* distinct and maximally diverse solutions from *P*. The solutions in *RefSet* are ordered according to quality, where the best solution is the first one in the list. The search is then initiated by assigning the value of TRUE to the Boolean variable *NewSolutions*. In step 3, *NewSubsets* is constructed and *NewSolutions* is switched to FALSE. We focus our attention in this basic illustrative scheme to subsets of size 2, and specify that the cardinality of *NewSubsets* corresponding to the initial reference set is given by $(b^2\text{-}b)/2$, which accounts for all pairs of solutions in *RefSet*. The pairs in *NewSubsets* are selected one at a time in lexicographical order and the Solution Combination Method is applied to generate one or more solutions in step 5. The *Improvement Method* is applied again to every newly created solution. If the final solution improves upon the worst solution currently in *RefSet*, the new solution replaces the worst and *RefSet* is reordered in step 6. The *NewSolutions* flag is switched to TRUE and the subset *s* that was just combined is deleted from *NewSubsets* in steps 7 and 8, respectively.

---

1. Start with $P = \emptyset$.
   - Use the Diversification Generation Method to construct a solution *y*.
   - Apply the Improvement Method to y. Let x be the final solution.
   - If $x \notin P$ then add *x* to *P* (i.e., $P = P \cup x$), otherwise, discard *x*.
   - Repeat this step until $|P| = PSize$. Build $RefSet = \{ x^1, \ldots, x^b \}$ with *b* diverse solutions in *P*.
2. Evaluate the solutions in *RefSet* and order them according to their objective function value such that x^1 is the best solution and x^b the worst. Make *NewSolutions* = TRUE.

while (*NewSolutions*) **do**

  3. Generate *NewSubsets*, which consists of all pairs of solutions in *RefSet* that include at least one new solution. Make *NewSolutions* = FALSE.

  while ($NewSubsets \neq \emptyset$) **do**

    4. Select the next subset *s* in *NewSubSets*.
    5. Apply the Solution Combination Method to *s* to obtain one or more new solutions *y*.
    6. Apply the Improvement Method to *y* and obtain *x*.

    if (*x* is not in *RefSet* and $f(x)<f(x^b)$) **then**

      6. Make $x^b = x$ and reorder *RefSet*.
      7. Make *NewSolutions* = TRUE.

end if
8. Delete *s* from *NewSubsets*.
end while
end while

Fig. 1.4. Outline of a simple scatter search approach

This illustrative procedure is very aggressive in trying to improve upon the quality of the solutions in the current reference set, to the extent that it sacrifices search diversity. In fact, the Diversification Generation Method is used only once to generate *PSize* different solutions at the beginning of the search and is never employed again. The initial *RefSet* is built by selecting a solution from *P* and then making *b*-1 more selections in order to maximize the minimum distance between the candidate solution and the solutions currently in *RefSet*. That is, for each candidate solution *x* in *P-RefSet* and reference set solution *y* in *RefSet*, we calculate a measure of distance or dissimilarity *d(x,y)*. We then select the candidate solution that maximizes $d_{\min}(x) = \min_{y \in RefSet} \{d(x, y)\}$.

The updating of the reference set is based on improving the quality of the worst solution and the search terminates when no new solutions are admitted to *RefSet*. The Subset Generation Method is also very simple and consists of generating all pairs of solutions in *RefSet* that contain at least one new solution. This means that the procedure does not allow for two solutions to be subjected to the Combination Method more than once. In the next section, several improvements to this basic scheme are proposed.

4.2.2 Path Relinking

Path relinking was originally suggested as an approach to integrate intensification and diversification strategies in the context of tabu search (Glover, 1994; Glover and Laguna, 1997). This approach generates new solutions by exploring trajectories that connect high-quality solutions, by starting from one of these solutions, called an *initiating solution*, and generating a path in the neighbourhood space that leads toward the other solutions, called *guiding solutions*. This is accomplished by selecting moves that introduce attributes contained in the guiding solutions.

PR can be considered an extension of the combination mechanisms of Scatter Search. Instead of directly producing a new solution when combining two or more original solutions, PR generates paths between and beyond the selected solutions in the neighborhood space. The character of such paths is easily specified by reference to solution attributes that are added, dropped or otherwise modified by the moves executed. Examples of such attributes include edges and nodes of a graph, sequence positions in a schedule, vectors contained in linear programming basic solutions, and values of variables and functions of variables.

To generate the desired paths, it is only necessary to select moves that perform the following role: upon starting from an *initiating solution,* the moves must progressively introduce attributes contributed by a *guiding solution* (or reduce the distance between attributes of the initiating and guiding solutions). The roles of the initiating and guiding solutions are interchangeable; each solution can also be induced to move simultaneously toward the other as a way of generating combinations. First consider the creation of paths that join two selected solutions *x*□ and *x*□, restricting attention to the part of the path that lies 'between' the solutions, producing a solution sequence *x*□ = $x(1)$, $x(2)$, ..., $x(r)$ = *x*□. To reduce the number of options to be considered, the solution $x(i + 1)$ may be created from $x(i)$ at each step by choosing a move that leaves a reduced number of moves remaining to reach x.

Instead of building a *Reference Set* as in SS, Path Relinking usually starts from a given set of elite solutions obtained during a search process. To simplify the terminology, we will also let *RefSet* (short for "Reference Set"), refer to this set of b solutions that have been selected during the application of the embedded search method. This method can be Tabu Search, as in Laguna et al. (1999), GRASP, as in Laguna and Martí (1999), or simply a diversification generator coupled with an improvement method as proposed in Scatter Search. From this point of view, SS and PR can be considered pool-oriented methods (Greistorfer and Voss, 2001) that operate on a set of reference solutions and basically differ in the way in which the Reference Set is constructed, maintained, updated and improved.

In the basic scheme of SS proposed above, all pairs of solutions in the RefSet are subjected to the combination method. Similarly, in this basic version of PR we will consider all pairs in the RefSet to perform a relinking phase. (For each pair (*x*□, *x*□) two paths are considered; one from *x*□ to *x*□ and the other from *x*□ to *x*□.)

Several studies have experimentally found that it is convenient to add a local search exploration from some of the generated solutions within the relinking path, as proposed in Glover (1994), in order to produce improved outcomes. We refer the reader to Laguna and Martí (1999), Piñana et al. (2001) or Laguna et al. (1999) for some examples. Note that two consecutive solutions after a relinking step are very similar and differ only in the attribute that was just introduced. Therefore, it is generally not efficient to apply an Improvement Method at every step of the relinking process. We introduce the parameter *NumImp* to control its application. In particular, the Improvement Method is applied every *NumImp* steps of the relinking process. (An alternative suggested in Glover (1994) is to keep track of a few "best solutions" generated during the path trace, or of a few best neighbors of the solutions generated, and then return to these preferred candidate solutions to initiate the improvement process.)

Figure 2.4 shows a simple PR procedure for a minimization problem. It starts with the creation of an initial set of b elite solutions (*RefSet*). As in the SS method, the solutions in *RefSet* are ordered according to quality, and the search is initiated by assigning the value of TRUE to *NewSolutions*. In step 3, *NewSubsets* is constructed with all the pairs of solutions in *RefSet,* and *NewSolutions* is switched to FALSE. The pairs in *NewSubsets* are selected one at a time in lexicographical order and the Relinking Method is applied to generate two paths of solutions in steps 5 and 7. The *Improvement Method* is applied every *NumImp* steps of the relinking process in each path (steps 6 and 8). Each of the generated solutions in each path (including also those obtained after the application of the Improvement Method), is checked to see whether it improves upon the worst solution currently in *RefSet*. If so, the new solution replaces the worst and *RefSet* is reordered in step 6. The *NewSolutions* flag is switched to TRUE and the pair (x□, x□) that was just combined is deleted from *NewSubsets* in step 11.

---

1. Obtain a RefSet of b elite solutions.
2. Evaluate the solutions in *RefSet* and order them according to their objective function value such that x^1 is the best solution and x^b the worst. Make *NewSolutions* = TRUE.

while (*NewSolutions*) **do**

  3. Generate *NewSubsets*, which consists of all pairs of solutions in *RefSet* that include at least one new solution. Make *NewSolutions* = FALSE.

  while (*NewSubsets* $\neq \varnothing$) **do**

    4. Select the next pair (x□, x□) in *NewSubSets*.

    5. Apply the Relinking Method to produce the sequence x□ = $x(1)$, $x(2)$, …, $x(r)$= x□

    for i=1 **to** i < r / *NumImp* **do**

      6. Apply the Improvement Method to $x(NumImp \cdot i)$.

    end for

    7. Apply the Relinking Method to produce the sequence x□= $y(1)$, $y(2)$, …, $y(s)$= x□

    for i=1 **to** i < s / *NumImp* **do**

      8. Apply the Improvement Method to $y(NumImp \cdot i)$.

    end for

    for (each generated solutions x) **if** (x is not in *RefSet* and $f(x)<f(x^b)$) **then**

      9. Make $x^b = x$ and reorder *RefSet*.

      10. Make *NewSolutions* = TRUE.

    end if, end for

    11. Delete (x□, x□) from *NewSubsets*.

  end while

end while

Fig. 2.4 Outline of a simple path relinking approach

As in the illustrative scheme proposed for SS, the updating of the reference set is based on improving the quality of the worst solution and the search terminates when no new solutions are admitted to *RefSet*. Similarly, the Subset Generation Method is also very simple and consists of generating all pairs of solutions in *RefSet* that contain at least one new solution. In the next section we propose strategies to overcome the limitations of this basic design.

4.3 Advanced Strategies

4.3.1 Scatter Search

A. REFSET REBUILDING

This update adds a mechanism to partially rebuild the reference set when no new solutions can be generated with the Combination Method. The update is performed when the inner "while-loop" in Figures 1 and 2 fails and the *NewSolutions* variable is FALSE. The *RefSet* is partially rebuilt with a diversification update, which works as follows. Solutions $x^{[b/2]+1}, \ldots, x^b$ are deleted from *RefSet*, where $[v]$ is the largest integer less than or equal to v. The frequency counts in the Diversification Generation Method are updated with the corresponding values from the solutions that remain in the reference set, that is, $x^1, \ldots, x^{[b/2]}$. Then, the generator is used to construct a set P of new solutions. Solutions $x^{[b/2]+1}, \ldots, x^b$ in *RefSet* are sequentially selected from P with the max-min criterion used in step 1 of Figure 1 (and described in the previous section). The min-max criterion is applied against solutions $x^1, \ldots, x^{[b/2]}$ when selecting solution $x^{[b/2]+1}$, then against solutions $x^1, \ldots, x^{[b/2]+1}$ when selecting solution $x^{[b/2]+2}$, and so on.

B. REFSET DYNAMIC UPDATE

In the basic design, the new solutions that become members of *RefSet* are not combined until all pairs in *NewSubsets* are subjected to the Combination Method. We call *Static Update* to this strategy. On the other hand, the Dynamic Update strategy applies the Combination Method to new solutions in a manner that is faster than in the basic design. That is, if a new solution is admitted to the reference set, the goal is to allow this new solution to be subjected to the Combination Method as quickly as possible.

A simple way to implement this strategy is by reducing *NewSubsets* to just one pair of solutions. Then, the procedure returns to the generation of subsets immediately after one application of the Combination Method. The downside of this intensification phase is that some solutions in *RefSet* could be replaced before being combined. To illustrate this, suppose that the initial *RefSet* = $\{ x^1, x^2, x^3, x^4 \}$ and that the combination of the pair (x^1, x^2) results in a solution y for which:

$$f(y) < f(x^3)$$

Then, the reference set is changed in such a way that the updated *RefSet* = $\{ x^1, x^2, y, x^3 \}$. The search continues by combining the members of the pair (x^1, y). Clearly, the quick updating of the reference set will eliminate the operation of combining members of the pair (x^1, x^4). Campos et al. (2001) compare this strategy with the basic design in the linear ordering problem.

C. MULTIPLE SOLUTION COMBINATIONS

The combination mechanism in SS is not limited in its general form in combining just two solutions. Glover (1997) proposes a procedure that generates subsets of the RefSet that have useful properties while avoiding the duplication of subsets previously generated. Specifically, four different collections of subsets of RefSet are introduced:

- SubsetType 1 : All 2- element subsets

- SubsetType 2: 3-element subsets derived from 2-element subsets by augmenting each 2-element subset to include the best solution not in this subset.
- SubsetType 3: 4-element subsets derived from 3-element subsets by augmenting each 3-element subset to include the best solution not in this subset.
- SubsetType 4: the subsets consisting of the best i elements, for i=5 to *b*.

These subsets interact with a dynamic update of the RefSet as described above. Algorithms for maintaining the RefSet with these Subsets and avoid duplications are detailed in Glover (1997).

D. HANDLING PROBLEMS WITH ZERO-ONE VARIABLES

When generating linear combinations of points it is important to be sure the results are dispersed appropriately, and to avoid calculations that simply duplicate other – particularly where the duplications occur after mapping continuous values into integer values. This caveat is especially to be heeded in the case of zero-one variables, since each continuous value can only map into one of the two alternatives 0 and 1, and a great deal of effort can be wasted by disregarding this obvious consequence.

Specifically, for the 0-1 case, a useful set of combinations and mappings (except for random variations) of r reference points consists simply of those that create a positive integer threshold $t \leq r$, and require that the offspring will have its ith component $x_i = 1$ if and only if at least t of the r parents have $x_i = 1$. (A preferred way to apply this rule is to subdivide the variables into categories, and to apply different thresholds to different categories – i.e., to different subvectors. Many problems offer natural criteria for subdividing variables into categories. However, in the absence of this, or in the case where greater diversity is desired, the categories can be arbitrarily generated and varied. At the extreme, for example, each variable can belong to its own category. This SS option gives the same set of alternative mappings as the so-called "Bernoulli crossover" introduced into GAs about a decade after the original SS proposals, except that Scatter Search does not select the 0 and 1 values at random as in the Bernoulli scheme, but instead selects these values by relying on frequency memory to achieve intensification and diversification goals.)

To illustrate the consequences of such a threshold mechanism applied to 2 parents (or to selected subvectors of two parents), t = 1 gives the offspring that is the union of the 1's in the parents and t = 2 gives the offspring that is the intersection of the 1's in the parents. But the threshold mechanism is not the only relevant one to consider. Two other kinds of combinations can be produced for r = 2, consisting of those equivalent to subtracting one vector (or subvector) from another and setting $x_i = 1$ if and only if the difference is positive. In this case it is not unreasonable to consider one more linear combination by summing these two offspring to get the symmetric difference between the parents.

To summarize: restricting attention to r = 2 yields the following options of interest:

(1) the intersection of 1's
(2) the union of 1's
(3) the 1's that belong to parent 1 and not to parent 2
(4) the 1's that belong to parent 2 and not to parent 1
(5) the symmetric difference (summing (3) and (4)).

In some contexts the last three options are not as useful as the first two, but in others they can be exceedingly important. An example of the importance of the symmetric difference is given in Glover and Laguna (1997).

An analogous set of options for r = 3 is given by:

(1) the intersection of 1's
(2) the union of 1's
(3) the 1's that belong to a majority of the parents
(4) the 1's that belong to exactly 1 parent but not more
(5) the 1's that belong to exactly 2 parents but not more
(6) the 1's that belong to the union of 2 parents, excluding those that belong to the 3rd parent.
(3 different cases)
(7) the 1's that belong to the intersection of 2 parents, excluding those that belong to the 3rd parent. (3 different cases)

Inclusion of the 6th and 7th options, each of which involves 3 different cases, entails a fair amount of effort, which may not be warranted in many situations. However, it would make sense to use (6) and (7) by defining the "3rd parent" to be the worst of the 3, hence the 1's that are excluded are those that belong to the worst parent.

In general, the relevant mappings for linear combinations involving 0-1 variables can be identified in advance by rules such as those indicated, applied either to full vectors or to subvectors. This results in a much more economical and effective process than separately generating a wide range of linear combinations and then performing a mapping operation. The latter can produce multiple duplications and even miss relevant alternatives.

4.3.2 Path Relinking

A. VARIATION AND TUNNELING

A variant of the Path Relinking approach starts with both endpoints *x*□ and *x*□ simultaneously, and produces two sequences *x*□ = *x*□(l), …, *x*□(*r*) and *x*□ = *x*□ (l), …, *x*□(*s*). The choices in this case are designed to yield *x*□(*r*) = x□(*s*), for final values of *r* and *s*. To progress toward this outcome when *x*□(*r*) = *x*□(*s*), either *x*□(*r*) is selected to create *x*□(*r* + 1), as by the criterion of minimizing reducing the number of moves remaining to reach $x\square(s)$, or $x\square(s)$ is chosen to create $x\square(s + 1)$, as by the criterion of minimizing (reducing) the number of moves remaining to reach $x\square(r)$. From these options, the move is selected that produces the smallest $c(x)$ value, thus also determining which of r or s is incremented on the next step. Basing the relinking process on more than one neighborhood also produces a useful variation.

The Path Relinking approach also benefits from a tunneling strategy that encourages a different neighborhood structure to be used than in the standard search phase. For example, moves for Path Relinking may be periodically allowed that normally would be excluded due to creating infeasibility. Such a practice is protected against the possibility of becoming 'lost' in an infeasible region, since feasibility evidently must be recovered by the time $x\square$ is reached. The tunneling effect therefore offers a chance to reach solutions that might otherwise be bypassed. In the variant that starts from both $x\square$ and $x\square$, priority may be given to keeping at least one of $x\square(r)$ and $x\square(s)$ feasible.

B. EXTRAPOLATED RELINKING

The Path Relinking approach goes beyond consideration of points 'between' x' and x'' in the same way that linear combinations extend beyond points that are expressed as convex combinations of two endpoints. In seeking a path that continues beyond x'' (starting from the point x') we once again invoke the Tabu Search concept of referring to sets of attributes associated with the solutions generated, as a basis for choosing a move that 'approximately' leaves the fewest moves remaining to reach x''. Let $A(\text{x})$ denote the set of solution attributes associated with ('contained in') x, and let A_drop denote the set of solution attributes that are dropped by moves performed to reach the current solution $x'(i)$, starting from x'. (Such attributes may be components of the x vectors themselves, or may be related to these components by appropriately defined mappings.)

Define a *to-attribute* of a move to be an attribute of the solution produced by the move, but not an attribute of the solution that initiates the move. Similarly, define a *from-attribute* to be an attribute of the initiating solution but not of the new solution produced. Then we seek a move at each step to maximize the number of *to-attributes* that belong to $A(x'') - A(x(i))$, and subject to this to minimize the number that belong to $A\_drop - A(x'')$. Such a rule generally can be implemented very efficiently by appropriate data structures.

Once $x(r) = x\square$ is reached, the process continues by modifying the choice rule as follows. The criterion now selects a move to maximize the number of its *to-attributes* not in A_drop minus the number of its *to-attributes* that are in A_drop, and subject to this to minimize the number of its *from-attributes* that belong to $A(x\square)$. The combination of these criteria establishes an effect analogous to that achieved by the standard algebraic formula for extending a line segment beyond an endpoint. (The secondary minimization criterion is probably less important in this determination.) The path then stops whenever no choice remains that permits the maximization criterion to be positive. The maximization goals of these two criteria are of course approximate, and can be relaxed.

C. MULTIPLE PARENTS

New points can be generated from multiple parents in Path Relinking by the following explicit process. Instead of moving from a point *x*□ to (or through) a second point *x*□, we replace *x*□ by a collection of solutions *X*□. Upon generating a point $x(i)$, the options for determining a next point $x(i + 1)$ are given by the union of the solutions in *X*□, or more precisely, by the union *A*□ of the attribute sets $A(x)$, for $x \in$ *X*□. *A*□ takes the role of $A(x)$ in the attribute-based approach previously described, with the added stipulation that each attribute is counted (weighted) in accordance with the number of times it appears in elements $A(x)$ of the collection. Still more generally, we may assign a weight to $A(x)$, which thus translates into a sum of weights over A'' applicable to each attribute, creating an effect analogous to that of creating a weighted linear combination in Euclidean space. Parallel processing can be applied to operate on an entire collection of points *x*□ $\in$ *X*□ relative to a second collection *x*□ $\in$ *X*□ by this approach. Further considerations that build on these ideas are detailed in Glover (1994), but they go beyond the scope of our present development.

This multiparent Path Relinking approach generates new elements by a process that emulates the strategies of the original Scatter Search approach at a higher level of generalization. The reference to neighborhood spaces makes it possible to preserve desirable solution properties (such as complex feasibility conditions in scheduling and routing), without requiring artificial mechanisms to recover these properties in situations where they may otherwise become lost.

Acknowledgments

Fred Glover's research is partially supported by the Office of Naval Research Contract N00014-01-1-0917 in connection with the Hearin Center of Enterprise Science at the University of Mississippi. Rafael Marti's research is partially supported by the Spanish Government (PR2002-0060 and TIC2000-1750-C06-01).

References

Campos, V., Glover, F., Laguna, M. and Martí, R. (2001), "An Experimental Evaluation of a Scatter Search for the Linear Ordering Problem" *Journal of Global Optimization* 21, 397-414

Glover F. (1994). "Tabu Search for Nonlinear and Parametric Optimization (with Links to Genetic Algorithms)," *Discrete Applied Mathematics*, vol. 49, 231-255.

Glover, F. (1997). "A Template for Scatter Search and Path Relinking," in *Lecture Notes in Computer Science*, 1363, J.K. Hao, E. Lutton, E. Ronald, M. Schoenauer, D. Snyers (Eds.), 1-53. (Expanded version available on request.)

Glover F. (1999). "Scatter Search and Path Relinking," In: D Corne, M Dorigo and F Glover (eds.) *New Ideas in Optimisation*, Wiley.

Glover, F. and M. Laguna (1997) *Tabu Search*, Kluwer Academic Publishers, Boston.

Glover, F., M. Laguna and R. Martí (2000), "Fundamentals of Scatter Search and Path Relinking" *Control and Cybernetics*, 29(3), 653-684

Glover, F., M. Laguna and R. Martí (2002). "Scatter Search," to appear in *Theory and Applications of Evolutionary Computation: Recent Trends*, A. Ghosh and S. Tsutsui (Eds.) Springer-Verlag.

Greistorfer, P. and Voss, S. (2001), "Controlled Pool Maintenance in Combinatorial Optimization" Conference on *Adaptive Memory and Evolution: Tabu Search and Scatter Search*, University of Mississippi.

Laguna, M. (2000) "Scatter Search," to appear in *Handbook of Applied Optimization*, P. M. Pardalos and M. G. C. Resende (Eds.), Oxford Academic Press.

Laguna, M. and R. Martí (1999). "GRASP and Path Relinking for 2-Layer Straight Line Crossing Minimization," *INFORMS Journal on Computing*, Vol. 11, No. 1, pp. 44-52.

Laguna, M. and R. Martí (2000a). "Experimental Testing of Advanced Scatter Search Designs for Global Optimization of Multi-modal Functions," Working paper, University of Colorado.

Laguna, M. and R. Martí (2000b). "The OptQuest Callable Library," *Optimization Software Class Libraries*, S. Voss and D. L. Woodruff (Eds.), Kluwer, Boston. pp. 193- 218

Laguna, M. and Martí, R. (2003), Scatter Search. Methodology and Implementations in C, Kluwer Academic Publishers, 312 pp.

Laguna, M., R. Martí and V. Campos (1999). "Intensification and Diversification with Elite Tabu Search Solutions for the Linear Ordering Problem," *Computers and Operations Research*, Vol. 26, pp. 1217-1230.

Piñana, E., Plana, I., Campos, V. and Martí, R. (2001). "GRASP and Path relinking for the matrix bandwidth minimization", *European Journal of Operational Research*, forthcoming.

5 Ant Colony Optimization

Vittorio Maniezzo, Luca Maria Gambardella, Fabio de Luigi

5.1 Introduction

Ant Colony Optimization (ACO) is a paradigm for designing metaheuristic algorithms for combinatorial optimization problems. The first algorithm which can be classified within this framework was presented in 1991 [21, 13] and, since then, many diverse variants of the basic principle have been reported in the literature. The essential trait of ACO algorithms is the combination of a priori information about the structure of a promising solution with a posteriori information about the structure of previously obtained good solutions.

Metaheuristic algorithms are algorithms which, in order to escape from local optima, drive some basic heuristic: either a constructive heuristic starting from a null solution and adding elements to build a good complete one, or a local search heuristic starting from a complete solution and iteratively modifying some of its elements in order to achieve a better one. The metaheuristic part permits the low-level heuristic to obtain solutions better than those it could have achieved alone, even if iterated. Usually, the controlling mechanism is achieved either by constraining or by randomizing the set of local neighbor solutions to consider in local search (as is the case of simulated annealing [46] or tabu search [33]), or by combining elements taken by different solutions (as is the case of evolution strategies [11] and genetic [40] or bionomic [56] algorithms).

The characteristic of ACO algorithms is their explicit use of elements of previous solutions. In fact, they drive a constructive low-level solution, as GRASP [30] does, but including it in a population framework and randomizing the construction in a Monte Carlo way. A Monte Carlo combination of different solution elements is suggested also by Genetic Algorithms [40], but in the case of ACO the probability distribution is explicitly defined by previously obtained solution components.

The particular way of defining components and associated probabilities is problem-specific, and can be designed in different ways, facing a trade-off between the specificity of the information used for the conditioning and the number of solutions which need to be constructed before effectively biasing the probability distribution to favor the emergence of good solutions. Different applications have favored either the use of conditioning at the level of decision variables, thus requiring a huge number of iterations before getting a precise distribution, or the computational efficiency, thus using very coarse conditioning information.

The chapter is structured as follows. Section 2 describes the common elements of the heuristics following the ACO paradigm and outlines some of the variants proposed. Section 3 presents the application of ACO algorithms to a number of different combinatorial optimization problems and it ends with a wider overview of the problem attacked by means of ACO up to now. Section 4 outlines the most significant theoretical results so far published about convergence properties of ACO variants.

5.2 Ant Colony Optimization

ACO [1, 24] is a class of algorithms, whose first member, called Ant System, was initially proposed by Colorni, Dorigo and Maniezzo [13, 21, 18]. The main underlying idea, loosely inspired by the behavior of real ants, is that of a parallel search over several constructive computational threads based on local problem data and on a dynamic memory structure containing information on the quality of previously obtained result. The collective behavior emerging from the interaction of the different search threads has proved effective in solving combinatorial optimization (CO) problems.

Following [50], we use the following notation. A combinatorial optimization problem is a problem defined over a set $\mathbf{C} = c_1, \ldots, c_n$ of basic *components*. A subset **S** of components represents a *solution* of the problem; $\mathbf{F} \subseteq 2^{\mathbf{C}}$ is the subset of *feasible solutions*, thus a solution **S** is feasible if and only if $\mathbf{S} \in \mathbf{F}$. A *cost function z* is defined over the solution domain, $z : 2^{\mathbf{C}} \rightarrow \mathbf{R}$, the objective being to find a minimum cost feasible solution S^*, i.e., to find S^*: $S^* \in \mathbf{F}$ and $z(S^*) \leq z(\mathbf{S})$, $\forall \mathbf{S} \in \mathbf{F}$.

Given this, the functioning of an ACO algorithm can be summarized as follows (see also [27]). A set of computational concurrent and asynchronous agents (a colony of ants) moves through states of the problem corresponding to partial solutions of the problem to solve. They move by applying a stochastic local decision policy based on two parameters, called *trails* and *attractiveness*. By moving, each ant incrementally constructs a solution to the problem. When an ant completes a solution, or during the construction phase, the ant evaluates the solution and modifies the trail value on the components used in its solution. This pheromone information will direct the search of the future ants.

Furthermore, an ACO algorithm includes two more mechanisms: *trail evaporation* and, optionally, *daemon actions*. Trail evaporation decreases all trail values over time, in order to avoid unlimited accumulation of trails over some component. Daemon actions can be used to implement centralized actions which cannot be performed by single ants, such as the invocation of a local optimization procedure, or the update of global information to be used to decide whether to bias the search process from a non-local perspective [27].

More specifically, an *ant* is a simple computational agent, which iteratively constructs a solution for the instance to solve. Partial problem solutions are seen as *states*. At the core of the ACO algorithm lies a loop, where at each iteration, each ant *moves* (performs a *step*) from a state ι to another one ψ, corresponding to a more complete partial solution. That is, at each step σ, each ant k computes a set $\mathbf{A}_k{}^{\sigma}(\iota)$ of feasible expansions to its current state, and moves to one of these in probability. The probability distribution is specified as follows. For ant k, the probability $p_{\iota\psi}{}^{k}$ of moving from state ι to state ψ depends on the combination of two values:

- the *attractiveness* $\eta_{\iota\psi}$ of the move, as computed by some heuristic indicating the *a priori* desirability of that move;
- the *trail level* $\tau_{\iota\psi}$ of the move, indicating how proficient it has been in the past to make that particular move: it represents therefore an *a posteriori* indication of the desirability of that move.

Trails are *updated* usually when all ants have completed their solution, increasing or decreasing the level of trails corresponding to moves that were part of "good" or "bad" solutions, respectively.

The general framework just presented has been specified in different ways by the authors working on the ACO approach. The remainder of Section 2 will outline some of these contributions.

5.2.1 Ant System

The importance of the original Ant System (AS) [13, 21, 18] resides mainly in being the prototype of a number of ant algorithms which collectively implement the ACO paradigm. AS already follows the outline presented in the previous subsection, specifying its elements as follows.

The move probability distribution defines probabilities $p\iota\psi k$ to be equal to 0 for all moves which are infeasible (i.e., they are in the tabu list of ant k, that is a list containing all moves which are infeasible for ants k starting from state ι), otherwise they are computed by means of formula (5.1), where α and β are user-defined parameters ($0 \leq \alpha,\beta \leq 1$):

$$
p_{\iota\psi}^{k} = \begin{cases} \dfrac{\tau_{\iota\psi}^{\alpha} + \eta_{\iota\psi}^{\beta}}{\sum\limits_{(\iota\zeta)\notin tabu_k} \left(\tau_{\iota\zeta}^{\alpha} + \eta_{\iota\zeta}^{\beta}\right)} & \text{if } (\iota\psi) \notin \text{tabu}_k \\ 0 & \text{otherwise} \end{cases} \tag{5.1}
$$

In formula 5.1, tabu_k is the tabu list of ant k, while parameters α and β specify the impact of trail and attractiveness, respectively.

After each iteration t of the algorithm, i.e., when all ants have completed a solution, trails are updated by means of formula (5.2):

$$\tau_{\iota\psi}(\tau) = \rho\, \tau_{\iota\psi}(\tau - 1) + \Delta\tau_{\iota\psi} \tag{5.2}$$

where $\Delta\tau_{\iota\psi}$ represents the sum of the contributions of all ants that used move $(\iota\psi)$ to construct their solution, ρ, $0 \le \rho \le 1$, is a user-defined parameter called *evaporation coefficient*, and $\Delta\tau_{\iota\psi}$ represents the sum of the contributions of all ants that used move $(\iota\psi)$ to construct their solution. The ants' contributions are proportional to the quality of the solutions achieved, i.e., the better solution is, the higher will be the trail contributions added to the moves it used.

For example, in the case of the TSP, moves correspond to arcs of the graph, thus state ι could correspond to a path ending in node i, the state ψ to the same path but with the arc (ij) added at the end and the move would be the traversal of arc (ij). The quality of the solution of ant k would be the length L_k of the tour found by the ant and formula (5.2) would become $\tau_{ij}(t)=\rho\, \tau_{ij}(t\text{-}1)+\Delta\tau_{ij}$, with

$$\Delta\tau_{ij} = \sum_{k=1}^{m} \Delta\tau_{ij}^{k} \tag{5.3}$$

where m is the number of ants and $\Delta\tau_{ij}^{k}$ is the amount of trail laid on edge (ij) by ant k, which can be computed as

$$\Delta\tau_{ij}^{k} = \begin{cases} \dfrac{Q}{L_k} & \text{if ant } k \text{ uses arc (ij) in its tour} \\ 0 & \text{otherwise} \end{cases} \tag{5.4}$$

Q being a constant parameter.

The ant system simply iterates a main loop where m ants construct in parallel their solutions, thereafter updating the trail levels. The performance of the algorithm depends on the correct tuning of several parameters, namely: α, β, relative importance of trail and attractiveness, ρ, trail persistence, $\tau_{ij}(0)$, initial trail level, m, number of ants, and Q, used for defining to be of high quality solutions with low cost. The algorithm is the following.

```
1. {Initialization}
   Initialize τ_ιψ and η_ιψ, ∀(ιψ).

2. {Construction}
   For each ant k (currently in state ι) do
       repeat
           choose in probability the state to move into.
```

```
            append the chosen move to the k-th ant's set tabu_k.
        until ant k has completed its solution.
    end for

3. {Trail update}
    For each ant move (ιψ) do
        compute Δτ_ιψ
        update the trail matrix.
    end for

4. {Terminating condition}
    If not(end test) go to step 2
```

5.2.2 Ant Colony System

AS was the first algorithm inspired by real ants behavior. AS was initially applied to the solution of the traveling salesman problem but was not able to compete against the state-of-the art algorithms in the field. On the other hand he has the merit to introduce ACO algorithms and to show the potentiality of using artificial pheromone and artificial ants to drive the search of always better solutions for complex optimization problems. The next researches were motivated by two goals: the first was to improve the performance of the algorithm and the second was to investigate and better explain its behavior. Gambardella and Dorigo proposed in 1995 the Ant-Q algorithm [35], an extension of AS which integrates some ideas from Q-learning [76], and in 1996 Ant Colony System (ACS) [36, 25] a simplified version of Ant-Q which maintained approximately the same level of performance, measured by algorithm complexity and by computational results. Since ACS is the base of many algorithms defined in the following years we focus the attention on ACS other than Ant-Q. ACS differs from the previous AS because of three main aspects:

5.2.2.1 Pheromone

In ACS once all ants have computed their tour (i.e. at the end of each iteration) AS updates the pheromone trail using all the solutions produced by the ant colony. Each edge belonging to one of the computed solutions is modified by an amount of pheromone proportional to its solution value. At the end of this phase the pheromone of the entire system evaporates and the process of construction and update is iterated. On the contrary, in ACS only the best solution computed since the beginning of the computation is used to *globally update* the pheromone. As was the case in AS, global updating is intended to increase the attractiveness of promising route but ACS mechanism is more effective since it avoids long convergence time by directly concentrate the search in a neighborhood of the best tour found up to the current iteration of the algorithm.

In ACS, the final evaporation phase is substituted by a *local updating* of the pheromone applied during the construction phase. Each time an ant moves from the current city to the next the pheromone associated to the edge is modified in the following way: $\tau_{ij}(t)=\rho\cdot\tau_{ij}(t-1)+(1-\rho)\cdot\tau_0$ where $0 \le \rho \le 1$ is a parameter (usually set at 0.9) and τ_0 is the initial pheromone value. τ_0 is defined as $\tau_0=(n\cdot L_{nn})^{-1}$, where L_{nn} is the tour length produced by the execution of one ACS iteration without the pheromone component (this is equivalent to a probabilistic nearest neighbor heuristic). The effect of local-updating is to make the desirability of edges change dynamically: every time an ant uses an edge this becomes slightly less desirable and only for the edges which never belonged to a global best tour the pheromone remains τ_0. An interesting property of these local and global updating mechanisms is that the pheromone $\tau_{ij}(t)$ of each edge is inferior limited by τ_0. A similar approach was proposed with the Max-Min-AS (MMAS, [70]) that explicitly introduces lower and upper bounds to the value of the pheromone trials.

5.2.2.2 State Transition Rule

During the construction of a new solution the state transition rule is the phase where each ant decides which is the next state to move to. In ACS a new state transition rule called *pseudo-random-proportional* is introduced. The *pseudo-random-proportional* rule is a compromise between the *pseudo-random* state choice rule typically used in Q-learning [76] and the *random-proportional* action choice rule typically used in Ant System. With the pseudo-random rule the chosen state is the best with probability q_0 (*exploitation*) while a random state is chosen with probability $1\text{-}q_0$ (*exploration*). Using the AS random-proportional rule the next state is chosen randomly with a probability distribution depending on η_{ij} and τ_{ij}. The ACS *pseudo-random-proportional* state transition rule provides a direct way to balance between exploration of new states and exploitation of a priori and accumulated knowledge. The best state is chosen with probability q_0 (that is a parameter $0 \le q_0 \le 1$ usually fixed to 0.9) and with probability $(1\text{-}q_0)$ the next state is chosen randomly with a probability distribution based on η_{ij} and τ_{ij} weighted by α (usually equal to 1) and β (usually equal to 2) .

$$s=\begin{cases}\arg\max\limits_{(ij)\notin tabu_k}\left\{\tau_{ij}^{\alpha}\cdot\eta_{ij}^{\beta}\right\} & \text{if } q\le q_0 \text{ (exploitation)}\\[2ex] \text{AS rule 5.1} & \text{otherwise (exploration)}\end{cases} \tag{5.5}$$

5.2.2.3 Hybridization and performance improvement

ACS was applied to the solution of big symmetric and asymmetric traveling salesman problems (TSP/ATSP) [36],[25]. For these purpose ACS incorporates an advanced data structure known as *candidate list* [60]. A candidate list is a static data structure of length *cl* which contains, for a given city *i*, the *cl* preferred cities to be visited. An ant in ACS first uses candidate list with the state transition rules to choose the city to move to. If none of the cities in the candidate list can be visited the ant chooses the nearest available city only using the heuristic value η_{ij}. ACS for TSP/ATSP has been improved by incorporating local optimization heuristic (*hybridization*): the idea is that each time a solution is generated by the ant it is taken to its local minimum by the application of a local optimization heuristic based on an edge exchange strategy, like 2-opt, 3-opt or Lin-Kernighan [48]. The new optimized solutions are considered as the final solutions produced in the current iteration by ants and are used to globally update the pheromone trails.

This ACS implementation combining a new pheromone management policy, a new state transition strategy and local search procedures was finally competitive with state-of-the-art algorithm for the solution of TSP/ATSP problems [5]. This opened a new frontier for ACO based algorithm. Following the same approach that combines a constructive phase driven by the pheromone and a local search phase that optimizes the computed solution, ACO algorithms were able to break several optimization records, including those for routing and scheduling problems that will be presented in the following paragraphs.

5.2.3 ANTS

ANTS is an extension of the AS proposed in [50], which specifies some underdefined elements of the general algorithm, such as the attractiveness function to use or the initialization of the trail distribution. This turns out to be a variation of the general ACO framework that makes the resulting algorithm similar in structure to tree search algorithms. In fact, the essential trait which distinguishes ANTS from a tree search algorithm is the lack of a complete backtracking mechanism, which is substituted by a probabilistic (*Non-deterministic*) choice of the state to move into and by an incomplete (*Approximate*) exploration of the search tree: this is the rationale behind the name ANTS, which is an acronym of *Approximated Non-deterministic Tree Search*. In the following, we will outline two distinctive elements of the ANTS algorithm within the ACO framework, namely the attractiveness function and the trail updating mechanism.

5.2.3.1 Attractiveness

The attractiveness of a move can be effectively estimated by means of lower bounds (upper bounds in the case of maximization problems) on the cost of the completion of a partial solution. In fact, if a state ι corresponds to a partial problem solution it is possible to compute a lower bound on the cost of a complete solution containing ι. Therefore, for each feasible move ι,ψ, it is possible to compute the lower bound on the cost of a complete solution containing ψ: the lower the bound the better the move. Since a large part of research in ACO is devoted to the identification of tight lower bounds for the different problems of interest, good lower bounds are usually available.

When the bound value becomes greater than the current upper bound, it is obvious that the considered move leads to a partial solution which cannot be completed into a solution better than the current best one. The move can therefore be discarded from further analysis. A further advantage of lower bounds is that in many cases the values of the decision variables, as appearing in the bound solution, can be used as an indication of whether each variable will appear in good solutions. This provides an effective way of initializing the trail values. For more details see [50].

5.2.3.2 Trail update

A good trail updating mechanism avoids stagnation, the undesirable situation in which all ants repeatedly construct the same solutions making any further exploration in the search process impossible. Stagnation derives from an excessive trail level on the moves of one solution, and can be observed in advanced phases of the search process, if parameters are not well tuned to the problem.

The trail updating procedure evaluates each solution against the last k solutions globally constructed by ANTS. As soon as k solutions are available, their moving average $\bar{z}$ is computed; each new solution z_{curr} is compared with $\bar{z}$ (and then used to compute the new moving average value). If z_{curr} is lower than $\bar{z}$, the trail level of the last solution's moves is increased, otherwise it is decreased. Formula (5.6) specifies how this is implemented:

$$\Delta\tau_{ij} = \tau_0 \cdot \left(1 - \frac{z_{curr} - LB}{\bar{z} - LB}\right) \tag{5.6}$$

where $\bar{z}$ is the average of the last k solutions and LB is a lower bound on the optimal problem solution cost. The use of a dynamic scaling procedure permits discrimination of a small achievement in the latest stage of search, while avoiding focusing the search only around good achievement in the earliest stages.

One of the most difficult aspects to be considered in metaheuristic algorithms is the trade-off between exploration and exploitation. To obtain good results, an agent should prefer actions that it has tried in the past and found to be effective in producing desirable solutions (exploitation); but to discover them, it has to try actions not previously selected (exploration). Neither exploration nor exploitation can be pursued exclusively without failing in the task: for this reason, the ANTS algorithm integrates the stagnation avoidance procedure to facilitate exploration with the probability definition mechanism based on attractiveness and trails to determine the desirability of moves.

Based on the elements described, the ANTS algorithm is as follows.

```
1.    Compute a (linear) lower bound LB to the problem
      Initialize τ_ιψ (∀ι,ψ) with the primal variable values

2.    For k=1,m (m= number of ants) do
        repeat
2.1       compute η_ιψ ∀(ιψ)
2.2       choose in probability the state to move into
2.3       append the chosen move to the k-th ant's tabu list
        until ant k has completed its solution
2.4     carry the solution to its local optimum
      end for

3.    For each ant move (ιψ),
      compute Δτ_ιψ and update trails by means of (5.6)

4.    If not(end_test) goto step 2.
```

It can be noted that the general structure of the ANTS algorithm is closely akin to that of a standard tree search procedure. At each stage we have in fact a partial solution which is expanded by branching on all possible offspring; a bound is then computed for each offspring, possibly fathoming dominated ones, and the current partial solution is selected from among those associated to the surviving offspring on the basis of lower bound considerations. By simply adding backtracking and eliminating the MonteCarlo choice of the node to move to, we revert to a standard branch and bound procedure. An ANTS code can therefore be easily turned into an exact procedure.

5.3 Significant problems

In the following of this section we will present applications of ACO algorithms to some significant combinatorial optimization problems. This is to give the reader an idea of what is involved by the use of an ACO algorithm for a problem: even though the last subsection presents an overview of recent application the list is by no means exhaustive, as it becomes readily evident by searching the web under the keywords "ant colony optimization".

5.3.1 Sequential ordering problem

The first ACO applications were devoted to solve the symmetric and the asymmetric traveling salesman problem. Given a set of cities $V = \{v_1, \ldots, v_n\}$, a set of edges $A = \{(i,j) : i,j \in V\}$ and a cost $d_{ij} = d_{ji}$ associated with edge $(i,j) \in A$, the TSP is the problem of finding a minimal length closed tour that visits each city once. In case $d_{ij} \neq d_{ji}$ for at least one edge *(i,j)* than the TSP becomes an Asymmetric TSP (ATSP). The first algorithm that applies an ACO based algorithm to a more general version of the ATSP problem is *Hybrid Ant System for the Sequential Ordering Problem* (HAS-SOP, [34]). HAS-SOP was intended to solve the sequential ordering problem with precedence constraints (SOP). The SOP in an *NP*-hard combinatorial optimization problem first formulated by Escudero [29] to design heuristics for a production planning system. The SOP models real-world problems like production planning [29], single-vehicle routing problems with pick-up and delivery constraints [64], and transportation problems in flexible manufacturing systems [2]. The SOP can be seen as a general case of both the ATSP and the pick-up and delivery problem [47]. It differs from ATSP because the first and the last nodes are fixed, and in the additional set of precedence constraints on the order in which nodes must be visited. It differs from the pick-up and delivery problem because this is usually based on symmetric TSPs, and because the pick-up and delivery problem includes a set of constraints between nodes with a unique predecessor defined for each node, in contrast to the SOP where multiple precedences can be defined.

HAS-SOP combines a constructive phase (ACS-SOP) based on the ACS algorithm [36] with a new local search procedure called SOP-3-exchange. SOP-3-exchange is based on a lexicographic search heuristic due to [64], on a new labeling procedure and on a new data structure called *don't push stack* inspired by the *don't look bit* [6] both introduced by the authors. SOP-3-exchange is the first local search able to handle multiple precedence constraints in constant time.

ACS-SOP implements the constructive phase of HAS-SOP but differs from ACS in the way the set of feasible nodes is computed and in the setting of one of the algorithm's parameters that is made dependent on the problem dimensions. ACS-SOP generates feasible solutions that does not violate the precedence constraints with a computational cost of order $O(n^2)$ like the traditional ACS heuristic.

A set of experiments based on the TSPLIB data shows that HAS-SOP algorithm is more effective than other existing methods for the SOP. HAS-SOP was compared against the two previous most effective algorithms: a branch-and-cut algorithm [2] that proposed a new class of valid inequalities and Maximum Partial Order/Arbitrary Insertion (MPO/AI), a genetic algorithm by Chen and Smith [17].

To better understand the role of the constructive ACS-SOP phase and the role of the SOP-3-exchange local search MPO/AI was also coupled with the SOP-3-exchange local search. Experimental results shows that MPO/AI alone is better than ACS-SOP due to the use of a simple local search embedded in its crossover operator. On the contrary, the combination between constructive phase and local search shows that HAS-SOP is better than both MPO/AI alone and MPO/AI + SOP-3-exchange. This is probably due to the fact that MPO/AI generates solutions that are already optimized and therefore the SOP-3-exchange procedure quickly gets stuck. On the contrary, ACS-SOP solution is a very effective starting point for the SOP-3-exchange local search therefore the HAS-SOP hybridization is very effective. Currently HAS-SOP is the best known method to solve the SOP and was able to improve 14 over 22 best known results in the TSPLIB data set.

5.3.2 Vehicle routing problems

A direct extension of the TSP, the first problem AS was applied to, are Vehicle routing problems (VRPs). These are problems where a set of vehicles stationed at a depot has to serve a set of customers before returning to the depot, and the objective is to minimize the number of vehicles used and the total distance traveled by the vehicles. Capacity constraints are imposed on vehicle trips, as well as possibly a number of other constraints deriving from real-world applications, such as time windows, backhauling, rear loading, vehicle objections, maximum tour length, etc. The basic VRP problem is the Capacitated VRP (CVRP): AS_{rank}, the rank-based version of AS, was applied to this problem by Bullnheimer, Hartl and Strauss [7, 8] with good results. These authors used various standard heuristics to improve the quality of VRP solutions and modified the construction of the tabu list considering constraints on the maximum total tour length of a vehicle and on its capacity.

Following these results, and the excellent ones obtained by ACS with TSP, SOP and QAP problems, ACS was applied to a VRP version more close to actual logistic practice, called VRPTW. VRPTW is defined as the problem of minimizing time and costs in case a fleet of vehicles has to distribute goods from a depot to a set of customers. The VRPTW minimizes a multiple, hierarchical objective function: the first objective is to minimize the number of tours (or vehicles) and the second objective is to minimize the total travel time. A solution with a lower number of tours is always preferred to a solution with a higher number of tours even if the travel time is higher. This hierarchical objectives VRPTW is very common in the literature and in case problem constraints are very tight (for example when the total capacity of the minimum number of vehicles is very close to the total volume to deliver or when customers time windows are narrow), both objectives can be antagonistic: the minimum travel time solution can include a number of vehicles higher than the solution with minimum number of vehicles (see e.g. Kohl et al., [45]).

To adapt ACS to a multiple objectives the Multiple Ant Colony System for the VRPTW (MACS-VRPTW [38]) has been defined. MACS-VRPTW is organized with a hierarchy of artificial ACS colonies designed to hierarchically optimize a multiple objective function: the first ACS colony (ACS-VEI) minimizes the number of vehicles while the second ACS colony (ACS-TIME) minimizes the traveled distances. Both colonies use independent pheromone trails but they collaborate by exchanging information through mutual pheromone updating. In the MACS-

VRPTW algorithm both objective functions are optimized simultaneously: ACS-VEI tries to diminish the number of vehicles searching for a feasible solution with always one vehicle less than the previous feasible solution.

ACS-VEI is therefore different from the traditional ACS applied to the TSP. In ACS-VEI the current best solution is the solution (usually unfeasible) with the highest number of visited customers, while in ACS the current best solution is the shortest one. On the contrary, ACS-TIME is a more traditional ACS colony: ACS-TIME, optimizes the travel time of the feasible solutions found by ACS-VEI. As in HAS-SOP, ACS-TIME is coupled with a local search procedure that improves the quality of the computed solutions. The local search uses data structure similar to the data structure implemented in HAS-SOP [36] and is based on the exchange of two sub-chains of customers. One of this sub-chain may eventually be empty, implementing a more traditional customer insertion.

```
0.  MACS-VRPTW algorithm

1.  {Initialization}
    Initialize ψ^gb the best feasible solution: lowest number
    of vehicles and shortest travel time.

2.  {Main loop}
    Repeat

      2.1  Vehicles ← #active_vehicles(ψ^gb)/* The number of used
                                               vehicles is computed */

      2.3  Activate ACS-VEI(Vehicles - 1) /* ACS-VEI searches for a
                                  feasible solution with always one
                                  vehicle less by maximising the num.
                                  of visited customers */

      2.4  Activate ACS-TIME(Vehicles)/* ACS-TIME is a traditional
                                  ACS colony that minimises the travel
                                  time*/

      While ACS-VEI and ACS-TIME are active

        Wait for an improved solution ψ from ACS-VEI or
             ACS-TIME

        2.5 ψ^gb ← ψ

        if #active_vehicles(ψ^gb) < Vehicles then

               2.6 kill ACS-TIME and ACS-VEI

      End While

    until a stopping criterion is met
```

Experimentally has been shown that the performance of the system increases in case the best solution ψ^{gb} calculated in ACS-TIME is used, in combination with the ACS-VEI best solution $\psi^{ACS-VEI}$, to update the pheromone in ACS-VEI equation (5.7).

$$\tau_{ij}(t) = \rho \cdot \tau_{ij}(t-1) + (1-\rho) / L_{\psi}^{ACS-VEI} \qquad \forall (i,j) \in \psi^{ACS-VEI} \tag{5.7}$$

$$\tau_{ij}(t) = \rho \cdot \tau_{ij}(t-1) + (1-\rho) / L_{\psi}^{gb} \qquad \forall (i,j) \in \psi^{gb}$$

MACS-VRPTW has been experimentally proved to be most effective than the best known algorithms in the field such as the the tabu search of Rochat and Taillard [61], the large neighbourhood search of Shaw [71] and the genetic algorithm of Potvin and Bengio [58]. MACS-VRPTW was also able to improve many results in the Solomon problem set both decreasing the number of vehicle or the travelled time.

MACS-VRPTW introduces a new methodology for optimising multiple objective functions. The basic idea is to coordinate the activity of different ant colonies, each of them optimizing a different objective. These colonies work by using independent pheromone trails but they collaborate by exchanging information. This is the first time a multi-objective function minimization problem is solved with a multiple ant colony optimization algorithm.

5.3.3 Quadratic Assignment Problem

The quadratic assignment problem (QAP) is the problem of assigning n facilities to n locations so that the assignment cost is minimized, where the cost is defined by a quadratic function. The QAP is considered one of the hardest CO problems, and can be solved to optimality only for small instances. Several ACO applications dealt with the QAP, starting using AS and then by means of several of the more advanced versions [54], [66]. The limited effectiveness of AS was in fact improved using a well-tuned local optimizer [53], but several other systems previously introduced were also adapted to the QAP. For example, two efficient techniques are the MMAS-QAP algorithm [69] and HAS-QAP [39]. For the testing of QAP solution algorithms, Taillard [74] proposed to categorize instances into four groups: (*i*) unstructured, uniform random (*ii*) unstructured, grid distance, (*iii*) real-world and (*iv*) real-world-like. Both MMAS-QAP and HAS-QAP have been applied to problem instances of type *i* and *iii*. The performances of these two heuristic approaches are strongly dependent on the type of problem. Comparisons with some of the best heuristics for the QAP have shown that HAS-QAP performs well as far as real-world, irregular and structured problems are concerned. On the other hand, on random, regular and unstructured problems the performance of this technique is less competitive.

This problem-dependency was not shown by ANTS, which was also applied to QAP. In order to apply ANTS to QAP (or any other problem), it is necessary to specify the lower bound to use and what is a move in the problem context (step 2.2). The application described in [50] made the following choices.

As for the lower bound, since there is currently no lower bound for QAP, which is both tight and efficient to compute, the LBD bound was used, which can be computed in $O(n)$ but which is unfortunately on the average quite far from the optimal solution.

As for the moves, it was declared that a move corresponds to the assignment of a facility to a location, thus adding a new component to the partial solution corresponding to the state from which the move originated. Some considerations on the move structure were used to improve the computational effectiveness of the resulting algorithm.

ANTS was tested on instances up to n=40 and showed to be effective on all instance types; moreover its direct transposition into an exact branch and bound was also effective when compared to other exact algorithms.

5.3.4 Other problems

This section outlines some of the more recent applications of ACO approaches to problems other than those listed in the previous ones. This variety is well represented in the many diverse conference with tracks entirely dedicated to ACO and most notably in ANTS conference series, entirely dedicated to algorithms inspired by the observation of ants' behavior (ANTS'98, ANTS'2000 and ANTS'2002). Many different applications have been presented: from plan merging to routing problems, from driver scheduling to search space sharing, from set covering to nurse scheduling, from graph coloring to dynamic multiple criteria balancing problems. A large part of the relevant literature can be accessed online from [1].

Moreover, several introductory overviews have been published. We refer the reader to [23], [24] and [52] for other overviews on ACO.

Among the problems not in the list above, a prominent role is played by the TSP. In fact, TSP has been and in many cases still is the first testbed for ACO variants, and more in general for most combinatorial optimization metaheuristics [68]. It was already on this problem that the limited effectiveness of the first variants emerged, and this fostered the design of improved approaches modifying some algorithm element and possibly hybridizing the framework with greedy local search or with other approaches, such as genetic algorithms or tabu seach [69], [42], [72]. These variants were then applied to other problems, for example MAX-MIN ant system was applied to the flow shop problem in [63], a problem then faced also with other ACO modifications [10], whereas in [8] a rank-based approach for the TSP is described or in [14] a so-called best-worst variant.

More recently, different authors ([75], [44]) have tackled the TSP with hybrid variants, mainly using tabu search, but also, in the case of large TSP instances, also with genetic evolution and nearest neighbor search, in order to improve both efficiency and efficacy. Moreover, variations of the basic TSP, such as the orienteering problem [49] or the probabilistic TSP where clients have to be visited with a certain probability [4] have also been studied.

Scheduling problems provide another common area for testing the effectiveness of ACO algorithms. An ACO approach for the job-shop scheduling is presented in [12], whereas applications to real-world scheduling cases have been recently described in [3] and [62].

More recently, the maturity of the field is showed by the fact that ACO approaches began to be proposed also for problems which are not standard combinatorial optimization testbed, but which are more directly connected to actual practice. For example, the problem of searching and clustering records of large databases is faced by means of ACO in [59], while an algorithm for document clustering is described in [78]. Even more theoretical problems linked to spatial data analysis were tackled with ACO techniques in [73] and [37].

Finally, a recent interesting research branch of ACO, not directly related to combinatorial optimization, is about telecommunication. In fact, the area of packet switching communications appear to be a promising field for ACO-related routing approaches [19, 20]. Whereas a standard optimization version of the frequency assignment problem was described in [51], an application to wavelength allocation was presented in [57], while techniques for path adaptive search are described in [77], [22], [9], [79] and an application to a satellite network in [67]. Moreover, applications directly related to communication Quality of Service (QoS) have been presented in [28], and more recently in [15], while an application which optimizes communication systems with GPS techniques is described in [16].

5.4 Convergence proofs

Recently, some works appeared which provide theoretical insight into the convergence properties of ant colony algorithms. All proofs refer to simplified versions of actually used systems, and do not provide direct guidelines for real-world usage, but they are of interest for the ascertainment of general properties of the systems used.

The first such proofs was proposed by Gutjahr [31], who worked on an ACO variant called Graph-Based Ant System (GBAS). The name derives from the analysis being carried on a so-called *construction graph*, which is a graph assigned to an instance of the optimization problem under consideration, encoding feasible solutions by "walks" on the graph. The objective function value of the walk is equal to the objective function value of the corresponding feasible solution of the original problem. It is always possible to design a construction graph for any given combinatorial optimization problem instance, with a number of nodes linear in the number of bits needed for the representation of a solution, and a number of arcs quadratic in this number of bits. Gutjahr proved that, under the conditions listed below, the solutions generated in each iteration of this Graph-based

Ant System converge with a probability that can be made arbitrarily close to 1 to the optimal solution of the given problem instance. Essential conditions are: (*i*) there is only one optimal walk in W, i.e., the optimal solution is unique, and it is encoded by only one walk in W; (*ii*) along the optimal walk w*, the desirability values satisfy $\eta_{kl}.(u) > 0$ for all arcs (*k,l*) of w* and the corresponding partial walks u of w*; (*iii*) a version of what is called *elitist strategy* is used, where only the best walks are rewarded: walks that are dominated by another already traversed walk do not get pheromone increments anymore. Especially the first of these conditions is quite restrictive.

Stützle and Dorigo [65] propose another convergence proof. They consider both the MAX-MIN Ant System and the ACS outlined in Section 3, and they show that in this case it is possible to prove that allowing more and more iterations the cost of the best solution found converges with probability equal to 1 to the optimal cost. This a property already guaranteed by random search alone, and it does not get lost imposing a minimum trail value. Moreover, the authors show that it is possible to compute a lower bound for the probability of the current best solution to be optimal.

Finally, Gutjahr [32] in a recent paper builds upon these results context of ACO and shows that for a particular ACO algorithm, a time-dependent modification of GBAS, the current solutions converge to an optimal solution with probability exactly one. More specifically, he shows that by using suitable parameter schemes, it can be guaranteed that the optimal paths get attractors of the stochastic dynamic process realized by the algorithm. This improves all previous results and proves a property of the same strength of the tightest one so far obtained in the whole metaheuristic area, which was that obtained by Hajek [41] for Simulated Annealing.

5.5 Conclusions

Ant Colony Optimization has been and continues to be a fruitful paradigm for designing effective combinatorial optimization solution algorithms. After more than ten years of studies, both its application effectiveness and its theoretical groundings have been demonstrated, making ACO one of the most successful paradigm in the metaheuristic area.

This overview tries to propose the reader both introductory elements and pointers to recent results, obtained in the different directions pursued by current research on ACO.

No doubt new results will both improve those outlined here and widen the area of applicability of the ACO paradigm.

References

1. M. Dorigo, Ant colony optimization web page, http://iridia.ulb.ac.be/mdorigo/ACO/ACO.html.
2. N. Ascheuer (1995) Hamiltonian path problems in the on-line optimization of flexible manufacturing systems. Ph.D.Thesis, Technische Universität Berlin, Germany
3. M. E. Bergen P.van Beek, T. Carchrae (2001) Constraint-based assembly line sequencing, *Lecture Notes in Computer Science*, 2056:88-99
4. L. Bianchi, L.M. Gambardella, M. Dorigo (2002) An ant colony optimization approach to the probabilistic traveling salesman problem. In *Proceedings of PPSN-VII, Seventh International Conference on Parallel Problem Solving from Nature*, Lecture Notes in Computer Science. Springer Verlag, Berlin, Germany, pp 883-892
5. E. Bonabeau, M. Dorigo, G. Theraulaz (2000) *Nature*, 406(6791):39-42
6. J.L. Bentley (1992) Fast algorithms for geometric traveling salesman problem, *ORSA Journal on Computing*, 4:387-411
7. B. Bullnheimer, R.F. Hartl, C. Strauss (1999) Applying the ant system to the vehicle routing problem, In: Voss S., Martello S., Osman I.H., Roucairol C. (eds.) *Meta-Heuristics: Advances and Trends in Local Search Paradigms for Optimization*, Kluwer, Boston, pp 285-296
8. B. Bullnheimer, R.F. Hartl, C. Strauss (1999) A new rank-based version of the ant system: a computational study, *Central European Journal of Operations Research* 7(1):25-38
9. C. N. Bendtsen, T. Krink (2002) Phone routing using the dynamic memory model, In *Proceedings of the 2002 congress on Evolutionary Computation*, Honolulu, USA, pp 992-997
10. C. Blum, M. Sampels (2002) Ant colony optimization for FOP shop scheduling: a case study on different pheromone representations, In *Proceedings of the 2002 congress on Evolutionary Computation*, Honolulu, USA, pp 1558-1563
11. T. Bäck, H.-P. Schwefel (1993) An overview of evolutionary algorithms for parameter optimization, *Evolutionary Computation* 1(1):1-23
12. M. den Besten, T. Stützle, M. Dorigo (2000) Ant colony optimization for the total weighted tardiness problem, In *Proceedings Parallel Problem Solving from Nature: 6th international conference*, Lecture Notes in Computer Science. Springer Verlag, Berlin, Germany, pp 611-620
13. A. Colorni, M. Dorigo, V. Maniezzo (1991) Distributed optimization by ant colonies, In *Proceedings of ECAL'91 European Conference on Artificial Life*, Elsevier Publishing, Amsterdam, The Netherlands, pp 134-142
14. O. Cordon, I. Fernandez de Viana, F. Herrera, L. Moreno (2000) A new ACO model integrating evolutionary computation concepts: the best-worst ant system, In *Proceedings of ANTS'2000 - From Ant Colonies to Artificial Ants: Second International Workshop on Ant Algorithms*, Brussels, Belgium, pp 22-29

15. C. Chao-Hsien, G. JunHua, H. Xiang Dan, G. Qijun (2002) A heuristic ant algorithm for solving QoS multicast routing problem, in *Proceedings of the 2002 congress on Evolutionary Computation*, Honolulu, USA, pp 1630-1635
16. D. Camara, A.A.F. Loureiro (2000) A GPS/ant-like routing algorithm for ad hoc networks, In *Proceeding of 2000 IEEE Wireless Communications and Networking Conference*, Chicago, USA, 3:1232-1236
17. S. Chen, S. Smith (1996) Commonality and genetic algorithms. *Technical Report CMU-RI-TR-96-27*, The Robotic Institute, Carnegie Mellon University, Pittsburgh, PA, USA
18. M. Dorigo (1992) Optimization, learning and natural algorithms, *Ph.D. Thesis*, Politecnico di Milano, Milano
19. G. di Caro, M. Dorigo (1998) Antnet: distributed stigmergetic control for communications networks, *Journal of Artificial Intelligence Research*, 9:317-365
20. G. di Caro, M. Dorigo (1998) Mobile agents for adaptive routing, In *Proceedings of the 31st Hawaii International Conference on System*, IEEE Computer Society Press, Los Alamitos, CA, pp 74-83
21. M. Dorigo, V. Maniezzo, A. Colorni (1991) The ant system: an autocatalytic optimizing process, *Technical Report TR91-016*, Politecnico di Milano
22. G. di Caro, M. Dorigo (1998) AntNet: distributed stigmergetic control for communications network, *JAIR, Journal of Artificial Intelligence Research*, 9:317-365
23. M. Dorigo, G. Di Caro (1999) The Ant Colony Optimization Meta-Heuristic. In D. Corne, M. Dorigo, F. Glover (eds) *New Ideas in Optimization*, McGraw-Hill, London, UK, pp 11-32
24. M. Dorigo, G. Di Caro, L.M. Gambardella (1999) Ant Algorithms for Discrete Optimization. *Artificial Life*, 5(2):137-172
25. M. Dorigo, L.M. Gambardella (1997) Ant colony system: a cooperative learning approach to the traveling salesman problem, *IEEE Transaction on Evolutionary Computation* 1:53-66
26. M. Dorigo, V. Maniezzo, A. Colorni (1996) The ant system: optimization by a colony of cooperating agents, *IEEE Transactions on Systems, Man, and Cybernetics-Part B* 26(1):29-41
27. M. Dorigo, T. Stützle (2002) The ant colony optimization metaheuristic: Algorithms, applications and advances. In F. Glover, G. Kochenberger (eds) *Handbook of Metaheuristics*. Kluwer Academic Publishers, Norwell, MA, pp 251-285
28. G. Di Caro, T. Vasilakos (2000) Ant-SELA: Ant-agents and stochastic automata learn adaptive routing tables for QoS routing in ATM networks, In *Proceedings of ANTS'2000 - From Ant Colonies to Artificial Ants: Second International Workshop on Ant Algorithms*, Brussels, Belgium, pp 43-46
29. L.F. Escudero (1988) An inexact algorithm for the sequential ordering problem. *European Journal of Operational Research* 37:232-253
30. T.A. Feo, M.G.C. Resende (1995) Greedy randomized adaptive search procedures, *Journal of Global Optimization* 6:109-133
31. W.J. Gutjahr (2000) A graph-based Ant System and its convergence. *Future Generation Computer Systems*. 16:873-888
32. W.J. Gutjahr (2002) ACO algorithms with guaranteed convergence to the optimal solution. *Information Processing Letters* 82(3):145-153
33. F. Glover (1989) Tabu search, *ORSA Journal on Computing* 1:190-206

34. L.M. Gambardella, M. Dorigo (2000) An ant colony system hybridized with a new local search for the sequential ordering problem, *INFORMS Journal on Computing* 12(3):237-255
35. L.M. Gambardella, M. Dorigo (1995) Ant-Q: a reinforcement learning approach to the travelling salesman problem, In *Proceedings of the Twelfth International Conference on Machine Learning*, Morgan Kaufmann, Palo Alto, California, USA, pp 252-260
36. L.M. Gambardella, M. Dorigo (1996) Solving Symmetric and Asymmetric TSPs by Ant Colonies, In *Proceedings of the IEEE Conference on Evolutionary Computation*, ICEC96, Nagoya, Japan, pp 622-627
37. Y. Gabriely, E. Rimon (2001) Spanning-tree based coverage of continuous areas by a mobile robot, *Annals of Mathematics and Artificial Intelligence* 31(1-4):77-98
38. L.M. Gambardella, E. Taillard, G. Agazzi (1999) MACS-VRPTW: A Multiple Ant Colony System for Vehicle Routing Problems with Time Windows, In D. Corne, M. Dorigo, F. Glover (eds) *New Ideas in Optimization*, McGraw-Hill, London, UK, pp 63-76
39. L.M. Gambardella, E. Taillard, M. Dorigo (1999) Ant colonies for the quadratic assignment problem, *Journal of the Operational Research Society* 50:167-176.
40. J.H. Holland (1975) Adaptation in natural and artificial systems, University of Michigan Press
41. B. Hajek (1988) Cooling schedules for optimal annealing, *Math. of OR*, 13:311-329
42. H.M. Botee, E. Bonabeau (1999) Evolving ant colony optimization, (SFI Working Paper Abstract)
43. C. Hurkens, S. Tiourine (1995) Upper and lower bounding techniques for frequency assignment problems, *Technical Report* 95-34, T.U. Eindhoven
44. T. Kaji (2001) Approach by ant tabu agents for Traveling Salesman Problem, In *Proceedings of the* IEEE *International Conference on System, Man and Cybernetics*
45. N. Kohl, J. Desrosiers, O. B. G. Madsen, M. M. Solomon, F. Soumis (1997) K-Path Cuts for the Vehicle Routing Problem with Time Windows, *Technical Report IMM-REP-1997*-12, Technical University of Denmark
46. S. Kirkpatrik, C.D. Gelatt, M.P. Vecchi (1983) Optimization by simulated annealing, *Science* 220:671-680
47. G.A.P. Kindervater, M.W.P. Savelsbergh (1997) Vehicle routing: handling edge exchanges, E. H. Aarts, J. K. Lenstra (eds) *Local Search in Combinatorial Optimization*. John Wiley & Sons, Chichester, UK, pp 311-336
48. S. Lin, B.W. Kernighan, (1973) An effective heuristic algorithm for the traveling salesman problem, *Operations Research*, 21:498–516
49. Yun-Chia Liang, S. Kulturel-Konak, A.E. Smith (2002) Meta heuristic for the orienteering problem, In *Proceedings of the 2002 Congress on Evolutionary Computation*, Honolulu, USA, pp 384-389
50. V. Maniezzo (1999) Exact and approximate nondeterministic tree-search procedures for the quadratic assignment problem, *INFORMS Journal of Computing* 11(4):358-369
51. V. Maniezzo, A. Carbonaro (2000) An ANTS Heuristic for the Frequency Assignment Problem, *Future Generation Computer Systems*; 16, North-Holland/Elsevier, Amsterdam, 927-935.
52. V. Maniezzo, A.Carbonaro (2001) Ant Colony Optimization: an overview, in C. Ribeiro (ed) *Essays and Surveys in Metaheuristics*, Kluwer, 21-44
53. V. Maniezzo, A. Colorni (1999) The ant system applied to the quadratic assignment problem, *IEEE Transactions on Knowledge and Data Engineering* 11(5):769-778

54. V. Maniezzo, A. Carbonaro (2000) A bionomic approach to the capacitated p-median problem, *Future Generation Computer Systems* 16(8):927-935
55. V. Maniezzo, R. Montemanni (1999) An exact algorithm for the radio link frequency assignment problem, *Technical Report* CSR99-02
56. V. Maniezzo, A. Mingozzi, R. Baldacci (1998) A bionomic approach to the capacitated p-median problem, *Journal of Heuristics* 4(3):263-280
57. G. Navarro Varela, M. C. Sinclair (1999) Ant Colony Optimization for virtual wavelength path routing and wavelength allocations, In *Proceeding of the Congress on Evolutionary Computation*, Washington DC, USA, pp 1809-1816
58. J.Y. Potvin, S. Bengio (1996) The vehicle routing problem with time windows – part II: genetic search, *Informs Journal of Computing* 8:165-172
59. R.S. Parpinelli, H.S. Lopes, A.A. Freitas (2002) Data mining with ant colony optimization algorithm, in *IEEE Transactions on Evolutionary Computation*, 6(4):321-332
60. G. Reinelt (1994) *The traveling salesman: computational solutions for TSP applications*. Springer-Verlag
61. Y. Rochat, E.D. Taillard (1995) Probabilistic diversification and intensification in local search for vehicle routing, *Journal of Heuristics* 1:147-167
62. H. Shyh-Jier (2001) Enhancement of hydroelectric generation scheduling using ant colony system based organization approach, in *IEEE Transactions of Energy Conversion*, Volume 16(3):296-301
63. T. Stützle (1998) An Ant Approach to the Flow Shop Problem, In *Proceedings of EUFIT'98* , Aachen, Germany, pp 1560-1564
64. M.W.P. Savelsbergh (1990) An efficient implementation of local search algorithms for constrained routing problems. *European Journal of Operational Research* 47:75-85
65. T. Stützle, M. Dorigo (2002) A Short Convergence Proof for a Class of ACO Algorithms, *IEEE Transactions on Evolutionary Computation*, 6(4):358-365
66. T. Stützle, M. Dorigo (1999) Aco algorithms for the quadratic assignment problem, In D. Corne, M. Dorigo, F. Glover (eds) *New Ideas in Optimization*, McGraw-Hill, London, pp 3-50
67. E. Siegel, B. Denby, S. Le Hégarat-Mascle (2000) Application of ant colony optimizatio to adaptive routing in a telecommunications satellite network, submitted *to IEEE Trasactions on Networks*
68. T. Stützle, A .Grün, S. Linke, M. Rüttger (2000) A comparison of nature inspired heuristic on the traveling salesman problem, In Deb et al. (eds) In *Proceeding* of PPSN-VI, *Sixth International Conference on Parallel Problem Solving from Nature*, 1917:661-670
69. T. Stützle, H. Hoos (2000) MAX-MIN ant system, *Future Generation Computer Systems*, Vol. 16(8):889-914
70. T. Stützle, H. Hoos (1997) Improvements on the ant system: Introducing MAX-MIN ant system, In *Proceeding of ICANNGA'97, International Conference on Artificial Neural Networks and Genetic Algorithms*, Springer Verlag, pp 245-249
71. P. Shaw (1998) Using Constraint Programming and Local Search Methods to Solve Vehicle Routing Problems, In *Proceeding of Proceedings of the Fourth International Conference on Principles and Practice of Constraint Programming*, M. Maher, J.-F. Puget (eds.), Springer-Verlag, pp 417-43
72. T. Stützle, H. Hoos (1998) The MAX-MIN Ant System and Local Search for Combinatorial Optimization Problems: Towards Adaptive Tools for Combinatorial Global Optimization In S. Voss, S. Martello, I.H. Osman, C. Roucairol (eds) *Meta-Heuristics:*

Advances and Trends in Local Search Paradigms for Optimization, Kluwer Academic Publishers, Boston, pp 313-329

73. M. Schreyer, G. R. Raidl, (2002) Letting ants labeling point feature, in *Proceedings of the 2002 congress on Evolutionary Computation*, Honolulu, USA, pp 1564-1569
74. E.D. Taillard (1995) Comparison of iterative searches for the quadratic assignment problem, *Location Science*, 3:87-105
75. C. Tsai, C. Tsai, C. Tseng (2002) A new approach for solving large traveling salesman problem, In *Proceeding of the 2002 Congress of Evolutionary Computation*, Honolulu, USA, pp 1636-1641
76. C.J. Watkins, P. Dayan (1992) Q-learning, *Machine Learning*, 8:279-292
77. O. Wittnr, B. E. Helvik, (2002) Cross-entropy guided ant-like agents finding dependable primary/backup path patterns in networks, In *Proceedings of the 2002 congress on Evolutionary Computation*, Honolulu, USA, pp 1528-1533
78. B. Wu, Y. Zheng, S. Liu, Z. Shi, (2002) CSIM: a document clustering algorithm based on swarm intelligence, In *Proceedings of the 2002 congress on Evolutionary Computation*, Honolulu, USA, pp 477-482
79. W. Ying, X. Jianying (2000) Ant colony optimization for multicast routing, In *Proceeding of the 2000 IEEE Asia-Pacific Conference on Circuits and Systems*, Tianjin, China

6 Differential Evolution

Jouni Lampinen and Rainer Storn

Abstract. This article discusses solving non-linear programming problems containing integer, discrete and continuous variables. A novel optimization method based on a recently introduced Evolutionary Algorithm called Differential Evolution is described. Three numerical examples are given to demonstrate the capabilities and practical use of the method. Since these classical examples have been used by a number of other researchers, it was possible to compare results between no less than 21 alternative optimization methods. The novel method was found easy to implement and use, effective, efficient and robust, which makes it an attractive and widely applicable approach for solving practical engineering design problems.

Keywords. evolutionary algorithms, differential evolution, non-linear optimization, mixed integer-discrete-continuous variables, penalty functions

6.1 Introduction

Most non-linear optimization methods assume that objective function variables are continuous. In practical engineering design work, however, problems are common in which some, or all, of the design parameters are discrete or integer variables. Discrete variables often occur because a design element is only available in a limited set of standard sizes. For example, the thickness of steel plate, the diameter of copper pipe, the size of a screw, the pitch of a gear tooth, the size of a roller bearing, the value of an electronic resistor, etc., are often limited to a set of commercially available standard sizes. Integer variables occur in problems with identical design elements. Examples of integer variables include the number of teeth on a gear, bolts or rivets needed to fix a structure, heat exchanger tubes, cooling fins on a heat sink, parallel V-belts in a transmission, coils of a spring, etc.

It is clear that a large number of engineering design optimization problems fall into the category of *mixed integer-discrete-continuous, non-linear programming* problems. Nevertheless, the most frequently discussed methods in the literature are for solving continuous problems. Generally, continuous problems are considered to be easier to solve than discrete ones, suggesting that the presence of any non-continuous variables considerably increase the difficulty of finding a solution. In practice, problems containing integer or discrete variables are usually solved as though they were continuous with the nearest available discrete values then being chosen. In most cases, this technique produces a result that is often quite far from optimal. The reason for this approach is that it is still commonly believed that there are no efficient, effective, robust, and easy-to-use non-linear programming methods currently available that are capable of handling mixed variables.

Another source of difficulty encountered in practical engineering design optimization involves constraint handling. Real-world limitations frequently impose multiple, non-linear and non-trivial constraints on a design. Constraints can limit the feasible solutions to only a small subset of the design space.

In response to these demands, a novel approach for solving mixed integer-discrete-continuous, non-linear engineering design optimization problems has been developed based on the recently introduced Differential Evolution (DE) algorithm [1,2,3] (see also [4]). This investigation describes the techniques needed to handle boundary constraints as well as those needed to deal with several non-linear and non-trivial constraint functions. After introducing these techniques, three illustrative and practical numerical examples are presented. The first example designs a gear train with a specified gear ratio, the second problem minimizes the manufacturing cost of a pressure vessel, and the third example uses DE to design a coil spring with a minimum amount of steel. The mixed-variable methods used to solve these problems are discussed in detail and compared with published results obtained using other optimization methods for the same problems. Although this investigation only focuses on engineering design applications, DE can be applied, in principle at least, to solve any mixed integer-discrete-continuous optimization task.

6.2 Mixed integer-discrete-continuous non-linear programming

A mixed integer-discrete-continuous, non-linear programming problem can be expressed as follows:

$$\begin{aligned}
&\text{Find} \\
&X = \{x_1, x_2, x_3, \ldots, x_D\} = \left[X^{(i)}, X^{(d)}, X^{(c)}\right]^T \\
&\text{to minimize} \\
&f(X) \\
&\text{subject to constraints} \\
&g_j(X) \leq 0 \qquad j = 1, \ldots, m \\
&\text{and subject to boundary constraints} \\
&x_i^{(L)} \leq x_i \leq x_i^{(U)} \qquad i = 1, \ldots, D \\
&\text{where} \\
&X^{(i)} \in \Re^i, \quad X^{(d)} \in \Re^d, \quad X^{(c)} \in \Re^c
\end{aligned} \tag{6.1}$$

$X^{(i)}$, $X^{(d)}$ and $X^{(c)}$ denote feasible subsets of integer, discrete and continuous variables, respectively. The above formulation is general and basically the same for all types of variables. Only the structure of the design domain distinguishes one problem from another. While both integer and discrete variables have a discrete nature, only discrete variables can assume floating-point values. In practice, it is not uncommon for the discrete values of a feasible set to be unevenly spaced. This is the main reason why integer and discrete variables are often handled differently.

Some researchers also define a fourth class of variables, so-called, "zero-one" (binary) variables that only assume the integer values, 0 or 1. This class of variables is important because it is capable of expressing many practically relevant machine states, like a switch that is on or off, a valve that is open or closed, a device that is connected or not, a clutch that is open or closed, etc. For this investigation, the class of binary-integer variables is treated as a special case of integer variables.

6.3 Differential Evolution

Price and Storn first introduced the Differential Evolution (DE) algorithm a few years ago [1,2]. DE can be classified as an *evolutionary optimization algorithm*. At present, the best-known representatives of this class are *genetic algorithms* [5] and *evolution strategies* [6] all of which belong to the class of stochastic population based optimization algorithms. Genetic algorithms (GAs) are less suitable for real-valued function optimization because they encode floating-point parameters into bitstrings. For most real-valued problems defined on the continuous domain, this parameter transformation has not proven beneficial [39,40]. On the other hand, encoding real parameters as floating-point numbers makes Evolution Strategies (ESs) well suited for real-parameter optimization. Like ES, DE also uses floating-point encoding. ESs, however, are more complicated and computationally expensive to use than DE [41,42]. This drawback mainly stems from the fact that ESs use predefined probability density functions (PDFs), usually gaussian or cauchy, for parameter perturbation. Since the proper variances and covariances for these PDFs are unknown in advance, they have to be adapted throughout the optimization by a process that imposes additional complexity to the optimization procedure. In contrast to classical ESs, however, DE is self-adjusting since it deduces-the perturbation information from the distances between the vectors that comprise the population itself. This feature automatically yields reasonably large vector perturbations at the beginning, i.e. the exploratory stage, of the optimization. Later on, when the vector population is closing in on the optimum, the distances between the vectors automatically get smaller. These smaller perturbations allow DE to conduct a fine-grained search for the optimum. The self-adjusting property of DE uses fewer control mechanisms when compared to other approaches, making DE both effective and easy to use.

Currently, there are several variants of DE [3]. The particular version used throughout this investigation is the *DE/rand/1/bin* scheme. The extension of DE to mixed-parameter optimization [8] will be described in detail. As the DE algorithm was originally designed to operate on continuous variables, the optimization of continuous problems is discussed first, while the manner in which this method has been adapted to handle integer and discrete variables is explained later.

Generally, the function to be optimized, f, is of the form:

$$f(X): \Re^D \to \Re \tag{6.2}$$

The optimization goal is to minimize the value of this *objective function f(X)*,

$$\min(f(X)) \tag{6.3}$$

by optimizing the values of its parameters:

$$X = (x_1, \ldots, x_D), \quad X \in \Re^D \tag{6.4}$$

where X denotes a vector composed of D objective function parameters. No generality is lost by the restriction to minimization since every maximization task can be cast into a minimization-problem by multiplying the objective function by −1. Usually, the parameters of the objective function are also subject to lower and upper boundary constraints, $x^{(L)}$ and $x^{(U)}$, respectively:

$$x_j^{(L)} \le x_j \le x_j^{(U)} \qquad j = 1, \ldots, D \tag{6.5}$$

Figure 6.1 illustrates a simple function to be minimized.

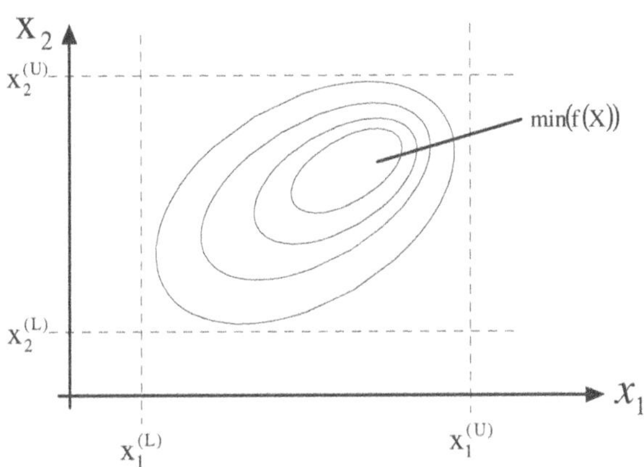

Fig. 6.1. Contour lines of a simple function *f(X)* the minimum of which has to be found.

As with all evolutionary optimization algorithms, DE operates on a *population*, P_g, of candidate solutions, not just a single solution. These candidate solutions are the *individuals* of the population. In particular, DE maintains a population of constant size that consists of *NP* real-valued vectors, $X_{i,g}$, where i indexes the population and g is the *generation* to which the population belongs.

$$P_g = X_{i,g} \qquad i = 1,\ldots,NP, \quad g = 1,\ldots,g_{max} \tag{6.6}$$

Each vector $X_{i,g}$ contains D real parameters (*chromosomes* of individuals):

$$X_{i,g} = x_{j,i,g} \qquad i = 1,\ldots,NP, \quad j = 1,\ldots,D \tag{6.7}$$

6.3.1 Initialization

In order to establish a starting point for optimum seeking, the population must be initialized. Often there is no more knowledge available about the location of a global optimum than the limits of the problem variables. In this case, a natural way to seed the initial population, $P_{g=0}$, is with random values chosen from within the given boundary constraints:

$$P_0 = x_{j,i,0} = rand_j[0,1] \cdot \left(x_j^{(U)} - x_j^{(L)}\right) + x_j^{(L)} \quad i = 1,\ldots,NP, \quad j = 1,\ldots,D \tag{6.8}$$

where $rand_j[0,1]$ denotes a uniformly distributed random value within the range: [0.0,1.0] that is chosen anew for each j. Equations 6.6 - 6.8 are illustrated in Fig. 6.2 for the current example, showing how an initial population might look like. The "x"-signs denote the endpoints of the population vectors.

6.3.2 Mutation and Crossover

DE's self-referential population reproduction scheme is different from other evolutionary algorithms. From the 1st generation on, vectors in the current population, P_g, are randomly sampled and combined to create candidate vectors for the subsequent generation, P_{g+1}. The population of candidate or "trial" vectors, $P'_g = U_{i,g} = u_{j,i,g}$, is generated as follows:

$$u_{j,i,g} = \begin{cases} v_{j,i,g} = x_{j,r3,g} + F \cdot \left(x_{j,r1,g} - x_{j,r2,g}\right) & \text{if} \quad rand_j[0,1) \leq CR \vee j = k \\ x_{j,i,g} & \text{otherwise} \end{cases} \tag{6.9}$$

where

$i = 1,\ldots,NP, \quad j = 1,\ldots,D$

$k \in \{1,\ldots,D\}$, random parameter index, chosen once for each i

$r_1, r_2, r_3 \in \{1,\ldots,NP\}$, randomly selected, except: $r_1 \neq r_2 \neq r_3 \neq i$

$CR \in [0,1], \quad F \in (0,1+]$

The randomly chosen indexes, r_1, r_2 and r_3 are different from each other and also different from the running index, i. New random integer values for r_1, r_2 and r_3 are chosen for each value of the index i, i.e., for each individual. The index k refers to a randomly chosen chromosome, which is used to ensure that each individual trial vector, $U_{i,g}$, differs from its counterpart in the previous generation, $X_{i,g}$, by at least one parameter. A new random integer value is assigned to k prior to the construction of each trial vector, i.e. for each value of the index i. The vector mutation scheme described by Eq. 6.9 is illustrated in Fig. 6.2.

F and CR are control parameters of DE. Like NP, both values remain constant during the search process. F is a real-valued factor in the range (0.0,1.0+] that scales the differential variations. The upper limit on F has been empirically determined. So far, it is considered that values of F greater than unity, while possible, do not appear to be productive. However, later on in this article it is demonstrated that values higher than F = 1.0, may still be useful.

CR is a real-valued crossover constant in the range [0.0,1.0] which controls the probability that a trial vector parameter will come from the randomly chosen, mutated vector, $V_{i,g}$, instead of from the current vector, $X_{i,g}$. Figure 6.3 gives a pictorial representation of DE's crossover operation.

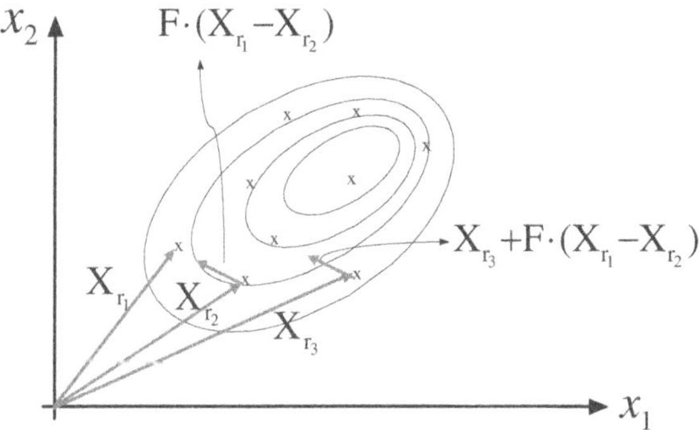

Fig. 6.2. Illustration of initial population and the mutation mechanism of DE.

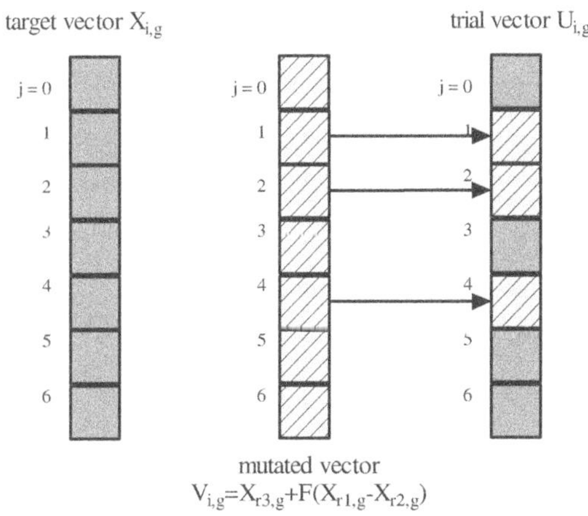

Fig. 6.3. Example for the binary crossover operation in DE.

Generally, both *F* and *CR* affect the convergence velocity and robustness of the search process. Their optimal values are dependent both on objective function characteristics and on the population size, *NP*. Usually, suitable values for *F*, *CR* and *NP* can be found by trial-and-error after a few tests using different values. Reasonable values to start with are $NP=5\times D...30\times D$, F=0.90, and CR=0.9. According our experiences, these settings are typically effective, while further fine-tuning of the control parameters often result in a considerably higher convergence rate. In order to demonstrate the usefulness of this advice, all the examples given later on in this article are solved with these particular settings. More practical advice on how to select control parameters *NP*, *F* and *CR* can be found in [1,2,3,7,9], for example.

6.3.3 Selection

DE's selection scheme also differs from other evolutionary algorithms. The population for the next generation, P_{g+1}, is selected from the current population, P_G, and the child population, P'_G, according to the following rule:

$$X_{i,g+1} = \begin{cases} U_{i,g} & \text{if } f(U_{i,g}) \leq f(X_{i,g}) \\ X_{i,g} & \text{otherwise} \end{cases} \tag{6.10}$$

Thus, each individual of the temporary population is compared with its counterpart in the current population. Assuming that the objective function is to be minimized, the vector with the lower or equal objective function value wins a place in the next generation's population. As a result, all the individuals of the next generation are as good or better than their counterparts in the current generation.

The interesting point concerning DE's selection scheme is that a trial vector is only compared to one individual, not to all the individuals in the current population. Such a scheme can be viewed as *binomial tournament selection*, since it's basic principle is similar with the general *tournament selection* rule [46] using tournament size 2. Here the binomial tournaments are kept between each trial individual and the corresponding current population member – the winners of a one-to-one competition enter the next stage of the optimization. An example of such a one-to-one competition is shown in Fig. 6.4.

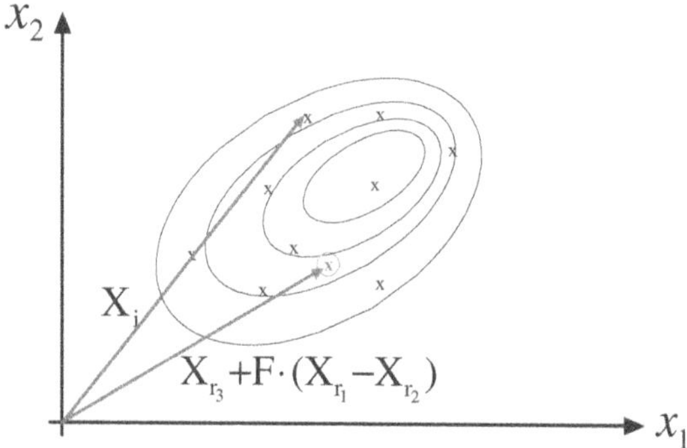

Fig. 6.4. Illustration of DE's standard selection scheme when two vectors compete. The winner point is encircled.

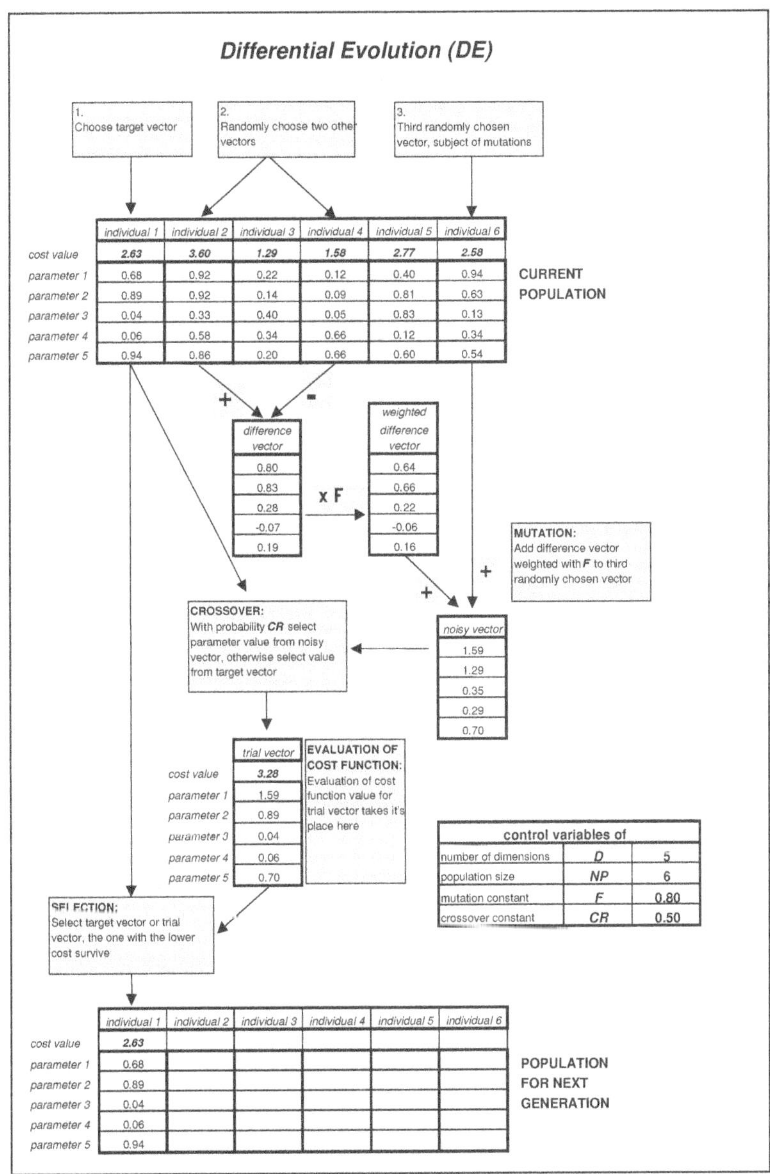

Fig. 6.5. Differential Evolution works directly with the floating-point valued variables of the objective function, not with their (binary) encoded equivalents. The functioning of DE/rand/1/bin is here illustrated in the case of a simple objective function $f(X) = x_1 + x_2 + x_3 + x_4 + x_5$.

The described binomial tournament selection (Eq. 6.10) is not the only selection scheme that allows DE to be effective. Different tournament sizes and alternative approaches for selecting the competing vectors for the tournaments have also been applied with success [1,2,3]. For example, "best of parents and children"-selection or ($\mu + \lambda$)-selection [41], chooses the best vectors from the combined parent and child populations to comprise the next generation. For DE, the current vectors are the $\mu = NP$ parents and new vectors are the $\lambda = NP$ children. In fact, ($\mu + \lambda$)-selection frequently offers faster convergence than binomial tournament selection since it is greedier in terms of movement towards lower objective function values. It is, however, also more prone to getting stuck in a local minimum and may require a larger *NP* to compensate.

Figure 6.5 shows that DE maintains two separate populations so that the current population is left intact while the population for the next generation is computed. Again, from a functional point of view this is not a necessity. It is also possible to immediately save the winning vectors in the array of the current population. This way the current population gradually changes into the new one. Keeping only one population requires less memory, so this method can be important for resource-limited computational environments. It is also evident that keeping only one population is greedier than DE's classical approach of maintaining two populations. With only one population, the population members have less time to participate as vector difference donors. As such, this scheme introduces asymmetry since the population members with higher indexes survive longer than those with lower indexes. This imbalance can be compensated for by altering the direction in which the population array is processed. For each new generation, the sequence in which the population vectors have to compete against the trial vectors should be reversed.

Figure 6.5 summarizes Mutation, Crossover, and Selection of *DE/rand/1/bin* in a data flow diagram.

6.3.4 DE dynamics

Understanding why DE works well as an optimization scheme is far from being trivial, but a few general statements can be made easily. One is that there are two opposing mechanisms that influence DE's population. The first mechanism is the population's tendency to expand and hence to explore the terrain on which it is working. There is a high probability that perturbations yielding acceptable new points will enlarge the region covered by the population. If the objective function surface is flat, the population will continue to expand, since a new point survives if its associated objective function value is (less than or) equal to that of its competitor. The new vector does not need to be better than the current one; equality suffices to supplant. The second mechanism is selection. By removing vectors, which are located in unproductive regions, selection can counteract the expansion of the population. The expansion mechanism prevents DE's population from converging prematurely, while the selection mechanism prevents the population from diverging into regions which are not of interest. Of course, these general mechanisms are not unique to DE; any evolution strategy adheres to these rules. The major difference between DE and other evolution strategies is the PDF that each employs.

One may wonder how the PDF generated by DE's mutation scheme compares to the gaussian and cauchy distributions upon which ESs rely. A simple investigation of the vector difference distribution reveals that NP population members have $NP(NP\text{-}1)/2$ interconnections if the connections between identical points are neglected. Since each connection has two directions, there are $NP(NP\text{-}1)$ difference vectors with a length greater than zero. Since a negative counterpart exists for each vector, the mean of the distribution will always be zero. Figure 6.6 shows a simple two-dimensional example to illustrate these findings. The distribution itself changes with and depends on the objective function surface being searched.

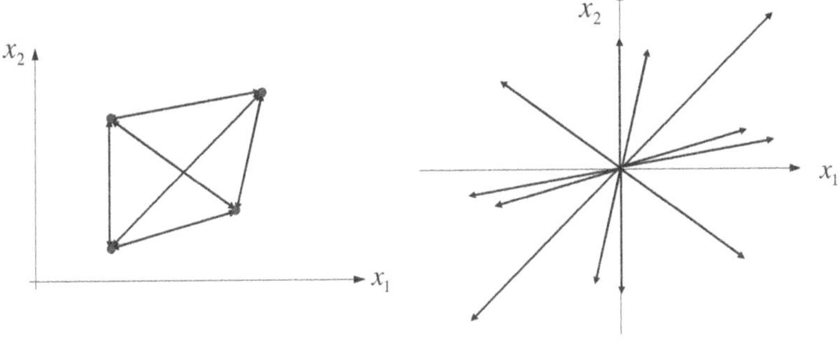

Fig. 6.6. Simple example to illustrate DE's difference vector distribution.

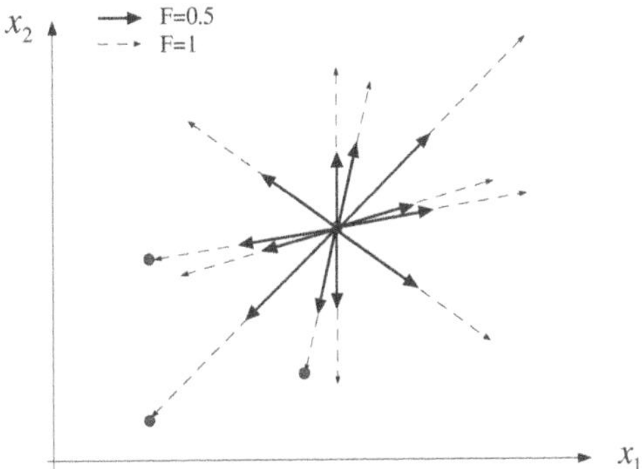

Fig. 6.7. Influence of the scaling factor F on the trial vectors.

The scaling factor, $F \neq 1.0$, attempts to dampen the expansion mechanism and to reduce the chance that trial points are too close to already existing points. The latter argument becomes more important as the population size, *NP*, gets smaller. The smaller the population size becomes, the greater the chances are that the population will stagnate [45]. Stagnation is a situation where none of the possible new vectors yields an improved solution. This lack of diversity in the set of vector differences makes it impossible to explore new territory. Stagnation is different from premature convergence because in the latter the population is still evolving. If stagnation occurs the population is basically frozen.

Figure 6.7 shows the potential points to be reached for F=1.0 and F=0.5 if the upper right point shown in Fig. 6.6 is used as a base point.

The crossover constant, $CR < 1.0$, may also help to prevent stagnation by increasing the number of possible perturbations beyond $NP(NP-1)$ thus allowing more potential points to be reached [45]. Figure 6.8 shows an example of the potential points that may be reached if the upper right point from Fig. 6.6 is the base vector, and crossover is introduced. Not only is the point defined by the addition of the base point and the weighted difference vector a potential trial point, but so are all points which mix the components of the target vector and the trial vector.

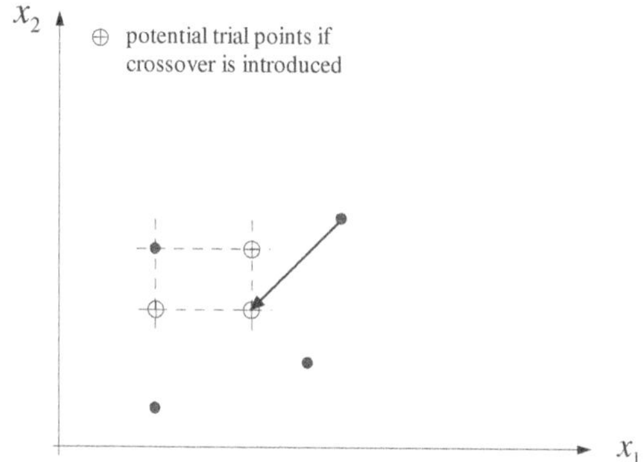

Fig. 6.8. Influence of the crossover constant *CR* on the trial vectors.

All these factors influence the outcome of the PDF used by DE and it is obvious that the shape of the PDF is difficult, if not impossible, to predict. The following sequence of pictures shows the evolution of both the vector population and the corresponding difference vector distribution when minimizing the two-dimensional function: $f(x_1, x_2) = 100 \cdot (x_2 - x_1^2)^2 + (1 - x_1)^2$. This function is the well-known "Rosenbrock's Saddle" [6] which has been used extensively as a test example in the optimization literature.

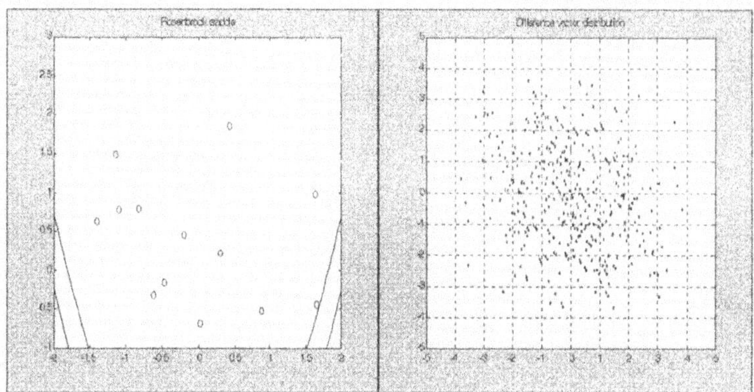

Fig. 6.9. Generation No. 1.

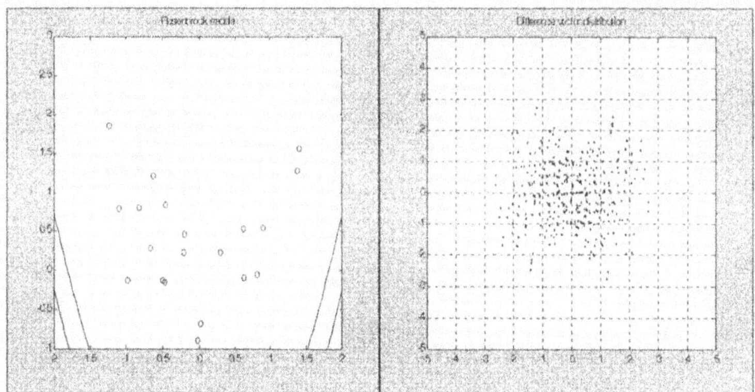

Fig. 6.10. Generation No. 6.

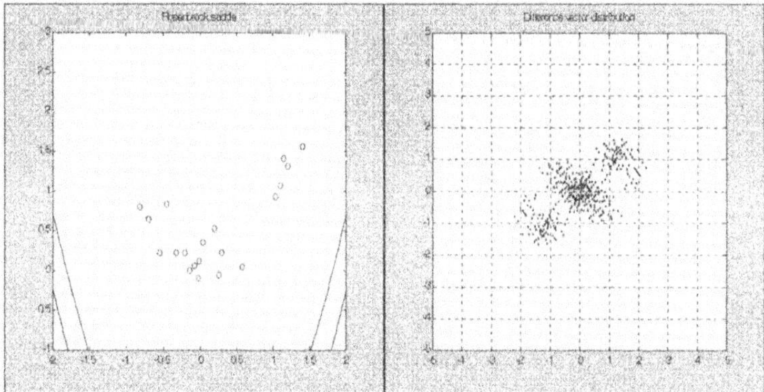

Fig. 6.11. Generation No. 12.

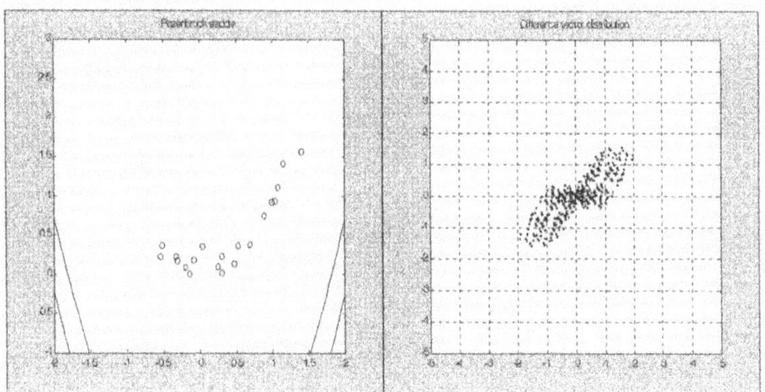

Fig. 6.12. Generation No. 18.

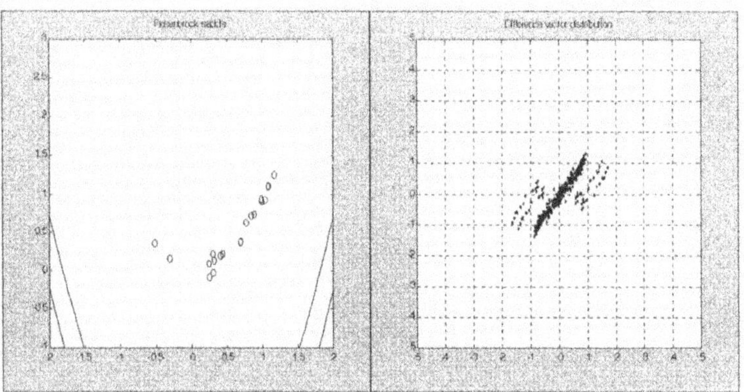

Fig. 6.13. Generation No. 24.

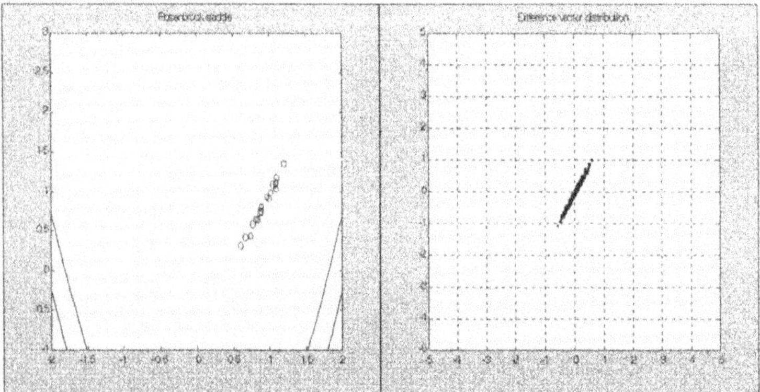

Fig. 6.14. Generation No. 30.

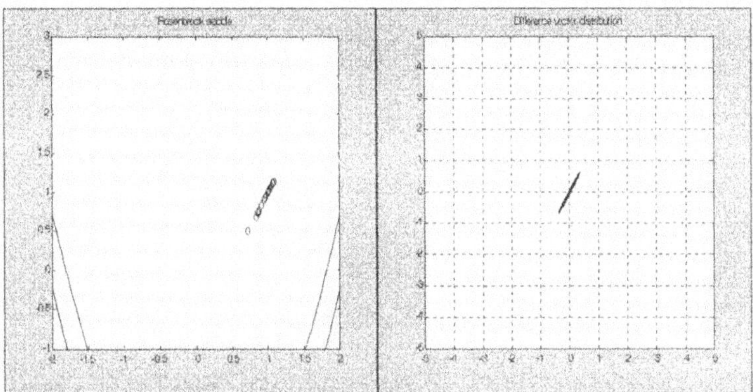

Fig. 6.15. Generation No. 36.

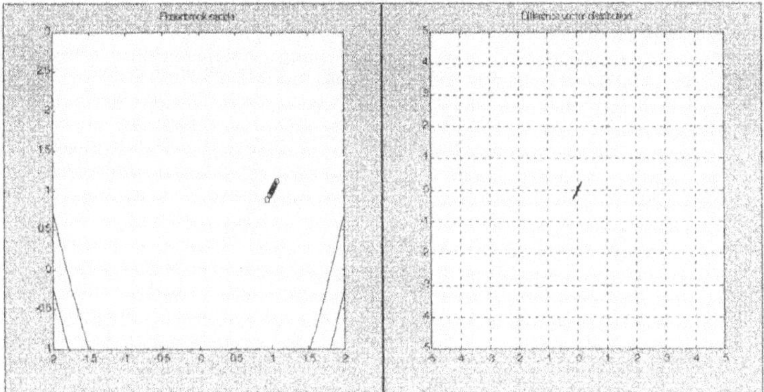

Fig. 6.16. Generation No. 42.

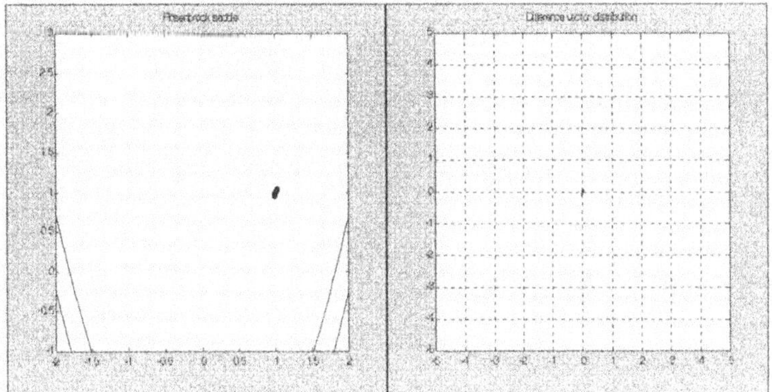

Fig. 6.17. Generation No. 48.

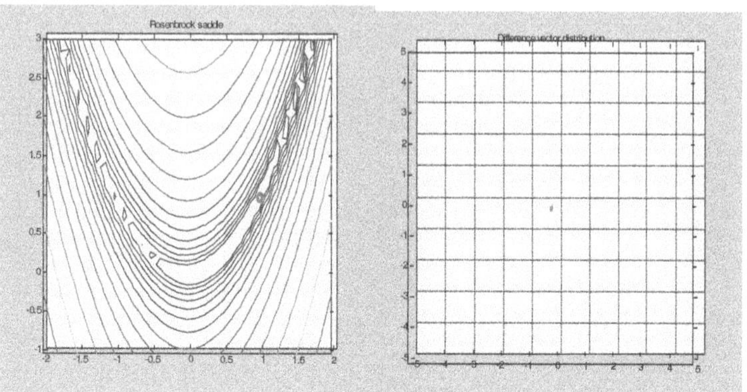

Fig. 6.18. Generation No. 54.

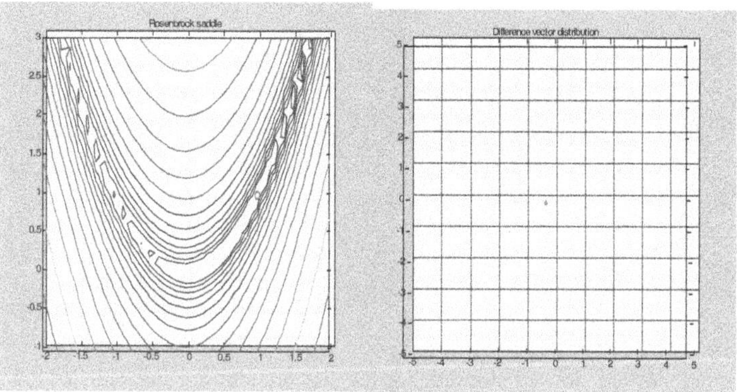

Fig. 6.19. Generation No. 60.

Figure 6.9 through Fig. 6.19 clearly show that the PDF used by DE very often has a shape different from either a gaussian or a cauchy distribution. It can also be seen that DE automatically adapts to the landscape it has to search and that its standard deviation diminishes as the population approaches the global optimum at the point (1,1).

6.4 Constraint handling

6.4.1 Boundary constraints

In boundary constrained problems, it is essential to ensure that parameter values lie inside their allowed ranges after performing the mutation and crossover operations.

A simple way to guarantee this is to replace parameter values that violate boundary constraints with random values generated within the feasible range:

$$u_{j,i,g} = \begin{cases} rand_j[0,1] \cdot \left(x_j^{(U)} - x_j^{(L)}\right) + x_j^{(L)} & \text{if } u_{j,i,g} < x_j^{(L)} \vee u_{j,i,g} > x_j^{(U)} \\ u_{j,i,g} & \text{otherwise} \end{cases} \quad (6.11)$$

where

$i = 1,...,NP, \quad j = 1,...,D$

This is the method that was used for this work. Another, less efficient method for keeping trial vectors within bounds is to regenerate the offending parameter value according Eq. 6.9 as many times as is necessary to satisfy the boundary constraint. It should also be mentioned that DE's generating scheme can extend its search beyond the space defined by initial parameter limits (Eqs. 6.8 and 6.9) if allowed to do so. This can be a beneficial property for unconstrained optimization problems because optima that are outside the initialized range can often be located.

6.4.2 Constraint functions

In the literature [4], mostly various penalty function methods have been applied with DE for handling constraint functions. In its simplest form, the function value $f'(X)$ to be minimized by DE can be computed by penalizing the objective function value with a weighted sum of constraint violations, for example as follows:

$$f'(X) = f(X) + \sum_{j=1}^{m} w_j \cdot \max\left(0, g_j(X)\right) \quad (6.12)$$

The penalty function approach effectively converts a constrained problem into an unconstrained one, and then $f'(X)$ is used instead of $f(X)$ as objective function.

Typically a penalty function method will establish one or more additional control parameters, penalty parameters or the weights like the ones, w_j, in Eq. 6.12. Appropriate values for the control parameters are expected to be set by the user *a priori*. In practice this stage always requires some additional efforts, and often finding suitable setting may be a laborious or even impossible task in practice. Quite often several trial runs of the DE are required for setting the penalty parameters by trial-and-error. The need for improved constraint handling techniques, that does not require user to set any extra search parameters, is obvious.

In [43,44] a new, more elegant approach for constraint handling is described, which does not establish any extra search parameters, like the weights, w_j, in Eq. 6.12, due to presence of the constraint functions. For constraint handling this novel approach applies a modified replacement rule of DE that is extended to handle also the involved constraint functions. In particular, DE's original replacement rule (Eq. 6.10) is substituted with the following one:

$$X_{i,g+1} = \begin{cases} U_{i,g} & \text{if} \begin{cases} \begin{cases} \forall j \in \{1,\ldots,m\}: g_j(U_{i,g}) \leq 0 \wedge g_j(X_{i,g}) \leq 0 \\ \wedge \\ f(U_{i,g}) \leq f(X_{i,g}) \end{cases} \\ \vee \\ \begin{cases} \forall j \in \{1,\ldots,m\}: g_j(U_{i,g}) \leq 0 \\ \wedge \\ \exists j \in \{1,\ldots,m\}: g_j(X_{i,g}) > 0 \end{cases} \\ \vee \\ \begin{cases} \exists j \in \{1,\ldots,m\}: g_j(U_{i,g}) > 0 \\ \wedge \\ \forall j \in \{1,\ldots,m\}: g'_j(U_{i,g}) \leq g'_j(X_{i,g}) \end{cases} \end{cases} \\ X_{i,g} & \text{otherwise} \end{cases} \tag{6.13}$$

where

$$g'_j(X_{i,g}) = \max\left(g_j(X_{i,g}), 0\right)$$
$$g'_j(U_{i,g}) = \max\left(g_j(U_{i,g}), 0\right)$$

Thus, when compared with the corresponding member, $X_{i,g}$, of the current population, the trial vector, $U_{i,g}$, will be selected if:

- it satisfies all constraints and provides a lower or equal objective function value, while both the compared solutions are feasible, or
- it is feasible while $X_{i,g}$ is infeasible, or
- it is infeasible, but provides a lower or equal constraint violation for all constraint functions.

Note that in case of an infeasible solution, the replacement rule does not compare the objective function values. Thus no selective pressure exists towards the search space regions providing low objective values, but infeasible solutions. Instead, a selective pressure towards regions, where constraint violation decreases, will be present in general. Because of that, also for finding the first feasible solution an effective selection pressure will be applied. This results in a fast convergence to the feasible regions of the search space.

Note also, that DE's replacement operation does not essentially require numerical objective function values for performing the selection, but only the result of the comparison; Which one of the compared solutions is better? Here it is considered that:

- If both the compared solutions are feasible, the one with lower objective function value is better.
- Feasible solution is better than infeasible.
- If both the compared solutions are infeasible, the situation is less obvious. Then the candidate vector can be considered less infeasible, and thus better than the current vector, if it does not violate any of the constraints more than the current vector and if it violates at least one of the constraints less.

The last mentioned concept for comparing two infeasible solutions is analogous with, and derived from, the concepts of Pareto-optimality. The selection is based on Pareto-dominance in the effective constraint function space, $g'(X) = \max(0,g(X))$. Note, that the Pareto-optimal front in the effective constraint function space is a single point ($g'(X) = 0$), if a feasible solution for the problem exists.

In addition, the candidate vectors considered equally good as the compared current population member are allowed to enter into the population in order to avoid stagnation at flat regions of the objective function surface.

This novel approach does not introduce any extra search parameters to be set by the user – simply, there are no constraint handling related, or any other, search parameters in Eq. 6.13.

An important special case should also be noted here. In case of an unconstrained problem ($m = 0$), Eq. 6.13 effectively reduces into the DE's original replacement rule (Eq. 6.10). Thus the new replacement rule does not change the DE from this point of view, except extending its capabilities for handling constraint functions.

In many real-world engineering design optimizations, the number of constraint functions is relatively high and the constraints are often non-trivial. It may well be, for example, that the feasible solutions comprise only a small subset of the search space. Feasible solutions may also divide the search space into isolated "islands". This discontinuity introduces stalling points for some genetic searches and also raises the possibility of new, locally optimal areas near the island borders. Furthermore, the user may easily define totally conflicting constraints so that no feasible solutions exist at all. Even so, if two or more constraints conflict so that no feasible solution exists, DE can find the nearest feasible solution. In the case of non-trivial constraints, the user is often able to judge which of the constraints are conflicting on the basis of the nearest feasible solution. It is then possible to reformulate the objective function or reconsider the problem setting itself to resolve the conflict.

Especially in cases for heavily constrained problems, where only a tiny fraction of the search space contains feasible solutions and where these feasible solutions are located in multiple separated islands around the search space, the novel constraint handling approach has been found particularly effective [43,44]. For further details, see [43,44].

6.5 Handling integer and discrete variables

6.5.1 Methods

In its canonical form, the Differential Evolution algorithm is only capable of handling continuous variables. Extending it to optimize integer variables is, however, very easy and requires only a couple of simple modifications. First, integer values should be used to evaluate the objective function, even though DE itself still works internally with continuous floating-point values. Thus,

$$f(y_i) \quad i = 1,\ldots,D \tag{6.14}$$

where

$$y_i = \begin{cases} x_i & \text{for continuous variables} \\ INT(x_i) & \text{for integer variables} \end{cases}$$

$$x_i \in X$$

INT() is a function for converting a real value to an integer value by truncation. Truncation is performed here to evaluate trial vectors and to handle boundary constraints. Truncated values are not assigned elsewhere. Thus, DE works with a population of continuous variables regardless of the corresponding object variable type. This is essential to maintain the diversity of the population and the robustness of the algorithm.

Instead of Eq. 6.8, integer variables should be initialized as follows:

$$P_0 = x_{j,i,0} = rand_j[0,1] \cdot \left(x_j^{(U)} - x_j^{(L)} + 1\right) + x_j^{(L)} \quad i = 1,\ldots,NP, \quad j = 1,\ldots,D \tag{6.15}$$

Additionally, instead of Eq. 6.11, boundary constraints for integer variables should be handled according to the prescription provided in Eq. 6.15 below:

$$u_{j,i,g} = \begin{cases} rand_j[0,1] \cdot \left(x_j^{(U)} - x_j^{(L)} + 1\right) + x_j^{(L)} \\ \quad \text{if } INT\left(u_{j,i,g}\right) < x_j^{(L)} \vee INT\left(u_{j,i,g}\right) > x_j^{(U)} \\ \\ u_{j,i,g} \\ \quad \text{otherwise} \end{cases} \tag{6.16}$$

where

$$i = 1,\ldots,NP, \quad j = 1,\ldots,D$$

Discrete values can also be handled in a straightforward manner. Suppose that the subset of discrete variables, $X^{(d)}$, contains l elements that can be assigned to variable x:

$$X^{(d)} = x_i^{(d)} \qquad i = 1,\ldots,l \tag{6.17}$$

where

$$x_i^{(d)} < x_{i+1}^{(d)}$$

Instead of the discrete value x_i itself, we may assign its index, i, to x. Now the discrete variable can be handled as an integer variable that is boundary constrained to the range $1,\ldots,l$. To evaluate the objective function, the discrete value, x_i, is used instead of its index i. In other words, instead of optimizing the value of the discrete variable directly, we optimize the value of its index, i. Only during evaluation is the indicated discrete value used. Once the discrete problem has been converted into an integer one, the previously described methods for handling integer variables can be applied (Eqs. 6.14–6.16).

6.5.2 An Illustrative Example

Consider minimization of a simple 2D objective function,

$$f(X) = |x_1| + |x_2| \tag{6.18}$$

subject to

$$-5 \le x_1, x_2 \le 5$$

In which the variables x_1 and x_2 are continuous. Then consider the same problem when the feasible set for x_1 and x_2 is limited to a set of integer values.

Figure 6.20 illustrates the application of the described integer and discrete variable handling techniques. The actual cost function surface from the DE's point of view is shown. Simply, the discrete problem is converted to a continuous one, in which flat regions (providing equal cost value) represent each alternative combination of discrete variables. The DE algorithm works internally with a continuous representation space, while the actual objective function evaluation space is discrete, since the truncation operation is applied before the objective function evaluation. Note that instead of truncation, a rounding operation may be applied as well. In this case, however, the Eqs. 6.14–6.16 should be modified respectively.

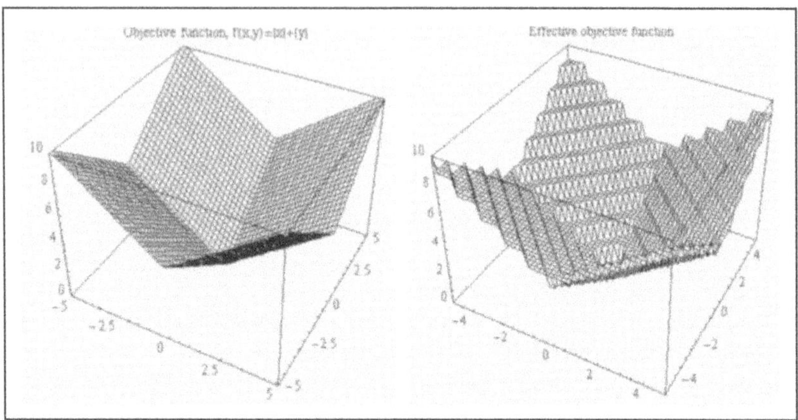

Fig. 6.20. The suggested method for handling integer and discrete variables is illustrated for the case of a simple 2D objective function $f(X) = |x1|+|x2|$. A continuous variable case (left) and a discrete case (right), in which only integer values are allowed is shown. In the discrete case the optimization algorithm still works with a continuous objective function surface, but the surface is modified according to Eqs. 6.14-6.17 as illustrated. A discrete problem is actually converted to a corresponding continuous one.

The modified cost-function surface does not actually make the problem more difficult to solve with DE. In fact, the modified cost function may be easier to solve than the original continuous problem. Note that continuous problems are often converted to a discrete one before problem solving, for example a widely applied binary encoded genetic algorithm follows this approach. Why convert a continuous problem to a discrete one, which is, in principle at least, more difficult to solve? The opposite approach considered here appears to be much more useful.

6.6 Numerical examples

To discover the effectiveness of the techniques proposed in Sects. 6.4 and 6.5, three numerical examples (Table 6.1) were optimized using DE and the described constraint handling approach (Eq. 6.13). These non-linear, engineering design optimization problems with discrete, integer and continuous variables were first investigated by Eric Sandgren [10] and subsequently by many other researchers who applied a variety of optimization techniques (Table 6.2). These problems represent optimization situations involving discrete, integer and continuous variables that are similar to those encountered in everyday mechanical engineering design tasks. Because the problems are clearly defined and relatively easy to understand, they form a suitable basis for comparing alternative optimization methods.

Table 6.1. Test problems.

Summary of Test Problems					
Example	*Description*	*Number of variables*			
		Total	*Discrete*	*Integer*	*Continuous*
1	Design of a Gear Train	4	0	4	0
2	Design of a Pressure Vessel	4	2	0	2
3	Design of a Coil Spring	3	1	1	1

Table 6.2. Alternative methods used to solve the test problems.

Compared Methods		
Reported by	*Solution technique*	*Reference*
1. Sandgren	Branch & Bound using Sequential Quadratic Programming	[10]
2. Fu, Fenton & Gleghorn	Integer-Discrete-Continuous Non-Linear Programming	[11]
3. Loh & Papalambros	Sequential Linearization Algorithm	[12,13]
4. Zhang & Wang	Simulated Annealing	[14]
5. Chen & Tsao	Genetic Algorithm	[15]
6. Li & Chou	Non-Linear Mixed Discrete Programming	[16]
7. Kannan & Kramer	Augmented Lagrange Method	[17]
8. Wu & Chow	Meta-Genetic Algorithm	[18]
9. Shih & Lai	Mixed-Discrete Fuzzy Programming	[19]
10. Lin, Zhang & Wang	Modified Genetic Algorithm	[20]
11. Deb & Goyal	Combination of Binary/Real-coded Genetic Algorithm	[21,22]
12. Cai & Thierauf	Two-level Parallel Evolution Strategy	[23,24,25]
13. Cao & Wu	Evolutionary Programming	[26,27]
14. Wang, Teo & Lee	Non-Linear Mixed Discrete Programming	[28]
15. Cao & Wu	Cellular Automata based Genetic Algorithm	[29]
16. Litinetski & Abrahamzon	Multistart Adaptive Random Search	[30]
17. Ndiritu & Daniell	Modified Genetic Algorithm	[31]
18. Lewis & Mistree	Foraging-Directed Adaptive Linear Programming	[32]
19. Coello Coello	Genetic Algorithm with self-adaptive penalties	[33,34]
20. Lin, Wang & Hwang	Hybrid Differential Evolution	[35]
21. This article	Differential Evolution	This article

In order to demonstrate that the DE is particularly easy to adapt for different problems, and finding suitable values for it's control parameters does not require dozens of trials-and-errors, the search parameters of DE were set to $F = 0.9$ and $CR = 0.9$ for all problems. These settings have been found effective for most of the problems, while they rarely provide the best efficiency. In other words, these settings are usually enough to solve the problem, but do not always provide the highest possible convergence rate. Thus, only the population size and the number of generations were coarsely varied to provide some problem specific adaptation. For each problem, the population size NP was selected among the multiples of 10 and the number of generations was selected from the multiples of 50. Any further settings, like penalty parameters, weights for each constraint function or other extra search parameters, were not required to be set by the user. No particular attempts were performed in order to optimize the search parameter settings to fit optimally to each test problem.

6.6.1 Example 1: Designing a gear train

The first example problem is to optimize the gear ratio for the compound gear train arrangement shown in Fig. 6.21. The gear ratio for a reduction gear train is defined as the ratio of the angular velocity of the output shaft to that of the input shaft. In order to produce the desired overall gear ratio, the compound gear train is constructed out of two pairs of gearwheels, *d-a* and *b-f*. The overall gear ratio, i_{tot}, between the input and output shafts can be expressed as:

$$i_{tot} = \frac{\omega_o}{\omega_i} = \frac{z_d z_b}{z_a z_f} \tag{6.19}$$

where ω_o and ω_i are the angular velocities of the output and input shafts, respectively, and z denotes the number of teeth on each gearwheel.

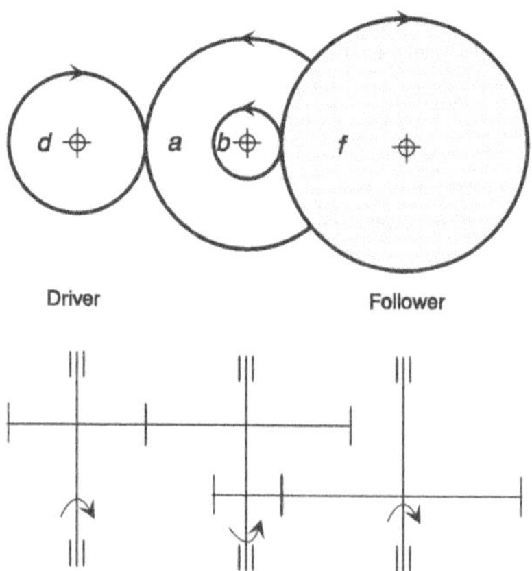

Fig. 6.21. Compound gear train for Example 1.

The optimization problem is to find the number of teeth for gearwheels *d, a, b* and *f* in order to produce a gear ratio, i_{tot}, as close as possible to the target ratio: i_{trg} = 1/6.931 (= 0.1443). For each gear, the minimum number of teeth is 12 and the maximum is 60.

The problem is formulated as follows:

Find

$$X = (x_1, x_2, x_3, x_4) = (z_d, z_b, z_a, z_f), \quad x \in \{12, 13, ..., 60\} \tag{6.20}$$

to minimize

$$f(X) = (i_{trg} - i_{tot})^2 = \left(\frac{1}{6.931} - \frac{x_1 x_2}{x_3 x_4}\right)^2$$

subject to

$$12 \le x_i \le 60 \ , \quad i = 1,2,3,4.$$

Thus, the goal is to find optimum values for four integer variables that will minimize the squared difference between the desired gear ratio, i_{trg}, and current gear ratio, i_{tot}. The objective (cost) function was simply defined as the squared error between the actual and the desired gear ratio. For this problem, each variable is subject only to upper and lower boundary constraints. The integer techniques described in Sect. 6.5.1 were invoked to handle boundary constraints.

The gear train problem was solved using the *DE/rand/1/bin* strategy with control settings of: *NP*=110, *F*=0.90 and *CR*=0.90.

Table 6.3 lists the various gear train solutions and compares DE's result with those reported in [10,11,13,14,17,18,20,21,22,26,27,30,32,35].

The solution found by DE was as good as the best solution in the literature. In fact, DE provided different results from run to run with the same objective function value (Table 6.4). In [21,22] it is reported that these solutions have been found to be globally optimal by applying explicit enumeration of all possible 48^4 (= $5.3 \cdot 10^6$) gear teeth combinations.

By inspecting Eq. 6.20, it is obvious that there are four global optima. Because DE can work with a population of solutions rather than just a single solution, it is capable of finding multiple global optima for this problem. By using a sufficiently large population, it is possible to obtain all four alternative solutions in a single run. Nevertheless, only one solution was extracted from the population of the last generation because the other three solutions can be found in a trivial way based on one solution and the symmetry of Eq. 6.20. In practice, however, optimization tasks exist with multiple global optima that cannot be detected so simply.

One may wonder why finding a single, globally optimal solution is not always sufficient when multiple global optima exist. One reason is that the sensitivity of the objective function to small changes in its variables may be different at the alternative optimal points. In practice, it is often important to select the most robust solution, i.e., the global optima with the least sensitivity to noise. For example, if a machine design is subject to optimization, it is possible that the optimized design variables cannot be manufactured accurately. Alternatively it may be that the design parameters change during the lifetime of the machine due to normal wear of its components, for example. In such cases, robust global optima are to be preferred over those that exhibit a high sensitivity to design implementation errors.

Table 6.3. Optimal solutions for the gear train problem.

Optimal solutions for the gear train problem							
Reported by	x_1 (z_d)	x_2 (z_b)	x_3 (z_a)	x_4 (z_f)	***f(x)***	*Gear Ratio*	*Error [%]*
Sandgren [10]	18	22	45	60	5.7×10^{-6}	0.146666	1.65
Fu et al. [11]	14	29	47	59	4.5×10^{-6}	0.146411	1.17
Loh & Papalambros [13]	19	16	42	50	0.233×10^{-6}	0.144762	0.334
Zhang & Wang [14]	30	15	52	60	2.36×10^{-9}	0.144231	0.034
Lin et al. [20]	19	16	49	43	$\mathbf{2.7\times10^{-12}}$	0.144281	**0.00114**
Kannan & Kramer [17]	13	15	33	41	2.12×10^{-8}	0.1441	0.11
Wu & Chow [18]	19	16	43	49	$\mathbf{2.7\times10^{-12}}$	0.144281	**0.00114**
Deb & Goyal [21,22]	19	16	49	43	$\mathbf{2.7\times10^{-12}}$	0.144281	**0.00114**
Litinetski & Abrahamzon [30] [1)]	19	16	43	49	$\mathbf{2.7\times10^{-12}}$	0.144281	**0.00114**
Cao & Wu [26,27]	30	15	52	60	2.36×10^{-9}	0.144231	0.034
Lewis & Mistree [32]	19	16	49	43	$\mathbf{2.7\times10^{-12}}$	0.144281	**0.00114**
Lin, Wang & Hwang [35]	19	16	43	49	$\mathbf{2.7\times10^{-12}}$	0.144281	**0.00114**
This article [2)]	16	19	43	49	$\mathbf{2.7\times10^{-12}}$	0.144281	**0.00114**

1) Only 6% out of 100 independent runs of the search algorithm found the optimum requiring averagely 5100 function evaluations. However, 95% of the runs found $f(X) = 2.7\times10^{-12}$ or a better solution.

2) Also alternative solutions with an equal target function value were obtained from run to run. All of the 100 independent runs of the search algorithm found the optimum after 110,000 function evaluations or earlier.

Table 6.4. Alternative solutions for the gear train problem found by DE. These solutions have been found to be globally optimal by applying explicit enumeration [21,22].

Alternative solutions for gear train problem				
Solution	z_d	z_b	z_a	z_f
1	19	16	43	49
2	16	19	43	49
3	19	16	49	43
4	16	19	49	43

Results for DE were recorded after 1000 generations, corresponding to 110,000 evaluations of the objective function, which is only 2.1% of all possible 48^4 (= $5.3 \cdot 10^6$) gear teeth combinations. The computation time required was 1.15 seconds using a PC with an AMD K6-233 MHz processor (SPECfp95 rating 5.5 [36]). To evaluate the robustness of the DE algorithm, 100 independent trials were performed. All runs yielded the reported value of $f(X)$. As mentioned, different solutions were obtained from run to run because of the existence of multiple global optima (Table 6.4).

6.6.2 Example 2: Designing a pressure vessel

The second test problem is to design a compressed air storage tank with a working pressure of 3,000 psi and a minimum volume of 750 ft^3. As Fig. 6.22 shows, the cylindrical pressure vessel is capped at both ends by hemispherical heads. Using a rolled steel plate, the shell is to be made out of two halves that are joined by two longitudinal welds to form a cylinder. Each head is forged and then welded to the shell.

The objective is to minimize the manufacturing cost of the pressure vessel. The cost is a combination of material cost, welding cost and forming cost. Refer to [10] for more details.

The design variables are shown in Fig. 6.22. Variables L and R are continuous while T_s and T_h are discrete. The thickness of the shell, T_s, and the head, T_h, are both required to be standard sizes. For this example, steel plate was available in multiples of 0.0625 inches.

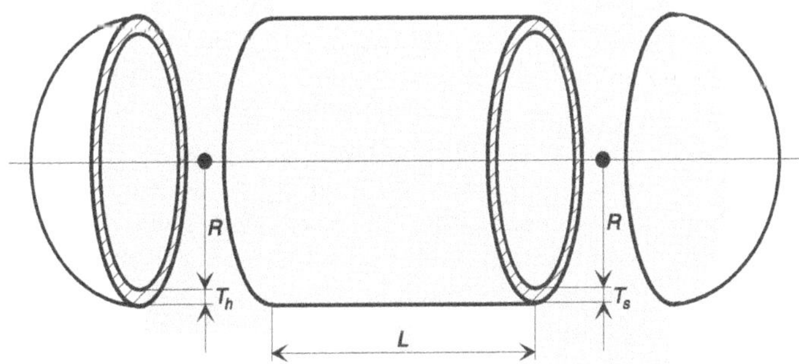

Fig. 6.22. Pressure vessel for Example 2.

The problem can be formulated as follows:

$$
\begin{aligned}
&\text{Find} \\
&X = (x_1, x_2, x_3, x_4) = (T_s, T_h, R, L) \\
&\text{to minimize} \\
&f(X) = 0.6224x_1x_3x_4 + 1.7781x_2x_3^2 + 3.1611x_1^2x_4 + 19.84x_1^2x_3 \\
&\text{subject to} \\
&g_1(X) = 0.0193x_3 - x_1 \le 0 \\
&g_2(X) = 0.00954x_3 - x_2 \le 0 \\
&g_3(X) = 750.0 \times 1728.0 - \pi x_3^2 x_4 - \frac{4}{3}\pi x_3^3 \le 0 \\
&g_4(X) = x_4 - 240.0 \le 0 \\
&g_5(X) = 1.1 - x_1 \le 0 \\
&g_6(X) = 0.6 - x_2 \le 0
\end{aligned}
\tag{6.21}
$$

The objective function, $f(X)$, represents the total manufacturing cost of the pressure vessel as a function of the design variables.

The constraints, $g_1,\ldots,g_6$, quantify the restrictions to which the pressure vessel design must adhere. These limits arise from a variety of sources. For example, the minimal wall thickness of the shell, T_s (g_1), and heads, T_h (g_2), with respect to the shell radius are limited by the pressure vessel design code. The volume of the vessel must be at least the specified 750 ft^3 (g_3). Available rolling equipment limits the length of the shell, L, to no more than 20 feet (g_4). According to the pressure vessel design code, the thickness of the shell, T_s, is not to be less than 1.1 inches (g_5) and the thickness of the head, T_h, is not to be less than 0.6 inches (g_6).

The DE control variables used to solve the pressure vessel design problem were: NP=30, F=0.90 and CR=0.90. The problem statements do not define the boundaries for the design variables, but the constraints, g_4, g_5 and g_6 are pure boundary constraints, so they were handled as lower boundary constraints for x_1 and x_2, and as an upper boundary constraint for x_4, respectively. The lower boundaries for x_3 and x_4 can be set to zero, since common sense demands that they must be non-negative values. The upper boundaries for x_1, x_2 and x_3, however, must still be specified in order to define the search space. Consequently, these bounds were set high enough to make it highly probable for the global optimum to lie inside of the defined search space. Since the possibility existed that the global optimum was outside of the initially defined search space, these estimated bounds were used only to initialize the population according to Eqs. 6.8 and 6.15. DE was then allowed to extend the search beyond these boundaries. The possibility of using this kind of "loose" boundary constraint for variables is one of the advantages of DE.

In practical engineering design work, it is not unusual for one or more boundaries to be unknown and the distance to the optimum cannot be reliably estimated. The boundary constraints used for each variable are shown in Table 6.5. The other constraints, g_1, g_2 and g_3 were handled as constraint functions.

Table 6.5. Boundary constraints used for the pressure vessel example.

Boundary constraints for pressure vessel example		
Lower limitation	*Constraint*	*Upper limitation*
constraint g_5	$1.1 \leq x_1 \leq 12.5$	Roughly guessed *
constraint g_6	$0.6 \leq x_2 \leq 12.5$	Roughly guessed *
non-negative value of x_3	$0.0 \leq x_3 \leq 240.0$	Roughly guessed *
non-negative value of x_4	$0.0 \leq x_4 \leq 240.0$	Constraint g_4

* The value of this boundary is not given among the problem statements. Thus, the value is estimated roughly and used only for initialization of the population. DE was allowed to extend the search beyond this limit.

Notice that it is not necessary to evaluate the constraint functions, g_4, g_5 and g_6, because they were handled as boundary constraints and DE was not allowed to generate a candidate vector that violated any of them.

The problem formulation given in Eq. 6.21 follows Sandgren's original problem statements [10]. In researching this problem, at least four different formulations were found in the literature. For some unknown reason, [16,23,24,25,28] have used a slightly reformulated constraint-function, g_5. Non-original problem statements have also been applied in [22,26,27,29,33,34,35], where the constraints g_5 and g_6 were ignored. Both these modifications extend the region of feasible solutions and also make it possible to obtain a significantly lower objective function value than with Sandgren's original problem statements [10]. For this reason, the results of [16,22,23,24,25,26,27,28,29,33,34,35] cannot be fairly compared with the results obtained using Sandgren's original problem statements [10]. To enable a more comprehensive comparison with the other methods, all four cases were solved using DE.

In Table 6.6, DE's results are compared to those reported in [10,11,13,17,18,19,28,30,31]. Since all these results are obtained using Sandgren's original problem statements [10], they are comparable with each other. Due to space limitations and the objectives for this introductory article, the results with the modified problem formulations are not included here. However, it can be briefly mentioned that the DE obtained continuously lower objective function values than any of these compared results [16,22,23,24,25,26,27,28,29,33,34,35], when identical problem statements were applied.

Table 6.6. Optimal solutions for the pressure vessel problem.

Optimal solutions for the pressure vessel design problem [1)]

Reported by	x_1 (T_s) [inch]	x_2 (T_h) [inch]	x_3 (R) [inch]	x_4 (L) [inch]	$g_1(X)$	$g_2(X)$	$g_3(X)$	$g_4(X)$	$f(X)$ [$]		
Sandgren [10]	1.125	0.625	48.97	106.72	0.179	0.1578	3.0	133.284	7982.5	100.0%	
Fu et al. [11]	1.125	0.625	48.38070	111.7449	0.191250	0.163449	75.8750	128.2551	8048.619	100.8%	
Loh & Papalambros [13]	1.125	0.625	58.2901	43.693	$1.068 \cdot 10^{6}$	0.068912	0.429283	196.3070	7197.734	90.2%	2)
Kannan & Kramer [17]	1.125	0.625	58.291	43.690	*violated*	–	–	–	*infeasible*	–	3)
Wu & Chow [18]	1.125	0.625	58.1978	44.2930	0.001782	0.069793	974.5829	195.7070	7207.497	90.3%	
Shih & Lai [19]	1.125	0.625	56.075	59.127	0.0427	0.090	26660.77	180.87	7462.1	93.5%	
Wang et al. [28]	1.125	0.625	58.290155	43.6927	$8.500 \cdot 10^{9}$	0.068912	0.441269	196.3073	7197.731	90.2%	4)
Litinetski & Abrahamzon [30]	1.125	0.625	58.29012	44.69286	$6.8400 \cdot 10^{7}$	0.068912	10674.40	195.3071	7242.547	90.7%	5)
Ndiritu & Daniell [31]	1.125	0.625	58.2127	44.0802	–	–	*violated*	–	*infeasible*	–	6)
This article	1.125	0.625	58.29016	43.69266	$4.8355 \cdot 10^{17}$	0.068912	$1.6087 \cdot 10^{11}$	196.3073	**7197.729**	**90.2%**	7)

1) In [14] and [20] it is reported that the value of $f(X) = 7197.7$ was reached. Because neither a more accurate result, nor details of the result were provided, they are not included in this comparison.

2) No values for constraint functions were reported in [13]. Also, the optimum solution was too inaccurately reported to reconstruct the results properly. The reported value, $x_3 = 58.290$, causes a violation of constraint g_3. For this reason, 58.2901 was used here to reconstruct the constraint functions and target function values. In [13], a value of $f(X) = 7197.7$ was originally reported.

3) In [17], a value of $f(X) = 7198.200$ was originally reported. Other than reported in [17], the solution results are in violation of constraint g_1. Since the violation is not caused by a simple rounding error of reported value for x_3, it was impossible to reconstruct a feasible result.

4) The optimum solution was too inaccurately reported in [28] to reconstruct the results properly. The reported value, $x_3 = 58.2902$, causes a violation of constraint g_1. For this reason, 58.290155 was used here to reconstruct the constraint functions and target function values. In [28], a value of $f(X) = 7198.005$ was originally reported, which is higher than the reconstructed value $f(X) = 7197.731$.

5) No values for constraint functions were reported in [30]. The reported $f(X) = 7197.729$ was considered to be erroneous. For this reason, both the $f(X)$ value and constraint function values, reported in this Table, are reconstructed.

6) Other than reported in [31] the solution is infeasible and violates clearly constraint g_3.

7) All of the 100 independent runs of the search algorithm found the reported optimum or better within 15,000 function evaluations.

As Table 6.6 shows, DE found a better solution for the pressure vessel design problem than the best solution found in the literature. Computations were carried out to 500 generations, corresponding to 15,000 evaluations of the cost function. Computation took 0.22 seconds on a PC with an AMD K6-233 MHz processor (SPECfp95 rating 5.5 [36]). In order to ensure robustness, one hundred independent runs were performed for each case. All trials yielded the reported value of $f(X)$ or (marginally) better. In addition, all solutions were within the feasible design domain.

6.6.3 Example 3: Designing a coil compression spring

The third example involves the design of a coil compression spring (Fig. 6.23). The spring is to be a helical compression spring to which a strictly axial and constant load will be applied.

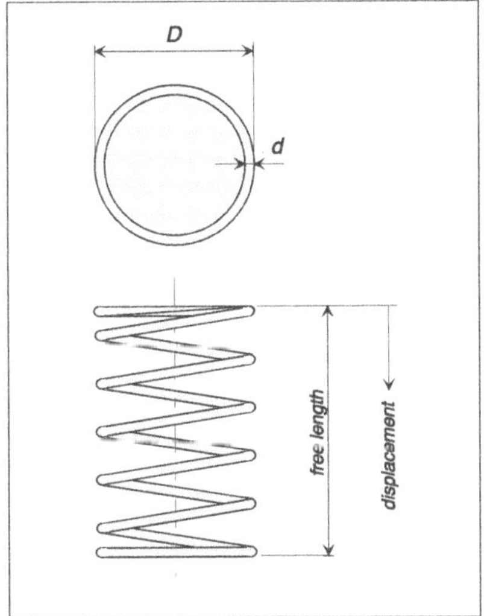

Fig. 6.23. Coil spring for Example 3.

The objective is to minimize the volume of spring steel wire needed to manufacture the spring. The design variables are the number of spring coils, N, the outside diameter of the spring, D, and the spring wire diameter, d. This example contains integer, discrete and continuous variables. The number of spring coils, N, is an integer variable and the outside diameter, D, is a continuous variable. Additionally, the spring wire diameter, d, is only available in the standard (discrete) sizes shown in Table 6.7.

The problem is formulated as follows:

$$
\begin{aligned}
&\text{Find}\\
&X = (x_1, x_2, x_3) = (N, D, d)\\
&\text{to minimize}\\
&f(X) = \frac{\pi^2 x_2 x_3^2 (x_1 + 2)}{4}\\
&\text{subject to}\\
&g_1(X) = \frac{8 C_f F_{\max} x_2}{\pi x_3^3} - S \le 0\\
&g_2(X) = l_f - l_{\max} \le 0\\
&g_3(X) = d_{\min} - x_3 \le 0\\
&g_4(X) = x_2 - D_{\max} \le 0\\
&g_5(X) = 3.0 - \frac{x_2}{x_3} \le 0\\
&g_6(X) = \sigma_p - \sigma_{pm} \le 0\\
&g_7(X) = \sigma_p + \frac{F_{\max} - F_p}{K} + 1.05(x_1 + 2)x_3 - l_f \le 0\\
&g_8(X) = \sigma_w - \frac{F_{\max} - F_p}{K} \le 0\\
&\text{where}\\
&C_f = \frac{4(x_2 / x_3) - 1}{4(x_2 / x_3) - 4} + \frac{0.615 x_3}{x_2} \le 0\\
&K = \frac{G x_3^4}{8 x_1 x_2^3}\\
&\sigma_p = \frac{F_p}{K}\\
&l_f = \frac{F_{\max}}{K} + 1.05(x_1 + 2)x_3
\end{aligned}
\tag{6.22}
$$

The objective function, $f(X)$, computes the volume of spring steel wire as a function of the design variables. The design constraints are specified as follows:

a) The maximum working load is: $F_{max} = 1000.0$ lb.
b) The allowable maximum shear stress is: $S = 189000.0$ psi (g_1).
c) The maximum free length is: $l_{max} = 14.0$ inch (g_2).

d) The minimum wire diameter is: $d_{min} = 0.2$ inch (g_3).
e) The maximum outside diameter of the spring is: $D_{max} = 3.0$ inch (g_4).
f) The pre-load compression force is: $F_p = 300.0$ lb.
g) The allowable maximum deflection under pre-load is: $\sigma_{pm} = 6.0$ inch (g_6).
h) The deflection from pre-load position to maximum load position is: $\sigma_w = 1.25$ inch (g_8).
i) The combined deflections must be consistent with the length, i.e., the spring coils should not touch each other under the maximum load at which the maximum spring deflection occurs (g_7).
j) The shear modulus of the material is: $G = 11.5\times10^6$.
k) The spring is guided, so the buckling constraint is bypassed.
l) The outside diameter of the spring, D, should be at least three times greater than the wire diameter, d, to avoid lightly wound coils (g_5).

A more detailed explanation of the coil spring design procedure can be found in [10,18] and in [37; pp.371–381].

Table 6.7. Allowable spring steel wire diameters for the coil spring design problem.

Allowable wire diameters [inch]					
0.009	0.0095	0.0104	0.0118	0.0128	0.0132
0.014	0.015	0.0162	0.0173	0.018	0.020
0.023	0.025	0.028	0.032	0.035	0.041
0.047	0.054	0.063	0.072	0.080	0.092
0.105	0.120	0.135	0.148	0.162	0.177
0.192	0.207	0.225	0.244	0.263	0.283
0.307	0.331	0.362	0.394	0.4375	0.500

Table 6.8. Boundary constraints used for the coil spring example.

Boundary constraints for coil spring example		
Lower limitation	*Constraint*	*Upper limitation*
At least one spring coil is required to form a spring.	$1 \le x_1 \le \frac{l_{max}}{d_{min}}$	Upper and lower surfaces of unloaded spring coils touch each other.
Constraints g_3 and g_5 together	$3d_{min} \le x_2 \le D_{max}$	constraint g_4
Constraint g_3	$d_{min} \le x_3 \le \frac{D_{max}}{3}$	constraints g_4 and g_5 together

The following DE control variable settings solved the coil spring problem: *NP*=50, *F*=0.90 and *CR*=0.90. Although the problem statements do not define the boundaries for design variables, the constraints, g_3 and g_4 are pure boundary constraints and were treated as a lower boundary constraint for x_3 and as an upper boundary constraint for x_2, respectively. Furthermore, $g5$ can also be handled as a boundary constraint. In order to define the search space, the other boundary constraints were chosen based on the problem statements and the simple geometric space limitations elaborated in Table 6.8. The remaining constraints were handled as soft-constraint functions.

Constraint functions: g_3, g_4 and g_5 did not have to be evaluated because they were handled as boundary constraints and DE was not allowed to generate a candidate vector that violated any of them.

Table 6.9. Optimal solutions for the coil spring problem.

Item	*Optimal solutions for the coil spring design problem* [2]						*Type of variable*
	Sandgren [10]	*Chen & Tsao [15]*	*Kannan & Kramer [17]*	*Wu & Chow [18]*	*Deb & Goyal [21,22]*	*This article* [3]	
x_1 *(N) [1]*	10	9	7	9	9	9	Integer
x_2 *(D) [inch]*	1.180701	1.2287	1.329	1.227411	1.226	1.2230410	Continuous
x_3 *(d) [inch]*	0.283	0.283	0.283	0.283	0.283	0.283	Discrete
$g_1(X)$	54309	415.969	*Violated* [1]	550.993	713.510	1008.8114	
$g_2(X)$	8.8187	8.9207	–	8.9264	8.933	8.94564	
$g_3(X)$	0.08298	0.08300	–	0.08300	0.083	0.083000	
$g_4(X)$	1.8193	1.7713	–	1.7726	1.491	1.77696	
$g_5(X)$	1.1723	1.3417	–	1.3371	1.337	1.32170	
$g_6(X)$	5.4643	5.4568	–	5.4585	5.461	5.46429	
$g_7(X)$	0.0	0.0	–	0.0	0.0	2.67581×10^{-16}	
$g_8(X)$	0.0	0.0174	–	0.0134	0.009	5.07515×10^{-16}	
$f(X)$ *[inch³]*	2.7995	2.6709	*Infeasible* [1]	2.6681	2.665	**2.65856**	
	100.0%	95.4%	–	95.3%	95.2%	**95.0%**	

1) In [17], a feasible solution with value of $f(X) = 2.365$ was originally reported. Other than reported in [17], the solution results in a severe violation of constraint g_1. Thus their solution is clearly infeasible – not a feasible high quality solution as they have reported.

2) In [32], a solution for the coil spring design problem was also reported. However, a problem formulation that is significantly different from [10] was applied in [32], thus their result can not be compared here at all.

3) All of the 100 independent runs of the search algorithm found the reported optimum or better within 12,500 function evaluations.

Table 6.9 compares DE's solution with results obtained by other researchers. DE obtained a better solution than the best solution found in the literature. In order to demonstrate the robustness of the DE algorithm, 100 independent optimization trials were performed. All trials yielded the reported value of $f(X)$ or better. All of the solutions were also within the feasible region of the design space. The number of generations was 250, corresponding to 12,500 function evaluations. The computation time was 0.22 seconds on a PC with an AMD K6-233 MHz processor (SPECfp95 rating 5.5 [36]).

6.7 DE's Sensitivity to Its Control Variables

In order to investigate DE's sensitivity to its control variables (*NP*, *F*, *CR*) and in order to investigate if any simple rules for selecting their values exists, the following experiments were performed. The focus was on the **convergence velocity** and on the **robustness** of the optimum seeking.

The coil spring design problem was used as a test problem. As a base point, the settings *NP*=50, *F*=0.90 and *CR*=0.90 were applied. One by one, each control variable was kept fixed, while the other two were varied incrementally as follows:

- *NP* from 4 to 100 with increments of 1
- *F* from 0.0 to 2.0 with increments of 0.1
- *CR* from 0.0 to 1.0 with increments of 0.1

100 independent optimization runs were performed with each individual set of control variable settings.

As a measure of the convergence velocity, the required number of cost function evaluations to reach the cost value $f(X) \leq 2.65856$ (the best value reported in Table 6.9) was used. The number of function evaluations was averaged over the above-mentioned 100 independent optimization runs. The maximum number of function evaluations was limited to 200,000. In cases when this stopping criterion was reached before the specified cost value, the optimization run was considered to have failed. The number of failures was used as a measure of the robustness. The results of the experiments are visualized in Figs. 6.24–6.26.

The results suggest that DE is not remarkably sensitive to its control parameters. Thus, their values are relatively easy to choose. It is a fact that generally, for an arbitrary objective function, a 0% failure risk within any finite number of function evaluations cannot be reached for any stochastic non-linear optimization algorithm. However, the results demonstrate that a low failure rate, say less than 1%, can be achieved with a finite number of function evaluations, in this case with a relatively low number of function evaluations.

Furthermore, Figs. 6.24–6.26 suggest that it should be relatively easy to find good initial values for *NP*, *F* and *CR* just by following the existing recommendations in [1,2,3,7,9] for the initial guess and by following the further advice given in this article. Actually, it is much more difficult to choose such values that do not result in any convergence at all.

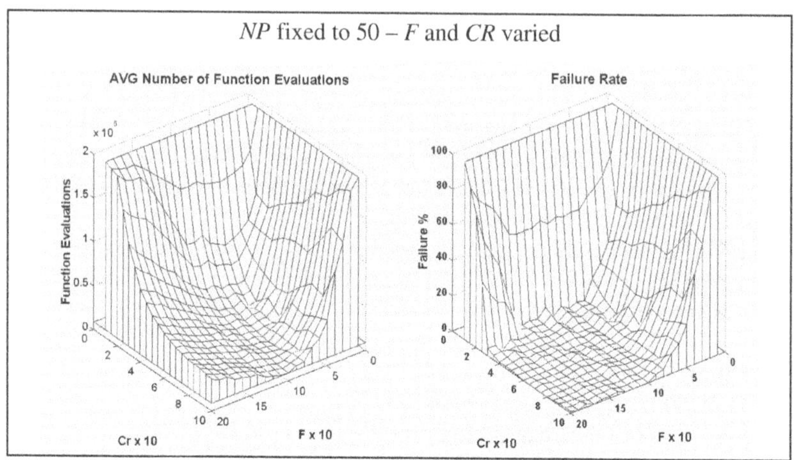

Fig. 6.24. Results of the 1st experiment. NP was kept fixed to 50 while *F* and *CR* were varied. The average number of function evaluations FE required to reach cost value $f(X) \leq 2.65856$ (left) and the number of failures (right), where the specified cost value was not reached within 200,000 function evaluations. Results are based on 100 independent trials at each studied point.

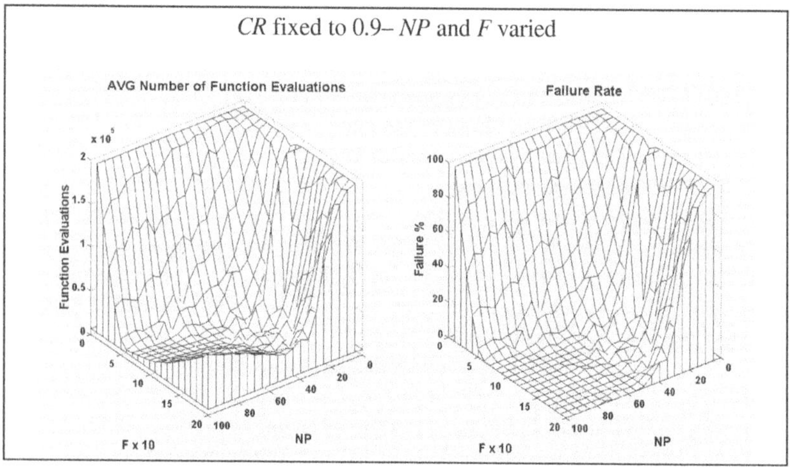

Fig. 6.25. Results of the 2nd experiment. CR was kept fixed to 0.9 while *F* and *NP* were varied. The average number of function evaluations FE required to reach cost value $f(X) \leq 2.65856$ (left) and the number of failures (right), where the specified cost value was not reached within 200,000 function evaluations. Results are based on 100 independent trials at each studied point.

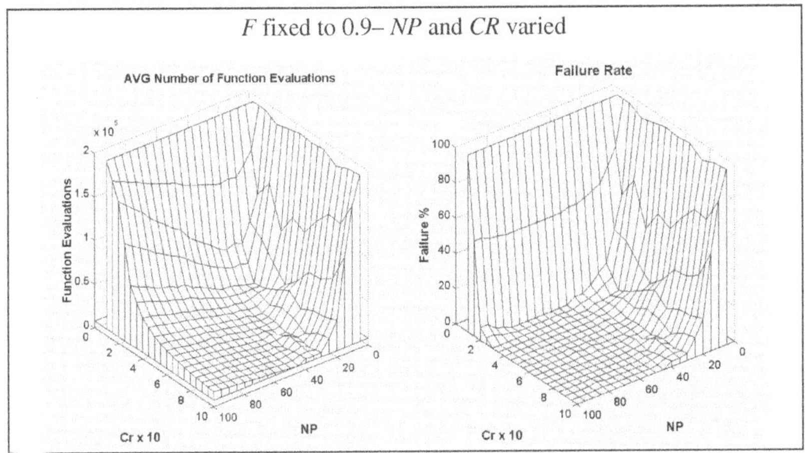

Fig. 6.26. Results of 3rd experiment. F was kept fixed to 0.9 while *NP* and *CR* were varied. The average number of function evaluations FE required to reach cost value $f(X) \leq 2.65856$ (left) and the number of failures (right), where the specified cost value was not reached within 200,000 function evaluations. Results are based on 100 independent trials at each studied point.

For *CR*, **generally, a relatively high value of *CR* should be the initial guess.** Only in the case that there is some *a priori* knowledge that the objective function is separable, or it is known that there is only a low degree of interaction between the function parameters, a low *CR*, say lower than 0.6, may result in a faster and/or more robust convergence. This is because *CR*=1.0 results in a rotationally invariant sampling of the search space, while *CR*=0.0 results in a search towards directions of the coordinate axes only. With *CR*=1.0 any search direction in the search space is equally likely, which is essential for solving common non-separable real-world problems efficiently. See [38] for further details. Note, how in this particular case of the spring design problem, with a high degree of non-linear interactions between the design parameters, higher values of *CR* are clearly the best ones, while low values result in a slow and failure prone convergence, despite whatever value *NP* or *F* was tried.

High values of both *F* and *NP* decrease the failure rate at the expense of the convergence velocity. However, both *F* and *NP* should be selected to be high enough, since when increasing them, both the convergence velocity and failure rate will improve until a certain point is reached, where the convergence velocity starts to decrease while the failure rate is still decreasing. One difficulty is the fact that all the control parameters *CR, F* and *NP* appeared to be dependent on each other. However, it seems that the nature of their interactions is not extremely complex. That explains why an intuitive tuning of the control parameters appeared to be so easy. Furthermore, it suggests that some further analysis in future may extract more accurate information on how to set the control parameters, or even select them automatically. In the current situation, selecting the control parameters without any previous experience, even randomly, within the recommended boundaries is highly likely to result in convergence. Nevertheless, finding near optimal values requires some hands-on experience and usually a few experiments with different values.

Actually, selecting *CR, F* and *NP* appears to be a non-linear and non-separable optimization problem itself. Indeed, one could apply an effective optimization method for optimizing itself, too. A future solution for a control parameter setting problem could be a *Meta-Differential Evolution* (as suggested by Kenneth Price), in which another higher level DE algorithm is applied to optimize the control parameters. The involved high computational costs have usually prevented using any *Meta-Evolutionary* approaches so far. However, it is likely that in the case of Meta-DE the computational cost would be much lower than with a *Meta-Genetic Algorithm*, for example. Figures 6.24-6.26 suggest that the objective function surface of the DE itself would not be too difficult – it appears to have an almost unimodal basic structure. This fact suggests that a relatively fast solution with a small population (in Meta-DE population) would be possible. Maybe it could be possible to solve a wide range of problems by Meta-DE using the same control parameters for all of the problems on outer DE, while the control parameters for the inner DE become optimized by the outer DE. At least, a Meta-Differential Evolution seems to be a subject worthy of further investigation.

6.8 Conclusions

Based on the results of this 3-problem test suite and comparisons with 20 other optimization methods, DE appears to be a promisingly efficient, effective and robust optimization algorithm. Additionally, DE is both easy to implement and easy to use. With the described extensions, DE was capable of optimizing all integer, discrete and continuous variables, and it was able to handle non-linear objective functions with multiple non-trivial constraints. Furthermore, high quality solutions were obtained in every case. In all test problems, the solution found by DE was better than or equal to the best solution found by any of the competing methods. Although the problems in this test set had generally been considered to be rather difficult, they were not sufficiently difficult to test DE's limits.

In order to evaluate and demonstrate DE's robustness, 100 independent trials were performed for each case studied. In every instance, all runs yielded the reported value of *f(X)* or better. In addition, all of the solutions were within the feasible region of the design space. Thus, it can be concluded that DE is a robust design algorithm for mixed parameter global optimization.

DE's ability to solve a problem is not particularly sensitive to the values of its control parameters *NP*, *F* and *CR*. Typically, it is sufficient to choose the values for the control parameters *F* and *CR* as multiples of 0.1 and the population size, *NP*, as a multiple of 10. Notice that exactly the same values of F and CR were effective for solving all the problems investigated here!

As demonstrated, the values of NP, F and CR, however, still affect the convergence velocity and robustness of the algorithm. In this investigation, values were selected roughly, and to favor robustness rather than convergence velocity. Because computationally inexpensive objective functions were used, the highest possible convergence velocity was not important. Despite this, convergence was still relatively fast in all cases. It is possible, however, to increase the convergence velocity significantly by favoring fast convergence over robustness.

A closer investigation of DE's sensitivity to the values of its control parameters provided further evidence supporting the above mentioned findings. Furthermore, more information about the contribution of each control parameter to the behavior of the DE algorithm was gathered. Understanding better the nature of each control parameter makes selecting their values easier and the resulting behavior can be estimated more reliably. However, the final target should be a DE without any user selected control parameters. This target can be reached only via a better understanding of the underlying behavior of the algorithm. The results of this investigation suggest that a Meta-DE could be a possibility to eliminate user-specified settings, for example. However, there are still a lot of open questions in this area.

It is sometimes argued that evolutionary optimizers require a large number of objective function evaluations. Before deciding whether or not this is true for DE, some further facts should be considered. Each of the real-world optimization problems discussed here can be solved on an ordinary PC using a couple of seconds for computation. Furthermore, this test problem set contains problems that are too difficult for most existing optimization methods. Unfortunately, insufficient data was available to accurately compare the number of objective function evaluations required by other methods with those required by DE. For some unknown reason, a large majority of the articles to which this investigation refers did not report the number of function evaluations properly. Besides, the number of objective function evaluations would be important only if the competing algorithm is, in fact, capable of finding the optimal solution.

Based on experience, DE seems to be better at locally fine-tuning a solution than traditional binary-encoded genetic algorithms. This property appeared especially when continuous parameters were the subject of optimization. This advantage appears not to decrease the capability of DE for global exploration, as is the case with many other methods. DE also seems capable of producing useful results with much lower population sizes than required by traditional genetic algorithms. Despite its small population sizes and greedy selection criterion, DE displayed a relatively low risk of premature convergence. As a result, DE required less than one tenth as many objective function evaluations as did a simple genetic algorithm.

Some methods for multi-constrained nonlinear optimization require a feasible initial solution as a starting point for a search. Preferably, this feasible solution should be rather close to a global optimum to ensure that the algorithm does not converge to a local minimum. If non-trivial constraints are imposed, it may be difficult or impossible to provide a feasible initial solution. The efficiency, effectiveness and robustness of many methods are often highly dependent on the quality of the starting point. The combination of DE with the described novel constraint handling approach does not require any initial solution, but it can still take advantage of a high quality initial solution if one is available. For example, initializing DE with normally distributed variations of a good solution creates a population that is biased towards the feasible region of the search space.

If there are no feasible solutions in the search space, as is the case for totally conflicting constraints, DE with the described constraint handling approach is still able to find the nearest feasible solution. This is important in practical engineering design work because often many non-trivial constraints are involved.

Another interesting topic is the fact that DE works with more than just one candidate solution. When multiple global optima exist, as was the case in the gear train example, multiple solutions can be found by performing only a single run of DE. In the case of multi-objective problems, the use of a population enables DE to locate multiple points from a Pareto-front instead of only a single Pareto-optimal solution.

It can be concluded, that the DE is a high potential novel alternative for practical engineering optimization. As this investigation demonstrates, DE has proven to be very effective for this purpose. DE has great potential to become a widely used, multipurpose optimization tool for solving a broad range of practical engineering design problems.

References

1. Storn R, Price KV (1995). *Differential evolution - a simple and efficient adaptive scheme for global optimization over continuous spaces*. Technical Report TR-95-012, ICSI, March 1995. Available via ftp://ftp.icsi. berkeley.edu/pub/techreports/1995/tr-95-012.ps.Z .
2. Storn R, Price KV (1997). *Differential Evolution – a Simple and Efficient Heuristic for Global Optimization over Continuous Spaces*. Journal of Global Optimization 11(4):341–359, December 1997. Kluwer Academic Publishers.
3. Price KV (1999). *An Introduction to Differential Evolution*. In: Corne D, Dorigo M, Glover F (eds) New Ideas in Optimization. McGraw-Hill, London (UK), pp 79–108. ISBN 007-709506-5.
4. Lampinen J (2002). *A Bibliography of Differential Evolution Algorithm*. Technical Report. Lappeenranta University of Technology, Lab. of Information Processing. Available via Internet: http://www.lut.fi/~jlampine/debiblio.htm . Cited 18th March 2002.
5. Goldberg DE (1989). *Genetic Algorithms in Search, Optimization and Machine Learning*. Addison-Wesley, Reading (MA). ISBN 0-201-15767-5.
6. Schwefel H-P (1995). *Evolution and Optimum Seeking*. John Wiley & Sons Inc., New York. ISBN 0-471-57148-2.
7. Price KV, Storn R (1997). *Differential Evolution – A simple evolution strategy for fast optimization*. Dr. Dobb's Journal, April 97, pp 18–24 and p 78.
8. Lampinen J, Zelinka I (1999). *Mechanical Engineering Design Optimization by Differential Evolution*. In: Corne D, Dorigo M, Glover F (eds). New Ideas in Optimization. McGraw-Hill, London (UK), pp 127–146. ISBN 007-709506-5.
9. Storn R (1996). *On the usage of differential evolution for function optimization*. In: 1996 Biennial Conference of the North American Fuzzy Information Processing Society (NAFIPS 1996), Berkeley, pp 519–523. IEEE, New York, NY, USA. Available via Internet http://www.icsi.berkeley.edu/~storn/litera.html .
10. Sandgren E (1990). *Nonlinear integer and discrete programming in mechanical design optimization*. Transactions of the ASME, Journal of Mechanical Design 112(2):223–229, June 1990. ISSN 0738-0666.
11. Fu J-F, Fenton RG, Cleghorn WL (1991). *A mixed integer-discrete-continuous programming method and its application to engineering design optimization*. Engineering Optimization 17(4):263–280. ISSN 0305-2154.
12. Loh H-T, Papalambros PY (1991). *A sequential linearization approach for solving mixed-discrete nonlinear design optimization problems*. Transactions of the ASME, Journal of Mechanical Design 113(3):325–334, September 1991.
13. Loh H-T, Papalambros PY (1991). *Computational implementation and tests of a sequential linearization algorithm for mixed-discrete nonlinear design optimization*.

Transactions of the ASME, Journal of Mechanical Design 113(3):335–345, September 1991.

14. Zhang C, Wang H-P (1993). *Mixed-discrete nonlinear optimization with simulated annealing*. Engineering Optimization 21(4):277–291. ISSN 0305-215X.
15. Chen JL, Tsao YC (1993). *Optimal design of machine elements using genetic algorithms*. Journal of the Chinese Society of Mechanical Engineers 14(2):193–199.
16. Li H-L, Chou C-T (1994). *A global approach for nonlinear mixed discrete programming in design optimization*. Engineering Optimization 22(2):109–122.
17. Kannan BK, Kramer SN (1994). *An Augmented Lagrange Multiplier Based Method for Discrete Continuous Optimization and Its Applications to Mechanical Design*. Transactions of the ASME, Journal of Mechanical Design 116(2):405–411, June 1994.
18. Wu S-J, Chow P-T (1995). *Genetic algorithms for nonlinear mixed discrete-integer optimization problems via meta-genetic parameter optimization*. Engineering Optimization 24(2):137–159. ISSN 0305-215X.
19. Shih CJ, Lai TK (1995). *Mixed-Discrete Fuzzy Programming for Nonlinear Engineering Optimization*. Engineering Optimization 23(3):187–199. ISSN 0305-215X.
20. Lin S-S, Zhang C, Wang H-P (1995). *On mixed-discrete nonlinear optimization problems: A comparative study*. Engineering Optimization 23(4):287–300. ISSN 0305-215X.
21. Deb K, Goyal M (1997). *Optimizing Engineering Designs Using a Combined Genetic Search*. In: Bäck T (ed) Proceedings of the 7th International Conference on Genetic Algorithms, pp 521–528.
22. Deb K, Goyal M (1998). *A Flexible Optimization Procedure for Mechanical Component Design Based on Genetic Adaptive Search*. Transactions of the ASME, Journal of Mechanical Design 120(2):162–164, June 1998.
23. Cai J, Thierauf G (1997). *Evolution Strategies in Engineering Optimization*. Engineering Optimization 29(1–4):177–199. ISSN 0305-215X.
24. Thierauf G, Cai J (1999). *Evolution Strategies – Parallelisation and Application in Engineering Optimization*. In: Topping BHV (ed) Parallel and Distributed Processing for Computational Mechanics: Systems and Tools, Saxe-Coburg Publications, Edinburgh (Scotland), pp 329–349. ISBN 1-874672-03-2.
25. Cai J, Thierauf G (2000). *Evolution Strategies and Genetic Algorithms and their Parallelisation for Structural Optimization: Part 2, Parallelisation and Applications*. In: Topping BHV, Lämmer L (eds). High Performance Computing for Computational Mechanics. Saxe-Coburg Publications, Edinburgh (Scotland), pp 195–205. ISBN 1-874672-06-7.
26. Cao YJ, Wu QH (1997). *Mechanical Design Optimization by Mixed-Variable Evolutionary Programming*. Proceedings of the 1997 IEEE Conference on Evolutionary Computation, IEEE Press, pp 443–446.
27. Cao YJ, Wu QH (1999). *A Mixed Variable Evolutionary Programming for Optimization of Mechanical Design*. International Journal of Engineering Intelligent Systems for Electrical Engineering and Communications 7(2):77–82, June 1999. CRL Publishing Ltd.
28. Wang S, Teo KL, Lee HWJ (1998). *A New Approach to Nonlinear Mixed Discrete Programming Problems*. Engineering Optimization 30(3–4):249–262. ISSN 0305-215X.

29. Cao YJ, Wu QH (1998). *A Cellular Automata based Genetic Algorithm and its Application in Mechanical Design Optimisation.* In: Proceedings of *the UKACC International* Conference on Control '98, 1.–4. September 1998. IEEE Conf. Publ. No. 455, vol. 2, pp. 1593–1598. ISBN 0-85296-708-X.
30. Litinetski VV, Abrahamzon BM (1998). *MARS – a multistart adaptive random search method for global constrained optimization in engineering applications.* Engineering Optimization 30(2):125–154. ISSN 0305-215X.
31. Ndiritu JG, Daniell TM (1999). *An Improved Genetic Algorithm for Continuous and Mixed Discrete-Continuous Optimization.* Engineering Optimization 31(5):589–614. ISSN 0305-215X.
32. Lewis K, Mistree F (1999). *Foraging-Directed Adaptive Linear Programming (FALP): A Hybrid Algorithm for Discrete/Continuous Design Problems.* Engineering Optimization 32(2):191–217. ISSN 0305-215X.
33. Coello Coello CA (1999). *Self-adaptive penalties for GA-based optimization.* In: Proceedings of the 1999 Congress on Evolutionary Computation, CEC 99, 6.–9. July 1999, Vol. 1, pp 573–580. ISBN 0-7803-5536-9.
34. Coello Coello CA (2000). *Use of a Self-Adaptive Penalty Approach for Engineering Optimization Problems.* Computers in Industry 41(2):113–127, March 2000.
35. Lin Y-C, Wang F-S, Hwang K-S (1999). *A hydrid method of evolutionary algorithms for mixed-integer nonlinear optimization problems.* In: Proceedings of the 1999 Congress on Evolutionary Computation, CEC'99, Vol. 3, pp 2159–2166. IEEE, Piscataway, NJ, USA. ISBN 0-7803-5536-9.
36. *SPECint95 and SPECfp95 computer benchmarks* (1998). The Standard Performance Evaluation Corporation. `http://www.specbench.org/` .
37. Siddall JN (1982). *Optimal engineering design: principles and applications.* Mechanical engineering series / 14. Marcel Dekker Inc. ISBN 0-8247-1633-7.
38. Salomon R (1996). *Reevaluating Genetic Algorithm Performance under Coordinate Rotation of Benchmark Functions.* BioSystems 39(3):263–278, Elsevier Science.
39. Bersini H, et.al. (1996) *Results of the First International Contest on Evolutionary Optimisation* (1st ICEO). In: Proceedings of the IEEE International Conference on Evolutionary Computation ICEC 96, pp 611-615.
40. Ingber L, Rosen B (1992). *Genetic Algorithms and Very Fast Simulated Reannealing: A Comparison.* Journal of Mathematical and Computer Modeling 16(11):87–100.
41. Bäck T, Hammel U, Schwefel H-P (1997). *Evolutionary Computation: Comments on the History and Current State.* IEEE Trans. Evol. Comp. 1(1):3–17, April 1997.
42. Sprave J (1995). *Evolutionäre Algorithmen zur Parameteroptimierung.* Automatisierungstechnik 43(3):110–117, Oldenbourg.
43. Lampinen J (2002). *A Constraint Handling Approach for the Differential Evolution Algorithm.* In: The 2002 IEEE World Congress on Computational Intelligence – WCCI 2002, 2002 Congress on Evolutionary Computation – CEC2002, Honolulu, Hawaii, May 12-17, 2002. 6 pages. IEEE. ISBN 0-7803-7281-6 (published as a CD-ROM).
44. Lampinen J (2002). *Multi-Constrained Nonlinear Optimization by the Differential Evolution Algorithm.* In: Roy R, Köppen M, Ovaska S, Furuhashi T, Hoffmann F (Eds) Soft Computing and Industry – Recent Advances, pp 305-318. Springer Verlag. ISBN 1-85233-539-4.
45. Lampinen J, Zelinka I (2000). *On Stagnation of the Differential Evolution Algorithm.* In: Ošmera P (ed) Proceedings of MENDEL 2000, 6th International Mendel Confer-

ence on Soft Computing, June 7.–9. 2000, Brno, Czech Republic. Brno University of Technology, Brno (Czech Republic), pp 76–83. ISBN 80-214-1609-2.

46. Goldberg DE, Deb K (1991). *A comparative analysis of selection schemes used in genetic algorithms*. In: Foundations of Genetic Algorithms, pp 69–93. Morgan Kaufmann.

7 SOMA - Self-Organizing Migrating Algorithm

Ivan Zelinka

7.1 Introduction

In recent years, a broad class of algorithms has been developed for stochastic optimization, i.e. for optimizing systems where the functional relationship between the independent input variables x and output (objective function) y of a system S is not known. Using stochastic optimization algorithms such as Genetic Algorithms (GA), Simulated Annealing (SA) and Differential Evolution (DE), a system is confronted with a random input vector and its response is measured. This response is then used by the algorithm to tune the input vector in such a way that the system produces the desired output or target value in an iterative process.

Most engineering problems can be defined as optimization problems, e.g. the finding of an optimal trajectory for a robot arm, the optimal thickness of steel in pressure vessels, the optimal set of parameters for controllers, optimal relations or fuzzy sets in fuzzy models, etc. Solutions to such problems are usually difficult to find their parameters usually include variables of different types, such as floating point or integer variables. Evolutionary algorithms (EAs), such as the Genetic Algorithms and Differential Evolutionary Algorithms, have been successfully used in the past for these engineering problems, because they can offer solutions to almost any problem in a simplified manner: they are able to handle optimizing tasks with mixed variables, including the appropriate constraints, and they do not rely on the existence of derivatives or auxiliary information about the system, e.g. its transfer function.

Evolutionary algorithms work on populations of candidate solutions that are evolved in generations in which only the best-suited - or fittest - individuals are likely to survive. This article introduces SOMA ('Self-Organizing Migrating Algorithm'), a new class of stochastic optimization algorithms. It explains the principles behind SOMA and demonstrates how this algorithm can assist in solving of various optimization problems. Functions on which SOMA have been tested can be found in this chapter.

SOMA, which can also works on a population of individuals, is based on the self-organizing behavior of groups of individuals in a "social environment". It can also be classified as an evolutionary algorithm, despite the fact that no new generations of individuals are created during the search (based on philosophy of this algorithm). Only the positions of the individuals in the search space are changed during a generation, called a 'migration loop'. Individuals are generated by random according to what is called the 'specimen of the individual' principle. The specimen is in a vector, which comprises an exact definition of all those parameters that together lead to the creation of such individuals, including the appropriate constraints of the given parameters.

SOMA is not based on the philosophy of evolution (two parents create one new individual – the offspring), but on the behavior of a social group of individuals, e.g. a herd of animals looking for food. One can classify SOMA as an evolutionary algorithm, because the final result, after one migration loop, is equivalent to the result from one generation derived by the classic EA algorithms - individuals hold new positions on the N dimensional hyper-plane. When the group of individuals is created, then the rule mentioned above governs the behavior of all individuals so that they demonstrate 'self-organization' behavior. Because no new individuals are created, and only existing ones are moving over the N-dimensional hyper-plane, this algorithm has been termed the Self-Organizing Migrating Algorithm, or SOMA for short. This algorithm was published in journals, books and presented at international conferences, symposiums, as well as in various invitational presentations, for example [1] – [11], [19], [20], [29].

In the following text the principle of the SOMA algorithm including its constraint handling and testing will be explained. The description is divided into short sections to increase the understandability of principles of the SOMA algorithm.

7.2 Function domain of SOMA

The set of functions on which a given algorithm shows good performance (i.e. its algorithm domain) should be clearly defined. A general definition can be found for example in [12]. This definition is, however, very general and hence not satisfactory for these purposes here. Based on experiences from artificial and real world test functions (for details see [11]) test functions can be classified like for example:

1. none-fractal type
2. defined at real, integer or discrete argument spaces
3. constrained, multiobjective, nonlinear
4. needle-in-haystack problems
5. NP problems

SOMA has been successfully tested on functions of type 1 - 3. Tests based on functions of type 4 and 5 are open for future research. Despite this fact, there is no rigorous presumption, which would eliminate functions 4 and 5 from the set defined above. Generally, the SOMA algorithm should be able to work on any system which provides an objective function, i.e. one that returns cost value. No auxiliary information, such as gradients etc. are needed.

7.3 Population

SOMA, as well as the other algorithms mentioned above, is working on a population of individuals. A population can be viewed as a matrix of size NxM (Table 7.1) where the columns represent individuals. Each individual in turn represents one candidate solution - or input vector - for the given problem or system, i.e. a set of arguments for the cost function. Associated with each individual is also a so-called cost value, i.e. the system response to the input vector. The cost value represents the fitness of the evaluated individual. It does not take part in the evolutionary process itself - it only guides the search process.

Table 7.1. Population – an example, P-parameter, I-individual, CV- cost value

	I1	I2	I3	I4	I5	I6	I7
CV	**204,92**	**93,75**	**163,80**	**121,73**	**107,53**	**121,06**	**120,21**
P 1	3,06	-46,64	5,02	38,72	35,82	0,07	23,76
P 2	2,51	54,04	85,10	0,29	24,11	4,29	20,38
P 3	46,75	51,28	11,35	3,08	24,66	60,24	33,44
P 4	72,49	15,08	2,92	3,67	5,81	4,54	4,05
P 5	6,32	57,16	58,83	26,61	12,44	23,89	4,23
P 6	73,79	-37,17	0,57	49,35	4,68	28,03	34,35

A population is usually randomly initialized at the beginning of the evolutionary process. Before that, a so-called Specimen (Eq.(7.1)) has to be defined on which the generating of the population is based.

$$\text{Specimen} = \{\{\text{Real}, \{\text{Low}, \text{High}\}\}, \{\text{Integer}, \{\text{Low}, \text{High}\}\}, \ldots\} \tag{7.1}$$

The Specimen defines for each parameter the type (e.g. integer, real, discrete, etc.) and its borders. For example, {Integer, {Low, High}} defines an integer parameter with an upper border *High* and a lower border *Low*. In other words, the borders define the allowed range of values for that particular cost function parameter. The careful selection of these borders is crucial for engineering applications, because without well-defined borders one can get solutions which are not applicable to the real physical system. For example one could get a negative thickness of the wall of a pressure vessel as an optimal result.

The borders are also important for the evolution process itself. Without them, the evolutionary process could go to infinity (author's experience with Schwefel's function – extremes are further and further away from the original. When a Specimen is properly defied then the population (Table 7.1) is generated as follows

$$P^{(0)} = x_{i,j}^{(0)} = rnd_{i,j}(x_j^{(High)} - x_j^{(Low)}) + x_j^{(Low)} \tag{7.2}$$

$$i = 1,\ldots,n_{pop} \quad , \quad j = 1,\ldots,n_{param}$$

Meaning of parameters is following – $P^{(0)}$ is the initial population and x is j^{th} parameter of individual which consist of n_{param} parameters. Population then consist of n_{pop} individuals. Eq. (7.1) ensures that the parameters of all individuals are randomly generated within the allowed borders, i.e. that the initial candidate solutions are chosen from that area within the search space that contains a feasible solution to the optimization problem, see Fig. 7.1.

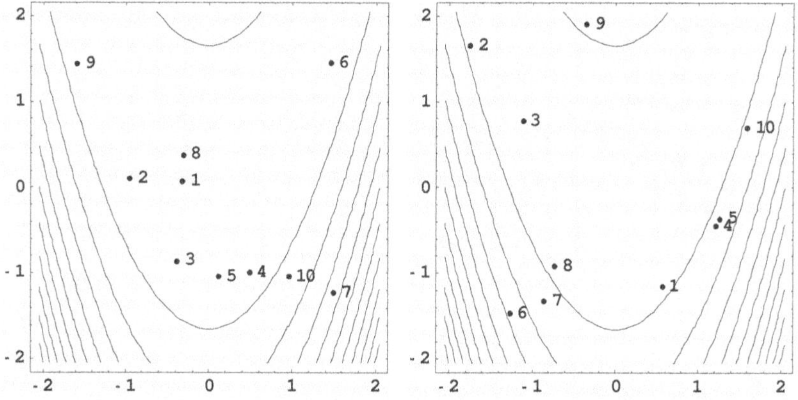

Fig. 7.1. Randomly generated population inside searched space, two examples

7.4 Mutation

Mutation, the random perturbation of individuals, is an important operation for EA strategies. It ensures the diversity amongst the individuals and it also provides the means to restore lost information in a population. Mutation in SOMA is different compared to other EA strategies. SOMA uses a PRT parameter to achieve perturbation. This parameter has the same effect for SOMA as mutation has for GA. It is defined in the range <0, 1> and is used to create a perturbation vector (PRTVector, see Table 7.2) as follows:

$$\text{if } rnd_j < PRT \text{ then } PRTVector_j = 1 \text{ else } 0, \quad j = 1,\ldots,n_{param} \tag{7.3}$$

Table 7.2. An example of perturbation vector for four parameter individual with PRT = 0.3

rnd_j	PRTVector
0.231	1
0.456	0
0.671	0
0.119	1

The "novelty" of this approach is that the PRTVector is created before an individual starts its journey over the search space (in standard EA terminology "before crossover"). The PRTVector defines the final movement of an active individual in *N-k* dimensional subspace (see next section).

7.5 Crossover

In standard EAs the 'Crossover' operator usually creates new individuals based on information from the previous generation. Geometrically speaking, new positions are selected from an *N* dimensional hyper-plane. In SOMA, which is based on the simulation of cooperative behavior of intelligent beings, sequences of new positions in the *N* dimensional hyper-plane are generated. They can be thought of as a series of new individuals obtained by the special crossover operation. This crossover operation determines the behavior of SOMA. The movement of an individual is thus given as follows:

$$\begin{aligned} &\vec{r} = \vec{r_0} + \vec{m}\, t\, \overrightarrow{PRTVector} \\ &\text{where} \quad t \in \langle 0, \text{by Step to, PathLength} \rangle \end{aligned} \tag{7.4}$$

or, more precisely:

$$\begin{aligned} &x_{i,j}^{ML new} = x_{i,j,start}^{ML} + (x_{L,j}^{ML} - x_{i,j,start}^{ML})\, t\, PRTVector_j \\ &\text{where} \quad t \in \langle 0, \text{by Step to, PathLength} \rangle \\ &\text{and} \quad ML \text{ is actual migration loop} \end{aligned} \tag{7.5}$$

It can be observed from Eq. (7.4, 7.5) that the PRTVector causes an individual to move toward the leading individual (the one with the best fitness) in *N-k* dimensional space. If all *N* elements of the PRTVector are set to 1, then the search process is carried out in an *N* dimensional hyper-plane (i.e. on a *N+1* fitness landscape). If some elements of the PRTVector are set to 0 (see 7.4 and 7.5) then the second terms on the right hand side of equation equal 0. This means those parameters of an individual that are related to 0 in the PRTVector are 'frozen', i.e. not changed during the search. The number of frozen parameters "*k*" is simply the number of dimensions which are not taking part in the actual search process. Therefore, the search process takes place in a N-*k* dimensional subspace.

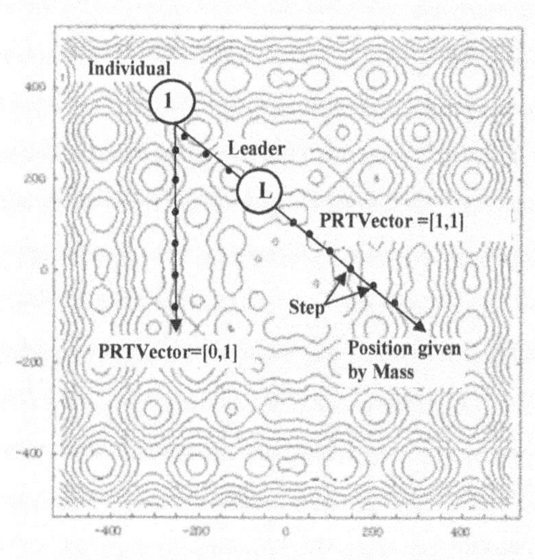

Fig. 7.2. PRTVector and its action on individual movement

7.6 Parameters and Terminology

SOMA, as other EAs, is controlled by a special set of parameters. Some of these parameters are used to stop the search process when one of two criteria are fulfilled; the others are responsible for the quality of the results of the optimization process. The parameters are shown in Table 7.3.

Table 7.3. SOMA parameters and their recommended domain

Parameter name	Recommended range	Remark
PathLength	<1.1, 3>	Controlling parameter
Step	<.11, PathLength>	Controlling parameter
PRT	<0, 1>	Controlling parameter
Dim	Given by problem	Number of arguments in cost function
PopSize	<10, up to user>	Controlling parameter
Migrations	<10, up to user>	Stopping parameter
MinDiv	<arbitrary negative, up to user >	Stopping parameter

A "disadvantage" of SOMA, as well as of other EAs, is that it has a slight dependence on the control parameter setting. During various tests it was found that SOMA is even more sensitive on the parameter setting than others algorithms. On the other side there was found setting that is almost universal, i.e. this setting was used almost in all simulations and experiments with very good performance of SOMA. The control parameters are described below and recommended values for the parameters, derived empirically from a great number of experiments, are given:

- **PathLength** ∈ <1.1, 3>. This parameter defines how far an individual stops behind the Leader (PathLength=1: stop at the leader's position, PathLength=2: stop behind the leader's position on the opposite side but at the same distance as at the starting point). If it is smaller than 1, then the Leader's position is not overshotted, which carries the risk of premature convergence. In that case SOMA may get trapped in a local optimum rather than finding the global optimum. Recommended value is 3.
- **Step** ∈ <.11, PathLength>. The step size defines the granularity with what the search space is sampled. In case of simple objective functions (convex, one or a few local extremes, etc.), it is possible to use a large Step size in order to speed up the search process. If prior information about the objective function is not known, then the recommended value should be used. For greater diversity of the population, it is better if the distance between the start position of an individual and the Leader is not a multiple of the Step parameter. That means that a Step size of 0.11 is better than a Step size of 0.1, because the active individual will not reach exactly the position of the Leader. Recommended value is 0.11.
- **PRT** ∈ <0, 1>. PRT stands for perturbation. This parameter determines whether an individual will travel directly towards the Leader, or not. It is one of the most sensitive control parameters. The optimal value is near 0.1. When the value for PRT is increased, the convergence speed of SOMA increases as well. In the case of low dimensional functions and a great number of individuals, it is possible to set PRT to 0.7-1.0. If PRT equals 1 then the stochastic component

of SOMA disappears and it performs only deterministic behavior suitable for local search.

- **Dim** - the dimensionality (number of optimised arguments of cost function) is given by the optimization problem. Its exact value is determined by the cost function and usually cannot be changed unless the user can reformulate the optimization problem.
- **PopSize** ∈ <10, up to the user>. This is the number of individuals in the population. It may be chosen to be 0.2 to 0.5 times of the dimensionality (Dim) of the given problem. For example, if the optimization function has 100 arguments, then the population should contain approximately 30-50 individuals. In the case of simple functions, a small number of individuals may be sufficient; otherwise larger values for PopSize should be chosen. It is recommended to use at least 10 individuals (two are theoretical minimum), because if the population size is smaller than that, SOMA will strongly degrade its performance to the level of simple and classic optimization methods.
- **Migrations** ∈ <10, up to user>. This parameter represents the maximum number of iterations. It is basically the same as generations for GA or DE. Here, it is called Migrations to refer to the nature of SOMA - individual creatures move over the landscape and search for an optimum solution. 'Migrations' is a stopping criterion, i.e. it tells the optimizing process when to stop.
- **MinDiv** ∈ <arbitrary negative, up to the user >. The MinDiv defines the largest allowed difference between the best and the worst individual from actual population. If the difference is too small then the optimizing process is will stop (see Fig. 7.3). It is recommended to use small values (see [11]). It is safe to use small values for the MinDiv, e.g. MinDiv = 1. In the worst case, the search will stop when the maximum number of migrations is reached. Negative values are also possible for the MinDiv. In this case, the stop condition for MinDiv will not be satisfied and thus SOMA will pass through all migrations.

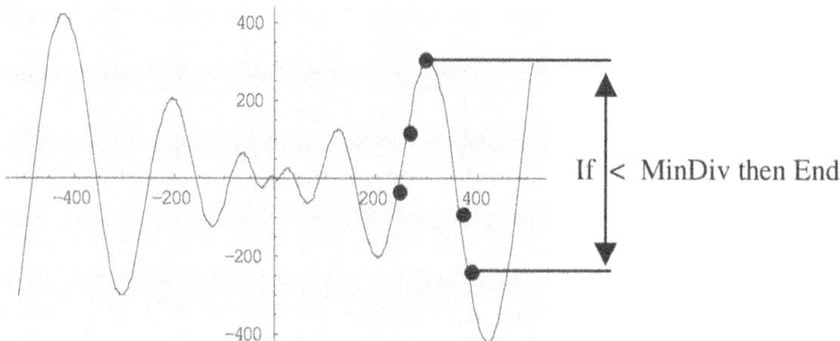

Fig. 7.3. Principle of MinDiv on population which consist of 5 individuals

When recommended values are taken into consideration like acceptable, then they can be included into algorithm or permanently set to be constant and number of control parameters will decrease from 6 to 1 (Migrations). The problem of deterministic finding suitable SOMA parameter settings for a given optimization problem is not absolutely solved and can be regarded as one of the future research activities.

7.7 Principles of SOMA

In the previous sections it was mentioned that SOMA was inspired by the competitive-cooperative behavior of intelligent creatures solving a common problem. Such a behavior can be observed anywhere in the world. A group of animals such as wolves or other predators may be a good example. If they are looking for food, they usually cooperate and compete so that if one member of the group is successful (it has found some food or shelter) then the other animals of the group change their trajectories towards the most successful member. If a member of this group is more successful than the previous best one (is has found more food, etc.) then again all members change their trajectories towards the new successful member. It is repeated until all members meet around one food source. This principle from the real world is of course strongly simplified. Yet even so, it can be said it is that competitive-cooperative behavior of intelligent agents that allows SOMA to carry out very successful searches. For the implementation of this approach, the following analogies are used:

- members of herd/pack ⇒ individuals of population, PopSize parameter of SOMA
- member with the best source of food ⇒ Leader, the best individual in population for actual migration loop
- food ⇒ fitness, local or global extreme on N dimensional hyper-plane
- landscape where pack is living ⇒ N dimensional hyper-plane given by cost function
- migrating of pack members over the landscape⇒ migrations in SOMA

The following section explains in a series of detailed steps how SOMA actually works. SOMA works in loops - so called Migration loops. These play the same role as Generations in classic EAs. The difference between SOMA's 'Migration loops' and EA's 'Generations' stems from the fact that during a Generations in classic EA's offspring is created by means of at least two or more parents (two in GA, four in DE). In the case of SOMA, there is no newly created offspring based on parents crossing. Instead, new positions are calculated for the individuals travelling towards the current Leader. The term 'Migrations' refers to their movement over the landscape-hyper-plane.

It can be demonstrated that SOMA can be viewed as an algorithm based on offspring creation. The leader plays the role of roe-buck (male), while other individuals play the role of roe (female); note that this has the characteristics of pack reproduction with one dominant male. Hence, GA, DE, etc. may be seen as a special case of SOMA and vice versa (see later SOMA strategy AllToAll). Because the original idea of SOMA is derived from competitive-cooperative behavior of intelligent beings, the authors suppose that this background is the most suitable one for its explanation.

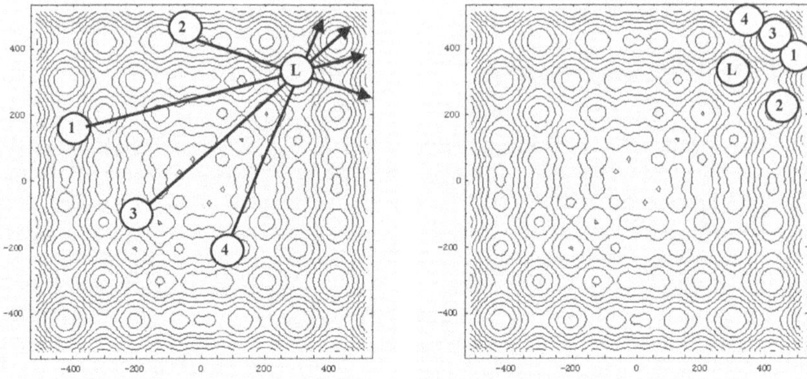

Fig. 7.4. Principle of basic version of SOMA – artificial example

The basic version of SOMA consists of the following steps:

1. **Parameter definition**. Before starting the algorithm, SOMA's parameters, e.g. Specimen, Eq. (7.1), Step, PathLength, PopSize, PRT, MinDiv, Migrations (see Table 7.3) and the cost function needs to be defined. Cost function is simply the function which returns a scalar that can directly serve as a measure of fitness. The cost function is then defined as a model of real world problems, (e.g. behavior of controller, quality of pressure vessel, behavior of reactor, etc.).
2. **Creation of Population**. A population of individuals is randomly generated. Each parameter for each individual has to be chosen randomly from the given range <Low, High> by using Eq. (7.1). The population (Table 7.1) then consists of columns - individuals which conform with the specimen, Eq. (7.1).
3. **Migrating loop**. Each individual is evaluated by cost function and the Leader (individual with the highest fitness) is chosen for the current migration loop. Then all other individuals begin to jump, (according to the Step definition) towards the Leader. Each individual is evaluated after each jump using the cost function. The jumping (Eq. (7.2 or 7.3)) continues, until a new position defined by the PathLength has been reached. The new position after each jump is calculated by Eq. (7.2) or Eq. (7.3). This is shown graphically in Fig. 7.2. The individual returns then to that position where it found the best fitness on its trajectory. Before an individual begins jumping towards the Leader, a random number is generated (for each individual's component), and then compared with PRT. If the generated random number is larger than PRT, then the associated component of the individual is set to 0 by means of the PRTVector (see Eq. (7.3) and Fig. 7.2). Hence, the individual moves in the *N-k* dimensional

subspace, which is perpendicular to the original space. This fact establishes a higher robustness of the algorithm. Earlier experiments have demonstrated that, without the use of PRT, SOMA tends to determine a local optimum rather than the global one. Migration can be also viewed as a competitive-cooperative phase. During the competitive phase each individual tries to find the best position on its way and also the best from all individuals. Thus during migration, all individuals compete among themselves. When all individuals are in new positions, they "release" information as to their cost value. This can be regarded as a cooperative phase. All individuals cooperate so that the best individual (Leader) is chosen. Competitive-cooperative behavior is one of the other important attributes typical for memetic algorithms [13].

4. **Test for stopping condition.** If the difference between Leader and the worst individual is not lower than the MinDiv and the maximum number of Migrations has not been reached, return to step 3 otherwise go to step 5 .
5. **Stop**. Recall the best solution(s) found during the search.

Steps 1 to 5 are graphically depicted in Table 7.4.

Table 7.4. Basic principle of SOMA

Parametry SOMA

Parameter	Value
Step	0,3
PathLength	3
PRT	0,1
MinDiv	0,1
Migrations	1000
PopSize	7

PRTVector, **different for each individual**

Rule	PRTVector
If rnd < PRT then 1 else 0	1
If rnd < PRT then 1 else 0	0
If rnd < PRT then 1 else 0	0
If rnd < PRT then 1 else 0	1
If rnd < PRT then 1 else 0	0
If rnd < PRT then 1 else 0	1

CostFunction Abs(Parameter 1)+ Abs(Parameter 2) +...+ Abs(Parameter 6)

Active individual (Individual 2) **Leader** (Individual 5)

	Individual 1	Individual 2	Individual 3	Individual 4	Individual 5	Individual 6	Individual 7
CV	**204,91528**	**261,3632**	**163,79679**	**121,73019**	**107,52784**	**121,06024**	**120,20974**
Parameter 1	3,0615753	-46,63569	5,0246553	38,723912	35,822343	0,0715185	23,761224
Parameter 2	2,5117282	54,036685	85,104704	0,2928606	24,111443	4,2879691	20,384665
Parameter 3	46,75014	51,282894	11,347164	3,0796963	24,657689	60,241731	33,437248
Parameter 4	72,486617	15,080129	2,916686	3,6713463	5,8142407	4,5385164	4,0482021
Parameter 5	6,316564	57,155744	58,829537	26,610056	12,43856	23,891907	4,2271271
Parameter 6	73,788657	-37,17206	0,5740442	49,352316	4,6835676	28,028598	34,351273

$$x_{i,j}^{MK+1} = x_{i,j,start}^{MK} + (x_{L,j}^{MK} - x_{i,j,start}^{MK})\,t\,PRTVector_j$$

$$t \in <0, \textit{by Step to}, PathLength>$$

New positions

	t = 0	t = 1	t = 2	...	t = 8	t = 9	t = 10
CV	**261,3632**	**221,28934**	**186,89373**	...	**384,17836**	**424,25222**	**464,32608**
	-46,63569	-21,89828	2,8391294	...	151,26359	176,001	200,73841
	54,036685	54,036685	54,036685	...	54,036685	54,036685	54,036685
	51,282894	51,282894	51,282894	...	51,282894	51,282894	51,282894
	15,080129	12,300362	9,5205959	...	-7,158003	-9,937769	-12,71754
	57,155744	57,155744	57,155744	...	57,155744	57,155744	57,155744
	-37,17206	-24,61537	-12,05868	...	63,281441	75,838128	88,394815

CV	Individual with lower CV	The best individual from newly calculated positions
	261,3632	**186,89373**
	-46,63569	2,8391294
	54,036685	54,036685
	51,282894	51,282894
	...	...

	Individual 1	Individual 2	Individual 3	Individual 4	Individual 5	Individual 6	Individual 7
CV	**204,91528**	**186,89373**					
Parameter 1	3,0615753	2,8391294					
Parameter 2	2,5117282	54,036685					
Parameter 3	46,75014	51,282894					
Parameter 4	72,486617	9,5205959					
Parameter 5	6,316564	57,155744					
Parameter 6	73,788657	-12,05868					

7.8 Variations of SOMA

Currently, a few variations - strategies of the SOMA algorithm exist. All versions are almost fully comparable with each other in the sense of finding of global optimum. These versions are:

1. 'AllToOne': This is the basic strategy, that was previously described. Strategy AllToOne means that all individuals move towards the Leader, except the Leader. The Leader remains at its position during a Migration loop. The principle of this strategy is shown in Fig. 7.4.
2. 'AllToAll': In this strategy, there is no Leader. All individuals move towards the other individuals. This strategy is computationally more demanding. Interestingly, this strategy often needs less cost function evaluations to reach the global optimum than the AllToOne strategy. This is caused by the fact that each individual visits a larger number of parts on the *N* dimensional hyper-plane during one Migration loop than the AllToOne strategy does. Fig. 7.5 shows the AllToAll strategy with PRT = 1.
3. 'AllToAll Adaptive': The difference between this and the previous version is, that individuals do not begin a new migration from the same old position (as in AllToAll), but from the last best position found during the last traveling to the previous individual.
4. 'AllToRand': This is a strategy, where all individuals move towards a randomly selected individual during the migration loop, no matter what cost value this individual has. It is up to the user to decide how many randomly selected individuals there should be. Here are two sub-strategies:

    - The number of randomly selected individuals is constant during the whole SOMA process.
    - For each migration loop, (in intervals of <1,PopSize>) the actual number of individuals is determined randomly. Thus, the number of randomly chosen individuals in the second sub-strategy is different in each migration loop.

5. 'Clusters': This version of SOMA with Clusters can be used in any of the above strategies. The word 'Cluster' refers to calculated clusters. Each individual from the population is tested for the cluster to which it belongs, according to Eq. (7.6) expressed below, where INDi is the *i*-th parameter of the individual; CCi is the *i*-th parameter of the leader (Cluster Center); HBi and LBi are the allowed bounds for the given parameter (see Specimen, Eq. (7.1)); and CD is the Cluster Distance given by the user. The result is that after a cluster calculation, clusters with their Leaders are derived, and each individual belongs to one cluster. In the case that all individuals create their own cluster (1 individual = 1 cluster), then each individual will jump toward all others, (this is identical with the 'AllToAll' strategy). Some clusters may be created or anihilated during migration loops.

$$CD > \sqrt{\sum_{i=1}^{\dim}\left(\frac{IND_i - CC_i}{HB_i - LB_i}\right)^2} \tag{7.6}$$

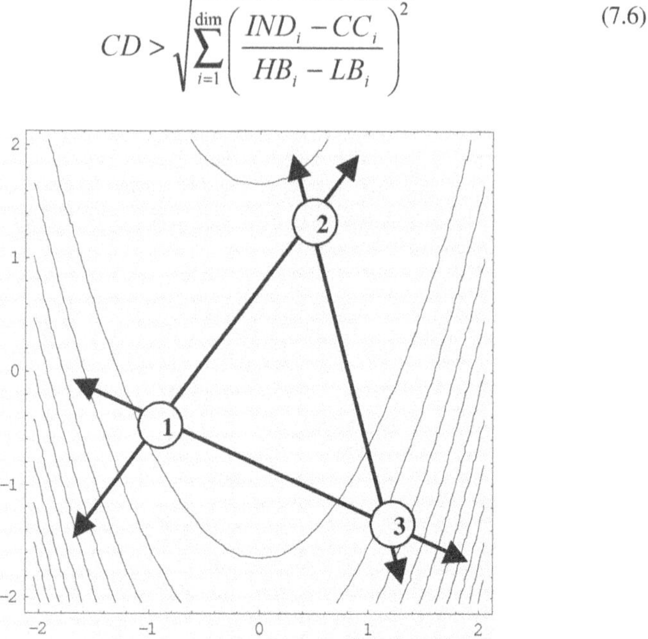

Fig. 7.5. AllToAll – an artificial example. Note that no Leader is present here.

By using SOMA with clusters, the user must define a so-called 'Cluster Distance' – the parameter, which says how large (how many of individuals) the cluster should be, and the domain of attraction of the local cluster Leader. Using this basic parameter, SOMA breaks itself up into more local SOMAs, each focusing on the contained Leaders. Therefore, independent groups of individuals are carrying out the search. These local SOMAs create clusters, which can join together or split into new clusters. This strategy has not been studied in detail yet, because of its increased complexity of computation compared with the low quality improvement of the optimization process. Other possible strategies or variations of SOMA are, for example, that individuals need not move along a straight line-vectors, but they can travel on curves, etc.

7.9 SOMA dependence on control parameters

As already mentioned above, the control parameters for SOMA are: PRT, PathLength, Step, MinDiv and PopSize. The quality of the results for the optimization partially dependant on the selection of these parameters. To demonstrate, how this influences the algorithm, some simulations were performed to show SOMA's dependence on them (Fig. 7.6).

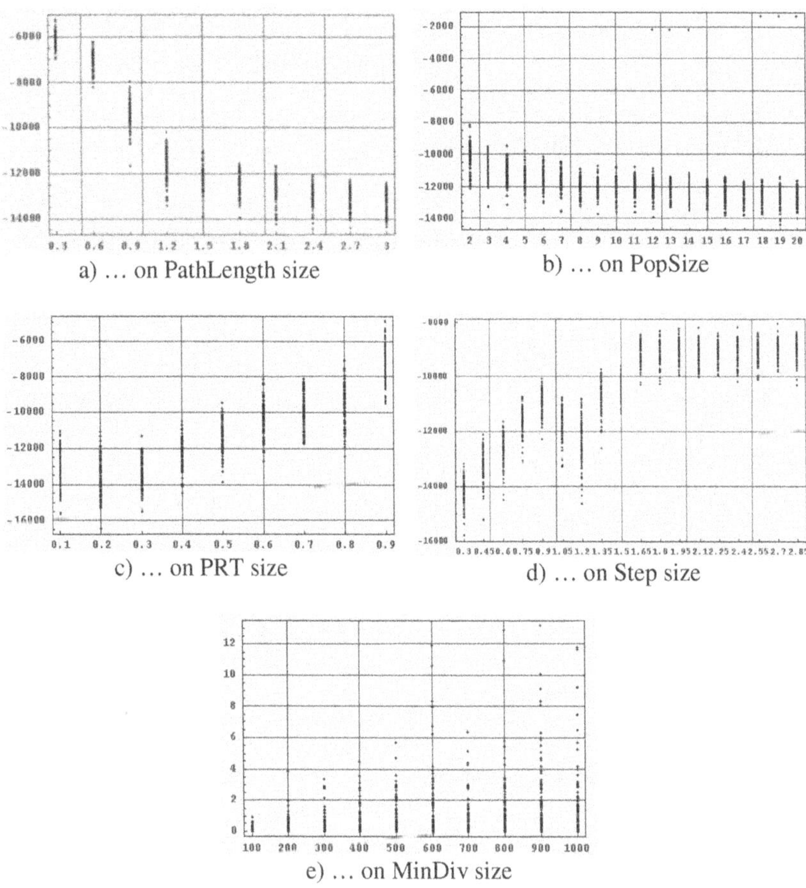

Fig. 7.6. Dependence SOMA on control parameters

These simulations demonstrate the dependency of the quality of optimization on the parameters PRT, PathLength, Step, MinDiv and PopSize. A total of 100 simulations were performed for each change of each parameter. The results are depicted as columns of points (point – the best value founded during actual simulation). Part a) shows the dependence on the PathLength parameter. It can be observed that an increases of the PathLength parameter resulted in deeper (better) extremes. This is logical since the search process passes through a lager search space. Part b) shows the dependence on the PopSize parameter. In addition, there is a small improvement. Part c) shows the dependence on PRT from which it is visible, that in the case of low PRT, SOMA's efficiency is very good. Part d) shows the dependence on the Step parameter. From this picture, it is clear that in the case of big steps, results are poor due to individuals performing large jumps and thus the searched space is poorly searched. The last part e) shows the dependence on the MinDiv parameter. From the picture it is visible that the MinDiv plays a role of something like a lens which determines the dispersion of final solutions in the case of repeated simulations (or/and solutions in the last population). MinDiv is the difference between the worst and the best individual and if it is quite large, then the process may stop before the optimal solution is found.

7.10 On the SOMA classification and some additional information

SOMA can be classified in two ways. Firstly, one can say that it is an evolutionary algorithm despite the fact that there are no new children created in the common 'evolutionary' way. The reason for such classification is simple. During a SOMA run, migration loops are performed, after which individuals are re-positioned, with the same effect as after one Generation in standard EAs. From this point of view, one can say that SOMA is an evolutionary algorithm.

Secondly, SOMA can be classified (based on the principles described – which is more logical), as a so-called memetic algorithm. This classification covers a wide class of meta-heuristic algorithms (see [13]). The key characteristic of these algorithms can be observed in various optimization algorithms, local search techniques, special recombination operators, etc. Basically speaking, memetic algorithms are classified as competitive-cooperative strategies [13] showing synergetic attributes.

SOMA algorithms show these attributes as well. Because of this, it is more appropriate to classify SOMA as a memetic algorithm (see also [14]). The word 'memetic' was coined by R. Dawkins in his book 'The Selfish Gene' [15]:

> *Examples of memes are tunes, ideas, catch phrases, clothes fashions, and ways of making pots or of building arches. Just as genes propagate themselves in the gene pool by leaping from body to body via sperm and eggs, so memes propagate themselves in the meme pool by leaping from brain to brain via a process that, in a broad sense, can be called imitation.*

In this sense, memetic algorithms can be taken to be like 'cultural, social or informational evolution' in comparison to GA, which stems from biological evolution. Dawkins defines memes as "...units of information in the brain...", where the phenotypes are gestures, communication and changes in behavior.

The synergetic effect of SOMA stems from the fact that during the migration loops individuals influence each others, and very often, one can observe the appearance and disappearance of time-space structures - sets consisting of individuals which travel over the search space. For a graphical view, see [11]. This is, of course, only a 'visual' effect.

The more important effect is that individuals - thanks to this mutual influence - change their trajectories and consequently find new and usually better positioning-solutions. Their final trajectory is not only given by deterministic rules but primarlily by their mutual interaction.

In Fig. 7.7 the movement of individuals for the simple version of SOMA is depicted (AllToOne, PRT=1, i.e. no perturbation) and also in 3D view in Fig. 7.8.

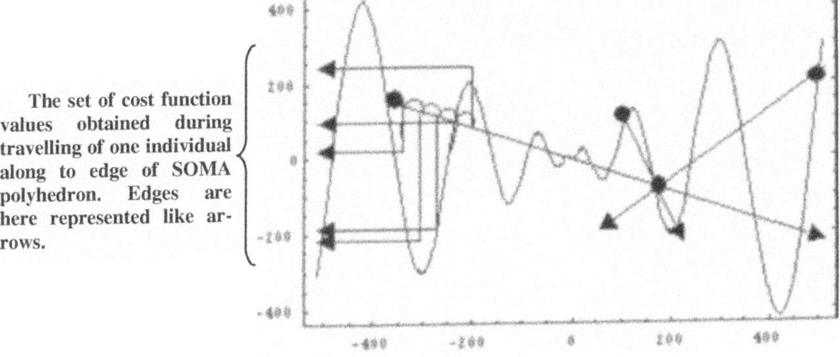

Fig. 7.7. Movement of individual for simplest version SOMA in 2D

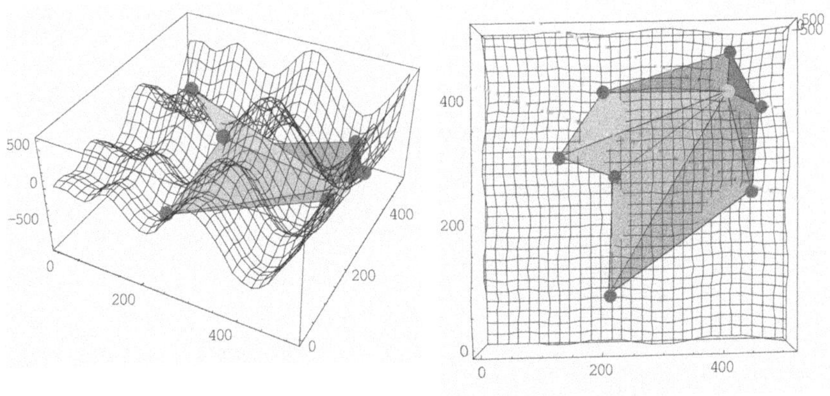

3D SOMA polyhedron with individuals at its vertices

3D SOMA polyhedron with individuals at its vertices. View from the top

Fig. 7.8. Movement of individual for simplest version SOMA in 3D along to edges of SOMA polyhedron

Because all individuals travel towards the Leader in straight lines (vectors of direction), their movement can be regarded as movement to the edges, thereby creating a polyhedron looking shape (Fig. 7.8). Movement is then determined not by the shape of the searched space but by the shape of polyhedron. Therefore, SOMA is able to overstep many local optima, which can usually cause problems to classic gradient-based methods. Fig. 7.8 shows only half of a real polyhedron. In the real SOMA process, individuals do not stop at the Leader position, but continue past the Leader to a position, which is given by the PathLength parameter.

7.11 Constraint Handling

SOMA, as well as other such evolutionary algorithms, can be used to solve optimization problems, sometimes called mixed integer-discrete-continuous, non-linear programming problems, etc. These can (see also [13]) be expressed as follows:

$$\begin{aligned}
&\text{Find} \\
&\mathrm{X} = \{\mathrm{x}_1, \mathrm{x}_2, \mathrm{x}_3, \ldots, \mathrm{x}_n\} = \left[X^{(i)}, X^{(d)}, X^{(c)}\right]^T \\
&\text{to minimize} \\
&f(X) \\
&\text{subject to constraints} \\
&g_j(X) \le 0 \qquad j = 1, \ldots, m \\
&\text{and subject to boundary constraints} \\
&x_i^{(Lo)} \le x_i \le x_i^{(Hi)} \qquad i = 1, \ldots, n \\
&\text{where} \\
&X^{(i)} \in \boldsymbol{R}^i, \quad X^{(d)} \in \boldsymbol{R}^d, \quad X^{(c)} \in \boldsymbol{R}^c
\end{aligned} \tag{7.7}$$

$X^{(i)}$, $X^{(d)}$ and $X^{(c)}$ denote feasible subsets of integer, discrete and continuous variables respectively. The above formulation is general and basically the same for all types of variables. Only the structure of the design domain distinguishes one problem from another. However, it is worth noticing here the principal differences between integer and discrete variables. While both integer and discrete variables have a discrete nature, only discrete variables can assume floating-point values. For example, Discrete = {-1, 2.5, 20, -3, -5.68…}. In practice, the discrete values of the feasible set are often unevenly spaced. These are the main reasons why integer and discrete variables require different handling. SOMA can be categorized as belonging to the class of floating-point encoded, 'memetic' optimization algorithms. Generally, the function to be optimized, f, is of the form:

$$f(X): \boldsymbol{R}^n \rightarrow \boldsymbol{R} \tag{7.8}$$

The optimization target is to minimize the value of this *objective function f(X),*

$$\min(f(X)) \tag{7.9}$$

by optimizing the values of its parameters:

$$X = (x_1, \ldots, x_{n_{param}}) \qquad x \in \boldsymbol{R} \tag{7.10}$$

where X denotes a vector composed of n_{param} objective function parameters. Usually, the parameters of the objective function are also subject to lower and upper boundary constraints, $x^{(Low)}$ and $x^{(High)}$, respectively:

$$x_j^{(Low)} \leq x_j \leq x_j^{(High)} \qquad j = 1, \ldots, n_{param} \tag{7.11}$$

7.11.1 Boundary constraints

With boundary-constrained problems, it is essential to ensure that the parameter values lie within their allowed ranges after recalculation. A simple way to guarantee this, is to replace the parameter values that violate boundary constraints with random values generated within the feasible range:

$$x'^{(ML+1)}_{i,j} = \begin{cases} r_{i,j}(x_j^{(High)} - x_j^{(Low)}) + x_j^{(Low)} \\ \text{if} \quad x'^{(ML+1)}_{i,j} < x_j^{(Low)} \vee x'^{(ML+1)}_{i,j} > x_j^{(High)} \\ \\ x'^{(ML+1)}_{i,j} \quad \text{otherwise} \end{cases} \tag{7.12}$$

where,

$i = 1, \ldots, n_{pop}$, $j = 1, \ldots, n_{param}$

7.11.2 Constraint functions

A soft-constraint (penalty) approach was applied for the handling of the constraint functions. The constraint function introduces a distance measure from the feasible region, but is not used to reject unfeasible solutions, as is the case with hard-constraints. One possible soft-constraint approach is to formulate the cost-function as follows:

$$
\begin{aligned}
& f_{cost}(X) = (f(X)+a)\cdot\prod_{i=1}^{m} c_i^{b_i} \\
& \text{where} \\
& c_i = \begin{cases} 1.0 + s_i \cdot g_i(X) & \text{if} \quad g_i(X) > 0 \\ 1 & \text{otherwise} \end{cases} \\
& s_i \geq 1 \\
& b_i \geq 1 \\
& \min(f(X)) + a > 0
\end{aligned}
\tag{7.13}
$$

The constant, a, is used to ensure that only non-negative values will be assigned to fcost. When the value of a is set high enough, it does not otherwise affect the search process. The constant, s, is used for appropriate scaling of the constraint function value. The exponent, b, modifies the shape of the optimization hyperplane. Generally, higher values of s and b are used when the range of the constraint function, g(X), is expected to be low. Often setting s=1 and b=1 works satisfactorily and only if one of the constraint functions, gi(X), remains violated after the optimization run, will it be necessary to use higher values for si and/or bi. In many real-world engineering optimization problems, the number of constraint functions is relatively high and the constraints are often non-trivial. It is possible that the feasible solutions are only a small subset of the search space. Feasible solutions may also be divided into separated islands around the search space, Eq. (7.9). Furthermore, the user may easily define totally conflicting constraints so that no feasible solutions exist at all.

For example, if two or more constraints conflict, so that no feasible solution exists, EAs are still able to find the nearest feasible solution. In the case of non-trivial constraints, the user is often able to judge which of the constraints are conflicting on the basis of the nearest feasible solution. It is then possible to reformulate the cost-function or reconsider the problem setting itself to resolve the conflict.

A further benefit of the soft-constraint approach is that the search space remains continuous. Multiple hard constraints often split the search space into many separated islands of feasible solutions. This discontinuity introduces stalling points for some genetic searches and also raises the possibility of new, locally optimal areas near the island borders. For these reasons, a soft-constraint approach is considered essential. It should be mentioned that many traditional optimization methods are only able to handle hard-constraints. For evolutionary optimization, the soft-constraint approach was found to be a natural approach.

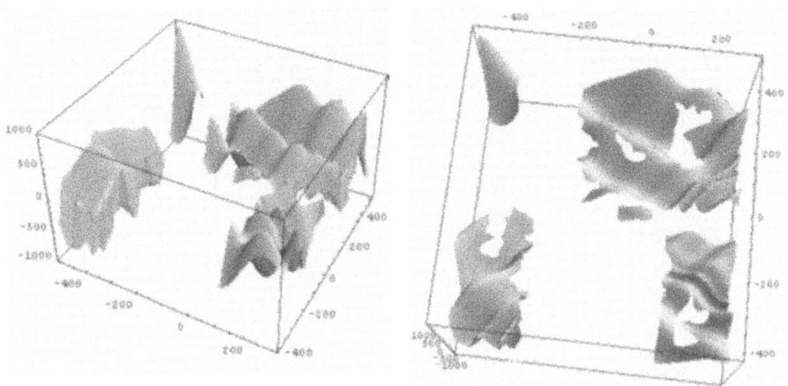

Fig. 7.9. Separated islands around the search space

7.11.3 Handling of Integer and Discrete Variables

In its canonical form, SOMA (as well as DE) is only capable of handling continuous variables. However extending it for optimization of integer variables is rather easy. Only a couple of simple modifications are required. First, for evaluation of cost-function, integer values should be used. Despite this, the SOMA algorithm itself may still work internally with continuous floating-point values. Thus,

$$\begin{aligned}
&f_{\text{cost}}(y_i) \qquad i = 1,\ldots,n_{param} \\
&\text{where} \\
&y_i = \begin{cases} x_i & \text{for continuous variables} \\ INT(x_i) & \text{for integer variables} \end{cases} \\
&x_i \in X
\end{aligned} \tag{7.14}$$

INT() is a function for converting a real value to an integer value by truncation. Truncation is performed here only for purposes of cost function value evaluation. Truncated values are not elsewhere assigned. Thus, EA works with a population of continuous variables regardless of the corresponding object variable type. This is essential for maintaining the diversity of the population and the robustness of the algorithm. Secondly, in case of integer variables, the population should be initialized as follows:

$$P^{(0)} = x_{i,j}^{(0)} = r_{i,j}(x_j^{(High)} - x_j^{(Low)} + 1) + x_j^{(Low)} \tag{7.15}$$
$$i = 1,\ldots,n_{pop} \quad , \quad j = 1,\ldots,n_{param}$$

Additionally, instead of Eq. (7.12), the boundary constraint handling for integer variables should be performed as follows:

$$x'^{(ML+1)}_{i,j} = \begin{cases} r_{i,j}(x_j^{(High)} - x_j^{(Low)} + 1) + x_j^{(Low)} \\ \text{if} \quad INT\left(x'^{(ML+1)}_{i,j}\right) < x_j^{(Low)} \vee INT\left(x'^{(ML+1)}_{i,j}\right) > x_j^{(High)} \\ \\ x'^{(ML+1)}_{i,j} \quad \text{otherwise} \end{cases} \tag{7.16}$$

where,

$$i = 1,\ldots,n_{pop} \quad , \quad j = 1,\ldots,n_{param}$$

Discrete values can also be handled in a straightforward manner. Suppose that the subset of discrete variables, X(d), contains l elements that can be assigned to variable x:

$$X^{(d)} = x_i^{(d)} \qquad i = 1,\ldots,l \quad \text{where} \quad x_i^{(d)} < x_{i+1}^{(d)} \tag{7.17}$$

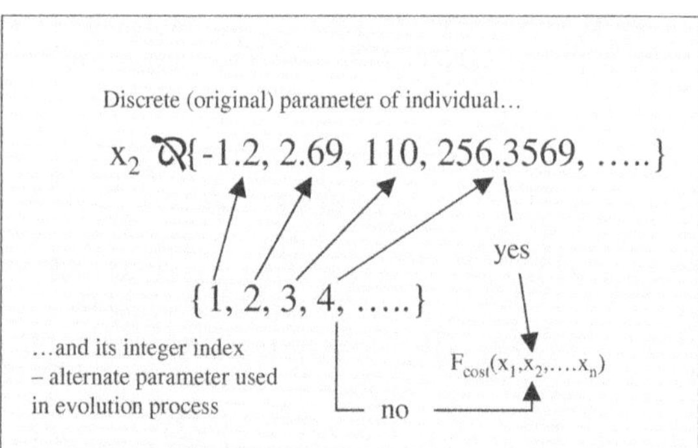

Fig. 7.10. Discrete parameter handling

Instead of the discrete value x_i itself, we may assign its index, *i*, to x. Now the discrete variable can be handled as an integer variable that is boundary constrained to range <1,2,3,...,N>. So as to evaluate the objective function, the discrete value, x_i , is used instead of its index *i*. In other words, instead of optimizing the value of the discrete variable directly, we optimize the value of its index *i*. Only during evaluation is the indicated discrete value used. Once the discrete problem has been converted into an integer one, the previously described methods for handling integer variables can be applied. The principle of discrete parameter handling is depicted in Fig. 7.10.

7.12 Selected Applications and Open Projects

The SOMA algorithm was not tested only on the mentioned 15 test functions. This algorithm was used on various examples usually from the engineering domain of interest. Some of examples are:

- **Active Compensation in RF-driven Plasmas**. Simulated annealing (SA), DE and SOMA were used in this experiment which was done at The Open University of Oxford, Oxford Research Unit. The aim of this experiment was to make a comparative study with previously published results (SA) and compare all three different evolutionary algorithms to deduce the fourteen Fourier terms in a radio-frequency (RF) waveform to tune a Langmuir probe inserted into plasma reactor (Fig. 7.11). Langmuir probes are diagnostic tools used to analyze the electron energy distribution in plasma processes. Two basic experiments were done, e.g. 1) SA and DE, 2) SA, DE and SOMA. The results of the first set of experiments (SA, DE) will be seen in a soon to be published book focusing on DE, and results of the second set of experiments will be published in a selected journal. For details please see [33].

Plasma reactor

Langmuir probe inside reactor chamber (needle in the middle)

Fig. 7.11. Plasma reactor at The Open University of Oxford, Oxford Research Unit

- **RESTORM (Radical Environmentally Sustainable Tannery Operations by Resource Management)** – a 5th framework program of the European Union (coordinated by BLC – British Leather Center, Northampton). In this project, based on the research cooperation of 15 European countries, is SOMA used in the frame of mathematical modeling for optimization of technological devices. The main aim of this optimization is to minimize operating costs and waste, which is usually produced during each technological process. In this research is going to be done a comparative study of SOMA with other well known algorithms like DE, GA, SA, etc.
- **Neural network learning**. The aim of this research was to show how evolutionary algorithms could be used as learning algorithms for various types of neural networks. The ways, in which differential evolution could be used for neural network learning, were examined. Various experiments were made and compared with standard neural algorithms [16] and [17].
- **Evolutionary identification of predictive models**. In this work it was shown how evolutionary algorithms could be used for retrieval of suitable predictive models. Two different algorithms were used. The first one was DE and the second was SOMA. Both algorithms were used in the same way to find the model whose response agreed closest with a given time series. The problem was formulated as an optimization problem where the cost function was based on the difference between the original time series and the time series produced by a candidate model [18], [19], [7] and [20].
- **Inverse fractal problem**. The aim of this project was to show how a new evolutionary algorithm could be used to solve inverse fractal problems (IFP). Two algorithms were used for mutual comparison i.e. DE [21], [22], [23] and [24] and SOMA [1].
- **Mechanical engineering problem optimization**. In this research, are described three illustrative and practical numerical examples. The first example was the design of a gear train with a specified gear ratio; the second problem is the minimization of manufacturing cost of a pressure vessel. A third example is to design a coil spring with a minimum amount of steel. The mixed-variable methods used to solve these problems are discussed in detail and compared with published results obtained with other optimization methods for the same problems. Even though this investigation only focuses on engineering design applications, SOMA as well as DE can be applied, in principle, to solve any mixed integer-discrete-continuous optimization task [13], [25], [26] and [1]. This case study is described in Chapter 26.
- **Statical optimization of chemical reactor**. In this ressearch, a chemical reactor was optimized by SOMA [6]. The reactor model was based on five interlocked ordinal differential equations (ODEs). The "speciality" of this example was that the optimization was based on this 5 ODEs model, not on its suitable analytical solution. Simply said, SOMA tried to search for such parameters, which resulted in a suitable behavior for the reactor produced [6]. Because there was no general analytic solution (from the point of view of reactor pa-

rameters), for each set of optimized parameters, numerical simulation of the 5 ODEs model had to be carried out.

- **Predictive control**. As in the previous example, the application of SOMA was used for predictive control of a given reactor from a different point of view [6].
- **Fuzzy logic setting**. In this research, both DE and SOMA were used to find suitable fuzzy logic setting. An input set, an output set and a rule base had to be optimized by SOMA and DE. The results were compared with expert and neural network settings [8].
- **Analytic programming.** In this application [29] of SOMA discrete variable handling was used to handle with arbitrary functions via integer index [13]. The result of analytic programming (AP) is thus the same as genetic programming. The output of AP are more or less complicated functions called "programs". The principles of AP are general enough to be used by any other evolutionary algorithms. In [29] the first results that demonstrated the ability of AP to solve selected examples from [30] were presented. Now is in process comparative study with examples from [31], [32]. Formulas in 7.18 - 7.21 demonstrate a few successful programs which were "generated" by analytic programming and which successfully solve so called "sextic polynomial" example [31].

$$-\frac{(K[1] - x^2 K[2] K[7]) \left(-x^2 - K[3] - \frac{K[4] K[8]}{x+K[9]}\right)}{x \left(-x + \frac{K[5]}{x} + K[6]\right)} \tag{7.18}$$

$$-\frac{x(x+K[1])}{K[2]\left(x + K[3] - \frac{K[4] K[5]}{x^2\left(2x - \frac{K[7]}{K[8]} + \frac{K[6]}{x}\right)}\right)} \tag{7.19}$$

$$\frac{2x^4}{K[3]} - 2x^2\left(\frac{x^2}{K[1]} + K[2]\right) \tag{7.20}$$

$$\frac{\frac{\left(2x - \frac{x}{K[5] + \frac{x K[10]}{K[9]}}\right)^x}{K[1]} + x}{-\frac{K[2] K[3]}{x} - \frac{(K[6]-x)(K[8]-K[7])}{K[4]}} \tag{7.21}$$

7.13 Gallery of test functions

Each new algorithm has to be theoretically analyzed and practically tested in order to verify how robust and powerful it is from a specific point of view (global extreme finding,...). For algorithm the SOMA a set of 15 test functions was used, which are depicted here, including their mathematical formulas. All these functions are a part of source C++ files which are accessible on [11]. On each contour plot, one or more black point sets are depicted, which represents a set of points which are "x%" (in the sense of 3D representation of appropriate function its and allowed boundaries) far from the global extreme from the cost value point of view. They are depicted here to demonstrate multiple global extremes and the complexity of test functions used for SOMA testing. It is important to note that these black points depend also on graphical resolution of used program.

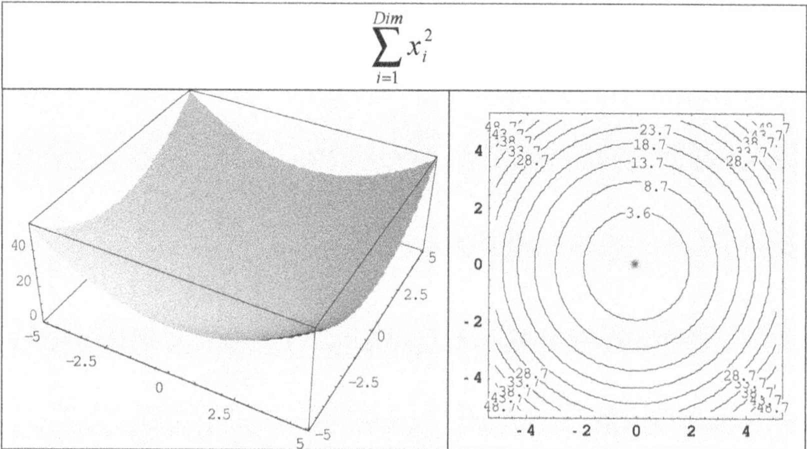

Fig. 7.12. 1st De Jong – black point represents a set of points which differ max. 0.01 % from global extreme in point of view of cost function value

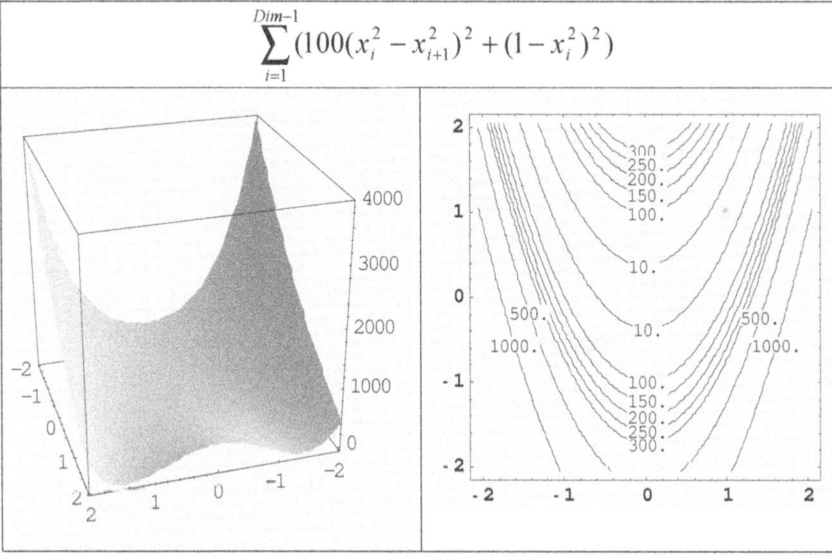

Fig. 7.13. Rosenbrock's saddle – black point represents a set of points which differ max. 0.0001 % from global extreme in point of view of cost function value

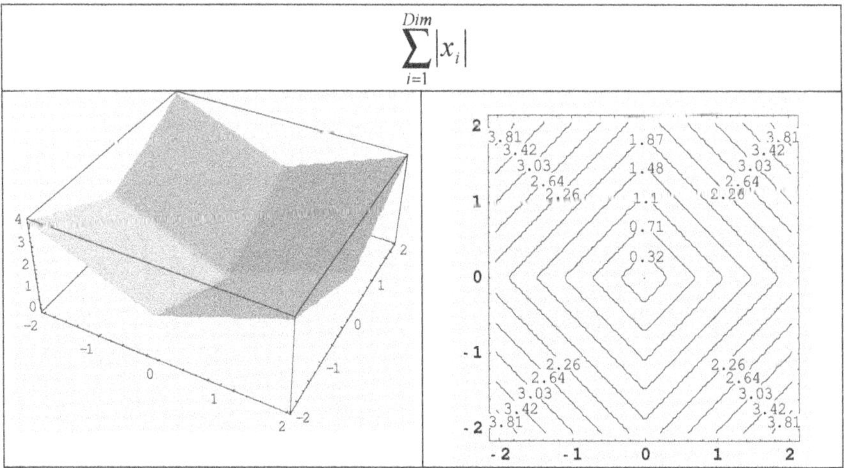

Fig. 7.14. 3rd De Jong – black point represents a set of points which differ max. 0.1 % from global extreme in point of view of cost function value

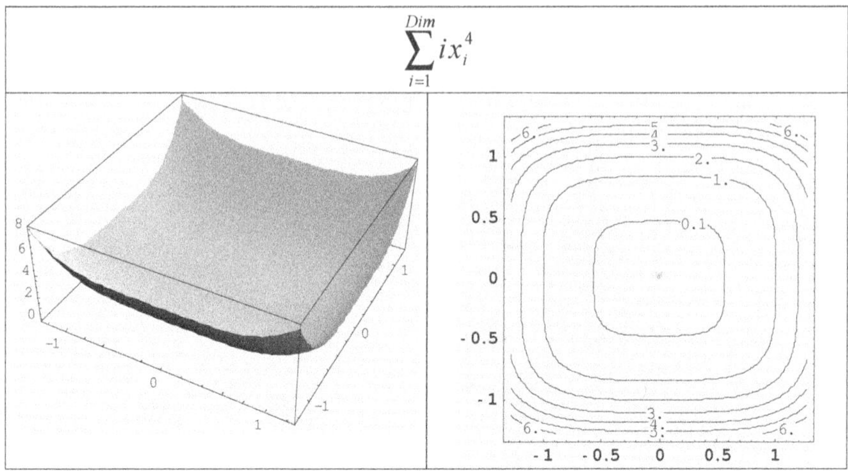

Fig. 7.15. 4th De Jong – black point represents a set of points which differ max. 0.000001 % from global extreme in point of view of cost function value

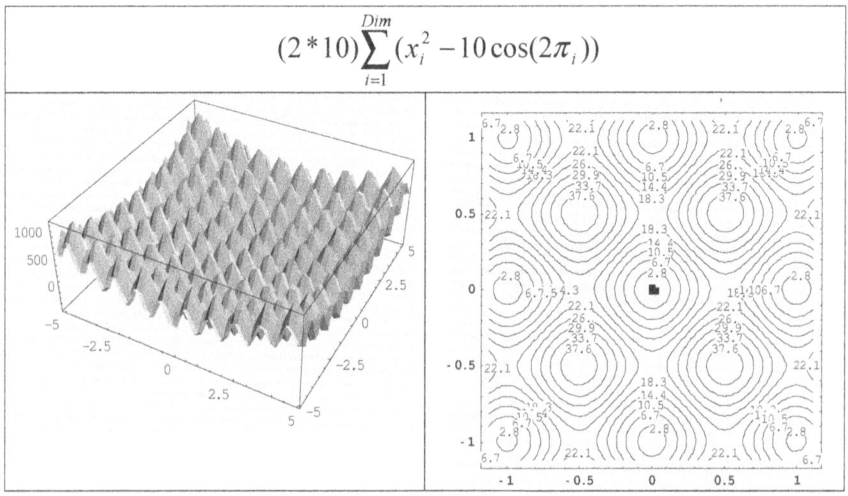

Fig. 7.16. Rastrigin's function - black point represents a set of points which differ max. 0.5 % from global extreme in point of view of cost function value

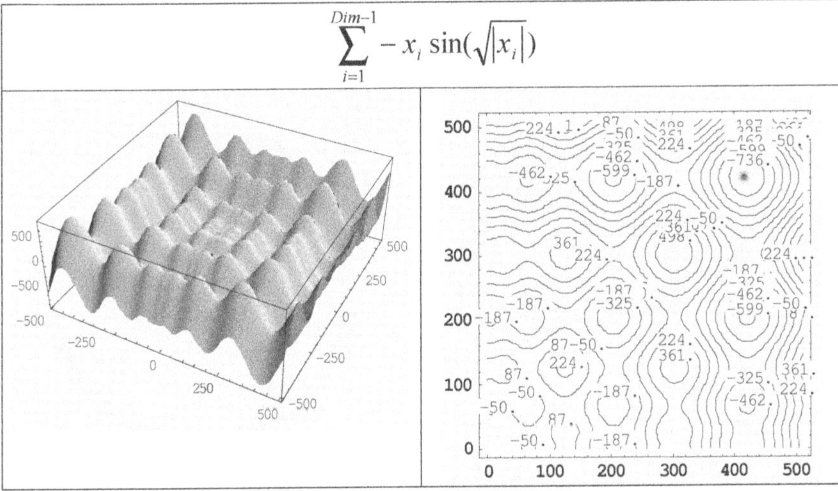

Fig. 7.17. Schwefel's function - black point represents a set of points which differ max. 0. 1 % from global extreme in point of view of cost function value

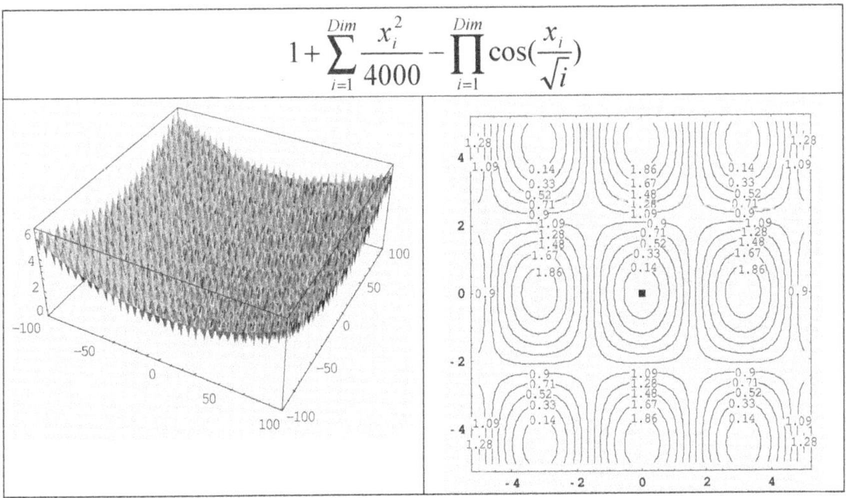

Fig. 7.18. Griewangk's function - black point represents a set of points which differ max. 0.01 % from global extreme in point of view of cost function value

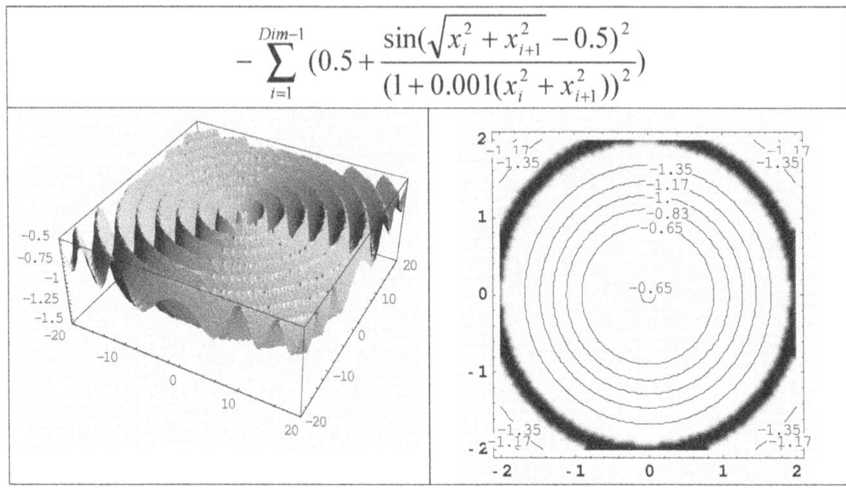

$$-\sum_{i=1}^{Dim-1}\left(0.5+\frac{\sin(\sqrt{x_i^2+x_{i+1}^2}-0.5)^2}{(1+0.001(x_i^2+x_{i+1}^2))^2}\right)$$

Fig. 7.19. Sine envelope sine wave function – black point represents a set of points which differ max. 1 % from global extreme in point of view of cost function value

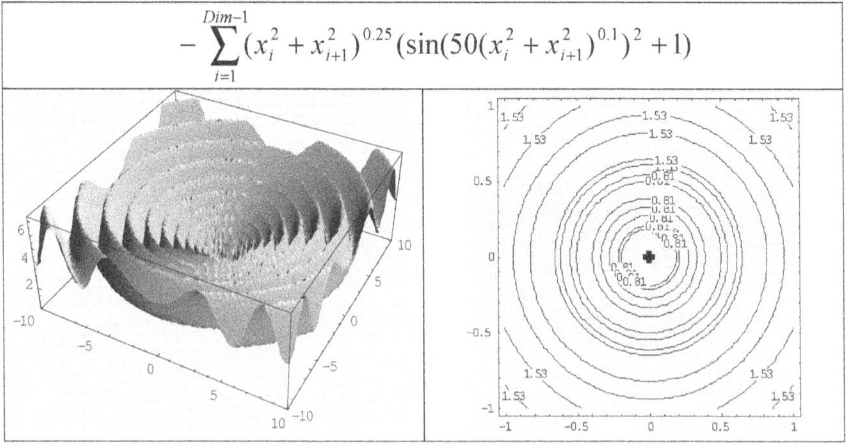

$$-\sum_{i=1}^{Dim-1}(x_i^2+x_{i+1}^2)^{0.25}(\sin(50(x_i^2+x_{i+1}^2)^{0.1})^2+1)$$

Fig. 7.20. Stretched V sine wave function – black point represents a set of points which differ max. 0. 1 % from global extreme in point of view of cost function value

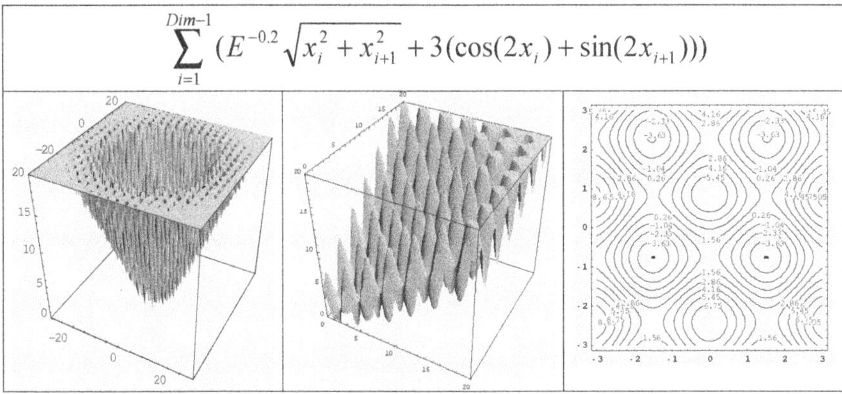

Fig. 7.21. Test function (Ackley) – black point represents a set of points which differ max. 0.1 % from global extreme in point of view of cost function value

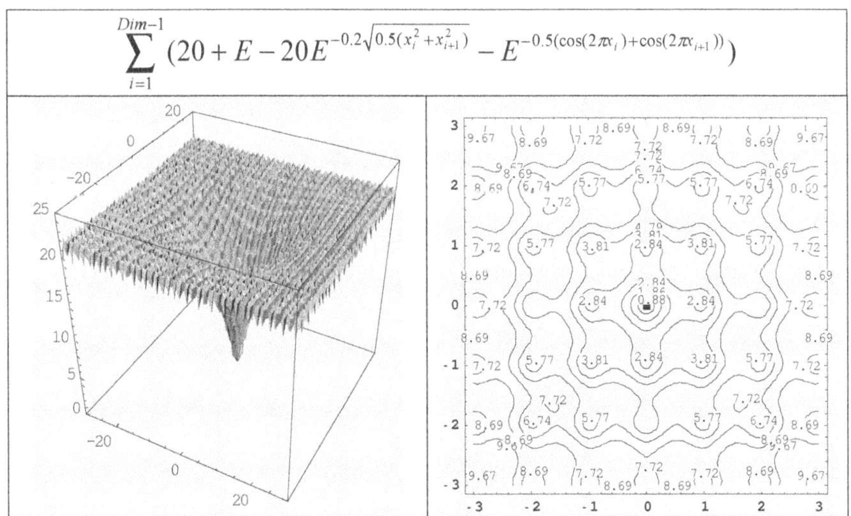

Fig. 7.22. Ackley's function – black point represents a set of points which differ max. 0.1 % from global extreme in point of view of cost function value

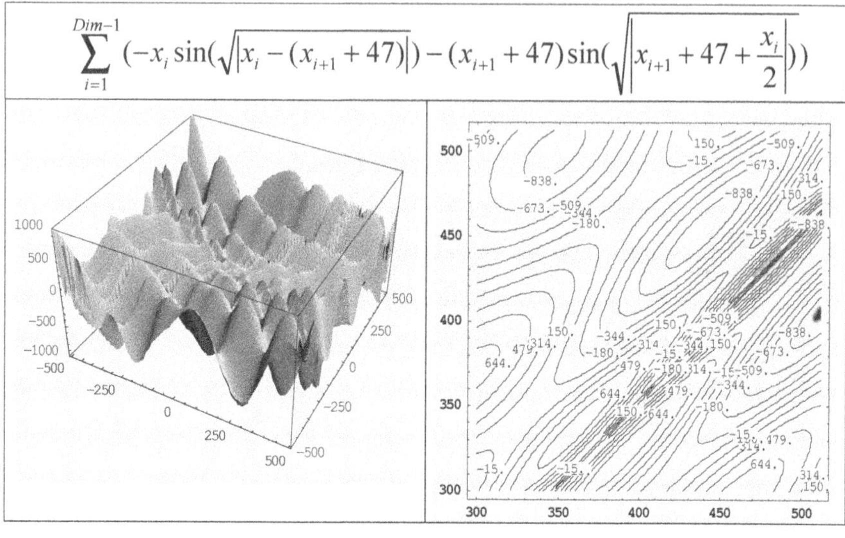

Fig. 7.23. Egg holder – black point represents a set of points which differ max. 1 % from global extreme in point of view of cost function value

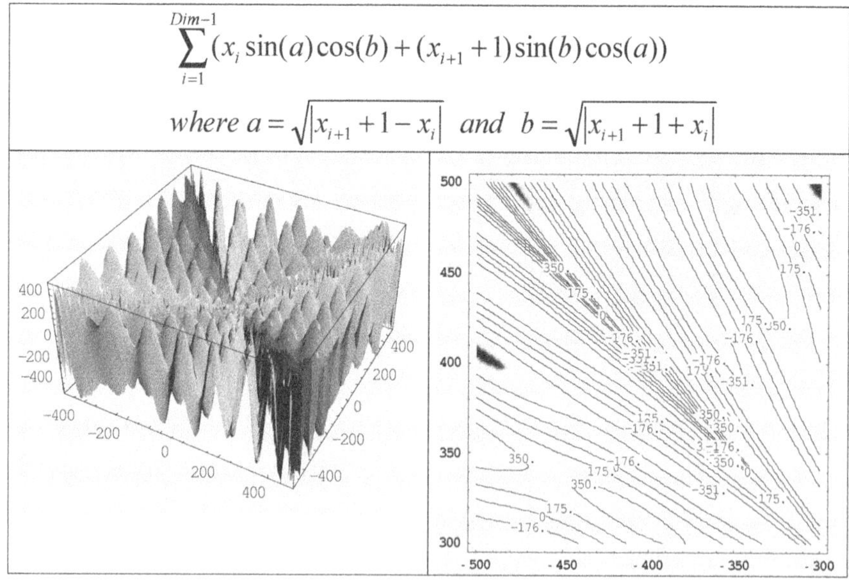

Fig. 7.24. Rana's function – black point represents a set of points which differ max. 2 % from global extreme in point of view of cost function value

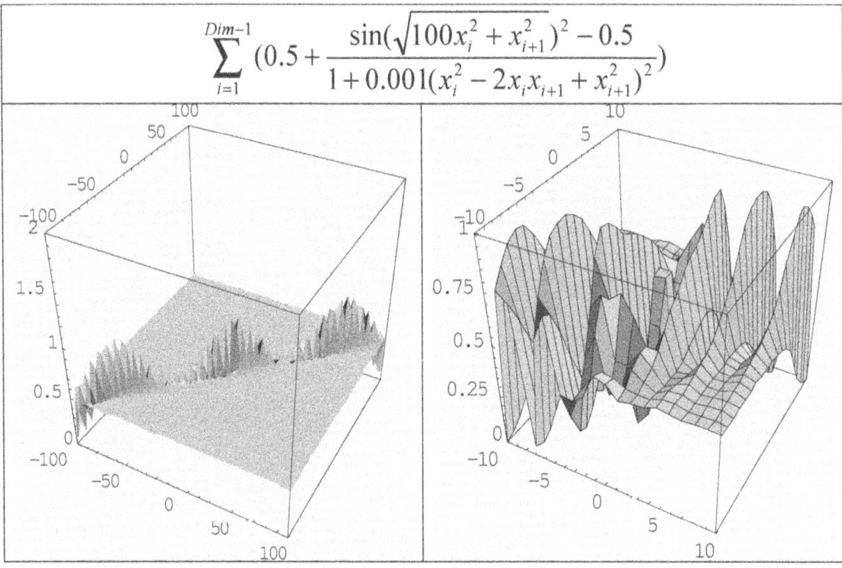

Fig. 7.25. Pathological function – detail is on the right. Due to extremely high complexity of this function, contour graph is not depicted here

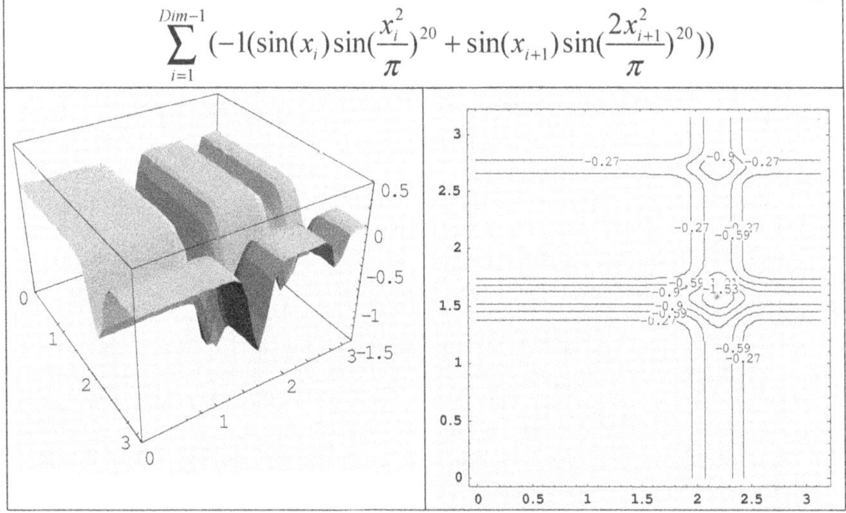

Fig. 7.26. Michalewicz's function – black point represents a set of points which differ max. 0.1 % from global extreme in point of view of cost function value

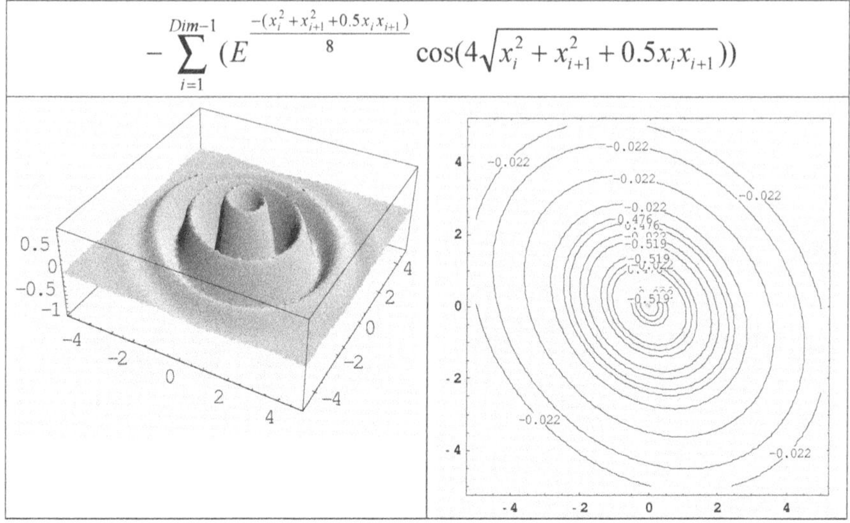

Fig. 7.27. Inverted cosine wave function (Masters) – black point represents a set of points which differ max. 0.1 % from global extreme in point of view of cost function value

7.14 SOMA on tested functions

The first testing of the SOMA algorithm was performed on 15 test functions, which are given and depicted in Fig. 7.12 – 7.27 and shortly in Table 7.5. In each case of the functions tested, the SOMA algorithm was searching for the global minimum of the given function. Each simulation was repeated 100 times in 100 dimensional space (i.e. the individuals consisted of 100 parameters). The left part of all figures below shows the descriptions for SOMA. All these results were compared with DE (at the right side) and are fully comparable for each function. The parameters setting for each simulation is given in the Table 7.7. One can use them to reproduce and verify the results by "ready to use" C++ code at [11].

Two figures represented results for each test function here. The first one shows average values of founded parameters and its deviations (based on all 100 simulations) and the second one (bar chart) shows cost value of the best individual after all migrations (SOMA) or in the last population (DE).

Each simulation was designed so that the same amount number of cost function evaluations was done. More details about DE (used here like comparative algorithm) can be found in [27], [28] and [13]. Number of cost function evaluation for SOMA and DE can be calculated quite easily. If principles of SOMA are taken into account, then number of cost function evaluations done during one individual run is given by

$$N_{eval\_1\_individual} = \frac{PathLength}{Step} \tag{7.22}$$

Because there is one Leader and thus (PopSize-1) individuals will run in one migration loop then for one migration is done number of cost function evaluations by

$$N_{eval\_1\_migration} = \frac{(PopSize - 1) * PathLength}{Step} \tag{7.23}$$

Finally, during all migrations is total number of cost function evaluations given by

$$N_{eval} = \frac{(PopSize - 1) * PathLength * Migrations}{Step} \tag{7.24}$$

For strategy AllToAll is situation similar. Because there is no Leader and all individuals run toward themselves, then nominator is multiplied by PopSize. Term (PopSize-1) means here the fact, that no one individual run toward itself. The number of cost function evaluations is for AllToAll strategy given by

$$N_{eval} = \frac{PopSize * (PopSize - 1) * PathLength * Migrations}{Step} \tag{7.25}$$

Under some assumptions can be also calculated probability that global extreme will be found. Main assumption is that searched space is discrete, i.e. can be covered by grid, fine or rough. This discretization can be done if there are no individuals, who are exactly at one point, i.e. there are no individuals with exactly the same coordinates. This is true every time, because probability that two or more individuals will share the same position is almost 0 (there is an infinite number of real numbers). Discretization can be done “a priori” by estimation or “a posteriori” so that after evolution is grid size based on minimal distance of individuals in the population. Thus each axe of searched space consist of L discrete elements and for cost function with “n” arguments (i.e. “n” axes) is probability done by

$$P_{GE} = \frac{N_{eval}}{L^n} \tag{7.26}$$

If grid size is constant, then probability of global extreme retrieval is bigger if N_{eval} increase and vice versa. Values represented by N_{eval} can increase only if PopSize, PathLength and Migrations are bigger or/and Step is lesser. All these theoretical conclusions are in good relation with numerical simulations described in Section 7.8

Table 7.5. SOMA and DE on test functions

Test function	Distance from global extreme			
	The worst case		The best case	
	SOMA	DE	SOMA	DE
Sphere model, 1st De Jong's function, Fig. 7.28	1.01	7.72	3.9 x10^{-7}	0.17
Rosenbrock's saddle, Fig. 7.29	138.35	222.39	93.27	94.08
3rd De Jong's function, Fig. 7.30	1.6 x10^{-9}	1.8 x10^{-5}	9.3 x10^{-11}	4.8 x10^{-6}
4th De Jong's function, Fig. 7.31	1.5 x10^{-19}	4.6 x10^{-7}	4.8 x10^{-22}	1.3 x10^{-8}
Rastrigin's function, Fig. 7.32	44.82	148.73	22.93	77.41
Schwefel's function, Fig. 7.33	-40928.8	-41405.1	-41995.7	-41997.8
Griewangk's function, Fig. 7.34	0.00785	0.14	6.7 x10^{-6}	1.1 x10^{-19}
Stretched V sine wave function, Fig. 7.35	0.80	0.13	0.035	1.2 x10^{-37}
Test function*, Fig. 7.36	-279.46	-283.44	-290.60	-290.60
Ackley's function, Fig. 7.37	5.15	0.022	-1.8 x10^{-4}	-1.8 x10^{-4}
Egg holder, Fig. 7.38	-44696.6	-29331.8	-76415.8	-33780.2
Rana's function, Fig. 7.39	- 25581.5	-17782.9	-38451	-19920.3
Pathological function, Fig. 7.40	17.30	29.80	2.34	27.06
Michalewicz's function, Fig. 7.41	-97.90	-98.69	-99.89	-99.89
Cosine wave function, Fig. 7.42	-77.40	-41.27	-92.65	-75.22

**Two equal global extremes are presented in this function*

Table 7.6. Overview of all simulations

Result	The worst		The best	
	SOMA	DE	SOMA	DE
Winner	10 x	6 x	9 x	4 x
Draw	0 x		3 x	

Table 7.7. Parameter settings

Test function	Algorithm setting							
	SOMA				DE			
Parameters with constant value for all simulations	PathLength = 3. D = 100, MinDiv = -1				D = 100			
Abbreviation (see below)	S*	PRT	PS**	M***	F	CR	NP	G****
Sphere model, 1st De Jong, Fig. 7.28	0.5	0.1	20	500	0.8	0.8	20	3300
Rosenbrock's saddle, Fig. 7.29	0.11	0.1	60	125	0.2	0.2	60	3450
3rd De Jong, Fig. 7.30	0.11	0.1	20	500	0.8	0.2	20	13301
4th De Jong, Fig. 7.31	0.11	0.1	20	500	0.8	0.2	20	13301
Rastrigin, Fig. 7.32	0.11	0.1	60	400	0.2	0.8	60	11015
Schwefel, Fig. 7.33	0.11	0.1	60	400	0.2	0.2	60	11015
Griewangk, Fig. 7.34	0.11	0.1	60	200	0.2	0.2	60	5508
Stretched V sine wave, Fig. 7.35	0.11	0.1	60	400	0.2	0.2	60	11015
Test function, Fig. 7.36	0.11	0.1	60	400	0.2	0.2	60	11015
Ackley's function, Fig. 7.37	0.11	0.1	60	400	0.2	0.2	60	11015
Egg holder, Fig. 7.38	0.11	0.1	60	800	0.2	0.2	60	22030
Rana's function, Fig. 7.39	0.11	0.1	60	800	0.2	0.2	60	22030
Pathological, Fig. 7.40	0.11	0.1	60	800	0.2	0.2	60	22030
Michalewicz, Fig. 7.41	0.11	0.1	60	200	0.2	0.2	60	5508
Cosine wave, Fig. 7.42	0.11	0.1	60	400	0.2	0.4	60	11015

*Step, **PopSize, ***Migrations, ****Generations

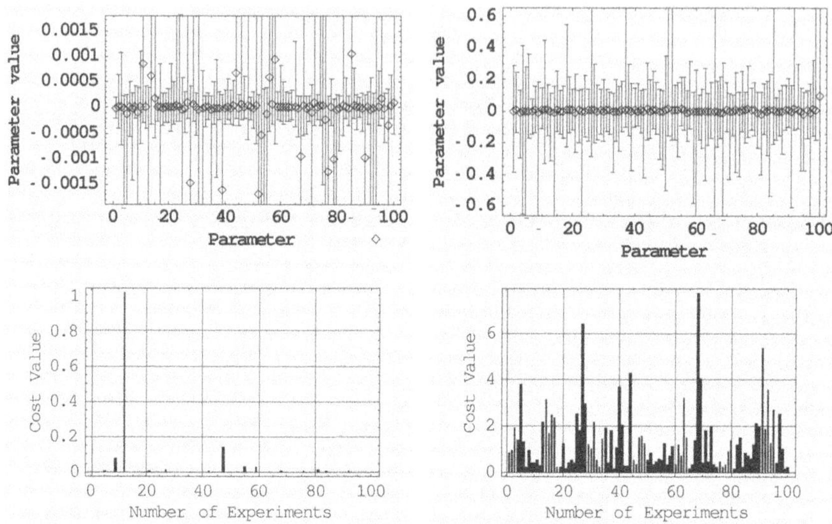

Fig. 7.28. 1st De Jong's function

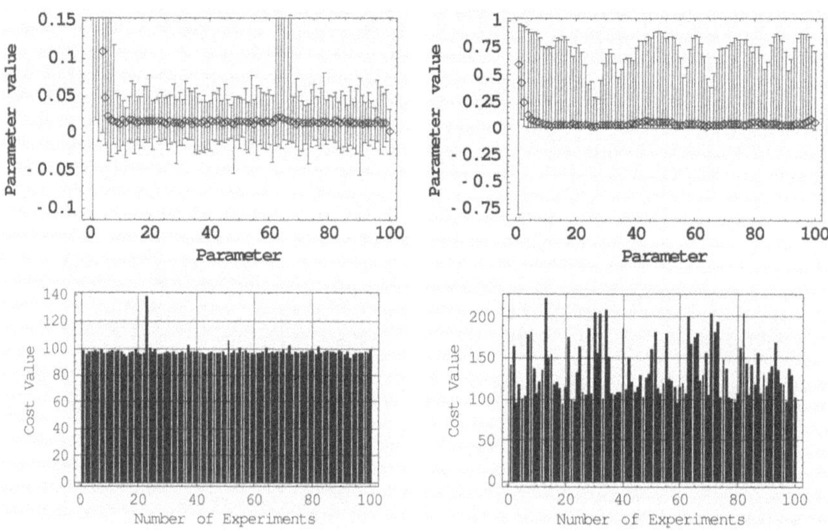

Fig. 7.29. Rosenbrock's saddle

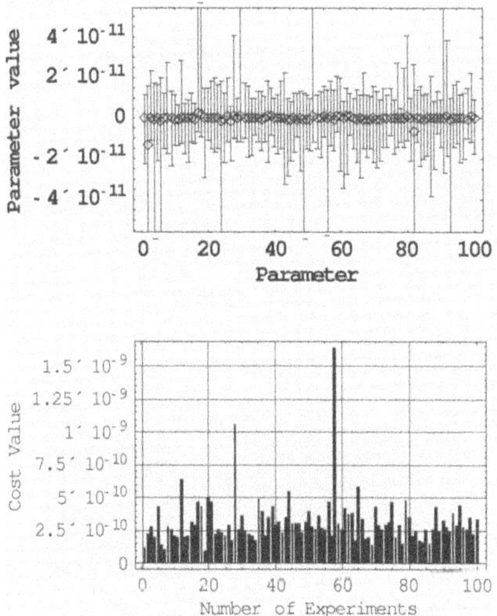

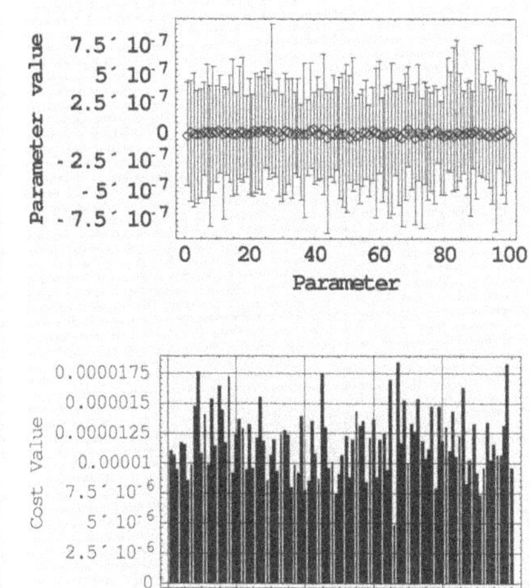

Fig. 7.30. 3rd De Jong's function

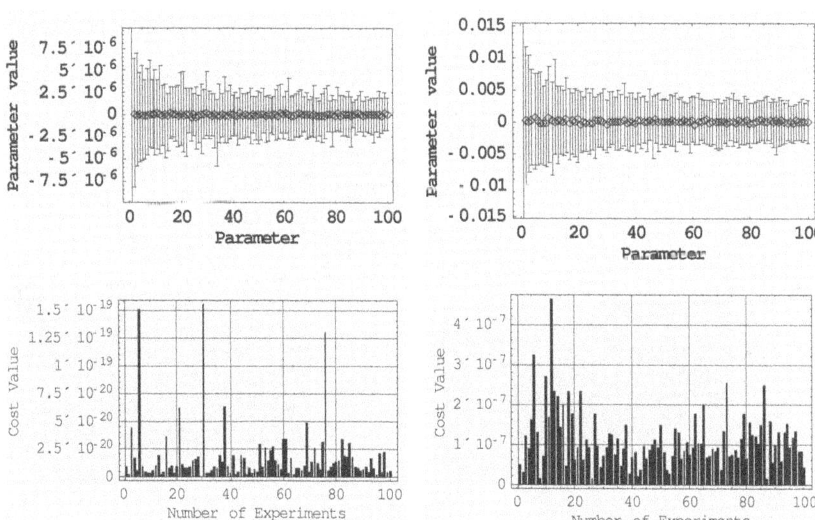

Fig. 7.31. 4th De Jong's function

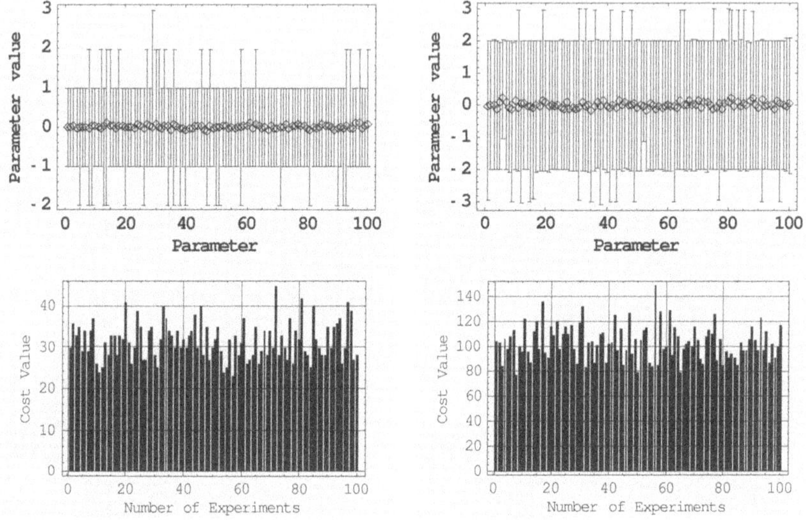

Fig. 7.32. Rastrigin's function

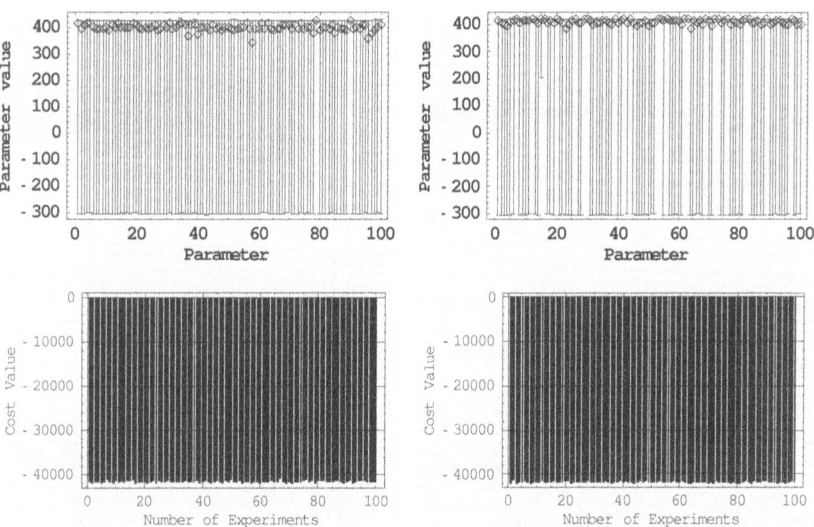

Fig. 7.33. Schwefel's function

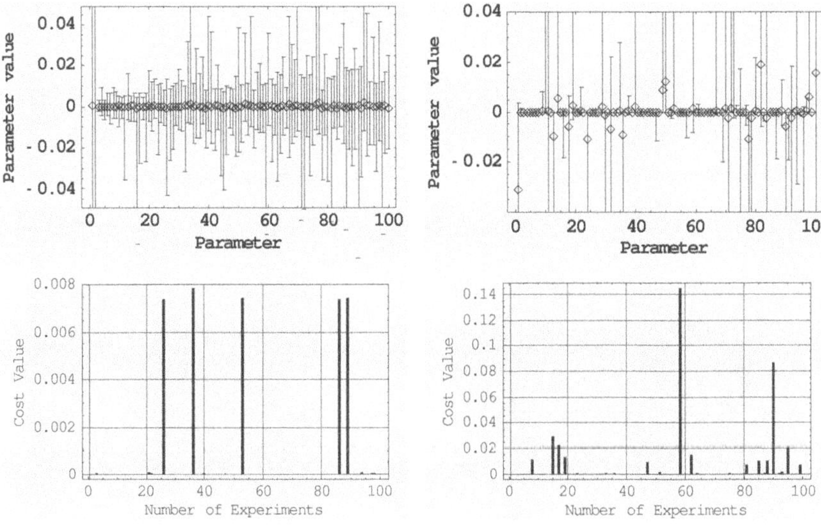

Fig. 7.34. Griewangk's function

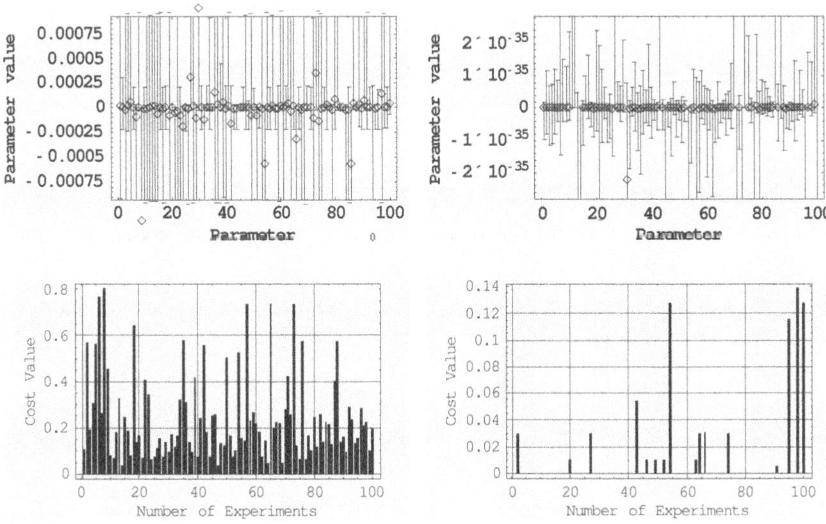

Fig. 7.35. Stretched V sine wave function (Ackley)

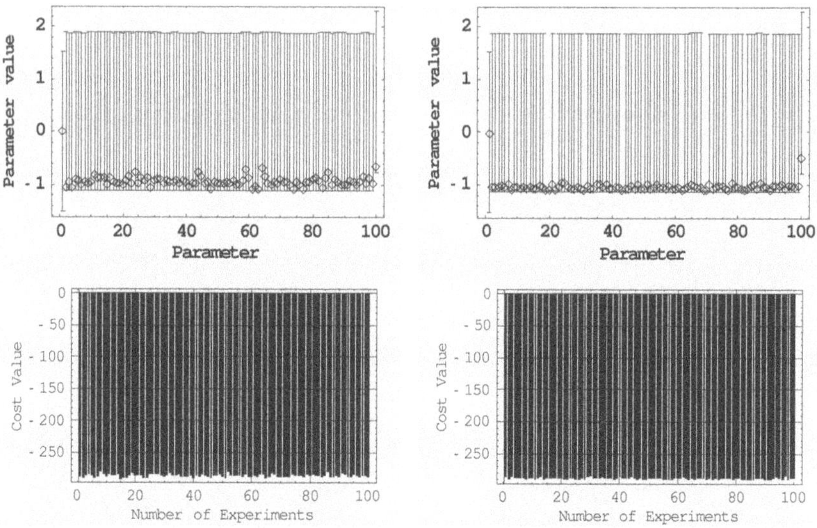

Fig. 7.36. Test function (Ackley)

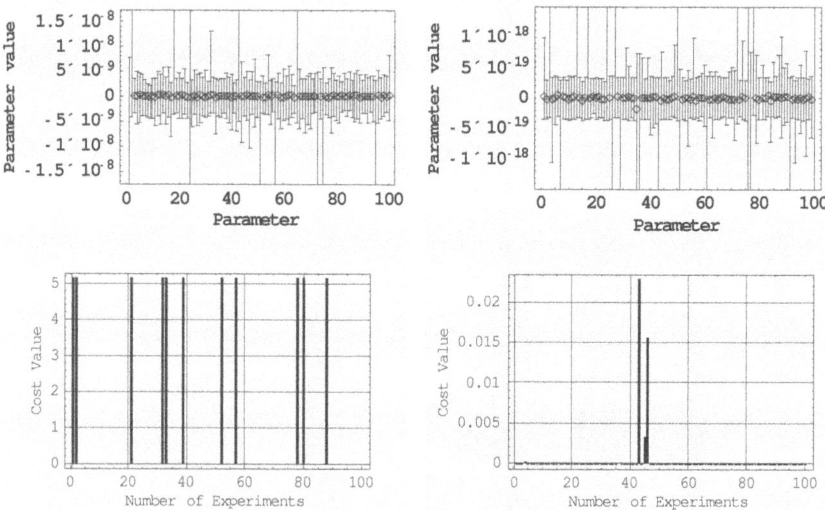

Fig. 7.37. Ackley's function

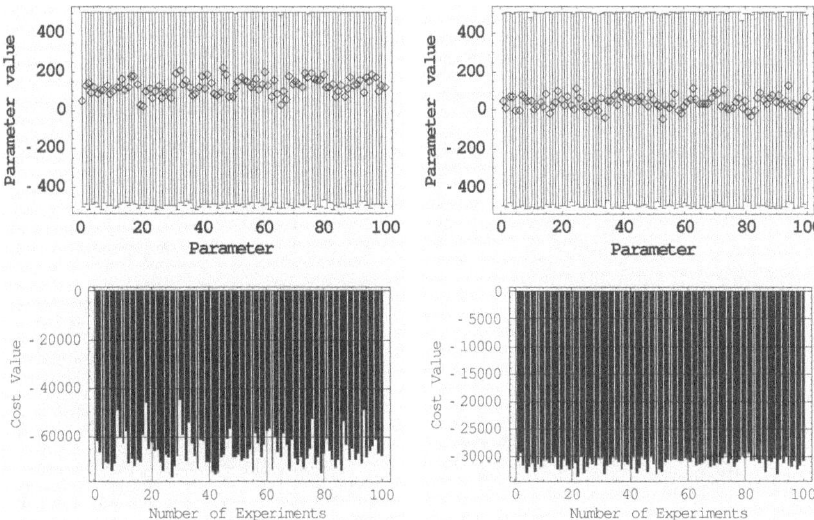

Fig. 7.38. Egg holder

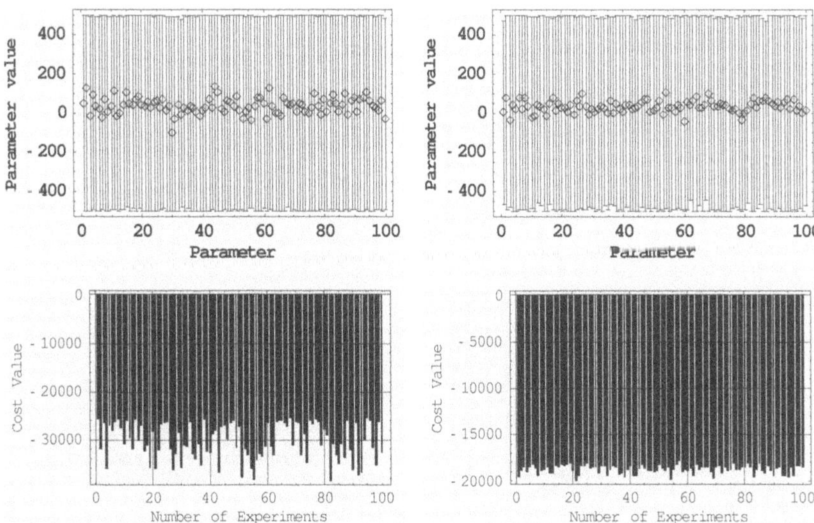

Fig. 7.39. Rana's function

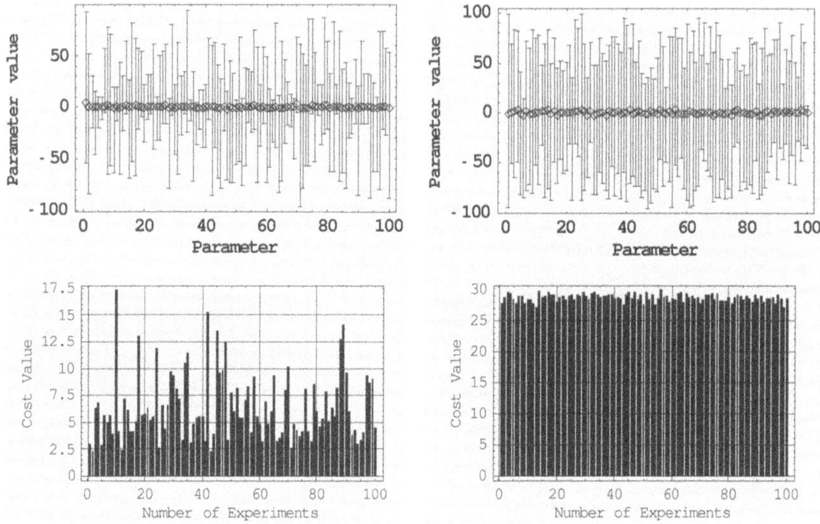

Fig. 7.40. Pathological function

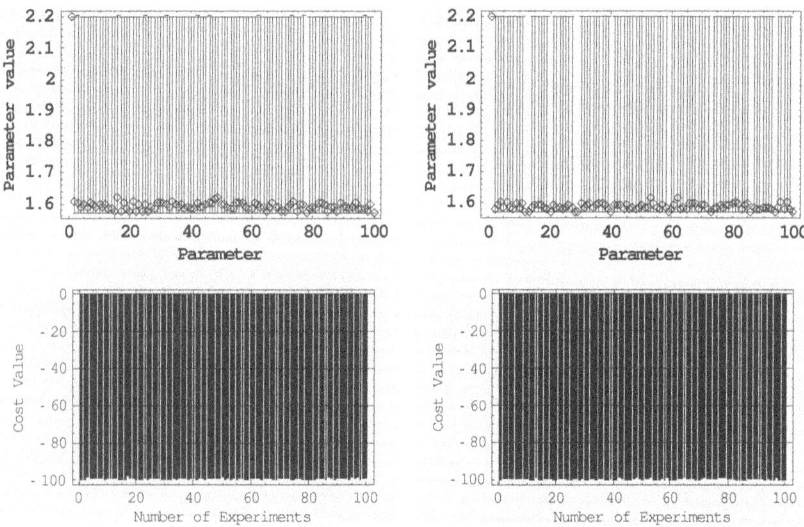

Fig. 7.41. Michalewicz's function

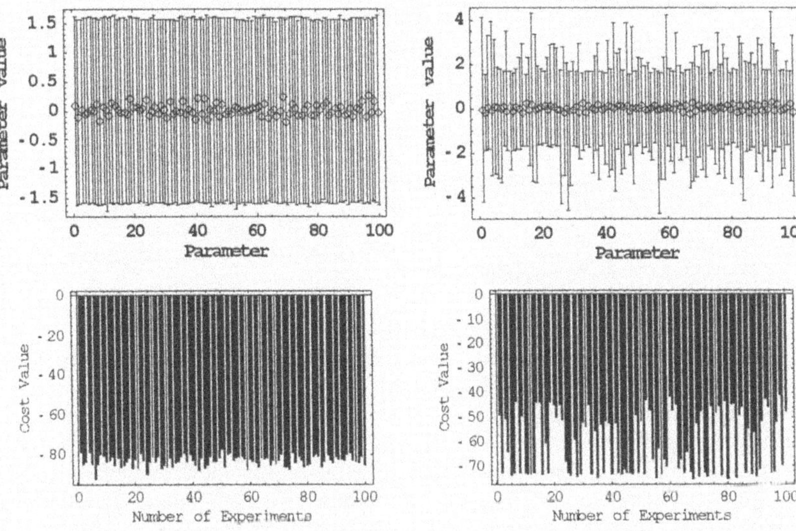

Fig. 7.42. Cosine wave function (Masters)

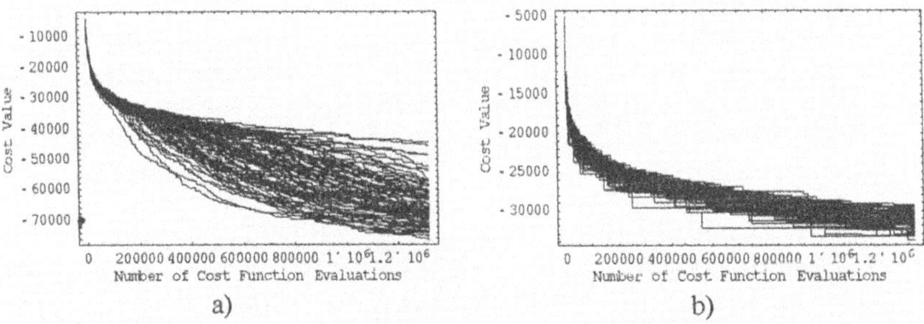

Fig. 7.43. SOMA (a) and DE (b) on Egg Holder. The flow of all solutions was for SOMA spread as is shown in a) only for three functions – Egg holder, Rana's function and Pathological function.

7.15 Conclusion

This article described SOMA, a new stochastic search algorithm for global optimization. In this chapter were introduced:

- Basic principles of SOMA algorithm
- Various strategies (versions) of the algorithm
- Selected applications
- Testing for robustness
- Handling of various constraints

The methods described for handling constraints are relatively simple, easy to implement and easy to use. Originally were introduced in [13] and used here because of their universality and easy implementation.

A soft-constraint (penalty) approach is applied for the handling of constraint functions. Some optimization methods require a feasible initial solution as a starting point for a search. Preferably, this solution should be rather close to a global optimum to ensure convergence to it instead of to a local optimum. If non-trivial constraints are imposed, it may be difficult or impossible to provide a feasible initial solution.

The efficiency, effectiveness and robustness of many methods are often highly dependent on the quality of the starting point. The combination of SOMA algorithm and the soft-constraint approach does not require an initial solution, but yet it can take advantage of a high quality initial solution if one is available.

For example, this initial solution can be used for initialization of the population in order to establish an initial population that is biased towards a feasible region of the search space. If there are no feasible solutions in the search space, as is the case for totally conflicting constraints, SOMA algorithms with the soft-constraint approach are still able to find the nearest feasible solution. This is often important in practical engineering optimization applications, because many non-trivial constraints are involved.

The test functions used for SOMA and DE testing had all 'negative' attributes mentioned in Section 7.2 Well known tested functions (Fig. 7.12 – Fig. 7.27) have been used and the results were depicted graphically (Fig. 7.28 – Fig. 7.42) and numerically (Table 7.5) including parameter setting for both algorithms (Table 7.7). Each test was designed so that the global extreme was searched in 101 dimensions, i.e. the cost functions had 100 arguments. For each cost function, the success for SOMA and DE for 100 simulations was given. These tests have demonstrated that SOMA is capable of solving hard optimization problems. However it is important to remember some important facts:

1. Two versions of both algorithms were compared here, one of the best versions of DE (DERand1Bin) with the basic SOMA (AllToOne). This approach was chosen for the following reasons: if two algorithms are mutually compared (one

of the the best DE versions and the worst version of SOMA) then it is almost sure that better versions of SOMA give better results (see Chapter 26).

2. The time needed to get results was relatively long – in the range from 40 seconds to 5 hours (on Intel Pentium 4, 2 GHz, 500 Mb RAM) for both algorithms (SOMA, DE). This surprisingly long time was caused by fact that each simulation was repeated 100 times. In reality, simulations took approx. from seconds to minutes, depending on how complicated the cost function was. This fact show that SOMA can be used for real time processes (see Section 7.11, "Active Compensation in RF - driven Plasmas"). Very often, it is difficult to find an analytical solution to a given engineering problem. It was shown that evolutionary algorithms usually find a solution faster than manual calculations.
3. Test functions (as well as the others problems) are sensitive to coordinates of the global extreme. This means that small differences in coordinates of the global extreme can cause a large change in the final cost values despite the fact that the evaluated position is not far from the position of the global extreme. This is especially important for high dimensional extremes. From this point of view, a difference of 10% or 39.56 (say from 0) does not means that the optimizing process is far from the global extreme (in sense of the cost function arguments - coordinates).
4. For some functions it is difficult to find the global optimum. Typical examples are Schwefel's function (Fig. 7.33) and Rosenbrock's saddle (Fig. 7.29). Schwefel's function being particularly difficult: The first problem is that the minimum is not in the origin as it usually is for other functions, but at the edge of the search space. The second problem is more interesting: in the case of symmetrical unfolding of the search space owing to the origin, the position of the global extreme cyclically changes its position (see Fig. 7.44 a)! A second tricky function is Rosenbrock's saddle (Fig. 7.44 b). There are almost two identical extremes but only one is the global one. Hence, this is a quite an unpleasant test function for any optimization algorithm. Both functions (and the others as well as) provide rather difficult test conditions for EAs as well as problems where global extreme change its position (that was problem of real example on plasma reactor, see Section 7.11, "Active Compensation in RF - driven Plasmas").

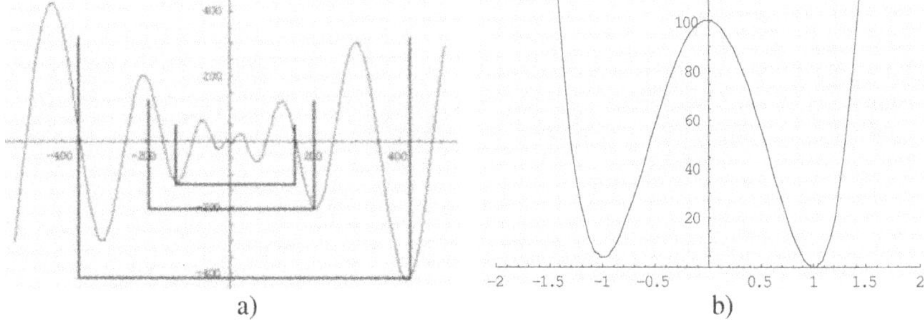

Fig. 7.44. Two tricky functions

5. The conditions for the optimization were set as highly difficult as possible i.e. a limited number of individuals was used, etc. Usually, 20 to 60 individuals were used for searching in 100 dimensional hyper-plane. It is for this reason that the global extreme was not exactly found in all cases. Different numbers of individuals, different parameter settings or different versions of the algorithms could improve this dramatically. Based on results can be declared that the SOMA performance was almost the same like DE. SOMA showed a very good, and sometimes better, ability to find extremes as DE (see Fig. 7.38, Fig. 7.39 and Fig. 7.40).
6. The number of cost function evaluations was preferred for "x" axis on graphs where history of cost value is depicted prior to Generations or Migrations (Fig. 7.43). The use of Generations or Migrations can in a graphical point of view cause that process with a lower convergence to global extreme to be viewed to the similarly to the faster, and vice versa.
7. Convergence speed. In the case of many test functions DE was faster in the final extreme (means here the best located extreme) than SOMA. As is visible from results described in the Chapter 26, SOMA was able to locate the same extreme as well as DE but no in all cases. Because examples in the Chapter 26 are low dimensional, it can be stated that SOMA gives better performance on problems which shows higher dimensionality and/or "irregularity" like functions on Fig. 7.23 – Fig. 7.25 and Fig. 7.38 – Fig. 7.39.

Despite comparing only the basic version of the SOMA algorithm with one of the best versions of DE algorithm, it is visible that SOMA's performance on the tested functions was very good. The SOMA algorithm works with minimum assumptions with respect to the objective function. The algorithm requires only the cost value returned from the objective function for guidance of its seeking for the optimum. No derivatives or other auxiliary information are needed. Including the algorithm's extensions discussed in this article, the SOMA algorithm can be applied to a wide range of optimization problems, which practitioners in the field of modern optimization would like to solve.

Acknowledgements

This work was supported by the grant No. MSM 26500014 of the Ministry of Education of the Czech Republic and by grants of the Grant Agency of Czech Republic GACR 102/03/0070 and GACR 102/02/0204.

References

[1] Zelinka Ivan, Artificial Intelligence in Problems of Global Optimization, Czech ed., BEN, ISBN 80-7300-069-5

[2] Zelinka Ivan, Vladimir Vasek, Jouni Lampinen, New Algorithms of Global Optimization, Automatizace (Journal of Automatization, Czech Ed.), 10/01, 628-634, ISSN 0005-125X

[3] Zelinka, I., Vašek, V., Lampinen, J.: SOMA and Differential Evolution - New Algorithms of Global Optimisation, Fine Mechanics and Optics, 4/2002, p. 112-117, ISSN 0447-6441

[4] Zelinka, I., Kolomazník, K., Lampinen, J., Nolle L.: SOMA – Selforganizing Migrating Algorithm and its application in the Mechanical Engineering. Part 1. – Theory of Algorithm, Journal of Engineering Mechanics, 21 p., ISSN 1210-2717

[5] Zelinka, I., Kolomazník, K., Lampinen, J., Nolle L.: SOMA – Selforganizing Migrating Algorithm and its application in the Mechanical Engineering. Part 1. – Testing and Applications, Journal of Engineering Mechanics, 23 p., ISSN 1210-2717

[6] Zelinka Ivan, Prediction and analysis of behavior of dynamical systems by means of artificial intelligence and synergetic, Ph.D. thesis, 2001, awarded by Siemens price like the best Ph.D thesis of year 2001 in the field of automation and cybernetics in Czech Rep.

[7] Kehar Jiri, Evolutionary algorithms in prediction and control, diploma thesis, 2000

[8] Dlapa Marek, Evolutionary algorithms in fuzzy logic and control, diploma thesis, 2000

[9] Zelinka Ivan, Lampinen Jouni, SOMA - Self-Organizing Migrating Algorithm, Mendel 2000, 6th International Conference on Soft Computing, Brno, Czech Republic, ISBN 80-214-1609-2

[10] Zelinka Ivan, Lampinen Jouni,SOMA - Self-Organizing Migrating Algorithm, Nostradamus 2000, 3rd International Conference on Prediction and Nonlinear Dynamic, Zlín, Czech Republic, ISBN

[11] Articles about SOMA algorithm (source codes, graphics animated gallery, bibliography, see http://www.ft.utb.cz/people/zelinka/soma

[12] Kvasnička V., Pospíchal J., Tiňo P. 2000, Evolučné algoritmy, STU Bratislava, ISBN 85-246-2000, 2000

[13] Lampinen Jouni, Zelinka, Ivan. New Ideas in Optimization - Mechanical Engineering Design Optimization by Differential Evolution. Volume 1. London : McGraw-Hill, 1999. 20 p.

ISBN 007-709506-5.
[14] http://www.densif.fee.unicamp.br/~moscato/memetic_home.html, http://www.geocities.com/ResearchTriangle/3123/ms2.html
[15] Dawkins Richard, The Selfish Gene, Oxford Univ Pr (Trade); ISBN: 0192860925

[16] Zelinka Ivan, DELA - An Evolutionary Learning Algorithms For Neural Networks. In Mendel '99, 5th International Conference on Soft Computing. Volume 1. Brno : PC-DIR, Brno, 1999, 1999, p. 410;414. ISBN 80-214-1131-7.
[17] Zelinka Ivan, DELA - An Evolutionary Learning Algorithms For Neural Networks. In Proces Controll '99. Volume 1. Slovak University of Technology : Vydavatelstvo STU, Bratislava, 1999, p. 346;351. ISBN 80-227-1228-0.
[18] Zelinka Ivan, Lampinen Jouni, Evolutionary Identification of Predictive Models. In Nostradamus 99. Volume 1. Zlín : Knihovna F.Bartoše, Zlín, 1999, p. 114;122. ISBN 80-214-1424-3.
[19] Zelinka Ivan, Lampinen Jouni,Evolutionary Identification of Predictive Models, ISF' 2000, The 20th International Symposium on Forecasting, Lisbon, Portugal, June 21-24, Editor - International Institute of Forecasters
[20] Zelinka Ivan, Lampinen Jouni, Evolutionary Identification of Predictive Models, EIS' 2000, International Symposium on Engineering of Intelligent Systems, Paisley, Scotland, UK, Editor - C. Fye, Publication by ICSC Academic Press International Computer Science Conventions, Canada, Switzerland, ISBN 3-906454-21-5
[21] Kenneth V. Price, Rainer Storn and Jouni Lampinen: Differential Evolution: Global Optimization for Scientists and Engineers
Ivan Zelinka, Inverse Fractal Problem, in print
[22] Zelinka Ivan, Fractal Sight by Means of Neural Network. In Mendel '98. Volume 1. Brno : PC-DIR, Brno, 1998, p. 305;308. ISBN 80-214-1199-6.
[23] Zelinka Ivan, Fraktální vidění pomocí neuronových sítí. In Proces controll '99. Volume 1. Slovak University of Technology, Bratislava, STU Bratislava, 1999, p. 318;322. ISBN 80-227-1228-0.
[24] Zelinka, Ivan, Inverse Fractal Problem By Means of Evolutionary Algorithms. In Mendel '99, 5th International Conference on Soft Computing. Volume 1. Brno : PC-DIR, Brno, 1999, p. 430;435. ISBN 80-214-1131-7.
[25] Lampinen Jouni, Zelinka Ivan, Mixed Integer-Discrete-Continuous optimalization by Differential Evolution 1. In Mendel '99, 5th International Conference on Soft Computing. Volume 1. Brno : PC-Dir, Brno, 1999, 1999, p. 71;76. ISBN 80-214-1131-7.
[26] Lampinen Jouni, Zelinka Ivan, Mixed Integer-Discrete-Continuous optimalization by Differential Evolution 1. In Mendel '99, 5th International Conference on Soft Computing. Volume 1. PC-DIR, Brno,

1999 : PC-Dir, Brno, 1999, p. 77;81. ISBN 80-214-1131-7.

[27] Lampinen Jouni (1999). A Bibliography of Differential Evolution Algorithm. Technical Report. Lappeenranta University of Technology, Department of Information Technology, Laboratory of Information Processing, 16th October 1999. Available via Internet http://www.lut.fi/~jlampine/debiblio.htm .

[28] Original homepage of DE http://www.icsi.berkeley.edu/~storn/code.html

[29] Zelinka I.: Analytic Programming by Means of Soma Algorithm. ICICIS'02, First International Conference on Intelligent Computing and Information Systems, Egypt, Cairo, 2002, ISBN 977-237-172-3

[30] Rektorys, Karel, Variational methods in Engineering Problems and Problems of Mathematical Physics. Czech Ed., 1999, ISBN 80-200-0714-8. English edition is also available on the Internet, see www.amazon.com

[31] Koza J.R. 1998, Genetic Programming, MIT Press, ISBN 0-262-11189-6, 1998

[32] Koza J.R.,Bennet F.H., Andre D., Keane M. 1999, Genetic Programming III, Morgan Kaufnamm pub., ISBN 1-55860-543-6, 1999

[33] Zelinka I., Nolle L., Active Compensation in RF-driven Plasmas by Means of Differential Evolution, in Price K., Storn R., Lampinen J., Differential Evolution: Global Optimisation for Scientists and Engineers, CRC Press, in print

8 Discrete Particle Swarm Optimization, illustrated by the Traveling Salesman Problem

Maurice Clerc

8.1 Introduction

The classical Particle Swarm Optimization is a powerful method to find the minimum of a numerical function, on a continuous definition domain. As some binary versions have already successfully been used, it seems quite natural to try to define a framework for a discrete PSO. In order to better understand both the power and the limits of this approach, we examine in detail how it can be used to solve the well known Traveling Salesman Problem, which is in principle very "bad" for this kind of optimization heuristic. Results show Discrete PSO is certainly not as powerful as some specific algorithms, but, on the other hand, it can easily be modified for any discrete/combinatorial problem for which we have no good specialized algorithm.

8.2 A few words about "classical" PSO

The basic principles in "classical" PSO are very simple. A set of moving particles (the swarm) is initially "thrown" inside the search space. Each particle has the following features.

1. It has a position and a velocity.
2. It knows its position, and the objective function value for this position.
3. It remembers its best previous position found so far.
4. It knows its neighbours' best previous position and objective function value (variant: current position and objective function value). See below for neighbour definition.

From now on, we will consider a particle to be one of its own neighbours, and so 3 and 4 can be combined.

There are many ways to define a "neighbourhood " [1], which is set of particles related to a given one, but we can distinguish two classes.

- "Physical" neighbourhood, which takes distances into account. In practice, distances are recomputed at each time step, which is quite costly, but some clustering techniques need this information.
- "Social" neighbourhood, which just takes "relationships" into account. In practice, for each particle, its neighbourhood is defined as a list of particles at the very beginning, and does not change. Note that, when the process converges, a social neighbourhood becomes a physical one.

At each time step, the behaviour of a given particle is a compromise between three possible choices:

- to follow its own way,
- to go towards its best previous position,
- to go towards the best neighbour's best previous position, or towards the best neighbour (variant).

This compromise is formalized by the following equations:

$$\begin{cases} v_{t+1} = c_1 v_t + c_2 \left(p_{i,t} - x_t\right) + c_3 \left(p_{g,t} - x_t\right) \\ x_{t+1} = x_t + v_{t+1} \end{cases} \tag{8.1}$$

where

v_t := velocity at time step t
x_t := position at time step t
$p_{i,t}$:= best previous position, found so far, at time step t
$p_{g,t}$:= best neighbour's previous best position, at time step t
c_1, c_2, c_3 := social/cognitive confidence coefficients

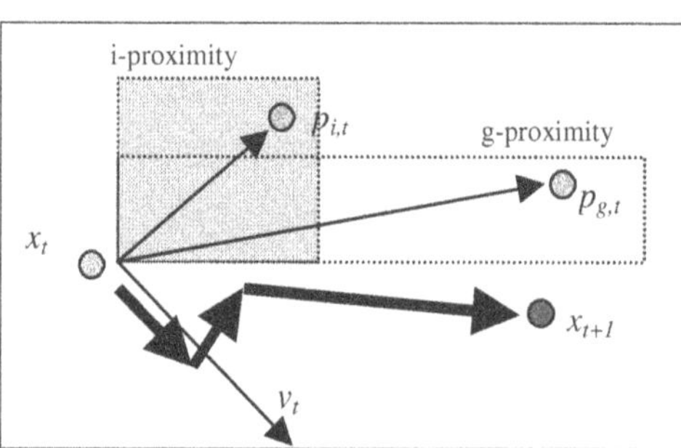

Fig. 1. Weighted combination of three possibles moves

The three social/cognitive coefficients respectively quantify:

- how much the particle trusts *itself now*,
- how much it trusts its *experience*,

- how much it trusts its *neighbours*.

It is important to note that the social/cognitive coefficients are usually randomly chosen within some given intervals, at each time step, and *for each velocity component*. It means a rule like "to go towards its best previous position" should be understood as "to go towards an area (proximity area) around of its best previous position". In classical PSO, these proximity areas are hyperparallelepids.

Of course, much works have been done to study and generalize this method [2, 3, 4, 5, 6, 7, 8]. In particular, in the following discussion, I use a "nohope/rehope" technique as defined in [9], and the convergence criterion proved in [10].

8.3 Discrete PSO

So, what do we really need for using PSO ?

- A search space of positions/states $S = \{s_i\}$
- A cost/objective function *f* on *S*, mapping *S* on a set of values $S \xrightarrow{f} C = \{c_i\}$, and the minimums of *f* are the solution states.
- An order on *C*, or, more generally, a semi-order, so that for every pair of elements of C (c_i, c_j), we can say we have either $c_i < c_j$ or $c_i \geq c_j$.
- If we want to use a physical neighbourhood, we also need a distance *d* in the search space.

Usually, *S* is a real space R^D, and *f* a numerical continuous function. However, Eq. 8.1 does not imply that this must be the case. In particular, *S* may be a finite set of states and *f* a discrete function, and, as soon as you are able to define the following basic mathematical objects and operations, you can use PSO:

- position of a particle
- velocity of a particle
- subtraction $(position, position) \xrightarrow{\text{minus}} velocity$
- external multiplication $(real\_number, velocity) \xrightarrow{\text{times}} velocity$
- addition $(velocity, velocity) \xrightarrow{\text{plus}} velocity$
- move $(position, velocity) \xrightarrow{\text{move (plus)}} position$

To illustrate this assertion, I have written yet-another-traveling-salesman-algorithm. This choice is intentionally very far from the "usual" use of PSO, so that we can have an idea of the power but also of the limits of this approach. The aim is certainly not to compete with powerful dedicated algorithms like LKH [11], but mainly to say something like "If you do not have a specific algorithm for your discrete optimization problem, use PSO: it works". We are in a finite case, so, to anticipate any remark concerning the NFL (No Free Lunch) theorem [12], let us say immediately it does not hold here for at least two reasons: the number of possible objective functions is infinite and the algorithm does use some specific information about the chosen objective function by modifying some parameters (and even, optionally, uses several different objective functions).

Note that some discrete PSO versions have already been defined and used successfully[13, 14], in particular an improved version of this "PSO for TSP" [15].

8.4 PSO elements for TSP

8.4.1 Positions and state space

Let $G = \{E_G, V_G\}$ be the weighted graph in which we are looking for Hamiltonian cycles. E_G is the set of weighted edges (arcs) and V_G the set of vertices (nodes). Graph nodes are numbered from 1 to N, so each element of V_G can be seen as just a label $i, i \in \{1, \ldots, N\}$. Each element of E_G is then a triplet $(i, j, w_{i,j}), i \in V_G, j \in V_G, w_{i,j} \in R^+$. As we are looking for cycles, we can consider just sequences of N+1 nodes, all different, except the last one equal to the first one. Such a sequence is here called a N-cycle and seen as a "position". So the search space is defined as follows: the finite set of all N-cycles.

8.4.2 Objective function

Let us consider a position like $x = (n_1, n_2, \ldots, n_N, n_{N+1}), n_i \in V_G, n_1 = n_{N+1}$. It is "acceptable" only if all arcs (n_i, n_{i+1}) do exist. In the graph, each existing arc has a value. In order to define the "cost" function, a classical way it to just complete the graph, and to create all non existent arcs with an arbitraty value l_{sup} large enough to ensure no solution could contain such a "virtual" arc, for example

$$\begin{cases} l_{sup} > l_{max} + (N-1)(l_{max} - l_{min}) \\ l_{max} = MAX(w_{i,j}) \\ l_{min} = MIN(w_{i,j}) \end{cases} \tag{8.2}$$

So, each arc (n_i, n_{i+1}) has a value, a real one or a "virtual" one. Now, at each position, a possible objective function can be simply defined by

$$f(x) = \sum_{i=1}^{N-1} w_{n_i, n_{i+1}} \tag{8.3}$$

This objective function has a finite number of values and its global minimum is indeed reached at a position (N-cycle) that is the best solution.

8.4.3 Velocity

We want to define an operator v which, when applied to a position during one time step, gives another position. So, here, it is a permutation of N elements, that is to say a list of transpositions. Let $\|v\|$ be the length of this list. A *velocity* is then defined by

$$v = ((i_k, j_k)), i_k \in V_G, j_k \in V_G, k = 1, \ldots, \|v\| \tag{8.4}$$

or, in short $v = ((i_k, j_k))$, which means "exchange nodes (i_1, j_1), then nodes (i_2, j_2), etc. and at last nodes $(i_{\|v\|}, j_{\|v\|})$". Note that we need two parentheses, for it is a list of pairs.

Two such different lists can generate the same result when applied independently to any position. We call such velocities *equivalent* and use the notation $\cong$ to express it. Thus, if v_1 and v_2 are equivalent, we say $v_1 \cong v_2$. For example, we have $((1,3),(2,5)) \cong ((2,5),(1,3))$. In fact, in this example, they are not only equivalent, they are opposite (see below): when using velocities to move on the search space, this one is like a « sphere »: you can reach the same point following two opposite paths. A null velocity is a velocity equivalent to $\varnothing$, the empty list.

8.4.4 Opposite of a velocity

It is defined by

$$\neg v = \left(\left(i_{\|v\|-k+1}, j_{\|v\|-k+1}\right)\right) \tag{8.5}$$

This formula means "to do the same transpositions as in v, but in reverse order". It is easy to verify that we have $\neg\neg v = v$ (and $v \oplus \neg v \cong \varnothing$, see below Addition "velocity plus velocity").

8.4.5 Move (addition) "position *plus* velocity"

Let x be a position and v a velocity. The position $x' = x + v$ is found by applying the first transposition of v to x, then the second one to the result etc.

8.4.5.1 Example

$$\begin{cases} x = (1,2,3,4,5,1) \\ v = ((1,2),(2,3)) \end{cases} \tag{8.6}$$

Applying v to x, we obtain successively

$$\begin{matrix}(2,1,3,4,5,2)\\(3,1,2,4,5,3)\end{matrix} \tag{8.7}$$

8.4.6 Subtraction "position *minus* position"

Let x_1 and x_2 be two positions. The difference $x_2 - x_1$ is defined as the velocity v, found by a given algorithm, so that applying v to x_1 gives x_2. The condition "found by a given algorithm" is necessary, for, as we have seen, two velocities can be equivalent, even when they have the same size. In particular, the algorithm is chosen so that we have $x_2 - x_1 = \neg(x_1 - x_2)$, and $x_1 = x_2 \Rightarrow v = x_2 - x_1 = \varnothing$.

8.4.7 Addition "velocity *plus* velocity"

Let v_1 and v_2 be two velocities. In order to compute $v_1 \oplus v_2$ we define the list of transpositions which contains first the elements (pairs, or transpositions) of v_1, followed by the elements of v_2. Optionally, we "contract" it to obtain a smaller equivalent velocity. In particular, this operation is defined so that $v \oplus \neg v = \varnothing$. Then, we have $\|v_1 \oplus v_2\| \leq \|v_1\| + \|v_2\|$, but we usually do *not* have $v_1 \oplus v_2 = v_2 \oplus v_1$.

8.4.8 Multiplication "coefficient *times* velocity"

Let c be a real coefficient and v be a velocity. There are several cases, depending on the value of c.

8.4.8.1 Case c = 0

We have $cv = \varnothing$.

8.4.8.2 Case c ∈]0,1]

We just "truncate" v. Let $\|cv\|$ be the greatest integer smaller than or equal to $c\|v\|$. So we define $cv = ((i_k, j_k)), k = 1, \ldots, \|cv\|$.

8.4.8.3 Case c > 1

It means we have $c = k + c', k \in \mathbf{N}^*, c' \in [0,1[$. So we can define $cv = \underbrace{v \oplus v \oplus \ldots \oplus v}_{k \text{ times}} \oplus c'v$.

8.4.8.4 Case c < 0

By writing $cv = (-c) \neg v$, we just have to consider one of the previous cases. Note that we have $v_1 \cong v_2 \Rightarrow cv_1 \cong cv_2$ if c is an integer, but it is usually not true in the general case.

8.4.9 Distance between two positions

Let x_1 and x_2 be two positions. The distance between these positions is defined by $d(x_1, x_2) = \|x_2 - x_1\|$. Note: it is a metric, for we do have (x_3 is any third position):

$$\|x_2 - x_1\| = \|x_1 - x_2\|$$
$$\|x_2 - x_1\| = 0 \Leftrightarrow x_1 = x_2$$
$$\|x_2 - x_1\| \leq \|x_2 - x_3\| + \|x_3 - x_1\|$$

8.5 The algorithm "PSO for TSP". Core and variations

8.5.1 Equations

We can now rewrite Eq. 8.1 as follows:

$$\begin{cases} v_{t+1} = c_1 v_t \oplus c_2 (p_{i,t} - x_t) \oplus c_3 (p_{g,t} - x_t) \\ x_{t+1} = x_t + v_{t+1} \end{cases} \tag{8.8}$$

In practice, we assume $c_3 = c_2$. No experiment has found that a different choice gives significantly better result on some class of problems. More precisely, for a given problem, you may find a better choice after a lot of trials, but there is no known rule to find it "in advance". Therefore, choosing $c_3 = c_2$ doesn't change the algorithm in any significant way. By defining an intermediate position $p_{ig,t} = p_{i,t} + (p_{g,t} - p_{i,t})/2$, we finally use the following system:

$$\begin{cases} v_{t+1} = c_1 v_t \oplus c_2' \left(p_{ig,t} - x_t\right) \\ x_{t+1} = x_t + v_{t+1} \end{cases} \tag{8.9}$$

The advantage of this equation is that it can be used as a guideline for choosing "good" coefficients [10], even if the proofs given in this article is applied to differentiable systems. It is indeed possible to use it just like that, but with such a straightforward and simple approach, the swarm size may need be quite big to avoid getting stuck at local minimums. Thus, some modifications of the core algorithm are quite helpful to improve performances. In particular, the No-Hope/ReHope process can be very powerful, which we describe in the next section.

8.5.2 NoHope tests

8.5.2.1 Criterion 0

If a particle has to move "towards" another one which is at distance 1, either it does not move at all, or it goes to exactly at the same position as the second one, depending on the social/confidence coefficients. When the swarm size gets smaller than $N(N-1)$, it may happen that all moves computed according to Eq. 8.9 are null. In this case there is absolutely no hope to improve the current best solution.

8.5.2.2 Criterion 1

The NoHope test defined in [9] is "Is the swarm to small?". This demand a computation of the swarm diameter at each time step, which is expensive. However, in a discrete case like the one being presented here, as soon as the distance between two particles tend to become "too small", the particles become identical (usually first by positions and then by velocities). So, at each time step, a "reduced" swarm is computed, in which all particles are different, which is not very expensive, and the NoHope test becomes "Is the swarm too reduced?", say by 50%.

8.5.2.3 Criterion 2

Another criterion has been added: "Is the swarm too slow?". This is done by comparing the velocities of all particles to a threshold, either individually or globally. In one version of the algorithm, this threshold is in fact modified at each time step, according to the best result obtained so far and to the statistical distribution of arc values.

8.5.2.4 Criterion 3

Another very simple criterion is defined: "No improvement for too many time steps". However, in practice, it appears that criteria 1 and 2 are sufficient.

8.5.3 ReHope process

As soon as there is "no hope", the swarm is re-expanded. The first two methods used here are inspired by the ones described in [9] and [16].

8.5.3.1 Lazy Descent Method (LDM)

Each particle goes back to its previous best position and, from there, moves randomly and slowly ($\|v\| = 1$) and stops *as soon as* it finds a better position or when a maximal number of moves M_{max} is reached (in the examples below $M_{max} = N$). If the current swarm is smaller than the initial one, it is completed by a new set of randomly chosen particles.

8.5.3.2 Energetic Descent Method (EDM)

Each particle goes back to its previous best position and, from there, moves slowly ($\|v\| = 1$) *as long as* it finds a better position in at most M_{max} moves. If the current swarm is smaller than the initial one, it is completed by a new set of particles randomly chosen. You may find, by chance, a solution, but it is more expensive than LDM. Ideally, it should be used only if the combination Eq. 8.9 + LDM seems to fail.

8.5.3.3 Local Iterative Levelling (LIL)

This method is more powerful … and more expensive. In practice, it should be used only when we *know* there is a better solution, but the combination Eq. 8.9 + EDM fails to find it.

The idea is as follows. There are infinitely possible objective functions with the same global minimum, so we can locally and temporarily use any of them to guide a particle. For each immediate physical neighbour y (at distance 1) of the particle x, a temporary objective function value $f_t(y)$is computed by using the following algorithm.

LIL algorithm

- Find all neighbours at distance 1.
- Find the best one, y_{min}.

- Assign to y the temporary objective function value $f_t(y) = (f(y_{\min}) + f(x))/2$ and, according to these temporary evaluations, x moves to its best immediate neighbour.

In TSP, the cost of such an algorithm is $O(N^2)$. If repeated too often, it considerably increases the total cost of the process. That is why it should be used only if there is really "no hope", and if you do want the best solution. In practice, the algorithm will usually find a good one, even without using LIL.

8.5.4 Adaptive ReHope Method (ARM)

The four methods described above (NoHope ReHope/LDM/EDM/LIL) can be used automatically in an adaptive way, depending on how many time steps has passed from the last improvement of the solution. A possible strategy is given in the following table.

Table 8.1. A possible ReHope strategy

Number of time steps without improvement	ReHope type
0-1	No ReHope
2-3	Lazy Descent Method
4	Energetic Descent Method
>4	Local Iterative Levelling

8.5.5 Queens

Instead of using the best neighour/neighbour's previous best for each particle, we can use an extra particle, which "summarizes" the neighbourhood. This method is a combination of the (unique) queen method defined in [9] and the multi-clustering method described in [17]. For each neighbourhood, we iteratively build a centroid and take it as the best neighbour ($p_{g,t}$ in Eq. 8.8).

8.5.6 Extra-best particle

In order to speed up the process, the algorithm can also use a special extra particle that stores the best position found so far. It is not absolutely necessary, for this position is also memorized as "previous best" in at least one particle, but it may avoid a whole sequence of iterations between two ReHope processes.

8.5.7 Parallel and sequential versions

The algorithm can run either in (simulated) parallel mode or in sequential mode. In the parallel mode, at each time step, new positions are computed for all particles and then the swarm is globally moved. In sequential mode, each particle is moved at a time: particle 1, then particle 2, ... then particle N, then again particle 1, etc. Note that Eq. 8.9 implicitly supposes a parallel mode, but in practice there is no clear difference in performances on a given sequential machine, and the second method is a bit less expensive.

8.6 Examples and results

8.6.1 Parameters choice

The purpose of this paper is mainly to show how PSO can be used, in principle, on a discrete problem. For that reason, we use some default values for the parameters. Of course, a better set of parameters can certainly be found by "optimizing the optimization" (in fact, PSO itself may be used to find an optimum in a parameter space).

8.6.1.1 Social/cognitive coefficients

If nothing else is explicitly mentioned, in all the examples we use $c_1 \in]0,1[$ (typically 0.5 if NoHope/ReHope is used or 0.999 if not). c_2 is randomly chosen at each time step in $[0,2]$. The convergence criterion defined in [10] is satisfied. This criterion is proved only for differentiable objective functions, but it is not unreasonable to think it should work here too. At least, in practice, it does.

8.6.1.2 Swarm size and neighbourhood size

Ideally, the best swarm and neighbourhood sizes depend on the number of local minimums of the objective function, say $n_{\text{local\_min}}$, so that the swarm can have sub-swarms around each of these minimums. Usually, we simply do not have this information, except the fact that there are at most N-1 such minimums. A possible way is to use a very small swarm at the beginning, just to have an idea of the landscape defined by the objective function (see the example below). Also, a statistical estimation of $n_{\text{local\_min}}$ can be made, using the arc values distribution. In practice, we usually use a swarm size S equal to N-1 (see below Structured Search Map).

Some considerations, not completely formalized, about the fact that each particle uses three velocities to compute a new one, indicate that a good neighbourhood size should be simply 4 (including the particle itself).

But all this is still "rules of thumb", and this is the main limitation of the approach: on a given example, we do not know in advance what are the best parameters and usually have to try different values. That is why, in particular, some adaptive PSO versions are now under development.

8.6.1.3 Performance criteria

Normally, the performance measures are reported in literature with too specific data. For example, "The program took 3 seconds on XYZ machine to solve instance xxxx from TSPLIB". Unfortunately, such data is not easy to compare with another performance measure which uses machine ZYX. Thus, we think it is better to have a measure like the following: "it needed 7432 tour evaluations to reach a solution".

However, completely compute the objective function value on a *N*-cycle or to update it after having swapped two nodes is not all the same: the first case is about *N*/4 times more expensive. So we use two criteria: the number of position evaluations *and* the number of arithmetic/logical operations. It is still not perfect, for the same algorithm can be written in different ways even in the same language, one more efficient, and the other less. However, if you know your machine is a xxx Mips computer, you can estimate how long it would take to run the example.

8.6.2 A toy example as illustration

In this example, we use here a very simple graph (17 nodes) from TSPLIB, called br17.atsp (cf. Appendix). The minimal objective function value is 39, and there are several solutions. Although it has only a few nodes, it is designed so that finding one of the best tours is not so easy, so it is nevertheless an interesting example.

8.6.2.1 What the landscape looks like

We define randomly a small swarm of, say, five particles, where p_{best} is the best one. For each other particle p_{i}, we now consider the linear sequence $p_k = p_i + k\left(p_{best} - p_1\right)/(N-1), k = 0,\ldots, N-1$, and plot the graph $\left(k/(N-1), f\left(p_{best}\right) - f\left(p_k\right)\right)$. It give us an idea of the landscape (see Fig. 2 and Fig. 3). The landscape is clearly quite chaotic, so we will surely need to use the NoHope/ReHope process quite often. Also, the Queen option is probably not efficient (a centroid does not give a better position).

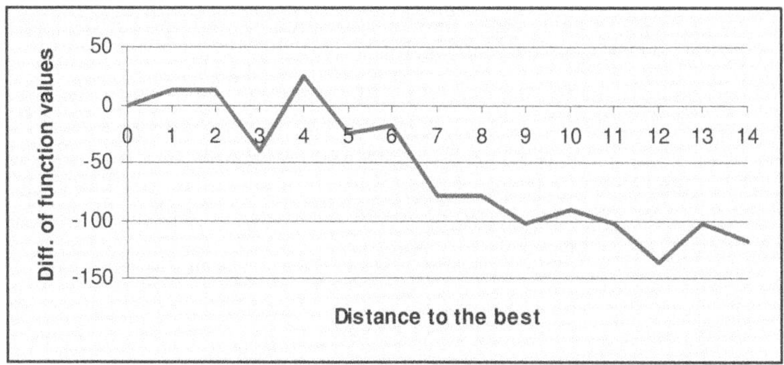

Fig. 2. One section of the search space landscape

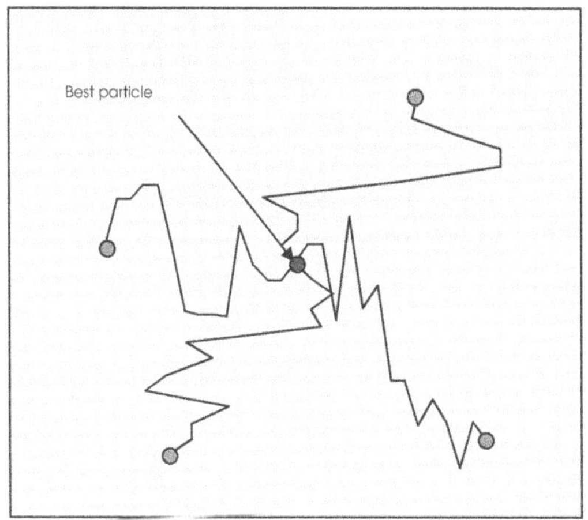

Fig. 3. Main sections for a 5-swarm

8.6.2.2 Structured Search Map. How the swarm moves

Let us suppose we know a solution x_{sol}. We consider all possible positions in the search space, and plot the map $(d(x_{sol},x),f(x))$. On this map, at each time step, we highlight the positions of the particles, so that we can see how it moves. We define some equivalence classes according to the equivalence relation $x \cong x' \Leftrightarrow d(x_{sol},x) = d(x_{sol},x')$. As we can see, not only the swarm globally moves towards the solution, but some particles also tend to move towards the minimum (in terms of objective function value) of the equivalence classes. It means that even if it does not find the best solution, it finds at least (and usually quite rapidly) interesting quasi-solutions. It also means that if the swarm is big enough, we may simultaneously find several solutions (this property is used in multiple objective optimization [18, 19]).

In our example, as the number of positions is quite big ($16! \approx 21^{12}$), we plot only a few of them (4700), as the "background" of the diagram sequence of Fig. 4 and 5. We clearly see that the example has been designed to be a bit difficult, for there is a "gap" between distances 4 and 8. Here, we use a swarm of 16 particles, and it finds three solutions in one "standard" run, and some others if we modify the parameters (see Appendix).

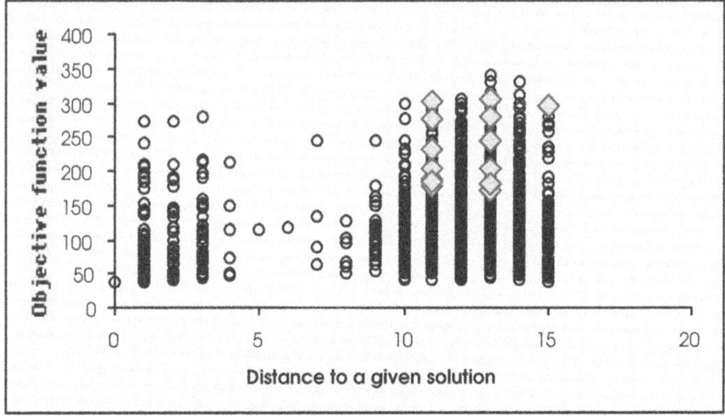

Fig. 4. Initial position on the Structured Search Map (swarm size = 16)

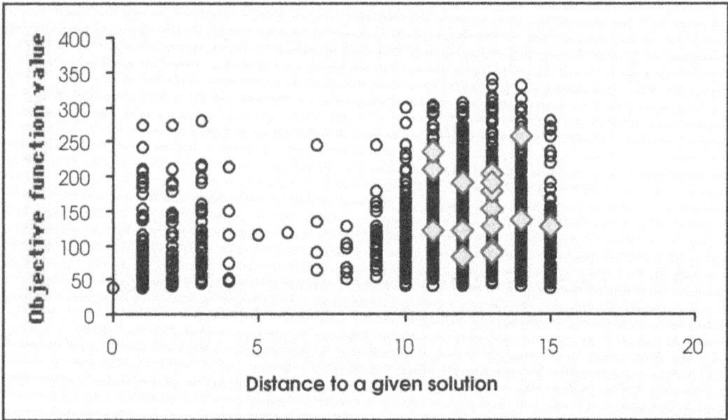

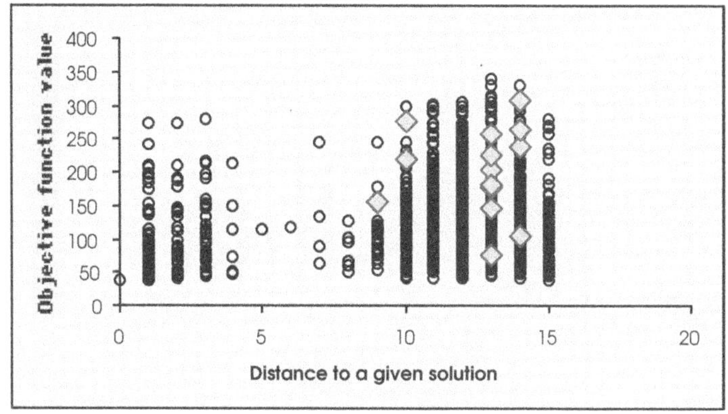
Objective function value
400
350
300
250
200
150
100
50
0
0
5
10
15
20
Distance to a given solution

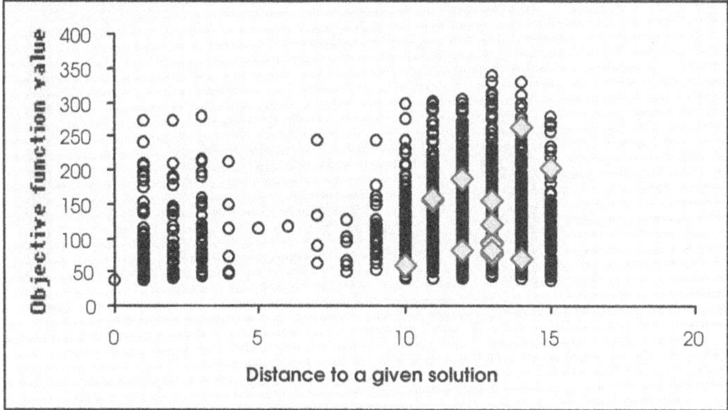
Objective function value
400
350
300
250
200
150
100
50
0
0
5
10
15
20
Distance to a given solution

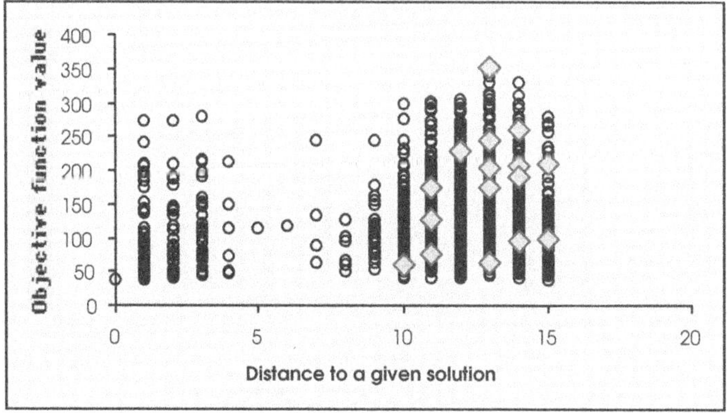
Objective function value
400
350
300
250
200
150
100
50
0
0
5
10
15
20
Distance to a given solution

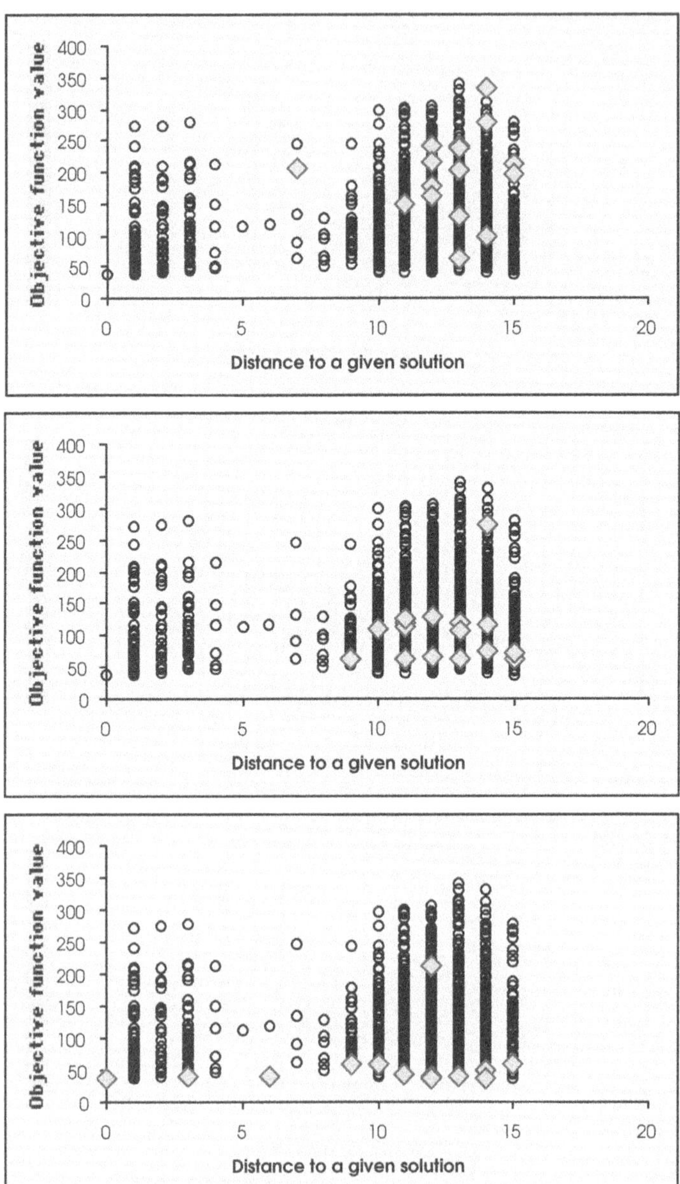

Fig. 5. How the swarm moves on the Structured Search Map (some times step between 1 and 480). It takes some times to jump the gap betwee distance 9 and 4

8.6.3 Some results, and discussion

Table 8.2. Some result using Discrete PSO on graph br17.atsp

Swarm size	Hood size	c_1	Random c_2	ReHope type	Best solution
16	4	0.500	]0,2]	ARM	39 (3 solutions)
16	4	0.500	]0,2]	ARM	39
16	4 with queens	0.500	]0,2]	ARM	39 (3 solutions)
8	4	0.500	]0,2]	ARM	39
1	1	0.500	]0,2]	ARM	44
128	4	0.999	]0,2]	no	47
32	4	0.999	]0,2]	no	66
16	4	0.999	]0,2]	no	86

Table 8.2 (cont.)

Hood type	Tour evaluations	Arithmetical/logical operations
social	7990	4.8 M
physical	7742	6.3 M
social or physical	9051	5.8 M
social	4701	2.8 M
social	926 to ∞	0.6 M
social	41837 to ∞	33.2 M to ∞
social	14351 to ∞	10.8 M to ∞
social	2880 to ∞	1.9 M to ∞

As we can see from Table 8.2:

- Using "physical" neighbourhood may need less tour evaluations, but is much more expensive than "social" option, if we consider the total number of arithmetical/logical operations (distances have to be recaculated at each time step). The fact that with social neighbourhood you find more solutions is not a general rule.
- Using only ReHope methods (in this case, using s particles for T time steps is equivalent to using just one particle for sT time steps) is not enough to reach the best solution.
- Using only core algorithm, with no ReHope at all, is not enough to reach the best solution, unless, probably, you use a huge swarm.
- Using queens is not interesting.

Results are not particularly good nor bad, mainly because we use a generic ReHope method which is not specifically designed for TSP. However, this is done on purpose. This way, by just changing the objective function, you could use Discrete PSO for different problems.

Appendix

Graph br17.atsp (from TSPLIB)

In the original data, diagonal values are equal to 9999. Here, they are equal to 0. For PSO algorithm, it changes nothing, and it simplifies the visualization program.
NAME: br17
TYPE: ATSP
COMMENT: 17_city_problem_(Repetto)_Opt.:_39
DIMENSION: 17
EDGE_WEIGHT_TYPE: EXPLICIT
EDGE_WEIGHT_FORMAT: FULL_MATRIX
EDGE_WEIGHT_SECTION

```
 0  3  5 48 48  8  8  5  5  3  3  0  3  5  8  8  5
 3  0  3 48 48  8  8  5  5  0  0  3  0  3  8  8  5
 5  3  0 72 72 48 48 24 24  3  3  5  3  0 48 48 24
48 48 74  0  0  6  6 12 12 48 48 48 48 74  6  6 12
48 48 74  0  0  6  6 12 12 48 48 48 48 74  6  6 12
 8  8 50  6  6  0  0  8  8  8  8  8  8 50  0  0  8
 8  8 50  6  6  0  0  8  8  8  8  8  8 50  0  0  8
 5  5 26 12 12  8  8  0  0  5  5  5  5 26  8  8  0
 5  5 26 12 12  8  8  0  0  5  5  5  5 26  8  8  0
 3  0  3 48 48  8  8  5  5  0  0  3  0  3  8  8  5
 3  0  3 48 48  8  8  5  5  0  0  3  0  3  8  8  5
 0  3  5 48 48  8  8  5  5  3  3  0  3  5  8  8  5
 3  0  3 48 48  8  8  5  5  0  0  3  0  3  8  8  5
 5  3  0 72 72 48 48 24 24  3  3  5  3  0 48 48 24
 8  8 50  6  6  0  0  8  8  8  8  8  8 50  0  0  8
 8  8 50  6  6  0  0  8  8  8  8  8  8 50  0  0  8
 5  5 26 12 12  8  8  0  0  5  5  5  5 26  8  8  0
```

Some solutions found by Discrete PSO

Each line of Table 8.3 gives the sequence of nodes of a minimum tour (common total weight=39). For example, according to the above matrix, the first tour gives the following weights: 5+0+3+0+0+0+8+0+0+0+6+0+12+0+0+5+0=39.

Table 8.3. Some solutions for br17.atsp

3	14	11	13	2	10	15	7	6	16	4	5	9	17	8	12	1
6	7	15	16	5	4	17	9	8	10	2	13	11	14	3	12	1
7	6	15	16	5	4	9	8	17	10	11	13	2	14	3	12	1
9	8	17	5	4	15	6	7	16	13	10	11	2	3	14	12	1
12	3	14	10	2	13	11	17	8	9	5	4	7	6	16	15	1
12	3	14	10	11	2	13	17	8	9	4	5	16	6	15	7	1
12	6	7	15	16	5	4	9	17	8	2	11	13	10	14	3	1
12	8	9	17	5	4	16	15	7	6	10	2	11	13	14	3	1
12	14	3	2	10	11	13	17	9	8	5	4	15	6	7	16	1
12	14	3	11	2	10	13	15	6	16	7	4	5	9	8	17	1
12	15	16	7	6	5	4	8	17	9	11	10	2	13	14	3	1
12	16	15	7	6	4	5	9	8	17	10	2	11	13	14	3	1
12	16	15	7	6	5	4	8	9	17	10	11	2	13	14	3	1
12	16	7	6	15	5	4	9	17	8	11	10	13	2	14	3	1
12	17	8	9	5	4	15	16	7	6	2	10	11	13	3	14	1
12	17	8	9	5	4	15	16	7	6	2	10	11	13	3	14	1
14	3	10	13	11	2	17	8	9	4	5	15	6	16	7	12	1
14	3	10	13	11	2	8	17	9	4	5	15	6	7	16	12	1
14	3	10	11	2	13	8	17	9	4	5	15	7	6	16	12	1
16	6	7	15	4	5	9	8	17	13	11	10	2	3	14	12	1
16	6	7	15	4	5	9	8	17	13	10	2	11	3	14	12	1
17	9	8	4	5	7	6	16	15	10	11	2	13	14	3	12	1
12	14	3	2	10	13	11	16	6	7	15	4	5	17	9	8	1

References

[1] Kennedy J., "Small Worlds and Mega-Minds: Effects of Neighborhood Topology on Particle Swarm Performance", *Congress on Evolutionary Computation*, Washington D.C., 1999, p. 1931-1938.

[2] Angeline P. J., "Using Selection to Improve Particle Swarm Optimization", *IEEE International Conference on Evolutionary Computation*, Anchorage, Alaska, May 4-9, 1998, p. 84-89.

[3] Eberhart R. C., Kennedy J., "A New Optimizer Using Particle Swarm Theory", *Sixth International Symposium on Micro Machine and Human Science*, Nagoya, Japan, 1995, p. 39-43.

[4] Kennedy J., Eberhart R. C., "Particle Swarm Optimization", *IEEE International Conference on Neural Networks*, Perth, Australia, 1995, p. 1942-1948.

[5] Kennedy J., "The Particle Swarm: Social Adaptation of Knowledge", *International Conference on Evolutionary Computation*, Indianapolis, Indiana, 1997, p. 303-308.

[6] Kennedy J., "The behavior of particles", *Evolutionary VII*, San Diego, CA, 1998, p. 581-589.

[7] Shi Y., Eberhart R. C., "Parameter Selection in Particle Swarm Optimization", *Evolutionary Programming VII*, 1998,

[8] Shi Y. H., Eberhart R. C., "A Modified Particle Swarm Optimizer", *International Conference on Evolutionary Computation*, Anchorage, Alaska, May 4-9, 1998, p. 69-73.

[9] Clerc M., "The Swarm and the Queen: Towards a Deterministic and Adaptive Particle Swarm Optimization", *Congress on Evolutionary Computation*, Washington DC, 1999, p. 1951-1955.

[10] Clerc M., Kennedy J., "The Particle Swarm-Explosion, Stability, and Convergence in a Multidimensional Complex Space", *IEEE Transactions on Evolutionary Computation*, vol. 6, 1, 2002, p. 58-73.

[11] Helsgaun K., An Effective Implementation of the Lin-Kernighan Traveling Salesman Heuristic, Department of Computer Science,Roskilde University, Denmark, 1997.

[12] Wolpert D. H., Macready W. G., No Free Lunch for Search, The Santa Fe Institute, 1995.

[13] Kennedy J., Eberhart R. C., "A discrete binary version of the particle swarm algorithm", *Conference on Systems, Man, and Cybernetics*, 1997, p. 4104-4109.

[14] Yoshida H., Kawata K., Fukuyama Y., "A Particle Swarm Optimization for Reactive Power and Voltage Control considering Voltage Security Assessment", *IEEE Trans. on Power Systems*, vol. 15, 4, 2001, p. 1232-1239.

[15] Secrest B. R., Lamont G. B., "Communication in Particle Swarm Optimization Illustrated by the Travelling Salesman Problem", *Workshop on Particle Swarm Optimization*, Indianapolis, IN: Purdue School of Engineering and Technology, 2001,

[16] He Z., Wei C., Jin B., Pei W., Yang L., "A New Population-based Incremental Learning Method for the Traveling Salesman Problem", *Congress on Evolutionary Computation*, Washington D.C., 1999, p. 1152-1156.

[17] Kennedy J., "Stereotyping: Improving Particle Swarm Performance With Cluster Analysis", *Congress on Evolutionary Computation,*, 2000, p. 1507-1512.

[18] Coello Coello C. A., Toscano Pulido G., Lechuga M. S., Handling Multiple Objectives with Particle Swarm Optimization, EVOCINV-02-2002, CINVESTAV, Evolutionary Computation Group, 2002.

[19] Hu X., Eberhart R. C., "Multiobjective Optimization Using Dynamic Neighborhood Particle Swarm Optimization", *Congress on Evolutionary Computation (CEC'2002)*, Piscataway, New Jersey, 2002, p. 1677-1681.

9 Applications in Heat Transfer

B V Babu

9.1 Introduction

In this chapter, we shall discuss two successful applications of Genetic Algorithms (GA) and Differential Evolution (DE) on heat transfer problems: (1) Estimation of heat transfer parameters in trickle bed reactors, and (2) Optimal design of shell-and-tube heat exchanger.

In the first problem, a new non-sequential technique is proposed for the estimation of *effective* heat transfer parameters using radial temperature profile measurements in a gas–liquid co-current downflow through packed bed reactors (often referred to as trickle bed reactors). Orthogonal collocation method combined with a new optimization technique, differential evolution (DE) is employed for estimation. DE is an exceptionally simple, fast and robust, population based search algorithm that is able to locate near-optimal solutions to difficult problems. The results obtained from this new technique are compared with that of radial temperature profile (RTP) method. Results indicate that orthogonal collocation augmented with DE offer a powerful alternative to other methods reported in the literature. The proposed technique takes less computational time to converge when compared to the existing techniques without compromising with the accuracy of the parameter estimates. This new technique takes on an average 10 s on a 90 MHz Pentium processor as compared to 30 s by the RTP method. This new technique also assures of convergence from any starting point and requires less number of function evaluations.

The second problem presents the application of Differential Evolution (DE) for the optimal design of shell-and-tube heat exchangers. The main objective in any heat exchanger design is the estimation of the minimum heat transfer area required for a given heat duty, as it governs the overall cost of the heat exchanger. Lakhs of configurations are possible with various design variables such as outer diameter, pitch, and length of the tubes; tube passes; baffle spacing; baffle cut etc. Hence the design engineer needs an efficient strategy in searching for the global minimum. In the present study for the first time DE, an improved version of Genetic Algorithms (GAs), has been successfully applied with different strategies for 1,61,280 design configurations using Bell's method to find the heat transfer area. In the application of DE 9680 combinations of the key parameters are considered. For comparison, GA is also applied for the same case study with 1080 combinations of its parameters. For this optimal design problem, it is found that DE, an exceptionally simple evolution strategy, is significantly faster compared to GA and yields the global optimum for a wide range of the key parameters.

Since their inception three decades ago, Genetic Algorithms (GA) have evolved like the species they try to mimic (Goldberg, 1989). Just as competition drives each species to adapt to a particular environmental niche, so too, has the pressure to find efficient solutions across the spectrum of real-world problems forced genetic algorithms to diversify and specialize (Davis, 1991; Moros *et al.*, 1996; Wolf and Moros, 1997; Chakraborti and Sastry, 1997, 1998; Babu and Mohiddin, 1999; Babu and Vivek, 1999). Differential Evolution (Price and Storn, 1997; Babu, 2001) is a search procedure similar to a GA applied on real variables, which is significantly fast at numerical optimization and is also more likely to find a function's true global optimum. DE is in similar to a real coded GA combined with an adaptive random search (ARS) (Boender and Romeijn, 1995; Maria, 1998) with a normal random generator. Among DE's advantages are its simple structure, ease of use, speed and robustness.

The various applications of DE include: digital filter design (Storn, 1995), fuzzy decision making problems of fuel ethanol production (Wang *et al.*, 1998), Design of fuzzy logic controller (Sastry *et al.*, 1998), batch fermentation process (Chiou and Wang, 1999; Wang and Cheng, 1999), multi sensor fusion (Joshi and Sanderson, 1999), dynamic optimization of continuous polymer reactor (Lee *et al.*, 1999), estimation of heat transfer parameters in trickle bed reactor (Babu and Sastry, 1999), optimization of alkylation reaction (Babu and Chaturvedi, 2000), optimal design of heat exchangers (Babu and Mohiddin, 1999; Babu and Munawar, 2000; Babu and Munawar, 2001), synthesis & optimization of heat integrated distillation system (Babu and Singh, 2000), optimization of non-linear functions (Babu and Angira, 2001a), scenario- integrated optimization of dynamic systems (Babu and Gautam, 2001), optimization of thermal cracking operation (Babu and Angira, 2001b), determining the number of components in mixtures of linear models (Dollena *et al.*, 2001), Identification of hysteretic systems using the differential evolution algorithm (Kyprianou *et al.*, 2001), optimization of Low Pressure Chemical Vapour Deposition Reactors Using Hybrid Differential Evolution (Lu and Wang, 2001), hybrid differential evolution for problems of Kinetic Parameter Estimation and Dynamic Optimization of an Ethanol Fermentation Process (Wang *et al.*, 2001), optimal design of auto-thermal ammonia synthesis reactor (Babu *et al.*, 2002), global optimization of MINLP problems (Babu and Angira, 2002a; Angira and Babu, 2003), optimization of non-linear chemical processes (2002b), etc. Upreti and Deb (1997) used GAs to optimize the length of ammonia reactor by solving coupled differential equations. They used binary strings to code the parameters; however this choice limits the resolution with which an optimum can be located to the precision set by the number of bits in the integer (Davis, 1991; Price and Storn, 1997; Wolf and Moros, 1997). Wolf and Moros (1997) encoded a floating point codes into mantissa, exponent and sign of exponent; however the string encoding can be completely circumvented by using the floating point codes directly. DE uses floating-point numbers that are more appropriate than integers for representing points in continuous space, which not only uses computer resources efficiently but also reduces the computational time which is very crucial for estimation problems. Upreti and Deb (1997) used bitwise mutation (logical exclusive OR, XOR). However addition is a more appropriate choice for mutation to search the continuum. Consider, for example, the consequences of using the XOR operator for mutation. To change a binary 15 (01111) into a binary 16 (10000) with an XOR operation requires inverting all the five bits. In most bit flipping schemes, a mutation of this magnitude is very rare even though the mutation operator is meant for fine tuning. The easiest way to restore the adjacency of neighboring points is addition. Using addition, 15 becomes 16 by just adding 1. Wolf and Moros (1997) divided the mutation probability into mantissa, exponent and the sign mutation probabilities and altered the selected segments randomly. The magnitude of alteration used by them was fixed, whereas, the magnitude of the mutation increments are automatically scaled by the simple adaptive scheme used by DE. The overall structure of the differential evolution (DE) algorithm developed by Price and Storn (1997) resembles that of most other population-based searches. Generally, a GA is used for maximizing a criterion, whereas DE is used for minimizing a criterion. DE utilizes NP parameter vectors of dimension D,

$$\mathbf{x}_{i,\Gamma}, \quad i = 0, 1, 2, \ldots\ldots\ldots, NP - 1 \tag{9.1}$$

as a population for each generation Γ. NP remains constant during the optimization process. The initial population is chosen randomly if nothing is known about the system. If a preliminary solution is available, the initial population is often generated by adding normally distributed random deviations to the nominal solution $x_{nom,0}$. The crucial idea behind DE is a new scheme for generating trial parameter vectors. DE generates new parameter vectors by adding the weighted difference vector between two population members to a third member. Extracting distance and direction information from the population to generate random deviations results in an adequate scheme with excellent convergence properties. The two operators used are *Mutation* and *Recombination*. For each vector, $\mathbf{x}_{i\Gamma}$, a trial vector $\mathbf{v}$ is generated according to,

$$\mathbf{v} = \mathbf{x}_{i,\Gamma} + \lambda \cdot \left(\mathbf{x}_{best,\Gamma} - \mathbf{x}_{i,\Gamma}\right) + F \cdot \left(\mathbf{x}_{r_1,\Gamma} - \mathbf{x}_{r_2,\Gamma}\right) \tag{9.2}$$

with

$$r_1,\ r_2 \in \left[0, NP - 1\right] \tag{9.3}$$

integer and mutually different, $\lambda > 0$, and $F > 0$.

The integers r_1 and r_2 are chosen randomly from the interval [0, NP-1] and are different from the running index i. F is a real and constant factor that controls the amplification of the differential variation. The idea behind using λ is to provide a means to enhance the greediness of the scheme by incorporating the current best value $x_{best,\Gamma}$. Recombination (crossover) provides an alternative and complementary means of creating viable vectors from the components of existing vectors. Previous studies on GA for solving PDEs (Upreti and Deb, 1997; Wolf and Moros, 1997) used a uniform crossover, however, a DE uses a nonuniform crossover that can take child vector parameters from one parent more often than it does from the other. In order to increase the diversity of the parameter vectors, the vector $\mathbf{u}_{i,\Gamma}$ with

$$\left(\mathbf{u}_{i,\Gamma}\right)_j = \begin{cases} v_j & \text{for } j = \langle n \rangle D, \langle n+1 \rangle D, ..., \langle n+M-1 \rangle D, \\ \left(x_{i,\Gamma}\right)_j & \text{otherwise} \end{cases} \tag{9.4}$$

is formed where the acute brackets $< >_p$ denote the modulo function with modulus number equal to dimension $\boldsymbol{D}$. A certain sequence of the vector elements of u is identical to the elements of **v**, the other elements of u acquire the original values of $\mathbf{x}_{i\Gamma}$. Choosing a subgroup of parameters for mutation is similar to a process known as *crossover* in evolution theory. The integer M is drawn from the interval $[0, \boldsymbol{D}\text{-}1]$ with the probability $Pr\ (\boldsymbol{M}=\gamma) = (CR)^{\gamma}$, $CR\ \varepsilon\ [0, 1]$ is the crossover probability and constitutes a control variable for the above-mentioned scheme. The random decisions for both n and $\boldsymbol{M}$ are made new for each trial vector **v**. Unlike many GAs, DE does not use proportional selection, ranking or even an annealing criterion that would allow occasional uphill moves. Instead the cost of each trial vector is compared to that of its parent target vector. The vector with the lower cost is rewarded by being selected to the next generation. This selection of the individuals to the next generation resembles *tournament selection* except that each child that is pitted against one of its parents, not against a randomly chosen competitor. If the resulting child vector yields a lower objective function value than a predetermined population member, the child vector replaces the parent vector with which it was compared. The comparison vector can, but need not, be a part of the generation process mentioned above. In addition, the best parameter vector $x_{best,\Gamma}$ is evaluated for every generation C in order to keep track of the progress that is made during the minimization process.

9.2 Heat Transfer Parameters in Trickle Bed Reactor

Trickle-bed reactors are widely used in petroleum and petro-chemical industries, and to a lesser extent in chemical and pharmaceutical industries. They also have a potential application in waste water treatment and in bio-chemical reactions. Various flow regimes such as trickle flow (at low liquid and low gas rates), pulse flow (intermediate gas and liquid rates), dispersed bubble flow (at high liquid and low gas rates), and spray flow (at low liquid and high gas rates) are encountered in a trickle-bed reactor depending upon the flowrates and physical properties of flowing phases and the packing geometry. The most common mathematical model for the non-adiabatic trickle-bed catalytic reactor is the two-dimensional pseudo-homogeneous model (Tsang *et al.*, 1976; Babu, 1993), which consists of coupled partial differential equations of the parabolic type. The inherent characteristic of the homogeneous model is that the system can be considered as a continuum; no distinction is made between the solid phase and the fluid phase. The assumption implies that the reactant and the product concentrations in the bulk fluid phase are same as that on the surface of the catalyst pellet. A similar implication holds for the bed temperature. The homogeneous model that describes the physical and chemical processes is as follows:

Mass transfer

$$u\frac{\partial C_A}{\partial z}+\rho_B r_A=\varepsilon D_{er}\left(\frac{\partial^2 C_A}{\partial r^2}+\frac{1}{r}\frac{\partial C_A}{\partial r}\right) \tag{9.5}$$

Heat transfer

$$u\rho_f C_p\frac{\partial T}{\partial z}+\rho_B(-\Delta\mathrm{H})r_A=k_{er}\left(\frac{\partial^2 T}{\partial r^2}+\frac{1}{r}\frac{\partial T}{\partial r}\right) \tag{9.6}$$

with boundary conditions

$$\begin{aligned}
&z=0,\quad 0\le r\le R &&C_A=C_0,\ T=T_0\\
&r=0,\quad 0\le z\le Z &&\frac{\partial C_A}{\partial r}=\frac{\partial T}{\partial r}=0,\\
&r=R,\quad 0\le z\le Z &&\frac{\partial C_A}{\partial r}=0,\ -k_{er}\frac{\partial T}{\partial r}=h_W(T-T_W)
\end{aligned} \tag{9.7}$$

In the typical operation of a trickle-bed reactor the heat transfer parameters, the effective radial thermal conductivity of the bed, k_{er} , and the effective wall-to-bed heat transfer coefficient, h_w, are unknown and need to be estimated. These parameters are extremely important in design and in process analysis. Once these parameters are estimated, the temperature profile can be generated numerically. The temperature profile in the reactor is important because it affects the selectivity, the yield and the stability of the reactor. Froment (1967) has demonstrated the sensitivity of the homogeneous model to the effective parameters. He concluded that changing the effective Peclet number of mass transfer had a negligible effect on the temperature and conversion, whereas a 10% increase either in k_{er} or h_w greatly changed the temperature and conversion pro-files in the reactor. Smith (1973) has analyzed the relative importance of the heat and mass transfer effects in the fixed-bed reactor and concluded that the radial temperature gradient is the most important heat transfer characteristic. Only a few studies have been reported on the heat transfer characteristics (Weekman and Myers, 1965; Hashimoto *et al.*, 1976; Muroyama *et al.*, 1978; Matsuura *et al.*, 1979a; Matsuura *et al.*, 1979b; Specchia and Baldi, 1979; Crine, 1982; Lamine *et al.*, 1996; Babu and Rao, 1998), which are essential for the proper design of a trickle-bed reactor, using only air–water and air–glycerol systems; although a lot of information is available on hydrodynamics and mass transfer for the same. Deviations of 30–40% were reported for the prediction of h_w with their own empirical correlations by different authors even for their own data. Moreover, the effect of gas rate on k_{er} and h_w is not well understood with respect to the flow regimes encountered in two-phase flow, especially in the pulse flow with packing geometry. Further, the pulse properties especially liquid holdup and pulse frequency could play an important role on the heat transfer parameters in pulse flow, and none of the earlier studies except Babu (1993) detailed their effects on heat transfer. These factors emphasize the importance of heat transfer and suggest the further need for study of the heat transfer phenomena. The study of heat transfer characteristics in a trickle bed can be simplified by testing the system with no reaction occurring. In this case only the heat transfer balance equation without the reaction term is needed, i.e.,

$$\left(LC_{pL} + GC^*_{pG}\right)\frac{\partial T}{\partial z} = k_{er}\left(\frac{\partial^2 T}{\partial r^2} + \frac{1}{r}\frac{\partial T}{\partial r}\right) \tag{9.8}$$

$$z = 0, \quad T = T_0, \tag{9.9}$$

$$r = 0, \quad \frac{\partial T}{\partial r} = 0, \tag{9.10}$$

$$r = R, \quad -k_{er}\frac{\partial T}{\partial r} = h_W\left(T - T_W\right) \tag{9.11}$$

Various methods can be used to integrate Eq. (9.8) subject to the boundary conditions. Among the most popular are analytical solution, finite difference techniques and the method of weighted residuals, the last two being numerical methods. Virtually all of the previous work on estimating the *effective* parameters has been based on the analytical solution, which is

$$\frac{T(r,z) - T_W}{T_0 - T_W} = 2\sum_{i=1}^{\infty}\frac{BiJ_o\left(b^r_{i,R}\right)\exp\left(-k_{er}b_i^2 z/\left(LC_{pL} + GC^*_{pG}\right)R^2\right)}{\left(Bi^2 + b_i^2\right)J_0\left(b_i\right)} \tag{9.12}$$

where

$$Bi = \frac{h_W R}{k_{er}} = b_i\frac{J_1\left(b_i\right)}{J_0\left(b_i\right)} \tag{9.13}$$

Since the eigenvalues b_i increase as i increases, there are certain ranges of parameters within which only the first few terms of the infinite series in Eqs. (9.12) and (9.13) is significant (Coberly and Marshall, 1951; Tsang *et al.*, 1976; Specchia and Baldi, 1979). Most of the previous studies, except Specchia and Baldi (1979) used graphical methods for estimating k_{er} and h_w considering only the first term of the infinite series in the analytical solution (Eq. (9.12)) of two-dimensional energy equation for simplicity leading to uncertainty in the estimated values. Babu (1993) concluded that the first seven terms of the infinite series would be sufficient for ensuring good convergence. Differing numbers and locations of temperature measurements made on the packed bed system have led to several types of parameter estimation methods using the analytical solution. The relative advantages and disadvantages have been accounted by Tsang *et al.* (1976). Tsang *et al.* (1976) proposed a technique using orthogonal collocation in an inverse problem. They used both gradient and gradient-free optimization schemes for parameter estimation. They showed that the results were accurate enough and the computational time required was less. They used Graeffe's method along with Newton's method for finding out the roots of the Jacobi polynomial. But the Newton's method is highly dependent on the initial guess, is less accurate and takes more computational time (Acton, 1970). Graeffe's method involves in squaring the polynomial and then taking the square root of the solution, which reduces the accuracy as the degree of the polynomial increases (Acton, 1970). It has been shown that the objective function is very flat near the minimum or the contours are long and narrow (Tsang *et al.*, 1976). For such kinds of problems the gradient-based optimization techniques fail and computational time taken is also large (Acton, 1970). Though gradient-free optimization solves some of the problems of the gradient methods, the global optimum is not assured and the computational time taken is still large. Thus, to develop a new technique for the estimation of the heat transfer parameters in a trickle bed, which is not only accurate but also guarantees faster convergence is the main objective of our study. The results obtained from DE in the present study are compared with those obtained from the radial temperature profile method employing Powell's method for optimization. This new technique is highly robust and is very fast in terms of computation time when compared to the radial temperature profile method. The results show that the Orthogonal Collocation augmented with the DE's offer a powerful alternative to the conventional estimation techniques while demonstrating the potential of Orthogonal Collation for solving boundary value problems.

9.2.1 Orthogonal collocation

The trickle-bed reactor model is not amenable to an analytical solution when the chemical reaction term is non-zero. In this case, a numerical integration method such as a finite difference technique must be used. However, the orthogonal collocation method (Villadsen and Stewart, 1967; Villadsen and Michelsen, 1978; Finlayson, 1972, 1980), a technique categorized as a method of weighted residuals, has been shown by Ferguson and Finlayson (1970), and Finlayson (1972, 1980) to be superior in some respects to the finite difference approaches. Furthermore, the orthogonal collocation method, besides obtaining the solution, gives the possibility of exploring the local stability within the system (Perlmutter, 1972; Bosch and Padmanabhan, 1974; Sorenson *et al.*, 1973). Young and Finlayson (1973) have used the orthogonal collocation technique to approximate the boundary condition at the entrance of a reactor and solved the coupled non-linear partial differential equations, which take both the axial and radial dispersions into account. Finlayson (1971) has also shown when the two-dimensional reactor model must be used instead of one-dimensional model. Karanth and Hughes (1974) used collocation to simulate the adiabatic packed bed reactor. Carey and Finlayson (1975) combined the orthogonal collocation method with finite element method to solve the catalyst pellet problem with large Thiele modulus. The orthogonal collocation method has been used in the above studies for the simulation, i.e., as a numerical method to solve the boundary value problems. Tsang *et al.*, (1976) used the orthogonal collocation method in an inverse problem for the estimation of the heat transfer parameters in a packed-bed reactor. Bosch and Hellinckx (1974) used Lobatto quadrature combined with the collocation method to estimate the parameters in the differential equations. This yielded a non-linear programming problem.

Polis *et al.* (1973) have used the Galerkin technique to estimate the parameters in distributed systems. The Galerkin method is also classified as one of the methods of weighted residuals, which can be used to reduce the partial differential equation to a set of ordinary differential equations. The set of ordinary differential equations can then be used to simulate the system response iteratively. However unlike all other weighted residual methods, which determine the undetermined coefficients of the trial function, the orthogonal collocation method gives the solution of the dependent variables at the collocation points directly. One can select the measurement locations to coincide with the collocation points. The various problems in estimating the parameters in a distributed system from the point of view of accurate parameter estimates have been discussed by Goodson and Polis (1974). The details of the use of the orthogonal collocation to estimate k_{er} and h_w are given below. By rewriting Eqs. (9.8)–(9.11) in dimensionless form, the parameter estimation problem becomes

Minimizing

$$\mathcal{F} = \varepsilon^T Q \varepsilon \tag{9.14}$$

subject to

$$\frac{\partial \tilde{T}}{\partial \tilde{z}} = P'_{er} \frac{1}{\tilde{r}} \frac{\partial}{\partial \tilde{r}} \left(\tilde{r} \frac{\partial \tilde{T}}{\partial \tilde{r}} \right) \tag{9.15}$$

$$\tilde{z} = 0, \quad \tilde{T} = \tilde{T}_d \tag{9.16}$$

$$\tilde{r} = 0, \quad \frac{\partial \tilde{T}}{\partial \tilde{r}} = 0 \tag{9.17}$$

$$\tilde{r} = 1, \quad -\frac{\partial \tilde{T}}{\partial \tilde{r}} = Bi\tilde{T} \tag{9.18}$$

where $\tilde{T} = T - T_W$, the modified Peclet number

$$P'_{er} = k_{er} Z / \left(GC^*_{pG} + LC_{pL}\right) R^2, \quad \tilde{r} = r/R, \quad \tilde{z} = z/Z \tag{9.19}$$

and ε is defined as an Nx1 column vector:

$$\varepsilon = \frac{\left(\tilde{T}_{\exp}(r_i, Z) - \tilde{T}_{cal}(r_i, Z)\right)}{\tilde{T}_d} \tag{9.19a}$$

N is the number of collocation or measurement points. Q is a NxN positive-definite weighting matrix, the identity matrix being used in the present study. The estimation criterion used, when Q is taken to be an identity matrix reduces from weighted least squares to a simple least-squares estimation. A detailed derivation of the solution of Eqs. (9.15)–(9.18) using orthogonal collocation is given in Babu and Sastry (1999). Using orthogonal collocation method, the estimation problem reduces to

Minimizing Eq. (9.15) subject to

$$\frac{d\tilde{\mathbf{T}}}{d\tilde{z}} = P'_{er}\left(\mathbf{B} - \frac{\mathscr{B}(\mathscr{A})^T}{Bi + A_{N+1,N+1}}\right)\tilde{\mathbf{T}} \tag{9.20}$$

$$\tilde{\mathbf{T}}(0) = \tilde{\mathbf{T}}_d \tag{9.21}$$

where **B**, $\mathscr{B}$ and $\mathscr{A}$ are the computational collocation matrix and vectors given by Villadsen and Stewart (1967). However, instead of using Graeffe's method and Newton's method, in the present study, the *Laguerre's* method (Ralston and Rabinowitz, 1978) for finding the roots of the Jacobi polynomial has been used. *Laguerre's* method is guaranteed to converge to all types of roots: real, complex, single or multiple from any starting point (Acton, 1970; Press *et al.*, 1996). In this study we used *LU-decomposition* and *LU-back-substitution* for finding out the inverse of a matrix required in calculating the collocation matrices (Villadsen and Stewart, 1967). The integration of the ordinary differential equation, Eq. (14), was computed using the fifth-order Runge–Kutta method with adaptive step size control (Cash and Karp, 1990; Press *et al.*, 1996) for greater accuracy.

9.2.2 Experimental setup and procedure

Experiments were carried out to obtain the data on radial temperature profile in a trickle-bed reactor (gas–liquid co-current downflow through packed beds). The schematic diagram of the setup is shown in Fig. 9.1.

The experimental setup mainly consists of a packed bed column of 50 mm diameter, comprising of air–liquid distributor, calming section, jacketed test section and air–liquid separator with other auxiliary parts. Air was drawn from 3.7 kW double piston–double action compressor, of maximum volumetric capacity of 14.98 m3/s at STP and a working pressure of 12 atm. The air drawn from the compressor was saturated with water in a saturator. The saturated air was introduced at the top of the column through a set of pre-calibrated rotameters to the air–liquid distributor at a constant pressure of 4 atm, monitored by a pneumatic pressure regulator. The air flowrates in the rotameter were controlled by needle valves. The air was passed through a filter before entering the distributor to remove traces of oil and dust, if any. Water was pumped through a 4.5 kW pump and was metered through a set of pre-calibrated rotameters to the air–liquid distributor at the top of the column. Water flowrates to the column were controlled by means of globe valves. Air and water were uniformly distributed through an air–liquid distributor at the top of the calming section. The air–liquid distributor essentially consists of two sets of openings, 21 copper tubes for distributing liquid and 16 nozzles for distributing air. The tubes and the nozzles were alternately arranged on a triangular pitch over the column cross-section. The number of liquid distribution tubes per unit area was approximately equal over the entire cross-section to ensure equal distribution of liquid. The calming section consisted of a long tube, which ensured a fully developed equilibrium gas–liquid flow before it entered the test section. The jacketed test section was designed for heat transfer studies. It consists of a jacket in order to circulate hot water at 60°C. Hot water was pumped with a 4.5 kW and metered through a set of pre-calibrated rotameters. Below the heat transfer section radial temperature profile measurement section

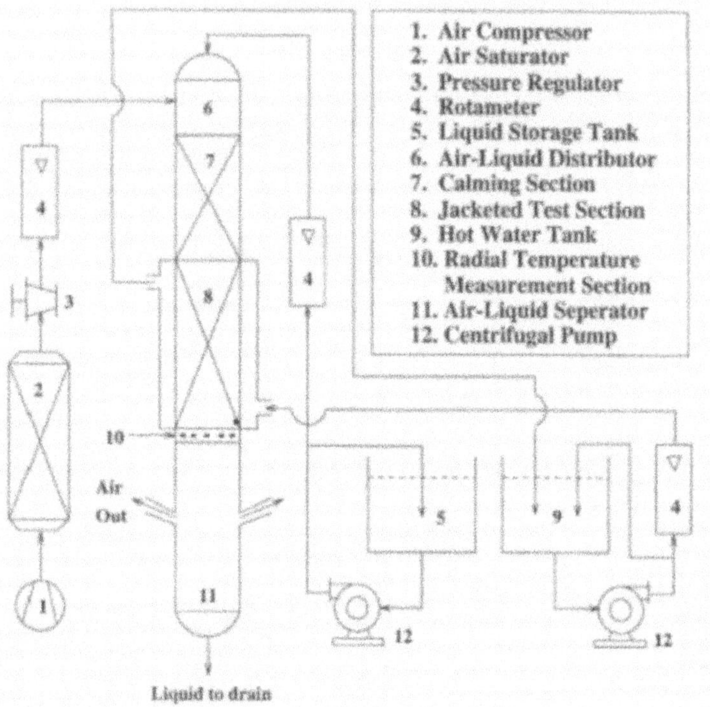

Fig. 9.1. Schematic diagram of experimental set-up (Babu and Sastry, 1999). Reprinted with permission from the Computers & Chemical Engineering.

was provided. An air–liquid separator was provided at the bottom of the column to separate the air and the liquid phases coming out of the test section. The radial temperature profile was obtained by measuring the temperatures at the bottom of the test section at three radial positions at r = 0.0, 0.4, and 0.8, and at three symmetric angular positions (120° apart) for each radial position. Wall temperature was measured by installing a thermocouple at the inside wall 3 mm above from bottom of the jacketed test section. Thermocouples were also installed at various locations to measure the inlet temperature of test liquid, and inlet and outlet temperatures of hot water. An INSREF constant temperature bath having an accuracy of 0.01°C and a MINCO platinum resistance thermometer bridge (MINCO RTB8078, Model No. S7929 Pail120C) with an accuracy of 0.025°C as a standard thermometer were used for calibrating all the chromel–alumel thermocouples used in the present study. All these thermocouples were connected to an APTEK multichannel digital temperature scanner for recording the temperatures. The detailed description of the experimental set-up, and the data collection and reduction procedures are reported elsewhere (Babu, 1993; Babu and Rao, 1998). Air and water were fed to the column from the top at the desired flowrates by means of pre-calibrated rotameters. Hot water was circulated through the jacket around the test section at sufficiently high flowrates (25–30 liters per minute) in order to maintain nearly constant wall temperature, and the minimum and maximum temperature difference between the inlet and the outlet hot water streams were 0.3°C at low flowrates to 2°C at high flowrates respectively of the flowing fluids. In general, it took 20–40 min for attaining the steady state. After steady state was attained, which was confirmed from the constant values of flowrates and temperatures, the flowrates of air and water and the temperatures were recorded. The average of the three angular positions was taken as the temperature at each radial position. This procedure was repeated for a wide range of air (0.01–0.898 kg/m2 s) and water flowrates (3.16– 71.05 kg/m2 s), covering trickle, pulse and dispersed bubble flow regimes. The length of the heat transfer test section used for heat transfer experiments was 0.715 m. The packing employed were 2.59 mm ceramic spheres, 4.05 and 6.75 mm glass spheres and 4.0 and 6.75 mm ceramic raschig rings.

9.2.3 Results and discussions

The psuedocode of the DE algorithm used in the present study is shown below:

- Initialize the values of D, F, CR, NP, λ and maximum number of generations MaxGen.
- Initialize all the vectors of the population randomly between a given lower bound LB, and upperbound UB
  for i =1 to NP
  for j =1 to D
  $(\mathbf{x}_{i,0})_j$ = LB + Random Number × (UB – LB)
- Evaluate the cost of each vector. Cost here is the value of the objective function to be minimized.
  for i =1 to NP
  $\mathcal{F}_i = \varepsilon^2\Big|_{\mathbf{x}_{i,0}}$
- Find out the vector with lowest cost i.e., the best vector so far
  $\mathcal{F}_{min} = \mathcal{F}_1$ and best = 1
  for i = 2 to NP
  if ($\mathcal{F}_i < \mathcal{F}_{min}$)
  then $\mathcal{F}_{min} = \mathcal{F}_i$ and best = i
- While the current generation is less than the maximum number of generations perform recombination, mutation, reproduction and evaluation of the objective function. While (Γ< MaxGen) do {

for i = 1 to NP {

- Select two distinct vectors randomly from the population other than the vector $\mathbf{x}_{i,r}$

do r_1 = Random Number × NP while (r_1 = i)
do r_2 = Random Number × NP while ((r_2 = i) or (r_1 = r_2))

- Perform D binomial trials, change at least one parameter of the trial vector $\mathbf{u}_{i,\,r}$ and perform mutation.
  j = Random Number × D
  for n =1 to D {
  if ((Random Number < CR) OR (n = (D – 1)))
  then $(\mathbf{u}_{i,\,r})_j = (\mathbf{x}_{i,\,r})_j + \lambda^* ((\mathbf{x}_{best,\,r})_j - (\mathbf{x}_{i,\,r})_j) + F * ((\mathbf{x}_{r1,\,r})_j - (\mathbf{x}_{r2,\,r})_j)$
  else $(\mathbf{u}_{i,\,r})_j = (\mathbf{x}_{i,\,r})_j$
  $j = \langle n+1 \rangle_D$ }
- Evaluate the cost of the trial vector.
  $$\mathcal{F}_{\text{trial}} = \varepsilon^2 \Big|_{\mathbf{u}_{i,r}}$$
- If the cost of the trial vector is less than the parent vector then select the trial vector to the next generation.
  If ($\mathcal{F}_{trial} \leq \mathcal{F}_i$) {
  $\mathcal{F}_i = \mathcal{F}_{trial}$
  If ($\mathcal{F}_{\text{trial}} < \mathcal{F}_{min}$)
  $\mathcal{F}_{min} = \mathcal{F}_{trial}$ and best = i}} /* for i = 1 to NP ends */
- Copy the new vectors $\mathbf{u}_{i,\,r}$ to $\mathbf{x}_{i,\,r}$ and increment Gamma, $\Gamma = \Gamma + 1$.
- Check for convergence and break if converged.} /* while Γ.... ends. */
- Print the results.

The collocation points or the measurement points were chosen to be that of the radial temperature profile measurements, i.e., r/R = 0, 0.4, 0.8. The values of *NP*, *CR*, λ and *F* are fixed empirically following certain heuristics (Price and Storn, 1997; Sastry *et al.*, 1998): (1) *F* and λ are usually equal and are between 0.5 and 1.0, (2) *CR* usually should be 0.3, 0.7, 0.9 or 1.0 to start with, (3) *NP* should be of the order of 10*D* and should be increased in case of misconvergence, and (4) if *NP* is increased then usually *F* has to be decreased. In the present study, the values of *D*, *NP*, *F*, λ and *CR* were taken as 2, 20, 0.7, 0.7 and 0.9, respectively. The maximum number of iterations was kept as 100; however, in all the runs the algorithm converged within 15 generations. The initial values of *Bi* and P_{er} were generated using Knuth's uniform random variate (Press *et al.*, 1996). The matrix inversions involved in computing the collocation matrices were achieved using LU decomposition and LU back-substitution. Runge–Kutta method with adaptive step size control was used for integrating the differential equation (Eq. (14)). Altogether 232 experimental data points were obtained covering a wide range of liquid and gas flowrates using five packings of different size and shape. The DE algorithm in conjunction with orthogonal collocation method was employed using the experimental temperature profile obtained for all the 232 data points. The typical radial temperature profile given by DE and RTP methods is shown in Table-9.1, which shows that the temperature profiles generated by both the methods are almost similar to the experimental profile. Similar trends were observed for all data points. The minimum and maximum sum square error for DE being $1.905052\text{x}10^{-6}$

and 7.57026x10^{-4}, respectively, and that for RTP being 1.905064x10^{-6} and 7.757032x10^{-4}, respectively with the present experimental data. The close agreement with the analytical solution and the experimental value shows that nonoptimal selection of collocation points, which are taken to be the measurement points and not the roots of the Jacobi polynomial, does not cause significant errors. The estimation errors (sum square error) given in Table-9.2 indicate that the temperature profile by DE is slightly better than that with RTP (DE's sum square error is 0.0001–0.001% less than RTP). Computational time taken by DE and RTP algorithm for randomly selected sample data points is compared in Fig. 9.2. Although the qualitative trends by both the methods is more or less the same, the average time taken, based on all data points, for an estimation by DE is 10 s as compared to 30 s by RTP on a 90 MHz Pentium processor. The function evaluations computed for sample data points by RTP and DE is shown in Fig. 9.3. DE also takes less number of function evaluations as compared to RTP (DE takes average of 800 function calls as compared to 2000 evaluations by RTP).

Table 9.1. Radial temperature profile generations: DE vs. RTP

$G = 0.0107,\ C^*_{pG} = 8952.44$			$G = 0.2041,\ C^*_{pG} = 6809.25$			$G = 0.5,\ C^*_{pG} = 9447.63$		
$\frac{LC_{pL}}{GC^*_{pG}} = 137.8,\ L = 3.16$			$\frac{LC_{pL}}{GC^*_{pG}} = 165.01,\ L = 54.89$			$\frac{LC_{pL}}{GC^*_{pG}} = 2.79,\ L = 11.11$		
$D_e = 8.1$, TF			$D_e = 4.89$, DBF			$D_e = 2.59$, PF		
EXP(^{0}C)	DE(^{0}C)	RTP(^{0}C)	EXP(^{0}C)	DE(^{0}C)	RTP(^{0}C)	EXP(^{0}C)	DE(^{0}C)	RTP(^{0}C)
56.48	56.605865	56.605843	38.00	38.075127	38.074937	57.30	57.276149	57.276099
56.95	56.776015	56.775936	38.98	38.876623	38.876634	57.46	57.493691	57.493661
57.20	57.248915	57.248689	41.09	41.118606	41.119144	58.08	58.069785	58.069799
58.91	58.91	58.91	49.85	49.85	49.85	59.12	59.12	59.12

The unit of D_e is mm.

Table 9.2. Estimation Characteristics: DE vs. RTP

Sum square error		h_W		k_{er}		Parameters			
DE	RTP	DE	RTP	DE	RTP	L	G	D_e	Flow
4.637816e^{-5}	4.637820e^{-5}	894.626343	894.51169	17.688709	17.693483	3.16	0.017	8.1	TF
2.333386e^{-4}	2.333388e^{-4}	2508.752686	2507.566406	27.927656	27.941372	9.57	0.2401	8.1	PF
1.764492e^{-4}	1.764498e^{-4}	3195.485596	3195.852539	53.098129	53.086544	19.83	0.5	8.1	PF
2.432793e^{-4}	2.432798e^{-4}	3999.604736	3999.616943	63.374241	63.372268	18.38	0.898	8.1	PF
1.568811e^{-4}	1.568813e^{-4}	1645.458130	1645.295044	26.286270	26.290203	6.52	0.0459	8.1	TF
3.369437e^{-5}	3.369470e^{-5}	4505.336914	4505.336914	99.590958	99.575050	54.89	0.2401	4.89	DBF
1.905052e^{-6}	1.905064e^{-6}	1761.488281	1761.581177	16.218536	16.217493	3.16	0.5	4.89	TF
1.340671e^{-4}	1.340675e^{-4}	4035.566406	4035.837646	65.492363	65.486969	36.87	0.0459	4.89	DBF
6.927241e^{-4}	6.927244e^{-4}	9719.313477	9718.518555	41.572712	41.573727	45.31	0.102	2.59	DBF
7.757026e^{-4}	7.757032e^{-4}	11872.214844	11868.007812	58.668274	58.678162	60.18	0.0459	2.59	DBF

The unit of D_e is mm.

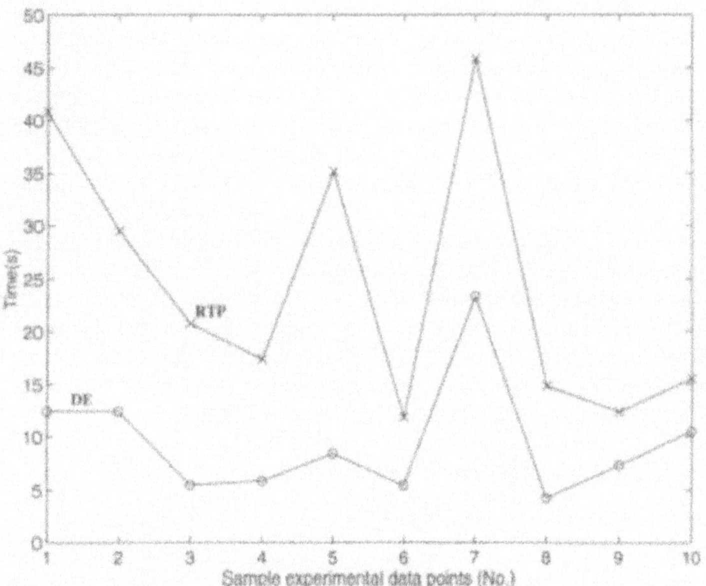

Fig. 9.2. Computational time: DE vs. RTP (Babu and Sastry, 1999). Reprinted with permission from the Computers & Chemical Engineering.

The estimation of values of h_w and k_{er} using both DE algorithm and RTP methods, are compared in Figs. 9.4 and 9.5 respectively, which shows that the estimation by DE is as accurate as the well-proven RTP algorithm. The estimations are also tabulated in Table 2, which show that the estimation error of DE is lower when compared to that of RTP algorithm. It also indicates that the estimation accuracy depends on the accuracy of the measured temperature profile, which in the present study is accurate only to two decimal places. Since a wide range of air and water flow rates was used covering the trickle, pulse and dispersed bubble flow for generalization of the estimation algorithm; the estimated parameters cover a wide range (894–11872 W/m^{2}K for h_w and 16-99 W/mK for k_{er}). The convergence criterion used for RTP is very stringent: the objective function, its relative change, the parameter values and their relative changes and the gradient of the objective function were all checked before the optimization scheme was terminated.

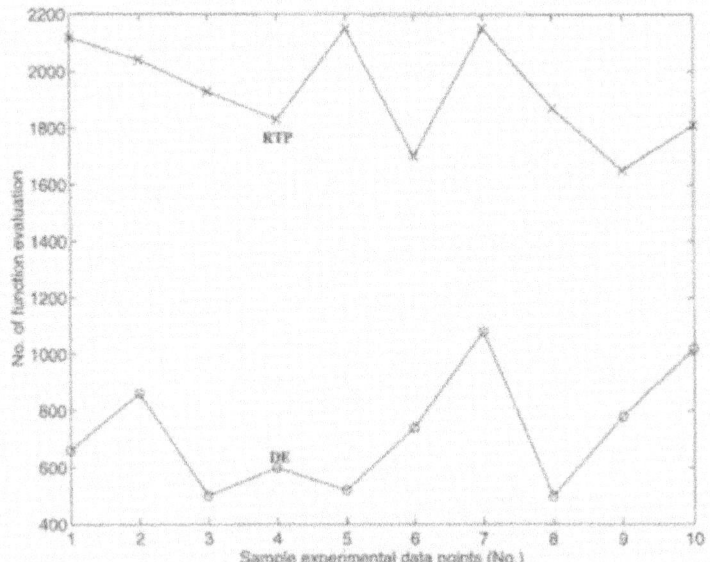

Fig. 9.3. Number of function evaluations: DE vs. RTP (Babu and Sastry, 1999). Reprinted with permission from the Computers & Chemical Engineering.

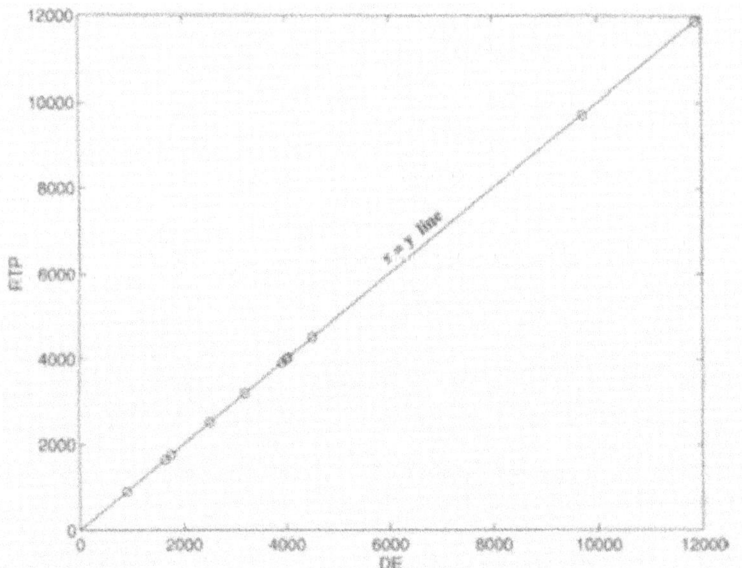

Fig. 9.4. Estimated value of h_w (W/m^2K): RTP vs. DE (Babu and Sastry, 1999). Reprinted with permission from the Computers & Chemical Engineering.

The RTP algorithm is said to have converged if it satisfies all the conditions given below (Eq. (9.22) - (9.27)):

$$\mathscr{F}_{\Gamma} < \delta_1 \tag{9.22}$$

$$\left|\mathscr{F}_{\Gamma} - \mathscr{F}_{\Gamma\text{-}1}\right| < \delta_2 \tag{9.23}$$

$$k_{er,min} \le k_{er,min} \le k_{er,max} \tag{9.24}$$

$$h_{W,min} \le h_{W,\Gamma} \le h_{W,max} \tag{9.25}$$

$$\left|k_{er,\Gamma} - k_{er,\Gamma\text{-}1}\right| < \delta_3 \tag{9.26}$$

$$\left|h_{W,\Gamma} - h_{W,\Gamma\text{-}1}\right| < \delta_4 \tag{9.27}$$

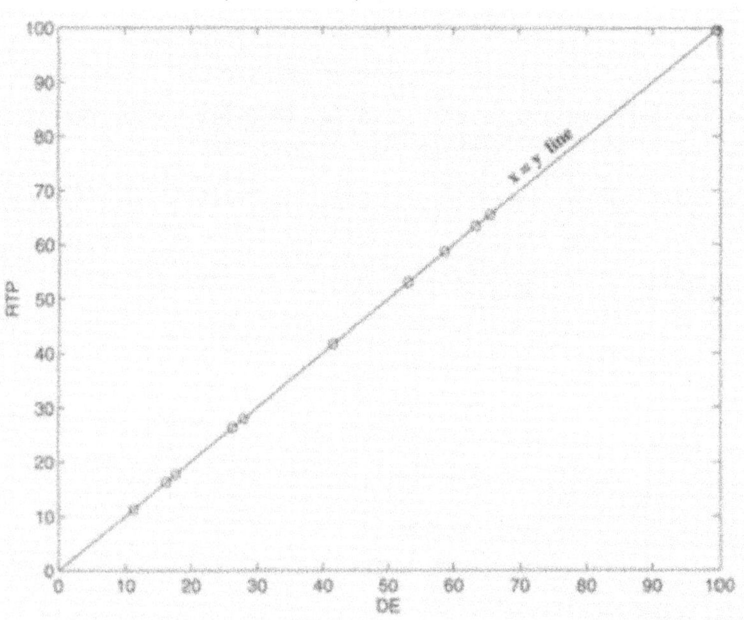

Fig. 9.5. Estimated value of k_{er} (W/m K): RTP vs. DE (Babu and Sastry, 1999). Reprinted with permission from the Computers & Chemical Engineering.

where δ_1, δ_2, δ_3 and δ_4 are constants. On the contrary, in case of DE the termination criterion is when 90–95% of the population has the same cost. The convergence criterion used for DE is:

$$\left|\sigma_{\Gamma}-\sigma_{\Gamma-1}\right|<\delta \tag{9.28}$$

where δ is a constant (in the present study $\delta = 1.0\times10^{-4}$) and σ_{Γ} is the cost variance given by

$$\sigma_{\Gamma}=\frac{\sum_{i=1}^{NP}\left(\mathscr{F}_{i,\Gamma}-\overline{\mathscr{F}}_{\Gamma}\right)^{2}}{NP-1} \tag{9.29}$$

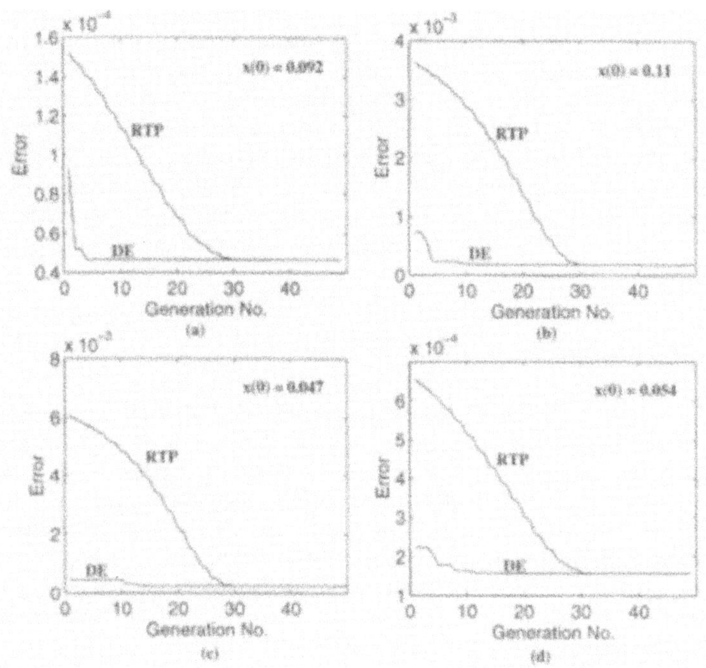

Fig. 9.6. Convergence of DE & RTP ($x(0)$ is RMS error of best initial guess of DE) (Babu and Sastry, 1999). Reprinted with permission from the Computers & Chemical Engineering.

RTP either failed or took a long time to converge when the initial guess of h_w and k_{er} was bad. Besides, DE initial guesses generated randomly were spread throughout the search space. Still the convergence of DE is faster than that of RTP. The convergence of DE and RTP is compared for sample data points in Fig. 9.6a–d, which clearly shows that DE converges much faster than RTP irrespective of the initial guesses. As depicted in Fig. 6a even though the initial estimation error of the best vector (the vector with least cost) of DE algorithm is 0.092, and in RTP method the initial estimation error is only 1.52×10^{-4}, DE converged in 10 generations whereas RTP took 32 iterations. Similarly, as shown in Fig. 6b–d, the initial estimation errors of the best vector of DE algorithm are 0.11, 0.047 and

0.54, respectively, and it took 10, 15 and 12 generations respectively to converge. On the other hand, even though the initial estimation errors are 3.6×10^{-3}, 6.0×10^{-3} and 6.5×10^{-4}, RTP method took 30, 30 and 31 iterations, respectively, to converge. The results clearly illustrate that DE is a design tool of great utility that is immediately accessible for practical applications.

The previous studies (Weekman and Myers, 1965; Hashimoto *et al.*, 1976; Muroyama *et al.*, 1978; Matsuura *et al.*, 1979a, b; Specchia and Baldi, 1979; Crine, 1982; Lamine *et al.*, 1996; Babu and Rao, 1997) on heat transfer effects in trickle-bed reactors have been conducted on non-reacting systems (air–water or air–glycerol). In the present study, the DE algorithm was applied on an air–water system as, to the best of our knowledge, no radial temperature profile data is available for trickle-bed reactors with reacting systems. As stated earlier, with the reaction term the pseudo-homogeneous model equations cannot be solved analytically. Since the proposed DE algorithm uses orthogonal collocation for solving the model equations, which can also be applied for systems with reaction. However, from the results shown above, it can be predicted that the DE algorithm will be equally effective in estimating heat transfer parameters in the presence of reaction effects.

9.2.4 Conclusions

The present study clearly shows the potential for using DE in estimating the heat transfer parameters in trickle bed reactors. In most of the studies carried out earlier in the estimation of heat transfer parameters in packed bed reactors, researchers have focused on using the first few terms of the analytical solution of the model equation and have used either gradient based or non-gradient based traditional optimization techniques for the estimation of h_w and k_{er}. Since the parameters are floating point and also due to its simple structure, ease of use, speed and robustness, it has been shown that a DE is the more appropriate choice for optimization. The results were compared with that of a RTP using analytical solution with Powell's method for estimation. In previous studies, DE has been used for design and control purposes, but in the present study DE is used as an estimator. DE algorithm is much faster, has less computational burden when compared to the RTP algorithm and the estimation is much more accurate. It is also observed that DE algorithm converges to the global optimum irrespective of its initial population, whereas the RTP Powell algorithm needed an initial guess nearer to the global optimum for convergence. The results presented in this study depict the scope of Differential Evolution in estimating the heat transfer parameters of a trickle-bed reactor. DE is more effective in terms of faster convergence, greater accuracy, lesser number of function evaluation and robustness. Based on these results, it is concluded that DE can be a very valuable resource for accurate and faster estimation of the heat transfer parameters, in multi-phase reactors such as trickle-bed reactors.

9.3 Design of Shell-and-Tube Heat Exchanger

This problem (Babu and Munawar, 2001) presents the application of Differential Evolution (DE) for the optimal design of shell-and-tube heat exchangers. The main objective in any heat exchanger design is the estimation of the minimum heat transfer area required for a given heat duty, as it governs the overall cost of the heat exchanger. Lakhs of configurations are possible with various design variables such as outer diameter, pitch, and length of the tubes; tube passes; baffle spacing; baffle cut etc. Hence the design engineer needs an efficient strategy in searching for the global minimum. In this study for the first time DE, an improved version of Genetic Algorithms (GAs), has been successfully applied with different strategies for 1,61,280 design configurations using Bell's method to find the heat transfer area. In the application of DE 9680 combinations of the key parameters are considered. For comparison, GA is also applied for the same case study with 1080 combinations of its parameters. For this optimal design problem, it is found that DE, an exceptionally simple evolution strategy, is significantly faster compared to GA and yields the global optimum for a wide range of the key parameters.

9.3.1 The Optimal HED problem

The proper use of basic heat transfer knowledge in the design of practical heat transfer equipment is an art. The designer must be constantly aware of the differences between the idealized conditions for which the basic knowledge was obtained versus the real conditions of the mechanical expression of his design and its environment. The result must satisfy process and operational requirements (such as availability, flexibility, and maintainability) and do so economically. Heat exchanger design is not a highly accurate art under the best of conditions (Perry & Green, 1993).

9.3.1.1. Generalized Design Procedure

The design of a process heat exchanger usually proceeds through the following steps (Perry & Green, 1993):

- Process conditions (stream compositions, flow rates, temperatures, pressures) must be specified.
- Required physical properties over the temperature and pressure ranges of interest must be obtained.
- The type of heat exchanger to be employed is chosen.
- A preliminary estimate of the size of the exchanger is made, using a heat transfer coefficient appropriate to the fluids, the process, and the equipment.
- A first design is chosen, complete in all details necessary to carryout the design calculations.

- The design chosen is now evaluated or rated, as to its ability to meet the process specifications with respect to both heat duty and pressure drop.
- Based on this result a new configuration is chosen if necessary and the above step is repeated. If the first design was inadequate to meet the required heat load, it is usually necessary to increase the size of the exchanger, while still remaining within specified or feasible limits of pressure drop, tube length, shell diameter, etc. This will sometimes mean going to multiple exchanger configurations. If the first design more than meets heat load requirements or does not use the entire allowable pressure drop, a less expensive exchanger can usually be designed to fulfill process requirements.
- The final design should meet process requirements (within the allowable error limits) at lowest cost. The lowest cost should include operation and maintenance costs and credit for ability to meet long-term process changes as well as installed (capital) cost. Exchangers should not be selected entirely on a lowest first cost basis, which frequently results in future penalties.

The flow chart given in Fig 9.7 (Sinnott, 1993) below gives the sequence of steps and the loops involved in the optimal design of a shell-and-tube heat exchanger.

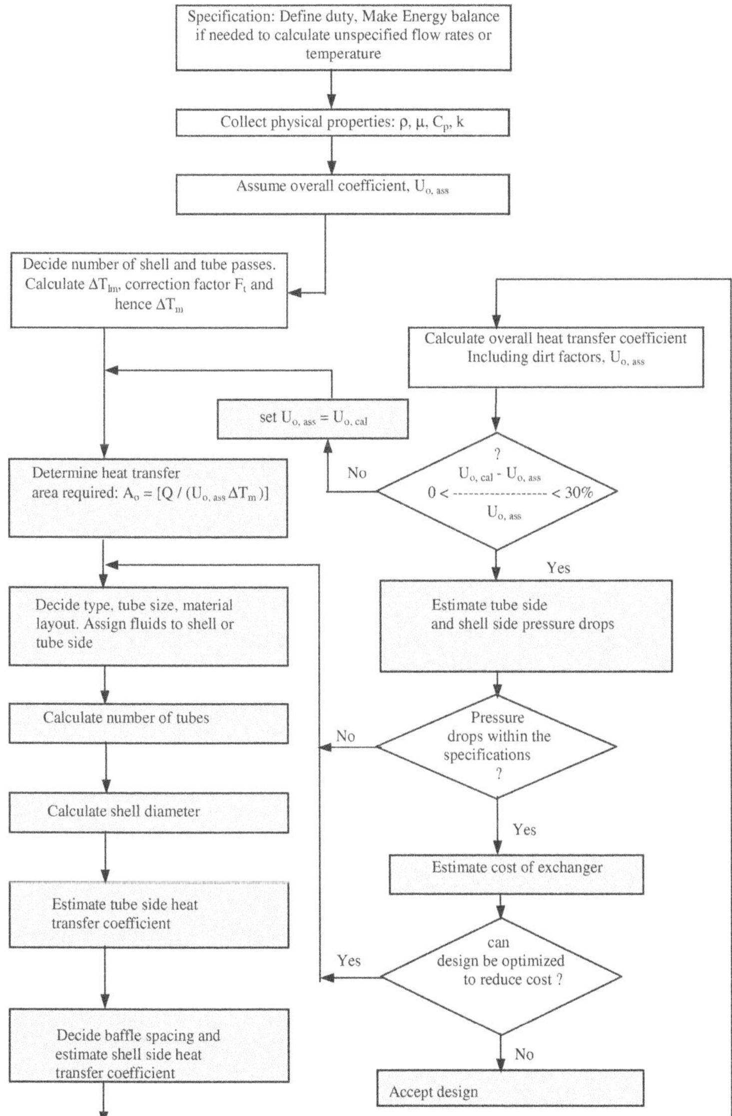

Fig. 9.7. Flow chart for optimal design of Shell-and-tube Heat Exchanger

9.3.2 Problem Formulation

The objective function and the optimal problem of shell-and-tube HED of this study are represented as shown in Table-9.3, similar to the problem formulation of Manish *et al.*, (1999).

Table 9.3. Problem formulation

$$\begin{array}{l}
\min C(\mathbf{X}) \text{ or } A(\mathbf{X}) \\
\mathbf{X} \in \{x_1, x_2, x_3, x_4, x_5, x_6, x_7\} \\
\text{where} \\
\quad x_1 = \{1,2,\ldots,12\} \\
\quad x_2 = \{1,2\} \\
\quad x_3 = \{1,2,3,4\} \\
\quad x_4 = \{1,2,\ldots,5\} \\
\quad x_5 = \{1,2,..,8\} \\
\quad x_6 = \{1,2,\ldots,6\} \\
\quad x_7 = \{1,2,\ldots,7\} \\
\text{subject to} \\
\quad \text{feasibility constraints [pressure-drop]}
\end{array}$$

The objective function can be minimization of HE cost C(**X**) or heat transfer area A(**X**) and **X** is a solution string representing a design configuration. The design variable x_1 takes 12 values for tube outer diameter in the range of 0.25" to 2.5" (0.25", 0.375", 0.5", 0.625", 0.75", 0.875", 1.0", 1.25", 1.5", 1.75", 2", 2.5"). x_2 represents the tube pitch - either square or triangular - taking two values represented by 1 and 2. x_3 takes the shell head types: floating head, fixed tube sheet, U tube, and pull through floating head represented by the numbers 1, 2, 3 and 4 respectively. x_4 takes number of tube passes 1-1, 1-2, 1-4, 1-6, 1-8 represented by numbers from 1 to 5. The variable x_5 takes eight values of the various tube lengths in the range 6' to 24' (6', 8', 10', 12', 16', 20', 22', 24') represented by numbers 1 to 8. x_6 takes six values for the variable baffle spacing, in the range 0.2 to 0.45 times the shell diameter (0.2, 0.25, 0.3, 0.35, 0.4, 0.45). x_7 takes seven values for the baffle cut in the range 15 to 45 percent (0.15, 0.2, 0.25, 0.3, 0.35, 0.4, 0.45).

In the present problem, the pressure drop on the fluids exchanging heat is considered to be the feasibility constraint. Generally a pressure drop of more than 1 bar is not desirable for the flow of fluid through a HE. For a given design configuration, whenever the pressure drop exceeds the specified limit, a high value for the heat transfer area is returned so that as an infeasible configuration it will be eliminated in the next iteration of the optimization routine. The total number of design combinations with these variables are 12 x 2 x 4 x 5 x 8 x 6 x 7 = 1,61,280. This means that if an exhaustive search is to be performed it will take at the maximum 1,61,280 function evaluations before arriving at the global minimum heat exchanger cost. So the strategy, which takes few function evaluations, is the best one. Considering minimization of heat transfer area as the objective function, differential evolution technique is applied to find the optimum design configuration with pressure drop as the constraint. For the case study considered, the performance of DE is compared with GA and the results are discussed in the fourth section.

9.3.3 Results & Discussions

The algorithm of Differential Evolution as given by Price and Storn (2002), is in general applicable for continuous function optimization. The upper and lower bounds of the design variables are initially specified. Then, after mutation because of the addition of the weighted random differential the parameter values may even go beyond the specified boundary limits. So, irrespective of the boundary limits initially specified, DE finds the global optimum by exploring beyond the limits. Hence, when applied to discrete function optimization the parameter values have to be limited to the specified bounds. In the present problem, since each design variable has a different upper bound when represented by means of integers, the same DE code given by Price and Storn (2002) cannot be used. We have used the normalized values for all the design variables and randomly initialized all the design variables between 0 and 1. Whenever it is required to find the heat transfer area using Bell's method for a given design configuration, these normalized values are converted back to their corresponding boundary limits. The pseudo code of the DE algorithm used in this study for the optimal heat exchanger design problem is given below:

- Choose a strategy and a seed for the random number generator.
- Initialize the values of D, NP, CR, F and MAXGEN.
- Initialize all the vectors of the population randomly. Since the upper bounds are all different for each variable in this problem, the variables are all normalized. Hence generate a random number between 0 and 1 for all the design variables for initialization.

```
for i = 1 to NP
    { for j = 1 to D
      x_i,j = random number
    }
```

- Evaluate the cost of each vector. Cost here is the area of the shell-and-tube heat exchanger for the given design configuration, calculated by a separate function cal_area () using Bell' method.

```
for i = 1 to NP
C_i = cal_area ()
```

- Find out the vector with the lowest cost i.e. the best vector so far.

```
C_min = C_1 and best =1
for i = 2 to NP
  { if (C_i < C_min)
    then C_min = C_i and best = i
  }
```

- Perform mutation, crossover, selection and evaluation of the objective function for a specified number of generations.

```
        While (gen < MAXGEN)
          { for i = 1 to NP
              {
```

- For each vector $\mathbf{X}_i$ (target vector), select three distinct vectors $\mathbf{X}_a$, $\mathbf{X}_b$ and $\mathbf{X}_c$ (select five, if two vector differences are to be used) randomly from the current population (primary array) other than the vector $\mathbf{X}_i$

```
    do
     {
          r1 = random number * NP
          r2 = random number * NP
          r3 = random number * NP
     }while (r1 = i) OR (r2 = i) OR (r3 = i) OR (r1 = r2) OR (r2 = r3) OR (r1 = r3)
```

- Perform crossover for each target vector $\mathbf{X}_i$ with its noisy vector $\mathbf{X}_{n,i}$ and create a trial vector, $\mathbf{X}_{t,i}$. The noisy vector is created by performing mutation. If CR = 0 inherit all the parameters from the target vector $\mathbf{X}_i$, except one which should be from $\mathbf{X}_{n,i}$.

```
   for exponential crossover
        { p = random number * 1
              r = random number * D
              n = 0
    do
   { Xn,i  = Xa,i  + F ( X b,i - X c,i )
          r = ( r+1 ) % D
              increment r by 1
    }  while ( ( p<CR ) and ( r<D ) )
          }
```

/* add two weighted vector differences for two vector perturbation. For best / random vector perturbation the weighted vector difference is added to the best / random vector of the current population. */

```
    for binomial crossover
        {      p = random number * 1
               r = random number * D
      for n = 1 to D
          {    if ( ( p<CR ) or ( p = D-1 ) )
                                     /* change at least one parameter if CR=0 */
               Xn,i  = Xa,i  + F ( X b,i - X c,i )
                   r = (r+1)%D
          }
        }
    if ( Xn,i > 1 )  Xn,i = 1
    if ( Xn,i < 0 )  Xn,i = 0
```

/* for discrete function optimization check the values to restrict to the limits */
/* 1 - normalized upper bound; 0 – normalized lower bound */

- Perform selection for each target vector, $\mathbf{X}_i$ by comparing its cost with that of the trial vector, $\mathbf{X}_{t,i}$; whichever has the lowest cost will survive for the next generation.

```
    C_t,i = cal_area()
    if ( C_t,i < C_i )    new X_i = X_t,i
    else                  new X_i = X_i  }          /* for i=1 to NP */
                          }
```

- Print the results.

The entire scheme of optimization of shell-and-tube heat exchanger design is performed by the DE algorithm, while intermittently it is required to evaluate the heat transfer area for a given design configuration. This task is accomplished through the separate function cal_area(), which employs Bell's method of heat exchanger design. Bell's method gives accurate estimates of the shell-side heat transfer coefficient and pressure drop compared to Kern's method, as it takes into account the factors for leakage, bypassing, flow in window zone etc. The various correction factors in Bell's method include: temperature correction factor, tube-side heat transfer and friction factor, shell-side heat transfer and friction factor, tube row correction factor, window correction factor for heat transfer and pressure drop, bypass correction factor for heat transfer and pressure drop, friction factor for cross-flow tube banks, baffle geometrical factors etc. These correction factors are reported in the literature in the form of monographs (Sinnott, 1993; Perry & Green, 1993). In this study, the data on these correction factors from the monographs are fitted into polynomial equations and incorporated in the computer program.

As a **case study** the following problem for the design of a shell-and-tube heat exchanger (Sinnott, 1993) is considered:

20,000 kg/hr of kerosene leaves the base of a side-stripping column at 200°C and is to be cooled to 90°C with 70,000 kg/hr light crude oil coming from storage at 40°C. The kerosene enters the exchanger at a pressure of 5 bar and the crude oil at 6.5 bar. A pressure drop of 0.8 bar is permissible on both the streams. Allowance should be made for fouling by including fouling factor of 0.00035 $(W/m^2\,{}^0C)^{-1}$ on the crude stream and 0.0002 $(W/m^2\,{}^0C)^{-1}$ on the kerosene side.

By performing the enthalpy balance, the heat duty for this case study is found to be 1509.4 kW and the outlet temperature of crude oil to be 78.6°C. The crude is dirtier than the kerosene and so is assigned through the tube-side and kerosene to the shell-side. Using a proprietary program (HTFS, STEP5) the lowest cost design meeting the above specifications is reported to be a heat transfer area of 55 m^2 based on outside diameter (Sinnott, 1993). Considering the result of the above program as the base case design, in this study DE is applied for the same problem with all the ten different strategies.

As a heuristic, the pressure drop in a HE normally should not exceed 1 bar. Hence, the DE strategies are applied for this case study separately with both 0.8 bar and 1 bar as the constraints. In both the cases, the same global minimum heat exchanger area of 34.44 m^2 is obtained using DE as against 55 m^2 reported by Sinnott, 1993. But, in the subsequent analysis, the results for 1 bar as the constraint are referred here in drawing the generalized conclusions.

A seed value for the pseudo random number generator should be selected by trial and error. In principle any positive integer can be taken. Integers 3, 5, 7, 10, 15 and 20 are tried with all the strategies for a NP value of 70 (10 times D). The F values are varied from 0.1 to 1.1 in steps of 0.1 and CR values from 0 to 1 in steps of 0.1, leading to 121 combinations of F and CR for each seed. When DE program is executed for all the above combinations, the global minimum HE area for the above heat duty is found to be 34.44 m^2 as against 55 m^2 for the base case design – indicating that DE is likely to converge to the true global optimum. For each seed, out of the 121 combinations of F and CR considered, the percentage of the combinations converging to this global minimum (C_{DE}) in less than 30 generations is listed. The average C_{DE} for each seed as well as for each strategy can be considered to be a measure of the likeliness in achieving the global minimum. It was found that the individual values of C_{DE} cover a wide range from 28.1 to 81.8 - indicating that DE is more likely to find the global optimum. Combining this above observation with the earlier one, it can be concluded that DE is more likely to find a function's true global optimum compared to the base case design. The average C_{DE} for each seed ranges from 40.7 to 64.8. In this range, the average C_{DE} for seeds 5, 7 and 10 is above 54 and so these seeds are good relative to others. With a benchmark of 40 for the individual C_{DE} values as well, seeds 5, 7 and 10 stand good from the rest. Thus with these seeds, there is more likeliness of achieving the global minimum. Whereas with seeds 3, 5 and 20 there are more of C_{DE} values below 40 and hence not considered to be good from likeliness point of view.

The average C_{DE} for different strategies varies from 44.2 to 59.4 and hence all are good if an average C_{DE} of 40 is considered to be the benchmark. With the same benchmark for C_{DE} values also is taken, then, excepting strategy numbers 2, 7 and 8 all other strategies are good. Considering 'speed ' as the other criteria, to further consolidate the effect of strategies on each seed and vice versa, the best combinations of F and CR - taking the minimum number of generations to converge to the global minimum (G_{min}) - are listed in Table-9.4.

The Criteria for choosing a good seed from 'speed' point view could be: (1) It should yield the global minimum in less number of generations, and (2) It should yield the same over a wide range of F and CR for most of the strategies. Following the first criteria with a benchmark of - achieving the global minimum in not more than two generations – seed 3 is eliminated. Satisfying the second criteria also, only seeds 7 and 10 remain. But comparing the overall performance of seeds 7 and 10, seed 10 yields the global minimum in two generations for more number of combinations of F and CR (18 times) than for seed 7 (11 times). Also, when random numbers are generated between 1 and 10 using seed 10, it is observed that it generates more 1's which is required in the design configuration leading to the global minimum. Though seed 5 was good from the 'more likeliness' point of view, but with 'speed' as the other criteria it got eliminated.

Table 9.4. Effect of seed on DE strategies w.r.t. G_{min}

S. No.	Strategy	NP = 70 & MAXGEN = 30																	
		Seed = 3			Seed = 5			Seed = 7			Seed = 10			Seed = 15			Seed = 20		
		F	CR	G_{min}	F	CR	G_{min}	F	CR	G_{min}	F	CR	G_{min}	F	CR	G_{min}	F	CR	G_{min}
1	DE/best/1/exp	0.7	0.9	3	0.9 1.0 1.1	1.0 1.0 1.0	3 3 3	0.4 0.5 0.6 0.7 0.8 0.9 0.9 1.0 1.1	0.4 0.4 0.4 0.4 0.4 0.4 0.7 0.4 0.4	5 5 5 5 5 5 5 5 5	0.8 0.9 0.9 1.0 1.1	0.8 0.8 1.0 1.0 1.0	2 2 2 2 2	0.6 0.7	0.3 0.3	4 4	0.8 0.9 1.0	0.4 0.4 0.4	3 3 3
2	DE/rand/1/exp	1.1	0.1	3	1.0 1.1	0.6 0.6	6 6	0.9 1.0	0.9 0.9	6 6	0.9	0.9	7	0.9 1.0 1.1	0.4 0.4 0.4	7 7 7	1.0 1.1	0.5 0.5	6 6
3	DE/rand-to-best/1/exp	1.0	0.9	4	1.0 1.1	0.2 0.2	5 5	0.9 1.0 1.1	1.0 1.0 1.0	2 2 2	0.6 0.8 0.9 1.0	0.8 0.9 1.0 0.7	4 4 4 4	0.7	0.9	3	1.0 1.0	0.7 0.9	4 4
4	DE/best/2/exp	0.4 0.7 0.9 1.1	0.9 0.9 1.0 0.7	4 4 4 4	1.1	1.0	2	0.8	0.7	2	1.0 1.1	0.6 0.6	4 4	0.8	0.8	4	0.7 0.8	1.0 1.0	2 2
5	DE/rand/2/exp	0.6 1.1	0.9 0.7	12 12	1.0 1.1	0.5 0.5	6 6	0.8	0.9	2	1.0	1.0	3	0.6 1.0	0.9 0.3	1 9	0.2	1.0	6
6	DE/best/1/bin	0.4 1.0	0.6 0.9	7 7	0.9 1.0 1.1	1.0 1.0 1.0	3 3 3	1.0	0.8	3	0.5 0.6 0.6 0.6 0.7 0.9 1.0 1.0 1.1 1.1 1.1	0.7 0.4 0.5 0.9 0.4 1.0 0.9 1.0 0.8 0.9 1.0	2 2 2 2 2 2 2 2 2 2 2	0.9	0.8	5	0.4 1.0	0.7 0.9	5 5
7	DE/rand/1/bin	1.0	0.5	9	0.1 0.7 0.8	0.3 0.5 0.8	7 7 7	0.9	0.9	5	0.4	0.5	7	0.7	0.9	3	0.6	0.7	10
8	DE/rand-to-best/1/bin	0.7 1.1	0.1 0.9	6 6	0.7 0.8 1.1	0.4 0.5 0.4	5 5 5	0.9 0.9 0.9 0.9 1.0 1.1	0.6 0.7 0.8 1.0 1.0 1.0	2 2 2 2 2 2	0.9 0.9	0.7 0.8	3 3	0.7 0.7	0.8 1.0	5 5	1.0	0.5	3
9	DE/best/2/bin	0.9	1.0	4	0.6 0.7 1.1	0.8 0.8 1.0	2 2 2	1.1	0.8	3	0.5 0.6	0.5 0.5	2 2	0.7	0.4	5	0.7 0.8	1.0 1.0	2 2
10	DE/rand/2/bin	0.6	0.8	6	0.4 1.1	0.8 0.5	5 5	0.9	1.0	6	1.0	1.0	3	1.1	0.4	9	0.2 0.8	0.9 0.9	4 4

Similarly with seed 15, using DE/rand/2/exp, with F = 0.6 and CR = 0.9, though the global minimum is achieved in one generation itself, still it is not considered as more number of generations are taken to converge with other strategies and other combinations of F and CR. From the 'speed' point of view it can be observed from Table-9.9 that the strategy numbers 2, 5, 7, and 10 are good. Hence, from both 'more likeliness' and 'speed' point of view, for different seeds, strategy numbers 1, 3, 4, 6, and 9 are good.

From the above results, it is also observed that, if for a given seed the random numbers generated are already good, then by using DE/best/1/...(strategy numbers 1 and 6) the global minimum is achieved in few generations itself. For example, with seeds 3 and 10 the global minimum is achieved in two generations itself with DE/best/1/... strategies. To summarize the effect of seed on each strategy, from Table-9.9, the best seeds and the number of combinations of F and CR in which the global minimum is achieved are found for each strategy along with G_{min}.

The seed value of 10 works well for four out of the ten strategies considered for a total of 19 combinations of F and CR. Seed 7 does well with 3 strategies, but for a total of 10 combinations of F and CR only. From this table also, it is clear that the seed value of 10 is better out of all the seeds considered.

Once a good seed is chosen, the next step is to study the effect of the key parameters of DE to find the best combinations of NP, F and CR for each strategy. For this, the DE algorithm for the optimal HED problem is executed for all the ten strategies for various values of NP, F and CR with MAXGEN = 15. The population size, NP is normally about five to ten times the number of parameters in a vector D (Price & Storn, 1997). To further explore the effect of the key parameters in detail, the NP values are varied from 10 to 100 in steps of 10 and F & CR values are varied again as before. The reason for taking MAXGEN = 15 can be explained as follows. As can be seen from Table-9.9, at the maximum 12 generations are taken to converge to the global minimum with NP=70. Hence, to study the effect of NP, the MAXGEN is restricted to only 15 so that with other NP values if more number of generations are taken then NP=70 itself can be claimed to be the best population size.

With MAXGEN = 15, and for the selected seed value of 10, the percentage of the combinations converging to the global minimum (C_{DE}) are found for each strategy for different NP values.

Though the NP values are varied from 10 to 100, the results for NP = 10, 20, and 30 are not considered in the subsequent analysis for the following reasons: (1) The global minimum is not achieved for any of the combinations of F and CR with some strategies for these NP values, and (2) To have the same basis for comparison with GA, for which the N values are varied from 32 to 100 (the reasons for taking the above range of N for GA is explained later). From these results, it is seen that the individual C_{DE} values cover a wide range from 0.8 to 66.1. The average C_{DE} for each NP as well as for each strategy are also computed. This average is again a measure of the likeliness of achieving the global minimum. The average C_{DE} for different NP varies from 6.3 to 31.3. With a benchmark of 10 for this average C_{DE}, it can be observed that, NP value of 70 and above stand good from the rest. From 'more likeliness' point of view, with a benchmark of 20 for C_{DE} values, again NP values of 70 and above are good. But for NP values of 100 and above it is observed that the likeliness decreases. The average C_{DE} for different strategies varies from 6.0 to 32.1. With a benchmark of 15 for this average C_{DE}, from 'more likeliness' point of view, strategy numbers 1, 3, 4, 6, 8, and 9 stand good from the rest.

Considering 'speed' again as the other criteria, the best combinations of F and CR – taking the least number of generations to converge to the global minimum (G_{min}) – are determined for various strategies with different NP values. It was found that for certain combinations of F and CR, DE is able to converge to the global minimum in a single generation also. From the 'speed' point of view also again it is evident that the NP values of 70 and above are good – indicating that at least a population size of 10 times D is essential to have more likeliness in achieving the global minimum. It is in agreement with the simple rules proposed by Price & Storn (1997). And strategy numbers 1, 4, 6, and 9 are good. Hence, from 'more likeliness' as well as 'speed' point of view, for different NP values, strategy numbers 1, 4, 6, and 9 are good.

Combining the results of variations in seed and NP, from 'more likeliness' as well as 'speed' point of view, it can be concluded that DE/best/...(strategy numbers 1, 4, 6, and 9) are good. Hence, for this optimal HED problem, the best vector perturbations either with a single or two vector differences are the best with either exponential or binomial crossovers.

The Number of Function Evaluations (NFE) are related to the population size, NP as NFE = NP * (G_{min} + 1) (plus one corresponds to the function evaluations of the initial population). It can be inferred from the simulated results (using NP and G_{min} values to compute NFE) that NFE varies from only 100 to 1300, out of the 1,61,280 possible function evaluations considered. Hence, the best combination corresponding to the least function evaluations is found to be for NP = 50 and DE/best/1/exp strategy, with 100 function evaluations as it converges in one generation itself. For this combination DE took 0.1 seconds of CPU time on a 266 MHz Pentium-II processor.

Now, to study the effect of F and CR on various strategies, the DE algorithm is executed for the present optimal problem, for different values of F and CR. For the selected seed value of 10 and NP = 70, the F values are varied from 0.5 to 1.0 in steps of 0.1 and CR values from 0 to 1 in steps of 0.1. For the ten different strategies considered in the above range of F and CR, 21 plots are made for number of generations versus CR values with F as a parameter; and 18 plots for the number of generations versus F values with CR as a parameter. Typical graphs are shown in Figs. 9.8 – 9.11. The numbers given in the legend correspond to the strategy. Whenever the global minimum is obtained, it is observed that DE converges in less than 70 generations for strategies 1, 4, 6, and 9; and in less than 100 generations for the rest of the strategies. This implies that for any particular combination of the key parameters, if convergence is not achieved in less than 100 generations, then it is never achieved. So the points at 100 generations in the graphs correspond to the misconvergence and hence represent local minima (other than 34.44 m^2 and obviously higher values of area).

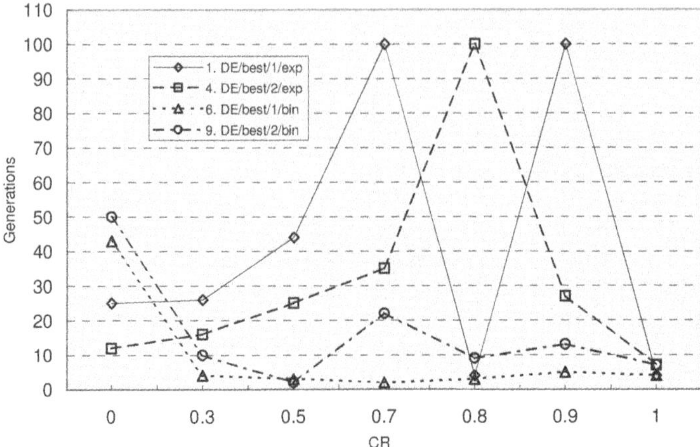

Fig. 9.8. Comparison of DE strategies for best vector perturbation with F as a parameter (F=0.5)

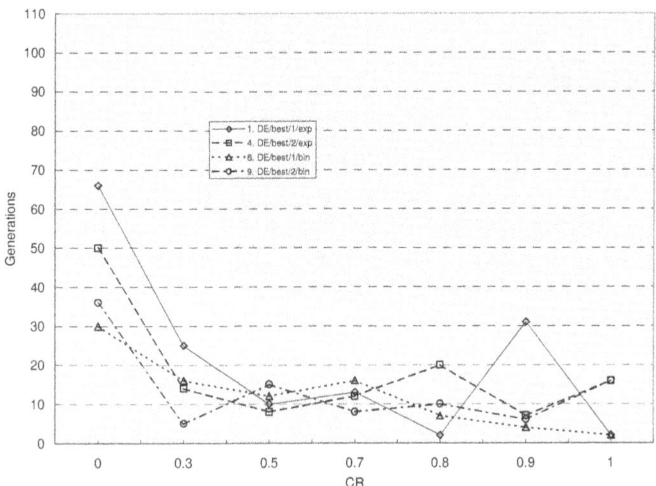

Fig. 9.9. Comparison of DE strategies for best vector perturbation with F as a parameter (F=0.9)

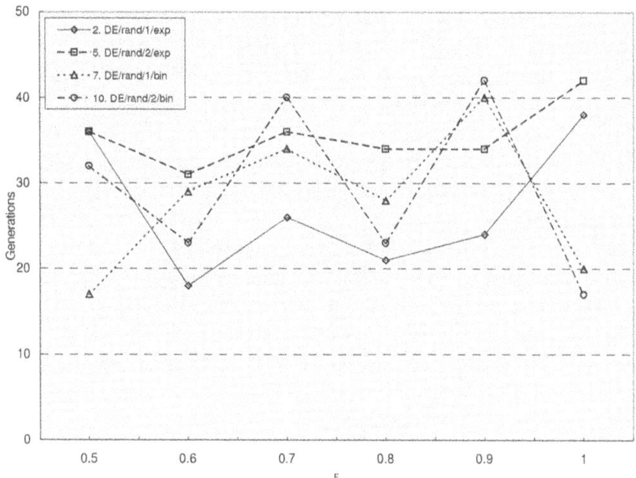

Fig. 9.10. Comparison of DE strategies for random vector perturbation with CR as a parameter (CR=0.5)

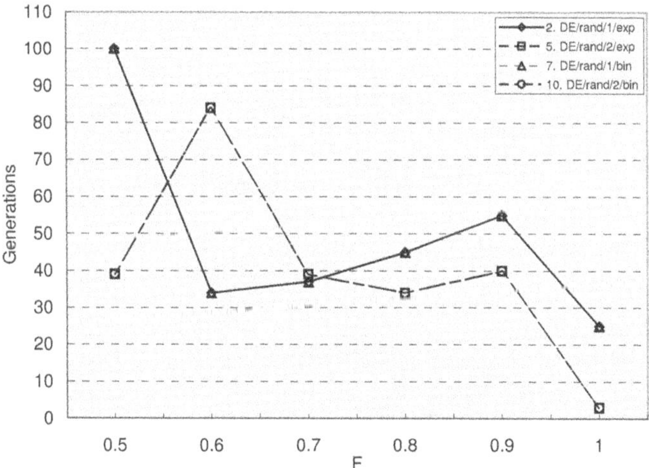

Fig. 9.11. Comparison of DE strategies for random vector perturbation with CR as a parameter (CR=1.0).

For studying the effect of F and CR on each strategy, the criterion considered is 'speed'. So, it should yield the global minimum in less number of generations. From the 'speed' point of view, it can be seen that with F as a parameter, in general (excluding the misconvergence points) DE/best/1/...(strategy numbers 1 and 6) are better than DE/best/2/.. (strategy numbers 4 and 9) unlike from 'more likeliness' point of view, where DE/best/2/.. is better than DE/best/1/.. as the former one takes less number of generations. Similar trends are observed for CR also as a parameter. It means that if best vector perturbation is to be tried then, from the 'speed' point of view, it is worth trying it with single vector perturbation first to see quickly if there is any convergence. If misconvergence occurs DE/best/2/.. can be tried as it has more likeliness of achieving the global minimum. With DE/best/1/.. it is observed that at low values of F binomial crossover is better than exponential crossover; and as F values are increased there is no marked difference in the performance of exponential and binomial crossover. From the 'speed' point of view this observation is again clearly evident from the results obtained. But for DE/best/2/.. binomial crossover seems to be better than that of exponential, as it yields the global minimum in less number of generations at medium to higher values of F. Similar trends are observed with CR also as a parameter.

If misconvergence occurs with DE/best/..., either DE/rand-to-best/1/... or DE/rand/...seem to be the immediate good options. From the 'speed' point of view DE/rand/2/... is seen to be better than DE/rand/1/... at high values of F and CR. This indicates that if random vector perturbation is to be used then, from the 'speed' point of view, it is better to use it with two vector differences. With DE/rand/1/.., exponential crossover seems to be a better option. For the optimal HED problem considered, it is concluded from the preceding discussions that for variations in seed and NP, from 'more likeliness' as well as 'speed' point of view DE/best/.. strategies are better than DE/rand/.. From these results it is also observed that the DE strategies are more sensitive to the values of CR than to F. Extreme values of CR are worth to be tried first. It was observed that there is no difference in the performance of exponential and binomial crossover at CR=1.0, which is quite obvious from the nature of these operators. The priority order of the strategies to be employed for a given optimal problem may vary from what is observed above. Selection of a good seed is indeed the first hurdle before investigating the right combination of the key parameters and the choice of the strategy. Some of these observations made in present the study form the supportive evidence of the recommendations and suggestions, made for other applications by Price and Storn (2002).

For comparison, Genetic Algorithms with binary coding for the design variables are also applied for the same case study with Roulette-wheel selection, single-point crossover, and bit-wise mutation as the operators for creating the new population. The GA algorithm is executed for various values of N - the population size, p_c – the crossover probability and p_m – the mutation probability. With a seed value of 10 for the pseudo random number generator, N is varied from 32 to 100 in steps of 4; p_c from 0.5 to 0.95 in steps of 0.05; and p_m from 0.05 to 0.3 in steps of 0.05, leading to a total of 1080 combinations. The reasons for not considering the N values of 30 and below are same as those already explained for NP while discussing the results with DE. N/2 has to be an even number for single-point crossover and hence the starting value of 32 and the step size of 4 are taken. The step size is smaller for GA compared to DE, as it can be seen later that GA has less likeliness so more search space is required. For each population size, 60 combinations of p_c and p_m are possible in this range. For the case study considered, the same global minimum heat transfer area is obtained ($34.44m^2$) by using GA also. The minimum number of generations required by GA to converge to the global minimum (G_{min}), in the above range of the key parameters is listed in Table-9.5 along with the Number of Function Evaluations (NFE).

Table 9.5. GA parameters converging to the global minimum

S. No.	Seed = 10 & MAXGEN = 100					
	N	p_c	p_m	C_{GA}	G_{min}	NFE
1	44	0.55	0.05 - 0.30	20	53	2376
		0.60			63	2816
2	48	0.75	0.05 - 0.30	10	5	288
3	52	0.75	0.05 - 0.30	20	67	3536
		0.90			19	1040
4	60	0.65	0.05 - 0.30	20	71	4320
		0.70			14	900
5	64	0.50	0.05 - 0.30	30	59	3840
		0.60			43	2816
		0.80			25	1664
6	68	0.65	0.05 - 0.30	20	64	4420
		0.85			6	476
7	72	0.50	0.05 - 0.30	10	47	3456
8	76	0.75	0.05 - 0.30	10	90	6916
9	80	0.60	0.05 - 0.30	50	23	1920
		0.75			18	1520
		0.80			39	3200
		0.85			35	2880
		0.95			46	3760
10	84	0.75	0.05 - 0.30	20	97	8232
		0.90			70	5964
11	100	0.60	0.05 - 0.30	20	13	1400
		0.75			4	500

For each combination of N and p_c listed in this table, GA is converging to the global minimum heat transfer area of 34.44 m^2 for all the six values of p_m from 0.05 to 0.3 in steps of 0.05. While executing the GA program, it is observed that more number of generations is taken by GA to converge and hence, the maximum number of generations (MAXGEN) is specified as 100. For a given N, the percentage of the combinations converging to the global minimum (C_{GA}) in less than 100 generations, out of the 60 possible combinations of p_c and p_m considered, is also listed in Table-9.5. As can be seen, C_{GA} ranges from 10 to 50 only as against the individual C_{DE}, which varies from 0.8 to 66.1. Comparing the individual C_{DE} range for one of the best strategy DE/best/2/bin, which varies 11.6 to 66.1 (except one that is 3.3), C_{GA} is relatively less. The average C_{GA} is calculated to be 20.9, whereas the average C_{DE} is above 22 for four out of the ten strategies (strategy numbers 1, 4, 6 and 9). It is interesting to note that, had the basis of MAXGEN been same (i.e. 100) for both GA and DE then, it is quite obvious that the C_{DE} values would have been very high (may be close to 100) – indicating that DE has 'more likeliness' of achieving the global optimum compared to GA – as it has a wide range of the individual C_{DE} values. And also DE has more strategies to choose from, which is an advantage over GA. As a measure of 'likeliness' another criteria is identified and defined – the percentage of the key parameter combinations converging to the global minimum, out of the total number of combinations considered (C_{tot}). In Table-9.10, out of the 1080 combinations of the key parameters considered with GA, only 138 combinations (i.e. C_{tot} = 12.8) are converging to the global minimum in less than 100 generations. Whereas in DE, out of the total of 9680 possible combination of key parameters considered 1395 combinations (i.e. C_{tot} = 14.4) are converging to global minimum in less than 15 generations itself.

The relation between NFE and the number of generations in GA also, remains the same as in DE. ($NFE = N * (G_{min} + 1)$). Using GA, with a seed value of 10, NFE varies from 288 to 8148 as against a small range of 100 to 1300 only for DE, which is an indication of the tremendous 'speed' of the DE algorithm. The above two observations clearly demonstrate that for the case study taken up, the DE algorithm is significantly faster, has more likeliness in achieving the global optimum and so is efficient compared to GA. It is also evident from Table-9.5 that, the best combination corresponding to the least function evaluations is for N = 48, p_c = 0.75, and p_m 0.05 to 0.3 (entire range of p_m), with 288 function evaluations as it converges in 5 generations itself. For this combination of parameters GA took 0.46 seconds of CPU time (on a 266 MHz Pentium-II processor) to obtain 34.44 m^2 area. The summary of the results from the preceding discussions, for the selected seed value of 10 is listed in Table-9.6.

The performance of DE and GA can be compared from this table with respect to the various criteria identified and defined. Out of the entire range of key parameter combinations considered, the range of the individual key parameter values converging to the global minimum heat exchanger area of 34.44 m^2, along with the best strategies recommended are listed for DE and GA. For comparison of DE and GA, the characteristic criteria identified are the 'likeliness' and the 'speed' in achieving the global minimum. As a measure of 'likeliness' the following parameters are defined: C_{DE} and average C_{DE} for DE; C_{GA} and average C_{GA} for GA; C_{tot} for both GA and DE. Similarly, as a measure of 'speed' the parameters defined are: G_{min}; NFE; and CPU time. From the range of the values of these identified parameters for the above criteria listed in Table-8, it is evident that DE has shown remarkable performance within 15 generations itself from both 'likeliness' and 'speed' point of view, which GA could not show even in 100 generations. Hence, it can be concluded that DE is significantly faster at optimization and has more likeliness in achieving the global optimum. The results shown in this comprehensive table consolidate all the observations made and the conclusions drawn from the preceding discussions. So, the authors recommend the above range of the key parameter values (as listed in Table-9.6) for the optimal design of a shell-and-tube heat exchanger using differential evolution.

Table 9.6. Comparison of DE and GA w.r.t. various criteria for the entire range of parameters

S. No.	Seed = 10			
	Criteria		DE	GA
1	MAXGEN		15	100
2	Global minimum heat exchanger area (m^2)		34.44	34.44
3	Parameter values converging to the global minimum	(a) Key parameters	NP: 50 – 100 F: 0.3 – 1.1 CR: 0.1 – 1.0	N: 44 – 100 P_m: 0.05 – 0.3 P_c: 0.5 – 0.95
		(b) Strategy	DE/best/1/exp DE/best/1/bin DE/best/2/exp DE/best/2/bin	-
4	Measure of 'likeliness' in achieving the global minimum		C_{DE}: 0.6–66.1 Avg. C_{DE}: 6.0-32.1 C_{tot}: 12.8	C_{GA}: 10 - 50 Avg. C_{GA}: 0.9 C_{tot}: 14.4
5	Measure of 'speed' in achieving the global minimum	(a) G_{min}	1 – 12	4 – 97
		(b) NFE	100 – 1300	288 – 8148
		(c) CPU time (s)	0.1 – 2.23	0.46 – 15.29

The performance of DE and GA is compared for the present problem in Table-9.7, with respect to the 'best' parameters – parameter values converging to the global minimum out of the entire range considered.

Table 9.7. Comparison of DE and GA w.r.t. various criteria for the best parameters

S. No.	Seed = 10			
	Criteria		DE	GA
1	Global minimum heat transfer area (m^2)		34.44	34.44
2	Best parameter values converging to the global minimum	(a) Key parameters	NP: 50 F: 0.8 – 1.1 CR: 0.7	N: 48 p_m : 0.05 – 0.30 p_c : 0.75
		(b) Strategy	DE/best/1/exp	-
3	G_{min}		1	5
4	NFE		100	288
5	CPU time (s)		0.1	0.46

For NP=50, with DE/best/1/exp strategy, CR=0.7 and F = 0.8 to 1.1 (any value in steps of 0.1), as already been mentioned DE took one generation, 100 function evaluations and 0.1 seconds of CPU time. But with GA, for N = 48, p_c = 0.75 and p_m = 0.05 to 0.3 (any value in steps of 0.05) it took 5 generations, 288 function evaluations and 0.46 seconds of CPU time.

From Table-9.7, it can be seen that DE is almost 4.6 times faster than GA. And by using DE, there is 78.3 % savings in the computational time compared to GA. Comparing the results of the proprietary program (HTFS, STEP5) with these algorithms (both DE and GA), there is 37.4 % saving in the heat transfer area for the case study considered. For the optimal shell-and-tube HED problem considered, the best population size using both DE and GA is seen to be around 7 times the number of design variables with about 70 percent crossover probability. Hence, for the heat duty of the case study taken up, the best design configuration of the shell-and-tube heat exchanger with respect to the design variables considered, corresponds to the global minimum heat exchanger area of 34.44 m^2. The best design variables are listed in Table-9.8 along with the best key parameters of the DE algorithm used for this optimization.

Table 9.8. Summary of the proposed final design for the case study

S. No.	Parameters		
1	Heat duty (kW)		1509.4
2	Best design variables	(a) tube outer diameter (inch)	½
		(b) tube pitch	5/8" triangular
		(c) shell head type	Fixed tube sheet
		(d) tube passes	Single
		(e) tube length(ft)	24
		(f) baffle spacing	20%
		(g) baffles cut	15%
3	Heat exchanger area (m^2)		34.44
4	Pressure drop (bar)	(a) tube-side	0.67
		(b) shell-side	0.31
5	DE parameters	(a) strategy	DE/best/1/exp
		(b) seed	10
		(c) NP	50
		(d) F	0.8 – 1.1
		(e) CR	0.7

9.3.4 Conclusions

This problem demonstrates the first successful application of Differential Evolution for the optimal design of shell-and-tube heat exchangers. A generalized procedure has been developed to run the DE algorithm coupled with a function that uses Bell's method of heat exchanger design, to find the global minimum heat exchanger area. For the case study taken up, application of all the ten different working strategies of DE is explored. The performance of DE and GA is compared. The following conclusions are drawn from this study:

1. The population-based algorithms such as GAs and DE provide significant improvement in the optimal designs, by achieving the global optimum, compared to the traditional designs.
2. For the present optimal shell-and-tube HED problem:
   - The best population size, using both DE and GA, is about 7 times the number of design variables

- From 'more likeliness' as well as 'speed' point of view, DE/best/... strategies are better than DE/rand/... for the selected seed value of 10
- DE/best/1/.. strategy is found to be the best out of the presently available ten strategies of DE

3. Differential Evolution, a simple evolution strategy is significantly faster compared to GA.
4. DE achieves the global minimum over a wide range of its key parameters – indicating the 'likeliness' of achieving the true global optimum.
5. And DE proves to be a potential source for accurate and faster optimization.

Nomenclature

$\overline{\mathscr{F}}$	average cost
$\tilde{z}$	dimensionless axial position
$\tilde{r}$	dimensionless radius, r/R
$\tilde{T}$	temperature vector, $(= T(r, z) - T_w)$ K
$\tilde{T}_d$	Temperature vector, $(= T_o - T_w)$, K
$x_{i,\Gamma}$	ith vector in generation Γ
$x_{best,\Gamma}$	vector with minimum cost in generation Γ
$x_{r_1,\Gamma}, x_{r_2,\Gamma}$	randomly selected vector
ΔH	heat of reaction, J/kg
ΔT_{lm}	log-mean temperature difference
ΔT_m	mean temperature difference
$\mathscr{A}$	collocation vector
A(X)	objective function heat transfer area, m^2
A_{ij}	element of the collocation matrix A
A_o	heat transfer area based on outer surface, m^2
B	collocation matrix
$\mathscr{B}$	collocation vector
Bi	biot number
b_i	coefficients of the analytical solutions
B_{ij}	element of the collocation matrix B
C(X)	objective function heat exchanger cost
C^*_{pG}	rate of change of enthalpy with respect to the rise in temperature of the gas phase, J/kgK

C_A	concentration of reactant A, kmol/m3
C_{DE}	percentage of the combinations converging to the global minimum of 34.44 m^2 of heat exchanger area, out of the 121 combinations of F and CR considered (Tables-2 & 5).
C_{GA}	percentage of the combinations converging to the global minimum of 34.44 m^2 of heat exchanger area, out of the 60 combinations of p_c and p_m considered (Tables-7 & 9).
C_o	inlet concentration, kmol/m^3
C_p	heat capacity at constant pressure, J/kg K
C_{pL}	specific heat of liquid, J/kgK
CPU time	the time taken on 266MHz Pentium-II processor
CR $(0<CR<1)$	crossover constant
C_{tot}	percentage of the key parameter combinations converging to the global minimum, out of the total number of combinations considered.
D	dimension
D	number of design variables or dimension of the parameter vector.
D_e	effective diameter of the packing, m
D_{er}	effective diffusivity, m^2/s
F $(0<F<1.2)$	scaling factor
$\mathscr{F}$	cost or the value of the objective function
F	weight applied to the random differential
F_t	LMTD correction factor
G	gas flow rate, kg/m^2 s
G_{min}	minimum number of generations required to converge to the global minimum
h_w	Wall heat transfer coefficient, W/m^2K
k	thermal conductivity, W/m K
k_{er}	effective thermal conductivity of the bed, W/mK¸
L	liquid flowrate, kg/m^2 s
M $(0<M<D-1)$	random integer
N	measurement points or number of collocation points
N	population size in GA
NP	population size in DE
P'_{er}	modified Peclet number, ("$k_{er}Z/(_{¸}C_{pL}/GC^{*}_{pg})R^2$)

p_c	crossover probability in GA
Pi	Jacobi polynomial of order i
p_m	mutation probability in DE
Pr	binomial probability function
Q	positive-definite weighting matrix
Q	heat duty, W
r	radial position in trickle bed, m
R	trickle-bed radius, m
r	random number
r_A	reaction rate
T	temperature, K
T_o	inlet fluid temperature, K
T_w	wall temperature, K
u	fluid velocity, m/s
u, v	trial vectors
$U_{0,cal}$	calculated value of overall heat transfer coefficient based on outside area, $W/m^2\,{}^0C$
$U_{o.ass}$	assumed value of overall heat transfer coefficient based on outside area, $W/m^2\,{}^0C$
X	a design configuration
x	a design variable
z	axial direction in the trickle bed, m
Z	length of the trickle bed, m

Greek letters

$\delta, \delta_1, \delta_2, \delta_3, \delta_4$	constants
$\in$	dimensionless temperature vector, $\tilde{T}(\tilde{r},1)/\tilde{T}_d$
ε	void fraction of the bed
γ	integer
Γ	generation number
λ $(0<\lambda 1.2)$	greediness scaling factor
ρ_B	catalyst bulk density, kg/m3
ρ_f	fluid density, kg/m3
σ_A	cost variance in generation "
ρ	density, kg/m^3
μ	viscosity, kg/m s

Abbreviations

ARS	adaptive random search
DBF	dispersed bubble flow
DE	differential evolution
GA	Genetic Algorithms
HE	Heat Exchanger
HED	Heat Exchanger Design
HTFS	Heat Transfer Flow Systems
HTRI	Heat Transfer Research Institute
LMTD	Log-Mean Temperature Difference
MAXGEN	Maximum Number of Generations specified
NFE	Number of Function Evaluations
PDE	Partial differential equation
PF	pulse flow
RTP	radial temperature profile method
TF	trickle flow

References

Acton, F. S. (1970). *Numerical methods that work.* Harper& Row, New York.

Angira, R. and Babu, B.V. (2003). "Evolutionary Computation for Global Optimization of Non-Linear Chemical Engineering Processes". *Proceedings of International Symposium on Process Systems Engineering and Control (ISPSEC'03) - For Productivity Enhancement through Design and Optimization*, IIT-Bombay, Mumbai, January 3-4, 2003, Paper No. FMA2, pp 87-91 (2003). *(Also available via Internet as .pdf file at http://bvbabu.50megs.com/custom.html/#56.*

Babu, B.V. (1993). *Hydrodynamics and heat transfer in single-phase liquid and two-phase gas-liquid co-current downflow through packed bed columns.* Ph.D. thesis. IIT, Bombay.

Babu, B.V. (2001). "Evolutionary Computation-At a Glance" NEXUS, Annual Magazine of Engineering Technology Association, BITS, Pilani, 3-7, 2001. Also available via Internet as .pdf file at http://bvbabu.50megs.com/custom.html/#36.

Babu, B.V. and Angira, R. (2001a). "Optimization of Non-Linear Functions Using Evolutionary Computation", *Proceedings of 12th ISME Conference on Mechanical Engineering*, Crescent Engineering College, Chennai, January 10-12, 2001, Paper No. CT07, 153-157. Also available via Internet as .pdf file at http://bvbabu.50megs.com/custom.html/#34.

Babu, B.V. and Angira, R. (2001b). "Optimization of Thermal Cracker Operation using Differential Evolution", *Proceedings of International Symposium & 54th Annual Session of IIChE (CHEMCON-2001)*, CLRI, Chennai, December 19-22, 2001. Also available via Internet as .pdf file at http://bvbabu.50megs.com/custom.html/#38 & Application No. 20, Homepage of Differential Evolution, the URL of which is: http://www.icsi.berkeley.edu/~storn/code.html

Babu, B.V. and Angira, R. (2002a). "A Differential Evolution Approach for Global Optimization of MINLP Problems", *Proceedings of 4th Asia-Pacific Conference on Simulated Evolution And Learning (SEAL-2002)*, Singapore, November 18 - 22, 2002 Paper No. 1033, Vol. 2, pp 880-884. *(Also available via Internet as .pdf file at http://bvbabu.50megs.com/custom.html/#46).*

Babu, B.V. and Angira, R. (2002b). "Optimization of Non-Linear Chemical Processes Using Evolutionary Algorithm". *Proceedings of International Symposium & 55th Annual Session of IIChE (CHEMCON-2002)*, OU, Hyderabad, December 19-22, 2002. *(Also available via Internet as .pdf file at http://bvbabu.50megs.com/custom.html/#54).*

Babu, B.V. and Chaturvedi, G. (2000). "Evolutionary Computation strategy for optimization of an Alkylation Reaction", *Proceedings of International Symposium & 53rd Annual Session of IIChE (CHEMCON-2000)*, Science City, Calcutta, December 18-21, 2000. Also available via Internet as .pdf file at http://bvbabu.50megs.com/custom.html/#31. & Application No. 19, Homepage of Differential Evolution, the URL of which is: http://www.icsi.berkeley.edu/~storn/code.html

Babu, B.V. and Gautam, K. (2001). "Evolutionary Computation for Scenario-Integrated Optimization of Dynamic Systems", *Proceedings of International Symposium & 54th Annual Session of IIChE (CHEMCON-2001)*, CLRI, Chennai, December 19-22, 2001. Also available via Internet as .pdf file at http://bvbabu.50megs.com/custom.html/#39 & Application No. 21, Homepage of Differential Evolution, the URL of which is: http://www.icsi.berkeley.edu/~storn/code.html.

Babu, B.V. and Mohiddin, S.B. (1999). "Automated Design of Heat Exchangers using Artificial Intelligence based Optimization", *Proceedings of International Symposium & 52nd Annual Session of IIChE (CHEMCON-1999)*, Panjab University, Chandigarh, December 20-23, 1999. Also available via Internet as .htm file at http://bvbabu.50megs.com/custom.html/#27.

Babu, B.V. and Munawar, S.A. (2000). "Differential Evolution for the Optimal Design of Heat Exchangers", *Proceedings of All India Seminar on Chemical Engineering Progress on Resource Development: A Vision 2010 and Beyond*, organized by IE (I), Orissa State Centre Bhuvaneshwar, March 13, 2000. Also available via Internet as .pdf file at http://bvbabu.50megs.com/custom.html/#28.

Babu, B.V. and Munawar, S.A. (2001). "Optimal Design of Shell-and-Tube Heat Exchangers using Different Strategies of Differential Evolution", PreJournal.com – The Faculty Lounge, Article No. 003873, posted on website Journal http://www.prejournal.com, March 03, 2001. *Also available via Internet as .pdf file at http://bvbabu.50megs.com/custom.html/#35* & Application No. 18, Homepage of Differential Evolution, the URL of which is: http://www.icsi.berkeley.edu/~storn/code.html.

Babu, B.V. and Rao, V.G. (1998). "Prediction of effective bed thermal conductivity and wall-to-bed heat transfer co-efficient in a packed bed under no flow conditions". *Journal of Energy, Heat and Mass Transfer*, **20**, 43-50. Also In: *Proceedings of the 1st ISHM1–ASME HM1 Conf.* (pp. 715–721), TMH Publishing Company Limited, New Delhi (1994).

Babu, B.V. and Sastry, K.K.N. (1999). "Estimation of Heat Transfer Parameters in a Trickle Bed Reactor using Differential Evolution and Orthogonal Collocation", *Computers & Chemical Engineering*, **23**, 327-339. Also available via Internet as .pdf file at http://bvbabu.50megs.com/custom.html/#24. & Application No. 13, Homepage of Differential Evolution, the URL of which is: http://www.icsi.berkeley.edu/~storn/code.html

Babu, B.V. and Singh, R.P. (2000). "Optimization and Synthesis of Heat Integrated Distillation Systems Using Differential Evolution", *Proceedings of All India Seminar on Chemical Engineering Progress on Resource Development: A Vision 2010 and Beyond*, organized by IE (I), Orissa State Centre Bhuvaneshwar, March 13, 2000.

Babu, B.V. and Vivek, N. (1999). "Genetic Algorithms for Estimating Heat Transfer Parameters in Trickle Bed Reactors", *Proceedings of International Symposium & 52nd Annual Session of IIChE (CHEMCON-99)*, Panjab University, Chandigarh, December 20-23, 1999. Also available via Internet as .htm file at http://bvbabu.50megs.com/custom.html/#26.

Babu, B.V., Angira, R. and Nilekar, A. (2002). "Differential Evolution for Optimal Design of an Auto-Thermal Ammonia Synthesis Reactor", *Computers & Chemical Engineering* (Communicated).

Boender, C.G.E. and Romeijn, H.E. (1995). "Stochastic methods". In: R. Horst, & P. M. Pardalos, (Eds.), *Handbook of global optimization* (pp. 829–869). Dordrecht: Kluwer Publishers.

Bosch, B.V.D. and Hellinckx, L. (1974). "A new method for the estimation of parameters in differential equations". *American Institute of Chemical Engineers Journal*, **20**, 250–255.

Bosch, B.V.D. and Padmanabhan, L. (1974). "Use of orthogonal collocation methods for the modeling of catalyst particles – II: Analysis of stability". *Chemical Engineering Science*, **29**, 805–810.

Carey, G.F. and Finlayson, B.A. (1975). "Orthogonal collocation on finite elements". *Chemical Engineering Science*, **30**, 587–596.

Cash, J.R. and Karp, A.H. (1990). "A variable order Runge–Kutta method for initial value problems with rapidly varying right–hand sides". *ACM Transactions in Mathematics and Software*, **16,** 201–222.

Chakraborti, C. and Sastry, K.K.N. (1997). "Genetic algorithm – an efficient alternative for proving logical arguments". *Evonews*, **5,** 17–18.

Chakraborti, C. and Sastry, K.K.N. (1998). "The genetic algorithms approach for proving logical arguments in natural language". In: J. R. Koza, W. Banzhaf, K. Chellapilla, K. Deb, M. Dorigo, D. B. Fogel, M. H. Garzon, D. E. Goldberg, H. Iba, & R. Riolo (Eds.), *Genetic programming 1998: Proceedings of the* 3rd *Annual Conference*. (pp. 463–470): San Mateo, CA. Morgan Kaufmann.

Chiou, J.P. and Wang, F.S. (1999). "Hybrid method of evolutionary algorithms for static and dynamic optimization problems with application to a fed-batch fermentation process", *Computers & Chemical Engineering*, **23**, 1277-1291.

Coberly, C.A. and Marshall, W.R. (1951). "Temperature gradients in gas streams flowing through fixed granular bed". *Chemical Engineering Progress*, **47** (3), 141–150.

Crine, M. (1982). "Heat transfer phenomena in trickle-bed reactors". *Chemical Engineering Communications*, **19**, 99–114.

Davis, L., (1991). *Handbook of genetic algorithms*. Van Nostrand Reinhold, New York.

Dollena S.H, David M.A. and Arnold J.S. (2001). "Determining the number of components in mixtures of linear models". *Computational Statistics & Data Analysis*, **38**, 15-48.

Ferguson, N.B. and Finlayson, B.A. (1970). "Transient chemical reaction analysis by orthogonal collocation". *Chemical Engineering Journal*, **1**, 327–335.

Finlayson, B.A. (1971). "Packed bed reactor analysis by orthogonal collocation". *Chemical Engineering Science*, **26**, 1081–1091.

Finlayson, B.A. (1972). *The method of weighted residuals and variational principles*. Academic Press, New York.

Finlayson, B.A. (1980). *Nonlinear analysis in chemical engineering*, Chemical Engineering Series McGraw-Hill International, New York.

Froment, G.F. (1967). Fixed bed catalytic reactors: Current design status. *Industrial & Engineering Chemistry - Research*, **59** (2), 18–27.

Goldberg, D.E. (1989). *Genetic algorithms in search, optimization, and machine learning*. Addison-Wesley, Reading, MA.

Goodson, R.E. and Polis, M.P. (1974). *Identification of parameters in a distributed systems*. A.S.M.E., New York.

Hashimoto, K., Muroyama, K., Fujiyoshi, K. and Nagata, S. (1976). "Effective radial thermal conductivity in co-current flow of a gas and liquid through a packed bed". *International Journal of Chemical Engineering*. **16**, 720–727.

Joshi, R. and Sanderson, A.C. (1999). "Minimal representation multi-sensor fusion using differential evolution". *IEEE Transactions on Systems, Man and Cybernetics, Part A* **29,** 63-76.

Karanth, N.G. and Hughes, R. (1974). "Simulation of an adiabatic packed bed reactor". *Chemical Engineering Science*, **29**, 197–205.

Kyprianou, A., Worden, K. and Panet, M. (2001). "Identification of hysteretic systems using the differential evolution algorithm". *Journal of Sound and Vibration,* **248** (2), 289-314.

Lamine, A.S., Gerth, L., Le Gall, H. and Wild, G. (1996). "Heat transfer in a packed bed reactor with co-current downflow of a gas and a liquid". *Chemical Engineering Science*, **51**, 3813–3827.

Lee, M.H., Han, C. and Chang, K.S. (1999). "Dynamic optimization of a continuous polymer reactor using a modified differential evolution". *Industrial & Engineering Chemistry Research,* **38** (12), 4825-4831.

Lu, J.C. and Wang, F.S. (2001). "Optimization of Low Pressure Chemical Vapour Deposition Reactors Using Hybrid Differential Evolution". *Canadian Journal of Chemical Engineering,* **79** (2), 246-254.

Manish, C.T., Yan Fu, and Urmila, M.D. (1999). "Optimal design of heat exchangers: A genetic algorithm framework" *Industrial & Engineering Chemistry Research,* **38**, 456-467.

Maria, G. (1998). "ARS and short-cut techniques for process model identification and monitoring". *FOCAPO98 Conference* Snowbird.

Matsuura, A., Hitaka, Y., Akehata, T. and Shirai, T. (1979a). "Apparent wall heat transfer coefficient in packed beds with downward cocurrent gas–liquid flow". *Heat Transfer – Japanese Research*, **8**, 53–60.

Matsuura, A., Hitaka, Y., Akehata, T. and Shirai, T. (1979b). "Effective radial thermal conductivity in packed beds with downward cocurrent gas–liquid flow". *Heat Transfer – Japanese Research*, **8,** 44–52.

Moros, R., Kalies, H., Rex, H.G. and Schaffarczyk, St. (1996). "A genetic algorithm for generating initial parameter estimations for kinetic models of catalytic processes". *Computers & Chemical Engineering*, **20**, 1257–1270.

Muroyama, K., Hashimoto, K. and Tomita, T. (1978). "Heat transfer from wall in gas–liquid cocurrent packed beds". *Heat Transfer – Japanese Research*, **7**, 87–93.

Perlmutter, D.D. (1972). *Stability of chemical reactors.* Prentice-Hall Publication, Englewood Cliffs, New Jersey.

Perry, R.H. and Green, D. (1993). *Perry's Chemical Engineers' Handbook*, 6th ed., McGraw Hill International Editions, New York.

Polis, M.P., Goodson, R.E. and Wozny, M.J. (1973). "Identification of parameters in a distributed systems". *Automatica*, **9**, 53–58.

Press, W.H., Teukolsky, S.A., Vellerling, W.T. and Flannery, B.P. (1996). *Numerical recipes in C: The art of scientific computing.* (2nd ed.), Cambridge University Press, Cambridge.

Price, K. and Storn, R. (1997). "Differential Evolution – A simple evolution strategy for fast optimization", *Dr. Dobb's Journal*, **22** (4), 18-24 & 78.

Price, K. and Storn, R. (2002). *Home Page on Differential Evolution as on July 2002.* http://www.ICSI.Berkeley.edu/~storn/code.html.

Ralston, A. and Rabinowitz, P. (1978). *A first course in numerical analysis.* McGraw-Hill, New York.

Sastry, K.K.N., Behera, L. and Nagrath, I.J. (1998). "Differential evolution based fuzzy logic controller for nonlinear process control", *Fundamenta Informaticae: Special Issue on Soft Computation.*

Sinnott, R.K. (1993). *Coulson & Richardson's Chemical Engineering (Design)*, **6**, 2nd ed., Pergamon Press, New York.

Smith, J.M. (1973). 'Heat & mass transfer effects in fixed bed reactors". *Chemical Engineering Journal*, **5**, 109–118.

Sorenson, J.P., Guertin, E.W. and Stewart, W.E. (1973). "Computational models for cylindrical catalyst particle". *American Institute of Chemical Engineers Journal*, **19**, 969–975.

Specchia, V. and Baldi, G. (1979). "Heat transfer in trickle bed reactors". *Chemical Engineering Communications*, **3**, 483–499.

Storn, R. (1995). "Differential evolution design of an IIR-filter with requirements for magnitude and group delay", *International Computer Science Institute*, TR-95-018.

Tsang, T.H., Edgar, T.F. and Hougen, J.O. (1976). "Estimation of heat transfer parameters in a packed bed". *Chemical Engineering Journal*, **11**, 57–66.

Upreti, S.R. and Deb, K. (1997). "Optimal design of an ammonia synthesis reactor using genetic algorithms". *Computers & Chemical Engineering*, **21**, 87–93.

Villadsen, J. and Michelsen, M.L. (1978). *Solution of differential equation models by polynomial approximation.* Prentice- Hall Publishers, Englewood Cliffs, New Jersey.

Villadsen, J.V. and Stewart, W.E. (1967). "Solution of boundary-value problems by orthogonal collocation". *Chemical Engineering Science*, **22**, 1483–1501.

Wang, F.S. and Cheng, W.M. (1999). "Simultaneous optimization of feeding rate and operation parameters for fed-batch fermentation processes", *Biotechnology Progress,* **15** (5), 949-952.

Wang, F.S., Jing, C.H. and Tsao, G.T. (1998). "Fuzzy-decision-making problems of fuel ethanol production using genetically engineered yeast", *Industrial & Engineering Chemistry Research,* **37**, 3434-3443.

Wang, F.S., Su, T.L. and Jang, H.J. (2001). "Hybrid Differential Evolution for Problems of Kinetic Parameter Estimation and Dynamic Optimization of an Ethanol Fermentation Process". *Industrial & Engineering Chemistry Research,* **40** (13), 2876-2885.

Weekman, V. W. and Myers, J. E. (1965). "Heat transfer characteristics of concurrent gas–liquid flow in packed beds". *American Institute of Chemical Engineers Journal*, **11**, 13–17.

Wolf, D. and Moros, R. (1997). "Estimating rate constants of heterogeneous catalytic reactions without supposition of rate determining surface steps – an application of genetic algorithm". *Chemical Engineering Science*, **52**, 1189–1199.

Young, L.C. and Finlayson, B. A. (1973). "Axial dispersion in non isothermal packed bed chemical reactors". *Industrial & Engineering Chemistry Fundamentals*, **12** (4), 412–422.

10 Applications in Mass Transfer

B V Babu

10.1 Introduction

In this chapter, we shall discuss two successful applications of Differential Evolution (DE) on mass transfer problems: (1) Optimization of Liquid Extraction Process, (2) Optimization of a Separation Train of Distillation Columns, and (3) Optimization and Synthesis of Heat Integrated Distillation Column Sequences.

In recent years, evolutionary algorithms (EAs) have been applied to the solution of NLP in many engineering applications. The best-known algorithms in this class include Genetic Algorithms (GA), Evolutionary Programming (EP), Evolution Strategies (ES) and Genetic Programming (GP). There are many hybrid systems, which incorporate various features of the above paradigms and consequently are hard to classify, which can be referred just as EC methods. They differ from the conventional algorithms since, in general, only the information regarding the objective function is required. In recent years, EC methods have been applied to a broad range of activities in process system engineering including modeling, optimization and control (Androulakis and Venkatasubramanian, 1991; McKay et al., 1997; Edwards et al., 1998; Chiou and Wang, 1999; Babu and Angira, 2001a; Babu and Angira, 2001b; Babu and Angira, 2002a; Babu and Angira, 2002b; Babu and Chaturvedi, 2000; Babu and Gautam, 2001; Babu et al., 2002; Angira and Babu, 2003). Differential Evolution (DE), developed by Storn & Price (1997), is one of the best EC methods. This method provides one of the best genetic algorithms for solving the real-valued test function. The convergent speed of the DE is very fast.

10.2 Optimization of Liquid Extraction Process

Liquid-Liquid extraction is a mass transfer operation in which a liquid solution (the feed) is contacted with an immiscible or nearly immiscible liquid (solvent) that exhibits preferential affinity or selectivity towards one or more of the components in the feed. Two streams result from this contact: the extract, which is the solvent rich solution containing the desired extracted solute, and the raffinate, the residual feed solution containing little solute.

Liquid-liquid extraction is a powerful separation technique which falls right behind distillation in the hierarchy of separation methods (Fig. 10.1). In situations where distillation is not feasible for reasons such as a complex process sequence, high investment or operating costs, heat sensitive materials, or low volatility, extraction is often the best technology to use.

LLE is carried out either (1) in a series of well-mixed vessels or stages (well mixed tanks or plate columns or (2) in a continuous process such as spray columns, packed columns, and rotating disk columns. This example illustrate the application of Evolutionary Computation method such as Differential Evolution (DE) to the optimization of a LLE system represented by a plug flow model.

Steady state continuous countercurrent liquid extraction can be modeled in a variety of ways, the most common of which are (1) a plug flow model and (2) an axial dispersion model. Jackson and Agnew (1980) demonstrated the effectiveness of an on-line model based steady-state optimization scheme in finding and holding the optimum operation conditions of a liquid extraction pilot plant in which acetic acid was extracted from amyl alcohol using water as the solvent. The equipment could be operated either manually or under computer control. Under automatic operation, the computer could maintain the interface level, feed and solvent flow rates, and the stirrer speed to their respective set points. The latter three set points were able to be changed by the optimization routine. The interface level was controlled via the extract flow. The measured variables, for both control and optimization purposes were the feed and solvent flow rates; the feed, raffinate and extract concentrations; the stirrer speed; the interface position; and the feed temperature. Further details are given in Jackson (1977).

The process operation was subject to upper and lower limits on feed and solvent phase superficial velocities and the stirrer speed, and to minimum throughput and flooding constraints. For use I an optimization scheme, a process model is required to be able to predict the steady-state process output for a given set of inputs. These predicted values can then be used to calculate the value of the performance function. To be of use in an on-line scheme, the model must be amenable to solution without excessive computational effort. They examined the accuracy of four models viz., a model based on plug flow of both phases, one based on axial dispersion superimposed on the flow of both phases, and two empirical models, one linear and one nonlinear) for a continuous pilot-scale extraction column in which water was used to extract acetic acid from amyl alcohol. The linear and non-linear models were direct correlations of experimentally obtained mass transfer data.

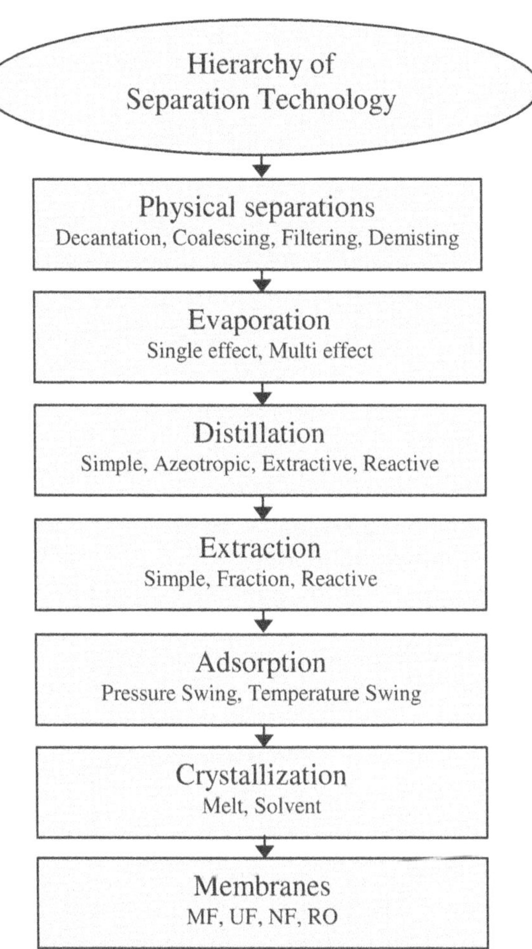

Fig. 10.1. Hierarchy of Separation Technologies

A statistical test (an F test or similar) on the model predictions of the process output variables provided a reasonable indication as to whether or not the model would be acceptable for use in a model-based optimization scheme. Two empirical models were rejected, as they were not fitting the data. Both plug-flow and axial-diffusion models enabled the correct optimum to be predicted in all cases, the differences in performance when these models were used were negligible, as there was little axial mixing of the phases in the Rotating Disc Contactor. Also, the marginally superior performance of the axial diffusion model was offset by the greater complexity of its solution compared to the plug flow model. Hence, the plug flow model offered the best compromise between predictive ability and complexity because of its greater simplicity. The same is used in the present study too. Once a model is specified, it can be used to determine the maximum extraction rate. A typical column is shown in Fig. 10.2. The process model, the objective function, and the constraints, are described below.

10.2.1 Process Model

Assuming that the concentrations are expressed on a solute free mole basis and that the equilibrium relation between Y and X is a straight, i.e., the phases are insoluble. The model is then given as below:

$$\frac{dX}{dZ} - N_{OX}\left(X - Y\right) = 0 \tag{10.1}$$

$$\frac{dY}{dZ} - FN_{OX}\left(X - Y\right) = 0 \tag{10.2}$$

where

F = extraction factor (mv_X / v_Y)
m = distribution coefficient ($m = 1.5$)
N_{OX} = number of transfer units
v_X, v_Y = superficial velocity in raffinate, extract phase
X = dimensionless raffinate phase concentration
Y = dimensionless extract phase concentration
Z = dimensionless contactor length

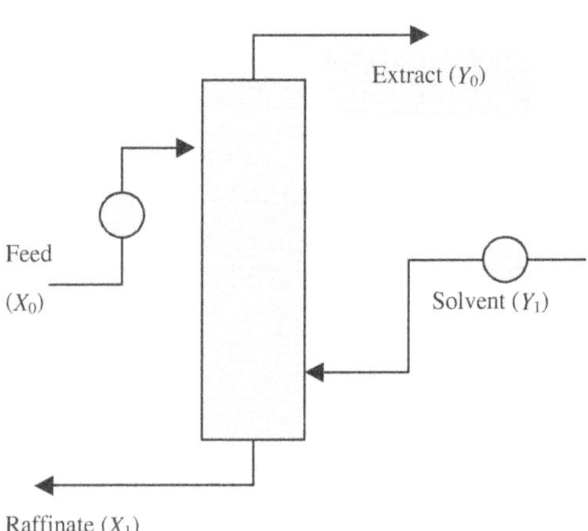

Fig. 10.2. Extraction Column, schematic

Fig. 10.2 shows the extraction column with boundary conditions X_0 and Y_1. A solution for Y_0 in terms of v_x and v_Y can be obtained, given the values for m, N_{OX}, and the length of the column.

Hartland and Mecklenburgh (1975) list the solution for the plug flow model (and also the axial dispersion model) for a linear equilibrium relationship, in terms of F:

$$Y_0 = \frac{F[1 - \exp\{N_{OX}(1 - F)\}]}{1 - F\exp[N_{OX}(1 - F)]} \quad (10.3)$$

For the plug flow and axial diffusion models, Jackson and Agnew (1980) summarized a number of correlations for N_{OX} given by an equation of the form:

$$N_{OX} = a\left(\frac{v_X}{v_Y}\right)^b (N)^c \quad (10.4)$$

The correlations obtained by non-linear least-squares regression were:

$$N_{OX} = 1.7\left(\frac{v_X}{v_Y}\right)^{0.24} (N)^{0.069} \text{ for } 2.08 \le N \le 4.2s^{-1} \quad (10.5a)$$

$$N_{OX} = 0.2\left(\frac{v_X}{v_Y}\right)^{0.24} (N)^{1.5} \text{ for } 4.2 \le N \le 8.33s^{-1} \quad (10.5b)$$

$$N_{OX} = 0.18\left(\frac{v_X}{v_Y}\right)^{0.18} (N)^{0.088} \text{ for } 2.08 \le N \le 4.3s^{-1} \quad (10.5c)$$

$$N_{OX} = 0.09\left(\frac{v_X}{v_Y}\right)^{0.24} (N)^{2.1} \quad \text{for } 4.3 \le N \le 8.33s^{-1} \quad (10.5d)$$

In the present study, equation (10.5b) is used with $N = 8.33s^{-1}$.

$$N_{OX} = 4.81\left(\frac{v_X}{v_Y}\right)^{0.24} \quad (10.6)$$

10.2.2 Objective function

We have used the same objective function as proposed by Jackson and Agnew (1980) i.e., to maximize the total extraction rate for constant disk rotation speed subject to the inequality constraints:

$$\text{Maximize } f = v_Y Y_0 \quad (10.7)$$

10.2.3 Inequality constraints

Implicit constraints exist because of the use of dimensionless variables

$$X_0 \le X \le X_1$$
$$Y_1 \le Y \le Y_0 \quad (10.8)$$

Constraints on v_X and v_Y would be upper and lower bounds such as:

$$0.05 < v_x < 0.25$$
$$0.05 < v_y < 0.30 \quad (10.9)$$

and flooding constraint is:

$$v_x + v_y \le 0.20 \quad (10.10)$$

10.2.4 Results & Discussion

Jackson and Agnew (1980) used a modified gradient-projection technique for linearly constrained optimization problems developed by Jackson (1977). Edgar & Himmelblau (1989) used GRG (generalized reduced gradient method) and obtained the same optimum (0.15, 0.05) as Jackson and Agnew. The objective function is 0.225 i.e. the true optimum lay on the flooding constraint.

In the present study, the DE (differential evolution), an evolutionary technique, is applied. Apart from the well known seventh strategy (DE-7) i.e., DE/rand/1/bin, the two new strategies (NS-1& NS-2) have been applied. The psuedo codes for DE-7, NS-1 and NS-2 are given below:

(a) For *DE-7*

```
{
    If (random number {0,1})<CR || k==D)
            {
                trial[j]=x1[c][j]+F*(x1[a][j]-x1[b][j]);
            }
                else trial[j]=x1[i][j];

            j=(j+1)%D;
}
```

(b) For *NS-1*

```
{
    If (random number {0,1})<CR || k==D)
            {
                trial[j]=x1[c][j]+F*(x1[a][j]-x1[b][j]);
            }
                else trial[j]=bestit[j]+F*(x1[a][j]-x1[b][j]);

            j=(j+1)%D;
}
```

(c) For *NS-2*:

```
{
    If (random number {0,1})<CR || k==D)
            {
             trial[j]=x1[c][j]+F*(x1[a][j]-x1[b][j])+(1-F)*(x1[d][j]-x1[e][j]);
            }
             else trial[j]=bestit[j]+F*(x1[a][j]-x1[b][j]);

            j=(j+1)%D;
}
```

Fig. 10.3 and Fig. 10.4 show the results obtained using DE-7, NS-1 and NS-2 and present the comparison of DE-7 with NS-1 & NS-2. The stopping criterion adopted is to terminate the search process when one of the following conditions is satisfied: (1) the maximum number of generations is reached (assumed 2000 generations). (2) $| f^k_{\max} - f^k_{\min} | < 10^{-5}$ where f is the value of objective function for k-th generation. After the mutation & recombination, if the bound (i.e. lower & upper limit of a variable) is violated then it is brought in the bound range (i.e. between lower & upper limit) by randomly assigning a value in the bound range (without forcing).

In Fig. 10.3 & Fig. 10.4, NRC & NFE represent respectively, the mean number of objective function evaluations and the percentage of runs converged to the global optimum in all the 10 executions (with different seed values). The code was run for all possible combination of *F* & *CR* (0.0<*F*<=1; and 0.0<*CR*<=1.0), where *F* is scaling factor & *CR* is crossover constant.

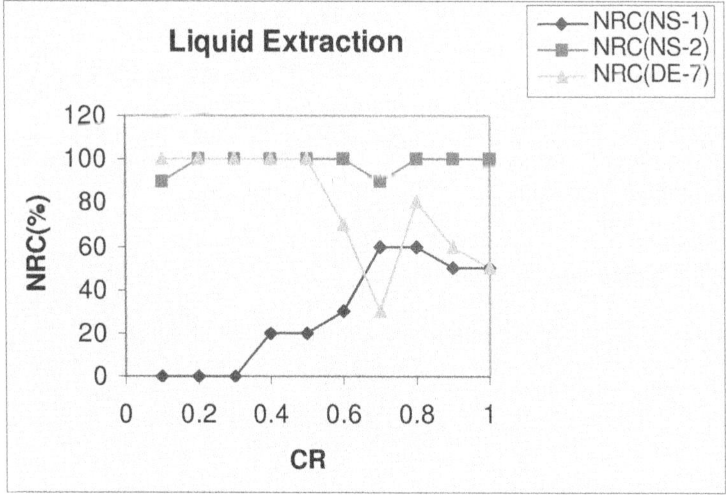

Fig. 10.3. NRC vs. CR (for F = 0.5)

Fig. 10.3 presents NRC vs. *CR* results obtained for $F = 0.5$ for all three strategies i.e., DE-7, NS-1 and NS-2. The NRC are more in case of new strategy (NS-2) as compared to DE-7 & NS-1. The new strategy NS-2 seems to be robust than DE-7 & NS-1 due to nearly 100% conversion to global optimum. Also, it is clear that till $CR = 0.3$ the NRC are nill for NS-1 while it is almost 100% for DE-7 & NS-2.

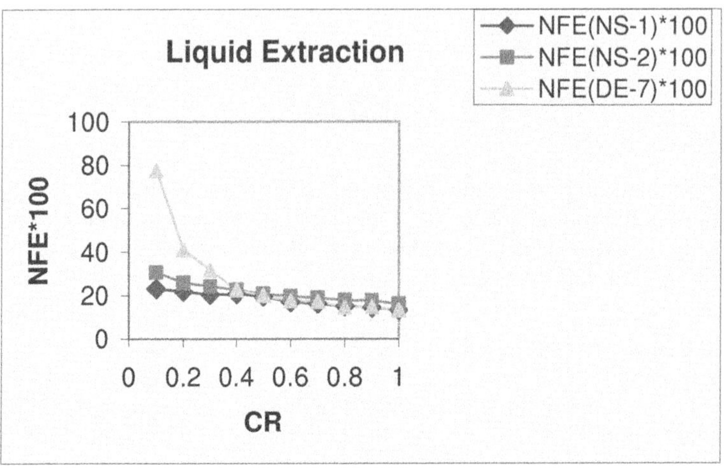

Fig. 10.4. NFE vs. CR (for F = 0.5)

From the Fig. 10.4, it is evident that NFE are least in case of NS-1. Also, difference of NFE between NS-1 & DE-7 becomes less after $CR = 0.5$ and both NS-1& DE-7 have same NFE at $CR = 1.0$. NFE are 3.55%, 6.56%, 1.24% more in case of DE-7 than the NS-1 at $CR = 0.6, 0.7, 0.9$ respectively.

From above Fig. 10.4 we can observe that till $CR = 0.4$, the NFE for NS-2 are less than that of DE-7. But for $CR > 0.4$ the NFE are slightly more than either DE-7 or NS-1. At $CR = 0.5$, the NFE are 6.65% and 4.41% more than that of NS-1 and DE-7 respectively. Although the NFE are slightly more for NS-2 but NRC is nearly 100% indicating the robustness of the strategy.

10.2.5 Conclusions

In the present study, the problem of optimization of liquid extraction has been solved using DE and two new DE strategies. It is found that that performance of DE is slightly better than that of NS-1. But NS-2 seems to be the best choice and certainly better than DE due to almost 100% NRC which indicate the robustness of this strategy.

10.3 Optimization of a Separation Train of Distillation Columns

In this example we illustrate use of DE (Differential Evolution) to optimize the operation of the solvent splitter column shown in figure-5. The feed is naphtha which has a value of \$42/bbl in its alternate use as a gasoline blending stock. The light ends sell at \$53/bbl while the bottoms are passed through a second distillation column to yield two solvents. A medium solvent containing 50 to 70% of the bottoms can be sold for \$68/bbl, while the remaining heavy solvent (containing 30 to 50% of the bottoms) can be sold for \$42/bbl. Another part of plant requires 200 bbl/day of medium solvent and an additional 200 bbl/day can be sold to an external market. The maximum feed that can be processed in first column is 2000 bbl/day. The operational costs (i.e., utilities) associated with each distillation column are \$1.25/bbl feed. The operating range for second column is given as the percentage split of medium and heavy solvent.

10.3.1 Problem formulation

The problem is formulated by developing the objective function and constraints. The five variables are identified (Fig. 10.5). The objective is to maximize the profit. The equality constraints arise from material balances and the inequality constraints from restrictions.

Let us define the operating variables as follows (all are in units of bbl/day):

x_1 = naphtha feed rate
x_2 = bottoms production rate
x_3 = light ends production rate
x_4 = medium solvent production rate
x_5 = heavy solvent production rate

10.3.1.1 Objective Function

The objective function in terms of profit is:
Maximize f = medium solvent sales + heavy solvent sales + light ends sales – feed cost – operational costs for each distillation column

$$\text{Max. } f = 68x_4 + 42x_5 + 53x_3 - 42x_1 - 1.25x_1 - 1.25x_2 \quad (10.11)$$

10.3.1.2 Inequality constraints

The inequality constraints for the ranges of production are:

$$\frac{x_4}{x_2} \geq 0.50 \quad \& \quad \frac{x_5}{x_2} \geq 0.30 .$$

Rearranging and substituting for x_2 yields:

$$x_4 - 0.5x_2 \geq 0 \Rightarrow x_4 - 0.5(x_4 + x_5) \geq 0 \Rightarrow x_4 - x_5 \geq 0 \tag{10.12a}$$

$$x_5 - 0.3x_2 \geq 0 \Rightarrow x_5 - 0.3(x_4 + x_5) \geq 0 \Rightarrow 0.7x_5 - 0.3x_4 \geq 0 \tag{10.12b}$$

Similarly,

$$\frac{x_4}{x_2} \leq 0.70 \quad \& \quad \frac{x_5}{x_2} \leq 0.50 \quad \text{yields}$$

$$0.3x_4 - 0.7x_5 \leq 0 \tag{10.12c}$$

$$x_5 - x_4 \leq 0 \tag{10.12d}$$

The constraints (10.12a) & (10.12d) and constraint (10.12b) & (10.12c) are the same. Therefore only two of the constraints are independent. The other inequality constraints (supply and operating constraints) are:

$$200 \leq x_4 \leq 400$$

$$x_1 \leq 2000 \quad \Rightarrow 1.667(x_4 + x_5) \leq 2000 \text{ Or } (x_4 + x_5) \leq 1200$$

Also,

$$x_5 \geq 0 .$$

Thus, total numbers of inequality constraints are:

$$x_4 - x_5 \geq 0 \tag{10.13}$$

$$0.7x_5 - 0.3x_4 \geq 0 \tag{10.14}$$

$$200 \leq x_4 \leq 400 \tag{10.15}$$

$$x_4 + x_5 \leq 1200 \tag{10.16}$$

$$x_5 \geq 0 \tag{10.17}$$

10.3.1.3 Equality constraints

(1). Total Material Balance:

$$x_1 = x_3 + x_4 + x_5 \tag{10.18}$$

(2). Material balance on column 2:

$$0.6x_1 = x_4 + x_5 \Rightarrow x_1 = 1.667(x_4 + x_5) \tag{10.19}$$

(3). Material balance on column 1:

$$x_3 = 0.4x_1 \qquad \Rightarrow x_3 = 0.667(x_4 + x_5) \tag{10.20}$$

$$x_2 = 0.6x_1 \qquad \Rightarrow x_2 = (x_4 + x_5) \tag{10.21}$$

10.3.1.4 Problem Reformulation

We simplify the objective function by substitution of the equality constraints (10.19), (10.20) & (10.21) into equation (10.11) so that it is expressed in terms of the two independent variables, x_4 and x_5. The final objective function is:

$$\text{Max. } f = 30x_4 + 4x_5 \tag{10.22}$$

The optimum values of x_4 and x_5 are:
x_4 = 400 bbl/day
x_5 = 800 bbl/day
The optimal profit is $13600/day.

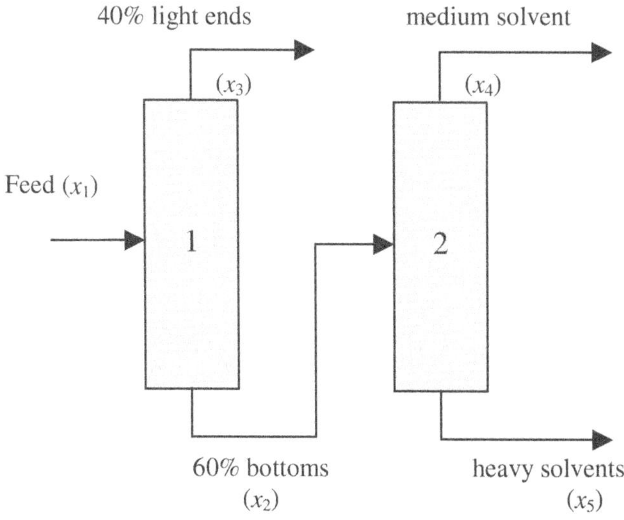

Fig. 10.5. Train of distillation columns

10.3.2 Results & Discussion

Edgar & Himmelblau (1989) used Linear programming Simplex method and obtained the same optimum as stated above. In the present study, the DE (differential evolution), an evolutionary technique, is applied. Apart from the well known seventh strategy (DE-7) i.e., DE/rand/1/bin, the two new strategies (NS-1& NS-2) have been applied.

Fig. 10.6 & Fig. 10.7 show the results obtained using DE-7, NS-1 and NS-2 and present the comparison of DE-7 with NS-1 & NS-2. The stopping criterion adopted is to terminate the search process when one of the following conditions is satisfied: (1) the maximum number of generations is reached (assumed 2000 generations). (2) $| f^k_{max} - f^k_{min} | < 10^{-5}$ where f is the value of objective function for k-th generation. After the mutation & recombination, if the bound (i.e. lower & upper limit of a variable) is violated then it is brought in the bound range (i.e. between lower & upper limit) by randomly assigning a value in the bound range (without forcing).

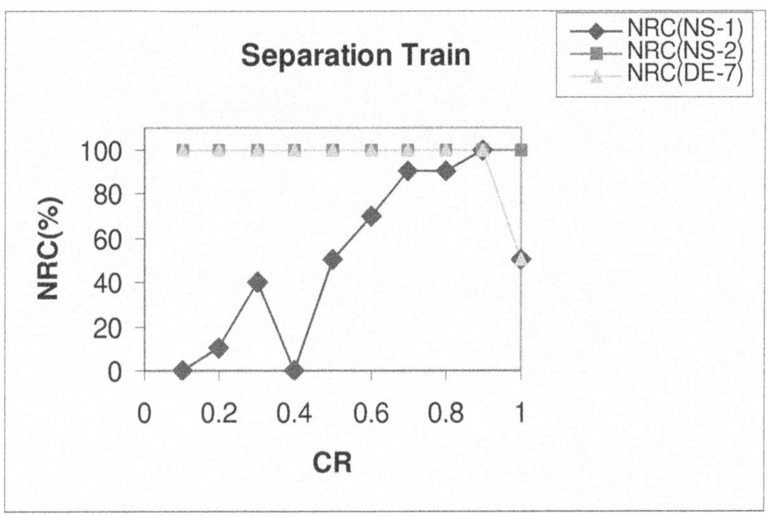

Fig. 10.6. NRC vs. CR (for F = 0.5)

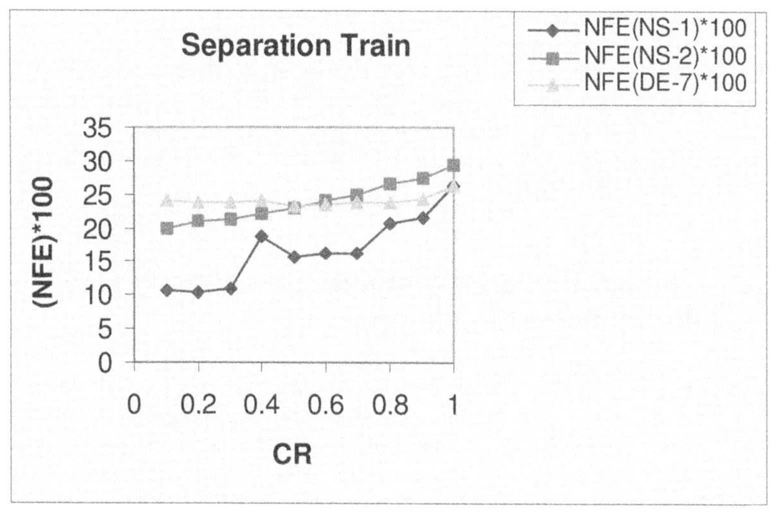

Fig. 10.7. NFE vs. CR (for F = 0.5)

In Fig. 10.6 & Fig. 10.7, NFE & NRC represent respectively, the mean number of objective function evaluations and the percentage of runs converged to the global optimum in all the 10 executions (with different seed values). The code was run for all possible combination of *F* & *CR* ($0.0<F<=1$; and $0.0<CR<=1.0$), where F is scaling factor & *CR* is crossover constant.

Fig.10.6 presents NRC vs. *CR* results obtained for $F = 0.5$ for all three strategies i.e., DE-7, NS-1 and NS-2. The NRC are more (100%) in case of new strategy (NS-2) as compared to DE-7 & NS-1. Also, it is clear that for $CR = 0.1$ & 0.4 the NRC are nill for NS-1 while it is 100% for DE-7 & NS-2. Also, the NRC at $CR = 1.0$ are 100% for NS-2 but it is only 50% for both DE-7 & NS-1. Evidently, the new strategy NS-2 seems to be robust than DE-7 & NS-1 due to 100% conversion to global optimum for all values of *CR* ($0.1 \leq CR \leq 1.0$)

From the Fig. 10.7, it is evident that NFE are least in case of NS-1. Also, difference of NFE between NS-1 & DE-7 becomes less after $CR = 0.5$ and both NS-1& DE-7 have same NFE at $CR = 1.0$. NFE are 45.8%, 46.91%, 15.24% & 13.29% more in case of DE-7 than the NS-1 at $CR = 0.6$, 0.7, 0.8 & 0.9 respectively.

We can observe from Fig. 10.7 that till $CR = 0.5$, the NFE for NS-2 are less than that of DE-7. But for $CR >0.5$ the NFE are slightly more than the DE-7. At $CR = 0.6$, the NFE in NS-2 are 49.44% and 2.46% more than that of NS-1 and DE-7 respectively. Although the NFE are slightly more for NS-2 but NRC is 100% for all values of *CR* ($0.1 \leq CR \leq 1.0$) indicating the robustness of this strategy.

10.3.3 Conclusions

In the present study, the problem of Optimization of a Separation Train of Distillation Columns has been solved using DE and two new DE strategies. It is found that that performance of DE is slightly better than that of NS-1. But NS-2 seems to be the best choice and certainly better than DE due to 100% NRC which indicate the robustness of this strategy.

10.4 Optimization and Synthesis of Heat Integrated Distillation Column Sequences

Distillation is most widely used in separation process in the chemical industry. It is also a highly energy intensive unit operation, with some process consuming one third or more of their energy in distillation alone. These systems are widely used in the manufacturing of petroleum and petrochemical products. Distillation systems consume large amount of energy to achieve separation. In view of increasing cost, it is important to design distillation systems, which consume less energy. Synthesis and design of these systems lies in the hands of human experts. The goal of energy conservation as an integral part of process synthesis is the main challenge in computer – aided design.

By definition, synthesis of heat integrated distillation systems will require optimization of following areas:

1. Determine the sequence of individual columns
2. Design of column sequence and their operating parameters e.g. pressure
3. Designing heat exchanger network

In last three decades, significant advances have been made in these areas. Heuristics methods are proposed by a number of researchers. These rules play important roles in reducing the huge search space when coupled with heat integration. Howevwr, lack of complete and systematic knowledge required often causes difficulty and leads to low quality solutions.

Decomposing approaches (Jaroslave, 1988; Rong, 1996) used pinch technology. Pinch technology, though has been proved to be very successful in engineering design, is useful for analysis of flow sheet. Many researchers (Andercovich, 1985; Kravanja and Grossman, 1990; Zorka et al., 1994; Zdravko, K., and Grossman, 1997) have suggested mixed integer linear or non-linear programming methods (MIL/NLP). Once one of the main solution techniques was found complex and computationally expensive for mathematical programming. In such techniques many simplification have to be assumed to make mathematical models manageable, which pose several limitations in practical applications.

All these advances show considerable degree of heat integration into model formulation, simultaneous optimization methods have not been developed in the way, which would allow efficient flow sheet optimization. Still there is a need to explore the alternative strategies for the synthesis and optimization of heat integrated – separation sequence and the heat exchanger network.

In synthesis and optimization of heat integrated columns variables are continuous and mixed integer search space persists. Mathematically, our goal is to search for minimum in such search space. Random search algorithms like genetic algorithm are capable of search in a multidimensional search space with many peaks and valleys. Improved genetic algorithm inherits its features from GA. Already using IGA; research has been reported (Wang et al, 1998). Differential evolution has shown its merits over GA and IGA in its robustness, ease of implementation and fast convergence. It has shown better result in an industrial case study for process simulation (Babu & Sastry, 1999). Differential evolution (Price and Storn, 1997) is another, yet relatively new, random search technique. It's faster in finding in global optimum for industrial applications with huge search space. Just 20 lines of C code are required for implementation. Inspired by their work we present and application of differential evolution to challenge the task.

10.4.1 Problem formulation

The problem to be addressed can be stated as follows:

Given a multi component feed stream of known conditions, it is to be separated into a number of products of specified compositions. The problem is to synthesize the optimal distillation sequence and its HEN, allowing use of sharp separators and minimize total annual cost. Basic assumptions made in this problem are as follows:

1. Heating and cooling requirement are not directly provided by hot and cold utilities. Heat integration should be taken into consideration for economical technological reasons. Pressures are selected from the range of allowable pressures. This allows column pressure with in an allowable range in order to carry out heat integration. Pressures of columns are treated as continuous variables.
2. Each distillation column performs a simple split (i.e. one feed and two products) and sharp splits. This excludes complex columns from the scope of this work, and is due to primary reasons first there are number of industrial examples there only simple columns (non sharp of sharp) are used. Second, I consider the approach in this problem as a step in direction understanding general separation sequences. Future work will attempt to incorporate complex columns.
3. Column feeds are the same as the outputs of the upstream columns. It simplifies the modal heat integration.

10.4.1.1 New method for representation of separation sequence

The superstructure for the separation sequence is of great importance, which is basis for modeling of synthesis and optimization. Cheng and Liu (1988) introduced component assignment diagram for representation of problem. Yee et al., (1990) proposed a network structure rather than tree representation. However, both of them need a large space and two products) superstructure would contain (N-1) columns for a feed stream containing N components. There are (2N-1) possible sequences in total. In this work I propose a (2N-1) dimensional array of integers *I* (2N-1) to show all possible sequences. The strategy is illustrated by an example of five components feed streams as follows:

1. N-1 integers are defined to show the decompositions point.

A	↑	B	↑	C	↑	D	↑	E
	1		2		3		4	0

shows one of components have been separated. A, B, C, D, E are refereed to as five components of a mixture which has been ranked according to physical properties as relative volatility.

2. *I* (2N-1) represent all possible sequences of feed stream, which is shown in the Table-10.1 below.

Table 10.1. Representation of I [2N-1] sequences for whole population

Sequences	Column 1	Column 2	Column 3	Column 4	I (2N-1)
1	ABCD/E	ABC/D	AB/C	A/B	{4,3,0,2,0,1,0,0,0}
2	ABCD/E	ABC/D	A/BC	B/C	{4,3,0,1,0,0,2,0,0}
3	ABCD/E	AB/CD	A/B	C/D	{4,2,0,1,3,0,0,0,0}
4	ABCD/E	A/BCD	B/CD	C/D	{4,1,0,0,2,0,3,0,0}
5	ABCD/E	A/BCD	BC/D	B/C	{4,1,0,0,3,2,0,0,0}
6	ABC/DE	A/BC	D/E	B/C	{3,1,4,0,2,0,0,0,0}
7	ABC/DE	AB/C	D/E	A/B	{3,2,4,1,0,0,0,0,0}
8	AB/CDE	A/B	CD/E	C/D	{2,1,4,0,0,3,0,0,0}
9	AB/CDE	A/B	C/DE	D/E	{2,1,3,0,0,0,4,0,0}
10	A/BCDE	BCD/E	BC/D	B/C	{1,0,4,3,0,2,0,0,0}
11	A/BCDE	BCD/E	B/CD	C/D	{1,0,4,2,0,0,3,0,0}
12	A/BCDE	BC/DE	B/C	D/E	{1,0,2,0,4,3,0,0,0}
13	A/BCDE	B/CDE	CD/E	C/D	{1,2,0,4,0,3,0,0,0}
14	A/BCDE	B/CDE	C/DE	D/E	{1,0,2,0,3,0,4,0,0}

3. *I* (2N-1) is generated automatically and randomly. Taking the example of ABCD/E, A/BCD, BC/D, B/C which is shown in the table, generation of *I* (2*5-1) is demonstrated as follows:

- Two integer variables a and b are defined to stand for the lower and upper limits of component number in the mixture, respectively is 0 while b is 5.
- For the first column *I* [1] = random (a, b) = 4, meaning top product is ABCD and bottom product is E for the first column.
- *I* [1] - a = 4, there are four components in the top product of the first column, which need further columns to separate. *I* [2] = random (a, *I* [1] = random (0, 4) = 1, which stands for the second column as top product A and bottom product BCD. Since b - *I* [1] = 1, bottom product of the first column is the pure product. Then *I* [3] = 0, standing for the pure product E, as one product has been separated, b <= b-1 = 4.

- I [2] - a = 1, standing for pure product in the top stream of the second column, then I [4] = 0. As A has been separated, a <= a + 1; for the bottom product is BC and the bottom product is D for the third column.
- I [5] - a = 2, there are two components in the top products of the third column. I [6] = random (a, I [2]) = 2; which stands for the fourth column with the top product B and bottom product C. for the bottom product of third column, as b - I [5] = 1, I [7] = 0, b <= b-1 = 3.
- I [6] - a = 1, the top product of the fourth column is a product so I [8] = 0, a <= a+1 = 2. For the bottom product of the fourth column, b- I [6] = 1, so I [9] = 0, b <= b-1 = 2.
- Since a = b, synthesis is finished, I = {4,1,0,0,3,2,0,0,0}.
- Merits of this coding method are obvious although the number of the feasible distillation sequences will increase exponentially with increase of the number of components, dimensions of array of integers I [2N-1] increase linearly; this new method provides a basis for DE to optimize the distillation sequence.

10.4.1.2 Heat integration strategy

For each sequence, compositions of feed streams, top products and bottom products of all N-1 columns are given. When operating conditions such as operating pressure are also given, the short cut simulation provides information of temperatures and condenser and reboiler, heat duties of top and bottom product streams. All units receive energy from environment and give off energy to environment. From the view point of process energy utilization's condensers and reboilers can be retreated as hot streams and cold streams, respectively. In this way the problem of heat integration is transformed into a synthesis problem of constrained HEN. This problem involves N-1 hot streams; N-1 cold streams and several levels of hot and cold utilities depending on temperatures of the units. Then the strategy to synthesize the HEN using DE can be used.

10.4.1.3 The cost model

The cost model involves a representation of cost and sizing requirement of distillation columns and HEN. The cost module method of Gutheri (1969) can determine cost of columns from operating pressure, shell size and material, type and numbers of trays (valve trays were used for all costing methods). The cost module method of Gutheri (1969) can also provide cost of HEN, which includes the investment and operating cost for heat exchangers.

10.4.2 Synthesis of Distillation system

Solutions are represented as strings of continuous numbers in order. Each string is composed of three parts:

1. The first part of the string is I [2N-1], representing all possible sequences of distillation. These are shown in the table.
2. The second part is O [m x (N-1)]. N-1 columns are needed for separation of N-components mixture. Each column is associated with m operating parameters.
3. The part is Q [(N-1)(N-1)], representing heat transferred in the heat exchanger between matched streams. A heat exchanger network involves N_H hot streams and N_C cold streams. If the network is acyclic, there are at most $Z = N_H \times N_C$ exchangers (Ponton and Donaldson, 1974).

As mentioned above, the problem of heat integration will be changed into a synthesis HEN, which involves N-1 hot streams, N-1 cold streams. For the structure including at most Z [(N-1)(N-1)] heat exchangers for an acyclic network, we can set up a matrix. The number of rows and columns of entry in the matrix denote the cold and hot streams, respectively. Instead of identifying each entry in the synthesis matrix by its row I and column j, a transformation is defined as follows:

$L = j + (N-1)(I-1)$, $I = 1,2,\ldots\ldots\ldots\ldots,$ N-1.

$J = 1, 2,\ldots\ldots\ldots\ldots\ldots\ldots\ldots\ldots,$ N-1.

This maps each (I, J) entry into a unique number L which ranges from 1 to Z. Each entry in the synthesis matrix may be denoted by its equivalent L number. Thus array of Q [(N-1)(N-1)] can represent all heat exchangers between process streams.

In order to generate a feasible population of the first generation, the value of Q [L] is taken randomly in a feasible range, from 0 to FQ [L] (the maximal permissible heat transferred of the entry L). For the solution to be feasible, FQ [L] must be known at first. They are determined as follows to satisfy the constraints of both heat balance and temperature approach. We refer to $T_H(I)$ as the source and target temperature of heat stream I, $T_C(j)$ the target and source temperature of cold stream j. Then

$FQ[L] = 0 \qquad (T_H(I) < T_C(J))$

$FQ[L] = \min(Q_H(I), Q_C(J)) \qquad (T_H(I) > T_C(J))$

Where $Q_H(I)$ is the residual heat duty of stream I; $Q_C(J)$ is the residual heat duty of cold stream j.

Note that the FQ [L] are determined not simultaneously, but successively. The order is determined by matches between I =1,...........,N-1; J =1,..........N-1. It is apparent that Q_H (I) and Q_C (J) will change from match to match. For example, if we refer the pressure as the operating parameters of each column, string for a distillation separation process diagram of five components, N = 5, three parts of the string are represented as follows:

{I [2x5-1]; O [1x(5-1); Q [(5-1)(5-1)]} = {I [9]; O [4]; Q [16]} = {4, 3, 0, 2, 0, 1, 0, 0, 0; p_1, p_2, p_3, p_4; Q_1, Q_2,......................., Q_{16}}.

The first part I [9] stands for distillation sequence {ABCD/E, ABC/D, AB/C, A/B}; the second part O [4] stands for pressure of each column. In the example, p_1, p_2, p_3, p_4 are pressure of columns ABCD/E, ABC/D, AB/C, A/B respectively. They are taken within the feasible range. There are two points, however, that should be noted.

1. Pressures are referred to as top pressure for each column, while bottom pressure of the column should plus an approach of the pressure depending on the number of trays.
2. The lower bounds of the pressure should be equal to the pressure of the previous column, since there is no constraint that temperature of the feed stream should be lower than that of bottom products.
3. This part is Q [16] representing heat transferred in heat exchangers between matched streams. If Q [16] are given randomly as {100, 0, 0, 40, 0, 0, 200, 0, 0, 0, 0, 0, 0, 160, 0, 0}, Q_H (I); Q_C (I) are given as {140, 320, 150, 160} and {100, 160, 240, 160} respectively. Then HEN is obtained, which is shown in Table-10.2 below.

Table 10.2. Heat Exchanger Network Obtained

	H1	H2	H3	H4	Hot Utility
C1	100	0	0	40	0
C2	0	0	200	0	120
C3	0	0	0	0	150
C4	0	160	0	0	0
Cold Utility	0	0	40	120	

10.4.2.1 Generation of First feasible generation and cost estimation

The main steps for generating the first feasible generations are as follows:
1. The first part of the string I [2N-1] is generated randomly.
2. The second part O [m x (N-1)] is generated randomly with in the given feasible range.
3. From operating conditions determined by the second part of the string, temperature and heat duty of the condenser and reboiler are calculated using the short cut method (Douglas, 1988). Condenser and reboiler are treated as hot streams and cold streams with fixed inlet and outlet temperature and flow heat capacity, and thus HEN is developed.
4. The third part of Q [16] is also taken randomly with in the calculated feasible range. Cost of columns is calculated of Gutherie (1969). From the first two parts of the strings. Heat duty of utility is calculated by heat balance of each stream from the third part of the strings. The total cost of operating and heat exchangers is calculated.

10.4.3 Results & Discussion

To illustrate the suitability of the proposed algorithm for the design of heat integrated distillation system following example is taken:

Example: Heat integration of propanol separation problem. The required data is presented in Table-10.3 below:

Table 10.3. Data for Example

Component	Feed fraction
A ethanol	0.25
B I-propanol	0.15
C n-propanol	0.10
D I-butanol	0.35
E n-butanol	0.15

Feed flow rate = 500.4 kmol/hours. Feed is saturated liquid in each column. Recovery of key components = 99%.

This is a typical five components separation problem which is discussed by many authors (Isia and Xenda, 1988; Schuttenhelm and Simmrock, 1992; Schembecker et al., 1994; Wang et al., 1998). A mixture of five components is to be separated into pure products. The system of rectification units should recover the process energy as soon as possible. The number of plates and reflux ratio of rectification columns are being calculated by Fenske-Gilland-Underwood method. The heat loads and the temperature of leaving streams are being calculated by short cut method given by those authors. Here we are supposed to do bubble point and dew point calculations to find the temperatures of the leaving streams. The heat loads are calculated by performing enthalpy or energy balance on column. Then heuristic for heat exchanger network is applied and found where heat exchangers can be put. After analyzing these all-possible heat transactions vectors are prepared. Pressures are also taken randomly between given ranges. Population of NP = 290 is chosen at first as D is 29 for these vectors. On various CR and F values results are generated for 500 iterations. As CR can be varied from 0 to 1.0, at every CR value between this range, F is incremented from 0.4 to 1.0.

After completing 500 iterations of above population results are found in this manner:
1, 0, 2, 0, 3, 0, 4, 0, 0, 23.3537, 35.6743, 30.2393, 39.0726, 4.99231e+006, 1.12053e+007, 690884, 4.47881e+006, 2.64236e+006, 1.74641e+006, 0, 0, 1.12667e+006, 0, 0, 0, 262561, 1.36789e+006, 0, 461565,

This result is for NP = 290, CR = 0.2 and F = 0.5 here optimum cost comes out to be $5770.17 per year.

In most of the generations it takes at most 30 to 50 iterations to find the minimum indicating efficiency of DE. After that DE tries to force all the vectors to come down to that minimum, and from all these generations and different parameter values we can notice that most of the generations with NP = 290 will have separation sequence as: 1, 0, 2, 0, 3, 0, 4, 0, 0, 0 that is A/BCDE, B/CDE, C/DE, D/E. The parameters pressures and heat loads will change according to operating pressure. Cost is also more less found to be between $ 50000 to $ 6000 which is much more improved then what Wang et al., (1998) could come to using genetic algorithm.

Final optimized vector for iteration number 500 for parameters NP = 290, D =29, CR = 0.2, F = 0.7 comes out to be:

1, 0, 2, 0, 3, 0, 4, 0, 0, 23.5175, 18.7515, 5.14835, 10.0602, 8.33306+006, 0, 0, 0, 0, 0, 0, 0, 0, 0, 0, 0, 0, 0, 0, 229255, cost = 5780.81.

Where pressures are in 10^6 kPa's and heat load's are in kCal's. Optimized heat exchanger network for this generation and final heat exchanger match are shown in Table-10.4 & Table-10.5 below.

Table 10.4. Heat Exchanger Network for Example

Streams	H1	H2	H3	H4
C1	8.3306e+006	0	0	0
C2	0	0	0	0
C3	0	0	0	0
C4	0	0	0	229255

Table 10.5. Final Heat Exchanger Match for Example

No. Of HE	Matched Streams	Heat loads in kCal
1	C1/H1	8.3306e+006
2	C4/H4	229255

Pressures in all the columns are listed in Table-10.6.

Table 10.6. Pressures for Example

Column No.	Pressure
1.	23.5175
2.	18.7515
3.	5.14835
4.	10.0602

These are results when we keep a population of NP = 145 i.e. (D*5) and CR = 0.2 and F = 0.7, D = 29.

After 500 iterations we following optimized vector.

1, 0, 2, 0, 3, 0, 4, 0, 0, 25.0035, 32.011, 7, 24.7422, 20.0442, 0, 0, 5.45524e+006, 0, 0, 0, 0, 5.45767e+007, 0, 0, 0, 0, 0, 1.26565e+006, 315948, 28994.6, cost = 5930.54

Here again we can notice that we get the same separation sequence, but the operating parameters are bit different and moreover we have 5 heat exchanger in HEN, the reason for this is we did not have sufficient number of vector's in our population compared to last time what we had. Hence we can notice that more the points in search space we get more closely to global optimum. In the case of 145 vector's population we slightly missed the global optimum vector which comes out when we increase our population to 290 that is 10 times of dimension of problem.

Parameter variation like in CR and F gives DE a faster and robust attitude when we increase or decrease them. Following heat exchanger network presented in Table-10.7 is achieved. The final match and the corresponding column pressures are shown in Table-10.8 & Table-10.9 respectively.

Table 10.7. Heat Exchanger Network for *NP* =145 for example

Stream	H1	H2	H3	H4
C1	0	0	0	0
C2	0	0	0	1.26565e+006
C3	5.45524e+006	0	0	315948
C4	0	5.45767e+007	0	28994.6

Table 8. Final match for example when *NP* = 145

No. of HE	Matched streams	Heat load (kcal)
1	C3/H1	5.45524e+006
2	C4/H2	5.45767e+007
3	C2/H4	1.26565e+006
4	C3/H4	315948
5	C4/H4	28994.6

Table 9. Pressures in the columns

Column	Pressure (10^6 kPa)
1	25.0035
2	25.0035
3	24.7422
4	20.0442

And, we can notice that cost of this particular vector, which is minimum in this population, is more than what we got in population of 290.

10.4.4 Conclusions

We proposed a differential evolution based synthesis and optimization strategy for heat integrated distillation systems. The strategy finds simultaneously the synthesized separation sequence, operating parameters and HEN for particular problem. One example has been taken to show the feasibility of DE for synthesis and design of heat integrated distillation systems. And performance evaluation on various *NP* values has been shown.

References

Andrecovich, M. J., and Westerberg, A. W. (1985). An MILP formulation for heat integrated distillation sequence synthesis. *A.I.Ch.E.J., 31*, 1461-1474.

Androulakis, I. P. & Venkatasubramanian, V. (1991). A genetic algorithm framework for process design and optimization. *Computers & Chemical Engineering, 15*(4), 217-228.

Angira, Rakesh and Babu, B.V. (2003). Evolutionary Computation for Global Optimization of Non-Linear Chemical Engineering Processes. *Proceedings of International Symposium on Process Systems Engineering and Control (ISPSEC '03)- For Productivity Enhancement through Design and Optimization*, IIT-Bombay, Mumbai, January 3-4, 2003, Paper No. FMA2, pp 87-91. (Also available via Internet as .pdf file at http://bvbabu.50megs.com/custom.html/#56).

Babu, B. V. and Angira, Rakesh (2001a). Optimization of Non-linear functions using Evolutionary Computation. *Proceedings of 12th ISME Conference*, Chennai, India, Jan, 10–12, 153-157. (Also available via Internet as .pdf file at http://bvbabu.50megs.com/custom.html/#34).

Babu, B. V. and Angira, Rakesh (2001b). Optimization of thermal cracker operation using Differential Evolution. *Proceedings of 54th Annual Session of IIChE (CHEMCON-2001), CLRI, Chennai, December 19-22, 2001.* (Also available via Internet as .pdf file at http://bvbabu.50megs.com/custom.html/#38) & Application No. 20, Homepage of Differential Evolution, the URL of which is: http://www.icsi.berkeley.edu/~storn/code.html

Babu, B. V., and Sastry, K.K.N. (1999). Estimation of heat transfer parameters in trickle bed reactors using differential evolution and orthogonal collection. *Computers and chemical engineering*, 23, 327-333. (Also available via Internet as .pdf file at http://bvbabu.50megs.com/custom.html/#24) & Application No. 13, Homepage of Differential Evolution, the URL of which is: http://www.icsi.berkeley.edu/~storn/code.html

Babu, B.V. and Angira, Rakesh (2002a). Optimization of Non-Linear Chemical Processes Using Evolutionary Algorithm. *Proceedings of International Symposium & 55th Annual Session of IIChE (CHEMCON-2002)*, OU, Hyderabad, December 19-22. (Also available via Internet as .pdf file at http://bvbabu.50megs.com/custom.html/#54).

Babu, B.V. and Angira, Rakesh (2002b). A Differential Evolution Approach for Global Optimization of MINLP Problems. *Proceedings of 4th Asia-Pacific Conference on Simulated Evolution And Learning (SEAL'02)*, Singapore, November 18-22, 2002, Vol. 2, pp 866-870. (Also available via Internet as .pdf file at http://bvbabu.50megs.com/custom.html/#46).

Babu, B.V. and Chaturvedi, Gaurav (2000). Evolutionary Computation strategy for optimization of an Alkylation Reaction. Proceedings of International Symposium & 53rd Annual Session of IIChE (CHEMCON-2000), Science City, Calcutta, December 18-

21, 2000. (Also available via Internet as .pdf file at http://bvbabu.50megs.com/custom.html/#31) & Application No. 19, Homepage of Differential Evolution, the URL of which is: http://www.icsi.berkeley.edu/~storn/code.html

Babu, B.V., and Gautam, K. (2001). Evolutionary Computation for Scenario-Integrated Optimization of Dynamic Systems. Proceedings of International Symposium & 54th Annual Session of IIChE (CHEMCON-2001), CLRI, Chennai, December 19-22, 2001. (Also available via Internet as .pdf file at http://bvbabu.50megs.com/custom.html/#39)

Babu, B.V., Angira, Rakesh, and Nilekar, Anand (2002). Differential Evolution for Optimal Design of an Auto-Thermal Ammonia Synthesis Reactor. Communicated to *Computers & Chemical Engineering,* 2002.

Cheng, S. H. and Liu, Y. A. (1988). Studies in chemical process design and synthesis. *Ind. Engng Chem. Res., 27,* 230-241.

Chiou, J. P. and Wang, F. S. (1999). Hybrid method of evolutionary algorithms for static and dynamic optimization problems with application to a fed-batch fermentation process. *Computers & Chemical Engineering, 23*(9), 1277-1291.

Douglas, J. M. (1988). *The conceptual design of chemical process.* New York. McGraw-Hill.

Edgar, T. F. & Himmelblau, D. M. (1989). *Optimization of Chemical Processes.* Singapore, McGraw-Hill, Inc., Page Nos. 534-539.

Edwards, K., Edgar, T. F., and Maousiouthakis, V. I. (1998). Kinetic model reduction using genetic algorithms. *Computers & Chemical Engineering, 22,* 239.

Gutheri, K.M. (1969). Capital cost estimating. *Chem. Engg.*, 24, 114-135.

Hartland, S., and Mecklenburgh, J. C. (1975). *The Theory of Backmixing*, Wiley, New York. Chap.10.

Isia, M. A., and Xerda, J. (1988). A heuristic method for the synthesis of heat integrated distillation system. *Chem. Engng J., 37*(3), 161-177.

Jackson, P. J. (1977). Ph.D. Thesis. Monash University, Victoria, Australia.

Jackson, P. J., and Agnew, J. B. (1980). A model based Scheme for the On-line Optimization of a Liquid Extraction Process. *Computers & Chemical Engineering, 4,* 241.

Jaroslav, J., and Radim Ptacnlk (1988). Synthesis of heat integrated rectification systems. *Computers and chemical engg., 12* (5), 427-432.

Kravanja, Z., and Grossman, I.E. (1990). PROSYN – an MINLP process synthesizer. *Computers and chemical engg., 14,* 1363-1378.

McKay, B., Willis, M., and Barton, G. (1997). Steady state modeling of chemical process systems using genetic programming. *Computers & Chemical Engineering, 21,* 981.

Ponton, J. W., and Donaldson, R. A. B. (1974). A fast method for the synthesis of optimal heat exchanger networks. *Chemical Engineering Science, 29,* 2375-2377.

Price, K. and Storn, R. (1997). Differential evolution. *Dr. Dobbs Journal,* April, 18-22.

Rong, B. G., Hu, Y. D., Zheng, Zhou, S. Q., and Han, F. Y. (1996). A hierarchical method to the synthesis of heat-integrated distillation flowsheets. Praha. Czech. Paper no. h5[508].

Schembecker, G., Schuttenhelm, W., and Simmrock, K. H. (1994). Cooperating knowledge integrating system for the systhesis of energy-integrated distillation process. *Computers & Chemical Engineering, 18*(Suppl), S131-S135.

Schuttenhelm, W., and Simmrock, K. H. (1992). Knowledge based synthesis of energy integrated distillation column and sequences. *Inst. Chem. Engng Symp., 28,* A461-A480.

Smith, B.D. (1967). Design of equilibrium stage processes. New York. McGraw-Hill.

Storn, R. and K. Price (1997). Differential Evolution: A simple and Efficient Heuristic for global optimization over continuous spaces. *Journal of Global Optimization*, *11*, 341.

Wang, Kefeng, Qian, yu, Yuan Yi and Ping Jing Yao (1998). Synthesis and optimization of heat integrated distillation systems using improved genetic algorithm. *Computers and chemical engg*. 23, 125-136.

Yee, T. F., Grossmann, I. E., and Kravanja, A. (1990). Simultaneous optimization model for multicomponent separation. *Computers & Chemical Engineering, 14*, 1185-1200.

Zdravko, K., and Grossmann, I. E. (1997). Multilevel-hierarchical MINLP synthesis of process flow sheets. *Computers and Chemical Engineering, 21* (Suppl), 421-426.

Zorka, N., Zdravko, K., & Grossmann, I. E. (1994). Simultaneous optimization model for multicomponent separation. *Computers and Chemical Engineering, 18* (Suppl), S125-S129.

11 Applications in Fluid Mechanics

B V Babu

11.1 Introduction

The optimization of non-linear constrained problems is relevant to chemical engineering practice (Salcedo, 1992; Floudas, 1995). In recent years, evolutionary algorithms (EAs) have been applied to the solution of NLP in many engineering applications. The best-known algorithms in this class include Genetic Algorithms (GA), Evolutionary Programming (EP), Evolution Strategies (ES) and Genetic Programming (GP). There are many hybrid systems, which incorporate various features of the above paradigms and consequently are hard to classify, which can be referred just as EC methods (Dasgupta and Michalewicz, 1997). They differ from the conventional algorithms since, in general, only the information regarding the objective function is required. EC methods have been applied to a broad range of activities in process system engineering including modeling, optimization and control. Differential Evolution (DE), developed by Price & Storn (1997), is one of the best EC methods. This method provides one of the best genetic algorithms for solving the real-valued test function. The convergence speed of DE is very high.

Some of the successful applications of DE include: digital filter design (Storn, 1995), fuzzy decision making problems of fuel ethanol production (Wang *et al.*, 1998), Design of fuzzy logic controller (Sastry *et al.*, 1998), batch fermentation process (Chiou and Wang, 1999; Wang and Cheng, 1999), multi sensor fusion (Joshi and Sanderson, 1999), dynamic optimization of continuous polymer reactor (Lee *et al.*, 1999), estimation of heat transfer parameters in trickle bed reactor (Babu and Sastry, 1999), optimization of alkylation reaction (Babu and Chaturvedi, 2000), optimal design of heat exchangers (Babu and Mohiddin, 1999; Babu and Munawar, 2000; Babu and Munawar, 2001), synthesis & optimization of heat integrated distillation system (Babu and Singh, 2000), optimization of non-linear functions (Babu and Angira, 2001a), scenario- integrated optimization of dynamic systems (Babu and Gautam, 2001), optimization of thermal cracking operation (Babu and Angira, 2001b), determining the number of components in mixtures of linear models (Dollena *et al.*, 2001), Identification of hysteretic systems using the differential evolution algorithm (Kyprianou *et al.*, 2001), optimization of Low Pressure Chemical Vapour Deposition Reactors Using Hybrid Differential Evolution (Lu and Wang, 2001), hybrid differential evolution for problems of Kinetic Parameter Estimation and Dynamic Optimization of an Ethanol Fermentation Process (Wang *et al.*, 2001), optimal design of auto-thermal ammonia synthesis reactor (Babu *et al.*, 2002), global optimization of MINLP problems (Babu and Angira, 2002a; Angira and Babu, 2003), optimization of non-linear chemical processes (Babu and Angira, 2002b), optimization of pyrolysis of biomass (Babu and Chaurasia, 2003) etc.

In this chapter, DE – an evolutionary computation method, is used to solve two classical problems in the area of fluid mechanics: (1) Gas transmission network, and (2) Water pumping system.

11.2 Gas Transmission Network

In this age of high competition in the several industries, it becomes necessary to cut down capital and operating costs as much as possible. Specifically in case of Chemical industries, the main focus is on reducing the processing costs, which include heating, cooling, transfer of various streams involved in any operational unit. The gas transmission network, which forms a considerable fraction of the operating cost, is also one of the focus areas. Broadly, a gas transmission system includes source of gas, delivery sites with the pipeline segments and compressor stations used to achieve desired pressure at the delivery site. As the design of an efficient and economical gas transmission network involves a lot of design parameters which directly/indirectly affect the capital and operating costs, this topic deserves special attention. A lot of work has been done in this area over the years. Larson and Wong (1968), Graham et al (1971), Martch and McCall (1972), Flanigan (1972), Mah and Shacham (1978), Cheesman (1971), Edgar et al. (1978) have discussed the various aspects of the problem. Larson and Wong determined the steady state optimal operating conditions of a straight natural line pipeline with compressors in series. They used dynamic programming to find the optimal suction and discharge pressures. The length and diameter of the pipeline segment were assumed to be constant because of limitations of dynamic programming. Martch and McCall (1972) modified the problem by adding branches too the pipeline segments. However, the transmission network was predetermined because of the limitations of the optimization technique used. Cheeseman introduced a computer optimizing code in addition to Martch and McCall (1972) problem they considered the length and diameters of the pipeline segments to be variables. But their problem formulation did not allow unbranched network, so complicated network systems couldn't be handled. Olorunniwo (1981) and Olorunniwo and Jensen (1982) provided further breakthrough by optimizing a gas transmission network including the following features:

1. The maximum no. of compressor stations that would ever be required during the specified time horizon.

2. The optimal location of these compressor stations.
3. The initial construction dates of the stations.
4. The optimal solutions of expansion for the compressor stations.
5. The optimal diameter sizes of the main pipes for each arc of the network.
6. The minimum recommended thickness of the main pipe.
7. The optimal diameter sizes, thicknesses and lengths of any required parallel pipe loops on each arc of the network
8. The timing of construction of the parallel pipe loops
9. The operating pressures of the compressors and the gas in the pipelines.

They used dynamic programming coupled with optimization logic to find the shortest route through the network.

Edgar & Himmelblau (1988) simplified the problem to make sure that the various factors involved in the design are clear. They assumed the gas quantity to be transferred along with the suction and discharge pressures to be given in the problem statement. They optimized the following variables:

1. The number of compressor stations
2. The length of pipeline segments between the compressors stations
3. The diameters of the pipeline segments
4. The suction and discharge pressures at each station.

They considered the minimization of the total cost of operation per year including the capital cost in their objective function against which the above parameters are to be optimized.

Edgar and Himmelblau (1988) also considered two possible scenarios:

1. The capital cost of the compressor stations is linear function of the horse power
2. The capital cost of the compressor stations is linear function of the horsepower with a fixed capital outlay for zero horsepower.

The first scenario is easy to solve as compared to the second one. They solved the second scenario using the branch and bound technique.

11.2.1 Problem Formulation

Babu et al. (2003a; 2003b) solved the above two possible scenarios of the problem using Differential Evolution (DE) as optimizer. The pipeline configuration is same as chosen by Edgar and Himmelblau (1988) and shown in Fig. 11.1.

Each of the compressor stations is represented by a node and each of the pipeline segments by an arc. Pressure is assumed to be increasing at a compressor and decreasing along the pipeline segment. The transmission system is presumed to be horizontal. This is a simple example chosen to illustrate a gas transmission system. However, a much more complicated network can be accommodated including various branches and loops at the cost of additional execution time.

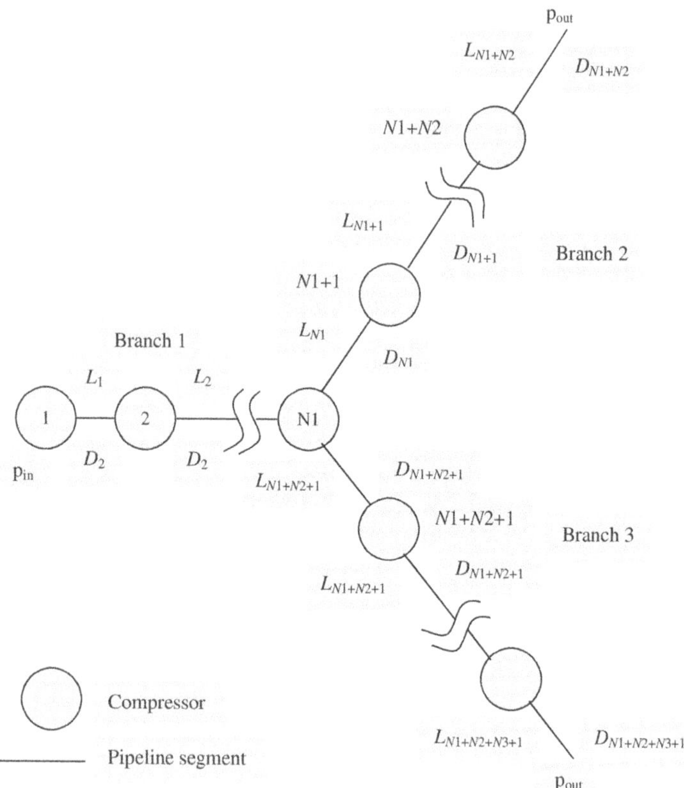

Fig. 11.1. Pipeline configuration with three branches.

Edgar and Himmelblau (1988) distinguished between two related problems (one is of a higher degree of difficulty than the other) before proceeding ahead with the details of the design problem. Fig. 11.2 shows the cost vs. horsepower. If the capital costs of the compressors are linear functions of horsepower as shown in line *A* of Fig. 11.2, the transmission line problem can be solved as a nonlinear programming problem by one of the methods discussed by Edgar and Himmelblau (1988). Alternately, if the capital costs are a linear function of horsepower with a fixed capital outlay for zero horsepower as shown by line *B* in Fig. 11.2, a condition that is more closely represents the practical problems, then the design problem becomes more difficult to solve and a branch-and-bound algorithm combined with a nonlinear programming algorithm has to be used. It would be clear the reason why branch and bound method is avoided for the case involving line *A* after the mathematical formulation of the objective function has been completed.

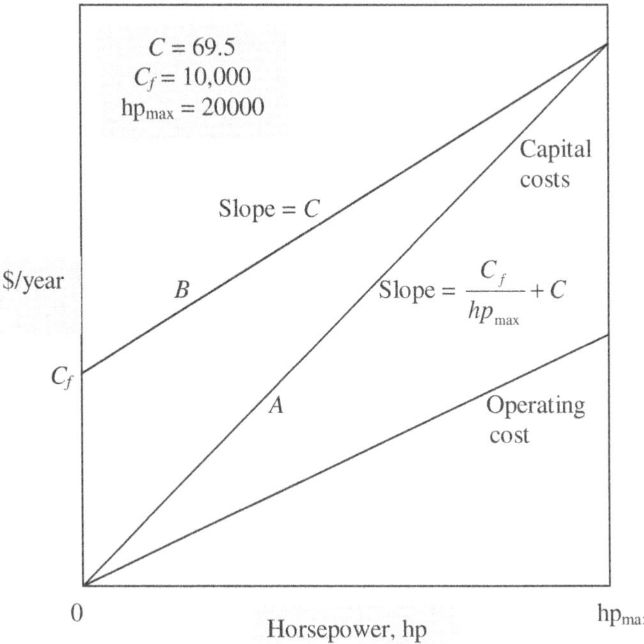

Fig. 11.2. Capital and operating costs of compressors

11.2.1.1 Number of Variables

Each node and each arc are labeled separately for a given pipeline configuration. The number of variables is as following:

- Total Compressors : N
- Suction Pressures : N-1
- Discharge Pressures : N
- Pipeline Lengths & Diameters: N+1

11.2.1.2 Variables

Each pipeline segments has the following variables associated with it:

1. The flow rate
2. The initial pressure
3. The outlet pressure
4. The pipe diameter
5. The pipeline segment length

It is assumed that each of the compressors has gas losses of one-half of one percent of the gas transmitted. As the mass flow rate is fixed, only the last four variables become important and need to be determined for each segment in the present problem.

11.2.1.3 Assumptions

The following assumptions are made:

1. Each compressor functions adiabatically with an inlet temperature equal to that of the surroundings.
2. Pipeline segment is long so that by the time gas reaches the next compressor it returns to the ambient temperature.
3. The annualized capital costs for each pipeline segment depend on pipe diameter and length, and have been taken as \$870/(inch)(mile)(year) as reported by Martch and McCall (1972).
4. The rate of work of one compressor is estimated using the following correlation:

$$W = (0.08531)Q\frac{k}{k-1}T_1\left[\left(\frac{p_d}{p_s}\right)^{z(k-1)/k} - 1\right] \tag{11.1}$$

   where
   $k = C_p/C_v$ for gas at suction conditions = 1.26 (Katz et al., 1959)
   z = compressibility factor of gas at suction conditions
   p_s = suction pressure, psia
   p_d = discharge pressure, psia
   T_i = suction temperature = 520^0R
   Q = flow rate into the compressor, MMCFD (million cubic feet per day)
   W = rate of work, horsepower
5. Total operating costs are linear function of compressor horsepower (Operation and maintenance costs per year can be related directly to horsepower (Cheesman, 1971b) and have been estimated to be between 8.00 and 14.0 \$/(hp)(year) (Martch and McCall, 1972))
6. Line A in Fig. 11.2 indicates that the cost is a linear function of horsepower (\$70.00/(hp)(year)) with the line passing through the origin.
7. Line B in Fig. 11.2 assumes a linear function of horsepower with a fixed initial capital outlay (\$70.00/(hp)(year) + \$10,000) which takes the installation costs, foundation, etc. into account.

11.2.1.4 Objective Function

As the objective in this study is to minimize the cost, the objective function comprises of the sum of the yearly operating and maintenance costs of the compressors in addition to the sum of the discounted capital costs of the pipeline segments and compressors over a period of 10 years. For line A, the objective function for the problem chosen in dollars per year is:

$$f=\sum_{i=1}^{n}(C_0+C_c)Q_i(0.08531)T_1\left(\frac{k}{k-1}\right)\left[\left(\frac{p_{d_i}}{p_{s_i}}\right)^{z(k-1/k)}-1\right]+\sum_{j=1}^{m}C_sL_jD_j \tag{11.2}$$

where,
n = number of compressors in the system
m = number of pipeline segments in the system (= n + 1)
C_0 = annual operating cost, \$/(hp)(year)
C_c = compressor capital cost, \$/(hp)(year)
C_s = pipe capital cost, \$/(in)(mile)(year)
L_j = length of pipeline segment j, mile
D_j = diameter of pipeline segment j, inch

As can be seen from the above expression, it is now clear why a branch and bound technique is not required to solve the design problem for line A. The term involving compressor i vanishes from the first summation in the objective function if the ratio (p_d/p_s) = 1, because of the way the objective function is formulated. This is equivalent to the deletion of compressor i in the execution of a branch and bound strategy. Nevertheless, the joined pipeline segments at node i may be of different diameters. But when line B represents the compressor costs, the fixed incremental cost for each compressor in the system at zero horsepower (C_f) would not be multiplied by the term in the square brackets of Eq. 11.2. C_f would be added in the sum of the costs whether or not compressor i is in the system, and a non-linear programming technique alone could not be used. Hence, a different solution procedure is required if line B applies.

However, DE has the capability of dealing with above complications as it is a population-based search algorithm, and hence, DE is used in this study (Babu et al., 2003a; 2003b) for both the possible scenarios.

11.2.1.5 Inequality Constraints

A constraint is there for operation of each compressor as the discharge pressure is always greater than or equal to the suction pressure:

$$\frac{p_{d_i}}{p_{s_i}}\geq 1 \qquad i=1,2,......,n \tag{11.3}$$

and the compressor ratio should not exceed some prespecified maximum limit K

$$\frac{p_{d_i}}{p_{s_i}}\leq K_i \qquad i=1,2,......,n \tag{11.4}$$

Also, the upper and lower bounds are placed on each of the four variables

$$p_{d_i}^{\min}\leq p_{d_i}\leq p_{d_i}^{\max} \tag{11.5}$$

$$p_{s_i}^{\min}\leq p_{s_i}\leq p_{s_i}^{\max} \tag{11.6}$$

$$L_i^{\min} \le L_i \le L_i^{\max} \tag{11.7}$$

$$D_i^{\min} \le D_i \le D_i^{\max} \tag{11.8}$$

11.2.1.6 Equality Constraints

For the gas transmission network chosen in the problem, there are two classes of equality constraints. First, as the length of the system is fixed, there would be two constraints for two branches as given below:

$$\sum_{j=1}^{N1-1} L_j + \sum_{j=N1}^{N1+N2} L_j = L_1^* \tag{11.9}$$

$$\sum_{j=1}^{N1-1} L_j + \sum_{j=N1+N2+1}^{N1+N2+N3+1} L_j = L_2^* \tag{11.10}$$

where L_k^* represents the length of a branch. Secondly, each pipeline segment must satisfy the Weymouth flow equation (GPSA, 1972):

$$Q_j = 871 D_j^{8/3} \left[\frac{p_d^2 - p_s^2}{L_j} \right]^{1/2} \tag{11.11}$$

Where Q_j is a fixed number, and p_d & p_s are the discharge pressure & suction pressure at the entrance and exit of the segment respectively.

11.2.2 Results & Discussion

The seventh of the ten DE strategies (DE/rand/1/bin) proposed by Price & Storn (2003) is used for solving this problem. The key parameters of DE are: NP, the population size; CR, the crossover constant; and F, scaling factor. The steps carried out in this strategy are:

Step 1: Generate NP random vectors as the initial population.

*Generally, NP is taken to be 5-10 times of the dimension (D) of the problem. In the present case the dimension of our problem is 4. Hence NP is chosen to be 40. Generate (NP*D) random numbers and linearize the range (0-1) to cover the entire range of the function. From these (NP*D) random numbers, generate NP random vectors, each of dimension D by mapping the random numbers over the range of the function.*

All the NP random vectors that are generated should satisfy the constraints (Eq. 11.3 – 11.11).

Step 2: Choose a target vector from population of size NP.

First generate a random number between 0-1. From the value of the random number, decide which population member is to be selected as the target vector (X_i). (a linear mapping rule can be used)

Step 3: Choose two vectors at random from the population and find the weighted difference.
Generate two random numbers. Decide which two population members are to be selected (X_a, X_b). *Find the vector difference between the two vectors* (X_a - X_b). *Multiply this difference with F to obtain the weighted difference. The value of F ranges between 0 and 1.2. But, the optimal range is 0.4-1.0 for many problems. As a guideline initially F may be chosen to be 0.5. If this leads to premature convergence, then it may be increased. In the present problem, F is chosen to be 0.75.*
Weighted difference = $F(X_a - X_b)$
Step 4: Find the noisy random vector.
Generate a random number. Choose a third random vector from the population (X_c). *Add this vector to the weighted difference to obtain the noisy random vector* (X_c').
Noisy Random Vector (X_c') = $X_c + F(X_a - X_b)$
Again, the NP noisy random vectors that are generated should satisfy the constraints (Eq. 11.3 – 11.11).
Step 5: Perform Crossover between X_i ***and*** X_c' ***to find*** X_t, ***the trial vector.***
Generate D random numbers. For each of the D dimensions:
If random no. > CR ; copy the value from X_i *into the trial vector.*
If random no. < CR ; copy the value from X_c' *into the trial vector.*
CR = 0.9 is a good first guess. Subsequently, CR may be chosen judging by the speed between 0 and 1. In this problem, CR is chosen to be 0.9.
Step 6: Calculate the cost of the trial vector and the target vector.
For a minimization problem, calculate the function value directly and this is the cost. For a maximization problem, transform the objective function f(x) using the rule, $F(x) = \dfrac{1}{1 + f(x)}$ *and calculate the value of the cost. Alternately, directly calculate the value of f(x) and this yields the profit.*

If the objective function is formulated in terms of cost, the vector that yields the lesser cost replaces the population member in the initial population. If the objective function is in terms of profit function, then the vector with greater profit replaces the population member in the initial population.

Steps 1 to 6 are continued till some stopping criterion is met. This may be of two kinds. One may be some convergence criterion that states that the error in the minimum or maximum between two previous generations should be less than some specified value (standard deviation may be used). The other may be an upper bound on the number of generations. The stopping criteria may be a combination of the two as well. Either way, once the stopping criterion is met, the computations are terminated.

Fig. 11.3 shows the design problem outlined. The maximum number of compressors in branches 1, 2, and 3 are set at 4, 3, and 3 respectively. The input pressure was fixed at 500 psia at a flow rate of 600 MMCFD, and the two output pressures are 600 and 300 psia respectively for branches 2 and 3. The total length of the branches 1 and 2 put together is constrained to be 175 miles, whereas the total length of the branches 1 and 3 put together is constrained to be 200 miles. The upper bound for diameter of pipeline segments in branch 1 is set at 36 inches, and those for branches 2 and 3 are set at 18 inches. The lower bound on the diameter of all pipeline segments is set at 4 inches. A lower bound of 2 miles is placed on each pipeline segment to ensure that the natural gas is at ambient conditions when it entered at subsequent compressor in the pipeline.

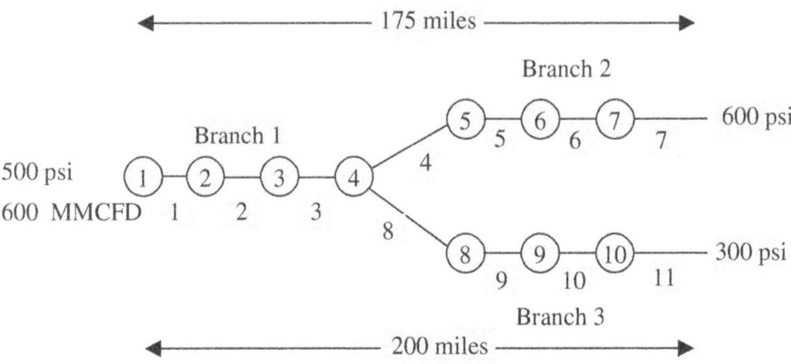

Fig. 11.3. Initial configuration of chosen gas transmission network

The resulting solution obtained by Edgar and Himmelblau (1988) to the example designed problem as shown in Fig. 11.3 using the cost relation of line A of Fig. 11.2 (using non-linear programming) and for line B of Fig. 11.2 (using branch and bound technique) are shown in Fig. 11.4 & Table-11.1, and Fig. 11.5 & Table-11.2 respectively. The optimum value of objective function for line A (Fig. 11.2) is reported to be 7.289×10^6 \$/yr. Based on the results listed the calculated optimum value of objective function for line B is found to be 7.389×10^6 \$/yr.

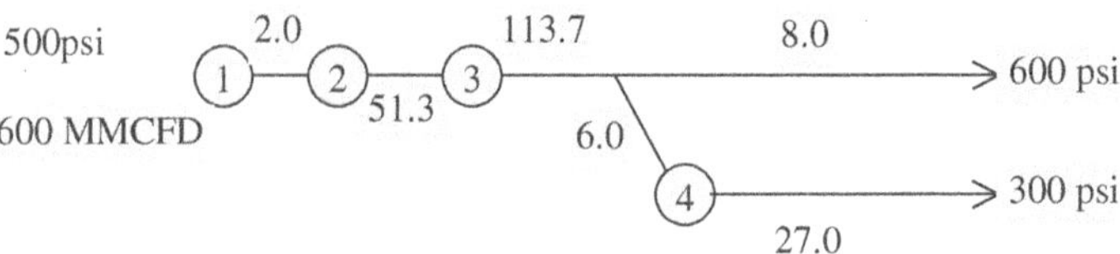

Fig. 11.4. Final optimal gas transmission network for line A of Fig. 11.2 (Edgar and Himmelblau, 1988)

Table 11.1. Values of operating variables for the optimal network configuration using the costs of line *A*, Fig. 11.2 (Edgar and Himmelblau, 1988)

Pipeline segment	Discharge Pressure (psia)	Suction Pressure (psia)	Pipe Diameter (inch)	Length (mile)	Flow rate (MMCFD)
1	719.1	715.4	35.0	2.0	597.0
2	1000.0	889.3	32.4	51.3	594.0
3	1000.0	735.8	32.4	113.7	591.0
4	735.7	703.8	18.0	2.0	294.0
5	703.8	670.6	18.0	2.0	292.6
6	670.6	636.1	18.0	2.0	291.1
7	636.1	600.0	18.0	2.0	289.7
8	735.8	703.8	18.0	2.0	294.0
9	685.2	859.1	18.0	2.0	292.6
10	859.1	832.5	18.0	2.0	291.1
11	832.5	300.0	18.0	27.0	289.7

Compressor station	Compression ratio	Capital cost ($/year)
1	1.44	70.00
2	1.40	70.00
3	1.12	70.00
4	1.00	70.00
5	1.00	70.00
6	1.00	70.00
7	1.00	70.00
8	1.26	70.00
9	1.00	70.00
10	1.00	70.00

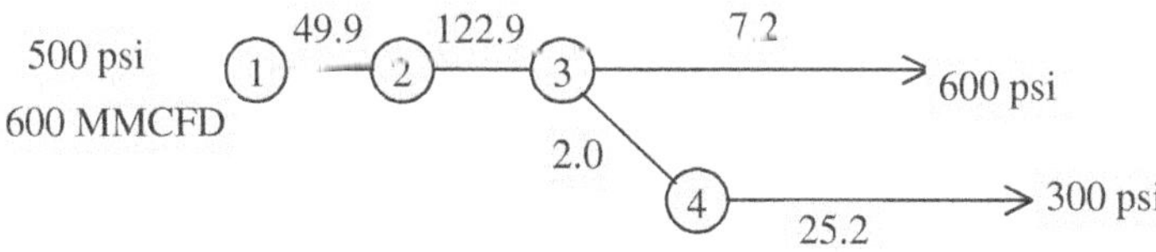

Fig. 11.5. Final optimal gas transmission network for line B of Fig. 11.2 (Edgar and Himmelblau, 1988)

Table 11.2. Values of operating variables for the optimal network configuration using the costs of line B, Fig. 11.2 (Edgar and Himmelblau, 1988)

Pipeline segment	Discharge Pressure (psia)	Suction Pressure (psia)	Pipe Diameter (inch)	Length (mile)	Flow rate (MMCFD)
1	954.5	837.2	32.3	49.9	597.0
2	1000.0	699.7	32.3	122.9	594.0
3	699.7	600.0	15.2	2.2	295.5
4	699.7	665.7	18.0	2.0	295.5
5	952.2	300.0	16.9	25.2	294.0

Compressor station	Compression ratio	Capital cost ($/year)
1	1.91	69.50
2	1.19	69.50
3	1.00	69.50
4	1.43	69.50

As can be seen from Table-11.1, the discharge pressure of pipeline segment 9 does not satisfy the second equality constraint (Eq. 11.11). The value of the compression ratio of third compressor station for line A (Fig. 11.2) does not match with the corresponding ratio of discharge and suction pressures shown in Table-11.1. Edgar and Himmelblau (1988) modified the second equality constraint (Eq. 11.11) to avoid problems in taking square roots by squaring it as given below:

$$(871)^2 D_j^{16/3}\left(p_d^2 - p_s^2\right) - L_j Q_j^2 = 0 \tag{11.12}$$

But, all the results shown in Table-11.1 do not satisfy the above constraint.

All problems mentioned above have been addressed and sorted out in this study using DE. The resulting solution obtained to the example designed problem as shown in Fig. 11.3 using the cost relation of line A of Fig. 11.2 and for line B of Fig. 11.2 are shown in Fig. 11.6 & Table-11.3.

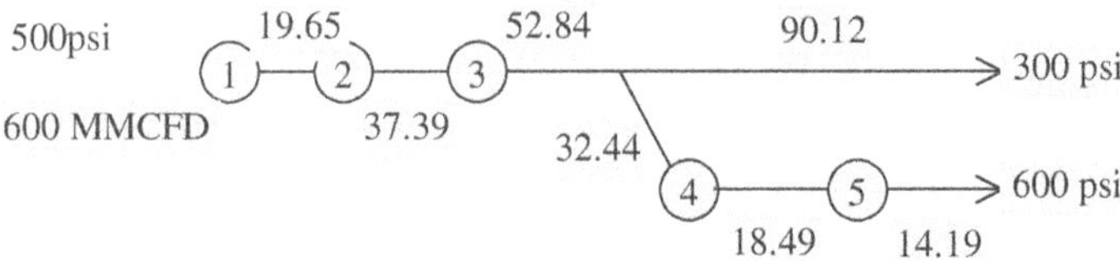

Fig. 11.6. Final optimal gas transmission network for both line A & line B of Fig. 11.2 (Babu et al., 2003a)

Table 11.3. Values of operating variables for the optimal network configuration using the costs of line *A* & line *B*, Fig. 11.2 (Babu et al., 2003a)

Pipeline segment	Discharge Pressure (psia)	Suction Pressure (psia)	Pipe Diameter (inch)	Length (mile)	Flow rate (MMCFD)
1	730.56	691.44	34.77	19.65	597.000
2	942.47	852.83	32.08	37.39	594.015
3	865.68	736.92	32.61	52.84	591.045
4	736.92	690.05	24.33	14.51	294.022
5	806.07	715.99	22.09	17.93	292.552
6	715.99	620.00	22.45	18.49	291.079
7	663.72	600.00	23.27	14.19	289.624
8	970.15	775.08	21.03	33.88	294.022
9	775.08	749.88	22.76	5.89	292.552
10	749.88	711.46	21.40	6.27	291.079
11	711.46	300.00	21.15	44.07	289.624

Compressor station	Compression ratio	Capital cost ($/year)
1	1.4611	70.00
2	1.3630	70.00
3	1.0150	70.00
4	1.0000	70.00
5	1.1681	70.00
6	1.0000	70.00
7	1.1062	70.00
8	1.0000	70.00
9	1.0000	70.00
10	1.0000	70.00

As can be seen from the above results, all the constraints are satisfied with this optimal gas transmission network. Also, a single network has been obtained for both the possible scenarios (line A & line B of Fig. 11.2) described earlier, and DE alone could give the solution for both scenarios. The optimum values of objective function for both the possible scenarios are obtained as 7.692×10^6 $/yr and 7.792×10^6 $/yr respectively. Though the objective function values are slightly higher than those reported by Edgar and Himmelblau (1988), these values correspond to satisfying all the constraints. In addition, by approximating the compression ratio of compressor station 3 (1.0150 in Table-11.4) to 1.00 (considering only two digits as Edgar and Himmelblau, 1988), it is possible to remove one more compressor from the proposed final network (Fig. 11.6), in which case the total number of compressors would be only 4 and the final objective function value would further reduce.

Babu et al. (2003b) further obtained an improved solution (better than that reported by above two studies) by forcing some of the parametric values (of pipe diameters and lengths) in optimization as shown in Table 11.3. The resulting final network for both line A and line B is shown in Fig. 11.7.

Table 11.4. Values of operating variables for the optimal network configuration using the costs of line *A* & line *B*, Fig. 11.2 (Babu et al., 2003b)

Pipeline segment	Discharge Pressure (psia)	Suction Pressure (psia)	Pipe Diameter (inch)	Length (mile)	Flow rate (MMCFD)
1	691.41	639.94	33.03	18.44	597.000
2	882.36	701.54	34.89	103.88	594.015
3	715.17	634.30	35.62	44.68	591.045
4	735.70	703.80	18.00	2.00	294.022
5	703.80	670.60	18.00	2.00	292.552
6	670.60	636.10	18.00	2.00	291.079
7	636.10	600.00	18.00	2.00	289.624
8	735.80	703.80	18.00	2.00	294.022
9	885.20	859.10	18.00	2.00	292.552
10	859.10	832.50	18.00	2.00	291.079
11	832.50	300.00	18.00	27.00	289.624

Compressor station	Compression ratio	Capital cost ($/year)
1	1.2877	70.00
2	1.2390	70.00
3	1.0398	70.00
4	1.0202	70.00
5	1.0000	70.00
6	1.0000	70.00
7	1.0000	70.00
8	1.2577	70.00
9	1.0000	70.00
10	1.0000	70.00

The optimum values of objective function for both the possible scenarios are obtained as 6.834×10^6 \$/yr and 6.934×10^6 \$/yr respectively. This shows a saving of close to 7.2 million dollars from its first feasible state and a savings of close to 0.5 million dollars that predicted by Edgar and Himmelblau (1988). It can be seen from Table-11.4 that out of the ten possible compressor stations only five remained in the final optimum network (see Fig. 11.7). As the compression ratio is unity for compressors 5, 6, 7, 9 and 10 they do not exist in the final optimal configuration.

DE converged to an optimal solution, and the final cost is less than what reported earlier in the literature.

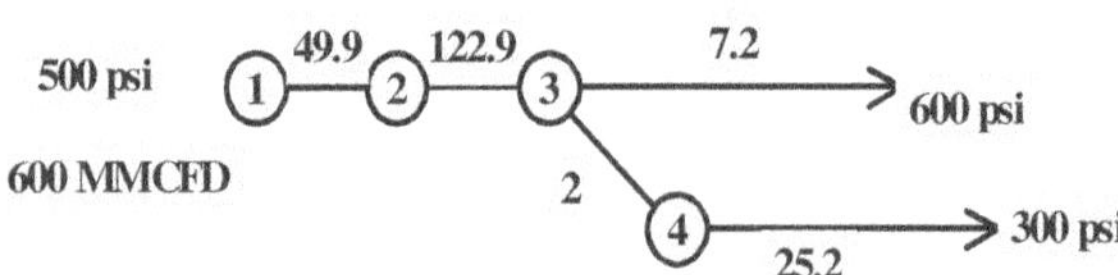

Fig. 11.7. Final optimal gas transmission network for both line A & line B of Fig. 11.2 (Babu et al., 2003b).

11.3 Water Pumping System

This problem arises from the area of chemical engineering, and represent difficult non-linear optimization problem, with equality constraints. Babu and Angira (2003) solved this problem using DE and compared with Branch & Reduce algorithm (Ryoo & Sahinidis, 1995). It is found that DE, an exceptionally simple evolutionary computation method, is significantly faster and yields the global optimum for a wide range of the key parameters.

11.3.1 Differential Evolution Strategies

Differential Evolution (Price & Storn, 1997) is an improved version of GA (Goldberg, 1989) for faster optimization. Unlike simple GA that uses binary coding for representing problem parameters, DE uses real coding of floating point numbers. Among the DE's advantages are its simple structure, ease of use, speed and robustness. Price & Storn (1997) gave the working principle of DE with single strategy. Later on, they suggested ten different strategies of DE (Price & Storn, 2002). A strategy that works out to be the best for a given problem may not work well when applied for a different problem. Also, the strategy and key parameters to be adopted for a problem are to be determined by trial & error. The key parameters of control are: NP - the population size, CR - the crossover constant, F - the weight applied to random differential (scaling factor). The Pseudo codes, for all ten DE strategies, used in the present study are given below:

```
/*----------------DE/rand/1/bin------------------------*/
        if (strategy = 1)
        {
            j = int (random number [0,1])*D);
            for (k = 1;k<=D;k++)
            {
                if ((random number [0,1]))<CR or| k=D)
                {
                    trial[j]=x1[c][j]+F*(x1[a][j]-x1[b][j]);
                }
                else trial[j]=x1[i][j];
                    j=(j+1)%D;
            }
        }

/*------------------------DE/best/1/bin----------------------*/
 else if (strategy = 2)
        {
            j=int (random number [0,1])*D);
            for (k=1;k<=D;k++)
```

```
                    {
                            if ((random number [0,1]))<CR  or k=D)
                            {
                            trial[j]=bestit[j]+F*(x1[a][j]-x1[b][j]);
                            }
                            else trial[j]=x1[i][j];
                      j=(j+1)%D;
                    }
              }

/*------------------DE/best/2/bin-----------------------*/
 else if (strategy = 3)
    {
                     j=int (random number [0,1])*D);
                    for (k=1;k<=D;k++)
                    {
                            if ((random number [0,1]))<CR  or k=D)
                            {
                    trial[j]=bestit[j]+F*(x1[a][j]+x1[b][j]-x1[c][j]-x1[d][j]);
                            }
                            else trial[j]=x1[i][j];
                      j=(j+1)%D;
                    }
    }

/*------------------DE/rand/2/bin----------------------*/
 else if (strategy = 4)
        {
                    j=int (random number [0,1])*D);
                    for (k=1;k<=D;k++)
                    {
                        if ((random number [0,1]))<CR  or k=D)
                            {
                                trial[j]=x1[e][j]+F*(x1[a][j]+x1[b][j]-x1[c][j]-
x1[d][j]);
                            }
                        else trial[j]=x1[i][j];
                      j=(j+1)%D;
                    }
        }

/*------------------DE/rand-to-best/1/bin----------------------*/
 else if (strategy = 5)
        {
                        assignd (trial,x1[i]);
                    j=int (random number [0,1])*D);
```

```
                for (k=1;k<=D;k++)
                {
                    if ((random number [0,1]))<CR  or k=D)
                          {
                            trial[j]=trial[j]+F*(bestit[j]-trial[j])+F*(x1[a][j]-
x1[b][j]);
                          }
                    else trial[j]=x1[i][j];
                  j=(j+1)%D;
                }
        }

/*------------------DE/rand/1/exp-----------------------*/
  else if (strategy = 6)
        {       assignd (trial,x1[i]);
                j=int (random number [0,1])*D);
                k=0;
                do
                  {
                    trial[j]=x1[c][j]+F*(x1[a][j]-x1[b][j]);
                    j=(j+1)%D;
                    k++;
                  }
                 while((random number [0,1] ))<CR and k < D);
        }

/*------------------DE/best/1/exp-----------------------*/
else if (strategy = 7)
        {       assignd (trial,x1[i]);
                j=int (random number [0,1])*D);
                k=0;
              do
                {
                    trial[j]=bestit[j]+F*(x1[a][j]-x1[b][j]);
                    j=(j+1)%D;
                    k++;
                }
              while ((random number [0,1]))<CR and k < D);
        }

/*------------------DE/best/2/exp-----------------------*/
 else if (strategy = 8)
        {       assignd (trial,x1[i]);
                j=int (random number [0,1])*D);
                k=0;
```

```
            do
              {
                  trial[j]=bestit[j]+F*(x1[a][j]+x1[b][j]-x1[c][j]-x1[d][j]);
                     j=(j+1)%D;
                     k++;
              }
            while ((random number [0,1]))<CR and k < D);
        }

/*-----------------DE/rand/2/exp----------------------*/
else if (strategy = 9)
        {         assignd (trial,x1[i]);
                  j=int (random number [0,1])*D);
                  k=0;
            do
              {
                  trial[j]=x1[e][j]+F*(x1[a][j]+x1[b][j]-x1[c][j]-x1[d][j]);
                     j=(j+1)%D;
                     k++;
              }
            while ((random number [0,1]))<CR and k < D);
        }

/*-----------------DE/rand-to-best/1/exp----------------------*/
 else if(strategy =10)
        {            assignd (trial,x1[i]);
                     j=int (random number [0,1])*D);
                     k=0;
              do
                {
                        trial[j]=trial[j]+F*(bestit[j]-trial[j])+F*(x1[a][j]-
x1[b][j]);

                          j=(j+1)%D;
                          k++;
                }
              while ((random number [0,1]))<CR and k < D);
        }
```

The crucial idea behind DE is a scheme for generating trial parameter vectors. Basically, DE adds the weighted difference between two population vectors to a third vector. Price & Storn (2002) have given some simple rules for choosing key parameters of DE for any given application. DE has been successfully applied in various fields.

11.3.2 Problem Formulation

A water pumping system (Stoecker, 1971) consists of two parallel pumps drawing water from a lower reservoir and delivering it to another that is 40 m higher, as shown in Fig. 11.8. In addition to overcoming the pressure difference due to the elevation, the friction in the pipe is $7.2w^2$ kPa, where w is the combined flow rate in kilograms per second. The pressure-flow-rate characteristics of the pumps are:

$$\text{Pump 1: } \Delta p\,(\text{kPa}) = 810 - 25w_1 - 3.75w_1^2 \tag{11.13}$$

$$\text{Pump 2: } \Delta p\,(\text{kPa}) = 900 - 65w_2 - 30w_2^2 \tag{11.14}$$

where w_1 and w_2 are the flow rates through pump 1 and pump 2, respectively.

The system can be represented by four simultaneous equations. The pressure difference due to elevation and friction is:

$$\Delta p = 7.2w^2 + \frac{(40\,\text{m})(1000\,\text{kg/m}^3)(9.807\,\text{m/s}^2)}{1000\,\text{Pa/kPa}} \tag{11.15}$$

$$\text{Pump 1: } \Delta p = 810 - 25w_1 - 3.75w_1^2 \tag{11.16}$$

$$\text{Pump 2: } \Delta p = 900 - 65w_2 - 30w_2^2 \tag{11.17}$$

$$\text{Mass balance: } \quad w = w_1 + w_2 \tag{11.18}$$

Stocker, (1971) used the method of successive substitution for solving this problem.

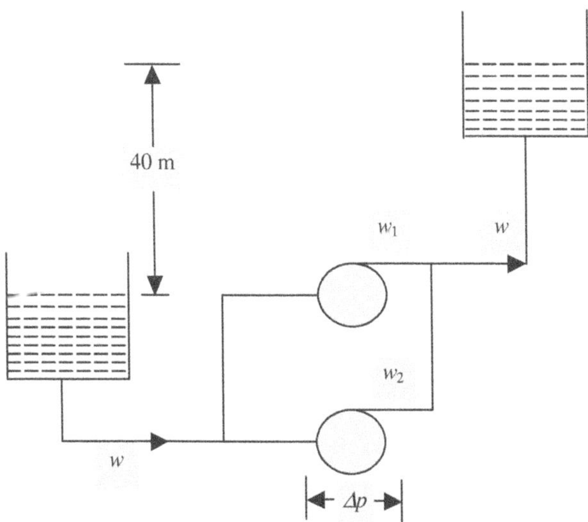

Fig. 11.8. Water Pumping System

11.3.2.1 Problem Modification

Liebman et al. (1986) modified the above problem as given below:

$$\text{Min.} \quad f = x_3 \tag{11.19}$$

such that

$$x_3 = 250 + 30x_1 - 6x_1^2 \tag{11.20}$$

$$x_3 = 300 + 20x_2 - 12x_2^2 \tag{11.21}$$

$$x_3 = 150 + 0.5(x_1 + x_2)^2 \tag{11.22}$$

$$0 \le \mathrm{x} \le (9.422, 5.903, 267.42) \tag{11.23}$$

Ryoo & Sahinidis (1995) solved this problem using Branch and Reduce algorithm. They used different strategies of Branch and Reduce algorithm. The CPU-time reported by them ranges from a minimum of 0.3 s to a maximum of 150 s for various strategies used on Sun SPARC station 2. The termination criterion used was an accuracy (ε) = 10^{-6}. The global optimum reported is (x; f) = (6.293429, 3.821839, 201.159334; 201.159334).

11.3.2.2 Problem Reformulation

When DE was applied to the above problem, it was found that the equality constraints were difficult to deal with. Hence, the problem was reformulated. The reformulated problem (Babu and Angira, 2003) is as follows:

$$\text{Min.} \ f = 150 + 0.5(x_1 + x_2)^2 \tag{11.24}$$

such that

$$6x_1^2 - 30x_1 - 249.9999999 + 150.0 + 0.5(x_1 + x_2)^2 \ge 0.0 \tag{11.25}$$

$$12x_2^2 - 20x_2 - 299.9999999 + 150.0 + 0.5(x_1 + x_2)^2 \ge 0.0 \tag{11.26}$$

$$0 \le \mathrm{x} \le (9.422, 5.903) \tag{11.27}$$

The global optimum obtained is: (x; f) = (6.293429, 3.821839; 201.159334).

11.3.3 Results & Discussion

Table 11.5 shows the results obtained by both DE and Branch & Reduce algorithm (Ryoo & Sahinidis, 1995). It may be noted that the global optimum is same as reported by Ryoo & Sahinidis (1995) i.e. f = 201.159334 and the flow rates through Pump 1 & Pump 2 are x_1 = 6.293430 & x_2 = 3.821839 respectively.

Table 11.5. Comparison of DE with Branch & Reduce

Parameters	Using DE	Using Branch & Reduce
x_1	6.293430	6.293429
x_2	3.821839	3.821839
x_3	201.159334	201.159334
CPU-time (s)	0.0714*	0.3$
Objective function (f)	201.159334	201.159334

* CPU-time on Pentium PIII, 500 MHz PC with strategy no. 10
$ CPU-time on Sun SPARC Station 2

Tables 11.6 & 11.7 present the comparison, in terms of the number of objective function evaluations, CPU-time and proportion of convergences to the optimum, between the different DE strategies. The termination criterion used is accuracy of 10^{-6} and 10^{-7} respectively. In these tables, *NFE, NRC* and CPU-time represents, respectively the mean number of objective function evaluations over all the 10 executions, the percentage of convergences to the global optimum and the average CPU time per execution (key parameters used are NP = 20, CR = 0.5, F = 0.8).

Table 11.6. Results of DE with all ten strategies (accuracy (ε)= 10^{-6}).

S. No.	Strategy	NFE	CPU-time	NRC
1	DE/rand/1/bin	3134	0.1319	100
2	DE/best/1/bin	2406	0.0879	100
3	DE/best/2/bin	4444	0.1758	100
4	DE/rand/2/bin	4644	0.1758	100
5	DE/rand-to-best/1/bin	2364	0.0879	100
6	DE/rand/1/exp	3214	0.1154	100
7	DE/best/1/exp	2372	0.0934	100
8	DE/best/2/exp	4506	0.1648	100
9	DE/rand/2/exp	4652	0.1868	100
10	DE/rand-to-best/1/exp	2162	0.0714	100

Table 11.7. Results of DE with all ten strategies [accuracy (ε)= 10^{-7}].

S. No.	Strategy	NFE	CPU- time (s)	NRC
1	DE/rand/1/bin	3524	0.1374	100
2	DE/best/1/bin	2624	0.1099	100
3	DE/best/2/bin	5016	0.1868	100
4	DE/rand/2/bin	5158	0.2033	100
5	DE/rand-to-best/1/bin	4146	0.1648	99
6	DE/rand/1/exp	3542	0.1319	100
7	DE/best/1/exp	2636	0.0989	100
8	DE/best/2/exp	5048	0.1923	100
9	DE/rand/2/exp	5206	0.1978	100
10	DE/rand-to-best/1/exp	2484	0.0989	100

The time taken by DE is much less than that of Branch & Reduce algorithm (Table-1). Of course the CPU-times cannot be compared directly because different computers are used. From the above table-2 & 3 it is evident that the strategy number 10 (DE/rand-to-best/1/exp) is the best strategy. It takes least average CPU-time, maximum *NRC* and minimum *NFE.*

11.4 Conclusions

DE is successfully applied to two classical fluid mechanics problems. In the first problem, all the constraints are satisfied with this optimal gas transmission network (Babu et al., 2003a). Also, a single network has been obtained for both the possible scenarios (line A & line B of Fig. 11.2), and DE alone could give the solution for both scenarios as against two different approaches adopted by earlier investigators. The optimum values of objective function for both the possible scenarios are obtained as 7.692×10^6 \$/yr and 7.792×10^6 \$/yr respectively. Though the objective function values are slightly higher than those reported by Edgar and Himmelblau (1988), these values correspond to satisfying all the constraints. In addition, by approximating the compression ratio of compressor station 3 (1.0150 in Table-11.4) to 1.00 (considering only two digits as Edgar and Himmelblau, 1988), it is possible to remove one more compressor from the proposed final network (Fig. 11.6), in which case the total number of compressors would be only 4 and the final objective function value would further reduce. Subsequently, Babu et al. (2003b) obtained an improved solution (better than that reported earlier studies) by forcing some of the parametric values (of pipe diameters and lengths) in optimization as shown in Table 11.3. The optimum values of objective function for both the possible scenarios are obtained as $6.834x10^6$ \$/yr and $6.934x10^6$ \$/yr respectively.

In the second problem, the optimization of water pumping system using Differential evolution (DE) has been presented. The key parameters used for the present problem are: NP = 40, CR = 0.8, F = 0.5. The strategy that took minimum CPU-time with highest NRC is strategy no. 10 (DE/rand-to-best/1/exp). The results obtained by two methods (viz. DE & Branch & Reduce algorithm) are same and matches with that reported in literature.

References

Angira, R. and Babu, B.V. (2003). Evolutionary Computation for Global Optimization of Non-Linear Chemical Engineering Processes". Proceedings of International Symposium on Process Systems Engineering and Control (ISPSEC' 03) - For Productivity Enhancement through Design and Optimization, IIT Bombay, Mumbai, January 3-4, 2003, Paper No. FMA2, pp 87-91 (2003). *(Also available via Internet as .pdf file at http://bvbabu.50megs.com/custom.html/#56).*

Babu, B.V. and Angira, R. (2001a). Optimization of Non-linear functions using Evolutionary Computation. *Proceedings of 12th ISME Conference*, India, January 10–12, 153-157 (2001). *(Also available via Internet as .pdf file at http://bvbabu.50megs.com/custom.html/#34).*

Babu, B.V. and Angira, R. (2001b). Optimization of thermal cracker operation using Differential Evolution. Proceedings of International Symposium & 54th Annual Session of IIChE (*CHEMCON-2001*), Chennai, December 19-22. (Also available via Internet as .pdf file at *http://bvbabu.50megs.com/custom.html/#38*) & Application No. 20, Homepage of Differential Evolution, the URL of which is: http://www.icsi.berkeley.edu/~storn/code.html

Babu, B.V. and Angira, R. (2002a). A Differential Evolution Approach for Global Optimization of MINLP Problems. Presented at *4th Asia-Pacific Conference on Simulated Evolution And Learning (SEAL'02)*, Singapore, November 18 – 22. (Also available via Internet as .pdf file at *http://bvbabu.50megs.com/custom.html/#46*).

Babu, B.V. and Angira, R. (2002b). Optimization of Non-Linear Chemical Processes Using Evolutionary Algorithm". Proceedings of International Symposium & 55th Annual Session of IIChE (CHEMCON-2002), OU, Hyderabad, December 19-22, 2002. *(Also available via Internet as .pdf file at http://bvbabu.50megs.com/custom.html/#54).*

Babu, B.V. and Angira, R. (2003). "Optimization of Water Pumping System Using Differential Evolution Strategies". To be presented at *The Second International Conference on Computational Intelligence, Robotics, and Autonomous Systems (CIRAS-2003)*, Singapore, December 15-18, 2003.

Babu, B.V. and Chaturvedi, G. (2000). Evolutionary Computation strategy for Optimization of an Alkylation Reaction. Proceedings of International Symposium & 53rd Annual Session of IIChE (*CHEMCON-2000*), Calcutta, December 18-21. (Also available via Internet as .pdf file at *http://bvbabu.50megs.com/custom.html/#31*) & Application No. 19, Homepage of Differential Evolution, the URL of which is: http://www.icsi.berkeley.edu/~storn/code.html

Babu, B.V. and Chaurasia, A.S. (2003). "Optimization of Pyrolysis of Biomass Using Differential Evolution Approach". To be presented at *The Second International Conference on Computational Intelligence, Robotics, and Autonomous Systems (CIRAS-2003)*, Singapore, December 15-18, 2003.

Babu, B.V. and Gautam, K. (2001). Evolutionary Computation for Scenario-Integrated optimization of Dynamic Systems. Proceedings of International Symposium & 54th Annual Session of IIChE (*CHEMCON-2001*), Chennai, December 19-22. (Also available via Internet as .pdf file at *http://bvbabu.50megs.com/custom.html/#39*) & Application No. 21, Homepage of Differential Evolution, the URL of which is: http://www.icsi.berkeley.edu/~storn/code.html

Babu, B.V. and Mohiddin, S.B. (1999). "Automated Design of Heat Exchangers using Artificial Intelligence based Optimization", *Proceedings of International Symposium & 52nd Annual Session of IIChE (CHEMCON-1999)*, Panjab University, Chandigarh, December 20-23, 1999. Also available via Internet as .htm file at http://bvbabu.50megs.com/custom.html/#27.

Babu, B.V. and Munawar, S.A. (2000). Differential Evolution for the optimal design of heat exchangers. *Proceedings of All-India seminar on Chemical Engineering Progress on Resource Development: A Vision 2010 and Beyond*, IE (I), Bhubaneswar, India, March 11, (2000). (Also available via Internet as .pdf file at *http://bvbabu.50megs.com/custom.html/#28*).

Babu, B.V. and Munawar, S.A. (2001). Optimal Design of Shell & Tube Heat Exchanger by Different strategies of Differential Evolution. *PreJournal.com - The Faculty Lounge, Article No. 003873, posted on website Journal http://www.prejournal.com. (Also available via Internet as .pdf file at http://bvbabu.50megs.com/custom.html/#35)* & Application No. 18, Homepage of Differential Evolution, the URL of which is: http://www.icsi.berkeley.edu/~storn/code.html

Babu, B.V. and Sastry, K.K.N. (1999). Estimation of heat-transfer parameters in a trickle-bed reactor using differential evolution and orthogonal collocation. *Computers & Chemical Engineering,* **23**, 327–339. *(Also available via Internet as .pdf file at http://bvbabu.50megs.com/custom.html/#24)* & Application No. 13, Homepage of Differential Evolution, the URL of which is: http://www.icsi.berkeley.edu/~storn/code.html

Babu, B.V. and Singh, R.P. (2000). Synthesis & optimization of Heat Integrated Distillation Systems Using Differential Evolution. *Proceedings of All-India seminar on Chemical Engineering Progress on Resource Development: A Vision 2010 and Beyond*, IE (I), Bhubaneswar, India, March 11, (2000).

Babu, B.V., Angira, R. and Nilekar, A. (2002). "Differential Evolution for Optimal Design of an Auto-Thermal Ammonia Synthesis Reactor", *Computers & Chemical Engineering* (Communicated).

Babu, B.V., Angira, R., Chakole, P.G. and Mubeen, J.H.S. (2003a), "Optimal Design of Gas Transmission Network using Differential Evolution", To be presented at *The Second International Conference on Computational Intelligence, Robotics, and Autonomous Systems (CIRAS-2003)*, Singapore, December 15-18, 2003.

Babu, B.V., Chakole, P.G. and Mubeen, J.H.S. (2003b). "Differential Evolution Strategy for Optimal Design of Gas Transmission Network", *European Journal of Operational Research* (Communicated).

Cheesman, A.P. (1971a). "How to Optimize Gas Pipeline Design by Computer". *Oil and Gas Journal,* **69** (51), December 20, 64.

Cheesman, A.P. (1971b). "Understanding Origin of Pressure is a Key to Better Well Planning". *Oil and Gas Journal,* **69** (46), November 15, 146.

Chiou, J.P. and Wang, F.S. (1999). Hybrid method of evolutionary algorithms for static and dynamic optimization problems with application to a fed-batch fermentation process. *Computers & Chemical Engineering,* **23**, 1277-1291.

Dasgupta, D. and Michalewicz, Z. (1997). *Evolutionary algorithms in Engineering Applications,* 3 - 23, Springer, Germany,.

Dollena S.H, David M.A. and Arnold J.S. (2001). "Determining the number of components in mixtures of linear models". *Computational Statistics & Data Analysis,* **38**, 15-48.

Edgar, T.F., Himmelblau, D.M. (1988). *Optimization of Chemical Processes,* McGraw Hill Book Company, New York.

Edgar, T.F., Himmelblau, D.M., and Bickel, T.C. (1978). "Optimal Design of Gas Transmission Network", *Society of Petroleum Engineering Journal,* **30**, 96.

Flanigan, O. (1972). "Constrained Derivatives in Natural Gas Pipeline System Optimization", *Journal of Petroleum Technology,* **May,** 549.

Floudas, C.A. (1995). *Nonlinear and mixed-integer optimization.* Oxford University Press, New York.

Floudas, C.A. and Pardalos, P.M. (1990). A Collection of Test Problems for Constrained Global Optimization Algorithms. *Lecture notes in computer Science,* **Vol. 455**. Springer, Germany.

Goldberg, D.E. (1989). *Genetic Algorithms in search, Optimization, and Machine learning,* Reading, MA, Addison-Wesley.

GPSA (1972). "Gas Processor Suppliers Association", *Engineering Data Book.*

Graham, G.E., Maxwell, D.A. and Vallone, A. (1971). "How to Optimize Gas Pipeline Networks", *Pipeline Industry,* **June,** 41-43.

Happel, J. and Jordan, D.G. (1975). *Chemical Process Economics.* 2nd Edition, Marcel Dekker, New York.

Joshi, R. and Sanderson, A.C. (1999). "Minimal representation multi-sensor fusion using differential evolution". *IEEE Transactions on Systems, Man and Cybernetics, Part A* **29,** 63-76.

Katz, D.L. (1959). *Handbook of Natural Gas Engineering,* McGraw Hill, New York.

Kyprianou, A., Worden, K. and Panet, M. (2001). "Identification of hysteretic systems using the differential evolution algorithm". *Journal of Sound and Vibration,* **248** (2), 289-314.

Larson, R.E. and Wong, P.J. (1968). "Optimization of Natural Gas System via Dynamic Programming", *Industrial and Engineering Chemistry,* **AC 12** (5), 475-481.

Lee, M.H., Han, C. and Chang, K.S. (1999). "Dynamic optimization of a continuous polymer reactor using a modified differential evolution". *Industrial & Engineering Chemistry Research,* **38** (12), 4825-4831.

Leibman J., Lasdon, L., Schrage, L. and Waren, A. (1986). Modeling and optimization with GINO. *The Scientific Press*, Palo Alto, CA.

Letterman, R.D. (1980). "Economic Analysis of Granular Bed Filtration", *Transactions of American Society of Civil Engineers (Journal of Environmental Engineering Division),* **106**, 279.

Lu, J.C. and Wang, F.S. (2001). "Optimization of Low Pressure Chemical Vapour Deposition Reactors Using Hybrid Differential Evolution". *Canadian Journal of Chemical Engineering,* **79** (2), 246-254.

Mah, R.S.H. and Schacham, M. (1978). "Pipeline Network Design and Synthesis", *Advances in Chemical Engineering,* **10**.

Martch, H.B. and McCall, N.J. (1972). "Optimization of the Design and Operation of Natural Gas Pipeline Systems", Paper No. SPE 4006, *Society of Petroleum Engineers of AIME, 1972.*

McCabe, W.L., Smith, J.C. and Harriot, P. (1985). *Unit Operations of Chemical Engineering*, 4th Edition, McGraw-Hill.

Olorunniwo, F.O. (1981). "A Methodology for Optimal Design and Capacity Expansion Planning of Natural Gas Transmission Networks", Ph.D. Dissertation, The University of Texas at Austin, May, 1981.

Olorunniwo, F.O. and Jensen, P.A. (1982). "Optimal Capacity Expansion Policy for Natural Gas Transmission Networks – A Decomposition Approach", *Engineering Optimization,* **6,** 95.

Price, K. and Storn, R. (1997). Differential Evolution - A simple evolution strategy for fast optimization. *Dr. Dobb's Journal,* **22** (4), 18 – 24 and 78.

Price, K. and Storn, R. (2002). *Web site of DE as on July, 2003,* the URL of which is: http://www.ICSI.Berkeley.edu/~storn/code.html

Ryoo, H.S. & Sahinidis, N.V. (1995). Global optimization of nonconvex NLPs and MINLPs with Applications in Process Design. *Computers& Chemical Engineering,* **19**(5), 551-566.

Salcedo, R.L. (1992). Solving Nonconvex Nonlinear Programming Problems with Adaptive Random Search. *Industrial & Engineering Chemistry Research,* **31**, 262.

Sastry, K.K.N., Behera, L. and Nagrath, I.J. (1998). "Differential evolution based fuzzy logic controller for nonlinear process control", *Fundamenta Informaticae*: *Special Issue on Soft Computation.*

Stoecker, W.F. (1971). *Design of Thermal Systems.* 3rd ed., McGraw-Hill International edition, Singapore, pp 117-121.

Storn, R. (1995). Differential Evolution design of an IIR-filter with requirements for magnitude and group delay. *International Computer Science Institute,* TR-95-026.

Wang, F.S. and Cheng, W.M. (1999). Simultaneous optimization of feeding rate and operation parameters for fed-batch fermentation processes. *Biotechnology Progress,* **15** (5), 949-952.

Wang, F.S., Jing, C.H. and Tsao, G.T. (1998). "Fuzzy-decision-making problems of fuel ethanol production using genetically engineered yeast", *Industrial & Engineering Chemistry Research,* **37**, 3434-3443.

Wang, F.S., Su, T.L. and Jang, H.J. (2001). "Hybrid Differential Evolution for Problems of Kinetic Parameter Estimation and Dynamic Optimization of an Ethanol Fermentation Process". *Industrial & Engineering Chemistry Research,* **40** (13), 2876-2885.

12 Applications in Reaction Engineering

B V Babu

12.1 Introduction

This chapter presents the application of Genetic Algorithms (GA) & Differential Evolution (DE) on two most important chemical engineering problems: (1) Optimal design of an auto-thermal ammonia synthesis reactor, and (2) Optimization of thermal cracking operation.

In the first problem, the Differential Evolution (DE), an evolutionary computation technique, is applied to the optimal design of an auto-thermal ammonia synthesis reactor. This paper also presents the new concept of "Nested" DE (DE is also used to find out the best combination of key parameters of DE itself). The main objective in the optimal design of an auto-thermal ammonia synthesis reactor is the estimation of the optimal length of reactor for different top temperatures with the constraints of energy and mass balance of reaction and feed gas temperature & mass flow rate of nitrogen for ammonia production. Thousands of combinations of feed gas temperature, nitrogen mass flow rate, reacting gas temperature and reactor length are possible. This section also presents the application of four methods, *viz.*, Euler's method, Runge-Kutta method (both variable & constant step size), & Gear's method in combination with DE, and compare the results reported using GA in earlier literature. A software package "POLYMATH" is also used to solve the three equality constraints i.e. three coupled differential equations. Apart from determining the optimal reactor length, the comparison of results obtained from different methods is presented. DE found to be a robust, fast and simple evolutionary computation technique for optimization problems.

The second paper presents the application of Differential Evolution (DE) for the optimization of Thermal Cracking operation. The objective in this problem is the estimation of optimal flow rates of different feeds to the cracking furnace under the restriction on ethylene and propylene production. Thousands of combinations of feeds are possible. Hence an efficient optimization strategy is essential in searching for the global optimum. In the present study LP Simplex method and DE, an improved version of Genetic Algorithms (GA), have been successfully applied with different strategies to find the optimum flow rates of different feeds. In the application of DE, various combinations of the key parameters are considered. It is found that DE, an exceptionally simple evolution strategy, is significantly faster and yields the global optimum for a wide range of the key parameters. The results obtained from DE are compared with that of LP Simplex method.

Price and Storn (1997) gave the working principle of DE with single strategy. Later on, they suggested ten different strategies of DE (Price and Storn, 2002). DE is an improved version of GA for faster optimization. Among the DE's advantages are its simple structure, ease of use, speed and robustness. The key parameters of control are: NP - the population size, CR - the crossover constant, F - the weight applied to random differential (scaling factor). Babu (2001) gave an overview of DE at a glance.

Differential Evolution (DE) is an improved version of GA. It is exceptionally simple, significantly faster & robust at numerical optimization and is more likely to find a function's true global optimum. DE has been successfully applied in various fields. The various applications of DE are: digital filter design (Storn, 1995), fuzzy decision making problems of fuel ethanol production (Wang *et al.*, 1998), Design of fuzzy logic controller (Sastry *et al.*, 1998), batch fermentation process (Chiou and Wang, 1999; Wang and Cheng, 1999), multi sensor fusion (Joshi and Sanderson, 1999), dynamic optimization of continuous polymer reactor (Lee *et al.*, 1999), estimation of heat transfer parameters in trickle bed reactor (Babu and Sastry, 1999; Babu and Vivek, 1999), optimization of alkylation reaction (Babu and Chaturvedi, 2000), optimal design of heat exchangers (Babu and Mohiddin, 1999; Babu and Munawar, 2000; Babu and Munawar, 2001), synthesis & optimization of heat integrated distillation system (Babu and Singh, 2000), optimization of non-linear functions (Babu and Angira, 2001a), scenario- integrated optimization of dynamic systems (Babu and Gautam, 2001), optimization of thermal cracking operation (Babu and Angira, 2001b), determining the number of components in mixtures of linear models (Dollena *et al.*, 2001), Identification of hysteretic systems using the differential evolution algorithm (Kyprianou *et al.*, 2001), optimization of Low Pressure Chemical Vapour Deposition Reactors Using Hybrid Differential Evolution (Lu and Wang, 2001), hybrid differential evolution for problems of Kinetic Parameter Estimation and Dynamic Optimization of an Ethanol Fermentation Process (Wang *et al.*, 2001), Global optimization of MINLP problems (Babu and Angira, 2002a), Optimal design of an auto-thermal ammonia synthesis reactor (Babu *et al.*, 2002b) Optimization of non-linear chemical processes (Babu and Angira, 2002c; Angira and Babu, 2003), etc.

12.2 Design of Auto-Thermal Ammonia Synthesis Reactor

12.2.1 Ammonia Synthesis Reactor

Ammonia is one of the most important chemicals produced as it enjoys the wide use in the manufacture of fertilizers. Hence modeling and simulation of ammonia manufacturing process has received considerable attention among the process industries. Simulation models for ammonia synthesis converters of different types have been developed for design & optimization (Annable, 1952; Eymery, 1964; Dyson, 1965; Murase *et al.*, 1970; Singh and Saraf, 1979; Upreti and Deb, 1997), and control (Shah, 1967) purposes.

Annable (1952) compared the performance of an autothermal ammonia synthesis reactor (Fig. 1) with the maximum yield that could be obtained if one had direct control of the temperature profile. He found that conversion could be increased by 14%. Obviously, one does not have direct control of the temperature profile, but it could be affected by the configuration of the heat transfer surface, *viz.*, added insulation and/or fins. Dyson (1965) considered the general problem of determining the heat transfer coefficient vs. length function that would maximize the yield in an autothermal reactor.

Murase *et al.* (1970) computed the optimum temperature trajectory along the reactor length applying the Pontryagin's maximum principle. Although their formulation was correct, the stated objective function was wrong. Edger and Himmelblau (1989) rectified the same and used Lasdon's generalized reduced-gradient method to arrive at an optimal reactor length corresponding to a particular reactor top temperature of 694 K. However they also ignored a term mentioned in Murase's formulation, pertaining to the cost of ammonia already present in the feed gas, in the objective function. Also the expressions of the partial pressures of nitrogen, hydrogen and ammonia, used to simulate the temperature and flow rate profiles across the length of the reactor, were not correct.

Upreti & Deb (1997) rectified above stated shortcomings. They used Murase's formulation with correct objective function and correct stoichiometric expressions of the partial pressures of N_2, H_2, and NH_3. They used GA in combination with Gear package of NAG library's subroutine, D02EBF, for the optimization of ammonia synthesis reactor. GA has the tendency to locate the near global optima but not necessarily the global optima. Also, Upreti & Deb (1997) have not tried all possible combinations of key parameters (p_c – crossover probability, p_m – mutation probability, N – population size). Moreover, there is a contradiction in the temperatures & gas flow rate profiles obtained. They reported the profiles that were not so smooth as in earlier literature. Also, they reported reverse reaction condition at the top temperature of 664 K, which was not found in literature earlier. Hence, the present study is carried out in order to take care of the above deficiencies.

A typical ammonia production process consists of production of the synthesis gas from the petroleum feed stock, compression of the gas to the required pressure and the synthesis loop in which the conversion to ammonia takes place. The ammonia converter is part of the synthesis loop, and its operation is quite crucial in the overall control strategy of the plant. In the converter the following catalytic reaction takes place at elevated temperatures and pressures releasing a large amount of heat.

$$N_2 + 3H_2 \Leftrightarrow 2NH_3; \Delta H = -22.0 \text{ kcal} \tag{12.1}$$

This heat has to be removed to obtain a reasonable conversion as well as to protect the catalyst life. At the same time, the released heat energy is utilized to heat the incoming feed-gas to proper reaction temperature.

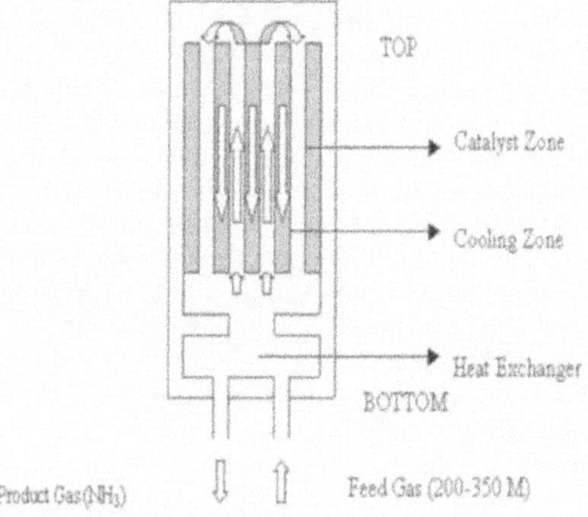

Fig. 12.1. Ammonia Synthesis Reactor

A typical ammonia synthesis reactor is as shown in the following Fig 12.1. The reaction zone (shaded) contains the catalyst. A number of cooling tubes are inserted vertically through the reaction zone. The feed gas comes in from the lower part of the reactor and flows up through the top of the reactor. Then, changing direction, it flows down through the reaction zone and heat exchanger to the outlet. As with all reversible exothermic reactions, the temperature, at which the reaction rate is maximum, decreases as the conversion increases. Even though one would like to maximize the reaction rate at each instant, it is impossible to obtain the ideal temperature profile by control of available design variables. The counter-current flow does, however, cause the temperature to decrease in the bottom part of the reactor because of heat transfer between the reacting and feed gas. Babu *et al.* (2002) carried out detailed simulation incorporating the optimization for the above problem.

12.2.2 Problem Formulation

The Problem formulation of Babu *et al.* (2002) is similar to that in Murase *et al.* (1970) including the modifications mentioned in Upreti and Deb (1997). Feed gas contains 21.75 mole% nitrogen, 65.25 mole% hydrogen, 5 mole% ammonia, 4 mole% methane and 4 mole% argon. In a typical ammonia reactor, feed gas enters the bottom of the reactor. The production of ammonia depends on the temperature of feed gas at the top of the reactor (henceforth called top temperature), the partial pressures of the reactants (nitrogen and hydrogen), and the reactor length.

12.2.2.1 Assumptions

The following assumptions are used in modeling the ammonia synthesis reactor (Murase *et al*, 1970):

1. The rate expression is valid.
2. The model is one dimensional, i.e., the temperature and concentration gradients in the radial direction are neglected.
3. Heat & mass diffusion in the longitudinal direction are negligible.
4. The temperature of the gas flowing through the catalyst zone is equal to the reacting gas and the catalyst particles.
5. The heat capacities of the reacting gas and the feed gas are constant.
6. The catalyst activity is uniform along the reactor and equal to unity.
7. Pressure drop across the reactor is negligible compared to the total pressure of the system.

12.2.2.2 Objective function

The objective function is the economic return based on the difference between the value of the product gas (heating value and the ammonia value) and the value of feed gas (as a source of heat only) less the amortization of reactor capital costs. Other operating costs are omitted. As shown by Upreti and Deb (1997), the final consolidation of the objective function terms is:

$$f(x, N_{N_2}, T_f, T_g) = 1.33563\times10^7 - 1.70843\times10^4 N_{N_2} + 704.09(T_g - T_0) \\ -699.27(T_f - T_0) - [3.45663\times10^7 + 1.98365\times10^9 x]^{1/2} \quad (12.2)$$

It is clear from the above expression that the objective function depends on four variables: the reactor length x, proportion of nitrogen N_2, the top temperature T_g, the feed gas temperature T_f.

12.2.2.3 Constraints

There are mistakes in modeling equations reported in Murase *et al.* (1970), which were corrected and some of them were reported by Upreti & Deb (1997). Hence for the sake of clarity, all the corrected modeling equations are presented below:

a. Energy-Balances

Feed Gas: Referring to Fig. 12.2, an energy balance on the feed stream yields the following equation:

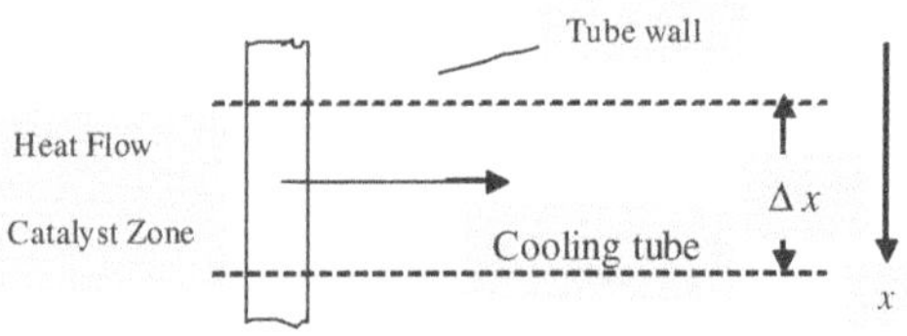

Fig. 12.2. Energy & Material Balance on control volume

$$\frac{dT_f}{dx} = \frac{US_1}{WC_{pf}}\left(T_g - T_f\right) \tag{12.3}$$

where,

U	= Overall heat transfer coefficient, Kcal/m^2.hr.K
S_1	= Surface area of cooling tubes per unit length of reactor, m
T_g	= Temperature of reacting gas, K
W	= Total mass flow rate, kg/hr
C_{pf}	= Specific heat of feed gas, kcal/kg. K
x	= Distance along axis, m

Reacting Gas: Similarly, for the reacting gas,

$$\frac{dT_g}{dx} = -\frac{US_1}{WC_{pg}}\left(T_g - T_f\right) + \frac{(-\Delta H)S_2}{WC_{pg}}\left(\frac{-dN_{N_2}}{dx}\right) \tag{12.4}$$

where,

ΔH	= Heat of reaction, kcal/kg mole of N_2
S_2	= Cross-sectional area of catalyst zone, m^2
$\frac{dN_{N_2}}{dx}$	= Reaction rate, kg moles of N_2 /hr.m^3
C_{pg}	= Heat capacity of reacting gas, kcal/kg. K

b. Material balance

Considering the incremental distance in the catalyst zone (Fig. 12.2) and performing a N_2 material balance yields:

$$\frac{dN_{N_2}}{dx} = -f\left[K_1 \frac{p_{N_2} p_{H_2}^{1.5}}{p_{NH_3}} - K_2 \frac{p_{NH_3}}{p_{H_2}^{1.5}}\right] \tag{12.5}$$

where,

f = Catalyst activity

$p_{N_2}, p_{H_2}, p_{NH_3}$ = Partial pressure of N_2, H_2, and NH_3

K_1, K_2 = Rate constants

$$K_1 = 1.78954 \times 10^4 \exp\left(\frac{-20800}{RT_g}\right) \tag{12.6}$$

$$K_2 = 2.5714 \times 10^{16} \exp\left(\frac{-47400}{RT_g}\right) \tag{12.7}$$

In order to maintain the energy & material balance of reaction, the above three coupled differential equations (Eqs. 12.3, 12.4, & 12.5) must be satisfied. It turns out that three of the above variables (T_f, T_g, N_{N_2}) can be eliminated by satisfying these energy & material balance equations. Thus, practically, there is only one design variable for given top temperature. The partial pressures appearing in the differential equations are computed as follows:

$$p_{N_2} = \frac{286 N_{N_2}}{2.598 N_{N_2}^0 + 2N_{N_2}} \tag{12.8}$$

$$p_{H_2} = 3 p_{N_2} \tag{12.9}$$

$$p_{NH_3} = \frac{286(2.23 N_{N_2}^0 - 2N_{N_2})}{2.598 N_{N_2}^0 + 2N_{N_2}} \tag{12.10}$$

The boundary conditions are:

$$T_f = T_0 \text{ at x=0} \tag{12.11}$$

$$T_g = T_f \text{ at x=0} \tag{12.12}$$

$$N_{N_2}^0 = 701.2 \text{ kmol/hr.m}^2 \text{ at x=0} \tag{12.13}$$

The three inequality constraints that limit the values of three of the design variables are as given below:

$$0.0 \le N_{N_2} \le 3220 \tag{12.14}$$

$$400 \le T_f \le 800 \tag{12.15}$$

$$0.0 \le x \le 10.0 \tag{12.16}$$

Since the reaction gas temperature (T_g) depends on the nitrogen mass flow rate (N_{N_2}), feed gas temperature (T_f) and reactor length (x), explicit bounds on T_g are not necessary. From the system model, we have three differential equations and four variables, making the degrees of freedom equal to one. We specify the length of the reactor, calculate the remaining variables using the system model and then pass these variables to the optimization algorithm. The computation procedure for the optimization carried out is shown in Fig 12.3.

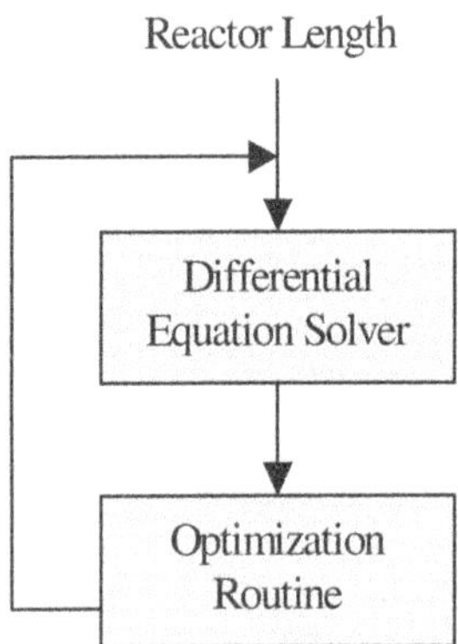

Fig. 12.3. Computation Procedure

12.2.3 Simulated Results & Discussion

12.2.3.1 Temperature & Flow rate Profiles

First, the system equations (12.3, 12.4, & 12.5) were solved using Runge-Kutta fixed step size method (RKFS) and Fig. 12.4 (a) shows the profiles obtained.

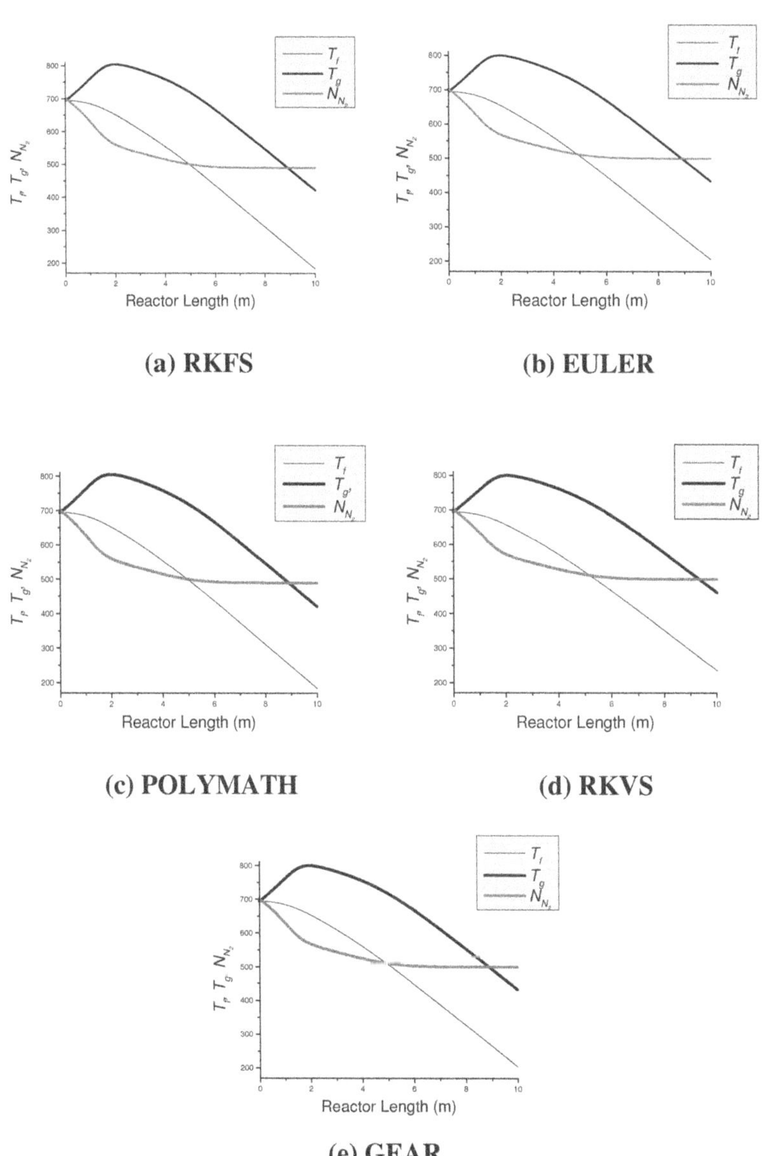

Fig. 12.4. The Profiles obtained using different Numerical Methods

From the graph it is clear that the profiles of N_{N2} & T_f intersect at a reactor length of 4.933 m. And the profiles of N_{N2} & T_g intersect at a reactor length of 8.915 m. It is evident from the graph that the profiles are very smooth and there are no such spikes as those reported by Upreti and Deb (1997). Therefore, in order to check if it is the limitation of RKFS because of which spikes are not obtained, the other numerical methods viz. Euler's method (EULER), Runge-Kutta variable step size method (RKVS) & Gear's method (GEAR) are also used for simulating the results. Apart from above mentioned numerical methods, the POLYMATH (a software package) is also used to simulate the results for profiles. Figs. 12.4 (b), (c), (d), & (e) show the profiles obtained using the above four methods. The profiles obtained are quite smooth without any spikes and similar to that of RKFS. The profiles are same qualitatively but they differ slightly quantitatively.

To illustrate the exact difference quantitatively, the above observations & comparison of various numerical methods used for simulating the results are presented in Table-12.1 for the reactor lengths of 10 m. Table-12.2 & Table-12.3 show the reactor length for which the variables N_{N2} & T_g and N_{N_2} & T_f intersect respectively.

12.2.3.2 Comparison of Results

From the Table-12.1, Table-12.2 and Table-12.3, it is evident that all the numerical methods (stated above) are equally good barring a few of the following differences. There is good agreement between RKFS & POLYMATH, which can be explained with the fact that POLYMATH software is based on RKFS algorithm. Similarly GEAR & EULER are giving same results though slightly different from RKFS & POLYMATH. The difference in the prediction is less than 2.3 % between GEAR/EULER & RKFS/POLYMATH respectively. Similarly, the difference in prediction of intersections is less than 5.0% between RKVS & GEAR/EULER method. RKVS is latest & has the ability to adopt its step size & varies as per requirement at every move. Based on the above observations, it can be substantiated that all the five methods are equally good both qualitatively giving the same results and quantitatively with slight difference. Hence, any one of the above numerical methods can be used for the solution of three coupled differential equations.

Upreti and Deb (1997) reported that the reverse reaction predominates the forward reaction at the top temperature of 664 K. In this study, RKVS, which is well accepted for many engineering problems, is used to generate temperature & flow rate profiles at that temperature. Surprisingly, there is no such trend observed in the profiles obtained [for which Upreti and Deb (1997) gave a very good physical explanation] as can be seen in Fig 12.5 (a). To see the presence of any reverse reaction effect, the program is executed for temperatures even below 664 K with interval of 10 K up to 600 K. But no such trend is observed. Typical results obtained at top temperature of 640 K and 600 K are shown in Fig. 12.5 (b) & (c) respectively.

Table 12.1. Comparison of different Numerical Methods at a reactor length of 10 m

Methods → / Parameters ↓	RKFS	EULER	POLYMATH	RKVS	GEAR
x	10.00	10.00	10.00	10.00	10.00
N_{N_2}	490.55	499.48	490.46	499.68	499.46
T_g	422.00	432.89	422.25	462.86	432.69
T_f	183.23	204.23	183.38	235.71	204.73

Table 12.2. Reactor Length at which variables N_{N_2} & T_g intersect

Methods → / Parameters ↓	RKFS	EULER	POLYMATH	RKVS	GEAR
x	8.915	8.915	8.916	9.355	8.895
N_{N_2}, T_g	490.48	499.56	490.56	499.68	499.57

Table12.3. Reactor Length at which variables N_{N_2} & T_f intersect

Methods → / Parameters ↓	RKFS	EULER	POLYMATH	RKVS	GEAR
x	4.933	4.900	4.935	5.156	4.900
N_{N_2}, T_f	501.525	511.50	501.64	511.75	511.55

As is evident from the figures there is no reverse reaction effect even at a top temperature as low as 600 K. Again, to check if it is the limitation of RKVS, the other methods are applied but none of them could show this effect.

Also, Upreti and Deb (1997) reported that the three differential equations (12.3, 12.4, & 12.5) become unstable at the top temperature of 706 K. However, using the above stated numerical methods, it is observed that the equations are not unstable even at a top temperature as high as 800 K. It may be noted that Upreti & Deb (1997) used NAG library's subroutine D02EBF (which is now replaced by D02EJF (NAG, 2002)). So it may be because of the error in the software package that they reported the reversible reaction effect.

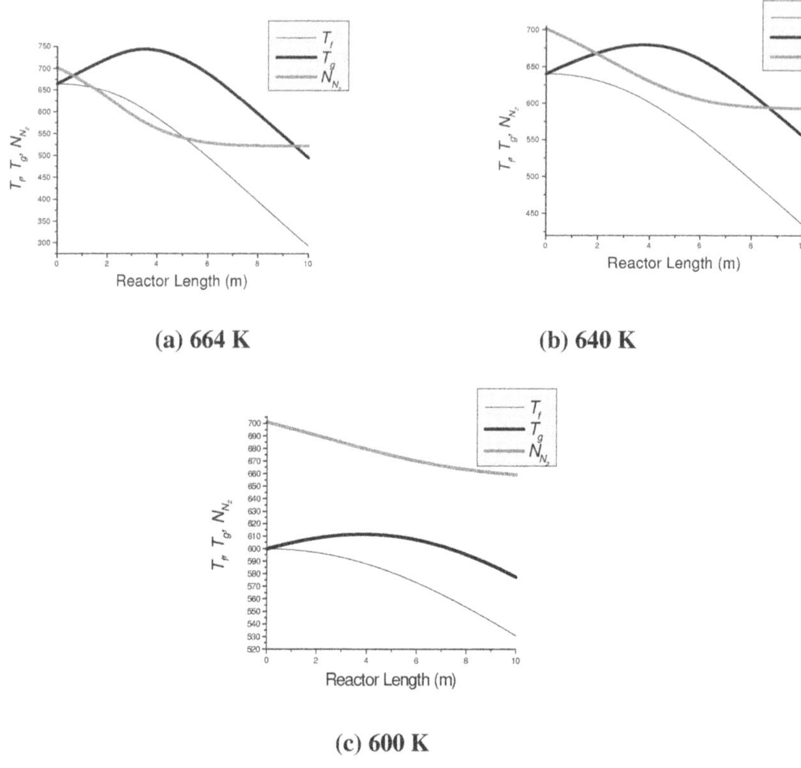

Fig. 12.5. The profiles obtained using RKVS for different top temperatures

12.2.4 Optimization

12.2.4.1 "Nested" DE – A new Concept

Choosing NP, F, and CR depends on the specific problem applied, and is often difficult. But some general guidelines are available. Normally, NP should be about 5 to 10 times the number of parameters in a vector. As for F, it lies in the range 0.4 to 1.0. Initially F = 0.5 can be tried then F and/or NP is increased if the population converges prematurely. A good first choice for CR is 0.1, but in general CR should be as large as possible (Price & Storn, 1997). The best combination of these key parameters of DE for each of the strategies mentioned earlier is again different. Price & Storn (2002) have mentioned some simple rules for choosing the best strategy as well as the ranges of corresponding key parameters.

12.2.4.2 Effect of DE Key Parameters

The key parameters are generally found by trial & error method. In this method one parameter is kept constant while varying the others in steps. In this process we may miss out optimum combination that gives global optima, as it is very difficult to cover the entire range of all the key parameters, however small the incremental step may be. The comparison of results obtained using different combination of DE key parameters (CR and F) are shown in Fig. 12.6 (a) and (b). These graphs show us how difficult it is to find out which combination can be the most suitable one for our problem as there is no specific trend observed.

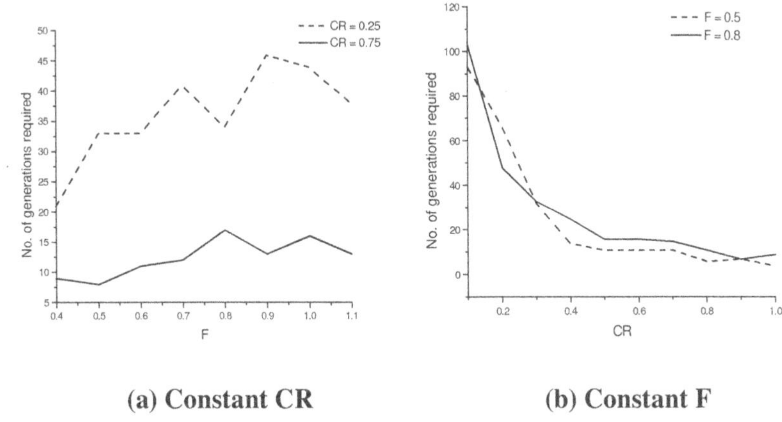

(a) Constant CR **(b) Constant F**

Fig. 12.6. The Effect of F and CR on Number of generations

By looking at the problems mentioned above and the results obtained in this study in choosing the right combination of the DE key parameters, it was felt that why not to use DE itself for finding the optimum key parameters along with the optimum variables of the actual problem formulation (i.e. using DE within DE wherein outer loop takes care of optimization of key parameters (NP, CR, F) with the objective of minimizing the number of generations required, while inner loop taking care of optimizing the problem variables). Yet complex objective can be one that takes care of minimizing the number of generations/function evaluations & the standard deviation (SD) in the set of solutions at the last generation/function evaluation, and try to maximize the robustness of the algorithm. The optimum reactor length obtained using GEAR with "Nested" DE is 6.79 m (at top temperature of 694) with objective function value of \$4848383.0 per year. Using RKFS with DE the values are 6.59 m (at top temperature of 694 K) and \$5006814.58 per year respectively. To find the best combination of these parameters for the present problem, we vary CR & F, but keep NP to the higher side (NP=10*D, for being on the safer side). The algorithm followed for the "Nested" DE operation in combination with GEAR is shown in Fig. 12.7.

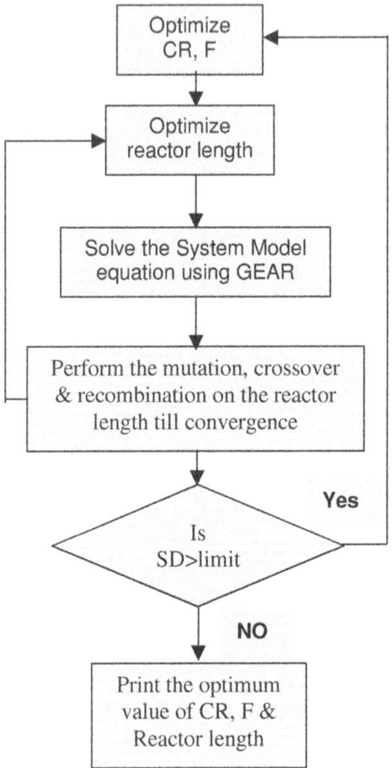

Fig. 12.7. The Solution Strategy For Nested DE

The optimum DE parameters using "Nested" DE (with the strategy DE/rand/1/bin) with gear method for various standard deviations are shown in Table-12.4. For population of x_i*'s (i=1,2,3...NP,)* we define

$$\text{Standard Deviation (SD)}=\sum_{i=1}^{NP}\frac{(\bar{x}-x_i)^2}{(NP-1)} \tag{12.17}$$

where,

$$\bar{x}=\sum_{i=1}^{NP}\frac{x_i}{NP} \tag{12.18}$$

From Table-12.4, it is clear that irrespective of the values of SD (1.0, 0.5, 0.25, 0.1 0.01 & 0.01) and the corresponding CR & F values, we obtained almost same values of the objective function (4848383.0 $/year) and reactor length (6.79 m). It consolidates the robustness of the DE algorithm. The more wider the range of values of CR, F, and SD for which same values of objective function and reactor length are obtained, the more robust is the algorithm.

Table 12.4. Optimum DE parameters Using Nested DE with GEAR for various SD values

Standard Deviation	Reactor length (m)	Objective function ($/year)	CR	F	G_{min}
1.0	6.79	4848382.5	0.745547	0.520328	1
0.5	6.79	4848382.5	0.939156	0.485891	2
0.25	6.79	4848382.5	0.995696	0.7012	4
0.1	6.79	4848383.0	0.8686	0.553093	6
0.01	6.79	4848383.0	0.902649	0.434454	10
0.001	6.79	4848383.0	0.793195	0.43707	18

Also, the code of "Nested" DE was run with another strategy DE/rand/1/exp, and found that the results are exactly same. This also proves DE's power and robustness. Now that we have already tried the effect of DE parameters and strategy on speed and accuracy and found that it has little or no effect on accuracy, we can say that the NP in a generation, if taken within the limits, would not have any effect on the result.

Table-12.5 shows the results obtained from different methods & its comparison with those obtained by Murase *et al.* (1970), Edgar and Himmelblau (1989) & Upreti and Deb (1997) using Pontryagin's maximum principle (PMP), Lasdon's generalized reduced-gradient method (LGRG) & Genetic Algorithm (GA) respectively.

Table 12.5. Results of various Numerical Methods and their comparison

Methods → Parameter ↓	PMP (Murase et al, 1970)	LGRG (Edgar & Himmelblau, 1989)	D02EBF With GA (Upreti & Deb, 1997)	EULER With DE	RKFS With DE	GEAR With DE	RKVS With DE
Optimum x (m)	5.18	2.58	5.33	6.8	6.59	6.79	7.16
Objective function ($/year)	Not reported	$1.29x10^6$	$4.23x10^6$	$4.84x10^6$	$5.00x10^6$	$4.84x10^6$	$4.84x10^6$

From Table-12.5, we observe that an optimum reactor length of 2.58 m is reported by Edgar and Himmelblau (1989) and 5.18 m by Murase *et al.* (1970), both of which are wrong due to the errors in their problem formulations as pointed out by Upreti and Deb (1997). An optimum reactor length of 5.33 m and the corresponding objective function value is 4.23 x 10^6 \$/year, reported by Upreti and Deb (1997) are also not correct as found in this study due to a possible error in the software they used (D02EBF – NAG Library sub-routine). This may also be wrong because simple GA does not ensure global optimum.

Among other methods, GEAR, EULER & RKVS have the same objective function value though the optimum reactor length is slightly different in each case. RKVS is an improved method over the other methods & well-accepted one. Also, this study establishes the accuracy and robustness of DE. Hence, the correct optimum reactor length can be considered as 7.16 m with an objective function value of 4.841x 10^6 \$/year.

12.2.5 Conclusions

In this problem, Differential Evolution (DE) is used for the optimal design of an auto-thermal ammonia synthesis reactor. The new concept of "Nested" DE is introduced. Results indicate that the profiles of temperatures & flow rate are smooth and there is no reverse reaction effect irrespective of numerical method used for the solution of differential equations. The optimum reactor length depends upon the top temperature. Also, the power & robustness of DE is brought out using new concept of nested DE. This successful application of DE for the optimal design of ammonia synthesis reactor indicates that DE has great potential and can be applied to advantage in all the highly non-linear & complex engineering problems.

12.3 Thermal Cracking Operation

This problem (Babu and Angira, 2001b) presents the application of Differential Evolution (DE) for the optimization of Thermal Cracking operation. The objective in this problem is the estimation of optimal flow rates of different feeds to the cracking furnace under the restriction on ethylene and propylene production. Thousands of combinations of feeds are possible. Hence an efficient optimization strategy is essential in searching for the global optimum. In this problem LP Simplex method and DE, an improved version of Genetic Algorithms (GA), have been successfully applied with different strategies to find the optimum flow rates of different feeds. In the application of DE, various combinations of the key parameters are considered.

12.3.1 Thermal Cracking

It is defined as the thermal decomposition, under pressure, of large hydrocarbon molecules to form smaller molecules. Lighter, more valuable hydrocarbons may thus be obtained from such relatively low value stocks as heavy fuel/gas oils (boiling up to 540^{0}C) and residues. This is conducted without any catalyst. Thermal cracking is normally carried out at temperatures varying from 450^{0}C to 750^{0}C and pressures ranging from atmospheric to 1000 psig (Hobson, 1975). The important reactions occurring are:

- Decomposition and destructive condensation.
- Hydrogenation and de hydrogenation.
- Polymerization.
- Cyclization.

The first two reactions are endothermic, while polymerization is exothermic. Coke formation is an additional reaction, which plays an important role in thermal cracking. Although the mechanism by which coke is formed are not entirely understood. It is thought, however, that coke results from extensive degradation of relatively heavy molecules to form increasing quantities of light hydrocarbon gases (dry gas) and polycyclic compounds having low hydrogen to carbon ratios. The rate at which hydrocarbon crack, is strongly dependent on temperature. Cracking reactions begin about 315-370^{0}C, depending on the type of material being cracked (Hobson, 1975).

Depending upon the pressure and temperature employed for the cracking and the characteristics of feed, there are various thermal cracking processes in which the product yields and characteristics are different (Gupta, 1994). Mainly there are four commercial processes employed for thermal cracking in oil refineries. They are:

- Dubbs thermal cracking process
- Pyrolysis.
- Visbreaking.
- Coking.

Pyrolysis or mild thermal cracking is done mainly for the production of lighter products mainly unsaturated like olefins (ethylene, propylene) and naphthene polymers, diolefins, benzene & toluene etc (Sourander *et al*, 1984). It is carried out at high temperature (650-700^{0}C) and low pressure.

12.3.2 Problem Description

This problem (Edgar and Himmelblau, 1989) deals with maximization of profit while operating within furnace and down stream process equipment constraints.

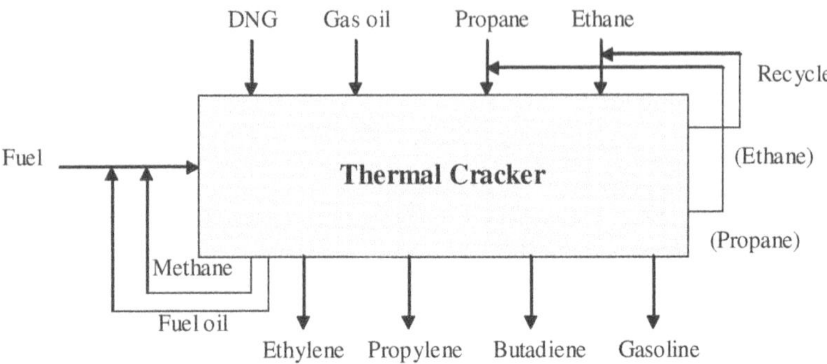

Fig. 12.8. Thermal Cracker

Fig 12.8 lists various feeds & corresponding product distribution for a thermal cracker, which produces olefins. The capacity to run gas feeds through the cracker is 200,000 lb/stream hr. (total flow based on an average mixture). Ethane uses the equivalent of 1.1 lb of capacity per pound of ethane; propane uses 0.9 lb of capacity per pound of propane; gas-oil uses 0.9 lb/lb; and DNG has a ratio of 1.0.

Based on the plant data, eight products are produced in varying proportions according to the following matrix as given in Table-12.6:

Table 12.6. Yield structure (Weight fraction):

Product	Feed			
	Ethane	Propane	Gas-oil	DNG
Methane	0.07	0.25	0.10	0.15
Ethane	0.40	0.06	0.04	0.05
Ethylene	0.50	0.35	0.20	0.25
Propane	-	0.10	0.01	0.01
Propylene	0.01	0.15	0.15	0.18
Butadiene	0.01	0.02	0.04	0.05
Gasoline	0.01	0.07	0.25	0.30
Fuel oil	-	-	0.21	0.01

Down stream processing limits exists of 50,000 lb/stream hr. on the ethylene and 20,000 lb/stream hr. on the propylene. The fuel requirements to run the cracking system for each feedstock type are as follows:

- Ethane 8364 Btu/lb.
- Propane 5016 Btu/lb.
- Gas oil 3900 Btu/lb.
- DNG 4553 Btu/lb.

Methane and fuel oil produced by the cracker are recycled as fuel. All the ethane and propane produced is recycled as feed. Heating values are as follows:

- Natural gas 21,520 Btu/lb.
- Methane 21,520 Btu/lb.
- Fuel oil 18,000 Btu/lb.

Because of heat losses and the energy requirements for pyrolysis, a fixed fuel requirement of 20.0×10^6 Btu/stream hr. occurs.

The price structure on the feeds and products and fuel costs is (all values are in cents per pound):

Feeds:	Ethane	6.55
	Propane	9.73
	Gas-oil	12.50
	DNG	10.14
Products:	Methane	5.38 (fuel value)
	Ethylene	17.75
	Propylene	13.79
	Butadiene	26.64
	Gasoline	9.93
	Fuel-oil	4.50 (fuel value)

Assume an energy (fuel) cost of $\$2.5/10^6$ Btu.

The variables to be optimized are the amounts of the four feeds (viz. Gas-oil, Propane, Ethane & Debutanized natural gasoline (DNG)). This problem was solved using both the Linear Programming Simplex method and Differential Evolution (DE). The assumption used in formulating the objective function and constraints are:

1. 20.0×10^6 Btu/hr. fixed fuel requirement (methane) to compensate for the heat-loss.
2. All propane and ethane are recycled with the feed, and all methane and fuel oil will be recycled as fuel.

12.3.2.1 Objective Function

Objective function for the profit is given by:

$$f = 2.84x_1 - 0.22x_2 - 3.33x_3 + 1.09x_4 + 9.39x_5 + 9.51x_6. \quad (12.19)$$

where,

f = Profit function (cents/ hr.)

x_1 = Fresh ethane feed (lb/hr.)

x_2 = Fresh propane feed (lb/hr.)

x_3 = Gas-oil feed (lb/hr.)

x_4 = DNG feed (lb/hr.)

x_5 = Ethane recycle (lb/hr.)

x_6 = Propane recycle (lb/hr.)

x_7 = Fuel added (lb/hr.)

12.3.2.2 Constraints

(a) Cracker capacity of 200,000 lb/hr,

$$1.1x_1 + 0.9x_2 + 0.9x_3 + 1.0x_4 + 1.1x_5 + 0.9x_6 \leq 200{,}000 \tag{12.20}$$

(b) Ethylene processing limitation of 50,000lb/hr,

$$0.5x_1 + 0.35x_2 + 0.2x_3 + 0.25x_4 + 0.5x_5 + 0.35x_6 \leq 50{,}000 \tag{12.21}$$

(c) Propylene processing limitation of 20,000lb/hr,

$$0.01x_1 + 0.15x_2 + 0.15x_3 0.18x_4 + 0.01x_5 + 0.15x_6 \leq 20{,}000 \tag{12.22}$$

(d) Ethane recycle

$$0.4x_1 + 0.06x_2 + 0.04x_3 + 0.05x_4 - 0.6x_5 + 0.06x_6 = 0 \tag{12.23}$$

(e) Propane recycle

$$0.1x_2 + 0.01x_3 + 0.01x_4 - 0.9x_6 = 0 \tag{12.24}$$

(f) Heat Constraints

$$\begin{aligned} &-6857.6x_1 + 364x_2 + 2032x_3 - 1145x_4 - 6857.6x_5 \\ &+ 364x_6 + 21{,}520x_7 = 20{,}000{,}000 \end{aligned} \tag{12.25}$$

12.3.3 Problem Reformulation

When DE was applied to the above problem, it was found that the equality constraints were difficult to deal with. Hence, the problem was reformulated by eliminating the equality constraints and incorporating them in inequality constraints thereby reducing the number of constraints and parameters. The reformulated problem is as follows:

12.3.3.1 Objective Function

$$\textbf{Max. } f = 9.1x_1 + 1.88x_2 - 2.5879x_3 + 1.9886x_4 \tag{12.26}$$

12.3.3.2 Constraints

(a) Cracker capacity of 200,000 lb/hr,

$$16.5x_1 + 10.1x_2 + 8.861x_3 + 9.926x_4 \leq 1800000 \tag{12.27}$$

(b) Ethylene processing limitation of 50,000lb/hr,

$$7.5x_1 + 4.0x_2 + 2.14x_3 + 2.665x_4 \leq 450000 \tag{12.28}$$

(c) Propylene processing limitation of 20,000lb/hr,

$$0.15x_1 + 1.51x_2 + 1.3711x_3 + 1.6426x_4 \leq 180000 \tag{12.29}$$

12.3.4 Simulated Results and Discussion

The reformulated problem was solved using both the Differential Evolution (DE) & LP Simplex method and the following results (Table 12.7) were obtained:

Table 12.7. Results of LP Simplex and DE

S No	Stream	Flow Rate (lb\hr.) using DE	Flow rate (lb\hr.) using LP Simplex
1	Fresh Ethane feed (x_1)	60,000	60,000
2	Fresh propane feed (x_2)	0	0
3	Gas-oil feed (x_3)	0	0
4	DNG feed (x_4)	0	0
5	Ethane recycle (x_5)	40,000	40,000
6	Propane recycle (x_6)	0	0
7	Fuel added (x_7)	32795.539	32795.539
8	Ethylene	50,000	50,000
9	Propylene	1000	1000
10	Butadiene	1000	1000
11	Gasoline	1000	1000
12	Methane	7000	7000
13	Fuel oil	0	0
14	Objective function (cents\hr.)	369560.00	369560.00

It may be noted that the maximum possible amount of ethylene is produced. As the ethylene production constraint is relaxed, the objective function value increases. Once the constraint is raised above 90,909.0909 lb\hr, the objective function remains constant at 676018.1875 cents/hr. LP simplex solution was cross-checked using a software package named *TORA*, (Taha, 1997), and the same results were obtained as shown in Table-12.7.

Table-12.8 presents the comparison, in terms of the number of objective function evaluations, CPU-time and proportion of convergences to the optimum, between the different DE strategies. In this table, F_A, *NRC* and CPU-time represents, respectively the mean number of objective function evaluations over all the 10 executions, the percentage of convergences to the global optimum and the average CPU time per execution (key parameters used are NP=40, CR=0.9, F=0.6, accuracy=0.0001%).

Table 12.8. Results of DE with all ten strategies

S No	Strategy	F_A	CPU- time	NRC
1	DE/rand/1/bin	6268	0.28	100
2	DE/best/1/bin	3168	0.145	100
3	DE/best/2/bin	9076	0.418	100
4	DE/rand/2/bin	11696	0.539	100
5	DE/rand-to-best/1/bin	6052	0.28	100
6	DE/rand/1/exp	5252	0.22	100
7	DE/best/1/exp	2796	0.126	100
8	DE/best/2/exp	10132	0.44	100
9	DE/rand/2/exp	12600	0.55	100
10	DE/rand-to-best/1/exp	6536	0.275	100

From the above Table-12.8 it is evident that the strategy number 7 is the best strategy. It takes least average CPU-time, maximum NRC and minimum F_A. However the best key parameters for strategy no.7 are NP=40, CR=0.8, F=0.5 giving CPU-time of 0.113 s, NRC=95 and F_A=2656.

12.3.5 Conclusions

The best key parameters for the present problem are: NP=40, CR=0.8, F=0.5. The strategy that took minimum CPU-time with highest NRC is strategy no. 7. The results obtained by two methods (viz. DE & LP Simplex) are same and matches with that reported in literature. Differential Evolution exhibits difficulties in dealing with equality constraint problems but in general, they are the most efficient in terms of function evaluations.

References

Angira, R. and Babu,B.V. (2003). "Evolutionary Computation for Global Optimization of Non-Linear Chemical Engineering Processes", *Proceedings of International Symposium on Process Systems Engineering and Control (ISPSEC '03)* - For Productivity Enhancement through Design and Optimization, IIT-Bombay, Mumbai, January 3-4, 2003, Paper No. FMA2, pp 87-91. Also available via Internet as .pdf file at http://bvbabu.50megs.com/custom.html/#56.

Annable, D. (1952). "Application of the Temkin kinetic equation to ammonia synthesis in large – scale reactors". *Chemical Engineering Science,* **1** (4), 145.

Babu, B.V. (2001). "Evolutionary Computation-At a Glance", NEXUS, Annual Magazine of Engineering Technology Association, BITS-Pilani, 3-7. Also available via Internet as .pdf file at http://bvbabu.50megs.com/custom.html/#36.

Babu, B.V. and Angira, R. (2001a). "Optimization of Non-Linear Functions Using Evolutionary Computation", *Proceedings of 12th ISME Conference on Mechanical Engineering*, Crescent Engineering College, Chennai, January 10-12, 2001, Paper No. CT07, 153-157. Also available via Internet as .pdf file at http://bvbabu.50megs.com/custom.html/#34.

Babu, B.V. and Angira, R. (2001b). "Optimization of Thermal Cracker Operation using Differential Evolution", *Proceedings of International Symposium & 54th Annual Session of IIChE (CHEMCON-2001)*, CLRI, Chennai, December 19-22, 2001. Also available via Internet as .pdf file at http://bvbabu.50megs.com/custom.html/#38 & Application No. 20, Homepage of Differential Evolution, the URL of which is: http://www.icsi.berkeley.edu/~storn/code.html

Babu, B.V. and Angira, R. (2002a). "A Differential Evolution Approach for Global Optimization of MINLP Problems", *Proceedings of 4th Asia-Pacific Conference on Simulated Evolution And Learning (SEAL-2002)*, Singapore, November 18 22, 2002, Paper No. 1033, Vol. 2, pp 880-884. Also available via Internet as .pdf file at http://bvbabu.50megs.com/custom.html/#46.

Babu, B.V. and Angira, R. (2002c). "Optimization of Non-Linear Chemical Processes Using Evolutionary Algorithm", *Proceedings of International Symposium & 55th Annual Session of IIChE (CHEMCON-2002)*, OU, Hyderabad, December 19-22, 2002. Also available via Internet as .pdf file at http://bvbabu.50megs.com/custom.html/#54.

Babu, B.V. and Chaturvedi, G. (2000). "Evolutionary Computation strategy for optimization of an Alkylation Reaction", *Proceedings of International Symposium & 53rd Annual Session of IIChE (CHEMCON-2000)*, Science City, Calcutta, December 18-21, 2000. Also available via Internet as .pdf file at http://bvbabu.50megs.com/custom.html/#31. & Application No. 19, Homepage of Differential Evolution, the URL of which is: http://www.icsi.berkeley.edu/~storn/code.html

Babu, B.V. and Gautam, K. (2001). "Evolutionary Computation for Scenario-Integrated Optimization of Dynamic Systems", *Proceedings of International Symposium & 54th Annual Session of IIChE (CHEMCON-2001)*, CLRI, Chennai, December 19-22, 2001. Also available via Internet as .pdf file at http://bvbabu.50megs.com/custom.html/#39 & Application No. 21, Homepage of Differential Evolution, the URL of which is: http://www.icsi.berkeley.edu/~storn/code.html.

Babu, B.V. and Mohiddin, S.B. (1999). "Automated Design of Heat Exchangers using Artificial Intelligence based Optimization", *Proceedings of International Symposium & 52nd Annual Session of IIChE (CHEMCON-1999)*, Panjab University, Chandigarh, December 20-23, 1999. Also available via Internet as .htm file at http://bvbabu.50megs.com/custom.html/#27.

Babu, B.V. and Munawar, S.A. (2000). "Differential Evolution for the Optimal Design of Heat Exchangers", *Proceedings of All India Seminar on Chemical Engineering Progress on Resource Development: A Vision 2010 and Beyond*, organized by IE (I), Orissa State Centre Bhuvaneshwar, March 13, 2000. Also available via Internet as .pdf file at http://bvbabu.50megs.com/custom.html/#28.

Babu, B.V. and Munawar, S.A. (2001). "Optimal Design of Shell-and-Tube Heat Exchangers using Different Strategies of Differential Evolution", PreJournal.com – The Faculty Lounge, Article No. 003873, posted on website Journal http://www.prejournal.com, March 03, 2001. *Also available via Internet as .pdf file at http://bvbabu.50megs.com/custom.html/#35* & Application No. 18, Homepage of Differential Evolution, the URL of which is: http://www.icsi.berkeley.edu/~storn/code.html.

Babu, B.V. and Sastry, K.K.N. (1999). "Estimation of Heat Transfer Parameters in a Trickle Bed Reactor using Differential Evolution and Orthogonal Collocation", *Computers & Chemical Engineering,* **23**, 327-339. Also available via Internet as .pdf file at http://bvbabu.50megs.com/custom.html/#24. & Application No. 13, Homepage of Differential Evolution, the URL of which is: http://www.icsi.berkeley.edu/~storn/code.html

Babu, B.V. and Singh, R.P. (2000). "Optimization and Synthesis of Heat Integrated Distillation Systems Using Differential Evolution", *Proceedings of All India Seminar on Chemical Engineering Progress on Resource Development: A Vision 2010 and Beyond*, organized by IE (I), Orissa State Centre Bhuvaneshwar, March 13, 2000.

Babu, B.V. and Vivek, N. (1999). "Genetic Algorithms for Estimating Heat Transfer Parameters in Trickle Bed Reactors", *Proceedings of International Symposium & 52nd Annual Session of IIChE (CHEMCON-99)*, Panjab University, Chandigarh, December 20-23, 1999. Also available via Internet as .htm file at http://bvbabu.50megs.com/custom.html/#26.

Babu, B.V., Angira, R. and Nilekar, A. (2002b). "Differential Evolution for Optimal Design of an Auto-Thermal Ammonia Synthesis Reactor", *Computers & Chemical Engineering* (Communicated).

Chiou, J.P. and Wang, F.S. (1999). "Hybrid method of evolutionary algorithms for static and dynamic optimization problems with application to a fed-batch fermentation process", *Computers & Chemical Engineering*, **23**, 1277-1291.

Dollena S. Hawkins, David M. Allen and Arnold J. Stromberg (2001). "Determining the number of components in mixtures of linear models". *Computational Statistics & Data Analysis,* **38**, 15-48.

Dyson, D. C. (1965). *Optimal Design of Reactors for Single Exothermic Reversible Reactions,* Ph.D. Thesis, London University.

Edgar, T.F. and Himmelblau, D.M. (1989). *Optimization of Chemical Processes*, McGraw-Hill Book Company, New York.

Eymery, J. (1964). *Dynamic Behavior of an Ammonia Synthesis Reactor,* D. Sc. Thesis, M.I.T.

Gupta, O.P. (1994). *Elements of Fuels, Furnaces and Refractories,* Khanna Publishers, New Delhi.

Hobson, G.D. (1975). *Modern Petroleum Technology,* Applied Science Publisher, Great Britain.

Joshi, R. and Sanderson, A. C. (1999). "Minimal representation multi-sensor fusion using differential evolution". *IEEE Transactions on Systems, Man and Cybernetics, Part A* **29,** 63-76.

Kyprianou, A., Worden, K. and Panet, M. (2001). "Identification of hysteretic systems using the differential evolution algorithm". *Journal of Sound and Vibration,* **248** (2), 289-314.

Lee, M. H., Han, C. and Chang, K. S. (1999). "Dynamic optimization of a continuous polymer reactor using a modified differential evolution". *Industrial & Engineering Chemistry Research,* **38** (12), 4825-4831.

Lu, J. C. and Wang, F. S. (2001). "Optimization of Low Pressure Chemical Vapour Deposition Reactors Using Hybrid Differential Evolution". *Canadian Journal of Chemical Engineering,* **79** (2), 246-254.

Murase, A., Roberts, H. L. and Converse, A. O. (1970). "Optimal Thermal Design of an Autothermal Ammonia Synthesis Reactor". *Industrial & Engineering Chemistry Research,* **9,** 503- 513.

NAG (2002). *Web site of Numerical Algorithm Group* as on February 2002. http://www.nag.com/numeric/FL/manual/html/genint/FLwithdrawn.asp

Price, K. and Storn, R. (1997). "Differential Evolution – A simple evolution strategy for fast optimization", *Dr. Dobb's Journal*, **22** (4), 18-24 & 78.

Price, K. and Storn, R. (2002). *Home Page on Differential Evolution as on July 2002.* http://www.ICSI.Berkeley.edu/~storn/code.html.

Sastry, K.K.N., Behera, L. and Nagrath, I.J. (1998). "Differential evolution based fuzzy logic controller for nonlinear process control", *Fundamenta Informaticae: Special Issue on Soft Computation.*

Shah, M. J. (1967). "Control simulation in ammonia production". *Industrial & Engineering Chemistry*, **59**, 72.

Singh, C. P. P., and Saraf, D. N. (1979). "Simulation of Ammonia Synthesis Reactors". *Industrial & Engineering Chemistry Process Design Development,* **18** (3), 364-370.

Sourander, M.L., Kolari, M., Cugini, J.C., Poje, J.B. and White, D.C. (1984). "Control and Optimization of Olefin-Cracking Heaters", *Hydrocarbon Processing,* **63,** 63-69.

Storn, R. (1995). "Differential evolution design of an IIR-filter with requirements for magnitude and group delay", *International Computer Science Institute*, TR-95-018.

Taha, H.A. (1997). *Operations Research – An Introduction,* Prentice-Hall of India Limited, New Delhi.

Upreti, S.R., and Deb, K. (1997). "Optimal design of an ammonia synthesis reactor using Genetic Algorithms". *Computers & Chemical Engineering,* **21**, 87 - 92.

Wang, F. S. and Cheng, W. M. (1999) "Simultaneous optimization of feeding rate and operation parameters for fed-batch fermentation processes". *Biotechnology Progress,* **15** (5), 949-952.

Wang, F. S., Su, T. L. and Jang, H. J. (2001). "Hybrid Differential Evolution for Problems of Kinetic Parameter Estimation and Dynamic Optimization of an Ethanol Fermentation Process". *Industrial and Engineering Chemistry Research,* **40** (13), 2876-2885.

Wang, F.S. and Cheng, W.M. (1999). "Simultaneous optimization of feeding rate and operation parameters for fed-batch fermentation processes", *Biotechnology Progress,* **15** (5), 949-952.

Wang, F.S., Jing, C.H. and Tsao, G.T. (1998). "Fuzzy-decision-making problems of fuel ethanol production using genetically engineered yeast", *Industrial & Engineering Chemistry Research,* **37**, 3434-3443.

13 New Ideas and Applications of Scatter Search and Path Relinking

Fred Glover, Manuel Laguna and Rafael Martí

Practical elements of Scatter Search and Path Relinking are illustrated by seven recent applications. The computational outcomes, based on comparative tests involving real world and experimental benchmark problems, demonstrate that these methods provide useful alternatives to more established search procedures. The designs in these applications are straightforward, and can be readily adapted to other optimization problems of varied structures.

13.1 Introduction

Scatter Search (SS) and Path Relinking (PR) have recently been investigated in a number of studies. In this chapter we disclose some of the practical performance aspects of these methods by examining the following seven recent applications:

a. Neural Network Training
b. Multi-Objective Routing Problem
c. OptQuest: A Commercial Implementation
d. A Context-Independent Method for Permutation Problems
e. Classical Vehicle Routing
f. Matrix Bandwidth Minimization
g. Arc Crossing Minimization

The designs in these applications are straightforward, and can be readily adapted to other optimization problems of similarly diverse structures.

SS and PR may be viewed as evolutionary algorithms that construct solutions by combining others, and derive their foundations from strategies originally proposed for combining decision rules and constraints (Glover, 1963, 1965). Chapter 4 describes the fundamental principles and processes underlying these methodologies. We limit our attention here to sketching specific applications that demonstrate the scope and impact of these procedures.

13.2 Scatter Search Applications

The descriptions that follow are edited versions of reports by researchers and practitioners who are responsible for the applications of SS cited in this section. Definitions of terms and basic procedural components employed are taken from Chapter 4.

13.2.1 Neural Network Training

A highly effective adaptation of Scatter Search to the neural network training problem has been developed by Laguna and Martí (2000). The improvement procedure embedded in the SS method consists in this case of the well known Nelder and Mead (1965) Simplex method for unconstrained nonlinear optimization. Given a set of weights *w*, the Simplex method starts by perturbing each weight to create an initial simplex from which to begin the local search. The algorithm uses the implementation of the Nelder-Mead method described in Press, et al. (1992).

The SS procedure itself begins by generating the appropriate data normalizations, and then creates an initial reference set (*RefSet*) of *b* solutions. A set *P* of *PSize* solutions (bounded between *wlow* and *whigh*) is built with the diversification method, based on a controlled randomization scheme, given in Glover, Laguna and Martí (1999). *RefSet* is filled with the best *b*/2 solutions in *P* that result by applying the improvement method.Then *b*/2 additional solutions are generated as perturbations of the first *b*/2 and are added to *RefSet*. The perturbation consists of multiplying each weight by 1 + U[-0.05,0.05], where U is the uniform distribution.

In step 2, the solutions in *RefSet* are ordered according to quality, where the best solution is the first one in the list. Then, the *NewPairs* set is constructed consisting of all the new pairs of solutions that can be obtained from *RefSet*, where a "new pair" contains at least one new solution. The pairs in *NewPairs* are selected one at a time to create linear combinations. The improvement method is applied to the best *b* solutions created as linear combinations. Each improved solution is then tested for admission into *RefSet*. If a newly created solution improves upon the worst solution currently in *RefSet*, the new solution replaces the worst and *RefSet* is reordered.

In step 3 the procedure intensifies the search around the best-known solution. At each intensification iteration, the best-known solution is perturbed (multiplied by 1 + U[-0.05,0.05]) and the improvement method is applied. The best solution is updated if the perturbation plus the improvement generates a better solution. After *IntLimit* intensification iterations without improving the best solution, the procedure abandons the intensification phase and returns to step 2. Previously, the improvement method is applied to the best *b*/2 solutions in *RefSet* and the worst *b*/2 solutions are replaced with perturbations of the best *b*/2 (now, each improved solution is multiplied by 1+U[-0.01,0.01]). The training procedure stops when the number of objective function evaluations reaches the total allowed. Preliminary experimentation determined that reasonable values for the parameters *wlow, whigh, b* and *IntLimit* are –2, 2, 10 and 20 respectively.

Computational experiments on the neural network training problem, applied to benchmark problems previously reported in the literature, show that the scatter search implementation compares very favorably with the best known methods for these problems (which include simulated annealing, tabu search, and genetic algorithms). SS reaches a prediction accuracy that that makes it possible to filter out potentially bad solutions generated during the optimization of a simulation, and does so within a computational time that is practical for on-line training.

13.2.2 Multi-Objective Routing Problem

Corberán et al. (2001) address the problem of routing school buses in a rural area. The authors approach this problem with a node routing model with multiple objectives that arise from conflicting viewpoints. From the point of view of cost, it is desirable to minimize the number of buses (m) used to transport students from their homes to school and back. And from the point of view of service, it is desirable to minimize the time that a given student spends in route. The current literature deals primarily with single-objective problems and the models with multiple objectives typically employ a weighted function to combine the objectives into a single one.

The solution procedure considers each objective separately and search for a set of efficient solutions instead of a single optimum. The SS approach for constructing, improving and then combining solutions consists of the following elements:

H1 and H2:	Two constructive heuristics to generate routes
SWAP:	An exchange procedure to find a local optimal value for the length of each route
INSERT:	An exchange procedure to improve upon the value of *tmax*, which identifies the maximum time in the bus.
COMBINE:	A mechanism to combine solutions in a reference set of solutions in order to generate new ones.

The overall procedure operates as follows. (See the outline in Figure 1.) The constructive heuristics H1 and H2 are applied with several values for *tmax* and the resulting solutions are stored in separate pools, one for each value of m. The larger the value of *tmax* the larger the frequency in which the heuristics construct solutions with a small number of routes. Conversely, solutions with a large number of routes are obtained when the value of *tmax* is decreased. The procedure then attempts to improve upon the solutions constructed by H1 and H2. The improvement consists of first applying SWAP to each route and then applying INSERT to the entire solution. If any route is changed during the application of INSERT then we apply SWAP one more time to all the changed routes. The procedure now iterates within a main loop, in which a search is launched for solutions with a common number of routes. The main loop terminates when all the m-values have been explored.

From all the solutions with m routes, the best b are chosen to initialize the reference set (*RefSet*). The criterion for ranking the solutions at this step is *tmax*, since all solutions have the same number of routes. The procedure performs iterations in an inner-loop that consists of searching for a solution with m routes with an improved t_{max} value. The combination procedure COMBINE is applied to all pairs of solutions in the current reference set *RefSet*. The combined solutions are improved in the same way as described above, that is, by applying SWAP then INSERT and finally SWAP to the routes that changed during the application of INSERT. We refer to the resulting set of distinct solutions as *ImpSet*. The reference set is then updated by selecting the best b solutions from the union of *RefSet* and *ImpSet*. Steps 5, 6 and 7 in the outline of Figure 1.13 are performed as long as at least one new solution is admitted in the reference set.

---

1. *Construct solutions* — Apply constructions heuristics H1 and H2 with several values of TMAX.
2. *Improve solutions* — Apply SWAP to each route in a solution and INSERT to the entire solution. Finally, apply SWAP o any route changed during the application of INSERT.
3. *Build solution pools* — Put all solutions with the same number of routes in the same pool.

for (each solution pool) **do**

4. *Build the reference set* — Choose the best b solutions in the pool to build the initial *RefSet*.

while (new solutions in *RefSet*) **do**

5. *Combine solutions* — Generate all the combined solutions from pairs of reference solutions where at least one solution in the pair is new.
6. *Improve solutions* — Apply SWAP to each route in a solution and INSERT to the entire solution. Finally, apply SWAP o any route changed during the application of INSERT.
7. *Update reference set* — Choose the best b solutions from the union of the current reference set and the combined-improved solutions to update the *RefSet*.

end while

end for

Fig. 13.1. SS for Multi-Objective Vehicle Routing

After the reference set is updated, the combination procedure may be applied to the same solution pairs more than once. Since the combination procedure includes some randomized elements, the combination of two solutions may result in a different outcome every time COMBINE is applied. Also, the size of the reference set is increased if the updating procedure fails to add at least one new solution. The additional solutions come from the original pool of solutions generated with the construction heuristics. The reference set size is increased up to 2*b, where b is the initial size.

The computational testing on a real-world problem with 42 primary (elementary) schools and 16 (middle) secondary schools in the Province of Burgos (Spain), reveals the procedure is capable of generating a highly effective approximation of the efficient frontier for each routing problem. Decision makers may use efficient solutions to estimate the best service level (given by the maximum route length) that can be obtained with each level of investment (given by the number of buses used). The results show that several of the solutions implemented in practice are not efficient and can be improved by the SS methodology of the study.

13.2.3 OptQuest: A Commercial Implementation

OptQuest, a registered trademark of OptTek Systems Inc., is commercial software designed for optimizing complex systems, such as those formulated as simulation models. Many real world optimization problems in business, engineering and science are too complex to be given tractable mathematical formulations. Multiple non-linearities, combinatorial relationships and uncertainties often render challenging practical problems inaccessible to modeling except by resorting to more comprehensive tools (like computer simulation). Classical optimization methods encounter grave difficulties when dealing with the optimization problems that arise in the context of complex systems. In some instances, recourse has been made to itemizing a series of scenarios in the hope that at least one will give an acceptable solution. Due to the limitations of this approach, a long-standing research goal has been to create a way to guide a series of complex evaluations to produce high quality solutions, in the absence of tractable mathematical structures. (In the context of optimizing simulations, a "complex evaluation" refers to the execution of a simulation model.)

The OptQuest Callable Library (OCL) is designed to search for optimal solutions to the following class of optimization problems:

Max or Min	$F(x,y)$	
Subject to	$Ax \leq b$	(Constraints)
	$g_l \leq G(x,y) \leq g_u$	(Requirements)
	$l \leq x \leq u$	(Bounds)
	y = alldifferent	

where x can be continuous or discrete with an arbitrary step size and y represents a permutation.

In a general-purpose optimizer such as OCL, it is desirable to separate the solution procedure from the complex system to be optimized. The disadvantage of this "black box" approach is that the optimization procedure is generic and has no knowledge of the process employed to perform evaluations inside of the box and therefore does not use any problem-specific information. The main advantage, on the other hand, is that the same optimizer can be applied to complex systems in many different settings. The optimization procedure uses the outputs from the system evaluator, which measures the merit of the inputs that were fed into the model. On the basis of both current and past evaluations, the optimization procedure decides upon a new set of input values (see Figure 2.13). The optimization procedure is designed to carry out a special "strategic search," where the successively generated inputs produce varying evaluations, not all of them improving, but which over time provide a highly efficient trajectory to the best solutions. The process continues until an appropriate termination criterion is satisfied (usually based on the user's preference for the amount of time to be devoted to the search).

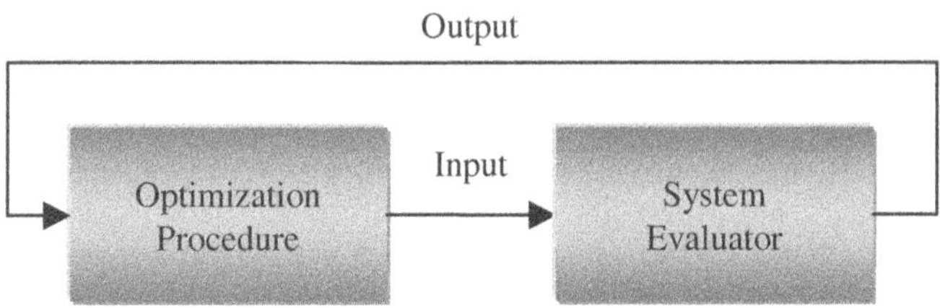

Fig. 13.2. Coordination between optimization and system evaluation

It is assumed that the user has a system evaluator that, given a set of input values, it returns a set of output values that can used to guide a search. For example, the evaluator may have the form of a computer simulation that, given the values of a set of decision variables, it returns the value of one or more performance measures (that define the objective function and possibly a set of requirements). The user-written application uses OCL functions to define an optimization problem and launch a search for the optimal values of the decision variables.

The scatter search method implemented in OCL begins by generating a starting set of diverse points. This is accomplished by dividing the range of each variable into 4 sub-ranges of equal size. Then, a solution is constructed in two steps. First, a sub-range is randomly selected. The probability of selecting a sub-range is inversely proportional to its frequency count (which keeps track of the number of times the sub-range has been selected). Second, a value is randomly chosen from the selected sub-range.

A subset of diverse points is chosen as members of the reference set. A set of points is considered diverse if its elements are "significantly" different from one another. OCL uses a Euclidean distance measure to determine how "close" a potential new point is from the points already in the reference set, in order to decide whether the point is included or discarded.

When the optimization model includes discrete variables, a rounding procedure is used to map fractional values to discrete values. When the model includes linear constraints newly created points are subjected to a feasibility test before they are sent to the evaluator (i.e., before the objective function value F(x) and the requirements G(x) are evaluated). If the solution is infeasible with respect to one or more constraints, OCL formulates and solves a linear programming (LP) problem. The LP (or mixed-integer program, when x contains discrete variables) has the goal of finding a feasible solution x^* that minimizes a deviation between x and x^*.

Once the reference set has been created, a combination method is applied to initiate the search for optimal solutions. The method consists of finding linear combinations of reference solutions. The number of solutions created from the linear combination of two reference solutions depends on the quality of the solutions being combined.

In the process of searching for a global optimum, the combination method may not be able to generate solutions of enough quality to become members of the reference set. If the reference set does not change and all the combinations of solutions have been explored, a diversification step is triggered. This step consists of rebuilding the reference set to create a balance between solution quality and diversity. To preserve quality, a small set of the best (*elite*) solutions in the current reference set is used to seed the new reference set. The remaining solutions are eliminated from the reference set. Then, the diversification generation method is used to repopulate the reference set with solutions that are diverse with respect to the elite set. This reference set is used as the starting point for a new round of combinations.

In Laguna and Martí (2002), the functionality of the library is illustrated with an example in the context of nonlinear optimization. The authors tested OCL by comparing its performance with Genocop III, a third-generation genetic algorithm. Experiments with 30 nonlinear optimization problems show that OCL is a search method that is both aggressive and robust, finding high-quality solutions early in the search and continuing to improve upon the best solution when allowed to search longer. The quality of solutions obtained by OCL uniformly dominated that of solutions obtained by Genocop III, with marked superiority on the more difficult problems. In addition, OCL obtained these improved solutions with speeds ranging from one to three orders of magnitude faster than the genetic algorithm approach. These characteristics make OCL especially useful for applications in which the evaluation of the objective function requires a non-trivial computational effort. OCL has now been used to solve complex optimization problems in more than 20,000 real world applications. More details can be found on the website www.opttek.com .

13.2.4 A Context-Independent Method for Permutation Problems

Campos, Laguna and Martí (2001) develop a context-independent method for solving problems whose solutions can be represented with a permutation. As in the case of OCL, described in the previous section, this general-purpose heuristic is based on a model that treats the objective function evaluation as a black box, making the search algorithm context-independent. The procedure is a scatter search/tabu search hybrid. The scatter search framework provides a means for diversifying the search throughout the exploration of the permutation solution space. Two improvement methods are used to intensify the search in promising regions of the solution space: a simple local search based on exchange moves and a short-term memory tabu search. Improved solutions are then used for combination purposes within the scatter search design.

The solver is designed in such a way that the user must specify whether the objective function evaluation is more sensitive to the "absolute" positioning of the elements in the permutation or to their "relative" positioning. Hence, we differentiate between two classes of problems:

A-permutation problems —	for which absolute positioning of the elements is more important
R-permutation problems —	for which relative positioning of the elements is more important

The distance between two permutations $p = (p_1, p_2, ..., p_n)$ and $q = (q_1, q_2, ..., q_n)$ depends on the type of problem being solved. For A-permutation problems, the distance is given by:

$$d(p,q) = \sum_{i=1}^{n} |p_i - q_i|.$$

The distance for R-permutation problems is defined as: $d(p,q)$ = number of times p_{i+1} does not immediately follow p_i in q, for $i = 1, \ldots, n\text{-}1$.

In order to design a context-independent combination methodology that performs well across a wide collection of different problems, the authors propose a set of 10 combination methods from which one is probabilistically selected according to its performance in previous iterations. Solutions in the *RefSet* are ordered according to their objective function value. So, the best solution is in the first one in *RefSet* and the worst is the last one. When a solution obtained with combination method i (referred to as cm$_i$) qualifies to be the jth member of the current *RefSet*, we add b-j+1 to *score*(cm$_i$). Therefore, combination methods that generate good solutions accumulate higher scores and increase proportionally their probability of being selected. Other SS elements in the method follow the standard description given in Chapter 4. Two different solvers are proposed, the first one implements a local search phase as the improvement method, the second one uses a short term memory tabu search as the improvement method.

The performance of the procedure has been assessed using 157 instances of four different permutations problems. Solutions to these problems are naturally represented as permutations: the bandwidth reduction problem (BRP), the linear ordering problem (LOP), the traveling salesman problem (TSP), and a single machine sequencing problem (SMS). Solutions obtained with the scatter search/tabu search method have been compared with the best-known solutions to each problem. The procedure has been shown competitive with methods specifically designed for the LOP and SMS problems. The method also provides reasonable results for TSP problems, although not competitive with those obtained by methods that are customized to exploit the special structure of the TSP. The method also is not highly appropriate for the BRP due to the min-max nature of the objective function calculation associated with this class of problems.

For the permutation problems considered, the method was shown superior to comparable procedures that are commercially available. (In all cases, the method produces substantial improvements). The experimentation shows that context-independent methods can be useful in the context of permutation problems, when the associated objective function is capable of discriminating among solutions in a given neighborhood

13.2.5 Classical Vehicle Routing

Rego and Leão (2002) identify a general design for solving vehicle routing problems using scatter search that has proved exceptionally effective. The vehicle routing problem (VRP) is a classic application in Combinatorial Optimization that can be defined as follows. Let $G = (V, A)$ be a graph where $V = \{v_0, v_1, \ldots, v_n\}$ is a vertex set, and $A = \{(v_i, v_j) | v_i, v_j \in V, i \neq j\}$ is an arc set. Vertex v_0 denotes a *depot*, where a fleet of m identical vehicles of capacity Q are based, and the remaining vertices $V' = V \setminus \{v_0\}$ represent n *cities* (or client locations). A nonnegative *cost* or distance matrix $C = (c_{ij})$ that satisfies the triangle inequality $(c_{ij} \leq c_{ik} + c_{kj})$ is defined on A. It is assumed that $m \in [\underline{m}, \overline{m}]$ with $\underline{m} = 1$ and $\overline{m} = n - 1$. The value of m can be a decision variable or can be fixed depending on the application.

Vehicles make pickups or deliveries but not both. With each vertex v_i is associated a quantity q_i $(q_0 = 0)$ of some goods to be delivered by a vehicle and a service time δ_i $(\delta_0 = 0)$ required by a vehicle to unload the quantity q_i at v_i. The VRP consists of determining a set of m vehicle routes of minimal total cost, starting and ending at a depot v_0, such that every vertex in $v_i \in V'$ is visited exactly once by one vehicle, the total quantity assigned to each route does not exceed the capacity Q and the total duration of any vehicle route does not surpass a given bound D.

The algorithm implementation is structured into five basic components, following the characteristic scatter search design:

13.2.5.1 Diversification Generation Method

To start with an initial set of trial solutions that differ significantly from each other, the generator of combinatorial objects described in Glover (1997) is used to generate permutations in n-vectors where components are all vertices $v_i \in V \setminus \{v_0\}$. For a given permutation P(h), each cluster of vertices in a route is obtained by successively assigning a vertex $v_i (i \in P(h))$ to a route R_{h_k} (initially k=1) until any of the cumulative values for $Q_k = \sum_{v_i \in R_{h_k}} q_i$ or $D_k = \sum_{(v_i, v_j) \in R_{h_k}} (c_{ij} + \delta_i)$ does not exceed Q or D, respectively, with the insertion of a new vertex v_{k_j}. As soon as such a cutoff limit is attained a new assignment is created by incrementing k by one unit, and the process goes on until all vertices have been assigned.

The result obtained can be viewed as a generalized assignment process that does not rely on the order in which clients are visited, though it ensures that all the initial solutions that can be created are feasible and different (since they derive from distinct permutations). Vehicle routes are then determined by using the stem-and-cycle ejection chain algorithm for the traveling salesman problem described in Rego (1998a).

13.2.5.2 Improvement Method

The improvement method is based on the Flower Ejection Chain (FEC) algorithm described in Rego (1998b). For the purpose of the proposed scatter search algorithm, the original FEC procedure has been modified as in the scatter search phase this improvement method is required to deal with infeasible solutions. The method works in two stages. The first stage is concerned with making the solution feasible while choosing the most favorable move and the second stage is the improvement process that operates only on feasible solutions. The method considers varying penalty factors associated with the problem constraints to drive the search toward the feasible region.

13.2.5.3 Reference Set Update Method

A set of reference solutions is created and maintained as follows. Intensification is achieved by the selection of high-quality solutions (in terms of the objective function value) and diversification is induced by including diverse solutions from the current candidate set *CS*. Thus the reference set *RS* is defined by two distinct subsets *B* and *D*, representing respectively the subsets of high-quality and diverse solutions, hence $RS = B \cup D$. A diversity measure, $d_{ij} = |(S_i \cup S_j) \setminus (S_i \cap S_j)|$ is used to express the distance between solutions S_i and S_j, identifying the number of edges by which the two solutions differ from each other. Candidate solutions are included in *RS* according to the *Maxmin* criterion that maximizes the minimum distance of each candidate solution to all the solutions currently in the reference set.

13.2.5.4 Subset Generation Method

Subsets of reference solutions are generated to create structured combinations in the next step. The method is typically designed to organize subsets of solutions to cover different promising regions of the solution space. In a spatial representation, the convex-hull of each subset delimits the solution space in subregions containing all possible convex combinations of solutions in the subset. In order to achieve a suitable intensification and diversification of the solution space, three types of subsets are required to be organized:

1. subsets containing only solutions in *B*,

2. subsets with only solutions in D, and
3. subsets mixing in solutions in B and D in different proportions.

Subsets defined by solutions of type 1 are conceived to intensify the search in regions of high-quality solutions while subsets of type 2 are created to diversify the search to unexplored regions. Finally, subsets of type 3 integrate both high-quality and diverse solutions with the aim of exploiting solutions across these two types of subregions.

13.2.5.5 Solution Combination Method

The solution combination method is designed to explore subregions within the convex-hull of the reference set. Solutions consist of vectors of variables x_{ij} representing edges (v_i, v_j). New solutions are generated by weighted linear combinations that are structured by the subsets defined in the last step. In order to restrict the number of solutions only one solution is generated in each subset by a convex linear combination. Nevertheless, the set of the generated edges does not necessarily (and usually does not) represent a feasible graph structure for a VRP solution insofar as it may produce a subgraph containing vertices with a degree different from two. Such subgraphs can be viewed as fragments of solutions (or partial routes). Structural feasibility is restored by either linking vertices of degree 1 directly to the depot or dropping edges with the smallest scores, from among those incident at vertices of degree greater than 2, until the degree of each vertex becomes equal to two. While the resulting subgraphs are feasible for the VRP routing structure, they may not yield a feasible solution in relation to the capacity or route length constraints. This latter form of infeasibility is handled by the improvement method as previously indicated.

Computational testing was performed on a set of 26 standard benchmark instances taken from Christofides, Mingozzi and Toth (1972) and Rochat and Taillard (1995). Comparisons with previous VRP algorithms in the literature show the scatter search algorithm not only is competitive with the best of them across a broad spectrum of problems but is highly robust. For example, in 7 out of the 14 instances from the Christofides, Mingozzi and Toth's testbed, the SS approach obtains a solution that succeeds in matching the best solution previously found by any method. For Rochat and Taillard's instances, the SS algorithm dominates all other methods in all instances. Moreover, the approach offers an additional important advantage. Because the problem constraints are handled separately from the solution generation procedures, and are therefore independent of the problem context, this scatter search design can be directly used to solve other classes of vehicle routing problems by applying any domain-specific (local search) heuristic that is able to start from infeasible solutions.

13.3 Path Relinking Applications

13.3.1 Matrix Bandwidth Minimization

The matrix bandwidth minimization problem (MBMP) has been the subject of study for at least 32 years, beginning with the Cuthill - McKee algorithm in 1969. The problem consists of finding a permutation of the rows and the columns of a matrix that keeps all the non-zero elements in a band that is as close as possible to the main diagonal. This problem has generated considerable interest over the years because of its practical relevance for a significant range of global optimization applications. They include preprocessing the coefficient matrix for solving the system of equations, finite element methods for approximating solutions of partial differential equations or large-scale power transmission systems.

Given a matrix $A=\{a_{ij}\}_{nxn}$ the problem can be stated in terms of graphs considering a vertex for each row (column) and an edge in E as long as either $a_{ij} \neq 0$ or $a_{ji} \neq 0$. The problem consists of finding a labeling *f* of the vertices that minimizes the maximum difference between labels of adjacent vertices. In mathematical terms, given a graph $G=(V,E)$ with vertex set V ($|V|=n$) and edge set E, we seek to minimize $B_f(G)=\max\{B_f(v): v \in V\}$ where $B_f(v)=\max\{|f(v)-f(u)|: u \in N(v)\}$.

In this expression, $N(v)$ is the set of vertices adjacent to v, $f(v)$ is the label of vertex v and $B_f(v)$ is the bandwidth of vertex v. A labeling *f* of *G* assigns the integers $\{1, 2, \ldots, n\}$ to the vertices of *G;* thus, it is simply a renumbering of these vertices. Then, the bandwidth of a graph is $B(G)$, the minimum $B_f(G)$ value over all possible labelings *f*. The MBMP consists of finding a labeling *f* that minimizes $B_f(G)$.

Piñana et al. (2001) propose a PR implementation for this problem consisting of two phases. The first phase uses a GRASP method to generate an initial set of elite (high quality) solutions. Instead of retaining only the best solution overall when running GRASP, this phase stores the 10 best solutions obtained with the method. In the second phase a relinking process is applied to each pair of solutions in the elite set. Given the pair (A,B), two paths are considered: from A to B (where A is the *initiating solution* and B the *guiding one*), and from B to A (where they interchange their roles).

The relinking process implemented in the search may be summarized as follows: Let C be the candidate list of vertices to be examined. At each step, a vertex v is chosen from C and labeled in the *initiating solution* with its label $g(v)$ in the *guiding solution*. To do this, we look in the *initiating solution* for the vertex u with label $g(v)$ and perform $move(u,v)$, then vertex v is removed from C. The candidate set C is initialized with a randomly selected vertex. In subsequent iterations, each time a vertex is selected and removed from C, its adjacent vertices are included in C.

In a primitive version, the method employs no improvement procedure, but simply operates on the initial elite set of solutions generated by GRASP method (first building a large set of solutions from which the *n*_best are included in the elite set). The relinking process is then applied to all pairs of solutions in the elite set. Each time the relinking process produces a solution that is better than the worst in the elite set, the worst solution is replaced by the new one. The procedure terminates when no new solutions are admitted to the elite set.

It is shown that in most cases this primitive version (which lacks an improvement method) does not produce better solutions than the initiating and guiding solutions. Upon adding a local search exploration from some of the visited solutions in order to produce improved outcomes, the results are in line with those reported in Laguna and Martí (1999) for the arc crossing problem. Specifically, a local search method is applied to some of the solutions generated in the path. Two consecutive solutions after a relinking step differ only in the label of two vertices and hence it is not efficient to apply the local search exploration at every step of the relinking process. The parameter *n_improves* controls the application of the exchange mechanism. In particular, the exchange mechanism is applied *n_improves* times in the relinking process.

Overall experiments with 211 instances were performed to assess the merit of the procedures. Three methods were considered: the acclaimed GPS approach, a tabu search procedure (Martí et al., 2001) that has previously obtained the best known results for this problem, and the proposed PR method. The experiments reveal that the performance of the GPS approach was clearly inferior, with average deviations several orders of magnitude larger than those obtained with the other methods. The PR procedure outperforms the TS method in small instances. In large instances, the TS method obtains better solutions than the PR, although it employs longer running times. The PR procedure has been shown to be robust in terms of solution quality within a reasonable computational effort.

13.3.2 Arc Crossing Minimization

Researches in the graph-drawing field have proposed several aesthetic criteria that attempt to capture the meaning of a "good" map of a graph. Although readability may depend on the context and the map's user, most authors agree that crossing reduction is a fundamental aesthetic criterion in graph drawing. In the context of a 2-layer graph and straight edges, the bipartite drawing problem or BDP consists of ordering the vertices in order to minimize the number of crossings.

A *bipartite graph* $G=(V,E)$ is a simple directed graph where the set of vertices V is partitioned into two subsets, V_1 (the left layer) and V_2 (the right layer) and where $E \subseteq V_1 \times V_2$. The direction of the arcs has no effect on crossings so G is considered to be an undirected graph, the arcs to be edges and denote G by the triple (V_1, V_2, E). Let $n_1 = |V_1|$, $n_2 = |V_2|$, $m = |E|$, and let $N(v) = \{w \in V \mid e = \{v, w\} \in E\}$ denote the set of neighbors of $v \in V$. A solution is completely specified by a permutation π_1 of V_1 and a permutation π_2 of V_2, where $\pi_1(v)$ or $\pi_2(v)$ is the position of v in its corresponding layer.

Laguna and Martí (1999) propose a PR procedure for arc crossing minimization in the context of GRASP. This is the first implementation of PR for the purpose of improving the performance of GRASP (as opposed to accompanying it, as in the study previously cited). In the proposed path relinking implementation, the procedure stores a small set of high quality (elite) solutions to be used for guiding purposes. Specifically, after each GRASP iteration, the resulting solution is compared to the best three solutions found during the search. If the new solution is better than any one in the elite set, the set is updated. Instead of using attributes of all the elite solutions for guiding purposes, one of the elite solutions is randomly selected to serve as a guiding solution during the relinking process. The relinking in this context consists of finding a path between a solution found after an improvement phase and the chosen elite solution. Therefore, the relinking concept has a different interpretation within GRASP, since the solutions found from one GRASP iteration to the next are not linked by a sequence of moves (as in the case of tabu search). The relinking process implemented in our search may be summarized as follows:

The set of elite solutions is constructed during the first three GRASP iterations. Starting with the fourth GRASP iteration, every solution after the improvement phase (called the *initiating solution*) is subject to a relinking process by performing moves that transform the initiating solution into the guiding solution (i.e., the elite solution selected at random). The transformation is relatively simple, at each step, a vertex v is chosen from the initiating solution and is placed in the position occupied by this vertex in the guiding solution. So, if $\pi_1^g(v)$ is the position of vertex v in the guiding solution, then the assignment $\pi_1^i(v) = \pi_1^g(v)$ is made. We assume that an updating of the positions of vertices in V_1 of the initiating solution occurs. After this is done, an expanded neighborhood from the current solution defined by $\pi_1^i(v)$ and $\pi_2^i(v)$ is examined. The expanded neighborhood consists of a sequence of position exchanges of vertices that are one position away from each other, which are performed until no more improvement (with respect to crossing minimization) can be found. Once the expanded neighborhood has been explored, the relinking continues from the solution defined by $\pi_1^i(v)$ and $\pi_2^i(v)$ before the exchanges were made. The relinking finishes when the initiating solution matches the guiding solution, which will occur after $n_1 + n_2$ relinking steps.

Two consecutive solutions after a relinking step differ only in the position of two vertices (after the assignment $\pi_1^i(v) = \pi_1^g(v)$ is made). Therefore, it is not efficient to apply the expanded neighborhood exploration (i.e., the exchange mechanism) at every step of the relinking process. The parameter β is used to control the application of the exchange mechanism, by applying the mechanism every β steps of the relinking process.

Overall, experiments with 3,200 graphs were performed to assess the merit of the procedure. The proposed method is shown competitive in a set of problem instances for which the optimal solutions are known. For a set of sparse instances, the method performed remarkably well outperforming the best procedures reported in the literature.

Acknowledgments

Fred Glover's research partially supported by the Office of Naval Research Contract N00014-01-1-0917 in connection with the Hearin Center of Enterprise Science at the University of Mississippi. Rafael Marti's research is partially supported by the Spanish Government (PR2002-0060 and TIC2000-1750-C06-01).

References

Campos, V., M. Laguna and R. Martí (2001) "Context-Independent Scatter and Tabu Search for permutation problems", technical report TR03-2001, University of Valencia. http://matheron.uv.es/investigar/tr03-01.ps.

Christofides, N., A. Mingozzi, P. Toth (1972) "The Vehicle Routing Problem," *Combinatorial Optimization*, Vol 11, pp 315-338.

Corberán, A., E. Fernández, M. Laguna and R. Martí (2001), "Heuristic Solutions to the Problem of Routing School Buses with Multiple Objectives" *Journal of the Operational Research Society*, 53 (4), 427-435.

Fleurent, C. and F. Glover (1997) "Improved Constructive Multistart Strategies for the Quadratic Assignment Problem Using Adaptive Memory," University of Colorado.

Glover, F. (1963), "Parametric combination of local job shop rules", chapter IV, ONR Research memorandum n. 117, GSIA, Carnegie Mellon University, Pittsburgh, PA.

Glover, F. (1965), "A Multi-Phase Dual algorithm for the zero-one integer programming problem", *Operations Research*, 13, (6), 879.

Glover, F. (1997) "A Template for Scatter Search and Path Relinking," in Lecture Notes in Computer Science, 1363, J.K. Hao, E. Lutton, E. Ronald, M. Schoenauer, D. Snyers (Eds.), pp 13-54.

Laguna, M. and Martí R. (2000) "Neural Network Prediction in a System for Optimizing Simulations," *IIE Transaction on Operations Engineering*, forthcoming.

Laguna, M. and Martí R. (2002) "The OptQuest Callable Library", Optimization Software Class Libraries, Stefan Voss and David L. Woodruff (Eds.) 193-218, Kluwer, Boston.

Laguna, M. and Martí, R. (1999), GRASP and Path Relinking for 2-Layer straight line crossing minimization", INFORMS Journal on Computing, vol. 11 (1), pp. 44 – 52.

Martí, R., Laguna, M., Glover, F. and Campos, V. (2001) "Reducing the Bandwidth of a Sparse Matrix with Tabu Search", *European Journal of Operational Research*, 135(2), pp. 211-220.

Nelder, J. A. and R. Mead (1965) "A Simplex Method for Function Minimization," *Computer Journal*, vol. 7., pp. 308-313.

Piñana, E., I. Plana, V. Campos and R. Martí (2001), "GRASP and Path Relinking for the Matrix Bandwidth Minimization", *European Journal of Operational Research*, forthcoming.

Press, W. H., S. A. Teukolsky, W. T. Vetterling and B. P. Flannery (1992) Numerical Recipes: The Art of Scientific Computing, Cambridge University Press (www.nr.com).

Rego, C. (1998a) "Relaxed Tours and Path Ejections for the Traveling Salesman Problem," in Tabu Search Methods for Optimization, *European Journal of Operational Research*, 106, pp 522-538.

Rego, C. (1998b) "A Subpath Ejection Method for the Vehicle Routing Problem, Management Science," Vol. 44, No 10, pp 1447-1459.

Rego, C. and P. Leão (2002) "A Scatter Search for the Vehicle Routing Problem" Research Report HCES-08-02, Hearin Center for Enterprise Science, School of Business Administration, University of Mississippi.

Rochat, Y, E. Taillard (1995) "Probabilistic Diversification and Intensification in Local Search for Vehicle Routing," *Journal of Heuristics*, Vol 1, pp 147-167.

14 Improvement of Search Process in Genetic Algorithms: An Application of PCB Assembly Sequencing Problem

Nguyen Van Hop and Mario T Tabucanon

14.1 Introduction

The challenge of modeling real-world problems, especially the class of NP-complete problems, is that the solution space is usually very large. It often requires very large time to examine all possible solutions to choose the best one. It is sometimes impossible to find the real optimum solution and we have to be satisfied with a certain pseudo-optimum one. There are many optimization and search methods to obtain the pseudo-optimum solutions. In general, these procedures start by checking some given number of variants, then change the search direction towards the more promising area, and finally pick up the best among the examined ones. In these methods, the key point is on how to guide the search process.

Among the existing search techniques, Genctic Algorithms (GAs) are popular and have been applied to many problems because of its ability in escaping local optimum. Therefore, GAs are very suitable to problems with heterogeneous (multi-hill) search space. A genetic algorithm is a search paradigm that mimics the natural evolutionary processes in living organisms. A genetic algorithm searches through an iterative process to find better solutions in the search space. It starts with a randomly initialized population of candidate solutions and implements probabilistic and parallel exploration in the search space using the domain-independent genetic operators of selection, crossover and mutation. A solution of the real-world problem is encoded as a string of symbols. The string is referred to as chromosomes and the symbols as genes. The selection process chooses individual chromosomes, probabilistically, according to their fitness, which measures the quality of a solution. The higher the fitness, the more likely it is for an individual to be selected. The genetic algorithm often manipulates crossover and mutation to produce new individuals. The crossover operator exchanges genetic information between their selected parents. Mutation randomly changes one gene value to the generated offspring. The process of one generation involving selection, crossover and mutation is called one cycle of iteration and is repeated until convergence is reached or the number of generations achieves the established limit. A general form of genetic algorithm is given as follows (see Figure 14.1).

1. Randomly initialize population
2. Evaluate initial population
3. *Repeat*
   - Select parents
   - Crossover
   - Mutation
   - Evaluate new population
   - Substitute old population.

   Until a stopping criterion is met

Fig. 14.1 Typical Genetic Algorithm

There are many variants of this general form of genetic algorithms. Basically, the effectiveness of genetic algorithm depends on the encoding scheme of solutions and the solution diversity. The problem structure which reflects the nature of decision variables and the relationship among them will decide the encoding scheme of solutions. For example, we should use integer values instead of binary numbers to indicate the traveling sequence in the encoding chromosome of the Traveling Salesman Problem's solution. The vital role of genetic operators is to increase the diversity, which allows the searching processes to find better solutions and avoid falling to the local optimum. However, these operators also partly destroy the structure of solutions. Therefore, keeping feasibility of chromosomes is an important issue when applying genetic operators (Gen, 1996). The search process in GAs tries to look for the solutions in a most efficient way to get closer and closer to the optimal solution. The GAs search for the optimal solution through the evolution processes in which only superior chromosome can survive from a set of random solutions in the solution space. Several schemes have proposed their own rule to define how to move to a better solution (Melanie et al., 1991; Salomon, 1997; Thierens and Goldberg, 1994). For instance, the GA performance is improved by transforming it into a "deterministic" GA with fixed number of nested loops of fixed sizes (Salomon, 1997). Through these attempts, the performance of GA has been improved for some specific classes of problems. However, in the general case when we do not know beforehand the class or the kind of the search space, which can be any, it is difficult to improve GA performance by some fixed scheme. On the other hand, some GAs often avoid local traps by diversifying the population of solutions, based on pure random search processes. However, keeping a rigid random search may increase searching efforts (Goldberg et al., 1990; Melanie et al., 1991, 1994; Justinian, 1997; Salomon, 1996, 1997; Schwefel, 1995; Srinivas and Patnaik, 1994) because we can not avoid revisiting the same or the similar variants. Thus, the computation time of

GA is usually long. To speed up the searching processes, it is better to "guide" the GA to look for the optimal solution.

In this chapter, we follow the approach of guiding the GA. However, unlike the mentioned methods we modify the GA in a more flexible way to guide the search instead of letting it go randomly. The search is not led by a rigid and fixed scheme but by an adaptive mechanism. The adaptation approach in GA is not new and there is a recent classification of works on this topic (Hinterding et al., 1997). However, some branches in the hierarchy of this classification are still empty. We adopt in this chapter the two-stage guiding strategy that is initially presented in recent work of VanHop and Hanh (2000) with the enhancement on the diversification and guiding capabilities. The search in the current Guided Genetic Algorithm (GGA) is still carried out in two stages: survey and evolution. The search process in GGA is not only more powerful by the mechanism which collects the data for adaptation evolution by a GA itself but also more diversified in the survey stage by efficient genetic operators (roulette wheel selection, weighted recombination crossover and shift mutation) and more oriented in the fine-tuning search of evolution stage with a new guiding drive. The purpose of the survey stage is to collect as much as possible the information of the solution space by diversifying the population. This information is then used to guide the search process in the evolution stage to find out the optimal solution. In the survey stage, a short searching life GA with a large population is built to investigate the behaviour of the solution space: what structure would probably fit to the given search space, what direction the search process could move to obtain good children (what solution/chromosome should be selected for the evolution stage). In order to collect complete information about the solution space, a large population could be used to search. This large search space could lead to increase the computation efforts. Fortunately, in this stage, we do not need to have the detailed information of the solution space. Only a rough evaluation is necessary to collect the good search direction in terms of good children. Thus the short iteration life GA is suitable to save the computation time. After each generation, the best chromosome is selected for the next stage. This chromosome indicates for a solution at a certain hill in search space. In order to force the searching process to move to another hill, the next population should be diversified to be as much different as possible with the current population. The diversification could be secured by the pure random selection and the shift mutation that modifies strongly the structure of the chromosomes to create a different population. In addition, the investigation could be strengthened with the introduction of a new crossover operation, weight recombination crossover. These operators, together with the fitness function, are used to filter the good solution. After the survey stage, we have information about the good hill in the form of selected chromosomes for initial population of the evolution stage. Thus, in the evolution stage, another GA is used to search for the optimal solution. The requirement is that the searching process should converge to the optimal solution as fast as possible. A smaller size population, longer life GA is directed by the schemes obtained from the first stage. In this stage, the search process is reinforced to increase the convergence speed by keeping a certain number of best chromosomes in the new generation. Only the best solutions are deterministically transmitted into the next generation. Thus, by spending some extra generations in the survey stage, we yield much more information in reducing the number of generations in the evolution stage by making the evolution faster and more oriented (see Figure 14.2)

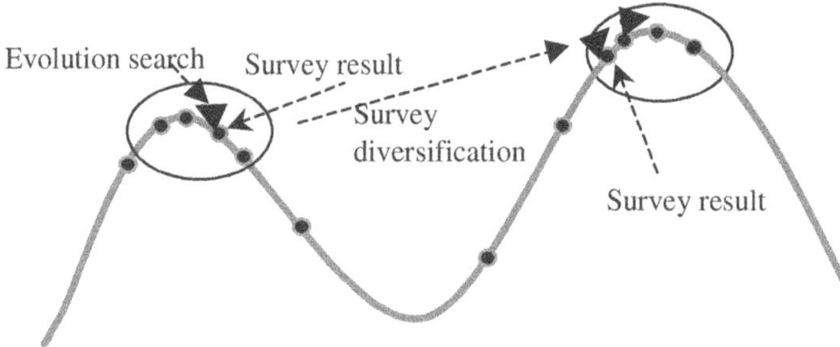

Fig. 14.2 Guiding the GA

This strategy could be used to solve any applied-GA problem such as grouping, facility layout and scheduling problems. For illustration, the printed circuit board (PCB) assembly sequencing problem on multiple non-identical parallel machines is solved by applying the GGA. The following section will describe the GGA in general. Then, Section 3 presents the PCB sequencing problem on multiple non-identical parallel machines in detail and its GGA solution. The final section presents the conclusion remarks and some future research recommendations.

14.2 Guided Genetic Algorithm (GGA)

The search in the Guided Genetic Algorithm (GGA) is carried out in two stages: the survey stage and the evolution stage. In the survey stage, a GA with a big population P_s and small number of generations G_s is created. The fitness function f_s is used to measure not the evolution of the whole population as usual but to investigate the behavior of the mating pairs, the pairs of chromosomes which are selected to generate the new children (chromosomes) of the next generation by applying GA operators. This is to spread members of the population over the search space by using the power of randomness. These members play the roles of agents who will help us to gather the information about the possible distribution of hills in the given search space. Therefore, the more random the population is generated and is reproduced, the more accurate the survey is. Hence, the goal of the first stage is not to make the population evolved but give them freedom and let it reproduce as it is so that we can examine and explore its nature which shows us the structure of the search space. Once we know about the search space we can always exploit this knowledge to find shorter ways to evolve the population in the evolution stage. In this stage, a GA with a smaller population size P_e ($P_e < P_s$) and longer life in terms of the number of generations G_e ($G_e > G_s$) is built. The fitness function f_e is used to measure the evolution of the whole population as usual while the search process is directed by the obtained solutions of the first stage representing the hills of the search space. In addition, while the survey GA tries to diversify the solution population as much as possible, the evolution GA guides the searching drive faster by keeping back a certain percentage of elite solutions αS in each generation, where α is the percentage of elite solutions being kept in each population of S chromosomes. This percentage is predetermined and serves as a guiding factor for the searching process. The remaining $(1-\alpha)S$ chromosomes are used for diversifying purpose. The pairs of chromosomes are randomly chosen from these chromosomes to create $(1-\alpha)S$ new children by using genetic operators. The αS elite solutions of the previous generation and $(1-\alpha)S$ obtained children combine to form S chromosomes of the next generation. The process is repeated until a close optimum solution is found. The main components of the GGA are described below.

14.2.1 Coding scheme

The solutions are often encoded as strings of genes. Each string is called as a chromosome (from now onward, the terms of solution and chromosome are used exchangeable). There are two kinds of chromosome, the binary and sequencing chromosomes corresponding to the binary or sequencing GAs, respectively. The binary chromosomes represent for the 0-1 decision problems' solutions where each gene stands for a binary decision variable. Each sequencing chromosome is a sequence of natural numbers where each gene indicates an assignment decision variable of the job order or the processing position. It is noticed that the chromosome representation should be feasible. It means that the structure of the chromosomes should satisfy constraints of the concerned problem. An example of binary chromosome is the solution of the grouping problem in which each grouping solution is presented as a string of (m x n) binary digits, which includes m segments representing for m groups. The length of each segment is n genes, which each gene stands for an object of the group. If object j ($j=1,\ldots,n$) belongs to group g then gene j has a value of 1 in segment g. Otherwise, this gene has a value of 0 (see Figure 14.3). A chromosome is feasible if all genes appear in a string only one value of 1 because each object can be assigned to one group only.

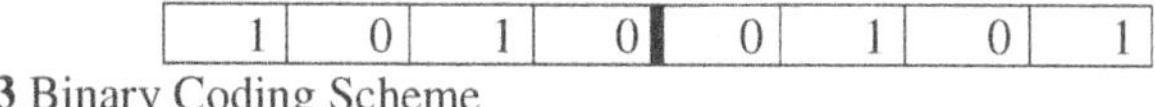

Fig. 14.3 Binary Coding Scheme

The example of sequencing is illustrated in Figure 14.4. The string "364198" represents for a job sequence in which job "3" is processed before job "6", job "6" is processed before job "4" and so on. This sequence is feasible if a job can not be appeared twice in the sequence

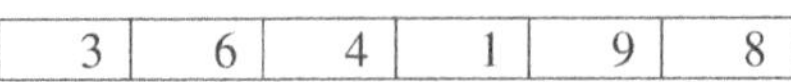

Fig. 14.4 Sequencing Coding Scheme

14.2.2 Fitness function

Fitness function will be the objective function of the concerned problem. Its value is determined from the structure of the corresponding chromosome. For example, the objective function of the grouping problem is the total dissimilarity between objects in the group. Thus the fitness value of the above chromosome is the total dissimilarity of objects 1 and 3 in group 1 plus the total dissimilarity of objects 2 and 4 in group 2. For the sequencing chromosome, the fitness value could be the total processing time of the job sequence "364198".

14.2.3 Genetic Operators

The genetic operators for Survey GA should guarantee to generate high degree of diversity solutions without destroying good solutions. Random selection based on roulette wheel principle is selected to diversify the population of solutions for the survey GA because the main purpose of this stage is to generate as much random population of solutions as possible to examine the characteristic search space. In roulette-wheel drawings, individuals are randomly chosen according to their fitness value, the high-fitted individuals being often selected. The roulette-wheel selection is represented by a pseudo-code as follows.

```
Procedure roulette_wheel_selection
Begin
sumfitness = 0;
for index1 = 0 to popsize do
     begin
       fitness[index1]=objfunc(chromosome[index1]);
       sumfitness=sumfitness+fitness[index1];
     end
 randvalue= random(sumfitness); index2 = 0; sum = 0;
do
    sum=sum+fitness[index2];
    index2 = index2 +1;
while((sum<randvalue)&&(index2<popsize-1));
select chromosome[index2]
End
```

In this procedure, the sum of fitness values is accumulated by variable *sum*. The variable *randvalue* contains the location where the wheel has landed after a random spin generated by the *random()* function. The do-while loop searches through the weighted roulette wheel until the searching variable index2 reaches the maximum index of the population *popsize* or the *sum* value is not less than the stopping point *randvalue*. The function returns the selected chromosome of the current population controlled by searching variable *index2*.

For evolution GA, the selection operator is used to guide the search process. This "guided" selection is known as the biased selection. In the biased selection, a certain fraction α of the best solutions at each generation will survive for the next generation. The remaining $(1\text{-}\alpha)$ solutions are randomly selected and placed in a mating pool, a term given to the place where the selected chromosomes are stored before genetic operators are applied.

The other genetic operators, crossover and mutation, should modify as much as possible the structure of the parent chromosomes to generate the new population of solutions for the next generation. There are a wide range of these operators, such as single point crossover, multi-point crossover, edge recombination crossover, random mutation, shift mutation, inversion mutation, etc. In fact, some operators are not efficient in terms of diversification. For instance, one point crossover operator often creates new similar chromosomes and the population of solutions tends to converge to a homogeneous state. This fact increases the danger of being trapped into sub-optimal solutions. In nature, genetic diversity is maintained by other processes besides mutation and crossover such as inversion, transformation, conjugation, translocation and transposition, etc. (Gould et al., 1996, reported by Simões and Costa, 1999). Among these operations, asexual transposition (crossover) seems to be a superior mechanism for binary genetic algorithm (Simões and Costa, 1999, 2000a, 2000b). The asexual transposition crossover (ATC) is directly inspired from the biology processes. After selecting one individual for reproduction, an asexual transposition is performed. Asexual transposition is characterized by the presence of mobile genetic units inside the chromosomes, moving themselves to new locations or duplicating and inserting themselves elsewhere. These mobile units are called transposons. One or several genes or just a control unit can form transposons. The movement can take place in the same chromosome or to a different one. The beginning of the transposon will be randomly chosen (gene *T*). The transposons within a chromosome are flanked by identical or inverse repeated sequences with a fixed length, *FSL*, defined as flanking sequences. According to the flanking sequence length, the *FSL* bits before the gene *T* make the first flanking sequence. The search for the second flanking sequence begins after gene *T* and stops when an equal or inverse sequence is found. The genes enclosed by gene *T* and the last gene of the second flanking sequence constitute the transposon. The insertion point is searched in the same chromosome and this process starts in the bit after the second flanking sequence. The insertion point is defined when an equal or inverse sequence of bits is found in the chromosome. Notice that the chromosome is viewed as having a circular form. Therefore after reaching the end of the chromosome the search continues in its first bit. When the insertion point is found, the transposon excises from its original position and will integrate in the insertion point. Figure 14.5 illustrates this operation as follows.

1. Selected Chromosome

0 0 1 1 1 0 0 0 0 0 | 1 1 0 0 1 1 1 1 1 1

2. Building the Transposon

0 0 1 **1** 1 0 0 0 0 0 | 1 1 0 0 1 1 1 1 1 1

□

gene **T** (random)

3. Finding the insertion point

0 0 1 **1** 1 0 0 0 0 0 | 1 1 0 0 1 **1** 1 1 1 1

□

insertion point

4. Transposon excision

0 0 1 | 1 0 0 1 **1** 1 1 1 1

1 1 0 0 0 0 0 | 1

5. Transposon integration

0 0 1 1 0 0 1 **1** 1 0 | 0 0 0 0 1 **1** 1 1 1 1

6. Obtained chromosome after asexual transposition

0 0 1 1 0 0 1 1 1 0 | 0 0 0 0 1 1 1 1 1 1

Fig. 14.5 Asexual Transposition

The Edge Recombination Crossover (ERC) is often used as an efficient crossover operator for scheduling problems (Cheng et al., 1999). ERC builds offspring using only the edges present in both parents. The best genes of parents in terms of number of edges are recombined into new off-springs. However, the number of edges in scheduling problems does not play such an important role, but the weights (processing time, setup time, costs, etc.) do. Therefore, a new crossover is developed that recombines the best genes of parents in terms of these weights instead of number of edges in the normal ERC.

Table 14.1 Weights of Jobs

Job	1	2	3	4	5	6	7	8	9	10
1	-	1	1	1	1	2	0	2	2	0
2	1	-	2	0	1	0	2	1	1	1
3	1	2	-	0	0	1	1	2	1	2
4	1	0	0	-	2	1	1	1	0	1
5	1	1	0	2	-	1	0	2	0	1
6	2	0	1	1	1	-	1	2	0	0
7	0	2	1	1	0	1	-	0	0	0
8	2	1	2	1	2	2	0	-	2	1
9	2	1	1	0	0	0	0	2	-	1
10	0	1	2	1	1	0	0	1	1	-

This new crossover is labeled as Weighted Recombination Crossover (WRC). The following section will describe the principle of WRC. Suppose the edges between jobs in the sequence are weighted by the job's processing time given in Table 14.1.

First, we calculate the total weights of each job in parents (see Figure 14.6 and Table 14.2).

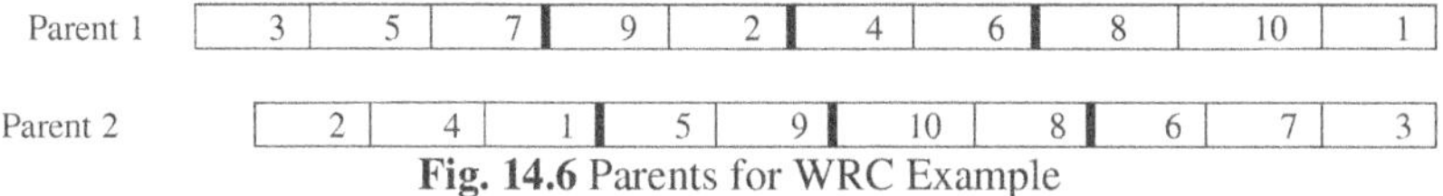

Fig. 14.6 Parents for WRC Example

Table 14.2 Computation of Weights

Job	Total weights	Job	Total weights
1	(1, 10) + (1,8) +(1,2) + (1,4) = 4	6	(6, 4) + (6,7) + (6, 3) = 3
2	(2, 9) + (2,4) + (2,1) = 2	7	(7,3) + (7, 5) + (7,6) = 2
3	(3, 5) + (3, 7) + (3,6) = 2	8	(8, 10) + (8,1) = 3
4	(4, 6) + (4,1) + (4,2) = 2	9	(9, 2) + (9,5) = 1
5	(5, 3) + (5, 7) + (5,9) = 0	10	(10, 8) + (10, 1) = 1

Pick a job that has smallest weight to add to the child. Check the feasibility constraints, if it is not violated then the job is selected as a member of the child. Otherwise, the job with the next least weight is picked. From the above table it can be seen that job 5 with a total weight of 0 has the minimum weight. Hence it is selected to be a member of next generation (Figure 14.7).

5									

Fig. 14.7 The First Node Selected in the WRC Method

For all the remaining jobs we remove the weights associated with the job selected above. The updated table is shown as Table 14.3.

Table 14.3 Weights after Removing the Selected Job and Its Weights

Job	Total Weights	Job	Total Weights
1	(1, 10)+(1,8)+(1,2) + (1,4) = 4	6	(6, 4) + (6,7) + (6, 3) = 3
2	(2, 9) + (2,4) + (2,1) = 2	7	(7,3)+ **(7, 5)**+(7,6) = 2 – 0 = 2
3	**(3, 5)**+(3, 7)+(3,6)= 2 – 0 = 2	8	(8, 10) + (8,1) = 3
4	(4, 6) + (4,1) + (4,2) = 2	9	(9, 2) + **(9,5)** = 1 – 0 = 1
5	(5, 3) + (5, 7) + (5,9) = 0	10	(10, 8) + (10, 1) = 1

Repeat the process until all of the jobs have been selected (Table 14.4). The Figure 14.8 shows the final child.

Table 14.4 Weights Calculation Process

Job	Total Weights	Job	Total Weights
1	**(1, 10)** + (1,8) +**(1,2)** + **(1,4)** = 4→3→3→2→0	6	(6, 4) + **(6,7)** + **(6, 3)** = 3→ 2→1→0
2	**(2, 9)** + (2,4) + (2,1) = 2→1→0	7	**(7,3)** + **(7, 5)** + (7,6) = 2 – 0 = 2→1→0
3	**(3, 5)** + (3, 7) + (3,6)= 2→ 0	8	**(8, 10)** + (8,1) = 3→2→0
4	**(4, 6)**+ (4,1) + **(4,2)** = 2→2→1→0	9	**(9, 2)** + **(9,5)** = 1→0
5	(5, 3) + (5, 7) + (5,9) = 0	10	(10, 8) + (10, 1) = 1→0

5	9	2	10	3	7	6	4	1	8

Fig. 14.8 Final Child of the WRC Process

Mutation operator brings random changes in a parent solution to generate an offspring. This change should satisfy the constraints of the problem. The inversion mutation is the most popular for binary GA because it is simple (Figure 14.9).

Parent	0	0	1	1	0	0	**1**	1	1	0	0	0	0	0	1	1

Child	0	0	1	1	0	0	**0**	1	1	0	0	0	0	0	1	1

Fig. 14.9 Binary Inversion Mutation

For sequencing, studies have shown that the shift mutation may give good results (Goldberg, 1989; Sikora, 1996.). Therefore, we will employ these operators to generate new offspring. In this mutation, a gene at one position is removed and replaced at another position as shown in Figure 14.10. The two positions are randomly chosen.

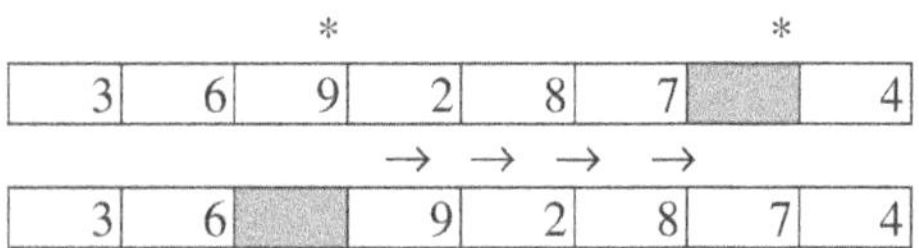

Fig 14.10 Sequencing Shift Mutation

14.2.4 Input parameters

Other input parameters of genetic algorithm are selected based on De Jong's (1975) suggestion: "...good GA performance requires the choice of a high crossover probability, a low mutation probability (inversely proportional to the population size), and a moderate population size..." (Also referred by Goldberg, 1989). The following parameter values are initially chosen:

- Survey population size $P_s = 100$

- Evolution population size $P_e = 20$
- Probability of mutation , $p_m = 0.04$
- Probability of crossover, $p_c = 0.8$
- Number of survey generations $G_s = 20$
- Number of evolution generations $G_e = 50$.
- Stopping condition: when the number of iterations reaches G_e

The remaining parameter, α, (0<α<1) will be selected based on the experiments that will be discussed in Section 3.4. The complete algorithm is described in the following form.

14.2.5 Guided Genetic Algorithm (GGA)

Step 1. Let $s = 0$; s is the generation count of survey stage.
Step 2. Initialize survey population with the size of P_s.
Step 3. Evaluate initial survey population
Repeat

Step 4. Generate new survey population using GA operators

- Roulette-wheel selection
- ATC for binary GA and WRC for sequencing GA with probability p_c.
- Inversion and shift mutations for binary and sequencing GA, respectively with mutation probability p_m.
- Perform the correction of the chromosome's structure for satisfying the feasibility conditions (if any).

Step 5. Evaluate new population using fitness function $f_s \Rightarrow$ determine the best children for evaluation population. If (number of selected children<P_e), increase G_s to ($G_s + 1$). Otherwise, break.
Step6. Substitute old population by new population. Increase s to ($s + 1$).

Until ($s = G_s$)
Step 6. Let $e = 0$, e is the generation count of evolution stage.
Step 7. Form the initial evaluation population from the results of survey stage with the size of P_e ($P_e < P_s$)
Step 8. Evaluate initial evolution population by the fitness function f_e
Repeat

Step 9. Generate the next evolution generation using genetic operators.

- Select the ($\alpha * P_e$) best solutions for next generation

Repeat

- Take pairs of chromosomes from the remaining (1-α)* P_e chromosomes of the old evolution population into the mating pool using the roulette rule.
- ATC for binary GA and WRC for sequencing GA take place for chromosomes in the mating pool with crossover probability p_c.
- Inversion and shift mutations for binary and sequencing GA, respectively with mutation probability p_m.

- Perform the correction of the chromosome's structure for satisfying the feasibility conditions (if any).

Until $(1\text{-}\alpha)^* P_e$ *children are generated*

Step 10. Combine (α^*P_e) best solutions and $(1\text{-}\alpha)^* P_e$ new generated children into new evolution population

Step 11. Evaluate new population using fitness function $f_e \Rightarrow$ determine the best child of the evaluation population.

Step 12. Substitute old population by new population. Increase e *to* $(e+1)$.

Until $(e = G_e)$

14.3 The GGA for the PCB Assembly Sequencing Problem

14.3.1 The PCB Sequencing Problem on Multiple Non-identical Parallel Machines

We consider the printed circuit board (PCB) scheduling problem on multiple non-identical parallel machines. The set of n PCBs and their component requirements are given in the form of incidence matrix. In the present case, each board is considered as a job. The jobs must be sequenced on a set of m parallel machines such that the total setup time and load variation between machines are minimized. The general requirement of the problem is to find the optimum grouping and board sequence for each machine in order to minimize total setup of the system. This requirement is often achieved through exploiting similarities between boards, or between board and current setup on magazine of a machine. In addition, balancing of the load is to be considered. When the assumption of identical parallel machines is relaxed to non-identical parallel machines, capacities of machines have to be included in the model. Hence a multiple criteria formulation becomes a necessity.

Each PCB j $(j = 1, .., n)$ requires $SIZE_j$ number of components or component types. The capacity C_t of machine t $(t = 1,...m)$ is presented as the number of slots on the magazine of that machine. It is assumed that components of a given type occupy only one slot on a magazine and that any machine capacity C_t is always greater than any $SIZE_j$ of a PCB j. Furthermore, the setup time of a board in the sequence cannot be predetermined as in a regular scheduling situation, as it depends on the previous jobs. That is, it depends on the current set of components on the magazine, and the additional requirements for the incoming job. For the assembly, total setup time on a machine would include the time to fix all PCBs on the machine, and the time to exchange components. It is expressed as

$$TS_t = n_t S + s\sum_{k=0}^{n_t-1} sw_{kt} ; t = 1,...m \tag{14.1}$$

Where TS_t is the total setup time for machine t; $t = 1,\ldots,m$

n_t is the number of boards allocated to machine t

sw_{kt} is the number of component changeovers between board $(k)^{th}$ to board $(k+1)^{th}$ in the board sequence for machine t. And so, sw_{0t} is the first loading.

s is the time to load and unload a component type on the feeder rack

S is the time to fix a PCB on machine table, PCB setup time.

The overall problem considers the three inter-related issues of board grouping, sequencing, and component switching and it is modeled as a multi-objective optimization problem, with three objectives. The first objective looks at the similarity between boards in a group, the second balances the load, and then the third one minimizes the setup times at individual groups. Appropriate priority weights are given to the objectives. The solution will simultaneously yield the groupings, sequencing and component switching for each individual group.

We formulate a composite model with all the related issues and objectives as

$$\text{Minimize } Z = \lambda_1 f_1 + \lambda_2 f_2 + \lambda_3 f_3 \tag{14.2}$$

Subject to:

$$\sum_{t=1}^{m} x_{jt} = 1;\ j = 1,\ldots,n \tag{14.3}$$

$$\sum_{i=1}^{p} w_{ij} x_{jt} = C_t;\ j = 1,\ldots,n; t = 1,\ldots,m \tag{14.4}$$

$$a_{ij} y_{jk} x_{jt} \le w_{ik};\ i = 1,\ldots,p;\ j,k = 1,\ldots,n; t = 1,\ldots,m \tag{14.5}$$

$$\sum_{j=1}^{n} y_{jk} x_{jt} = 1;\ k = 1,\ldots,n; t = 1,\ldots,m \tag{14.6}$$

$$\sum_{k=1}^{n} y_{jk} x_{jt} = 1;\ j = 1,\ldots.,n; t = 1,\ldots,m \tag{14.7}$$

$$b_{ik} \le w_{i(k+1)} x_{(k+1)t} - w_{ik} x_{kt};\ k = 1,\ldots,n-1; i = 1,\ldots,p; t = 1,\ldots,m \tag{14.8}$$

$$sw_{0t} + sw_{kt} = \left(\sum_{i=1}^{p} a_{ij}\right) y_{j1} x_{jt} + 2 * \sum_{i=1}^{p} b_{ik};\ k = 1,\ldots,n-1;\ j = 1,\ldots,n; t = 1,\ldots,m \tag{14.9}$$

$$\lambda_1 + \lambda_2 + \lambda_3 = 1 \tag{14.10}$$

Where

n	Total number of PCBs
m	Number of machines
p	Total number of required components.
i	Component index, $i = 1,\ldots, p$
j	Board index, $j = 1,\ldots, n$
k	Sequence position index, $k = 1, 2,\ldots,n$
t	Machine (or group) index, $t = 1, \ldots,m$
C_t	Capacity of machine t, as number of slots in the magazine.

$A=\{a_{ij}\}$=PCB component incidence matrix.

S	Time to fix one board on the machine table

s Component setup time (for loading or unloading of one component)
sw_{kt} Number of component changeovers between board $(k)^{th}$ to board $(k+1)^{th}$ in the board sequence for machine t, and sw_{0t} is the first loading.
TS Total set-up time
$SIZE_j$ Size of PCB j (number of component types required).
λ_1 Weight for similarities criterion
λ_2 Weight for load balancing criterion
λ_3 Weight for setup time criterion
f_1 Similarities function
f_2 Load balancing function
f_3 Setup time function
x_{jt} = 1, if board j belongs to group t; 0, otherwise.
w_{ij} = 1, if setup j contains component type i; 0, otherwise.
y_{jk} = 1, if board j is assembled in the kth job in the sequence; 0, otherwise.
b_{ik} = 1, if component type i has to load on the machine right after k^{th} job; 0, otherwise.

Note:

k = sequence position is a relative position in each group. The real position will be converted based on this relative position. For example, if the solution for group $t = 1$ is $x_{21}=1$, $y_{25}=1$; $x_{61}=1$, $y_{62}=1$; $x_{71}=1$, $y_{79}=1$ then the board sequence for boards in group 1 is 6-2-7 because its relative positions are 2-5-9, respectively.

Each of the objectives can be developed as per its requirements. For f_1, the similarities function which addresses the grouping problem, the computation is based on the Jarcard's similarity coefficient. The coefficient is defined as

$$SIM_{jk} = \frac{COM_{jk}}{TOTAL_{jk}} \tag{14.11}$$

Where COM_{jk} is the number of common components between board j and board k
$TOTAL_{jk}$ is the number of components in both board j and board k
So, the total similarities between boards in group t is

$$f_s^t = \sum_{j=1}^{n} \sum_{k=1}^{n} SIM_{jk} x_{jt} x_{kt} \tag{14.12}$$

Where

x_{jt} equals to 1 if board j belongs to group t; otherwise $x_{jt} = 0$
x_{kt} equals to 1 if board k belongs to group t; otherwise $x_{kt} = 0$

The similarities function is given by

$$f_1 = BIG - \min\{ f_s^t ; t = 1,...,m\} \tag{14.13}$$

Where BIG is a very large positive number
The second component of (14.2), representing the load balancing criterion, will be

$$f_2 = \frac{\max\{f_l^t; t=1,...,m\} - \min\{f_l^t; t=1,...,m\}}{\max\{f_l^t; t=1,...,m\}} \tag{14.14}$$
$$= \frac{\left(\max\{\sum_{j=1}^{n} SIZE_j x_{jt}; t=1,...,m\} - \min\{\sum_{j=1}^{n} SIZE_j x_{jt}; t=1,...,m\}\right)}{\max\{\sum_{j=1}^{n} SIZE_j x_{jt}; t=1,...,m\}}$$

Where $f_l^t = \sum_{j=1}^{n} SIZE_j x_{jt}$ is the total number of components loaded on machine t.
This function expresses the difference between maximum and minimum loads for the machines. This loading variance should be minimized.

Now, the total setup time is expressed (based on (14.1)) as follows:

$$f_3 = TS = \max\{TS_t; t=1,...,m\} = \max\{n_t * S + s\sum_{k=0}^{n_t-1} sw_k; t=1,..m\} \tag{14.15}$$
$$= \max\left\{\left(\sum_{j=1}^{n} x_{jt}\right) * S + \left(\left(\sum_{i=1}^{p} a_{ij}\right) y_{j1} x_{jt} + 2 * \sum_{i=1}^{p} b_{ik}\right) * s\right\}; k = 1,..,n-1; j = 1,..,n; t = 1,..,m$$

The first component of (14.15) indicates total board fixing time of machine t. Hence, it relies on the total number of boards assigned to that machine. The second element is constraint (14.9). This constraint determines the total component exchange time, which depends on the total number of component switches. Function (14.15) needs to be minimized.

The remaining elements of the objective function are the weights of the criteria. The weights are given subjectively, based on their relative importance. These weights also need to satisfy condition (14.10). As our main concern is the total setup time, λ_3 dominates the other objectives. We take $\lambda_1 = 0.2$, $\lambda_2 = 0.2$, and $\lambda_3 = 0.6$.

Other equations are explained as follows. Constraint (14.3) expresses the grouping condition, in which each board belongs to only one group. Constraint (14.4) assures that exactly C_t component types are placed on machine t for any PCB of group t. Constraint (14.5) indicates that if PCB j is assembled as the kth job on the machine t, then all the component types needed for PCB j must be set up on that machine. Constraints (14.6) and (14.7) assign exactly one PCB to exactly one position in the sequence for each machine. Constraint (14.8) expresses the loading components relationship with the setup status. The number of component changeovers is counted in the next constraint (14.9). The last constraint (14.10) is a normalizing constraint for the priority weights of the multiple objectives.

14.3.2 Related works

The PCB scheduling problems have also been considered in several works, under different types of machine settings. Most of the current research works emphasize on a single machine configuration. Since the PCB assembly is characterized by small batches and high speed assembly, the main concern even at scheduling is the setup time including setup of the boards, and loading and unloading of components. Hence, most of the related research works in PCB assembly select setup time as the main criterion (Maimon and Braha, 1998; Smed et al., 1999; Rajkumar and Narendran, 1998; etc.).

Tang and Denardo (1988) propose a greedy heuristic to determine job sequence in a flexible manufacturing systems (FMS) environment, which can be applied for the case of PCB assembly. Rajkumar and Narendran (1998) present another greedy heuristic, which exploits the similarities between boards and current set up on the magazine of the machine. Some local search heuristics (2-opt, tabu search, genetic algorithm, simulated annealing) are also used to solve the problem. Maimon and Braha (1998) develop a genetic algorithm and compare it with a TSP-based spanning tree algorithm. Gronalt et al. (1997) focus on the component switching problem and the feeder assignment problem with the assumption that the sequence of board types to be processed on a single machine is given. They model the combined set-up (loading) and feeder assignment problem, called a component-switching problem, as a mixed-integer linear program. A recursive heuristic was proposed to solve this combined problem. Gunther et al. (1998) extend the PCB assembly setup problem for single machine when a feeder requires more slots in the magazine. They solve three related sub-problems of board sequence, component switching, and feeder assignment, sequentially by a TSP (Traveling Salesman Problem) -based heuristics. Their heuristic first constructs the board sequence using the upper bound on component changeovers between two consecutive boards. The "keep tool needed soon" (KTNS) rule developed by Tang and Denardo (1988) is implemented to evaluate the performance of the board sequence. Then an iterative procedure with a heuristic to improve the solution is developed. A drawback of this approach is that it does not take into account the current status of the magazine.

For the single problem of component switching, Tang and Denardo (1988) propose a heuristic, called KTNS (Keep Tool Needed Soon) rule, and provide a proof of its optimality. According to KTNS rule, the tool magazine is kept in full and at any instance g tool u is

- Inserted if $L(u,g) = g$; $J_u = 0$;
- Kept if : $L(u,g) = g$; $J_u = 1$;
- Removed if $u = \max\left\{ \underset{q:Jq=1}{L(q,g)} \right\}$

Where $L(u,g)$ is the first instant at or after instant g at which tool u is needed.

$J_u = 1\ (0)$ if tool u is (is not) on the machine at a given instant g

There are considerable extensions for the case of non-uniform tool sizes including the works of Crama et al. (1994); Privault and Finke (1995); Gunther et al. (1998); Matzliach and Tzur (1998); Djellab et al. (2000); Matzliach and Tzur (2000). Some beginning attempts have been made to handle the non-uniform tool sizes issue using different formulations such as an integer programming (Cramma et al., 1994) and network flows (Privault and Finke, 1995) which are solved by greedy algorithms. Recently, Matzliach and Tzur (2000) have shown the complexity of this case to be NP-complete. They also propose two constructive heuristics, which provide solutions that are extremely close to the optimal solution (within less than two percent). Matzliach and Tzur (1998) analyze another aspect of the tool switching problem, when parts that need to be processed on the machine arrive randomly and tool sizes are non-uniform. Djelab et al. (2000) present an iterative best insertion procedure (IBI) using hypergraph representation for the tool-switching problem which gives quite good solutions. In another direction, Rupe and Kuo (1997) relax the assumption of tool magazine capacity where jobs may require more tools than the magazine's capacity can hold while the uniform tool-sizes assumption is still maintained. Hence, job splitting is taken into account. When jobs are split into sub-jobs to allow tool changes at any instant between tool uses, the KTNS policy still yields an optimal solution. When jobs are split completely into sub-jobs such that each sub-job requires one tool and tools are permitted to change concurrently with job changes, the "Get Tool Needed Soon (GTNS)" policy developed by the authors is proved to give optimal solution.

For the case of multiple machines setting, Ben Arieh and Maimon (1992) use a Simulated Annealing algorithm to sequence PCBs on two serial machines. Van Hop and Tabucanon (2001) attempt to solve PCB assembly setup problem for a line of machines from a multiple criteria viewpoint. They propose an approach using both multi-attribute decision-making and multi-objective decision-making technique together to solve the problem. So far, there has not been any literature addressing the scheduling problem of PCBs for multiple non-identical machines. A similar work of Rajkumar and Narendran (1997) tries to allocate PCBs to machines but their work was developed for the case of multiple identical parallel machines only. Moreover, they focus on assigning PCBs to a machine and assume that the boards sequence is already given.

The present case is an effort to consider the combined problem of scheduling n PCBs for m non-identical parallel machines, which includes all three issues of board grouping, board sequencing and component switching. The following section will describe GGA solution for this problem.

14.3.3 The GGA Solution

Solving the integrated problem described above is difficult since its sub-problems are complex even when tackled in isolation. The board grouping problem belongs to the class of partitioning problems which are NP-hard in strong sense. Once the grouping is done, the optimal board sequence for each machine is still hard to be determined as the setup time for each board is sequence dependent. The component switching problem is even more complicated as the loading needs to be determined by the current and future requirements. The computational time increases exponentially as the problem size gets larger. Hence, searching techniques like genetic algorithms are attractive. In this case, solving the integrated problem is used as an example to illustrate the efficiency of the Guided Genetic Algorithm

(GGA). The main components of the GGA do not change as described above but the following items should be adapted to the characteristics of the problem.

14.3.3.1 Coding scheme

In this case, a solution is presented as a string of paired numbers. The string contains m sections, each section representing a group of boards allocated to a specific machine. A section is further divided into n genes indicating n boards. Each gene has a pair of numbers. The first number in gene k of section t is a binary number with a "*1*" if board k belongs to group t (corresponding to machine t), and "*0*" otherwise. The second value in gene k represents the relative position of board k in the sequence of its group. Figure 14.11 illustrates a sample chromosome. For this chromosome structure, we have 7 boards and 2 machines (groups). The first group includes board 1, 3, 6 with the relative positions 4, 6, 2, respectively. Ordering the boards according to their relative positions, the sequence for the first group is obtained as 6-1-3. Similarly, in the second group the board sequence is 4(1)-5(3)-2(4) -7(7).

1,4	0,3	1,6	0,1	0,5	1,2	0,7	0,5	1,4	0,2	1,1	1,3	0,6	1,7

Fig. 14.11 A Sample Chromosome

14.3.3.2 Fitness function

The fitness value is calculated based on the objective function (14.2) and definitions (14.11) – (14.15). The only exception is, instead of calculating number of component changeovers as in (14.15), we apply KTNS principle to determine total number of component switches for each machine t.

14.3.3.3 Genetic operators

In this problem, the genetic operators of the GGA are implemented for both grouping and sequencing at the same time. In addition, since the nature of the problem is different, the weight of the kth job in the PCB sequences is calculated as the total setup time when job k is loaded on the machine plus the total setup time when job k is unloaded from the machine (see figure 14.12) because the loading depends on the current (job k) and future component (job $k+1$) requirements as well as the current status of the magazine.

$$\text{Parent 1} \quad \cdots \longrightarrow J^1_{k-1} \xrightarrow{w^1_{k-1,k}} J^1_k \xrightarrow{w^1_{k,k+1}} J^1_{k+1} \longrightarrow \cdots$$

$$\text{Parent 2} \quad \cdots \longrightarrow J^2_{k-1} \xrightarrow{w^2_{k-1,k}} J^2_k \xrightarrow{w^2_{k,k+1}} J^2_{k+1} \longrightarrow \cdots$$

Where $w^{(*)}_{k-1,k} = \text{TS}(*, \text{k}) - \text{TS}(*, \text{k-1})$ and $w_k = w^1_{k-1,k} + w^2_{k-1,k} + w^1_{k,k+1} + w^2_{k,k+1}$

TS(*, k) = total setup time of the sequence (*) until job k

Fig. 14.12 Weight Calculation for Job *k* in the Parent Chromosomes

14.3.4 Experimental Results

The proposed GGA is encoded in C++ language. Several data sets have been used to test the performance of our GGA. Six groups of data sets have been used to verify influence of parameters on the performance of the GGA as presented in Table 14.7. The first group of data (Random 1) is randomly generated to verify the influence of searching drive α (percentage of best solutions). The second set of random data (Random 2) is then used to investigate the performance of our genetic algorithms with the varying of population size of survey and evolution stages. The third group of random data set (Random 3) checks the effect of the number of generation steps in each stage on the performance of the GGA. Using the results of the first three investigations, we may have conclusion about the best compromise between parameters of the GGA. Then, the last three groups of data set are used to investigate the performance of GGA when the number of components is randomly generated by a uniform distribution function with a mean value of 100 and three levels of low, medium and high variances, respectively. In order to secure the assumption that magazine's capacity of the machines can hold all component types for any PCB, the magazine's capacities C_1, C_2, C_3 are set to be the maximum component requirement for a board plus two. The ratio between the time to fix a board on machine table and time to load/unload one component (S/s) is assumed to be equal to three. For each case, the experiments are run for ten times, with number of board types ranging from 30 to 100.

Table 14.7 Test Problems

Data set	*Name*	P_s/P_e	G_e/G_s	α	p	m	n	S/s
1	Random 1	100/20	50/20	[0,1]	[80,120]	3	30	3
2	Random 2	[1,10]	50/20	0.3	[80,120]	3	30	3
3	Random 3	100/20	[1,10]	0.3	[80,120]	3	30	3
4	Random 4	100/20	100/20	0.3	[80,120]	3	[30,100]	3
5	Random 5	100/20	100/20	0.3	[70,130]	3	[30,100]	3
6	Random 6	100/20	100/20	0.3	[60,140]	3	[30,100]	3

For evaluating the GGA, we use the most important objective that is, minimizing of total setup time. The evaluation is carried out by comparing the solution of each problem to the possible upper and lower bounds on the setup time for that problem. The lower bound is obtained by relaxing the constraint on the capacity of a machine. In such case, all the required components can be loaded simultaneously. Thus the total number of component changeovers is *p*, and the total number of PCBs set up would be *n*. Therefore, the average setup time on each machine would be given by $(n*S+p*s)/m$. The upper bound is attained from the case when each PCBs processing is completed, all components are removed from the magazine and remounted for another PCBs. In such case, the number of component switches is the total number of required components for all PCBs. Thus, the average setup time on each machine would be $(n*S + \sum_{j=1}^{n} SIZE_j *s)/m$.

The performance of the algorithm is now computed as Quality Index (QI), in terms of the total setup time (TS) of the obtained solution. The Quality Index is measured as the percentage ratio of the difference between the upper bound and the obtained solution, and the difference between upper and lower bounds.

$$QI = \frac{UB - TS}{UB - LB} \times 100\% \tag{14.16}$$

First, the algorithm is tested with the first group of random data to analyse the influence of the searching drive α (fraction of best solutions carried over to next generation). The first set of random data has been used. For illustration, Random1 has 30 board types (n = 30), survey population size P_s = 100, evolution population size P_e = 20, number of survey generations G_s = 20, number of evolution generations G_e = 50, and the distribution of component requirement of the boards in the range of (80, 120). We need to choose an appropriate value for searching drive α. If this value is too small, then the search procedure will be closer to pure random search. It may reach global optimum, but may take very long time to converge on the solution. If α is larger, then the solution may move in the direction of global optimum. However if α is too close to 1, the search procedure may get stuck at a local optimum, as some good solutions may dominate the population. They will be used several times to form new offsprings. The entities of the population gradually become more and more similar. The algorithm will converge to a generation that is completely homogeneous, and therefore, the search engine of GA falls into a local trap. Therefore, it is better to investigate a suitable value for α. We run the program on the same data set (Random1) with different values of α. Other input parameters were kept unchanged. Figure 14.13 illustrates the influence of α on the optimality. As we can see, with the value of α in the range of 0.05- 0.35, better solutions can be achieved. For the rest of problem sets, we select α = 0.3.

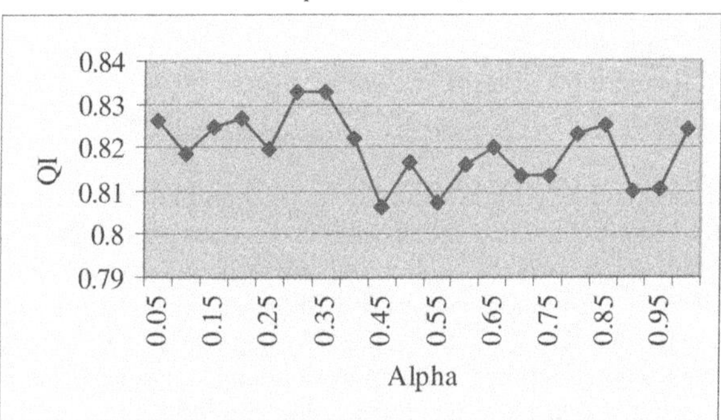

Fig. 14.13 The Effect of Searching Drive α

Next, we investigate the performance of the GGA by varying the P_s/P_e ratio while keeping the value of $\alpha = 0.3$ and other parameters as constants as in the random data set Random 2 (see Table 14.7). As the results presented in Figure 14.14, when the population size of the survey stage is closed to the evolution stage, the survey GA do not have enough information of the solution stage. Thus the search procedure may easily get stuck at local optimal hill. The searching mechanism can not move further to other hills of the solution space. When the value of P_s is larger relative to the evolution population size, P_e, the survey GA has a large enough searching space. It can collect enough information of the solution hills of the solution space. Thus, the quality of the solution is better. However, the searching time may increase because the population space is larger but the number of generations of survey GA is still limited. Thus, it is better to select a reasonable value of P_s/P_e ratio to secure the ability of survey without increasing the computation time. The value of $P_s/P_e = 5$ could be appropriate.

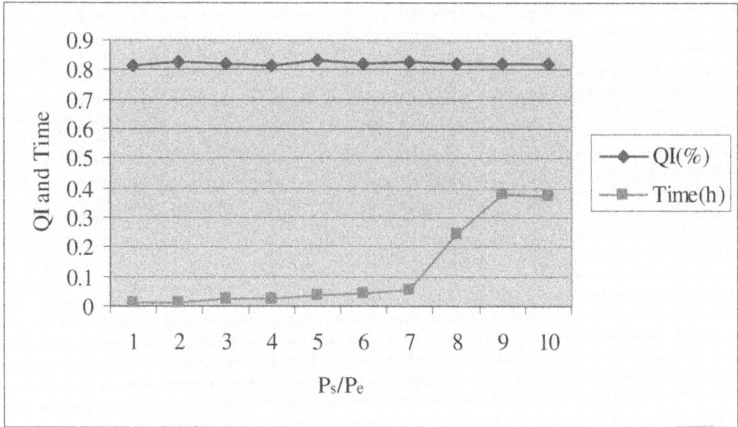

Fig. 14.14 The Effect of P_s/P_e Ratio

With the selection of searching drive factor and population size ratio P_s/P_e in hand, we move forward to test the influence of G_e/G_s ratio by using the random data set Random 3. In this case, if the life of the survey search is as large as evolution search, the searching time will be longer although the quality of the solution may be better. Otherwise, if the life of the survey GA is very short compared with the evolution search, the survey GA does not have enough time to collect enough information of the solution space. Thus the quality of the solution may be reduced. Figure 14.15 shows these results. Therefore, the ratio of 5 could be an appropriate value of G_e/G_s

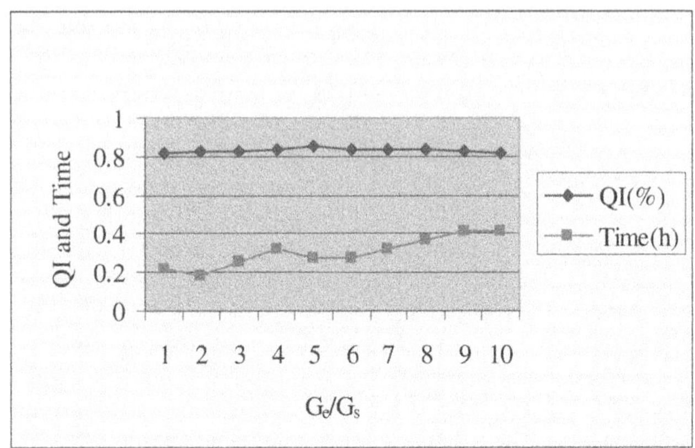

Fig. 14.15 The Effect of G_e/G_s Ratio

After selecting the suitable input parameter for the GGA, we investigate the performance of the GGA for the considered PCB sequencing problem on multiple non-identical parallel machines with three random data set: Random 4, Random 5 and Random 6 corresponding to three varying level of component requirements: low, medium and high variations, respectively. The results are presented in Figure 14.16. In all cases, the quality of the GAs is quite good in terms of proximity of convergence to optimum solutions. As we can see, the overall performance of our GGA is QI > 80%.

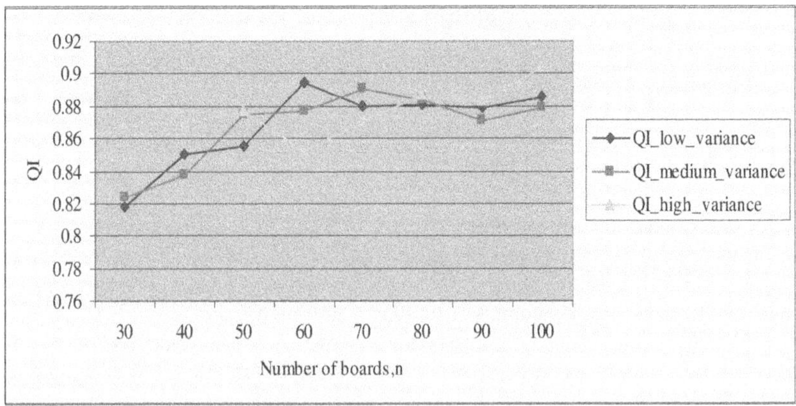

Fig. 14.16 Performance of GGA with Selected Parameters

These results have shown the advantage of the proposed approach in solving PCB scheduling problem in specific and the NP-hard problems in general. On an average, the proposed approach provides a good quality solution in terms of set up time in a reasonable computation time. In case a company has a computerized database containing information about daily production plans, the algorithm can quickly generate daily grouping and schedules to help the planners. The savings in total setup time would increase the value added for such high flexibility production systems, adding productivity to flexibility.

14.4 Concluding Remarks

In the present chapter, the guided genetic algorithm (GGA) has been presented. In this algorithm, the search process is carried out in two stages: the survey and evolution stages. The survey stage collects information of the search space by a large population and short generation life GA. This GA explores the structure of the search space in terms of selected good children after each generation. In order to have complete information of the search space, the search process should visit as many hills as possible by diversifying the next population as much as possible. This is done by the efficient genetic operators such as roulette wheel selection, weighted recombination crossover and shift mutation. After the survey stage, the information about the good hill in the form of selected chromosomes is used to form the initial population for the evolution stage. In the evolution stage, another GA is used to search for the optimal solution. The search process in the evolution GA should converge to the optimal solution as fast as possible with smaller size populations, longer generation life GA by keeping a certain number of best chromosomes in the new generation. Thus, by spending some extra generations in the survey stage we yield much more information to reduce the number of generations in the evolution stage by making the evolution faster and more oriented. This GGA can be used to solve any kind of NP-hard problems. For illustration, the PCB scheduling problem on multiple non-identical parallel machines is solved by the GGA. The obtained results are quite promising in terms of good quality solution with a reasonable computation time. However, in order to improve further the search process, the similar guiding framework could be used but the problem specific genetic operators should be developed in future applications. In addition, large-scale numerical experiments are needed to be conducted to determine the appropriate parameters for the GGA on the other applications. The effect of other parameters such as rates and probabilities of operators should also be investigated because a drawback for GA in general and GGA in specific is the difficulty in selection of parameters and to combine these operators in an efficient way (see Muhlenbein, 1997 for further analysis on GAs).

References

Ben-Arieh D, Maimon O (1992) Annealing method for PCB assembly scheduling on two sequential machines. IJ Computer Integrated Manufacturing 5: 361-367.

Cheng R, Gen M, Tsujimura Y (1999) A tutorial survey of job-shop scheduling problems using genetic algorithms, part II: hybrid genetic search strategies Computers and Industrial Engineering 36: 343-364.

Crama Y, Kolen AWJ, Oerlemans AG, Spieksma FCR (1994) Minimizing the number of tool switches on a flexible machine. IJ Flexible Manufacturing Systems 6: 33-54.

DeJong KA (1975) An analysis of the behavior of a class of genetic adaptive systems. Ph.D. dissertation, University of Michigan, USA.

Djellab H, Djellab K, Gourgand M (2000) A new heuristic based on a hypergraph representation for the tool switching problem. IJ Production Economics 64: 165-176.

Gen M, Tsujimura Y, Li Y (1996) Fuzzy assembly line balancing using genetic algorithms. Computers and Industrial Engineering 31: 631-634.

Goldberg DE (1989) Genetic algorithms in search, optimization and machine learning. Wiley, Massachusetts.

Goldberg DE, Deb K, Korb B (1990) Messy genetic algorithms revisited: studies in mixed size and scale. Complex Systems 4:415--444.

Gould JL, Keeton WT (1996) Biological Science. W.W. Norton & Company

Gronalt M, Grunow M, Gunther HO, Zeller R (1997) A heuristic for component switching on SMT placement machines. IJ Production Economics 53: 181-190.

Gunther HO, Gronalt M, Zeller R (1998) Job sequencing and component set-up on a surface mount placement machine. Production Planning and Control 9: 201-211.

Hinterding R, Michalewicz Z, Eiben AE (1997) Adaptation in evolutionary computation: a survey. In: proceedings of the 4th IEEE international conference on evolutionary computation. Indianapolis, pp 65-69.

Justinian R (1997) Analysis of complexity drift in genetic programming, In: genetic programming 1997: proceedings of the second annual conference. pp 28-31.

Maimon OZ, Braha D (1998) A genetic algorithm approach to scheduling PCBs on a single machine. IJ Production Research 36: 761-784.

Matzliach B, Tzur M (1998) The online tool switching problem with non-uniform tool size. IJ Production Research 36: 3407 –3420.

Matzliach B, Tzur M (2000) Storage management of items in two levels of availability. European Journal of Operational Research 121: 363 –3379.

Melanie M, Forrest S, Holland J (1991) The Royal Road for genetic algorithms: fitness landscapes and GA performance. In: proceedings of the first European conference on artificial life. MIT Press, Cambridge.

Melanie M, Holland J, Forrest S (1994) When will a genetic algorithm outperform hill climbing ? In: Cowan J, Tesauro G, Alspector J (eds) Advances in neural information processing systems. Morgan Kauffman, San Francisco.

Muhlenbein H. (1997) Genetic algorithms. In: Aarts A, Lenstra JK (Eds.) Local search in combinatorial optimization. Wiley, Chichester, pp 137-171.

Privault C, Finke G (1995) Modeling a tool switching problem on a single NC- machine. J Intelligent Manufacturing 6: 87-94.

Rajkumar K, Narendran TT (1997) A bi-criteria model for loading on PCB assembly machines. Production Planning & Control 8: 743-752.

Rajkumar K, Narendran TT (1998) A heuristic for sequencing PCB assembly to minimize set-up times. Production Planning and Control 9: 465-476.

Rupe J, Kuo W (1997) Solutions to a modified tool loading problem for a single FMM. IJ Production Research 35: 2253 –2268.

Salomon R (1996) Reevaluating genetic algorithm performance under coordinate rotation of benchmark functions; a survey of some theoretical and practical aspects of genetic algorithms. BioSystems 39: 263-278.

Salomon R (1997) Improving the performance of genetic algorithms through derandomization. In: software concept and tools, Springer, Berlin.

Schwefel HP (1995) Evolution and optimum seeking. Wiley, New York.

Sikora R (1996) A genetic algorithm for integrating lot-sizing and sequencing in scheduling a capacitated flow line. Computers and Industrial Engineering 30: 969-981.

Simões AB, Costa E (1999) Transposition versus crossover: an empirical study. In: proceedings of the genetic and evolutionary computation conference, GECCO 1999, Florida, USA.

Simões AB, Costa E (2000a) Using genetic algorithms with sexual or asexual transposition: a comparative study. In: proceedings of the congress on evolutionary computation, CEC 2000, San Diego, USA.

Simões AB, Costa E (2000b) Using genetic algorithms with asexual transposition. In: proceedings of the genetic and evolutionary computation conference, GECCO 2000, Las Vegas, USA.

Smed J, Johnsson M, Puranen M, Leipala T, Nevalainen O (1999) Job grouping in surface mounted component printing. Robotics and Computer-Integrated Manufacturing 15: 39-49.

Srinivas M, Patnaik L (1994) Genetic algorithms: a survey. IEEE Press.

Tang CS, Denardo EV (1988) Models arising from a flexible manufacturing machine, part I: minimization of the number of tool switches. Operations Research 36: 767-777.

Thierens D, Goldberg DE (1994) Convergence models of genetic algorithm selection schemes. In: Davidor Y, Schwefel HP, Manner R (eds.) Proceedings of parallel problem solving from nature . Springer, Berlin, pp 119-129.

VanHop N, Hanh PH (2000) A self- guiding derandomized genetic algorithm for setup problem in PCB assembly. In: the 2000 international conference on artificial intelligence, IC-AI'2000. Las Vegas, Nevada, USA.

VanHop N, Tabucanon MT (2001) Set-up problem for a line of machines in PCB assembly process planning: a multiple criteria decision making approach. IJ Computer Integrated Manufacturing 14: 343-352.

15 An ANTS Heuristic for the Long - Term Car Pooling Problem

Vittorio Maniezzo, Antonella Carbonaro, Hanno Hildmann

Abstract

The rising auto usage deriving from growth in jobs and residential population is making traffic congestion less tolerable in urban and suburban areas. This results in air pollution, energy waste and unproductive and unpleasant consumption of people's time. Public transport cannot be the only answer to this increasing transport demand. Car pooling has emerged to be a viable possibility for reducing private car usage in congested areas. Its actual practice requires a suitable information system support and, most important, the capability of effectively solving the underlying combinatorial optimization problem. This paper presents an application of the ANTS approach, one of the approaches which follow the Ant Colony Optimization (ACO) paradigm, to the car pooling optimization problem. Computational results are presented both on datasets derived from the literature about problems similar to car pooling and on real-world car pooling instances.

15.1 Introduction

The rising auto usage deriving from growth in jobs and residential population is making traffic congestion less tolerable in urban and suburban areas. In fact, traffic results in air pollution, energy waste and unproductive and unpleasant consumption of people's time. Public transport cannot be the only answer to this increasing transport demand, and it becomes necessary to develop alternative mobility systems which can provide intermediate solutions, in terms of costs and flexibility, between public transport and private cars.

Among the most promising solutions so far tested are the car pooling and car sharing services, which have shown to have the potential to attack a mobility segment badly satisfied by public transport.

Car pooling is a mobility service proposed and controlled by large organizations, such as large companies or public administrations, which encourage their citizens or employees to pick up colleagues while driving to/from work in order to minimize the number of private cars traveling to/from a common destination site. The benefits which can be obtained are relevant both in terms of reduction of the use of private vehicles and of the parking space required. In particular, for individuals the reduced traveling expenses and need for a second car are evident; for the community there are reductions in vehicle emissions, traffic volumes, noise and congestion; there are fuel savings and fewer accidents. Moreover, the benefits for companies that adopt a car pooling service are multiple: this service permits to maximize the use of employee parking, to encourage sociability between the employees, to reduce stress in driving to work and to improve the company image.

To foster the acceptance of a car pooling service among employees, a benefit policy is usually in order. Benefits can be very diverse and can span from reserved parking areas to reserved road lanes. A carpool lane, or high occupancy lane (HOV - High Occupancy Vehicle lane), is in fact a lane reserved to carpoolers. HOV lanes are common in the USA and Asia, less so in Europe.

Car pooling appears in two different forms: it can be either a *Daily Car Pooling Problem* (DCPP) or a *Long - term Car Pooling Problem* (LCPP).

In the case of DCPP on each day a number of users (*servers*) declare their availability for picking up and later bringing back colleagues (*clients*) on that particular day. The problem asks to assign clients to servers and to identify the routes to be driven by the servers in order to minimize a penalty due to unassigned clients, subject to time window and car capacity constraints.

In the case of LCPP each user is available to act both as a server and as a client and the objective is to define crews - or *user pools* - where each user will in turn, on different days, pick up the remaining pool members. The objective becomes that of maximizing pool sizes and minimizing the total distance traveled by all users when acting as servers, again subject to car capacity and time window constraints.

As mentioned, car pooling is enjoying a rising interest as an alternative transportation service throughout the world. For example, in Australia the New South Wales Government [8] is pursuing a range of initiatives in accordance with its integrated transport plan. One such initiative is car pooling, particularly for journeys to work in areas where public transport is limited. Also in Europe initiatives are starting. For example, Zurich and surrounding towns are involved in CARPLUS [7], a European research project on carpooling. The aim of CARPLUS is to find methods for setting up services which promote organized carpooling services.

The U.S.A. have probably the most relevant carpooling applications. In fact there are experiences, such as those in [1] or [17], and even companies, such as Trapeze software [18] or ATC [3], which propose specialized software for carpooling management support.

Despite the increasing actual interest of the application, the car pooling problem has so far received little attention from the optimization community. To the best of our knowledge, the only mathematically founded contributions are in [2] and [15] for the DCPP and in [5] or the LCPP.

This paper presents an ANTS heuristics for the LCPP, where ants construct complete problem solutions. In particular, Section 2 describes the problem and its mathematical formulation, the reduction rules we used to simplify the overall complexity, the used cost function and the derived lower bounds; Section 3 reviews the ANTS approach, while Section 4 presents at the adaptation of ANTS to the LCPP, specifying lower bounds and local search procedures. Section 5 describes some elements of the actual decision support system embedding the ANTS heuristic, that we implemented for solving real world LCPP instances. Finally, Section 6 contains the obtained computational results.

15.2 Problem Definition and Formulation

The LCPP can be defined as follows. A number n of users must reach their common destination and later on the day get back home. The problem objective is to partition the set of users into subsets, or *pools*, such that each pool member in turns will pick up the remaining members in order to drive together to the workplace and back.

Each user i enlisted in the car pooling program specifies:

- the maximal driving extratime T_i user i is willing to accept, when picking up colleagues, in addition to the time needed to drive directly from home to the workplace or back;
- the minimum time e_i acceptable for leaving home;
- the maximum time u_i acceptable for arriving at work;
- the minimum time e_i' acceptable for leaving work;
- the maximum time u_i' acceptable for getting back home.
- the capacity Q_i of his[1] car.

One further parameter associated with each user is the penalty p_i incurred when the user cannot be pooled with anyone else.

The objective is to define user pools such that as few cars as possible are used and that the routes to be driven by the drivers are as short as possible, subject to the operational constraints listed above. Note that pools are supposed to be stable over a period of time and will not change every day. This entails that the number of people in a pool will be at most equal to the capacity of the smallest car among those owned by pool members, since each member will eventually pick up all other ones.

The LCPP problem can be modeled by means of a directed graph G = (V,A), where V = {0, … , n} is the set of nodes and A the set of arcs.

[1] Here and in the following we use the convention of denoting the employee by the pronoun "he". This is only because the notation "he/she" or "his/her" seems awkward to us, and any other denotation is as arbitrary as "he".

The set V is partitioned as V = {0} ∪ V′, where 0 is the node associated with the workplace, and V′ is the subset of nodes associated with the houses of the employees.

The set A is a set of directed weighted arcs *(ij)*, where each arc *(ij)* ∈ A is associated with a non-negative cost c_{ij} and a travel time t_{ij}.

Figure 15.1 shows a part of a feasible solution of problem instance LCPP50 (see section 15.6 for a description of the instances), where the optimal pools are shown only for upper-right employees.

The LCPP problem can be formulated as an integer program in different ways, each of which permits to make evident different characteristics of the problem itself. In the following we will present two of them.

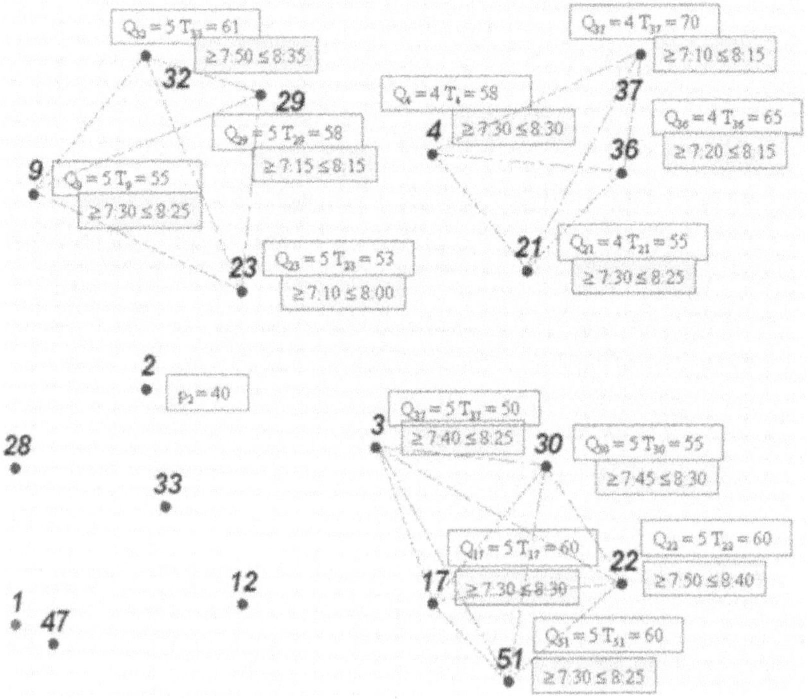

Fig. 15.1. Example of user data and of a part of a LCPP solution

15.2.1 The objective function

The LCPP as defined above is actually a multiobjective problem, requiring to:

1. maximize car usage, thereby minimizing the number of cars travelling to/from work;
2. minimize the length of the path to be driven by each employee, when acting as a driver.

However, the analysis of the problem structure suggests that it is possible to combine these two objectives in a single objective function, as follows.

Let k be a pool of clients. Each of them, on different days, will use his car to pick up the other pool members and go to work (and later come back), thus he has to define an *Hamiltonian path* on the partial subgraph of G identified by k, starting from the node associated to his house, passing through all other nodes and ending at the workplace.

Let *Hpath*(i,k) be such an Hamiltonian path, starting from $i \in k$, connecting all $j \in k \backslash \{i\}$ and ending in 0.

Hpath(i,k) is a *feasible path* iff $|k| \le Q_j$, $\forall j \in k$, and all user constraints are met.

The *minimum path*, min_path(i,k), for $i \in k$ is the shortest feasible Hamiltonian path for i.

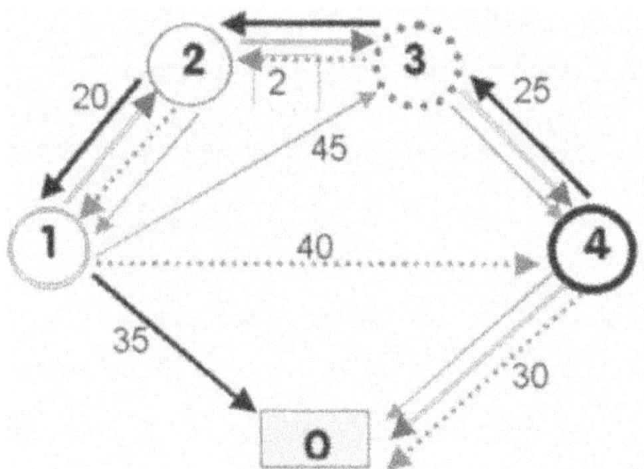

Fig. 15.2. Elements defining the cost of a pool

Computing the example in Figure 15.2:

$k = \{1,2,3,4\}$
min_path(1,k) = (20+25+25+30) = 100
min_path(2,k) = (20+45+25+30) = 120
min_path(3,k) = (25+20+40+30) = 115
min_path(4,k) = (25+25+20+35) = 105
cost(k) = (100+120+115+105)/$|k|$ = 110

In the computation of the cost of a pool, it is assumed that the shortest paths are chosen. The cost of a pool k is then defined to be:

$$cost(k) = \begin{cases} \sum_{i \in k} min\_path(i,k)/|k| & if\ |k| > 1 \\ c_{i0} + p_i & otherwise \end{cases} \quad (15.1)$$

Notice that a pool of one single person has a cost increased by a penalty, whose amount is associated with him. For example, in Figure 15.1 the pool comprising the single employee 2 has a penalty value equal to 40.

The cost of a complete LCPP solution σ is then defined to be the sum of the costs of the pools in it, that is, cost(σ) = $\sum_{k \in \sigma}$ cost(k). This view, originally proposed in [13], optimizes at the same time both objective functions. In fact, provided that the penalty p_i of a client is sufficiently bigger than 0, it is more convenient to pool clients together than to leave them alone (see subsection 15.2.3 for a discussion). Moreover, pool costs are directly minimized by the objective function.

In Figure 15.2 we graphically show an example of minimum paths and we indicate how to compute the cost of the presented pool.

15.2.2 A four-indices mathematical formulation

The first proposed formulation makes use of the following variables:

- x_{ij}^{hk} : binary variable equal to 1 iff arc $(ij) \in$ A is in the shortest Hamiltonian path of a server h of a pool k.
- y_{ik}: binary variable equal to 1 iff client i is in pool k.
- ξ_i: binary variable equal to 1 iff client i is not pooled with any other client;
- s_i^h : non negative integer variable denoting the pick-up time of client i by server h.
- f^h: non negative integer variable denoting the arrival time of server h at the workplace.
- l_i: latest arrival time of client i.
- K: index set of all pools.
- C: index set of all clients.

$$\text{(LCPP)} \quad z_{LCPP} = min \sum_{k \in \mathbf{K}} \left(\sum_{h \in \mathbf{C}} \sum_{(ij) \in \mathbf{A}} c_{ij} x_{ij}^{hk} \Big/ \sum_{i \in \mathbf{C}} y_{ik} \right) + \sum_{i \in \mathbf{C}} p_i \xi_i \quad (15.2)$$

$$s.t. \quad \sum_{j \in \mathbf{C} \setminus \{h\}} x_{ij}^{hk} = y_{ik} \qquad i,h \in \mathrm{C},\ k \in \mathrm{K} \quad (15.3)$$

$$\sum_{i \in C \setminus \{0\}} x_{ij}^{hk} = y_{jk} \qquad j,h \in C,\ k \in K \qquad (15.4)$$

$$\sum_{j \in C} x_{ji}^{hk} = \sum_{j \in C} x_{ij}^{hk} \qquad i,h \in C,\ k \in K \qquad (15.5)$$

$$\sum_{k \in K} y_{ik} + \xi_i = 1 \qquad i \in C \qquad (15.6)$$

$$\sum_{(ij) \in A} x_{ij}^{hk} \le Q_h \qquad h \in C,\ k \in K \qquad (15.7)$$

$$\sum_{(ij) \in A} t_{ij} x_{ij}^{hk} \le T_h \qquad h \in C,\ k \in K \qquad (15.8)$$

$$s_j^h - s_i^h \le t_{ij} + M\left(1 - \sum_{k \in K} x_{ij}^{hk}\right) \qquad (ij) \in A,\ h \in C \qquad (15.9)$$

$$f^h \ge s_i^h + t_{i0} - M\left(1 - \sum_{k \in K} x_{i0}^{hk}\right) \qquad i,h \in C \qquad (15.10)$$

$$f^h \le l_i + M\left(1 - \sum_{k \in K} \sum_{j \in \Gamma_i} x_{ij}^{hk}\right) \qquad i,h \in C \qquad (15.11)$$

$$s_i^h \ge e_i \qquad i,h \in C \qquad (15.12)$$

$$x_{ij}^{hk} \in \{0,1\} \qquad h \in C,\ k \in K,\ (i,j) \in A \qquad (15.13)$$

$$y_{ik} \in \{0,1\} \qquad i \in C,\ k \in K \qquad (15.14)$$

$$\xi_i \in \{0,1\} \qquad i \in C \qquad (15.15)$$

$$f^h \ge 0 \qquad h \in C \qquad (15.16)$$

where Γ_i denotes the adjacency set of vertex $i \in C$.

Equations (15.3) force a client i to be declared to be in pool k, if there is a path originated in h going from i to j, equations (15.4) do the same for client j and equations (15.5) are continuity constraints. Equations (15.6) force each client to be assigned to a pool or to be penalized, while constraints (15.7) and (15.8) are the car capacity and the maximum travel time constraints, respectively.

Constraints (15.9) and (15.12), where M is a big constant, collectively set feasible pick-up times, compatible with the paths denoted by the x and y variables, while constraints (15.10) and (15.11) set minimum and maximum values of feasible arrival times, respectively.

Constraints (15.13), (15.14) and (15.15) are binary integrality constraints while constraints (15.16) are the non-negativity constraints for the f^h variables.

15.2.3 A set partitioning formulation

A second formulation can be designed using binary variables which represent whole pools. In this case, assuming it is possible to a priori enumerate all feasible pools, the problem is that of selecting the set of pools that collectively satisfy the transportation request the best: the problem turns into a Set Partitioning problem.

Let P be the index set of all feasible pools, that is of all pools that meet capacity and time windows constraints. Each feasible pool k contains a number of nodes of V' ranging from 1 to $min_{j\in k}Q_j$. Let $P_i \subset P$ be the subset of pools containing node i, $i \in V'$, and let c_l be the cost of pool $l \in P$.

The cost c_l is computed as follows. For each member of pool l we compute both the minimum cost feasible path starting from his house, connecting all other members and reaching the workplace, and the minimum cost return path leading the same members to their houses. The average, over the number of members of pool l, of the sum of the costs of the outward and return trips gives the cost c_l.

Let x_l be a binary variable that is equal to 1 if and only if pool l is in the solution. A set partitioning formulation of the problem is as follows.

$$\text{(SP)} \qquad z(SP) = \min \sum_{l \in P} c_l x_l \tag{15.17}$$

$$\text{s.t.} \quad \sum_{l \in P_i} x_l = 1 \qquad \forall i \in V' \tag{15.18}$$

$$x_l \in \{0,1\} \qquad \forall l \in P \tag{15.19}$$

Notice that single - member pools are considered, as well as pools of any feasible cardinality. Notice moreover that the used objective function implicitly maximizes pool sizes, since a pool in a solution satisfies a number of constraints equal to the number of its members, therefore the marginal cost of a user is lower in a big size pool than in a smaller one as long as intra-pool distances are smaller than user penalties.

15.2.4 Reduction rules

It is possible to remove from the graph representing an LCPP instance a number of arcs which cannot belong to any feasible solution.

15.2.4.1 Definition

An arc $(ij) \in A$ can be part of a feasible LCPP solution (is a *feasible arc* iff:

$$(e_i + t_{ij} + t_{j0}) \leq u_i \tag{15.20}$$

$$(e_j + t_{j0}) \leq u_i \tag{15.21}$$

$$(t_{ij} + t_{j0}) \leq t_{i0}+T_i \tag{15.22}$$

Since the problem graph is potentially fully connected, the number of arcs grows as n^2, with n the number of nodes. Many of the arcs are however not feasible and should not be considered when trying to construct a solution. To filter those out and thus reduce the complexity of the problem we used the three constrains on feasibility presented above, which results in the following reduction rules for arc (ij), $\forall(ij)\in$ A.

15.2.4.2 Reduction 1

Unless at least one path connecting node i with node j and the workplace can meet the latest arrival time of i or of j, arc (ij) can be removed from A.

15.2.4.3 Reduction 2

Unless at least one path connecting node i with node j and the workplace can meet the maximum driving time of i, arc (ij) can be removed from A.

15.2.5 Lower bounds

Different lower bounds can be derived from the presented formulations.

Two simple bounds can be derived from formulation LCPP following the observation that

$$min \sum_{k\in \mathbf{K}} \left(\frac{\sum_{h\in \mathbf{C}} \sum_{(ij)\in A} c_{ij} x_{ij}^{hk}}{\sum_{i\in \mathbf{C}} y_{ik}} \right) \geq min \sum_{k\in \mathbf{K}} \left(\min_{h\in \mathbf{C}} \left[\sum_{(ij)\in A} c_{ij} x_{ij}^{hk} \right] \right) \tag{15.23}$$

By making this substitution in the objective function, two computationally efficient bounds can be derived as follows.

15.2.5.1 Lower bound LB1

We relax path time constraints (15.7) (thus all constraints (15.8) through (15.11)), capacity (15.6) and connectivity (15.4).

LB1 is the cost of the least cost partial graph where:

- at most 2 arcs are incident to each client;
- at most n arcs are incident to the workplace.

This problem can be solved, with minor modifications, as a balanced transportation problem.

15.2.5.2 .Lower bound LB2

We relax time constraints (15.7) (thus all constraints (15.8) through (15.12)), capacity (15.6) and max-degree constraints (15.3). Degree constraints can be later reintroduced in a Lagrangean fashion.
Lower bound LB2 is the cost of a minimum spanning arborescence rooted at the workplace, obtained using as cost associated with the arcs the reduced costs produced by lower bound LB1.

15.3 The ANTS metaheuristic

ANTS (an acronym of *Approximated Non-deterministic Tree Search*) is an extension of the Ant System proposed in [10], and described in the ACO chapter of this collection, which specifies some underdefined elements of the original algorithm, such as the attractiveness function to use or the initialization of the trail distribution. This turns out to be a variation of the general ACO framework that makes the resulting algorithm similar in structure to tree search algorithms. In this paper we assume that the reader is already familiar with the general ACO structure, and in particular, with the Ant System algorithm. We refer the reader to the ACO chapter and to [4] and [12] for details on these algorithms. In fact, the essential trait which distinguishes ANTS from a tree search algorithm is the lack of a complete backtracking mechanism, which is substituted by a probabilistic one, hence the name. In the following, we will outline two distinctive elements of the ANTS paradigm within the ACO framework, namely the attractiveness function and the trail updating mechanism. See [11] and [12] for a more detailed discussion.
In ACO and ANTS alike an ant is defined to be a simple computational agent, which iteratively constructs a solution for the problem to solve. Partial problem solutions are seen as states; each ant moves from state ι to state ψ, corresponding to a more complete partial solution. At each step σ, each ant k computes a set A_{σ}^{k} of feasible expansions to its current state, and moves to one of these in probability. For ant k, the probability $p_{\iota\psi}^{k}$ of moving from state ι to state ψ depends on the combination of two values:

i) the *attractiveness* η of the move, as computed by some heuristic indicating the *a priori* desirability of that move. Specifically, a variable $\eta_{\iota\psi}$ quantifies the attractiveness of moving from state ι to state ψ.

ii) the *trail level* τ of the move, indicating how proficient it has been in the past to make that particular move: it represents therefore an *a posteriori* indication of the desirability of that move. Specifically, a variable $\tau_{\iota\psi}$ quantifies the past proficiency of moving from state ι to state ψ.

The elements where ANTS essentially differs from general ACO are in how to compute attractiveness and trail levels.

15.3.1 Attractiveness

The attractiveness of a move can be effectively estimated by means of lower bounds (upper bounds in case of maximization problems) to the cost of the completion of a partial solution. In fact, if a state corresponds to a partial problem solution it is possible to compute a lower bound to the cost of a complete solution containing it. Therefore, for each feasible move $(\iota\psi)$, it is possible to compute the lower bound to the cost of a complete solution containing ψ: the lower the bound the better the move. Since large part of research in combinatorial optimization is devoted to the identification of tight lower bounds for the different problems of interest, good lower bounds are usually available.

A further advantage of lower bounds is that in many cases the values of the decision variables, as appearing in the bound solution, can be used as an indication of whether each variable will appear in good solutions. This provides an effective way for initializing the trail values. For more details see [12].

Notice finally that the attractiveness computed this way corresponds to a look-ahead procedure, where the cost of the incumbent partial solution is increased by a quantity depending on the possible future expansions.

15.3.2 Trail update

A good trail updating mechanism avoids stagnation, the undesirable situation in which all ants repeatedly construct the same solutions making any further exploration in the search process impossible. Stagnation derives from an excessive trail level on the moves of one solution, and can be observed in advanced phases of the search process, if parameters are not well tuned to the problem.

The trail updating procedure evaluates each solution against the last k ones globally constructed by ANTS. As soon as k solutions are available, their moving average $\bar{z}$ is computed; each new solution z_{curr} is compared with $\bar{z}$ (and then used to compute the new moving average value). If z_{curr} is lower than $\bar{z}$, the trail level of the last solution's moves is increased, otherwise it is decreased. Since, as in the case of the TSP, a state ι can be represented by the last node i visited by the ant, and the next state ψ is defined by the solution containing i plus the arc connecting i to the new visited node, say j, it is possible to represent states with the nodes corresponding to the last visited node of the associated partial solution. This observation is used in formula (15.24), which specifies how trail update is implemented:

$$\Delta\tau_{ij} = \tau_0 \cdot \left(1 - \frac{z_{curr} - LB}{\bar{z} - LB}\right) \tag{15.24}$$

where τ_0 is a user-defined parameter.

Based on the described elements, the ANTS metaheuristic is the following.

```
1. compute a (linear) lower bound LB to the problem to solve;
   initialize τij (∀i,j) with the optimal bound variable values;
2. For k=1,m (m= number of ants) do
        repeat
2.1          compute ηij ∀(ij);
2.2          choose in probability the state to move into;
2.3          append the chosen move to the k-th ant's tabu list;
        until ant k has completed its solution;
2.4     carry the solution to its local optimum;
     end for
3. For each ant move (ij)
        compute Δτij and update the trace matrix;
4. If not(end_test) goto step 2.
```

It can be noticed that the general structure of the ANTS algorithm is closely akin to that of a standard tree-search procedure. At each stage we have in fact a partial solution which is expanded by branching on all possible offspring; a bound is then computed for each offspring, possibly fathoming dominated ones. The current partial solution is selected on the basis of lower bound considerations among those associated to the surviving offspring.

By simply adding backtracking and eliminating the Montecarlo choice of the node to move to, we revert to a standard branch and bound procedure. An ANTS code can therefore be easily turned into an exact procedure.

15.4 ANTS approaches for the LCPP

In this section we describe an ANTS procedure for solving the LCPP. To this end, it is necessary to adapt the general ANTS scheme to the LCPP, specifying two components: the lower bound to use to compute the η values and the local search to use at the end of the loop at step 2.

15.4.1 Attractiveness quantification

A valid lower bound for LCPP can be computed in different ways, for example as the solution of a linear assignment problem or of a minimal spanning tree problem suitably defined over graph G, as introduced in Section 2. Using different bounds guarantees the possibility of analyzing different aspects of the problem, thus obtaining diverse contributions. In the case of linear lower bounds, as the two mentioned ones, it is possible to integrate the specific contribution of each single bound in an additive framework [6], where each successive bound contributes in refining the estimate of the dual costs associated to the original problem constraints, thus of the reduced costs of the problem variables.

The complete attractiveness quantification procedure becomes therefore a procedure that computes the $\eta_{\iota\psi}$ as follows. First, given the partial solution corresponding to state ψ, the corresponding cost $z\psi$ is calculated. Then, all nodes already in ψ and of all arcs incident into them are deleted from the problem graph G. Finally, a lower bound is computed on the resulting partial subgraph, obtaining a cost z_{LB}. The attractiveness of the move $(\iota\psi)$ is then defined to be $\eta_{\iota\psi} = z\psi + z_{LB}$.

15.4.2 Local optimization

The local optimization module makes use of five different neighborhoods of the current solution, thus it implements a *Variable Neighborhood Search* procedure, as proposed in [16]. The overall structure is very simple, as it consists of a main loop considering each neighborhood in turns. Each neighborhood is used to obtain its local optimum, then the next neighborhood is considered. The optimization stops when no neighborhood is capable of improving the current solution.

The five implemented neighborhoods are the following.

1. *split(pool1)*:
   Calculate the resulting cost when pooling any number of members of *pool1* into a new pool. If the cost of the two new pools decreases the overall solution cost, then accept the new solution.

2. *merge(pool1, pool2)*:
   Calculate the cost of a pool that contains all members of *pool1* and *pool2*. If the resulting cost is lower than the sum of the costs of *pool1* and *pool2*, then accept the new solution.

3. *insert(pool1)*:
   Calculate the cost of inserting any pool with only one member into *pool1*. If the overall cost decreases, then accept the new solution.

4. *swap_1vs1(pool1, pool2)*:
   Calculate the cost of swapping any two members of *pool1* and *pool2*. If the overall cost decreases, then accept the new solution.

5. *swap_1vs0(pool1, pool2)*:

Calculate the resulting cost when moving any member from *pool1* to *pool2*. If the overall cost decreases, then accept the new solution.

Insert is a particular case of merge, but it is considered separately for computational reasons.

15.5 A DSS for the LCPP

The method described in this paper was designed for actual use in real world settings. To make use practical, the ANTS algorithm was embedded in a decision support system, whose general architecture is presented in figure 15.3. The architecture is a standard DSS architecture [14] [19], consisting of a graphical user interface (GUI), a data management subsystem and a model management subsystem.

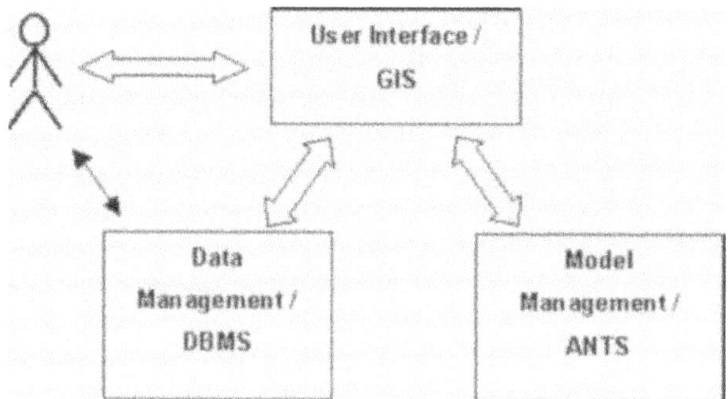

Fig. 15.3. DSS architecture

In the case of our specific application, the three subsystem were implemented as follows.

The Model Management subsystem contains two main algorithms: ANTS and a bucket-based implementation of the Dijkstra's shortest path algorithm.
Dijkstra is used, once the set of all clients to pool is defined, to identify the shortest time path between each pair of clients (subject to minor operational constraints, such as which types of road to avoid, if any, or maximum global travel time or distance), and between each client and the workplace. The result of this computation is stored in a time/distance matrix which will then be used by ANTS.
ANTS uses the matrix defined above, and data about user constraints, in order to generate the pools.

The Data Management subsystem is implemented by means of a standard relational DBMS and of spatial management structures. The relevant functionalities

were ensured by making use of a *Geographic Information System*, specifically ESRI's ArcView 3.1, which maintains relational tables in .dbf format and spatial data in shapefile format.
Relational data contains all elements relevant to specify user constraints, plus some accounting statistics. Spatial data consists of a detailed road map of the region of interest (a graph of about 150,000 arcs and 120,000 nodes) and of the location of clients' houses and of the workplace.

The Graphic User Interface was implemented as a customization of ArcView's GUI. It permits to insert all relevant data (or to download them from a company legacy system, in case they are already present), to access the DBMS and to export data and import results from the Model Management subsystem. A screenshot of our GUI is shown in figure 15.4.

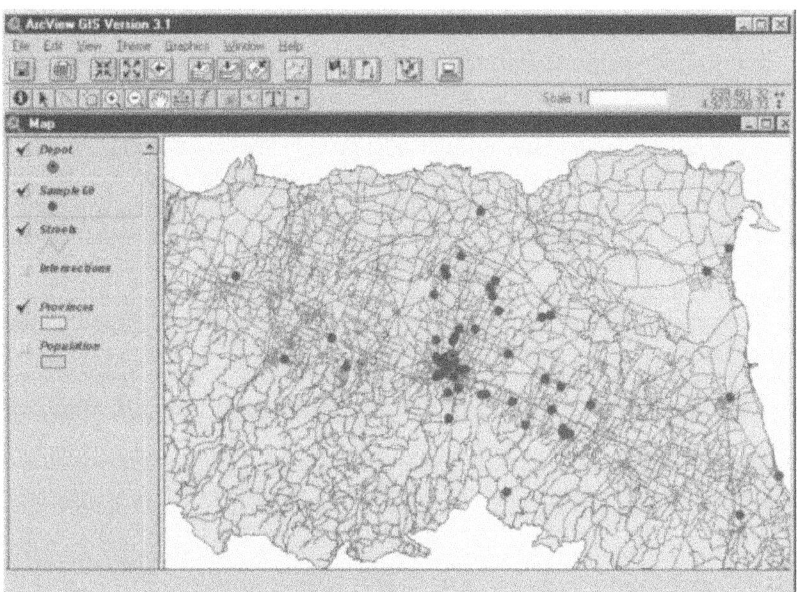

Fig. 15.4. The user interface

The only nonstandard characteristic worth mentioning is about data flow. As shown in Figure 15.3 in fact, there is no direct link between the DBMS and the optimization modules, but every communication is mediated by the GUI. This is to accommodate for data type conversions which increase computational efficiency. Moreover, there is a (thin) direct link between the user and the DBMS. This is because the user is assumed to be the system administrator, therefore it is possible for him to bypass the GUI and work directly on the stored relations.

15.6 Computational results

This section reports about the computational results obtained by applying ANTS to different sets of test problems. ANTS was implemented in C, under Microsoft Visual Studio 6, and all results were obtained running the code on a Windows ME Pentium III 733 MHz machine equipped with 512 Mb of RAM.

The computational testing was carried out on three sets of structurally different problem instances, named LCPP, B and I, respectively.

The LCPP instances were originally derived from VRP hard instances presented in the literature and adapted as DCPP instances [2]. We could use the same data files used in [2], simply ignoring the additional elements that differentiate DCPP from LCPP.

The I instances were also used in [2]. These are real-world instances, in that they are derived by taking random subsets from a big-size case study proposed by a company in northern Italy.

The B instances were generated by us as follows. We studied the space and time distribution of clients in the I instances, as a function of the distance from the workplace and of the distribution of the density of the population in the region, and designed a sample generator following a gravitational model which produces instances having the same aggregate distribution found in the I instances. We then applied this generator to the region around Bologna, in northern Italy. This permits to obtain instances where the clients are scattered all around the workplace, which is not the case in the I instances because there the workplace is located close to a major lake, thus the clients are all on one side of the workplace.

To test the algorithm, we underwent a parameters setting stage. The parameters to define (see an ACO introduction for those undefined in this paper) are: m, number of ants in the population, □, the evaporation coefficient, and □, the relative importance of visibility versus trail. To set them, we arranged a *coeteris paribus* test campaign, where the tested values were: $m \in \{\frac{1}{20}n, \frac{1}{6}n, \frac{1}{3}n, n\}$, □ $\in \{0, 0.05, 0.1\}$ and □ $\in \{0, 0.25, 0.5, 0.75, 1\}$. The tests, whose details can be found in [9] showed that a promising setting is $m = \frac{1}{6}n$, □ $= 0$ and □ $= 0.5$. It could seem surprising that the best value for □ is 0, but consider that the dynamic scaling implemented by formula (15.24) prevents unlimited accumulation of trail, and in fact makes evaporation non essential.

The algorithm was run three times on each problem instance, each run lasting 1200 seconds. The following tables report the best results obtained by these runs. In particular, the tables show the following columns (Table-15.1):

Table 15.1. The columns in the tables.

prob	problem name: the number at the end of the name indicates the number of clients in the instance;
reduct	number of arcs removed by the reduction rules;
Lb1	lower bound LB1;
Lb2	lower bound LB2;
lb_{zub}	upper bound provided by the LB2 heuristic;
T_{zub}	cpu time to compute the lower bounds and the LB2 heuristic;
Zub	best result produced by ANTS;
Avg	average of the best results found in the three runs;
Avg_t	average of the three CPU times taken to get the results in avg;
Bks	best known solution value.

Notice that the results presented in columns *bks* were themselves obtained by ANTS. In fact for each problem we kept track of the overall best solution found obtained trying to solve it (we actually ran the algorithm many times on each problem: for parameters setting, to test the impact of longer CPU times etc. Most of the results presented in column *bks* were obtained by 7200 sec. long runs).

Table 15.2 shows the results obtained on the LCPP instances. On these fictitious problems, the reduction rules often remove about 80% of the arcs, the actual data ranging from 80.91% in LCPP150 to 55.93% in LCPP120. The lower bounds are quite poor, but a significant improvement is expected by their combination in an additive framework, as introduced in Section 15.2.4. However, the indication they provide is essential for the correct functioning of ANTS. Moreover, they provide a means to quickly initialize an upper bound, which unfortunately proves to be of mediocre quality. ANTS itself on the contrary is capable of consistently producing good quality solutions.

Table 15.2. ANTS results on the set of LCPP test problems

Prob	reduct	lb1	lb2	lb zub	t zub	zub	avg	avg t	bks
Lcpp50	2015	691	939	2458	0.06	1327.50	1327.50	1	1327.50
Lcpp75	4460	734	1209	3718	0.00	1354.00	1354.00	372	1354.00
Lcpp100	8034	770	1425	5098	0.06	1564.43	1567.20	487	1528.27
Lcpp120	8054	469	989	12354	0.06	2352.98	2388.32	520	1882.72
Lcpp150	18204	921	1733	7524	0.16	2130.00	2136.89	493	1984.92
Lcpp170a	17475	843	1724	15738	0.22	3348.58	3388.20	161	2863.08
Lcpp170b	19966	920	1687	12652	0.16	3032.88	3075.47	377	2703.07
Lcpp195	22916	956	1921	17966	0.22	3725.47	3768.68	707	3140.93
Lcpp199	31787	1032	1995	9822	0.22	2657.92	2672.64	155	2507.30
Lcpp225	40828	1137	2145	11258	0.28	3007.75	3028.11	518	2841.15

Table 15.3 shows the results on what turned out to be the hardest instances. Already the reduction rules are much less effective than on the LCPP set, being able to remove from 19.68% of the arcs of B50a to 52.44% of the arcs of B400. This trend is maintained on all columns, where the lower bounds provide comparatively lower values and the upper bounds comparatively worse values than in the LCPP

case. Also ANTS suffers on some instances, even though usually the distance between best known solution and best ANTS solution (often obtained in one fourth of the cpu time) is not great.

Table 15.3. ANTS results on the set of B test problems

prob	reduct	lb1	lb2	lb zub	t zub	zub	avg	avg t	bks
B50a	492	685	901	3754	0.00	1128.32	1136.39	353	1106.25
B50b	569	679	857	3790	0.05	1222.43	1222.92	404	1181.42
B50c	577	606	690	3944	0.00	1140.10	1143.27	242	1072.38
B75a	2268	719	948	5608	0.00	1476.67	1499.04	450	1426.40
B75b	2342	589	1012	5682	0.00	1518.35	1530.93	432	1398.95
B75c	2535	989	1230	5732	0.05	1745.00	1750.42	432	1658.43
B100a	4561	901	1384	7434	0.06	2153.07	2153.89	201	2012.67
B100b	4578	1020	1287	7376	0.06	2083.47	2091.90	632	1920.13
B100c	5060	1034	1286	7474	0.11	2246.50	2255.87	372	2090.77
B150a	11252	1440	1746	11288	0.11	3144.28	3153.75	134	2838.42
B150b	10381	1001	1569	11018	0.11	2895.38	2903.94	602	2528.67
B200a	20950	1966	2441	15530	0.22	4669.90	4676.71	822	4161.33
B200b	19633	1852	2292	15218	0.27	4218.73	4266.25	558	3795.98
B250	32703	2121	2854	19124	0.39	5233.03	5259.63	550	3852.60
B300	45870	1927	2996	22492	0.49	5702.13	5712.11	894	4848.98
B400	83903	2456	3410	30342	0.98	7906.56	7906.56	78	6642.48

Table 15.4 shows the results on the only real-world data we could use. It appears that the difficulty of these instances is more similar to that of the LCPP set than to that of the B set.

Table 15.4. ANTS results on the set of I test problems

prob	reduct	lb1	lb2	lb zub	t zub	zub	avg	avg t	bks
I50a	1507	331	388	1134	0.00	532.95	532.95	432	532.95
I50b	1581	296	365	1138	0.00	458.65	458.67	248	458.65
I50c	1541	335	407	1152	0.06	569.23	569.23	277	569.23
I75a	4319	474	514	1500	0.05	696.97	696.97	405	696.97
I75b	4409	529	556	1504	0.06	814.18	814.18	56	805.28
I75c	4278	469	544	1530	0.00	657.37	657.37	614	657.37
I100a	8321	609	685	2078	0.06	855.35	858.88	714	849.10
I100b	8300	578	652	2058	0.11	901.20	904.29	224	891.17
I100c	8065	581	715	2094	0.05	878.20	881.65	696	877.43
I150A	18933	856	1046	3352	0.17	3144.28	3144.28	291	1320.68
I150B	19023	863	1010	3314	0.16	1360.50	1360.50	3	1288.60
I200A	34170	995	1354	4560	0.27	1737.46	1737.46	6	1609.52
I200B	34328	1112	1430	4568	0.22	1893.98	1893.98	7	1779.02
I250	53759	1367	1898	6202	0.44	2479.92	2479.92	10	2286.57
I300	76360	1498	2110	7564	0.61	2754.07	2754.07	19	2636.50
I400	136101	1868	2666	10024	1.26	3603.48	3603.48	32	3253.35

Again, the reduction rules are able to often remove more than 80% of the arcs (85.43 in the case of I200a) and the quality of lower and upper bounds improves.

Most significantly, the quality of the best ANTS solution is never too far from the best known solution. Notice how the best ANTS solution for the bigger instance is usually obtained in the earliest stages of the search process. This indicates that the allowed CPU time was not sufficient to permit an identification of an effective trail distribution, thus the randomization mechanism implemented within ANTS still worked as a diversification procedure, preventing the (premature) focusing on a specific subspace.

15.7 Conclusions

This paper presented an application of an ANTS metaheuristic to the Long-term Car Pooling Problem, with the objective of maximizing the size of user pools and of minimizing the cost of the paths to be driven by the drivers.

We presented two different mathematical formulations of the problem, and we used them to derive lower bounds to the cost of an optimal solution of the problem. These lower bounds, albeit rather loose, provide the basis for the definition of an ANTS approach to the LCPP. Further components of the algorithm include an effective local search procedure, in this case implemented as a variable neighborhood local search, and reduction rules to a priori reduce the search space.

The algorithm presented represents the first attempt to use acknowledged optimization techniques for solving the LCPP, thus we had to define also a representative problem base for testing. To this end, we adapted two testsets originally proposed for a similar problem, the Daily Car Pooling Problem, consisting of some real world instances and of adaptations of VRP instances very well-known in the literature. Moreover, we generated some new instances using a detailed road map and a gravitational model for user distribution defined after the real world case.

The computational results are interesting, even though we can reliably compare the ANTS results only against themselves: the lower bounds are in fact still loose and the initialization heuristic is not effective.

Our research effort on LCPP is however still ongoing. We are currently working trying to combine the two lower bounds presented in an additive framework, and to use the resulting bound as an initialization of a dual-based procedure for formulation SP. This should provide an effective lower bound. Moreover, we are implementing an exact procedure to assess the distance from optimality, at least for the smallest problems, and we are modifying the local search procedure in order to obtain a tabu search and other local descent based heuristics to be used for comparison.

References

1. Valley Metro Regional Public Transportation Authority. http://www.valleymetro.maricopa.gov/Rideshare/index.html, 2000.
2. R. Baldacci, V. Maniezzo, and A. Mingozzi. An exact algorithm for the car pooling problem. In Proceedings of CASPT - 2000, 8th International Conference "Computer-Aided Scheduling of Public Transport", 2000.
3. ATC Corporate. http://www.intelitran.com, 2000.
4. M. Dorigo, V. Maniezzo, and A. Colorni. The ant system: optimization by a colony of cooperating agents. IEEE Transactions on Systems, Man, and Cybernetics-Part B, 26(1):29–41, 1996.
5. D.Vigo. Heuristic approaches for the car pooling problem. Technical Report DEIS - OR - 00 - 02, University of Bologna, Italy, 2000.
6. M. Fischetti and P. Toth. An additive bounding procedure for combinatorial optimization problems. Operations Research, 37, 1989.
7. Telematics for car pooling. http://195.65.25.24, 1999.
8. The New South Wales Government. http://www.rta.nsw.gov.au./index.htm, 1999.
9. H. Hildmann. An ants metaheuristic to solve car pooling problems. M.Sc. AI thesis, University of Amsterdam, Faculty of Science, Department of Artificial Intelligence, 2001.
10. V. Maniezzo. Exact and approximate nondeterministic tree-search procedures for the quadratic assignment problem. INFORMS Journal on Computing, 11(4):358 – 369, 1999.
11. V. Maniezzo and A. Carbonaro. An ants heuristic for the frequency assignment problem. In Proceedings of MIC'99, 1999.
12. V. Maniezzo and A. Carbonaro. Ant Colony Optimization: an Overview, volume Recent Advances in Metaheuristics. Kluwer, 2001.
13. V. Maniezzo, A. Carbonaro, D. Vigo, and H. Hildmann. An ants algorithm for the car pooling problem. In Proceedings of ANTS'2000, 2nd International Conference "Ant Colony Optimization", 2000.
14. V. Maniezzo, Mendes I., and Paruccini M. Decision support for siting problems. Int. Journal of Decision Support Systems, 23:273 – 284, 1998.
15. A. Mingozzi, R. Baldacci, and V. Maniezzo. Lagrangean column generation for the car pooling problem. Technical Report WP-CO0002, University of Bologna, S.I., Cesena, Italy, 2000 (*Operations Research*, to appear).
16. N. Mladenovic and P. Hansen. Variable Neighborhood Search, volume Recent Advances in Metaheuristics. Kluwer, 2001.
17. Land of Sky Regional Council. Commute connections project report., 1999.
18. Inc. Trapeze Software Group. http://www.trapezesoftware.com, 2001.
19. E. Turban. Decision support and expert systems. Macmillan Publishing Company, 1990.

16 Genetic Algorithms in Irrigation Planning: A Case Study of Sri Ram Sagar Project, India

K Srinivasa Raju and D Nagesh Kumar

Abstract

The present study deals with application of Genetic Algorithms (GA) in the field of irrigation planning. The GA technique is used to achieve efficient operating policy with the objective of maximum net benefits for the case study of Sri Ram Sagar Project, Andhra Pradesh, India. Constraints include continuity equation, land and water requirements, crop diversification considerations, and restrictions on storage capacities. Penalty function approach is used to convert constrained problem into unconstrained one. For fixing GA parameters, namely, crossover and mutation probabilities, the model is run for 7 values of crossover and 6 values of mutation probabilities. It is found that appropriate parameters such as number of generations, population size, crossover probability, and mutation probability are 200, 50, 0.6 and 0.01 respectively for the present study. Maximum benefits obtained by LP solution is 2.4893 Billion Rupees where as these are 2.3903 Billion Rupees by GA (with a fitness function value of 2.3678 Billion Rupees). Results obtained by GA are compared with Linear Programming solution and found to be reasonably close.

Keywords: Genetic Algorithms, Irrigation Planning, Linear Programming, Sri Ram Sagar Project, India

16.1 Introduction

Many real world problems involve two types of problem difficulties i.e., multiple, conflicting objectives (instead of a single optimal solution, competing goals give rise to a set of compromise solutions denoted as pareto-optimal) and a highly complex search space which is difficult to be solved by exact methods. Thus efficient optimization strategies are required that can deal with both difficulties. *Genetic Algorithms* (GA) possess several characteristics that are desirable for this type of problems and make them preferable to classical optimization methods. Goldberg (1989) described the nature of genetic algorithms of choice by combining a *Darwinian survival of the fittest* procedure with a structured, but randomized, information exchange to form a canonical search procedure that is capable of addressing a broad spectrum of problems. Genetic Algorithms are search procedures based on the mechanics of natural genetics and natural selection. They combine the concept of artificial survival of fittest with genetic operators abstracted from nature to form a robust search mechanism. Goldberg (1989) identified the following significant differences between GAs and more traditional optimization methods:

- GAs work with a coding of the parameter set, not with the parameters themselves.
- GAs search from a population of points, not a single point.
- GAs use objective function information, not derivatives or other auxiliary knowledge.
- GAs use probabilistic transitions rules, not deterministic rules.

16.1.1 Working Principle of Genetic Algorithms

Any nonlinear optimization problem without constraints is solved using Genetic Algorithms involving basically three tasks, namely, Coding, Fitness evaluation and Genetic operation. First of all, decision variables are identified for the given optimization problem. These decision variables are coded into some string like structures. For coding the decision variables, binary coding is used. This coded string is called Chromosome. The length of chromosome depends on the desired accuracy of the solution. It is not necessary to code all the decision variables in the same sub string length.

Fitness function is first derived from objective function and is used in successive genetic operations. Genetic operators require that the fitness function should be nonnegative. If the problem is of maximization, fitness function is taken as directly proportional to the objective function. The fitness function value of a string is known as the string's fitness.

Once the fitness of each string is evaluated, the population is operated by three common operators for creating new population of points. They are reproduction, crossover and mutation. Reproduction selects good strings in a population and forms a mating pool. In this paper Roulette wheel simulation is used for the selection of good strings. In crossover operator, two strings are picked from the mating pool at random and some portions of the strings are exchanged between the strings. A single point crossover operation is performed by randomly choosing a crossing site along the string and by exchanging all bits on the right side of the crossing site. The mutation operator changes 1 to 0 and 0 to 1 with a small mutation probability, p_m, within the string. Mutation creates points in the neighborhood of the current point, which help in local search around the current solution. It is also used to maintain diversity in the population.

The newly created population using the above operators is further evaluated and tested for termination. If the termination criterion is not met, the population is iteratively operated by the above three mentioned GA operators and evaluated. This process is continued until termination criterion is met. One cycle of these operations and the subsequent evaluation procedure is known as a generation. The constrained problem, if any, is converted into unconstrained problem by using penalty function method. In this process, the solution falling outside the restricted solution region is considered at a very high penalty. This penalty forces the solution to adjust itself in such a way that after some generations it will fall into restricted solution space. In penalty function method a penalty term, corresponding to the constraint violation is added to the objective function. Generally bracket operator penalty term is used.

$$F_i = f(x) + \in \sum_{j=1}^{k} \delta_j (\phi_j)^2 \qquad (16.1)$$

Where F_i is fitness value, $f(x)$ is objective function value, k is total numbers of constraints, $\in$ is -1 for maximization and +1 for minimization, δ_j is penalty coefficient and ϕ_j is amount of violation. Once the problem is converted into unconstrained problem, the rest of the procedure remains the same. Excellent description of Genetic Algorithms is given by Deb (1995, 1999).

16.1.2 Necessity of Mathematical Modeling in Irrigation Planning

Need for efficient integrated management of an irrigation system is keenly felt due to growing demand for agricultural products, escalating cost of supplying water to farmer's fields and stochastic nature of water resources. Due to dwindling supply of irrigation water the profit conscious irrigators wish to so allocate the water as to maximize the net benefits with competing alternative crops. Investor's choice is further complicated by the fact that the allocation of water is required to be optimized over time, among the crops and also among the competing units of the same crop simultaneously. To meet the requirements, mathematical models and irrigation management methodologies are essential for optimum command area planning. In the present study, the concept of Genetic algorithms and irrigation planning problem are integrated for the case study of Sri Ram Sagar Project, Andhra Pradesh, India. The study is divided into four sections. Section 16.2 gives a brief literature review. Section 16.3 describes the case study followed by mathematical modeling. Section 16.4 analyses results obtained from mathematical model. Section 16.5 gives conclusions followed by references.

16.2 Literature Review

Various authors reported applications of various models in irrigation planning which are explained in brief. Lakshmi Narayana and Rajagopalan (1977) used Linear Programming (LP) model for maximizing the irrigation benefits for Bari

Doab basin in North India. Sensitivity analysis on the tube well capacity, the area available for irrigation, the operation costs for canals and tube wells etc., are also carried out. Loucks et al. (1981) discussed in detail the micro level irrigation planning with a detailed example. Multiobjective analysis is also reported in their studies. Irrigation planning studies using LP are reported by Maji and Heady (1980), Tandaveswara et al. (1992), Garg and Ali (1998). Srinivasa Raju and Nagesh Kumar (1999) proposed a crop-planning model with the objective of maximizing irrigation benefits for a typical irrigation system. Vedula and Nagesh Kumar (1996) developed an integrated model for irrigation planning for a reservoir system in Karnataka, India. Sabu Paul et al. (2000) proposed a multilevel approach based on stochastic dynamic programming for irrigation planning in Punjab, India. Kuo et al (2000) used Genetic Algorithm based model for irrigation project planning for a case study of Delta, Utah with the objective of maximization of net economic benefits for a culturable command area of 394.6 ha. They identified the optimum number of generations, population, crossover and mutation probabilities as 200, 50, 0.6 and 0.02 respectively. The present study deviates at least in two aspects (1) this study uses GA model for irrigation planning in Indian context. (2) A comparison is made between the solution obtained from GA model and that obtained by Linear Programming model.

16.3 Irrigation System and Mathematical Modeling

Sri Ram Sagar Project (SRSP) is a state sector major irrigation project located in Godavari River basin in Andhra Pradesh, India. Its head works are located in Pochampadu village in Nizamabad district of Andhra Pradesh at $18^0 58'$ N latitude and $70^0 20'$ E longitude. Salient features of Sri Ram Sagar project are presented in Table 16.1. Location map of SRSP is presented in Fig 16.1.

Table 16.1. Salient Features of Sri Ram Sagar Project

Item	Value
Type of Dam	Gravity
Length of Earth Dam	13.640 km
Length of Masonry Dam	0.958 km
Total Length of Dam	14.598 km
Maximum height of Masonry Dam	42.67 m
Gross Storage Capacity	3173 Mm^3
Full Reservoir Level (FRL)	332.5 m
Water Spread Area at FRL	434.8 Mm^3
Design flood Discharge	45300 cumecs
Culturable Command Area (stage 1)	1,78,100 ha

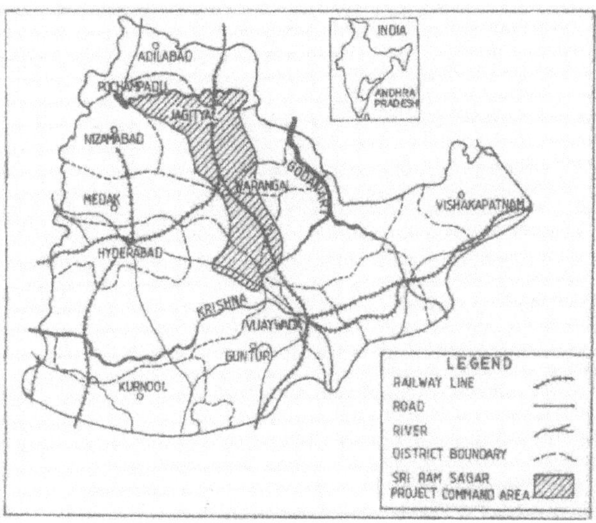

Fig 16.1. Location Map of Sri Ram Sagar Project

The climate of the area is subtropical and semi-arid. There is an extreme variation in temperature with average maximum and minimum values of 42.2^0 C and 28.6^0 C. The average relative humidity for the period from July to September remains above 80% whereas for April to June it is 65%. The evaporation loss varies from 124.3 mm in October to 386.3 mm in April. Rainfall is the primary source of water. The average rainfall of the study area is 944 mm out of which 800 mm falls during June to October. The culturable command area (CCA) of the project (stage 1) is 178,100 ha. Crops grown in the command area are Paddy (rice), Sorghum, Maize, Groundnut, Chillies and Sugarcane in both summer (Kharif) and winter (Rabi) seasons.

Mathematical modelling of the objective function with the corresponding constraints is explained below. The net benefits (BE) under different crops from command area of SRSP are to be maximized. Mathematically it can be expressed as

$$BE = \sum_{i=1}^{10} B_i A_i \tag{16.2}$$

Where i = Crop index [1 = Paddy (S), 2=Maize (S), 3= Sorghum (S), 4=Groundnut (S), 5=Paddy (W), 6= Maize (W), 7=Sorghum (W), 8=Groundnut (W), 9= Chilies (TS), 10=Sugarcane (P)].

S = Summer, W = Winter, TS =Two season, P = Perennial, t = Time index (1=January, ..., 12=December). BE = Net benefits from the whole planning region (Indian Rupees); B_i = Net benefits from the crops (excluding costs of water, fertilizers, labour employment etc) in Indian Rupees per hectare; A_i = Area of crop i grown in the command area (ha).

The model is subject to the following constraints.

16.3.1 Continuity equation

Reservoir operation includes water transfer, storage, inflow and spillage activities. Water transfer activities consider transport of water from the reservoir to the producing areas through canals to meet the water needs. A monthly continuity equation for the reservoir storage (Mm^3) (neglecting evaporation and other losses) can be expressed as

$$S_{t+1} = S_t + I_t - R_t ; t = 1,2,\ldots\ldots 12 \tag{16.3}$$

Where S_{t+1} = Reservoir storage in the reservoir at the end of month t (Mm^3); I_t = Monthly net inflows into the reservoir (Mm^3); R_t = Monthly releases from reservoir (Mm^3).

16.3.2 Crop area restrictions

The total cropped area allocated for different crops in the command area in a particular season should be less than or equal to the Culturable Command Area (CCA).

$$\sum_i A_i \leq CCA \quad \text{i = 1, 2, 3, 4, 9, 10} \qquad \text{Summer season} \tag{16.4}$$

$$\sum_i A_i \leq CCA ; \text{i = 5, 6, 7, 8, 9, 10} \qquad \text{Winter season} \tag{16.5}$$

Where CCA = Culturable Command area (Ha). Crops of two seasons, namely, Chillies and Sugarcane (indices 9 and 10) are included in both the equations because they occupy the land in both the seasons.

16.3.3 Crop water diversions

Crop water diversions for crop i in month t (CWR_{it}) are obtained from the project reports. During the absence of any crop activity, CWR_{it} is taken as zero. Total water releases from Sri Ram Sagar reservoir must satisfy the irrigation demands of the region.

$$\sum_{t=1}^{12} \sum_{i=1}^{10} CWR_{it} A_i - R_t \leq 0; t = 1,2;\ldots\ldots 12 \tag{16.6}$$

Where CWR_{it}= Crop water diversions for crop i in month t (meters)

16.3.4 Canal capacity restrictions

Releases from reservoirs cannot exceed the canal capacity.

$$R_t \leq CC; t = 1,2,.........12 \quad (16.7)$$

The maximum volume of water the canal can transport each month is calculated as CC = 0.0864 x 30.4 x (canal capacity in cumecs)

16.3.5 Live storage restrictions

Reservoir storage volume S_t in any month t must be less than or equal to the maximum live storage capacity of the reservoir.

$$S_t \leq \mathrm{LSP} \; ; \mathrm{t} = 1,2,..........., 12 \quad (16.8)$$

Where *LSP*= Maximum live storage capacity of Sri Ram Sagar reservoir (Mm^3)

16.3.6 Crop diversification considerations

Since the command area lies in a region, which predominantly depends on agricultural economy, the planners want to ensure production of certain cash crops in addition to food crops. The targets are based on the existing cropping pattern.

$$A_i \geq A_{i.\min}; i = 1,2,.........10 \quad (16.9)$$

$$A_i < A_{i.\max}; i = 1,2,.........10 \quad (16.10)$$

Where $A_{i.min}$ and $A_{i.max}$ are minimum and maximum allowable limits of the area under crops. All the above information including inflows are obtained from reports and from discussion with officials of the project. Additional information is also obtained from agricultural department and Marketing society etc.

16.4 Results and Discussion

Penalty function approach is used to convert constrained problem into unconstrained problem with a reasonable penalty function value. For fixing the GA parameters, namely, crossover and mutation probabilities, the model is run for different values of crossover and mutation probabilities. Seven values of crossover probability i.e., 0.6, 0.75, 0.8, 0.85, 0.9, 0.95, 1.0 and six values of mutation probabilities i.e., 0.01, 0.03, 0.05, 0.07, 0.1, 0.12 are chosen with a population size of 50 and maximum number of generations 200. Fig 16.2 presents maximum fitness

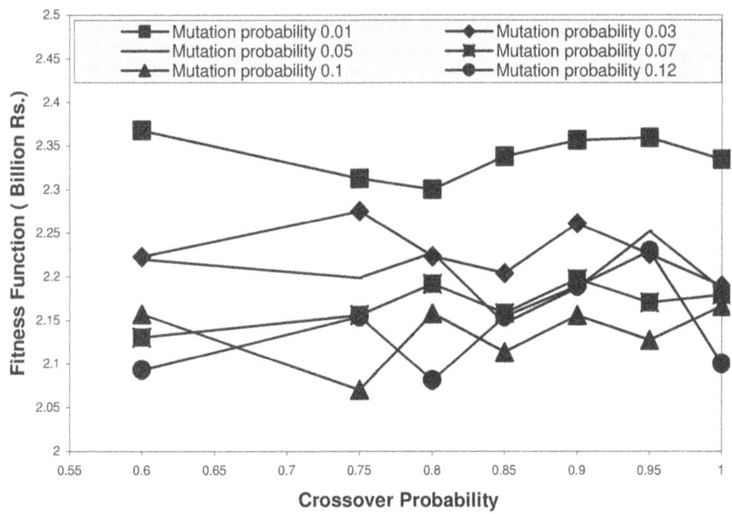

Fig 16.2. Comparison of Fitness Function Values for Various Crossover and Mutation Probabilities

function values for above mutation and crossover probabilities. Results obtained are compared in terms of total fitness function values in Fig 16.2 and number of generations in Fig. 16.3. It is observed from Fig 16.2 that for mutation probability value of 0.01 and for various crossover probabilities, each solution maintains its identity by deviating from other sets of solutions. Among these, the maximum fitness function value of 2.3678 Billion Rupees is achieved for crossover probability value of 0.6 and mutation probability value of 0.01 and this combination is used for further analysis.

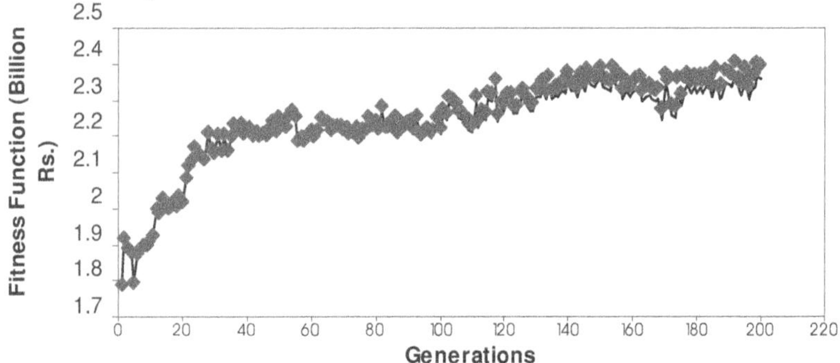

Fig 16.3. Comparison of Fitness Function Values for various Generations.

Fig 16.3 presents number of generations and corresponding fitness function values for the above selected probabilities. Maximum fitness function value occurred at generation 192 with a population of 50.

Efforts are also made to compare the solution of Genetic Algorithm (GA) with Linear Programming (LP) algorithm. Table 16.2 presents cropping pattern obtained by both the methods, which are self-explanatory.

Table 16.2. Crop Plans from the Planning Model for Maximization of Net Benefits

S .No.	Crops and related parameters	Unit	Solution from		% deviation between GA and LP
			GA	LP	
1	Paddy (s)	1000ha	29.43	30.00	1.90
2	Maize (s)	1000ha	29.37	30.00	2.10
3	Sorghum (s)	1000ha	49.71	50.00	0.58
4	Groundnut (s)	1000ha	8.950	9.000	0.55
5	Paddy (w)	1000ha	21.71	22.00	1.32
6	Maize (w)	1000ha	27.38	30.00	8.73
7	Sorghum (w)	1000ha	42.11	50.00	15.7
8	Groundnut (w)	1000ha	9.831	10.00	1.70
9	Chillies (ts)	1000ha	6.130	6.200	1.13
10	Sugarcane (ts)	1000ha	8.140	8.200	0.73
	Irrigated area	1000ha	247.0	259.8	4.92
	Net Benefits	Billion Rs.	2.3903	2.4893	3.97

s = Summer; w = Winter; ts=Two season

It is observed from Table 16.2 that maximum percentage crop acreage deviation of 15.78 and 8.73 occurs for Sorghum (w) and Maize (w) when comparing solution of LP with GA where as these are 1.9 % and 2.1% for Paddy (s) and Maize (s). Maximum benefits obtained by LP solution is 2.4893 Billion Rupees where as these are 2.3903 Billion Rupees by GA (with a fitness function value of 2.3678 Billion Rupees). Irrigated area and net benefits are deviated by 4.92% and 3.97% as compared to LP solution. Fig 16.4 presents graphical representation of crop acreages. Fig 16.5 presents release policy obtained by both the methods. It is observed that LP solution yields more releases in the months of January, February, May, August and September and releases obtained by the solution of GA is more for other months.

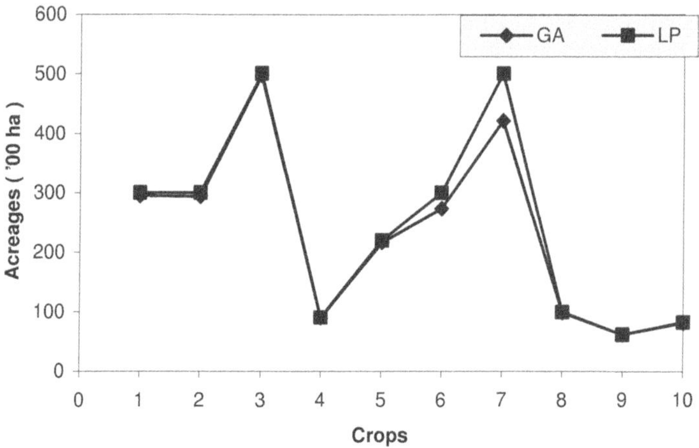

Fig 16.4. Comparison of Cropping Pattern

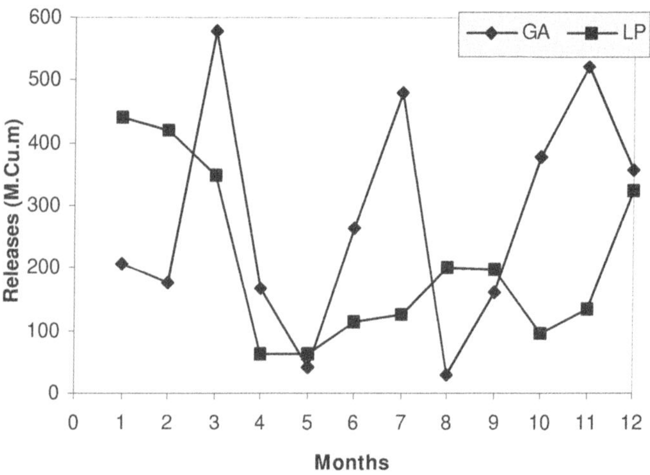

Fig 16.5. Comparison of Monthly Release

Fig 16.6 presents storage policy for the reservoir as obtained by both the methods. It is observed that reservoir reaches zero storage in the month of June and July as observed from the solution of LP whereas these are 49.46 and 782.40 as observed with GA solution. It is observed from above analysis and results that solutions obtained by both GA and LP are reasonably close. However, the solution obtained by GA for irrigation planning problem is to be further refined and investigated for number of factors such as penalty function values, efficient selection of mutation and crossover probabilities, generation and population which are targeted for further study.

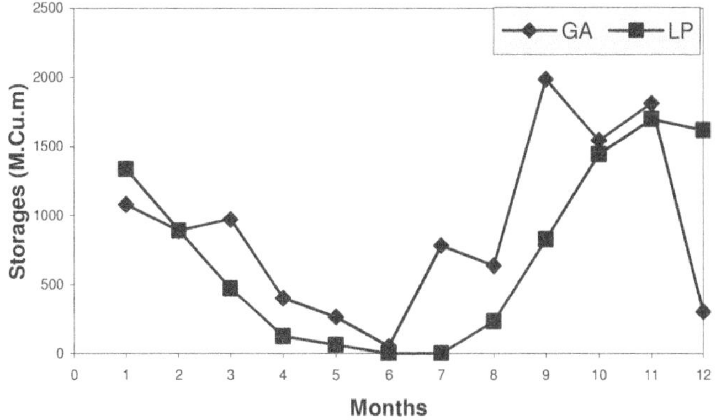

Fig 16.6. Comparison of Monthly Storage

16.5 Conclusions

The present study develops a GA based mathematical model for thecase study of Sri Ram Sagar Project, Andhra Pradesh, India for optimising net benefits from an irrigation system with the constraints such as continuity equation, land and water requirements, crop diversification considerations, canal capacity and storage restrictions. The results obtained from the GA model are compared with those obtained from Linear Programming optimisation model. The observations from the above study are as follows.

1. Maximum benefits obtained by LP solution is 2.4893 Billion Rupees where as these are 2.3903 Billion Rupees by GA (with a fitness function value of 2.3678 Billion Rupees).

2. It is observed that solutions obtained by both GA and LP are reasonably close. Irrigated area and net benefits obtained by GA are deviated by 4.92% and 3.97% as compared to LP solution.
3. Appropriate GA parameters identified from this study are: Number of generations =200, Population size=50, Crossover probability =0.6 and Mutation Probability=0.01.

References

Deb K (1996) Optimization for engineering design: algorithms and examples, Prentice Hall of India, New Delhi.

Deb K (1999) An introduction to genetic Algorithms. Sadhana 24: 293-315

Garg NK, Ali A (1998) Two level optimization model for lower Indus basin. Agricultural Water Management 36: 1-20

Goldberg DE (1989) Genetic algorithms in search, optimization, and machine learning, Addison-Wesley.

Kuo SF, Merkley GP, Liu CW (2000) Decision support for irrigation project planning using a genetic algorithm. Agricultural Water Management 45: 243-266

Lakshminarayana V, Rajagopalan P (1977) Optimal cropping pattern for basin in India. Journal of Irrigation and Drainage Engineering, ASCE 103 : 53-70

Loucks DP, Stedinger JR, Haith DA (1981) Water resources systems planning and analysis, Prentice-Hall, New Jersey.

Maji CC, Heady EO (1980) Optimal reservoir management and crop planning using deterministic and stochastic inflows. Water Resources Bulletin 16: 438-443

Sabu Paul, Panda SN, Nagesh Kumar D (2000) Optimal irrigation planning - A multilevel approach. Journal of Irrigation and Drainage Engineering, ASCE 126 : 149-156

Srinivasa Raju K, Nagesh Kumar D (1999) Multicriterion decision making in irrigation development strategies. Journal of Agricultural Systems 62: 117-129

Tandaveswara BS, Srinivasan K, Amarendra Babu N, Ramesh, S.K. (1992) Modelling an over developed irrigation systems in south India. International Journal of Water Resources Development 8: 17-29

Vedula S, Nagesh Kumar D (1996) An integrated model for optimal reservoir operation for irrigation of multiple crops. Water Resources Research 34 : 1101-1108

17 Optimization of Helical Antenna Electromagnetic Pattern Field

Ivan Zelinka

17.1 Introduction

Antennas and antenna systems play important role in today's communication technologies. Their use is thus vitally important for communication as well as for everything related to communication. These systems contain a rich set of antennas such as parabolic, linear, helical, etc. Each type of antenna usually has specific properties, which predestine the given antenna to the specific problems in the communication. This chapter will focus attention on helical antenna electromagnetic pattern fields designed to demonstrate the use and performance of the SOMA algorithm. It will also compare results obtained from a differential evolution (DE).

17.2 Problem description

This application of the SOMA algorithm is focused on the optimization of the helical antenna electromagnetic pattern field. Examples of problems about antennas, construction and their use can be for example found in [1] – [8]. The main aim of optimizations here is to find such parameters of construction, which satisfy "attributes" the electromagnetic field of a helical antenna is focused in one direction with a minimum of secondary "dew-lap". The principle of the helical antenna is depicted in Fig. 17.1. The parameters of the antenna, which were optimized here, are described in 17.1.

Table 17.1. Optimized parameters

Parameters	Interval of optimization	Description
▯	<1°, 80°>	raising angle
n	<1, 100>	number of coils

Both parameters determine the quality of the final shape of the optimized electro-magnetic field.

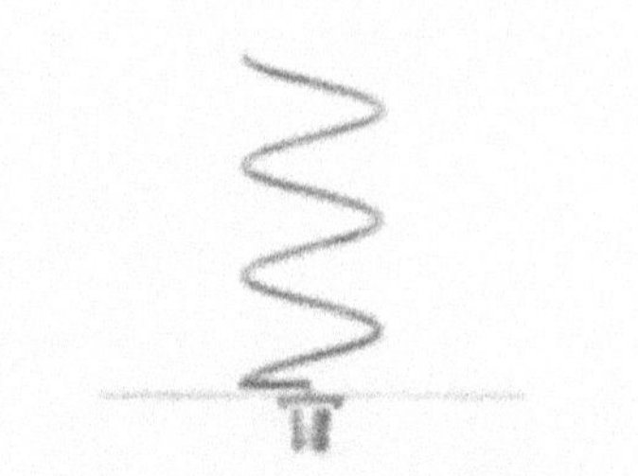

Principle of helical antenna…

…and its real example

Fig. 17.1. Helical antenna – basic principle and its realization (© - [3]). The main dew-lap of the electromagnetic field is along the main spiral axe

The cost function, which determines the final shape of the antenna electromagnetic field, is given by 17.1 [9].

$$f_{\cos t} = \sum_{\theta=.5\approx 28^\circ}^{5.78\approx 338^\circ} \left| \frac{(-1)^{n+1}\sin(\frac{\pi}{2n})\cos(\theta)\sin(\frac{1}{2}n(2\pi\tan(\alpha)(\cos(\theta)-1)-2\pi-\frac{\pi}{2}))}{\sin(\frac{1}{2}(2\pi\tan(\alpha)(\cos(\theta)-1)-2\pi-\frac{\pi}{2}))} \right| \quad (16.1)$$

where $\theta \in <0, 2\pi>$

Its minimization should lead to the optimal parameters "n" and "" (see Table 17.1). In fact, the optimization of equation (17.1) represents a search for the global extreme on the hyperplane in 3D. This hyperplane is depicted in Fig. 16.3. The cost function was designed so that its minimization should produce an antenna pattern as optimal as possible. Optimal here means one "dew-lap" (if possible) in the main direction of the antenna radiation.

The process of optimization was thus searching for a global extreme on 2D hyperplane in 3D space (third dimension was costvalue as it is graphically depicted in Fig. 17.3. the main aim during this optimization was to minimize the plane of the radiation field in the interval 0.5-5.78 (28°-338°), see Fig. 16.2. Intervals <0, 0.5> and <5.78, 6.28> were excluded from the minimization – maximal values were allowed in this interval. The main axe of antenna radiation is in this interval, which in this case is identical with axe "x" (see Fig. 16.2 and Fig. 16.4). The maximum in this interval <5.78, 0.5> then guarantees a helical antenna with a narrow main dew-lap.

The cost function (16.1) was two-parameters: the parameter "n" (the number of spirals) and „□“ (the raising angle). The cost function then provided the size of gray area from Fig. 17.2. These three numbers can be used to depict the cost function surface as it is done in Fig 17.3. The optimization process in this example was focused on a global minimum localization. Its coordinates were then the optimal parameters of the helical antenna.

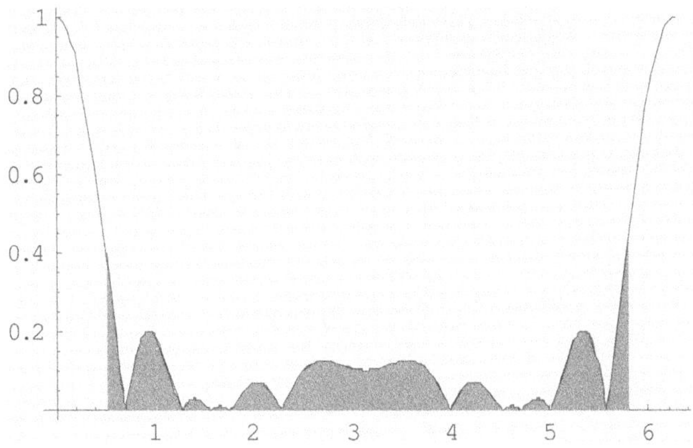

Fig. 17.2. Cost function principle. Every point of "z" axis of plane in Fig. 16.3 is the gray plane depicted here.

The used cost function (16.1) including the original parameter setting was taken from „Electrical Engineering Toolbox for Mathematica" which is a specialized tool for electrical engineering in the Mathematica® environment. The shape of the radiation field is depicted in Fig. 16.4.

From Fig. 16.4 it is clearly visible that the antenna field of radiation is quite wide with some small and medium sized dew-laps in others quadrants. These secondary dew-laps (especially in quadrant II and III) do not promote the main dew-lap and thus are useless. They just consume energy and money. For the minimization and contraction (from the sides, the "amplitude" of main dew-lap was conserved) of the main dew-lap two algorithms- SOMA and DE were used.

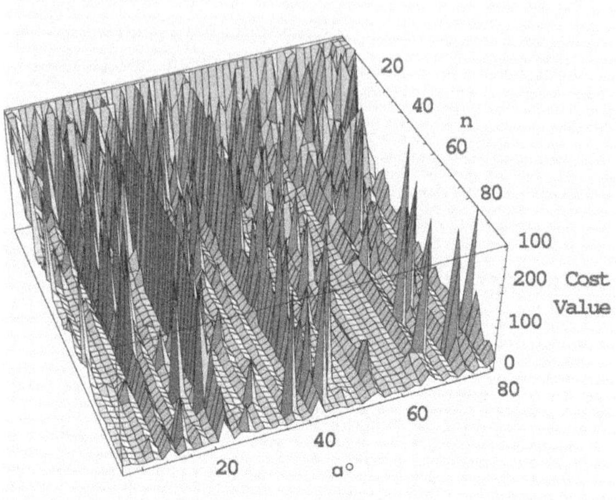

Fig. 17.3. Surface of all possible solutions.

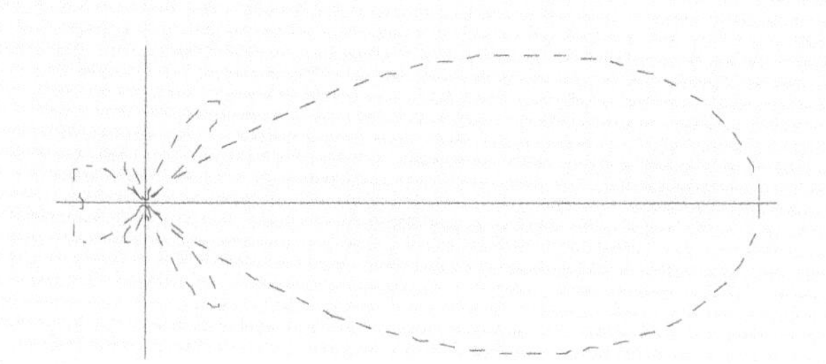

Fig. 17.4. Radiation antenna with original parameters n=12 coils and raising angle =10°, see also Fig. 17.2.

17.3 Simulations

Simulations were done in the Mathematica® environment with C++ code linked into the Mathematica® like an internal library. On [10] the same code for the DOS environment is accessible. The optimization was repeated 100 times and was done by means of algorithms SOMA and DE.

When the SOMA algorithm was used (see all 100 histories on Fig. 16.6), a configuration was found, which is depicted in Fig. 17.5. According to Fig. 16.5, which is based on optimal parameters, is visible that the main dew-lap including the secondary dew-laps were narrow, focused in the same direction as the primary one. The small dew-laps in quadrants II and III can be ignored.

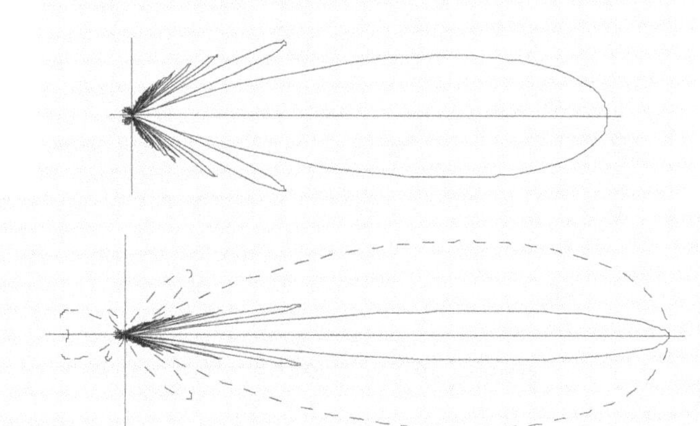

Fig. 17.5. Antenna radiation field optimized by algorithm SOMA (n = 99 coils and raising angle □ = 20.5°) and comparison with original dew-lap from Fig. 17.4 with n = 12 coils and raising angle □ = 10°

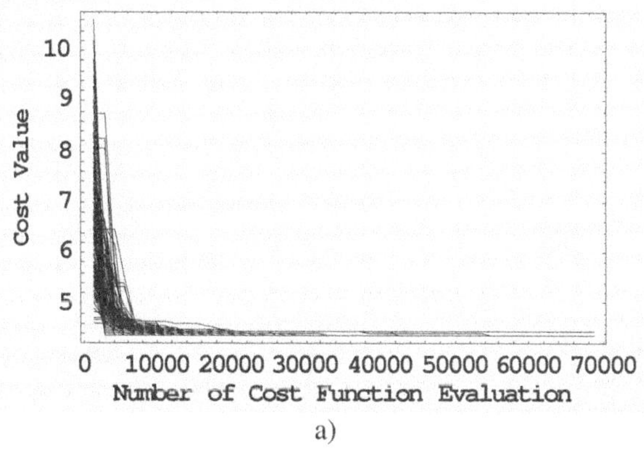

a)

Fig. 17.6. History of 100 times repeated optimization by SOMA.

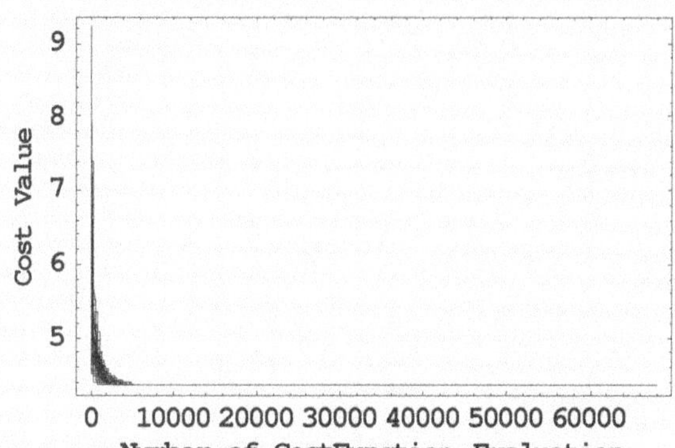

Fig. 17.7. History of 100 times repeated optimization by DE.

In the case of differential evolution the same results were obtained as in the case of the SOMA algorithm, (see 17.7, 17.8) with only mirror differences (10^{-2}). From 17.7 it is visible that DE has found the global extreme sooner than SOMA; however, the faster convergence also means a higher probability that the algorithm will stop in the local extreme. In contrast the slower convergence means that the hyperplane was more properly searched (see for example Chapter 7, especially „Test function – egg holder", „Rana's function" and „Pathological function".).

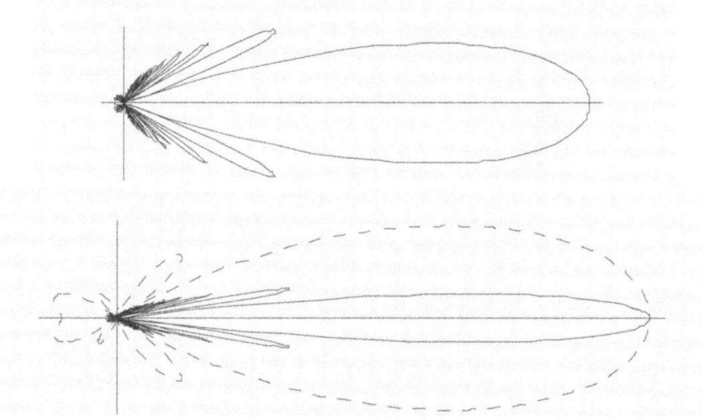

Fig. 17.8. Antenna radiation field optimized by DE (n = 99 coils and raising angle [illegible] = 20.5°, the same as in the case of SOMA) and comparison with original dew-lap from Fig. 16.4.

Each simulation of all 100 took a few seconds on a PC with Pentium III 800 MHz with the parameters described in Table 17.2.

17.4 Software support

To allow any reader to repeat these simulations including its modifications, a source code in C++ of this problem was released on the Internet pages, see [10]. The cost function in C++ is described in Table 17.3. The increment "0.01" in the loop is a sampling of the minimized plane (Fig.17.2) for 0.5°. Basically this loop is the "Sum" from (17.1). The cost function is in file "Antena.h" (Table 17.3) and the configuration file (Table 17.2) is „deantena.txt" or „soantena.txt" according to which algorithm was used.

Optimization can be started by DOS command

1. „Aasoma soantena" – SOMA, strategy AllToAll
2. „Aorsoma soantena" – SOMA, strategy AllToOne Randomly
3. „Aosoma soantena" – SOMA, strategy AllToOne
4. „Arsoma soantena" – SOMA, strategy AllToAll Adaptive
5. „Der1b deantena" – differential evolution, version DeRand1Bin

After optimization has finished two files will be created with extensions *.HST and *.OUT. "Soantena" or "deantena" will be found rather than *. In the HST file, the history of the cost value evolution with its dependence on migration loops (or generations for DE) is found, and in the OUT file there is a detailed report about the optimization and its results.

Table 17.2. Initial file „??antena.txt"

deantena.txt	Description	soantena.txt	Description
50	PopSize	3.	PathLength
2	Dim	.11	Step
1336	Generations	.1	PRT
.9	F	50	PopSize
.3	CR	2	D
0.	Lo	50	Migrations
100.	Hi	-1.	MinDiv
0.	Typ (real)	0.	Lo
0.	Lo	100.	Hi
6.28	Hi	0.	Typ (real)
0.	Typ (real)	0.	Lo
		6.28	Hi
		0.	Typ (real)

Table 17.3. Cost function in C++ (Antena.h)

```
//Antenna Field Patterns
//For more please see http://www.ft.utb.cz/people/zelinka/soma.htm

double Pi,theta,s,temp,temp1;
int     n;

n = getIntPopulation(0,Individual);
s = getPopulation(1,Individual);

temp=temp1=0;
Pi = 3.14159;

for(theta=.5;theta<=5.78;theta=theta+0.01)
        {
        temp = ((2*Pi)*s*(cos(theta)-1)-2*Pi-Pi/n)/2;
  temp1 = temp1 + fabs((pow(-1,(n+1))*sin(Pi/(2*n))*cos(theta)*sin(n*temp))/sin(temp));
  }

CostValue = temp1;
```

More detailed information about SOMA software support on different platforms (DOS, Unix, Mathematica®, Matlab®) and its use is easily accessible on [10].

17.5 Conclusion

The helical antenna pattern field optimization dealt with here was a rather simple task. However despite this simplicity, it is visible from Fig. 16.3 that the hyperplane, which represents a given problem is quite complicated, and thus the use of such algorithms as SOMA, DE etc. is well founded. A more important application of the antenna pattern field design would be in what is called a "dipoles field" [11]. In this case the antenna is created by dipoles organized into a matrix field of dimension N x M. Each dipole can have its own size, orientation, and currency with different phases and amplitudes. Such a system usually contains hundreds of variables whose settings influence the optimality of a given antenna system. Such problems are clearly predestined for evolutionary algorithms.

Acknowledgements

This work was supported by the grant No. MSM 26500014 of the Ministry of Education of the Czech Republic and by grants of Grant Agency of Czech Republic GACR 102/03/0070 and GACR 102/02/0204.

Special thanks belong to Jason Hecker [3] for his agreement allowing me to use pictures in Fig. 16.1 and for additional information about helical antennas.

References

[1] Bob Atkins KA1GT, Helical Antena Design, The ARRL UHF/Microwave Experimenters Handbook, ISBN 0-87259-312-6

[2] Karl Rothammel Y21BK, Die Helical-Antenne, Antennenbuch, Telekosmos Verlag, ISBN 3-440-04791-1

[3] Jason Hecker, How to Make a Simple 2.425GHz Helical Antenna, **http://www.wireless.org.au/~jhecker/helix/helical.html**

[4] Salema, C.; Fernandes, C. A.; Jha, R., Solid Dielectric Horn Antennas, Artech House, Boston, Mar 1998

[5] Fernandes, C. A., "Lens Antennas for mm-W applications", Contribution in Wireless Flexible Personalised Communications, L. M. Correia (Ed.), John Wiley & Sons, Mar 2001.

[6] Fernandes, C. A., "Shaped-Beam Antennas", Chapter in Handbook of Antennas in Wireless Communications, L. Godara (Ed.), CRC Press, August 2001

[7] Fernandes, C. A., "Complex modes on an Helical Antenna fed by a Current Disk", (in Portuguese), Master Thesis, Dep. Electrical and Computer Engineering, IST, Feb. 1985.

[8] Fernandes, C. A., "Radiation from an Helical Antenna fed by Circular Waveguide", (in Portuguese), Doctor Thesis, Dep. Electrical and Computer Engineering, IST, Nov. 1990.

[9] Electrical Engineering Toolbox for Mathematica, sw tool for Mathematica® environment, see also www.wolfram.com

[10] Articles about SOMA algorithm (source codes, graphics animated gallery, bibliography, see **http://www.ft.utb.cz/people/zelinka/soma**

[11] Prokop J., Vokurka J., Šíření elektromagnetických vln a antény, SNTL/Alfa, Praha 1980, ISBN 04-521-80

18 VLSI design: gate matrix layout problem

Pablo Moscato, Alexandre Mendes and Alexandre Linhares

18.1 Introduction

With applications ranging from fields as distinct as fuzzy modeling (Xiong 2001), autonomous robot behavior (Luk et al. 2001), learning with backpropagation (Foo et al. 1999), and multicriteria optimization (Viennet et al. 1996), evolutionary methods have become an indispensable tool for systems scientists. Although already studied in the past, an interesting emerging issue is the use of multiple populations, which is gaining increased momentum from the conjunction of two technologies. On the hardware side, computer networks, multi-processor computers and distributed processing systems (such as workstations clusters) are increasingly becoming widespread. Regarding the software issues, the introduction of Parallel Virtual Machine[1] (PVM), and later the Message Passing Interface Standard[2] (MPI), as well as web-enabled, object-oriented languages (such as Java[3]) have also had their role. Most *Evolutionary Algorithms* (EAs) are methods that are easy to parallelize as well as naturally suitable for heterogeneous systems. For most EAs the distribution of the tasks is relatively easy for most applications. The workload can be distributed at an individual or a population level; the final choice depending on the complexity of the computations involved.

The application presented in this chapter is motivated by the availability of these computer environments. However, here we do not report results on the use of parallel computers, or networks of workstations. The proposed *memetic algorithm* (MA) runs in a sequential way on a single processor, but a set of populations evolve separately, an approach that can be easily mapped to a parallel environment.

Species evolve naturally grouped in sub-populations, with boundaries defined by some specific features like distance or geographical barriers. The role of closed (or nearly closed) populations in biological evolution is extremely important. Con-

---

[1] `http://www.csm.ornl.gov/pvm/pvm_home.html`

[2] `http://www-unix.mcs.anl.gov/mpi/`

[3] JAVA is a general programming language developed by SUN Microsystems.

sider for instance the Galapagos Islands, an example of notorious inspirational role for Darwin's ideas when aboard the HMS Beagle (Darwin 1993). A set of islands separated by several kilometers of water can be colonized by a single species of birds. In the beginning of such colonization, all animals will share the same characteristics (and genetic pool), but as evolution takes place, the groups concentrated in each island will start to differentiate by adapting themselves to the particular characteristics present in each island (Weiner 1995). This independent adaptation may eventually lead to having different species after a sufficient number of generations, given very little or no migration at all exists between the islands. Even if the islands share the same characteristics, different species might arise, due to the relative isolation and the *genetic drift* phenomenon (Weiner 1995).

The genetic drift concept states that if two identical populations are separated and submitted to equal conditions, due to the random nature of the processes involved in evolution, they may still follow different evolutionary paths and become different species after a large number of generations. Analogously, it is quite common in EAs to find that, due to the random nature of the approach, if the same algorithm is run twice it may generate different final solutions. Usually viewed as a setback, this characteristic can actually be very useful when multiple populations are used. With several populations evolving in parallel, larger portions of the search space can be sampled, and any particularly important new features found could be spread among them through migration of individuals. This mechanism makes the parallel search potentially more powerful than single population approaches (Cantú-Paz 1997, 1999).

18.2 The gate matrix layout problem

The Gate Matrix Layout problem (GMLP) belongs to the NP-hard class (Lengauer 1990; Linhares 1999; Nakatani et al. 1986) and arises in the context of the physical layout of *Very Large Scale Integration* (VLSI) systems. Formally we can state the problem as:

Instance: A matrix M having Boolean coefficients.

Parameter: $p > 0$, integer.

Question: Is there a permutation of the columns of M so that if in each row we change to '*' every 0 lying between the row's leftmost and rightmost 1, then no column contains more than p 1's and *'s?

This problem has appeared in the literature in a number of combinatorially equivalent forms. The form given above was defined in (Lopez and Law 1980). It has been known as 'Weinberger arrays' or 'PLAs with multiple folding'. The problem is solvable in $O(n)$ for fixed p (Bodlaender 1993, 1996; Fellows and Langston 1989).

We can now state the problem in the terms of the familiar VLSI setting. Suppose that there are g gates (they can be described as vertical wires holding transistors at specific positions) and n nets (horizontal wires interconnecting all the distinct gates that share transistors at the same position). As described above, an instance of the GMLP can be represented as 0-1 matrix, with g columns and n rows. An entry '1' in the position (i, j) means a transistor must be implemented at gate i and net j; 0 means that no such connection is to be made. The VLSI architecture requires that all transistors in the same net must be interconnected. Given a specific gate sequence, whenever two nets overlap, their implementation needs two separated physical tracks. This superposition of interconnections defines the number of tracks needed to build the circuit.

Both types of formulation are equivalent. The objective in the GMLP is to find the minimal p. In other words, is to find a permutation of the g columns so that the superposition of interconnections is minimal, thus minimizing the number of tracks and the overall circuit area. The Fig. 18.1 shows a possible solution for an instance composed of seven gates and five nets, and how to go from the binary matrix representation to the circuit itself.

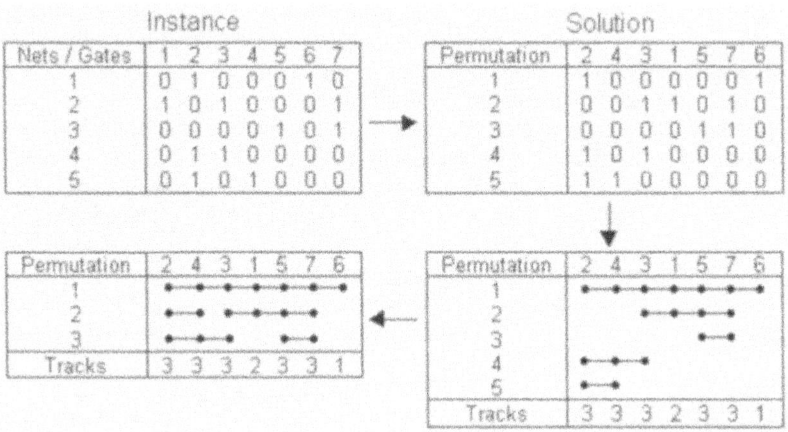

Fig. 18.1. The figure shows in clockwise order beginning in the upper-left: An instance matrix; a permutation of the matrix columns (a feasible solution); the correspondent circuit and the real circuit with only the necessary tracks.

The Fig. 18.1 represents a circuit with seven gates and five nets. Usually, real circuit matrices are very sparse, like the one in the example. Based on the instance, we created a feasible solution, represented by the columns permutation (2-4-3-1-5-7-6). The translation from the 0-1 representation into the circuit diagram is very simple. All the values '1' in the same net must be interconnected (represented by the horizontal lines). In the lower part of this diagram there is the sum of the transistors in each gate. Finally, the real circuit with only three tracks remaining, after the transistor-grouping routine was applied, is shown in the lower-left part of the Fig. 18.1.

For more information on this problem, including other industrial settings in which it arises, please refer to (Wong et al. 1988; Yanasse 1997; Linhares and Yanasse 2002). The reader should be aware that this is not just an ordinary NP-hard problem: it was in fact one of the first problems identified as being fixed-parameter tractable, and this result eventually led to the creation of a new, large class of problems under the label FPT, for fixed parameter tractability (Downey and Fellows 1995; Fellows and Langston 1987). In the next section we introduce a new MA for the GMLP.

18.3 The memetic algorithm

Since the publication of John Holland's book, 'Adaptation in Natural and Artificial Systems' (Holland 1975), the field of *Genetic Algorithms* (GAs), and the broader field of *Evolutionary Computation* (EC), were clearly established as new research areas. However, other pioneering works could also be cited, as they became increasingly conspicuous in many engineering fields and in industrial problems. In the mid 80's, a new class of *knowledge-augmented GAs*, also called *hybrid GAs*, began to appear in the computer science and engineering literature. The main idea supporting these methods is that of making use of other forms of *knowledge*, i.e., other solution methods already available for the problem at hand. As a consequence, the resulting algorithms had little resemblance with biological evolution analogies. Recognizing important differences and similarities with other population-based approaches, some of them were categorized as *Memetic Algorithms* in 1989 (Moscato 1989; Moscato and Norman 1992), a denomination inspired from the term *meme* introduced by Dawkins (Dawkins 1976). The field of *cultural evolution* was suggested as being more relevant, as a working metaphor, to understand the performance and find inspiration sources to improve these new methods. Next we will describe the main aspects of the MA implemented.

18.3.1 Population structure

The use of structured populations in evolutionary algorithms is still quite rare. Nevertheless, previous tests (França et al. 2001; Mendes et al. 2001) in different types of NP optimization problems have shown that this feature can generate a big leap in the method's performance. It was verified that the number of individuals utilized by the method is dramatically reduced when population structures such as binary or ternary trees are employed. The consequence is a reduction in the computational effort and a considerable performance improvement, when compared to ordinary non-structured population approaches (França et al. 2001; Mendes 1999).

It is illustrative to show how some MAs resemble more the cooperative problem solving techniques that can be found in some organizations. For instance, in the approach being described, we use a hierarchically structured population based on a complete ternary tree. In contrast with a non-structured population, the complete ternary tree can also be understood as a set of overlapping sub-populations (which we will refer to as *clusters*). The choice of the ternary tree structure was based mainly on empirical aspects. The first is motivated by the fact that any hierarchical tree behaves like a set of overlapping clusters, as said before. Therefore, the dynamics is similar to several populations evolving in parallel – each cluster acts as an independent population – and exchanging individuals at a given rate. This exchange of individuals comes as a consequence of the tree-restructuring phase, carried out to maintain a specific hierarchical consistence. Now, consider the use of trees with other degrees of complexity. A binary tree-based population, for instance, would be formed by 3-individual clusters only, with only two recombinations possibilities. This would degrade the 'multiple population' character of the tree structure. Trees with a greater order – quaternary or more – increase the multiple population character, but initial tests indicated that the performance does not improve at all and moreover, the number of individuals rapidly jumps to prohibiting levels in terms of computational effort requirements. The best trade-off points to the selected ternary tree structure.

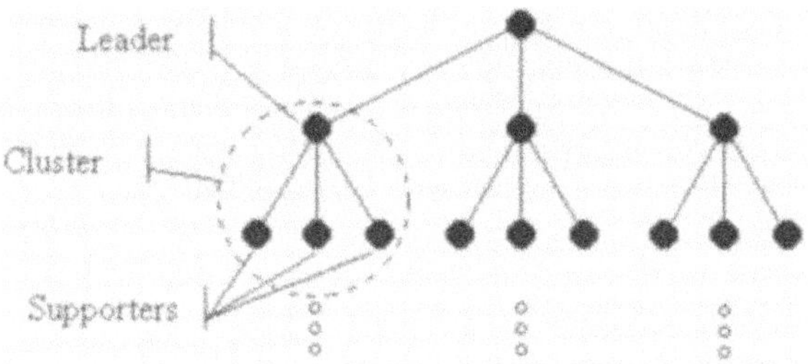

Fig. 18.2. Diagram of the ternary tree hierarchical structure utilized by the MA populations

In Fig. 18.2, we can see that each cluster consists of one single *leader* and three *supporter* individuals. Any leader individual in an intermediate layer has both leader and supporter roles. The hierarchy states that the leader individual must always contain the best solution – considering the number of tracks – present in the cluster. The number of individuals in the population is equal to the number of nodes in the ternary tree, i.e., we need 13 individuals to make a ternary tree with three levels and 40 individuals to have four levels. For this experiment, we fixed the population size to be 13. This value might seem too low at a first glance, but after several tests with 40 and 121 individuals, we concluded that 13 individuals are sufficient to make the algorithm keep its convergence speed under control. The use of 40 or more individuals does not deteriorate the algorithm's behavior, but the computational effort increases considerably, as well as the CPU time. Moreover, the MA with the 13-individual configuration had a very competitive performance with other types of problems (França et al. 2001; Mendes et al. 1999).

18.3.2 Representation and crossover

The success of a MA relies very much on the representation chosen to encode for feasible solutions, and to follow the terminology of genetic algorithms we will call them "chromosomes". It is preferable to use representations that are *compact*, *complete* and *stable*. A compact representation requires very few variables to represent a solution in a unique way. A complete representation must be able to represent all the possible solutions of the problem, including, of course, the optimal one. A stable representation is one in which small changes in the chromosome structure would result in solutions with small modifications in the individual's fitness. If small disturbances generate large fitness modifications, the MA might have trouble to evolve the population because important information learned during the evolutionary process is continually lost. The representation chosen for the problem is quite intuitive. A solution is represented as a *chromosome* in which the *genes*[4] assume different integer values in the $[1, g]$ interval, where g is the number of columns of the associated binary matrix. These values will define the sequence (permutation) of the gates.

The recombination algorithm used to address the GMLP is a variant of the well-known *Order Crossover* (OX) (Goldberg 1989) called *Block Order Crossover* (BOX). The OX is very simple and easy to implement, but lacks an essential feature for the GMLP. It begins by copying a piece of the chromosome from one of the parents into the offspring. This can already be a major limitation if the good features of the individuals are not grouped together, but separated in many blocks. In this case, a better-tailored crossover should be able to copy several parts of the chromosome into the offspring. This is the motivation for the BOX, where instead of just one piece, several pieces of the parent are copied. The BOX resembles the second variant of the OX crossover presented in (Syswerda 1991).

The rest of the BOX crossover is the same of the OX: The *loci*[5] in the offspring that were not filled with the information of the first parent are sequentially filled out with the alleles inherited from the second parent. Both procedures – OX and BOX – tend to perpetuate the relative order of the gates, but the BOX can better communicate separated blocks to the offspring. Nevertheless, given the stochastic characteristic of the crossover process, undesirable alterations might – and will – appear during the evolutionary process, as it happens in nature.

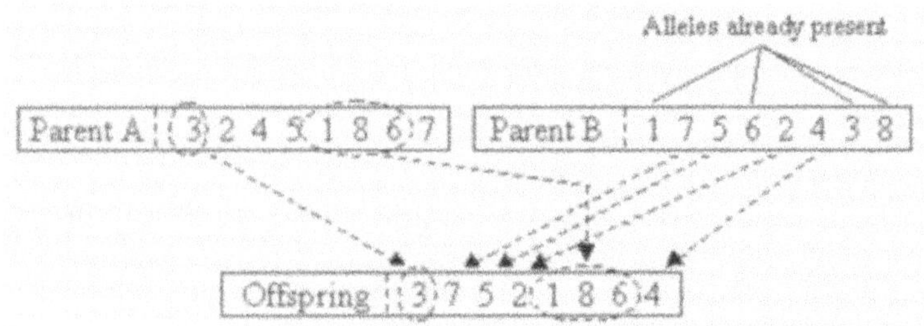

Fig. 18.3. Block Order Crossover (BOX) example

[4] Chromosomes are composed of ordered sequences of genes.

[5] Loci (plural of locus) is the name given to the positions of the chromosome where the alleles are placed.

In Fig. 18.3, *Parent A* contributes with two pieces of its chromosome to the offspring – one with a single allele and the other with three alleles. These parts are copied in the same positions that they occupy in the parent. The blank spaces are then filled with the information from *Parent B*, going from left to right. The values in *Parent B* already present in the offspring are skipped; being copied only the new ones. In the structured population approach, *Parent A* is always a leader and *Parent B* a supporter belonging to the same cluster.

The contribution from each parent to the offspring is another important issue. In nature, parents contribute in different proportions to the offspring's genetic material, given the randomized characteristics of the genetic recombination. It is somewhat common to see GA/MA work where the genetic inheritance is proportional to the parent's fitness. In other words, parents better adapted will influence more their offspring's genetic conformation. In this implementation we decided that each parent contributes with a fixed 50% of its genetic material. The decision of not using the fitness was based on the little influence it had on the results. This has probably occurred due to the use of a very strong local search[6]. The local search capacity to improve the individuals' fitness has made any little modification in the crossover characteristics be a negligible factor in the algorithm's performance. Nevertheless, we must emphasize that the crossover is a crucial step of the evolutionary process; a situation confirmed by the considerable difference of performance between the OX and BOX crossovers in our preliminary tests.

The number of descendants created every generation is another important issue and can strongly influence the quality of the algorithm. Generally the number of descendants has a close relation to the percentage of them that will be included in the population, replacing the old ones. If all new individuals are accepted, the crossover rate shall be set at low levels, otherwise good information already present in the gene pool might be lost. If the offspring's insertion policy is more restrictive, more individuals should be created, since only a few of them will survive. As the policy adopted is indeed very restrictive[7], we assumed that the crossover rates should be higher than normal. In fact, after tests with values ranging from 0.5 up to 2.5 times the population size, the best value found for the crossover rate was 2.0. This number corroborates the relation described before, characterizing a high offspring mortality selection process.

18.3.3 Mutation

Mutation plays a critical role in evolutionary algorithms and in memetic algorithms design in particular (Merz and Freisleben 1999). It consists of a random change in a part, or parts, of the individual's genetic code. This random nature makes the mutation be much more destructive than constructive. Indeed, the great majority of mutations are destructive, creating worse or non-viable individuals. Such individuals are usually eliminated through natural selection and the results of

[6] The local search is described in Sect. 18.3.4.

[7] The offspring's insertion policy is described in Sect. 18.3.6.

disastrous mutations are not perpetuated. But the strong point is that, when successful, mutation can create good, complex features, or even make the individual jump out of *evolution traps* by improving its fitness. When a population is very similar in terms of genetic characteristics, making any further improvement almost impossible, we can say it became stuck in an evolution trap. A good mutation movement can put an end to this situation by creating a new better individual that could be virtually unreachable through recombination alone.

Two mutation strategies are utilized in the MA. The first is a *light mutation*, which makes slight changes in the chromosome. It is based on the swapping of gates. Two positions of the chromosome are selected uniformly at random and the values of their alleles are swapped. This mutation procedure is applied (in average) on 10% of all new individuals every generation (see Fig. 18.4).

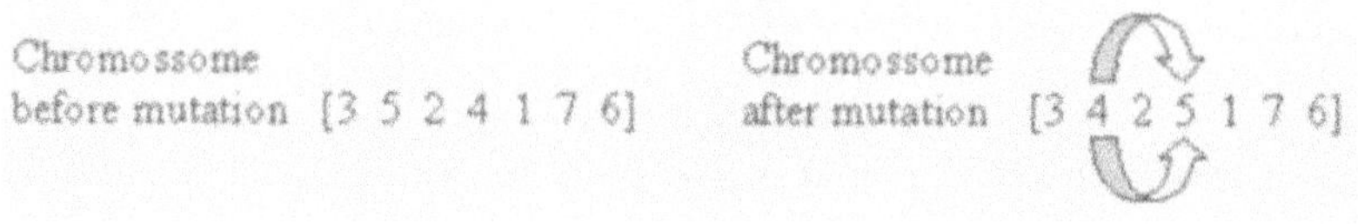

Chromossome before mutation [3 5 2 4 1 7 6]

Chromossome after mutation [3 4 2 5 1 7 6]

Fig. 18.4. Swap mutation example. Two positions of the chromosome are selected uniformly at random and their alleles are swapped

The second mutation strategy is called *heavy mutation*. Since the population continuously loses diversity during the evolutionary process, after a sufficient number of generations all individuals will share almost the same genetic code. When that happens, we say that the population has converged to a low-diversity situation[8]. In such a situation, any further improvement is too costly. A heavy mutation procedure makes the population be filled with new individuals and a diversified genetic load, allowing the MA to continue its search. The procedure adopted executes a light mutation move exactly 10.*g* times in each individual, except the best one[9]. The resulting population will thus be composed of several randomized, low-quality individuals, and a highly adapted one.

18.3.4 Local search

The local search is one of the main differences between this chapter's MA and an ordinary GA. Its influence on the algorithm's performance is overwhelming, being quite hard to think of a method that does not utilize a neighborhood-based search and can deal with problems as complex as the GMLP.

[8] The definition of population convergence adopted this work is based on the individual's replacement rate, rather than on chromosome similarities. Both criteria are related but the first is less restrictive. Please refer to Sect. 18.3.6.

[9] The preservation of the best individual is a kind of *elitism*. It guarantees that the valuable genetic information present in its chromosome will not be lost during the evolutionary process.

The design of local searches strategies is a rather complicated job. On one hand, we try to get as much search power as possible and on the other we must keep an eye on the increasing computational complexity. Next there is a step-by-step description of the local search construction, highlighting the mistakes and successes during the process.

Local search algorithms for combinatorial optimization problems generally rely on a neighborhood definition that establishes a relationship between solutions in the configuration space. Two neighborhood definitions were utilized in this specific local search. The first one was the *all-pairs*. It consists of swapping pairs of columns/gates from a given solution. A hill-climbing algorithm can be defined by reference to this neighborhood; i.e., starting with an initial permutation of all columns, every time a proposed swap reduces the number of tracks utilized, it is confirmed and another cycle of swaps takes place, until no further improvement can be achieved. The second neighborhood implemented is named *insertion*. It consists of removing a column from one position and inserting it in another position (which could include any point between a pair of gates, or the left/right extremes of the permutation). The hill-climbing iterative procedure is the same regardless the neighborhood definition. Both neighborhoods have an $O(n^2)$ complexity[10]. For small instances, with less than 30 gates, it is possible to evaluate every swap or insertion possibility, but since the goal was to develop a local search capable of running on a wide spectrum of instances, the use of neighborhood reductions became imperative.

Reduction techniques are designed to work with smaller neighborhood sizes without considerable loss of performance. Most reductions are based on general empirical conclusions and special characteristics of the problem. The first type utilizes less problem-specific information and its performance tends to be worse. Nevertheless, it is easier to implement them and can serve as an initial step to more complex reductions. That is the reason why we started with a simple reduction rule allowing the swap and insertion moves only between columns that are close enough to each other. The motivation for such approach is that the crossover and mutation operators are relatively successful in finding the region where each gate should be placed. No local search must be applied to make an allele swap its position with another allele on the opposite side of the chromosome. The evolutionary operators will sometime make such a change. But when a fine-tuning is required, crossover and mutation are useless. Local search must be employed in this case and the search can be concentrated only in the region being adjusted. In fact, the distance that characterizes such fine-tuning region is a critical parameter and should vary proportionally to the instance's size. For instance, suppose there is an instance with 50 gates. A good choice observed in our preliminary tests was to fix the distance at k = 10, meaning that every column should be tested for swap/insertion with the 10 nearest columns to the right and the left, totaling 20 tests. Of course, larger values for k, like 20 or 30, should yield to better performances, but the computational complexity increases proportionally. This tradeoff

[10] The $O(n^2)$ means that the computational complexity of visiting the entire neighborhood is proportional to the square of the instance's size.

analysis is a very stressing and time-consuming issue but also a necessary step. During this analysis, we experienced some problems with the largest instance, with 141 gates. The minimum k values necessary to reach optimal solutions for it were still very high, making the search process too sluggish.

After the definition of the first reduction strategy, it became clear that a somewhat 'smarter' technique, utilizing specific GMLP information should be created. The resulting reduction is named *Critical Column Based Local Search* (CCBLS). This new local search reduction scheme prohibits useless swaps and insertion tries by using information of the so-called *critical columns*. The critical columns define the maximum number of tracks required by the solution. In any given solution, every gate will have to implement a specific number of transistors. This number is given by the sum of the transistor-interconnecting horizontal lines that pass through the gate. After calculating this value for every column, we can define that column i is critical if it satisfies the equation:

$$n_t(i) = \max_{\forall k \in G} [n_t(k)] \tag{18.1}$$

where $n_t(\cdot)$ denotes the number of tracks in gate $(\cdot)$ and G is the set of gates.

In other words, the critical columns define the solution's maximum number of tracks. The Fig. 18.5 shows an example of such columns in a given instance.

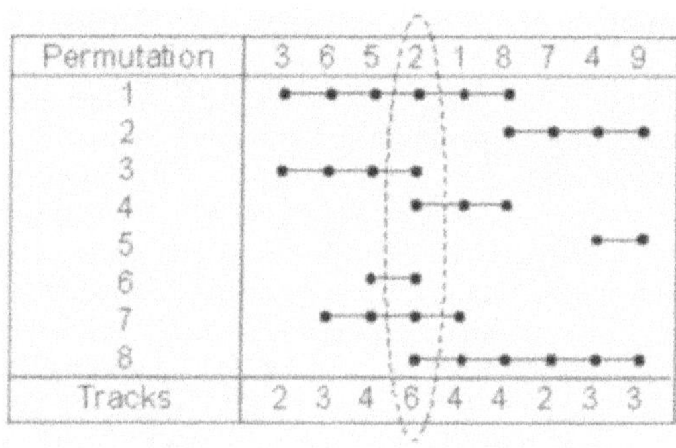

Fig. 18.5. Illustration of a *critical column* utilized in the CCBLS to reduce the overall computational effort

In the Fig. 18.5, the critical column is given by gate number two, which requires six tracks to be implemented. The local search reduction policy works by prohibiting any swap or insertion that cannot affect the critical column(s). In the example shown, movements involving any pair of gates extracted from the set {1, 8, 7, 4, 9} are not allowed, since they cannot decrease the number of tracks required by gate 2. Likewise, movements between gates belonging to the set {3, 6, 5} are also prohibited. If the movement does not affect the critical column, any reduction in the number of required tracks will be only a local improvement, not global. For instance, suppose the gates 8 and 9 are swapped in Fig. 5. The number of tracks needed to implement the gates 8, 7, 4 and 9 might be changed, but the value corresponding to the critical column will continue to be six, making the individual require at least the same number of tracks.

The reduction in the number of swap and insertion tries when the CCBLS is utilized is considerable. In the example described in Fig. 18.5, the number of possible swaps and insertions without any reduction is 108. With the CCBLS reduction scheme only, this number drops to 49. The case in which there is more than one critical column is analogous. Since the movements must affect all critical columns, the prohibited movements are the ones in which both columns belong to the following regions:

- Before the leftmost critical column.
- After the rightmost critical column.

As said before, since the computational complexity of the local search neighborhoods is $O(n^2)$, the reduction will become particularly more appealing in the larger instances. Concerning the MA performance in practice, the reduction strongly improved it, making the algorithm reach better values with a much lower number of individual's evaluations. The use of critical columns information has also allowed an increase in the number of nearest neighbors to be tested – the k-value. In other words, as the use of the CCBLS concentrated the computational effort only on movements that could improve the individual's fitness, it opened up space to increase the search horizon while maintaining the computational complexity at a low level.

After a well-suited local search is developed, the next step is to determine how to better use it. There are several possibilities, such as applying the local search operator in:

- All new individuals created during the generation.
- A percentage of the new individuals.
- The best individual of the population.

The first possibility is the most natural, since the offspring generally evolve in equal conditions. The problem is that depending on the complexity of the individual evaluation, the number of local searches becomes prohibitively large. In fact, we dropped this possibility after the first test, where the algorithm took more than five minutes to execute just one generation using the 141-gate instance. This might be a good example where distributed processing could be successfully employed, since the local search in an independent process.

The use of local search in a percentage of the new individuals appeared to be more realistic. Nevertheless, we had the same problem of wasting too much computational time applying local searches in individuals that did not return satisfactory results. The results were still disappointing, regardless of the percentage utilized (tests ranged from 10% to 90% of the new individuals, with 10% steps).

The very best results came when we concentrated the local search on the best individual only. There is only one local search per evolutionary cycle[11] and moreover, it is always concentrated on the most promising individual, what increases the chances to find higher quality solutions in a shorter CPU time. However, we

[11] An evolutionary cycle begins with the initial population creation and ends when it converges to a low-diversity state.

should emphasize that this choice was strongly influenced by the problem structure, as well as the fitness function. It is not recommended to make a direct relation between the local search complexity and its level of application into the individuals. This is a much more complex issue, being always necessary thorough tests considering a broad set of possibilities.

The use of the two neighborhood definitions – swap and insertion – together with the two reduction schemes, contributed to obtain a general superior performance, continuously surpassing the best previous approach to the problem[12]. Moreover, the use of local search only in the best individual also helped to reduce the number of fitness evaluations, a second criterion utilized to compare the results.

18.3.5 Selection for recombination

The selection for recombination is one of the two processes that play the role of natural selection in the MA – the other one is the *offspring insertion*. There are many ways of selecting parents to give birth to an offspring. Among them, we must cite:

- Uniformly at random selection.
- Biased random selection (e.g., *roulette wheel* or *tournament selections*).
- Restrictive selection operators are usually used in structured populations (or breed-separated populations) or when there is a limit in the number of recombinations a given individual can take part in.

The unbiased random selection was the first approach utilized by a GA, in the original algorithm proposed by Holland (Holland 1975). It is very simple to be implemented, but lacks an important nature-related feature. Nature always privileges the best-fitted individuals to generate descendants. The reason is that such individuals tend to live longer, having a more numerous offspring than worse fitted ones. The biased random selections are successful in simulating such conditions. For instance, the roulette wheel approach selects parents with a frequency that is proportional to their fitness, meaning that the best individuals will have their genetic information transmitted to a larger number of descendants. This is a more realistic scenario, and usually leads to better results compared to the first option.

The restrictive selection operators are better suited for special reproduction scenarios. The tree-structure separates the population in different clusters and reproduction occurs only between individuals belonging the same cluster. In contrast with more static schemes, our clustered environment is more flexible, since individuals from different clusters can swap places if one becomes better than the other. This adds a different dynamics to the population.

In the hierarchically tree-structured population, the reproduction algorithm can only involve a leader and one of its supporters within the same cluster. The recombination procedure selects any leader uniformly at random and then chooses – also uniformly at random – one of the three supporters. The adoption of a uni-

[12] The best previous results for the GMLP were obtained by (Linhares et al. 1999).

formly-at-random cluster selection is related to the 'multiple population' character of the hierarchical tree approach. By choosing the clusters in an unbiased fashion, we put the same 'evolutionary pressure' in all clusters. The use of biased selection operators is not necessary, since the evolutionary pressure on the best individuals is carried out by the population hierarchy adopted and by the offspring-acceptance policy. After an individual was created and went through – or not – a mutation process, the algorithm decides whether or not it should be discarded. This phase is described next, in Sect. 18.3.6.

18.3.6 Offspring insertion

The offspring insertion into the population is the second mechanism that plays the role of natural selection in the MA[13]. Among the several existing insertion policies, at least three should be highlighted:

- Always accept the offspring.
- Accept the offspring with a probability that is proportional to its fitness.
- Accept the offspring only if is better fitted than one of the parents.

The *always accept* policy has both strong and weak points. The strong one is that diversity is preserved for much longer and premature convergence is not a problem. The weak side is that since the population has a fixed size and all offspring are accepted, a situation where a good individual is replaced by a worse one might be possible, and even common. This could be partially overcome by setting the crossover rate at low levels and replacing individuals sequentially from the worse to the best one, in an *elitist* fashion.

The second policy is a more realistic representation of what happens in nature. In many species, the offspring has very limited parental support to survive its early days. The image of thousands of little turtles swarming over the beach heading to the sea just after being born is a somewhat suitable example. Predators are everywhere, choosing their preys among literally thousands of possibilities, in a strongly randomized process. The turtles that shall survive are the best fitted to reach the water as soon as possible and, of course, the lucky ones. The policy of accepting an offspring with a probability proportional to its fitness represents exactly this situation and shall improve the algorithm's performance in comparison to the first one.

The third policy, which was utilized in the MA tests, states that the offspring will only be inserted if it is better than one of its parents, replacing the leader of the supporter that took part in the recombination. If the new individual is better than the leader of the cluster, it takes the leader's place; otherwise it takes the supporter's. This is an extremely elitist and restrictive policy, which generates a very fast loss of diversity. The positive side is that the algorithm becomes more "focused" early on and evolves faster. This characteristic was especially beneficial when no local search was employed. Although our tests are concentrated in the MA, we decided to make a few tests with a population-structured GA version, i.e.,

[13] The first mechanism is the selection for recombination, described in Sect. 18.3.5.

the same MA without the use of local search. The use of this third policy strongly improved the performance of the algorithm, compared to the use of the two previous ones. This improvement was also noticeable in the MA, but with a lower intensity.

In order to deal with the accelerated loss of diversity, a more sensitive population-convergence checking had to be developed. Generally, population convergence is evaluated by the similarity degree of the individuals' chromosomes and/or fitness. In this work we adopted a criterion stating that if during an entire generation no individual was accepted for insertion, we conclude that the population has converged. In such case, the heavy mutation procedure is applied thereafter.

During the recombination phase, after each generation, the population goes through a restructuring procedure. The hierarchy described in Sect. 18.3.1 states that the fitness of the leader of a cluster must be lower than the fitness of the leader of the cluster just above it. Following this rule, the higher clusters will have leaders with better fitness and the best solution will be the leader of the root cluster. The adjustment is done comparing the supporters of each cluster with the leader. If any supporter turns out to be better than its respective leader, they swap their places. Considering the GMLP, the higher is the position that an individual occupies in the tree, the fewer is the number of tracks it requires.

18.3.7 Pseudo-code of the MA

The Fig. 18.6 shows the pseudo-code of the MA described in the previous sections. The internal *repeat-loop* is responsible for the population evolution. Just before it starts, there is an initialization step, where all individuals are created and evaluated. Then, the processing enters the recombination *for-loop* itself, where initially, parents are selected and an offspring is created from these parents. The offspring is then mutated, evaluated and inserted into the population. This process is repeated until the last offspring is created. The population is then restructured and, if a convergence criterion is not satisfied, the process continues, otherwise the processing exits the internal repeat-loop. After the internal loop finishes, the local search is applied on the best individual and another population cycle begins. After all the populations were processed, the individual-migration phase takes place and afterwards, the restart of all populations. This process is repeated until a time limit is reached and the best solution ever found is reported.

Method MultiPopMemeticAlg;
begin
    repeat
        for $i = 1$ **to** *numberOfPopulations* **do**
            initializePopulation(*pop*(*i*));
            evaluatePopFitness(*pop*(*i*));
            restructurePop(*pop*(*i*));
            repeat

```
            for j = 1 to numberOfRecombinations do
                selectParents(individualA, individualB) ⊆ pop(i);
                newInd = recombine(individualA, individualB);
                if (makeMutation newInd) then newInd = mutation(newInd);
                evaluteIndFitness(newInd);
                insertIntoPop(newInd, pop(i));
            end
            restructurePop(pop(i));
        until (populationHasConverged pop(i));
    end
    for i = 1 to numberOfPopulations do
        makeMigration(pop(i));
    end
  until(stopCriterion);
end
```

Fig. 18.6. Diagram of the implemented MA's pseudo-code

18.3.8 Migration policies

The MA uses an *island*-type migration model. Consider several populations evolving separately in parallel, with little or no migration at all between them. Due to the genetic drift, explained in the Sect. 18.1, it is reasonable to conclude that after several generations, the final populations will differ a lot. This differentiation can be very useful for the MA search process. Since the populations will be composed of high quality, different individuals, the migration can create a precious synergy, leading the algorithm to solutions otherwise unreachable (Gordon and Whitley 1993; Levine 1994).

The island model needs a definition of a communication topology between islands, according to which individuals will migrate. One of the most studied and used topologies is the so-called *ring structure*. The populations are located at nodes over a ring, restricting migration to adjacent populations only. This approach apparently works better than the topology where all populations are connected. This second choice reduces the action of the genetic drift, worsening overall performance.

The next factor to be defined is the migration rate. It determines how many individuals will migrate after the population convergence. This is also a critical parameter since a too high migration rate might influence the genetic drift effect in the same way the topology does. We defined three options for it.

- **0-Migrate**: No migration is used and all populations evolve in parallel without any kind of individual exchange.
- **1-Migrate**: Migration occurs in all populations and a copy of the best individual from each one migrates to the population right next to it, replacing a randomly chosen individual – except the best one. Every population receives only one new individual.

- **2-Migrate**: Migration does also occur in all populations, but two copies of each one's best individual migrate to the adjacent populations, replacing randomly chosen individuals – except the best one. Every population receives two new individuals.

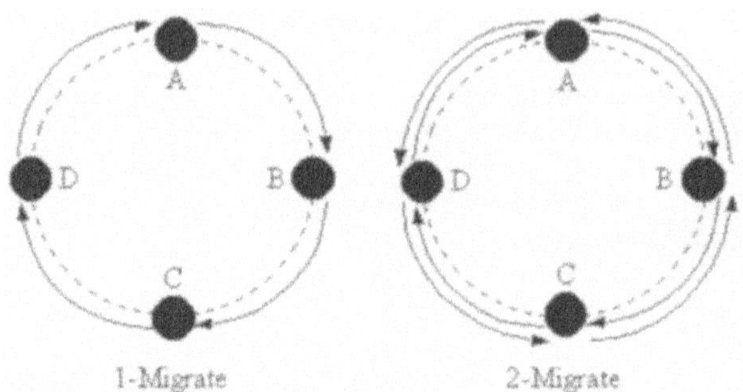

Fig. 18.7. The figure displays the diagrams of the 1-Migrate and 2-Migrate policies in an example with four populations arranged in a ring structure. The arrows indicate migratory movements.

The Fig. 18.7 shows the diagrams of two migration policies. The four populations are placed in a ring structure. In *1-Migrate* we have only one external individual being received by each population. In *2-Migrate*, this number rises to two individuals. The migration phase occurs always after all populations have converged and comparing the three policies is, in fact, a comparison between none, weak and strong migration, given the difference of communication intensity.

The last step is to determine the individuals to migrate. There are at least two choices: a randomly selected individual or the best one. Preliminary tests indicated that selecting a random individual is worse. The population's best individual carries the best genetic information available and shall be more valuable in a future recombination with individuals from another population. Therefore, the migration was set to exchange only the best individuals among the populations. After migration occurs, all populations will go through a heavy mutation process that will restart them, leaving only the best individuals untouched, including the ones that were migrated. Therefore, after the migration and heavy mutation phases, the conformation of the populations will be as follows:

- In the 0-Migrate every population will restart from the heavy mutation with just one high quality individual - the best one before the heavy mutation process.
- In the 1-Migrate policy, every population will restart with two high quality individuals. The original best individual and the one received from the adjacent population.
- In the 2-Migrate, the number of high quality individuals jumps to three. The original one and two received from the neighbor populations.

18.4 Computational experiments

The computational tests covered two aspects of the migration policies. The first was the influence of the number of populations on the overall performance. For this evaluation the number of populations varied from one up to five. The second aspect was the influence of migration intensity on the algorithm's performance: for each number of populations, the three migration policies were tested, totaling 13 configurations – with only one population its not necessary to test 1-Migrate and 2-Migrate. The tests were applied into five instances, for which we tested the whole set of configurations, ten times each one. In Table 18.1 we show some information about the instances utilized in this work.

Table 18.1. Information on the instances utilized in the computational tests

Instance name	Number of gates	Number of nets	Best known solution
W2	33	48	14
V4470	47	37	9
X0	48	40	11
W3	70	84	18
W4	141	202	27

There is one small instance, three medium-sized, and a large one. The work with the most extensive tests available in the literature (Linhares et al. 1999) presented 25 instances in total. However, most of them were too small and easy to solve either to optimality or to the best-known solution with the MA. Considering the instances' sizes, only V4470, X0, W2, W3 and W4 had more than 30 gates and for this reason we concentrated the study on them. The stop criterion for the MA was a time limit, fixed as follows:

- 10 seconds for the instance W2.
- 30 seconds for the instances V4470 and X0.
- 90 seconds for the instance W3.
- 40 minutes for the instance W4.

The difference between CPU times is due to the dimension of the instances and takes into account the average time to find high quality solutions. The increase in the CPU time from W3 to W4 reflects an explosive growth in the computational complexity. Although the instance size has just doubled, the time required to solve it was multiplied by more than 25. All tests were executed in a Pentium 366 MHz Celeron[14] computer, using Sun JDK 2.0 Java language running under Windows 98[15] environment. For comparison reasons, we estimate that the Pentium Celeron processor's performance lies somewhere between a Pentium and a Pentium II processor. The Java 2.0 is a *just-in-time* native compiler, much faster than previous Java versions.

[14] Pentium and Celeron are trademarks of Intel Microsystems.

[15] Windows is a trademark of Microsoft Inc.

The experimental results are shown next. Four figures describe the results for each configuration (see Fig. 18.8). In clockwise order we have in boldface the best number of tracks found for that instance. Next in the sequence we display the number of times this solution was found out of ten tries, the worst value found, and finally, in the lower-left part of the cell, is the average number of tracks.

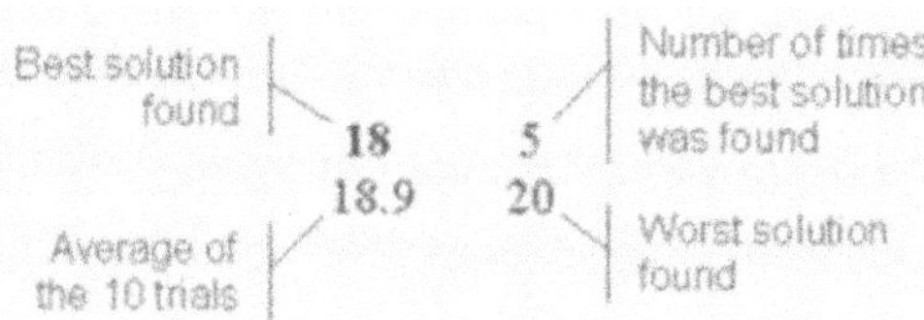

Fig. 18.8. Data fields for each configuration of number of populations and migration type

The instance W2 has reached the presumed optimal value of 14 tracks in all configurations of migration policy and number of populations. However, we could still conclude that the best configuration was the 1-Migrate with four populations because it required the lowest number of individual's evaluations. The other instances have shown more noticeable differences in performance and their results were thus detailed in the Tables 18.2 to 18.5.

Since the tests take into account two parameters – migration policy and number of populations – we will analyze them separately. At first, two aspects of randomized search algorithms must be pointed out: *exploitation* and *exploration*. Exploitation is the property of the algorithm to thoroughly explore a specific region of the search space, looking for any improvement in the current best available solution(s). Exploration is the property to explore wide portions of the search space, looking for promising regions, where exploitation procedures should be employed. In general, the island model of migration privileges exploration, with each population searching in a different part of the solution's space. Moreover, when an individual is migrated, the intention is to explore the region between the ones covered by the populations. The exploitation job, carried out mainly by the local search operator, is thus simultaneous with the exploration.

Table 18.2. Results for the V4470 instance

	Number of populations									
	1		2		3		4		5	
0-Migrate	**9**	2	**9**	1	**9**	4	**9**	1	**9**	5
	10.1	11	10.1	11	9.6	10	10.0	11	9.5	10
1-Migrate			**9**	2	**9**	6	**9**	5	**9**	4
			9.9	11	9.5	11	9.6	11	9.6	10
2-Migrate			**9**	5	**9**	4	**9**	2	**9**	5
			9.5	10	9.7	11	9.8	10	9.5	10

Table 18.3. Results for the X0 instance

	Number of populations									
	1		2		3		4		5	
0-Migrate	**11** 11.6	6 13	**11** 11.1	9 12	**11** 11.2	8 12	**11** 11.1	9 12	**11** 11.1	9 12
1-Migrate			**11** 11.0	10 11	**11** 11.0	10 11	**11** 11.0	10 11	**11** 11.1	9 12
2-Migrate			**11** 11.2	8 12	**11** 11.1	9 12	**11** 11.2	8 12	**11** 11.1	9 12

Table 18.4. Results for the W3 instance

	Number of populations									
	1		2		3		4		5	
0-Migrate	**18** 20.0	3 23	**18** 20.0	3 23	**18** 19.1	5 20	**18** 18.4	7 20	**18** 18.8	5 20
1-Migrate			**18** 20.0	3 22	**18** 18.9	5 20	**18** 18.6	6 21	**18** 18.1	9 19
2-Migrate			**18** 20.1	4 23	**18** 18.6	7 21	**18** 18.5	6 20	**18** 18.9	4 22

Table 18.5. Results for the W4 instance

	Number of populations									
	1		2		3		4		5	
0-Migrate	**29** 30.5	4 34	**28** 30.9	1 36	**28** 30.6	1 34	**28** 30.4	2 35	**28** 30.3	2 34
1-Migrate			**29** 30.9	2 34	**28** 30.6	2 35	**27** 29.4	2 34	**27** 30.2	1 34
2-Migrate			**28** 31.2	1 36	**28** 30.0	4 34	**28** 29.7	2 32	**28** 29.4	3 30

The first parameter – migration policy – has a trifling influence on the algorithm's performance. Nevertheless, the 1-Migrate appears to be the logical choice, since the best configuration for each instance utilized it. The other two policies have also had a good performance but were slightly worse. The 1-Migrate apparently better-balanced exploitation and exploration, returning the W4 presumed optimal solution – 27 tracks – in two different configurations. The other instance where migration has impacted the results was the V4470, which revealed to be a rather difficult problem. The 0-Migrate option was clearly worse than the other two, on average.

The second aspect to be analyzed is the number of populations. Although it is also not clear which configuration was the best, the use of only one is surely not the best choice since several multi-population configurations gave better final values. Based on the results, the conclusion is that when multiple populations are utilized, at least three of them should be employed. With only two populations, the algorithm does not seem to take advantage of the genetic drift effect. The use of more populations allows the exploration of much larger portions of the search space, being necessary at least three of them to make this feature noticeable.

The other characteristic of the method was its high efficiency. The number of individual's evaluations was very low. Compared to the best previous work (Linhares et al. 1999), the improvement is striking. Initially, we shall briefly explain that approach. The *Microcanonical Optimization* (MO) presented in (Linhares et al. 1999) is based on a fast variant of the well-known *Simulated Annealing* (SA) (Kirkpatrick et al. 1983) approach, which divides the search into two alternating phases, initiation and sampling. These phases have dual objectives: in initiation, the system strives to rapidly obtain a new locally optimum solution, while at the sampling phase the system moves out of the local optimum while retaining similar cost values (as controlled by parameters analogous to the temperature in SA). The proposed algorithm was able to outperform five previous approaches in all previously tested instances. The MO is also fully described in (Linhares et al. 1999). The Table 18.6 shows the results for both methods.

Table 18.6. Number of individual's evaluations required by the *memetic algorithm* and by the *microcanonical optimization* approaches to reach the presumed optimal solutions

	Number of evaluations					
	Memetic Algorithm			Microcanonical Optimization		
	Minimum	Average	Maximum	Minimum	Average	Maximum
W2	3,125	3,523	5,398	12,892	19,839	26,541
V4470	32,509	176,631	451,377	102,976	1,109,036	2,714,220
X0	18,136	43,033	117,384	52,126	95,253	187,335
W3	79,089	203,892	495,306	1,700,667	7,143,872	21,289,846
W4	3,213,532	9,428,591	15,643,651	24,192,291	167,986,282	405,324,093

Analyzing the results presented in Table 18.6, we conclude that the MA obtained much better values than the MO approach, in all instances. Taking the average results, the reduction is better than 80%, depending on the instance size. For the largest instance, W4, we had an outstanding 94% average reduction in the number of evaluations. Such results were absolutely unexpected, and thus we decided to check for any bugs in the software code. After discarding this possibility, we ran both algorithms – MA and MO – on the same machine. The result was finally confirmed by the difference of CPU time. For instance, the MO required over six hours of CPU time to find the W4 optimal value, while the MA found it within less than 20 minutes.

The comparison of the number of evaluations has also brought some light into the V4470 instance results. Although the V4470 is smaller than the X0, the MA had some difficulties in finding the optimal value. This was reflected in the number of evaluations, much larger than for the X0 instance. Initially, we thought this was a problem of the MA, but after checking the MO results we concluded this instance is intrinsically difficult, deserving a more careful study on its structure.

There are at least three reasons for these outstanding results. First, the local search embedded on the MA does not consider all possible movements, as a neighborhood reduction enables considerable computational gains (without major loss of search power). Second, the idea of discarding all possible movements that do not affect *critical columns* (and hence cannot improve solution quality) also is a major step towards greater efficiency. Finally, the application of local search is carried out only to the best individual of each population. These factors, taken in combination, help to explain the superior performance of this algorithm. Of course, all other factors – migration policy, crossover strategy, mutation, population structure and offspring's insertion policy – also played an important role. Nevertheless, the local search is by far, the most performance-influencing feature of the MA. Now, in order to give more information about the speed of the algorithm, the Table 18.7 presents some MA-related statistics.

Table 18.7. Statistics on the MA related to the number of individual's evaluations – the figures represent average values

Instance name	k-value	Evaluations per local search	Evaluations per second
W2	10	379	6,906
V4470	20	548	5,702
X0	20	585	5,557
W3	30	749	3,374
W4	60	3,242	662

The k-values presented in Table 18.7 are the same used in the computational tests. Next to them we have the number of evaluations per local search. This number does already take into account the two local search reductions – the k-based limit and the CCBLS. Finally, the fourth column shows the number of evaluations per second. The evaluation procedure requires the entire 0-1 matrix to be scanned and its complexity order is $O(g.n)$. There is little space for optimization and thus two different implementations of that procedure cannot differ too much in terms of complexity. We believe this value might be a good performance measure between different computer systems, facilitating future comparisons.

18.5 Discussion

This chapter presented a memetic algorithm (MA) application to the gate matrix layout problem (GMLP), which belongs to the NP-hard class. The main features of the MA were described, including:

- A hierarchically structured population and its specific selection for recombination procedure.
- The BOX crossover – a variant of the well-known OX crossover.
- The local search and two neighborhood-reduction policies.
- An island-model migration.

The items cited above were critical for the better performance of the MA, compared to the previous best method for the GMLP. Many questions might arise at the end of this chapter. Is the ternary-tree structure really the best choice? Can the local search be improved even further? Should the ring structure of the island-model be replaced by another one?

Every time a general method is specified and used on a given problem, questions related to the finally specification naturally start to pop-up. Nevertheless, the main goal of this implementation was to join new ideas, like the hierarchical population, the BOX crossover and the CCBLS local search, with "consecrated" ideas, such the island-model migration, with the intention to solve a complex combinatorial problem.

A conclusion derived from this implementation, which is worth emphasizing, is that local search *is very important.* No matter how hard is the problem being solved, if a GA was already developed for it, it seems reasonable to attempt a MA adding an *ad hoc* local search procedure to the GA.

The use of a standard set of "industrial-sized" real-world instances was very important in our study. Their sizes varied from 33 up to 141 gates. The solutions obtained have rivaled with the ones from previous methods', matching all the best-known solutions, while dramatically decreasing the computational effort. In all cases, the number of evaluations was reduced by at least a factor of eight.

This algorithm is included in a framework for general optimization, called NP-Opt (Mendes et al. 2001). This framework is an object-oriented, Java-based software. At present, it includes five different NP-problems: Single Machine Scheduling, Parallel Machine Scheduling, Flowshop Scheduling, Gate Matrix Layout and Gene Ordering. The framework is updated and improved continuously by a team of collaborators. For more information, please refer to the NP-Opt Homepage[16], where the latest version is always available for download, as well as the software guide and a set of test instances. This guide describes technical aspects of the framework including the classes' structure. It will also help the user to change the software code, adjusting it to different needs. More complex issues like how to add new methods, problems, and changing the graphical interfaces are also discussed in the user manual.

Although the MA used to address the GMLP contains several specially tailored features, it runs in this general optimization environment. The use of the NP-Opt framework facilitates the programming of optimization methods. We believe that further development and study of such frameworks may ultimately provide an invaluable tool to the field of systems science.

[16] http://www.densis.fee.unicamp.br/~smendes/NP-Opt.html

References

Bodlaender HL (1986) Classes of graphs with bounded tree-width. Technical Report RUU-CS-86-22. Department of Computer Science, University of Utrecht

Bodlaender HL (1993) A Tourist Guide through Treewidth. Acta Cybernetica 11:1–21

Cantú-Paz E (1997) A Survey of Parallel Genetic Algorithms. Technical Report 97003, Illinois Genetic Algorithms Laboratory, University of Illinois at Urbana-Champaign

Cantú-Paz E (1999) Topologies, Migration Rates, and Multi--Population Parallel Genetic Algorithms. Technical Report 97007, Illinois Genetic Algorithms Laboratory, University of Illinois at Urbana-Champaign

Darwin CR (1993) The origin of the species. Random House, New York

Dawkins R (1976) The selfish gene. Oxford University Press, Oxford

Downey RG, Fellows MR (1995) Fixed parameter tractability and completeness 1. Basic results. SIAM Journal on Computing 24:873–921

Fellows MR, Langston MA (1987) Non-constructive advances in polynomial--time complexity. Information Processing Letters 26:157–162

Fellows MR, Langston MA (1989) An analog of the Myhill-Nerode theorem and its use in computing finite basis characterizations. Proceedings of the 30th Annual Symposium on Foundations of Computer Science, pp 520–525

Foo SK, Saratchandran P, Sundararajan N (1999) An evolutionary algorithm for parallel mapping of backpropagation learning on heterogeneous processors. International Journal of Systems Science 30:309–321

França PM, Mendes AS, Moscato P (2001) A Memetic Algorithm for the total tardiness Single Machine Scheduling Problem. European Journal of Operational Research 132: 224–242

Goldberg DE (1989) Genetic Algorithms in Search, Optimization, and Machine Learning Addison-Wesley

Gordon VS, Whitley D (1993) Serial and Parallel Genetic Algorithms as Function Optimizers. Proceedings of the ICGA'93 – 5th International Conference on Genetic Algorithms, pp 177–183

Holland J (1975) Adaptation in Natural and Artificial Systems. The University of Michigan Press

Kirkpatrick S, Gellat DC, Vecchi M (1983) Optimization by simulated annealing. Science 220: 671–680

Lengauer T (1990) Combinatorial algorithms for integrated circuit layout. John Wiley & Sons, New York

Levine D (1994) A Parallel Genetic Algorithm for the Set Partitioning Problem. Technical Report ANL-94/23. Illinois Institute of Technology

Linhares A (1999) Synthesizing a Predatory Search Strategy for VLSI Layouts. IEEE Transactions on Evolutionary Computation 3:147–152

Linhares A, Yanasse H, Torreão J (1999) Linear Gate Assignment: a Fast Statistical Mechanics approach. IEEE Transactions on Computer-Aided Design on Integrated Circuits and Systems 18:1750–1758

Linhares A, Yanasse HH (2002) Connections between cutting-pattern sequencing, VLSI design, and flexible machines. Computers & Operations Research 29: 1759–1772

Lopez A, Law H (1980) A dense gate matrix layout method for MOS VLSI. IEEE Transactions Electron. Devices 27:1671–1675

Luk BL, Galt S, Chen S (2001) Using genetic algorithms to establish efficient walking gaits for an eight-legged robot. International Journal of Systems Science 32:703–713

Mendes AS (1999) Algoritmos Meméticos Aplicados aos Problemas de Sequenciamento em Máquinas (in Portuguese). Master Thesis. State University of Campinas, Brazil

Mendes AS, Muller FM, França PM, Moscato P (1999) Comparing Metaheuristic Approaches for Parallel Machine Scheduling Problems with Sequence-Dependent Setup Times. Proceedings of the CARS & FOF'99 – 15th International Conference on CAD/CAM Robotics & Factories of the Future, pp 1–6

Mendes AS, França PM, Moscato P (2001) NP-Opt: An Optimization Framework for NP Problems. Proceedings of the POM2001 – International Conference of the Production and Operations Management Society, pp 82–89

Merz P, Freisleben B (1999) Fitness landscapes and memetic algorithm design. In: New Ideas in Optimization. McGraw-Hill, pp 245–260

Moscato P (1989) On evolution, search, optimization, genetic algorithms and martial arts: towards memetic algorithms. Technical Report C3P 826, Caltech Concurrent Computation Program, California Institute of Technology, Pasadena, USA

Moscato P, Norman M (1992) A Memetic Approach for the Traveling Salesman Problem. Implementation of a Computational Ecology for Combinatorial Optimization on Message-Passing Systems. In: Parallel Computing and Transputer Applications. IOS Press, pp 177–186

Nakatani K, Fujii T, Kikuno T, Yoshida N (1986) A heuristic algorithm for gate matrix layout. Proceedings of the International Conference of Computer-Aided Design, pp 324–327

Syswerda G (1991) Schedule optimization using genetic algorithms. In: Handbook of Genetic Algorithms. Van Nostrand Reinhold, New York, pp 332–349

Viennet R, Fonteix C, Marc I (1996) Multicriteria optimisation using a genetic algorithm for determining a Pareto set. International Journal of Systems Science 27:255–260

Weiner J (1995) The beak of the finch. Vintage Books, New York

Wong DF, Leong HW, Liu CL (1988) Simulated Annealing for VLSI Design. Kluwer, Norwell

Xiong N (2001) Evolutionary learning of rule promises for fuzzy modelling. International Journal of Systems Science 32:1109–1118

Yanasse HH (1997) On a pattern-sequencing problem to minimize the number of open stacks. European Journal of Operational Research 100:454–463

19 Parametric Optimization of a Fuzzy Logic Controller for Nonlinear Dynamical Systems using Evolutionary Computation

Laxmidhar Behera

Abstract

Fuzzy logic controllers (FLC) have long been successfully implemented for effective tracking control and regulation of nonlinear dynamical systems. The optimization of parameters of a fuzzy logic controller is the focus of research in the domain of evolutionary computation (EC). The parameters include membership functions and rule sets. In this chapter we solve this optimization problem using three different algorithms. These are simple genetic algorithm (SGA), differential evolution (DE) algorithm and univariate marginal distribution algorithm (UMDA). Like simple genetic algorithm, differential evolution is an exceptionally simple, fast, and robust population based search algorithm that is able to locate near-optimal solutions to difficult problems. In contrast, univariate marginal distribution algorithm is a purely probabilistic search strategy over the possible solution space. We have selected two nonlinear control system problems for application. In the domain of process control, control of pH poses a difficult problem because of inherent nonlinearities and frequently changing process dynamics. The efficacy of DE over SGA is shown through the design of FLC for a pH neutralization process. This result is also verified through successful implementation on a laboratory scale pH plant setup. The next problem is the control of an one-link robot manipulator. The fuzzy model of the robot inverse dynamics in conjunction with a fuzzy PD (proportional plus derivative) controller is optimized using univariate marginal distribution algorithm and compared with SGA. Fuzzy UMDA model is found to be more accurate as compared to fuzzy SGA model. However, fuzzy SGA and UMDA controllers do fare well on equal footing and their performances are accurate and robust to model uncertainties.

Key words: Adaptive Control, Differential Evolution, Fuzzy Logic Control, Genetic Algorithms, Nonlinear Control, pH Control, Process Control

19.1 Introduction

An evolutionary algorithm is a non-gradient method and a very promising direction for global search in nonlinear optimization problems. These algorithms do not need the gradient and differentiable information to update the parameters. We find many applications of evolutionary algorithms to find optimal parameters (or weights) of fuzzy logic controllers (FLC)[Karr 93,Nordvik 91,Park 94], neural networks[Meeden 96,Whitley 90,Yang 00], and fuzzy neural networks [Tetta 01] in the field of system modeling and intelligent control. Although evolutionary algorithms became popular with the success of simple genetic algorithm (SGA) [Goldberg 89], the field is maturing with the inventions of novel evolutionary paradigms. Just as competition drives each species to adapt to a particular environmental niche, so too, has the pressure to find efficient solutions across the spectrum of real-world problems forced genetic algorithms to diversify and specialize. A GA that is well adapted to solving a combinatorial task like the traveling salesman problem may fail miserably when used to minimize functions with real variables and many local minima.

Evolutionary computation [Fogel95] and other variants [Back96] use real valued representation and focus on self-adaptive Gaussian mutation. Recently, a new evolutionary optimisation technique, called Particle Swam, has been developed by [Kenn01]. This technique is inspired by the population dynamics of "bird flocks" and "fish schools" and it is able to work within disjoint search spaces and on non-differentiable objective functions, since it uses only function values and no derivatives. Given the enormous scope of evolutionary computation, we limit our attention to differential evolution [Price 97] and univariate marginal distribution algorithm (UMDA) [Heinz98, Heinz01] apart from SGA to optimize the parameters of a fuzzy logic controller and fuzzy inverse dynamics in the context of nonlinear control.

Differential Evolution is a search procedure similar to a genetic algorithm applied on real variables which is significantly faster at numerical optimization than the traditional GA. DE is also more likely to find a function's true global optimum. DE is a design tool of great utility that is immediately accessible for practical applications. Among DE's advantages are its simple structure, ease of use, speed and robustness. DE has been used to design several complex digital filters. [Price 97].

In contrast, univariate marginal distribution algorithm has been proposed by [Heinz 98] where GA has been extended to population based search methods using probability distributions instead of string recombination/crossover.

Since the introduction of the basic methods of fuzzy reasoning by Zadeh [Zadeh 73], Fuzzy logic controllers (FLC's) [Lee 90, Mamdani 74] are being used successfully in an increasing number of application areas - such as in washing machines, elevators and automobiles. There is a considerable interest in applying fuzzy logic systems to process control [Karr 93, Qin 94, Morgan 96] including cement kiln control, task scheduling and robot arm manipulation. One such application is in the control of pH, which exemplifies the need for an adaptive controller [Gustaffson 92, Saraf 94]. [Qin 94] proposed a multiregion FLC for pH control, in which the process to be controlled is divided into fuzzy regions such as high-gain, low-gain, large-time-constant, and small-time-constant based on prior knowledge of the process. But this technique suffers miserably in the constantly changing pH systems and prior knowledge of the system is not always possible. We have proposed a novel control algorithm for pH control using fuzzy neural networks [Behera 99]. Fuzzy neural networks can solve the process of parameter optimization since these networks can be trained using back propagation algorithms. However, evolutionary algorithms also provide an excellent opportunity to optimize parameters in an efficient manner.

Karr & Gentry developed a technique wherein they used GA to generate the membership functions for the pH control process [Karr 93]. But since membership functions and rule sets are interrelated, using a hand-designed rule set with a GA designed membership function does not use the GA to the fullest extent. Previous work using GA for designing FLC's [Karr 93,Nordvik 91,Park 94] has focused on the development of rule sets or high performance membership functions, when GA's have been used to develop both, it has been done serially. However, the interdependence between these two components suggests a simultaneous design procedure would be more appropriate methodology [Homaifar 95]. Homaifar & McCormick [Homaifar 95] proposed a technique for simultaneous design of membership functions and the rule sets. This technique suffers from the fact that convergence is not fast enough for real-time implementation and it is susceptible for getting trapped in a local optimum. It has been proved that by using a search smoothing function, the convergence becomes faster and global optimization is assured [Gu 94]. The search space smoothing technique can be a powerful aid for a SGA FLC for application in real-time.

We have selected two nonlinear control system problems for application. In the domain of process control, control of pH poses a difficult problem because of inherent nonlinearities and frequently changing process dynamics. The efficacy of DE over SGA is shown through the design of FLC for a pH neutralization process. This result is also verified through successful implementation on a laboratory scale pH plant setup.

The design of FLC had been a laborious task done by human experts or by trial-and-error or iteratively. The use of DE, in simultaneous design of the membership functions and the rule sets of a FLC for the real-time control of pH, was the main objective of our present work. This technique was used to produce an adaptive DE FLC for a laboratory pH control system. Nonlinearities in the system are associated with the logarithmic pH scale, and changing process dynamics are introduced by altering the system parameters such as the desired set point and the concentration and the buffering capacity of input solutions. The results are compared with a GA based FLC developed by Homaifar & McCormick [Homaifar 95], with a slight modification that a search space smoothing function [Gu 94] is used for faster convergence. After search space smoothing, some local minimum points are temporarily filled and thus the problem of getting trapped in these points is solved. The results show that FLCs augmented with the DE offer a powerful alternative to conventional adaptive control techniques. They demonstrate the potential of adaptive FLC's in the volatile environment associated with pH systems.

The next problem is the control of an one-link robot manipulator. The fuzzy model of the robot inverse dynamics in conjunction with a fuzzy PD (proportional plus derivative) controller is optimized using univariate marginal distribution algorithm and compared with SGA. Fuzzy UMDA model is found to be more accurate as compared to fuzzy SGA model. However, fuzzy SGA and UMDA controllers do fare well on equal footing and their performances are accurate and robust to model uncertainties.

This chapter is organized as follows. In Section 2, we briefly discuss the basics of Differential Evolution. An account of Genetic Algorithms with Search Space Smoothing is given in Section 3. Section 4 compares DE and GA incorporated with a search space smoothing technique. The physical setup on which the FLCs were implemented and tested is described in Section 5. Simulation and the algorithm description of both DE FLC and GA FLC are described in Section 6. Section 7 contains a brief note on the experiments conducted and the results obtained. The univariate marginal distribution algorithm is presented in section 8 and strategies for robot control using UMDA and SGA are described in section 9. Finally, Section 10 concludes with a brief summary.

19.2 Differential Evolution

The overall structure of the Differential Evolution (DE) algorithm developed by Price & Storn (1997) resembles that of most other population-based searches. It utilizes N parameter vectors of dimension D,

$$\boldsymbol{x}_{i,G},\ i=0,1,2,....,N-1 \tag{19.1}$$

as a population for each generation G. N remains constant during the optimization process. The initial population is chosen randomly if nothing is known about the system. In case a preliminary solution is available, the initial population is often generated by adding normally distributed random deviations to the nominal solution $\boldsymbol{x}_{nom,0}$. The crucial idea behind DE is a new scheme for generating trial parameter vectors. DE generates new parameter vectors by adding the weighted difference vector between two population members to a third member. If the resulting vector yields a lower objective function value than a predetermined population member, the newly generated vector replaces the vector with which it was compared. The comparison vector can, but need not be part of the generation process mentioned above. In addition the best parameter vector $\boldsymbol{x}_{best,G}$ is evaluated for every generation G in order to keep track of the progress that is made during the minimization process.

Extracting distance and direction information from the population to generate random deviations results in an adequate scheme with excellent convergence properties. The two operators used are *mutation*, and *recombination*. The mutation step size is a function of the parameter to which it is being applied and a function of time. Any source of mutating noise must also adapt to a vector populations evolving shape in solution space. Such a source is the population itself and the vector differential of every pair of randomly chosen vectors can be used to perturb another vector. By mutating vectors with population-derived noise, DE ensures that the solution space will be efficiently searched in each dimension.

For each vector, $\boldsymbol{x}_{i,G}$, a trial vector $\boldsymbol{v}$ is generated according to,

$$\boldsymbol{V} = \boldsymbol{x}_{i,G} + \lambda \cdot \left(\boldsymbol{x}_{best,G} - \boldsymbol{x}_{i,G}\right) + \boldsymbol{F} \cdot \left(\boldsymbol{x}_{r_1,G} - \boldsymbol{x}_{r_2,G}\right) \tag{19.2}$$

with

$$r_1, r_2 \in [0, N-1], \textit{ integer and mutually different}, \ \lambda > 0, \text{ and } \boldsymbol{F} > 0 \tag{19.3}$$

The integers r_1 and r_2 are chosen randomly from the interval [0, N-1] and are different from the running index i. $\boldsymbol{F}$ is a real and constant factor that controls the amplification of the differential variation. The idea behind λ is to provide a means to enhance the greediness of the scheme by incorporating the current best value $\boldsymbol{x}_{best,G}$.

In order to increase the diversity of the parameter vectors, the vector

$$\boldsymbol{u} = (u_1, u_2, \ldots, u_D)^T \tag{19.4}$$

with

$$u_j = \begin{cases} v_j & j = \langle n \rangle_D, \langle n+1 \rangle_D, \ldots, \langle n+M-1 \rangle_D \\ \left(\boldsymbol{x}_{i,G}\right)_j & \textit{otherwise} \end{cases} \tag{19.5}$$

is formed where the acute brackets $\langle \ \rangle_D$ denote the modulo function with modulus D. A certain sequence of the vector elements of $\boldsymbol{u}$ are identical to the elements of $\boldsymbol{v}$, the other elements of $\boldsymbol{u}$ acquire the original values of $\boldsymbol{x}_{i,G}$. Choosing a subgroup of parameters for mutation is similar to a process known as *cross-over* in evolution theory. The integer M is drawn from the interval [0,D-1] with the probability P(M= γ)= $\delta^\gamma, \delta \in [0,1]$is the crossover probability and constitutes a control variable for the above mentioned scheme. The random decisions for both n and M are made anew for each trial vector $\boldsymbol{v}$. Unlike many GA's, DE does not use proportional selection, ranking or even an annealing criterion that would allow occasional uphill moves. Instead the cost of each trial vector is compared to that of its parent target vector. The vector with the lower cost is rewarded by being selected to the next generation.

19.3 Simple Genetic Algorithm with Search Space Smoothing

Genetic algorithms are powerful search and optimization algorithms, based on semblance of natural genetics. They ensure the proliferation of quality solutions while investigating new solutions via a systematic information exchange that utilizes probabilistic decisions. GAs require the problem of maximization (or minimization) to be stated in the form of an objective function. In GAs, a set of variables is encoded into a binary string, analogous to a chromosome in nature. Each string therefore contains the solution to the problem and has to be broken into individual substrings to evaluate the cost function, which yields the *fitness* of that string. GAs select parents from a pool of strings (population) according to the criteria of *survival of the fittest*. They create new strings by recombining parts of the selected parents in a random manner.

The re-population of the next generation is done using three methods: reproduction, crossover, and mutation [Goldberg 89]. Through reproduction, strings with high fitness duplicate into multiple copies in the next generation while strings with low fitness receive fewer copies or none. Crossover refers to taking a string, splitting it into two parts at a randomly generated point and recombining it with another string, which has also been split at the same crossover point. This operation brings about change in the strings and exchanges information between them. Mutation is the random alteration of a bit in the string, which assists in keeping diversity in the population.

The criteria of survival of the fittest can some times lead to premature convergence when used to optimize multi-modal convergence, as the string, which represents a local minimum, might duplicate. Simple GAs also suffer from the problem that convergence is too slow to apply it to optimize a controller (adaptation) for real-time control. This drawback can be eliminated to some extent by using a search space smoothing technique [Gu 94]. The smoothing function used has different levels of strength, resulting in a search space with varying degrees of smoothness. We used a *smoothing factor*, α, to characterize the degree of a smoothing operation and the smoothness of the resulting search space. If $\alpha = 1$, no smoothing is done and the search space is same as the original search space. If $\alpha > 1$, a smoothing operation is applied and the smoothed search space is flatter than the original search space. If $\alpha >> 1$, the smoothing effect is very strong and gives rise to a nearly flat search space as shown in Figure 19.1. The smoothing procedure is dealt with in Section 6 Simulation.

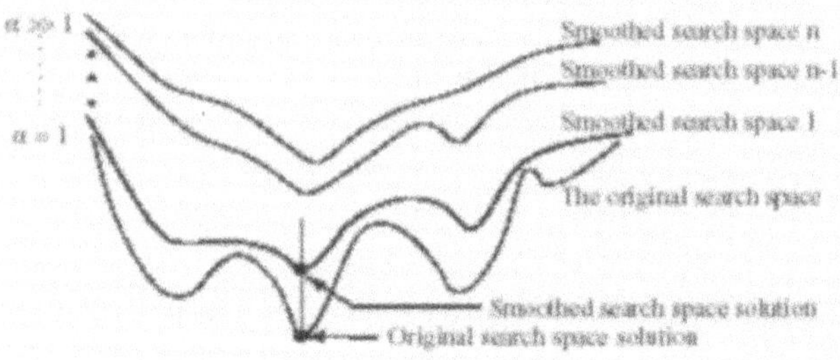

Fig. 19.1. A series of smoothed search spaces is generated. The smoother search spaces are used to guide the search of those of the original search spaces

19.4 Simple Genetic Algorithm Vs Differential Evolution

A SGA uses binary strings to code the parameters; however this choice limits the resolution with which an optimum can be located to the precision set by the number of bits in the integer. DE uses Floating-point numbers, which not only uses computer resources efficiently but also reduces the computational time, which is very crucial for control applications. while use of search space smoothing aids faster convergence of a SGA, it still takes substantially more time to converge as compared to a DE. Figure 19.2 plots the average cost/fitness in each generation. The algorithm is said to have converged if

$$\left|\sigma_G - \sigma_{G-1}\right| < \in \tag{19.6}$$

where $\in$ is a constant (in the present study $\in = 1.0e^{-4}$) and σ_G is the cost/fitness variance given by,

$$\sigma_G = \frac{\sum_{i=1}^{N} \left(F_{i,G} - \overline{F_G}\right)^2}{N-1} \tag{19.7}$$

As clearly depicted the DE converges almost in three generations and also to a much lower value.

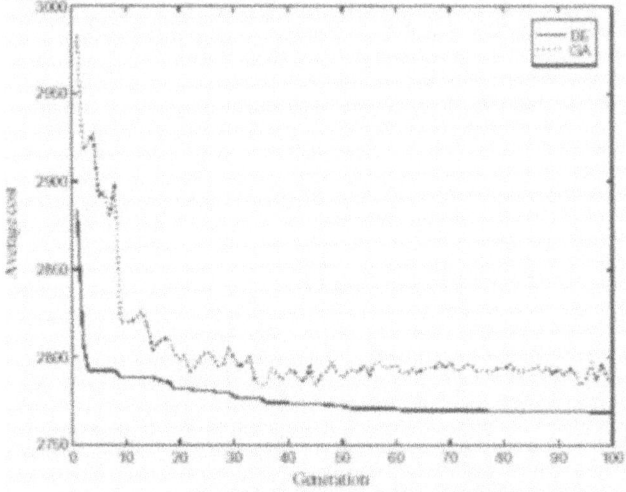

Fig. 19.2. The convergence of DE is much faster than a SGA

Just as floating-point numbers are more appropriate than integers for representing points in continuous space, addition is more appropriate than random bit flipping as used for mutation in a SGA for searching the continuum. Consider for example the consequences of using the logical exclusive OR (XOR) operator for mutation integers. To change a binary 15 (01111) into a binary 16 (10000) with an

XOR operation requires inverting all the five bits. In most bit flipping schemes, a mutation of this magnitude is very rare even though the mutation operator is meant for fine-tuning. The easiest way to restore the adjacency of neighboring points is addition. Using addition, 15 becomes 16 by just adding 1. The magnitude of the mutation increments is automatically scaled by the simple adaptive scheme used by DE.

Recombination (crossover) provides an alternative and complementary means of creating viable vectors from the components of existing vectors. The SGA uses a uniform crossover. However, a DE uses a nonuniform crossover that can take child vector parameters from one parent more often than it does from the other. The selection of the individuals to the next generation resembles *tournament selection* except that each child that is pitted against one of its parents, not against a randomly chosen competitor. Only the fitter of the two is then allowed to advance into the next generation

19.5 pH Neutralization Process

The laboratory pH system considered here was selected to be representative of pH systems present in industries and is schematically depicted in Figure 19.3. It contains both nonlinearities and changing process dynamics. The nonlinearities are due to the fact that the output of the pH sensor is proportional to the logarithm of concentration. While the changing process dynamics are brought about by introducing a buffer into the system, which significantly alters the response of the system, by changing the concentration of the control reagents and by changing the set point.

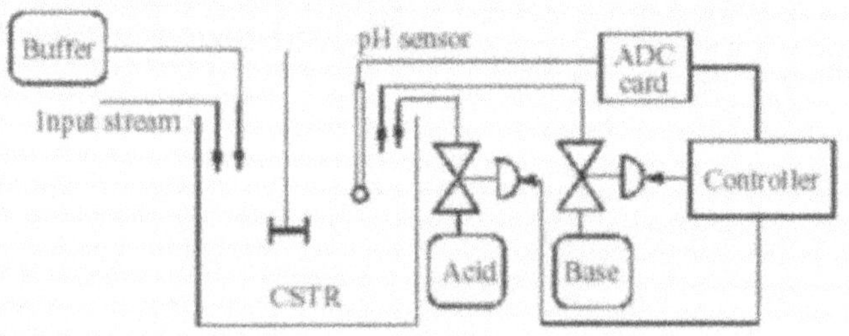

Fig. 19.3. Schematic of the physical pH control setup

The system consists of a continuous stirred tank reactor (CSTR) of 20 L volume; two control reagents, acid (HCl) and base (NaOH), which are manipulated by stepper motor driven needle valves, one input stream, whose pH has to be controlled; and another stream through which a buffer (acetic acid) can be introduced. The metering range of the control valves is 0.0-0.02 L/s with an increment of 0.0002 L/s. The objective of the control problem is to drive the pH of the system to the desired set point in the shortest time possible by adjusting the valves of the control reagents. Three industrial pH electrodes were used to account for spatial non-uniformity within the reactor. The pH value of the process stream was taken to be the median of the three sensor signals. The pH sensor signals were transmitted through a 12 bit add-on ADC card (Dynalog Microsystems, PCL 207) to a 33 MHz 386 IBM PC, which controlled the entire process. A sampling time of 30 ms was used throughout the study.

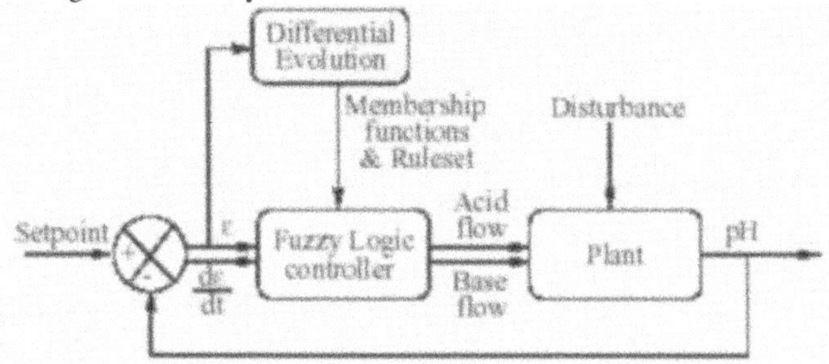

Fig. 19.4. Block Diagram of DE FLC structure

To develop an adaptive FLC using DE, a computer model of the physical system is required. Figure 19.4 shows a schematic of the basic design of an adaptive FLC that uses a DE for membership function and rule set selection. Fortunately, the dynamics of pH system are well understood and can be modeled by using conventional techniques for buffered reactions. The mathematical equations of the CSTR can be described as follows [McAvoy 72]:

$$\frac{Vd\xi}{dt} = F_{in}C_{in} - F_{out}\xi \tag{19.8}$$

$$\frac{Vd\left[Na^{+}\right]}{dt} = F_{b}C_{b,in} - F_{out}\left[Na^{+}\right] \tag{19.9}$$

$$\frac{Vd\left[Cl^{-}\right]}{dt} = F_{a}C_{a,in} - F_{out}\left[Cl^{-}\right] \tag{19.10}$$

$$\left[H^{+}\right]^{3} + (K_{a} + \varsigma)\left[H^{+}\right]^{2} + (K_{a}(\varsigma - \xi) - K_{w})\left[H^{+}\right] - K_{w}K_{a} = 0 \tag{19.11}$$

$$pH = -\log_{10}\left[H^{+}\right] \tag{19.12}$$

where

$$F_{out} = F_{in} + F_{a} + F_{b} \tag{19.13}$$

$$\xi = \left[HAc\right] + \left[Ac^{-}\right] \tag{19.14}$$

$$\varsigma = \left[Na^{+}\right] - \left[Cl^{-}\right] \tag{19.15}$$

The cubic equation must be solved for $\left[H^{+}\right]$ions, which directly yields the pH of the solution. The symbols are defined in the Nomenclature section (Appendix B). One of the strong points of fuzzy controllers is the fact that they do not require mathematical models. However, to obtain the cost for a given controller, the DE must have a method to evaluate the controller's performance. In this sense we have negated one of the fuzzy controller advantages in order to use the power of DE's to optimize this controller.

19.6 Simulation

The input parameters for the FLC viz the *error* ε , and *derivative of error* $\frac{\delta\varepsilon}{\delta t}$ had their base lengths determined by the DE, while the location of the peaks were kept constant. The rules used were of the form, **IF (** ε **is {NH, NM, NS, NZ, ZE, PZ, PS, PM, PH}) and (** $\frac{\delta\varepsilon}{\delta t}$ **is {NM, NS, ZE, PS, PM}) THEN {** ***output*** **}**, where *output* is the acid or base flow rate. The acronyms are defined in the Abbreviations section (Appendix A). The two input spaces use a total of fourteen triangles, so the dimension D of each parameter vector is (9+5+9×5) = 59. The outputs were singletons to ease the computational burden, and no additional parameters were required for them. The first forty five vectors of each parameter vector form the rule set and the next nine vectors form the base lengths of error membership functions and the rest the base lengths of the derivative of error membership functions. The rule set consisted of the flowrates of the control reagents, which were classified as **{VLA, LA, MA, SA, ZF, SB, MB, LB, VLB}**. The minimum overlap between two adjacent membership functions was chosen to be 0.2 and the maximum to be 0.8 to ensure that at least two membership functions exist through out state space. Thus the two main ingredients of a fuzzy controller, the rule set and the membership functions, are incorporated into a single individual which the DE will seek to optimize.

The task of defining a cost function is always application specific; it always comes down to accurately describing the goal of the controller. In this case, the objective of the controller is to drive the pH to the desired set point in the shortest possible time and to maintain the pH at the desired set point. The ability of the FLC to achieve these objectives can be represented by a cost

function that specifies how well the controller has reduced the error over some finite time period. To ensure that a robust FLC is obtained, the cost function should reflect the controller's ability to reach the set point from a number of initial condition cases. Mathematically, this fitness function is expressed as

$$F=\sum_{i=case1}^{case4}\sum_{j=0s}^{30s}\varepsilon^{2} \tag{19.16}$$

The initial condition cases were selected from different regions of the state space of the pH control system.

The coding of the membership functions was same in the case of GA except that the base lengths were coded as binary strings. The fitness function used was also same, but was incorporated with a search space smoothing technique. The smoothing function applied to the GA is expressed mathematically as,

$$F_i(\alpha)=\begin{cases}\overline{F}+\left(F_i-\overline{F}\right)^{\alpha} & \text{if } F_i \geq \overline{F}\\ \overline{F}-\left(F_i-\overline{F}\right)^{\alpha} & \text{if } F_i < \overline{F}\end{cases} \tag{19.17}$$

where F_i is the fitness of the ith individual, F is the average fitness, and $\alpha \geq 1$. α is decreased gradually from a large number, for eg. from 10 to 1. A search space generated by a larger α exhibits a smoother terrain surface, and that generated by a smaller α exhibits a more rugged terrain surface. The two extreme cases of the function are: 1) If α >> 1, then $F_i(\alpha)\to\overline{F}$, this is a trivial case; 2) If $\alpha=1$, then $F_i(\alpha)=F_i$, which is the original problem. The above-mentioned algorithm of DE FLC produced the membership functions as shown in Figure 19.5 and the simulation result is shown in Figure 19.6.

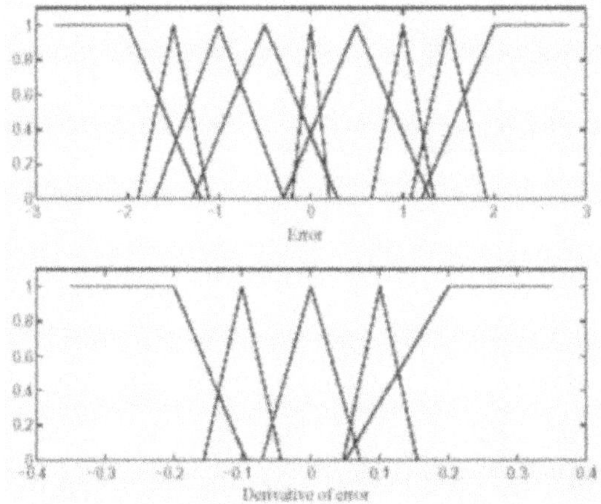

Fig. 19.5. Final Membership Function at which the DE converged

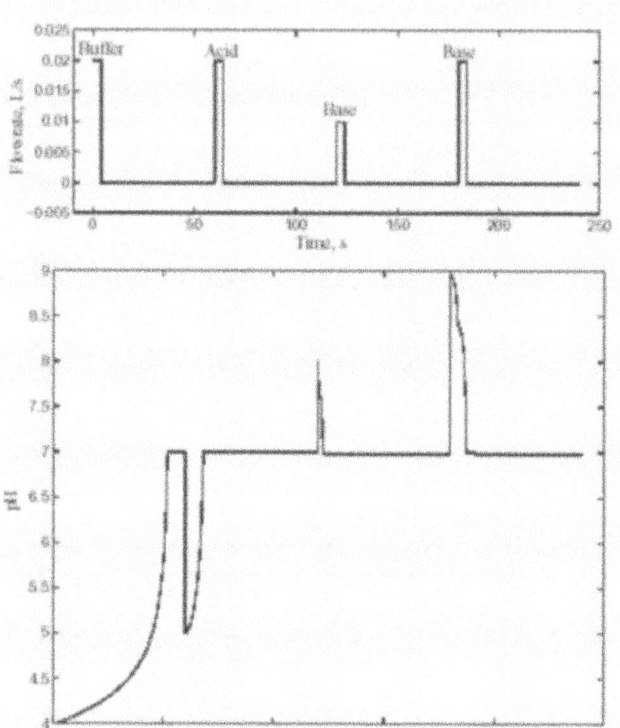

Fig. 19.6. Simulation of the response of DE FLC for perturbances like addition of buffer, acid and base

19.7 Experiments & Results

Using the aforementioned algorithm, an adaptive DE FLC was developed. The DE was used for off-line adaptation of the FLC. The performance of this adaptive DE FLC is demonstrated for three different situations, which provide a challenging test-bed for any control system. First the pH system is perturbed with the external addition of an acid, a base and a buffer along with a continuous input stream of pH 8.0. In this case the process dynamics are dramatically altered by the addition of buffer. Second, the desired set point is altered, which for all intents and purposes changes the objective of the controller. Third, the concentrations of the control reagents are changed, which causes the system to handle differently.

Consider the first case where the pH system is perturbed with a buffer, acid and base. Figure 19.7 compares the performance of a FLC developed by a DE in the third iteration with that of a FLC, which is designed by a GA after hundred iterations. The results show that the DE FLC is more efficient in terms of faster settling, lesser overshoot and steady-state error. Next consider a situation where the set point is changed, which implies that the objective of the controller is being changed. As shown in Figure 19.8 the DE FLC outperforms the GA FLC. Finally consider the case wherein the concentrations of the control reagents are changed. This is the most disruptive perturbance since it changes the system response completely. The DE FLC is able to maintain a high degree of control over the process despite the drastic change in the environment, which is clearly depicted in Figure 19.9. The results presented in this paper depict the scope of Differential Evolution in designing a FLC. The FLC augmented with DE is able to maintain a high degree of control over a bench scale pH setup despite the nonlinearities and the dramatic perturbations.

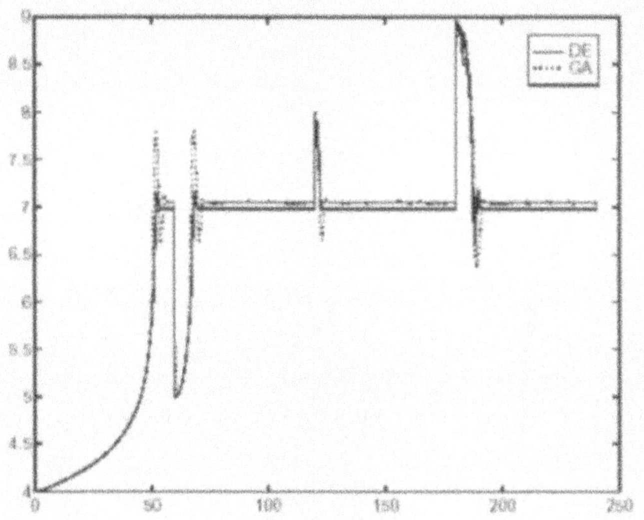

Fig. 19.7. Perturbation by the addition of buffer, acid and base: DE FLC shows better performance than the SGA FLC

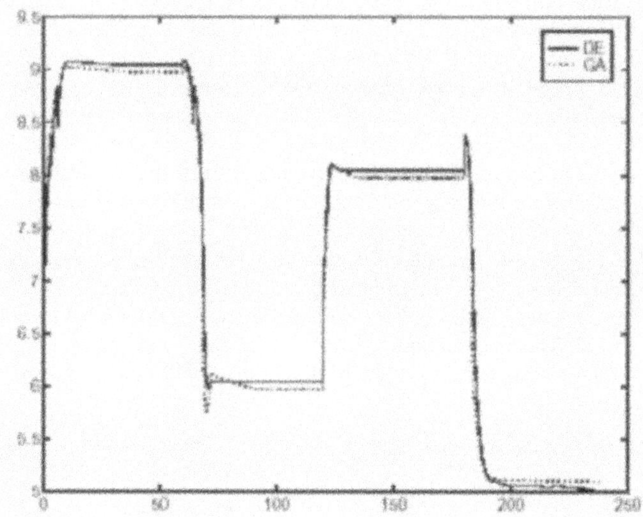

Fig. 19.8. Change of set point: DE FLC performs effeciently compared to SGA FLC

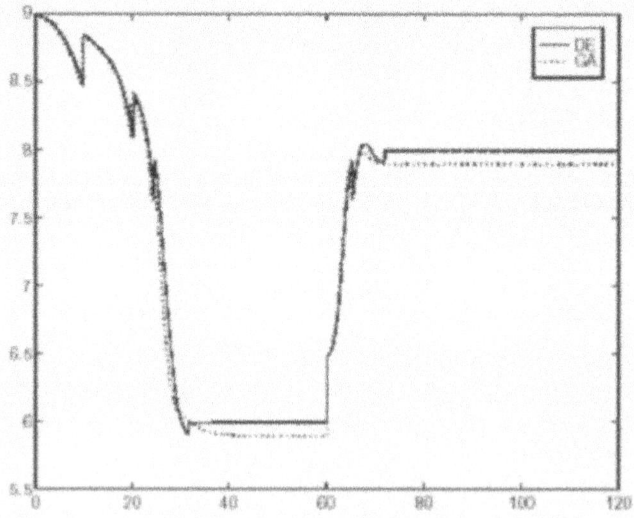

Fig. 19.9. Change of concentration of control reagents from 0.5N to 0.1N: DE FLC has less effect of parameter changes as compared to SGA FLC

19.8 The Univariate Marginal Distribution Algorithm

The univariate marginal distribution algorithm UMDA estimates the distribution of gene frequencies using mean-field approximation. Each string in the population is represented by a binary vector $\boldsymbol{x}$. The algorithm generates new points according to following distribution:

$$p(\boldsymbol{x},t)=\prod_{i=1}^{n} p_i^s(x_i,t) \tag{19.18}$$

The UMDA algorithm is given as follows:

- Step 1: Set $t = 1$, Generate $N >> 0$ binary strings randomly.
- Step 2: Select $M<N$ strings according to a selection method.
- Step 3: Compute the marginal frequencies $p_i^s(x_i,t)$ from the selected strings.
- Step 4: Generate new N points according to the distribution $p(\boldsymbol{x},t)=\prod_{i=1}^{n} p_i^s(x_i,t)$.
- Set $t=t+1$. If the termination criteria are not met, go to Step 2.

For infinite populations and proportionate selection, it has been shown [Mhlenbein and Mahnig 2001] that average fitness never decreases for maximization problem (increases for minimization problem).

19.9 Robot arm control

The multi-link robot dynamics have been represented by single link and double link robot equations to study the effects of nonlinearity in fuzzy and neuro-fuzzy learning models. We focus on a single link manipulator whose dynamics is governed by the following equation [Behera et al., 1994]:

$$ml^2\ddot{q}+mgl\cos q=\tau \tag{19.19}$$

where q and $\ddot{q}$ are respectively link position and link acceleration, m is the link mass, l is the link length, g is the gravity and τ the applied torque. The parameters are given as m=11.36 and l=0.432.

19.9.1 Control Architecture

The general approach to control a nonlinear system like robot arm is to actuate a feed-forward torque which is computed based on inverse dynamics model of the robot arm and a feedback torque that is computed using position and velocity feed-back terms [*Behera, 1995*]. The schematic diagram for such control architecture is given in Figure 10.

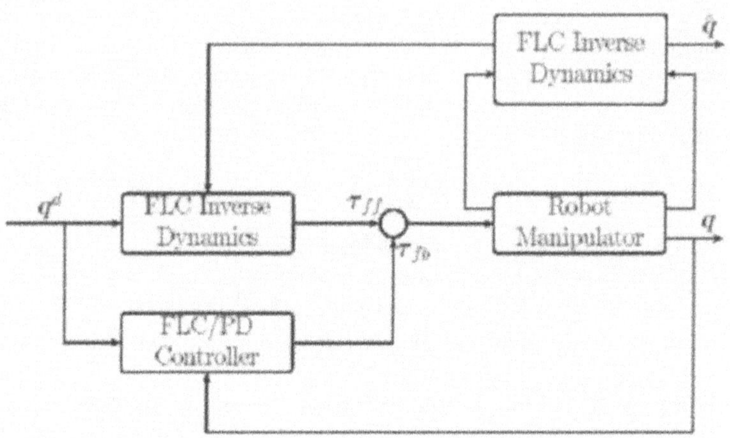

Fig. 19.10. A general control architecture for a robot manipulator

19.9.2 Inverse Dynamics Model

There are various methods available in the literature to compute the inverse dynamics of a robot arm. In the absence of dynamic and parametric uncertainties, the inverse model can be directly computed in terms of desired link position, velocity and acceleration. However, the inherent uncertainties compel a control engineer to estimate the model. The generic form of the inverse dynamics for a robot manipulator is given as follows:

$$\tau = f(q, \dot{q}, \ddot{q}) \tag{19.20}$$

where τ is the required torque, and q, $\dot{q}$, $\ddot{q}$ are respectively the link position, velocity and acceleration. Neural networks [Behera, 1995], fuzzy neural networks [Behera, 1999] and fuzzy models have become popular tools to estimate the inverse dynamic model (19). In this section, we derive a fuzzy model of the inverse dynamics using SGA and UMDA. Since the objective is to evaluate the performance of evolutionary computational approaches, we keep the model very simple. The generic form of the fuzzy model is given as follows:

$$\text{if } x_1 \text{ is } A_1, x_2 \text{ is } A_2, \ldots, x_n \text{ is } A_n, \text{ then } y \text{ is } B$$

where x_i is the fuzzy input variables for the model, A_i is the fuzzy attribute of x_i, y is the fuzzy output and B is the fuzzy attribute of y. We have only two fuzzy input variables, q and $\ddot{q}$, and one fuzzy output variable, τ, required to model the inverse dynamics for a single link manipulator. Each input variable is fuzzy partitioned into 6 regions: NB (negative big), NM (negative medium), NS (negative small), PS (positive small), PM (positive medium) and PB (positive big). Thus we have maximum possible 36 rules to describe the dynamics. For simplicity, we assume that the output fuzzy variable is a fuzzy singleton. The output of the fuzzy model is computed using center of gravity method:

$$y = \frac{\sum_{r=1}^{R} \mu_r B_r(y)}{\sum_r \mu_r} \tag{19.21}$$

where r is the index for a rule, R is the total number of rules and $\mu_r = \min(A_1(x_1), A_2(x_2), \ldots A_n(x_n))$.

Fuzzy parameters: Given that we have two input variables, each is fuzzy partitioned into 6 regions, and each region is characterized by two parameters, mean and variance, the total number of parameters in the input space are 24. There are 36 singleton parameters in the output space, one for each of the 36 rules. Thus it is required to estimate 60 parameters. Since the output is a nonlinear function in terms of these 60 parameters, the resulting nonlinear optimization problem is a good candidate within the evolutionary computation approaches. We select both SGA and UMDA to estimate these parameters.

We select binary string. Each parameter is represented by 8 bits. Initial population consists of 1000 strings. In SGA, proportional selection was adopted and multi-point crossover was done. The mutation rate was kept at 0.01.

In UMDA, 20% of the population is selected according to fitness. The univariate frequency of each bit is computed over the selected strings and a new population is generated according to these univariate frequencies.

Data for the inverse dynamics are generated using two following position trajectories: $q = \cos 3t$ and $q = \frac{1}{3}(\cos t + \cos 2t + \cos 3t)$. The data set has 500 data points. Both SGA and UMDA are evolved over 100 generations. We show the fitness of the best string in Figure 11. Here the fitness is the squared error between the desired output and predicted output. Our result shows that UMDA converges much faster compared to SGA with far better accuracy.

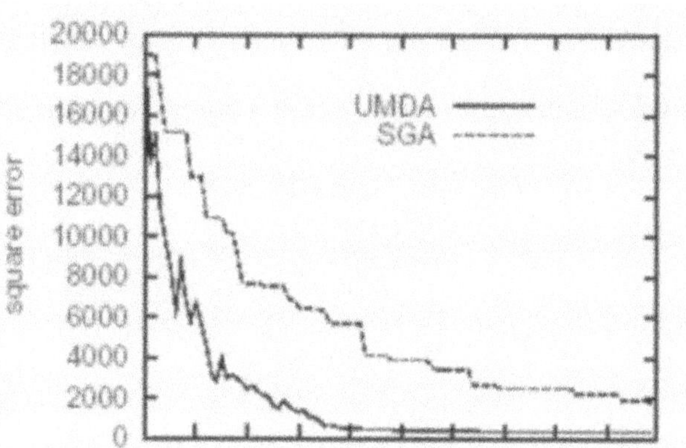

Fig. 19.11. Error convergence in parametric evolution using UMDA and SGA while modeling the inverse dynamics;UMDA is faster than SGA.

The accuracy of the predicted model is tested by comparing with the desired torque corresponding to two different trajectories. The model prediction for UMDA is shown in Figure 12. The model prediction for SGA is shown in Figure 13. The figures clearly show that UMDA does better as compared to SGA.

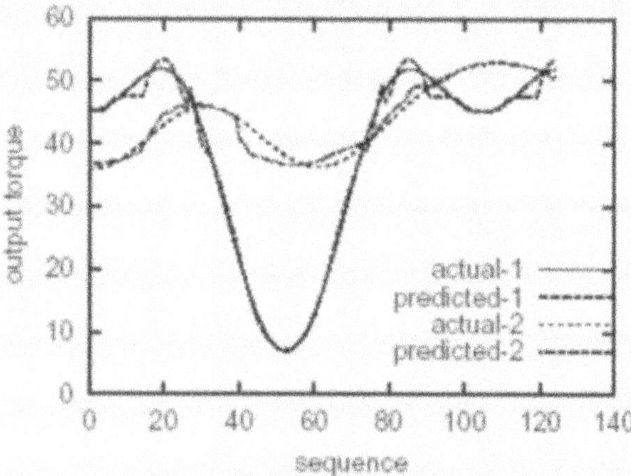

Fig. 19.12. Inverse dynamics model prediction in fuzzy parametric evolution using UDMA, actual-1 refers to the desired output corresponding to the trajectory $q = \cos 3t$ and similarly the output actual-2 corresponds to the trajectory $q = \frac{1}{3}(\cos t + \cos 2t + \cos 3t)$.

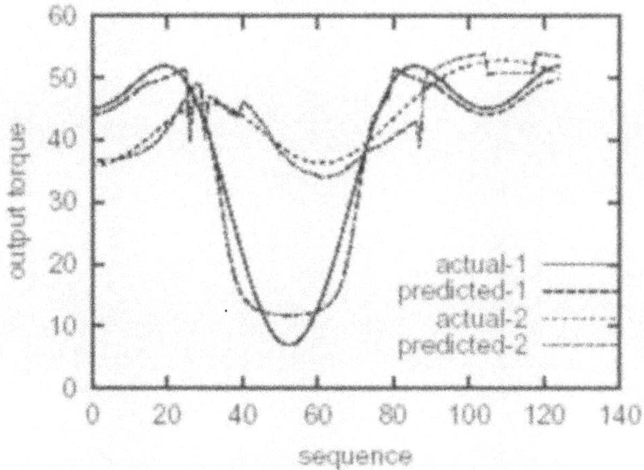

Fig. 19.13. Inverse dynamics model prediction in fuzzy parametric evolution using SGA, actual-1 refers to the desired output corresponding to the trajectory $q = \cos 3t$ and similarly the output actual-2 corresponds to the trajectory $q = \frac{1}{3}(\cos t + \cos 2t + \cos 3t)$.

19.9.3 Feedback fuzzy PD Controller

With reference to Figure 10, the fuzzy PD controller or a simple PD controller is necessary to guarantee the stability of the closed loop control system. The generic form of a rule in a fuzzy logic controller is as follows:

if x_1 is A_1, x_2 is $A_2, \ldots$ and x_n is A_n, then the control action τ_{fb} is B

The final control action is computed using center of gravity method. In case of single link manipulator, two input variables are link position and link velocity and are fuzzy partitioned into 6 regions each (NB, NM, NS, PS, PM, PB). So also the only control variable is fuzzy partitioned into 8 regions (NC, NB, NM, NS, PS, PM, PB, PC) where NC stands for negative critical and PC stands for positive critical. We represent each variable by two parameters, its mean and variance. Thus we have 40 variables. However, by fixing mean of some fuzzy attributes such as NS and PS at zero, we reduced the parameter size to 30. As before each parameter is represented by 8 bits. We show the results of error convergence as parameters are optimized in Figure 14. The figure shows that UMDA converges after 6 generations while SGA takes 25 generations to reach the same level of performance.

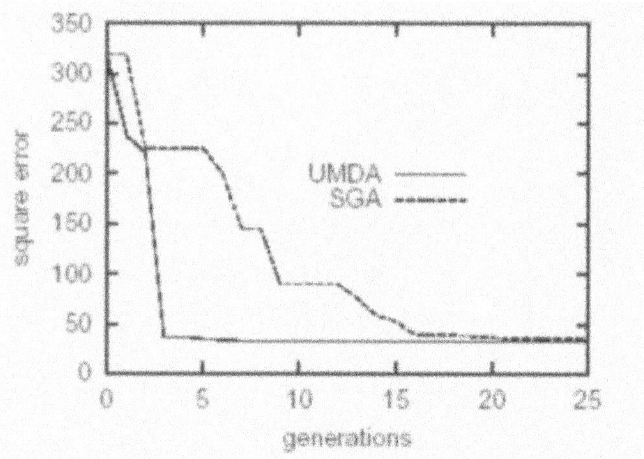

Fig. 19.14. Error convergence in parametric evolution using UMDA and SGA while designing fuzzy PD controller

We run the closed loop system using fuzzy UMDA inverse dynamics and fuzzy UMDA PD controller since UMDA performs better than SGA. In the first case, we used exact inverse dynamics along with fuzzy PD controller for a set-point tracking. The response was shown in curve 1 of Figure 15. Then we assumed 20% model uncertainties in the actual model and the control scheme was implemented using fuzzy PD control and a conventional PD controller. The respective responses are shown by curve 2 and 3 in Figure 15. The results show that FLC is robust to parametric uncertainties.

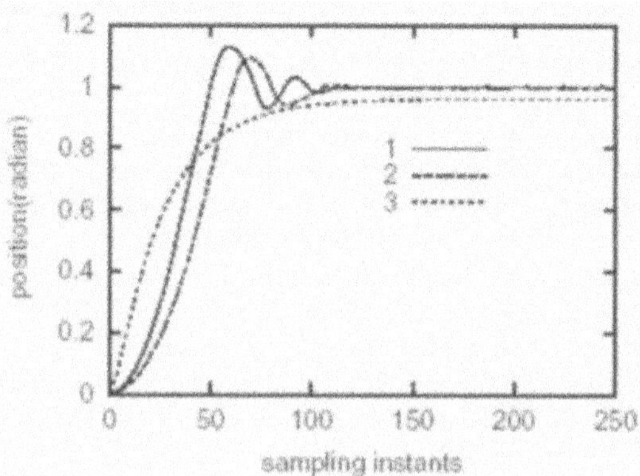

Fig. 19.15. Set point response (1) FLC without model uncertainty, (2) FLC with model uncertainty, (3) PD with model uncertainty.

19.10 Conclusions

This chapter shows the efficacy of using DE, UMDA and SGA in designing controllers without the previously needed human expert. We have earlier implemented DE for designing FLC's for the first time [Sastry, 1999]. In most of the works done earlier in the design of a self-organizing FLC, researchers have focused on using a SGA for altering the rule-set or the membership functions. Since the parameters are floating point and also due to its simple structure, ease of use, speed and robustness, it has been shown that a DE is the more appropriate choice. This methodology allows the complete design of both the membership functions and the rule-set, leading to high performance controllers. The technique has been applied to a bench scale pH setup in which the process dynamics change to a large extent. The system was subjected to perturbations like the introduction of a buffer, changes in the desired set point, and the alterations in the concentrations of the control reagents. In all instances, the adaptive DE FLC was able to successfully drive the system pH to the desired set point in a reasonable time. The results were compared with that of a SGA based FLC with a search space smoothing function. Though the smoothing function ensures faster convergence and global optimization, the results presented demonstrate that a DE FLC is more effective in terms of faster convergence, accurate control and robustness. We extended our work in this chapter by implementing fuzzy UMDA inverse dynamics and fuzzy PD controller for effective control of a robot arm. Like DE FLC, UMDA FLC and UMDA inverse dynamics achieved faster convergence in comparison to SGA FLC and SGA inverse dynamics.

References

Back T (1996), Evolutionary algorithms in Theory and Practice, Oxford University Press, New York, USA

Behera L, Gopal M, Chaudhury S (1994), Trajectory Tracking of a Robot Manipulator using Gaussian Networks, Robotics and Autonomous Systems, 13:107-115

Behera L (1995), Neural Controllers for Robot Manipulators, PhD Thesis, Indian Institute of Technology, Delhi

Behera L, Anand KK (1999), Guaranteed Tracking and Regulatory Performance of Nonlinear Dynamic Systems using Fuzzy Neural Networks, IEE Proc. Control Theory and Applications, 146:484-491

Fogel DB (1995), Evolutionary Computation: Toward a new Philosophy of Machine Intelligence, IEEE Press, Piscataway, NJ, USA

Gu J, Huang X (1994), Efficient Local Search with Search Space Smoothing: A Case Study of the Travelling Salesman Problem (TSP), IEEE Trans. on Systems, Man and Cybernetics, 24:728-735

Goldberg DE (1989), Genetic Algorithms in Search, Optimization, and Machine Learning, Reading, MA: Addison-Wesley

Gustaffson TK, Waller KV (1992), Nonlinear and Adaptive Control of pH, Ind. Eng. Chem. Res., 31:2681-2693

Mühlenbein H, Mahnig T (2001), Evolutionary Computation and Beyond, In: Foundations of Real-world Intelligence, ed. Uesaka et al, CSLI Publications, 123-188

Muhlenbein H (1998), The equation of response to selection and its use for Prediction, Evolutionary Computation, 5:303-346

Homaifar A, McCormick E (1995), Simultaneous Design of Membership Functions and Rule Sets for Fuzzy Controllers Using Genetic Algorithms, IEEE Trans. Fuzzy System, 3:129-138

Karr CL, Gentry EJ (1993), Fuzzy Control of pH Using Genetic Algorithms, IEEE Trans. Fuzzy Systems, 1:46-53

Kennedy J, Eberhart RC (2001), Swarm Intelligence, Morgan Kauffman

Lee CC (1990), Fuzzy Logic in Control Systems: Fuzzy Logic Controller, parts I and II, IEEE Trans. Syst. Man Cybern.,20:404-435

Maiti SN, Kapoor N, Saraf DN (1994), Adaptive Dynamic Matrix Control of pH, Ind. Eng. Chem. Res., 33:641-646

Mamdani EH (1974), Application of Fuzzy Algorithms for the Control of a Dynamic Plant, Proc. IEEE, 121:1585-1588

McAvoy TJ (1972), Dynamics of pH in Controlled Stirred Tank Reactor, Ind. Eng. Chem. Res. Process Des. Dev., 11:1254-1259

Meeden LA (1996), An increamental approach to developing inteligent neural network controllers for robots, IEEE Trans. on Syst. Man Cybern., 26: 474-485

Morgan P (1996), A Clear Look at Fuzzy PI Control, InTech, 50-54

Nordvik JP, Renders JM (1991), Genetic Algorithms and their Potential for Use in Process Control: A Case Study, In: Proc. of the Fourth Int. Conf. Genetic Algorithms, 260-265

Park D, et al (1994), Genetic-Based New Fuzzy Reasoning Models with Application to Fuzzy Control, IEEE Trans. on Syst. Man Cybern., 24:39-47

Price K, Storn R (1997), Differential Evolution, Dr. Dobb's Journal}, April:18-24

Qin SJ, Borders G (1994), A Multiregion Fuzzy Logic Controller for Nonlinear Process Control, IEEE Trans. Fuzzy Systems, 2:74-81

Sastry KKN, Behera L, Nagrath IJ (1999), Differential evolution based fuzzy logic controller for nonlinear process control, Fundamenta Informaticae, 37:121-136

Tettamanzi A, Tomassini M (2001), Soft Computing: Integrating Evolutionary, Neural, and Fuzzy Systems, Springer-Verlag.

Whitley D, Starkweather T, Bogart C (1990), Genetic algorithms and neural networks: Optimizing connections and connectivity, Parallel Computing, 14:347-361

Yang JM, Horng JT, Kao CY (2000), A genetic algorithm with adaptive mutations and family competition for training neural networks, International Journal of Neural Systems, 10:333-352

Zadeh LA (1973), Outline of A New Approach to the Analysis of Complex Systems and Decision Processes, IEEE Trans. Syst. Man Cybern., 3:28-44

20 DNA Coded GA: Rule Base Optimization of FLC for Mobile Robot

Prahlad Vadakkepat, Xiao Peng and Lee Tong Heng

20.1 Introduction

In recent years, the new concept of DNA (deoxyribonucleic acid) computing has drawn intensive research interests. The idea of DNA computing, proposed by Leonard Adleman (Leonard 1994) in 1994, is to express a problem in the form of DNA molecules and to realize the computation by operating on those DNA molecules. There are two major advantages of DNA computing: the great parallel computation power and the mega information storage ability. DNA computing is quick, as it can perform many calculations *simultaneously* or *in parallel* (Boneh et al 1995; Winfree 1995). Some of the very complex problems which are hard even for supercomputers can be solved by DNA computing (Lipton 1995; Boneh et al 1995). DNA computing also provides a huge storage media since it stores the information in DNA molecules (Baum 1995). DNA computing is such a novel idea that its future applications still remain unknown. However, it seems that DNA computing will make great changes in the fields of computer science, biology, chemistry and medicine (Leonard 1996; Beaver 1995).

Compared to the novel DNA computing, Fuzzy logic controllers (FLCs) have already been successfully applied in many industrial control systems. FLCs can exhibit better performance than the conventional controllers. Experience and knowledge about the plant are usually used in the design of FLCs. FLCs are especially suitable when the plant dynamics is uncertain or too complex to analyze.

The kernel of the FLC is a linguistic control rule base. The rule base determines the control decisions for the plant and plays a key role. Conventionally, fuzzy control rules are established based on the engineering knowledge and experience of experts and/or skilled operators. The design and fine-tuning work of rule base are usually accomplished by "trail and error" approach. There are some drawbacks associated with such an approach. As the method is human dependent, the resulted rule base can be far from optimal when the expert's knowledge and/or experience are not so reliable. Furthermore, the method is time consuming and even more inefficient when the design work must be accomplished on-line as in some adaptive systems. Due to these reasons, the recent research has focused on fuzzy-evolutionary systems (Karr 199; Thrift 1991), which can improve the FLC's performance by the automatic and efficient design of rule base. The evolutionary method mostly used in such systems is the genetic algorithm (GA).

Genetic algorithms are optimum searching algorithms that are based on concepts of natural selections and genetics (Holland 1975). Genetic algorithm is inspired by observing the natural phenomenon of evolution. Darwinian rule of "the survival of the fittest" is the spirit inside the mechanism behind GA. For problems that cannot be handled by traditional methods due to the lack of some specific knowledge, GA is useful as a black-box method. It is even more suitable in dynamic situations where the goal and constraints are changing all the time. Additionally, it is convenient to incorporate GA with other algorithms in the artificial intelligence field. Studies (Park et al 1994; Wong and Feng 1995; Homaifar and McCormick 1995; Surmann 1996) have shown that genetic algorithms have the ability to optimize the rule base of the conventional FLC.

The performance of GA is closely associated with the population size. The larger the population, the better is the chance to find a global optimum. However, a large population size increases the computational load. A trade-off must be made between the population size and computation time. The parallel computation potential and high information density of DNA computing may enable to handle a population size that is too big for the conventional computers to handle. It may then greatly increase the performance of GAs. Meanwhile, GA may be tolerant to the errors, which are prone to occur in the biological operations of DNA computing, by considering them as a kind of mutation. As a result, genetic algorithms are likely to provide a promising line of research in DNA computing.

In this article, a fuzzy logic controller is designed for Khepera robots to perform the obstacle avoidance task. A DNA coded GA is proposed in (Xiao et al 2001, 2002) to generate and optimize the FLC rule base. In the proposed algorithm, every individual is coded as a DNA chromosome. Simulation result shows that the GA optimized FLC works well on the Khepera robot. It seems that this method can be applied to robots with more complex tasks such as, "navigation and mapping" and "predator and evader" directions of study.

The organization of this article is as follows. The basics of DNA computing are provided in Section 2. The Khepera robot and the associated Webots simulation software are outlined in Section 3. The basic building blocks of a fuzzy logic controller and the block diagram of the designed FLC for the Khepera robots are described in Section 4. Section 5 deals with the DNA coded GA. The coding method and the correspondence between the amino acids and the linguistic variable of the FLC are discussed. Simulation results are included in Section 6 followed by discussions in Section 7.

20.2 DNA Computing

The idea of DNA computing is to use strands of DNA to encode the problem, and to manipulate the strands using techniques commonly available in any molecular laboratory, in order to simulate the operations that lead to the solution.

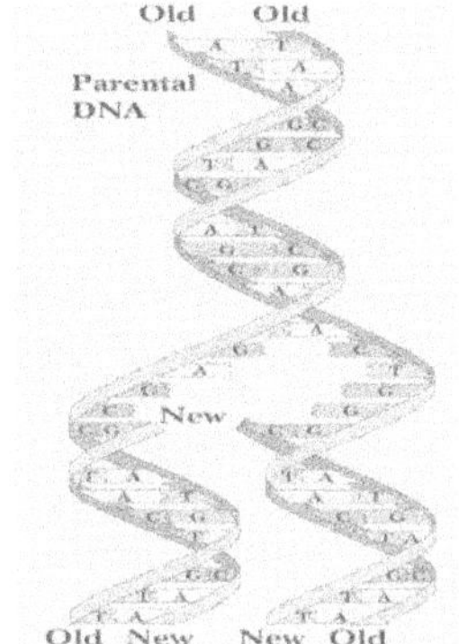

Fig. 20.1. Autoreplication of DNA

The name of DNA computing is easy to confuse with biocomputing. Usually, biocomputing refers to everything that the computer scientists can do, to help the biologists in the study of genes. In DNA computing, instead, molecular biology is suggested to solve the problems that computer scientists face. The DNA computing is different from the Genetic Algorithm as well. GA simulates the rules of the nature of evolution in computation. In this way, GA searches for the optimal solution of a problem. As to DNA computing, it dose not "simulate" anything in molecular biology, but actually uses DNA strands to perform the computation. The basis of DNA computing lies in the fact that DNA molecule is a natural born information container. The genetic information of life is encoded in DNA.

The special feature of DNA is determined by its chemical structure. A single strand of DNA is a concatenation of four kinds of nucleotides, differing only in the nitrogenous base: adenine (A), guanine (G), cytosine (C) and thymine (T). These nucleotides are also known as bases. The chemical structure of DNA, the famous double helix, consists of a particular bond of two single strands. This bond occurs between bases and follows the property of complementarity: adenine (A) bonds with thymine (T) and vice versa; cytosine (C) bonds with guanine (G) and vice versa. This is known as the *Watson-Crick complementarity*, denoted as: $\overline{A}=T$; $A=\overline{T}$; $\overline{G}=C$; $G=\overline{C}$. This complementarity enables the storage of information in DNA and thus forms a basis to perform the computations.

Single nucleotides are linked together, end to end, to form the DNA strands in a process called polymerization. This linking occurs via the reaction between the 5' phosphate of one nucleotide and the 3' hydroxyl of another. Every DNA strand has two distinct ends: one with a free 5' PO_4 group and the other with a free 3' OH group, referred to as the 5' and 3' ends, respectively. The 3' and the 5' ends determine the strand's polarity. Only two complementary strands of opposite polarity (also known as, in *antiparallel* fashion) can bond together to form the final double helix. Fig. 20.1 shows how a DNA autoreplicates itself.

Ordinary biological operations are employed to realize the computation on DNA strands. The DNA strand can be separated into two single chains at first and fused together later by heating up and cooling down the DNA solution through the process named "*melting and annealing*". DNA strands can be linked, cut or modified with the help of special enzymes. The length of a DNA strand can be measured using *gel electrophoresis* and its content can be read by *sequencing*. Just like the mathematical and logical operators ("ADD", "MINUS", "AND" and "OR") in the conventional computation, these biological operations (together with others not mentioned) constitute the operators handling the information encoded in DNA strands.

While a biological operation is performed, it works on all the DNA molecules in a container at the same time. Since there may be trillions of DNA molecules, DNA computing is massively parallel. With the current biological technology, it is possible to reach a speed of 10^{18} operations per second in DNA computing (Leonard 1996). It is approximately 1,200,000 times faster than a digital super computer. As the DNA molecules are smaller and information is stored in the form of nucleotides, an information density of the order of 1 bit per nm^2 is attainable, while the existing storage media can store at a density of about 1 bit per 10^{12} nm^2 (Leonard 1994).

As DNA computing is in a state of infancy, there are several problems unsolved. The errors in the biological operations and the decay of DNA strands are among the major difficulties. The improvements in fields including biology, chemistry, computer science, engineering, mathematics, and physics are all valuable for the appearance of a feasible DNA computer in the future.

20.3 The Khepera Robot and Webots Software

20.3.1 The Khepera Robot

The Khepera robot (K-team 1999) is one of the most popular robots used for research in universities and research centers. It is a miniature mobile robot with two wheels, each controlled by a DC motor with an incremental encoder (Fig. 20.2a). With the program downloaded to its memory, the Khepera can run autonomously. With 2 radio modem turrets or via a serial cable, the robot can communicate with a host computer. The basic Khepera has a diameter of 55 mm, height of 30 mm and weight of 70g. It supports a large number of hardware extension modules, such as gripper, vision turret and radio turret. Included in the associated Webots software is a useful library of on-board applications. The programs for Khepera can be developed within the standard and well known tools, such as C/C++ and Matlab.

(a) The robot

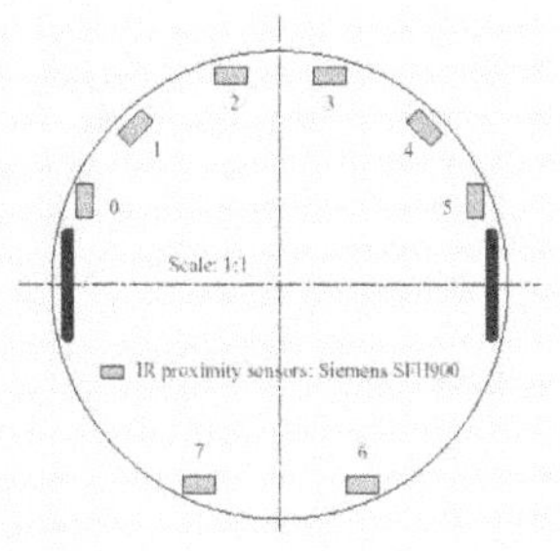

(b) Eight infrared sensors

Fig. 20.2. The Khepera Robot

Khepera has eight infrared proximity and ambient light sensors to detect its surroundings (Fig. 20.2b). Each sensor returns a value from 0 to 512 (a higher value means darker) for ambient light and, from 0 to 1023 (a higher value means nearer to obstacles) for proximity to obstacles. The object can be detected within the range of 5 cm.

Although the maximum speed of the Khepera robot is 60 cm/s, the speed range can be set from -20 units to 20 units, with each unit referring to 8 mm/s. The speed of the left and right wheels can be set separately.

20.3.2 The Webots Software

Khepera supports many simulation packages. Webots is the most powerful 3D mobile robot simulator (Cyberbotics 1999; Cyberbotics2 1999). Developed by Cyberbotics, Webots supports Khepera robots as well as other robots such as Alice and Koala robots. The user can program virtual robots using the C/C++ library. Webots also provides facility to control the real robot through a serial port communication.

A 3D environment editor (Fig. 20.3) allows the user to customize the robot world. It is possible to add objects such as walls, balls, cans and lamps into the world. The object properties such as size, color, position, orientation, etc are also definable.

Webots provides three kinds of Application Programming Interfaces (API) for users to construct the programs. The Khepera API provides control functions on robots. These functions are used to read the proximity and light measurements of the infrared sensors, to read and set the speeds of the motors, and so on. The supervisor API can control experiments, record experimental data such as robot trajectories, and implement inter-robot communications. The controller API enables the users to design a graphic user interface (GUI).

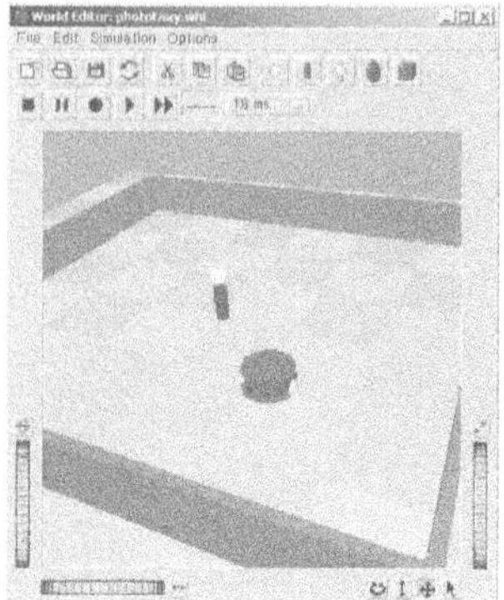

Fig. 20.3. World editor of the Webots software

In this work, a training environment is constructed in Webots for the optimization of fuzzy logic controller.

20.4 The Fuzzy logic controller

Unlike Boolean logic, fuzzy logic can deal with uncertain and imprecise situations. Linguistic variables (SMALL, MEDIUM, LARGE, etc.) are used to represent the domain knowledge, with their membership values lying between 0 and 1. Basically, a fuzzy logic controller (FLC) consists of the following components (Lee 1990).

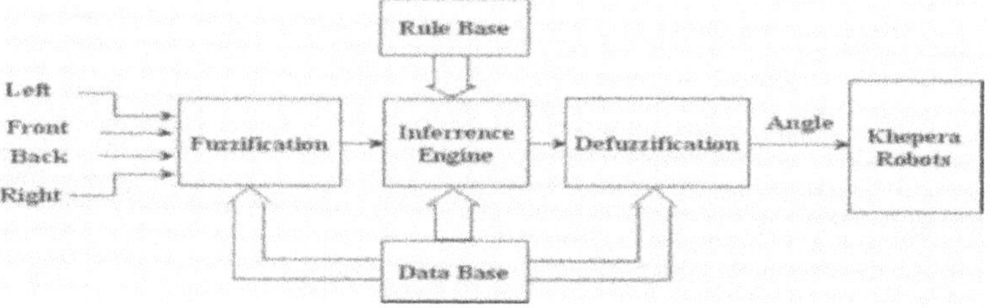

Fig. 20.4. Structure of fuzzy logic controller for Khepera robot

- A fuzzification interface, to scale and map the measured variables to suitable linguistic variables.
- A knowledge base, comprising of the linguistic control rule base.
- A decision making logic, to infer the fuzzy logic control action(s) based on the measured variables, which is much akin to the human decision making.

- A defuzzification interface, to scale and map the linguistic control actions inferred, to yield a non-fuzzy control input to the plant/process being controlled.

The fuzzy logic controller is designed to control the moving direction of the Khepera robot (Fig. 20.4). The proximity values received from the 8 sensors are used to generate inputs to the FLC. These 8 sensors, marked from 0 to 7, are divided into 4 groups to detect obstacles in 4 directions, which are left, front, right and back. The grouping is as follows:

Left = Sensor0 + Sensor1 Front= Sensor2 + Sensor3
Right = Sensor4 + Sensor5 Back = Sensor6 + Sensor7

For instance, the proximity value of an obstacle on the left side is the sum of the reading values from sensors 0 and 1. The "Left", "Front", "Right" and "Back" are the inputs to the FLC and their values range from 0 to 1024. Based on the input values, three linguistic variables, "Large", "Medium" and "Small", are used to represent each input. The input linguistic variables form the *antecedents* of the fuzzy rules. There are four output linguistic variables: "Forward", "Small Turn", "Large Turn" and "Backward". These variables form the *consequents* of the rules. The logic operation used is "AND". A typical fuzzy rule is:

If Left is Large AND Front is Large AND Right is Large AND Back is Small, the action is Backward.

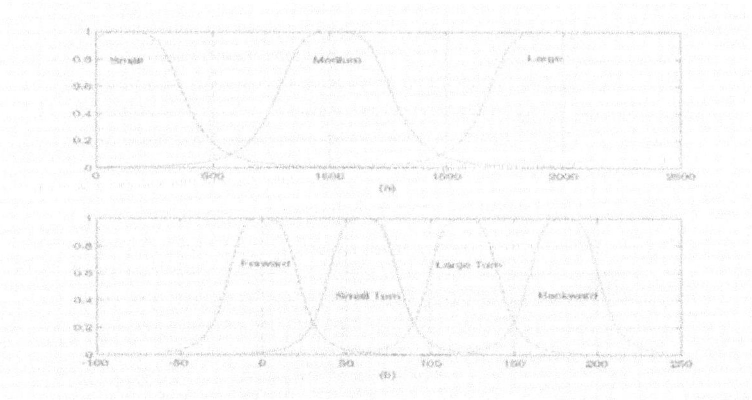

Fig. 20.5. Membership functions for the input and output variables

From the combination of three variables for four inputs, there are $3^4 = 81$ states for the robot to encounter. The FLC rule base thus consists of 81 such rules.

The consequent variable indicates the angle by which the robot should turn to. This angle, with a scope of 0 to 180 degree, is the final output of the FLC. The direction the robot to turn to is determined as follows. Readings from sensors on the left (sensor 0, 1, 2) and right (sensor 3, 4, 5) are summed up separately. If the sum value from the left group is bigger than the right, the robot turns to the right side and vice versa.

Bell-shaped membership functions are used to fuzzify the inputs into linguistic variables and to defuzzify the consequent variable. In fuzzification the four input variables utilize the same membership functions (Fig. 20.5a). The membership functions in Fig. 20.5b are employed in the defuzzification stage. In this work, the parameters of the membership functions are simple and are manually tuned.

20.5 DNA coded Genetic Algorithm for FLC

DNA computing is a newly developed computation method that has potential massive parallelism and power. A typical test tube can contain a large number of DNA molecules, and biological operations can be performed simultaneously. Meanwhile, the population size is an important factor, which affects the performance of GAs. Bigger population sizes always lead to better results. However, usually a trade-off between the size and computing time has to be made. On the other hand, with the application of DNA computing to GAs, the computation time will not explode with the increase in population size.

Table 20.1. The genetic code of amino acids

	U	C	A	G	
U	Phenylalaine	Serine	Tyrosine	Cysteine	U
	(Phe)	(Ser)	(Tyr)	(Cys)	C
	Leucine		Terminator	Ter	A
			(Ter)	Tryptophan(Trp)	G
C	Leu	Proline	Histidine	Arginie	U
		(Pro)	(His)	(Arg)	C
			Clutamine		A
			(Gln)		G
A	Iseleusine)	Threonine	Asparagine	Ser	U
	(Ile)	(Thr)	(Asn)		C
			Lysine	Arg	A
	Methionine(Met)		(Lys)		G
G	Valine	Alanine	Aspartic acid	Glycine	U
	(Val)	(Ala)	(Asp)	(Gly)	C
			Glutamic acid		A
			(Glu)		G

In this article the DNA coding method to encode the rule base of a FLC is discussed and then the DNA coded GA is used to optimize the FLC rule base (Xiao 2001, 2002).

It is known that the messenger Rribonucleic acid (mRNA) is first synthesized from DNA during the synthesis of proteins. In the mRNA three successive bases called codons are allocated sequentially. These codons are the codes for amino acids. Sixty four (64) kinds of codons correspond to 20 kinds of amino acids. The codons are expressed as Leu, Arg, Thr, etc., which are the abbreviations for the related amino acids. For example, Leu stands for leucine, Arg for arginine and Thr for threonine. Table 20.1 shows the correspondence between the codons and amino acids. Based on the three bases of a codon, the corresponding amino acid can be identified. The "AUG" along the left-top-right direction in Table 20.1 corresponds to "Met". The special Terminator codon (Ter) is the stop codon. The transcription, which is the process of synthesizing the RNA based on the DNA strand, is terminated when the "Ter" is found. Amino acids can be synthesized artificially. The meaning of each codon can be defined as desired, such as a variable or in the form of a function. In this way, problems under consideration can be encoded as DNA strand (Furuhashi 1997; Yoshikawa and Uchikawa 1996; Lee 2000).

The coding scheme differs depending on the application needs. A typical fuzzy rule often consists of antecedents (or premise variables), consequent variables (or conclusion) and fuzzy relations. All these parts can be encoded as DNA strands. For instance, some coding methods (Furuhashi 1997) encode the whole "IF--THEN" rule into chromosomes. Such a method is similar to the "broadcast language" (Holland 1975). DNA codons are used as alphabets and symbols to construct a presentation string. A parser scans the string to extract the fuzzy rules. However, in this work, the antecedent and fuzzy relation of a rule are fixed. Only the consequent part for each rule is needed to develop and the associated computational load is rather low. There are 4 consequent linguistic variables for each rule: Forward, Small Turn, Large Turn and Backward. Four amino acids are mapped to these variables as follows:

$$Phe \longleftrightarrow Forward,\ Leu \longleftrightarrow SmallTurn,$$
$$Ile \longleftrightarrow LargeTurn,\ Val \longleftrightarrow Backward$$

There are totally 81 rules in the rule table. The 81 consequent linguistic variables are combined into one chromosome as an individual of GA. The 81 rules are sorted in a certain sequence and thus each rule has its associated consequent part at a fixed position of the chromosome. Furthermore, the rule table has a kind of symmetry, resulting in part of the rules to have the same consequent part. For example, if the two rules have antecedents as,

Left is Large AND Front is Large AND Right is Small AND Back is Small ...
Left is Small AND Front is Large AND Right is Large AND Back is Small ...,

then the rules should have the same consequent (the same degree value to turn). Due the symmetry, the length of the chromosome is shortened to 54 simplifying the computations. One of the DNA chromosomes can be denoted as:

PheLeuIleValValPhePheIleIleLeu...
...LeuIleLeuValPheValIlePhePheLeu

Another issue worth mention is that when there are no obstacles around, the robot should always move forward. Then the consequent part of the rule for such a situation is "Forward" which means turning by 0 degree. This prevents the robot from spinning about its axis unnecessarily.

The fitness of the chromosome is initialized to zero. In each step, based on the sensor readings, the robot decides the angle to turn to by the FLC and then moves along this direction for a certain distance. The fitness is decreased by unity if an obstacle is detected in that step; otherwise the fitness is increased by unity. If the robot keeps approaching the obstacle and receives a very large proximity value from any of its sensors, it is considered as a crash and the fitness is decreased by 10. In addition, there is a penalty function associated with the angle turned by the robot. The fitness is decreased by a value, which is k times of the degree of angle. Based on simulation results, it seems that the suitable range of k is from 0.0055 to 0.0111. This function encourages the robot to act more efficiently, i.e. just turn by the necessary angle to avoid obstacles.

The evolution operations on the chromosomes are similar to those of the standard GA. Replication, cross-over, and mutation operations are used. With the help of enzymes (such as restriction enzymes and ligase), these operations can be carried out in a biological lab. In each generation, individuals are sorted by their fitness. Those with higher fitness values are selected as elite and are candidates for the evolution operations. The evolution stops when the best individual does not receive any improvements for a certain time.

20.6 Simulation Results

The 3D simulation software, Webots, is used in this work. The program is coded using MS C/C++. The Khepera robot is placed in the same position and orientation in the training world as shown in Fig. 20.6a. The fitness of an individual DNA strand is evaluated for 400 steps. In each step, the robot makes a turn and then moves forward for 64ms. The turn angle is decided by the FLC, which uses the rule base represented by the individual under evaluation. The speed of forward motion is set to 12 units (about 96 mm/s).

A small population size of 60 is used and it turns out to be enough for the "obstacle avoidance" task. The elite population size is 20. Out of the 40 children generated by cross-over, 20 of them are mutated. The rather high mutation rate avoids the "premature convergence" which often happens with a smaller population size. The searching is terminated when no improvement is reported in 20 successive generations.

The trajectories of the robot in the evolution process are shown and compared in Fig. 20.6b, 20.6c, and 20.6d. At the beginning, the individuals are generated by arbitrarily assigning the four consequent variables to each rule. As a result, the robot exhibited random behaviors and crashed frequently. It also leaned to linger at the corners of the obstacles for a long time before moving out. As the evolution continued, the robot gained the ability to avoid the obstacle "effectively" and crashes seldom took place. At the end of the evolution process, the best individual enabled the robot to avoid obstacles elegantly and escape from trap points quickly.

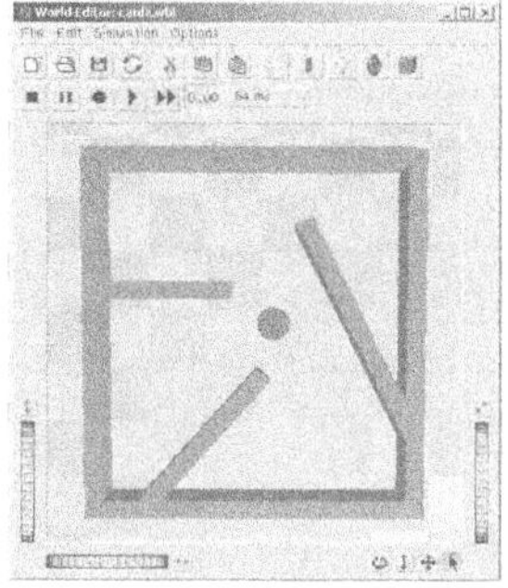

(a) Training world

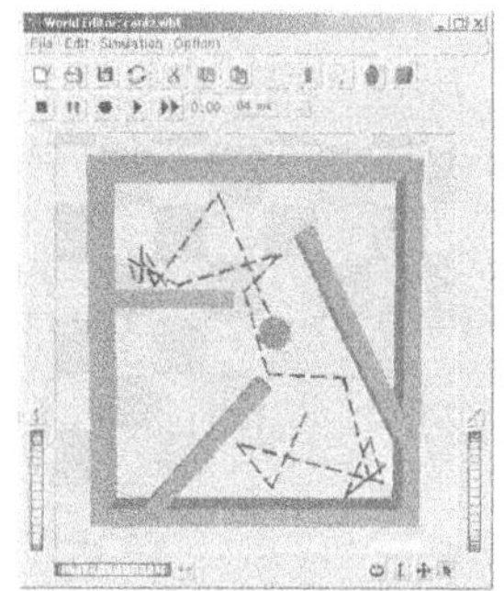

(b) Trajectory before evolution

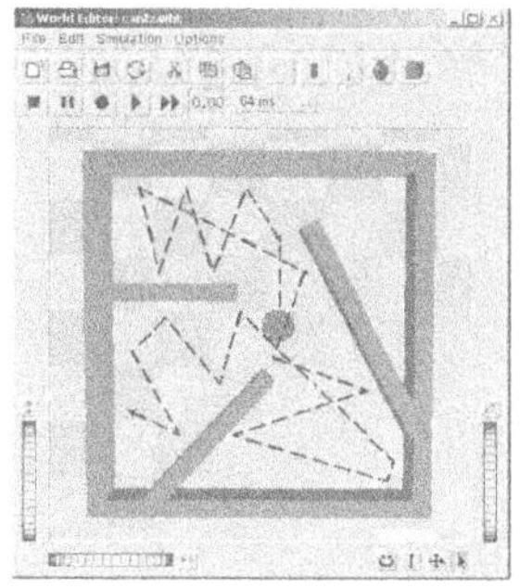

(c) Trajectory after 50 generations

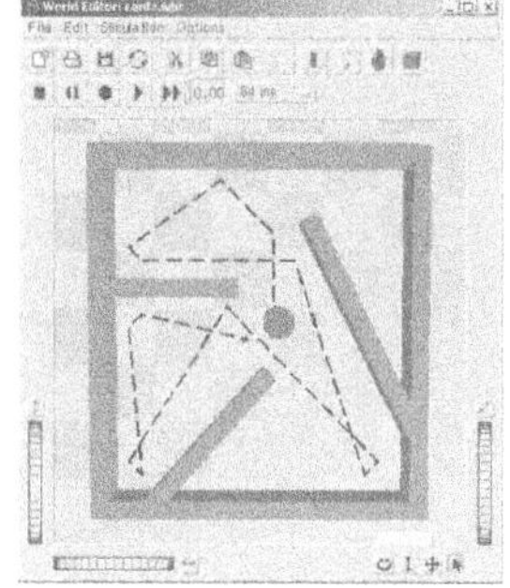

(d) Trajectory after evolution

Fig. 20.6. Simulation Results on Webots software

The optimized rule base resulted from simulation training world is tested in a real experimental scenario which is much different from the training world. In the experiment, the robot successfully navigated around and avoided the obstacles. However, the robot could not move around so elegantly as in the simulation. Sometimes it just turned by an angle that is more than necessary. The decrease in efficiency is due to errors coming from the real world environment. For instance, the environment's color and lighting condition can disturb the sensor reading and the inertia and friction can affect robot's mechanic motion. However, the FLC's performance is not tampered that much by the disturbances, depicting some grade of robustness and noise tolerance.

20.7 Discussion

The discussed DNA coded genetic algorithm constructed an optimized rule base of the FLC for the Khepera robot. The FLC enables the robot to perform well in the "obstacle avoidance" task. The optimized FLC enabled the robot to move around in an unknown world with full of obstacles without being trapped or stuck. Since the "obstacle avoidance" is a fairly simple task, the FLC designed is rather simple. Further work cases on more complex problems and further combination of algorithm with DNA computation to exploit the power of evolutionary method is suggested.

References

Baum, E. (1995) Building an Associative Memory Vastly Larger Than the Brain. Science, 268:583--585.

Beaver, D. (1995) A universal molecular computer. Technical Report CSE-95-01, Penn State University.

Boneh, D., Dunworth, C., Lipton, R.J. and Sgall, J. (1995) On the computational power of DNA. Technical Report CS-TR-499-95, Princeton University.

Boneh, D., Dunworth, C. and Lipton, R.J. (1995) Breaking DES using a molecular computer. Technical Report CS-TR-489-95, Princeton University.

Cyberbotics. (1999) Webots 2.0 User Guide.

Cyberbotics. (1999) Webots 2.0 Reference Manaul.

Furuhashi, T. (1997) Development of IF□HEN rules with the use of dna coding. in: W. Pedrycz (Ed.), Fuzzy Evolutionary Computation, Kluwer Academic Publishers, Boston.

Karr, C.L. (1997) Fuzzy-evolutionary Systems. In T. Back, D.B.Fogel, and Z.Michalewicz (eds) Handbook of Evolutionary Computation, Oxford University Press.

Holland, J. (1975) Adaptation in Natural and Artificial Systems. University of Michigan Press.

Homaifar, A. and McCormick. E. (1995) Simultaneous design of membership functions and rule sets for fuzzy controllers using genetic algorithms. IEEE Trans. Fuzzy Syst., vol.3, pp.129--139.

K-team. (1999) Khepera User Manual. Version 5.02.

Lee, C.C. (1990) Fuzzy logic control systems: fuzzy logic controller--Part 1. IEEE Trans. System Man Cybernet. SMC-20 404--418.

Lee, K.Y., Lee, D.W. and Sim,. K.B. (2000) Evolutionary neural networks for time series prediction based on l-system and dna coding method. in: IEEE. Proc. of the 2000 Congress on Evolutionary Computation, Vol. 1.2, pp. 1467--1474.

Leonard, A. (1994) Molecular Computation of Solutions to Combinatorial Problems. Science, 266, 1021-1024.

Leonard, A. (1996) On constructing a molecular computer. In proceedings of the first DIMACS workshop on DNA computing.

Lipton, R.J. (1995) Using DNA to solve NP-complete problems. Science, 268:542--545, Apr.28.

Park, D., Kandel, A. and G. Langholz. (1994) Genetic-based new fuzzy reasoning models with application to fuzzy control. IEEE Trans. Syst., Man, Cybern., vol. 24, pp. 39--47.

Surmann, H. (1996) Genetic optimization of a fuzzy system for charging batteries. IEEE Trans. Ind. Electron., vol. 43, pp. 541--548.

Thrift, P. (1991) Fuzzy logic synthesis with genetic algorithms. In proceedings of 4th Int. Conf. on Genetic Algorithms pp509--513.

Winfree, E. (1995) On The Computational Power of DNA Annealing and Ligation. Technical report, California Institute of Technology, USA.

Wong, C.C. and Feng, S.M. (1995) Switching type fuzzy controller design by genetic algorithm. Fuzzy Sets Syst., vol.74, no. 2, pp. 175C185.

Xiao, P., Prahald, V. and Tong H. L. (2001) DNA coded GA for the rule base optimization of a fuzzy logic controller, in: Proceedings of the 2001 Congress on Evolutionary Computation CEC2001, IEEE Press, COEX, Seoul, Korea, 2001, pp. 1191--1196.

Xiao P., Prahlad, V. and Tong H. L. (2002) Mobile robot obstacle avoidance: DNA coded GA for FLC optimization, in: Proceedings of the Congress on FIRA Robot World Cup 2002, Seoul, Korea, 2002.

Yoshikawa, T. and Uchikawa, Y. (1996) Effect of new mechanism of development from artificial dna and discovery of fuzzy control rules. in: Proc. of IIZUKA'96, pp. 498--501.

21 TRIBES application to the flow shop scheduling problem

Godfrey C Onwubolu

Hyper-spheres instead of hyper-parallelepipeds for proximity areas, and adaptation of the swarm size as well as the relationships between the particles offer an elegant and powerful theoretical framework for design and analysis of autonomous adaptive search heuristics. This chapter describes an autonomous adaptive search heuristic known as TRIBES, which in its current formulation is biased toward solving problems defined as rational points, and the transformation processes involved in realizing a version for solving combinatorial optimization problems. We apply the TRIBES methodology to the well-known flow shop-scheduling problem, and report simulation results. To demonstrate its effectiveness, we compare the solutions of the TRIBES with other emerging optimization techniques. The results show that TRIBES is promising for solving combinatorial optimization problems, which are of significant importance in the manufacturing sector.

21.1 Introduction

The concepts of *migration* in which birds fly together in flocks, and *assemblages* or *communities* in which a whole variety of birds gather together in one area/region have been of much interest to researchers in optimization. An analogy with the way birds flock has suggested the definition of a new computational paradigm, which is known as particle swarm optimization (Kennedy and Eberhart 1999). The main characteristics of this model are cognitive influence, social influence, and the use of constriction parameters. Cognitive influence accounts for a particle moving towards its best previous position, social influence accounts for a particle moving towards the best neighbor, and constriction parameters control the explosion of the systems' velocities and positions. Particle swarm optimization (PSO) heuristics usually need some predefined parameters, such as numerical coefficients, swarm size, neighborhood size and topology. Adaptive PSO is not a new concept; some adaptive PSO versions have already been designed (Carlisle and Dozier 1998; Clerc 1999; Kennedy and Eberhart 1999; Shi and Eberhart 2001; Hu and Eberhart 2002; Angeline 1998). Some use selection, some modify a coefficient during the process, some use clustering techniques or fuzzy rules, but all of them do need some predefined parameters, such as swarm size, weighting coefficient, or neighborhood size and topology. A framework for more complete adaptive versions has been defined in Clerc (2002), but there was still the problem of an initial weighting coefficient, and no neighborhood adaptation was suggested.

This chapter presents an autonomous version of PSO referred to as TRIBES (Clerc 2003) which does not require any parameter at all. TRIBES uses two techniques, namely *hyper-spheres* instead of hyper-parallelepipeds for proximity areas, and *adaptation* of the swarm size, as well as the relationships between the particles. First, we introduce the concepts of *tribes* and *informer groups* (i-groups) and then each of the key elements is then described in some details in the subsequent sections. Further, we show how to remove weighting coefficients by using uniform random distribution in hyper-spheres. An important point is that this first improvement can be used even for non-adaptive methods. In the adaptation part we present some rules to generate/remove not only particles themselves but also relationships between particles.

In its current formulation, TRIBES is biased toward solving problems defined as rational points. Consequently, we present the transformation processes involved in realizing a version for solving combinatorial optimization problems. The contribution of the current article consists of the strategies which have been devised for inclusion in TRIBES, for transforming particle positions normally in the form of rational numbers into a *sequence*, which is made up of discrete numbers. Moreover, a local search engine is included to improve the solution quality. It should be noted that in the original TRIBES, no local search is needed to improve differentiable functions since the particle position, which is a rational form is exceptionally adequate to obtain very good solution quality, but which cannot be used for combinatorial optimization problems. The overall strategy, the individual components of which are then discussed, can be summarized as follows:

- after initialization, transform rational position into discrete position;
- evaluate the objective function of the position;
- access the local search routine to obtain better solution; and
- memorize the best solution so far.

Finally, we apply the TRIBES methodology to the well-known flow shop-scheduling problem, and report simulation results. To demonstrate its effectiveness, we compare the solutions of the particle swarm optimization with other emerging optimization techniques.

21.2 Flow-shop scheduling problem (FSP)

A flow-shop problem (FSP) is one in which all jobs must visit machines or work centers in the same sequence. The flow shop can be formatted generally by the sequencing of n jobs on m machines under the precedence condition. Typical objective functions include the minimizing of average flow time, minimizing the time required to complete all jobs or makespan, minimizing maximum tardiness, and minimizing the number of tardy jobs. If the number of jobs is relatively small, then the problem can be solved without using any generic optimizing algorithm. Every possibility can be checked to obtain results and then sequentially compared to capture the optimum value. But, more often, the number of jobs to be processed is large, which leads to big-O order of $n!$ Consequently, some kind of algorithm is essential in this type of problem to avoid combinatorial explosion. We define one basic objective function for minimization of the completion time for the flowshop problem using PSO.

The minimization of completion time for a flow shop schedule is equivalent to minimizing the objective function F.

$$\Im = \sum_{j=1}^{n} C_{m,j} \tag{21.1}$$

Where, $C_{m,j}$ = the completion time of job j. To calculate $C_{m,j}$ the recursion procedure is followed for any ith machine jth job as follow:

$$C_{i,j} = \max\left(C_{i-1,j}, C_{i,j-1}\right) + P_{i,j} \tag{21.2}$$

Where, $C_{1,1} = k$ (any given value) and $C_{i,j} = \sum_{k=1}^{j} C_{1,k}$; $C_{j,i} = \sum_{k=1}^{i} C_{k,1}$

$i \Rightarrow$ machine number, $j \Rightarrow$ job in sequence, $P_{i,j} \Rightarrow$ processing time of job j on machine i.

21.3 TRIBES approach

A tribe is just a sub-swarm (Clerc 2003). Any particle belongs to one and only one tribe. An informer for a given particle P is a particle Q that can "give P" some information. Typically this information includes the best position of Q found so far, and the function value at this best position. The key elements of TRIBES, which are hereafter discussed after terminology is given, are:

- Informers;
- Hyper-spheres for defining promising search area; and
- Adaptations.

21.3.1 Terminology and concepts

At the very beginning, there are n particles $\{0, 1, 2,...,n\}$, $n = 3$ for the complete adaptive version. So we have at the same time: the swarm, a tribe T0, and three i-groups *I0*, *I1*, and *I2* (informers group). Each particle belongs to the tribe T0. For each particle, its *i*-group contains (i) the particle itself; and (ii) the best particle of the tribe.

At each time-step each particle *moves* towards a more promising area, using its knowledge $(x,\ p\_i,\ p\_g)$. Let us define the following:

- (H_i) be the hyper-sphere radius = norm$(p_i - p_g)$, center (p_i)
- (H_g) be the hyper-sphere radius = norm$(p_i - p_g)$, center (p_g)
- a point p(p_i) is randomly chosen in (H_i)
- a point p(p_g) is randomly chosen in (H_g)
- x(t+1) is a weighted combination of p(p_i) and p(p_g), according to the *f* values on (p_i) and (p_g).

If we do not use adaptive version, it means we have a global PSO (the neighborhood of each particle is the whole swarm). For each particle, we have

$$x(t+1) = x(t) + (w(t),\ p\_i(t),\ p\_g(t)) \tag{21.3}$$

where $x(t)$ = position at time *t*, $p\_i(t)$ = previous best position of the particle (at time t), $p\_g(t)$ = previous best position found so far in the i-group (informer group) of the particle (at time t), including the particle itself, and $w(x_1,\ x_2,\ x_3)$ = a vectorial function of three positions.

Unlike the classical PSO, TRIBES has no explicit velocity and *w* is not at all depending on some more or less arbitrary "weights" cognitive or social coefficients (often called alpha, phi_1, phi_2). A promising area is defined using hyperspheres.

21.3.2 Informers

The set of informers of a particle, its *i*-group, contains, but is not necessarily limited to, its tribe. Also, the relation "to be an informer of" does not need to be symmetric. The present version of TRIBES is symmetric.

21.3.3 Hyper-spheres, and promising areas

21.3.3.1 Hyper-spheres

A hyper-sphere is the set of all points in n-dimensional space ($\Re^n$) which are equidistant from a given point (for our purposes, the origin). Clearly circles and spheres fall into this definition as the two- and three-dimensional hyper-spheres. Interestingly, a 1-sphere takes the form of two points and encloses a line segment. Though hyper-spheres of dimension greater than four cannot be visualized directly by inhabitants of our three-dimensional space, they can be imaged by cross-sections. The planar cross-sections of a sphere are circles whose radii progress from *0* to r, the radius of the sphere, and back to *0*. Similarly, the cross-sections of a glome (the 4-sphere) are *3* spheres whose radii progress from *0* to r and back to *0*. This method of constructing hyper-spheres lends itself well to answering the natural questions that arise from such considerations. Such considerations include the formulae for the enclosed hyper-volume by (or "content" of) and surface "hyper-area" of hyper-spheres in terms of their dimension n and radius r. Relevant literature on hyper-sphere are well documented (Conway 1993; Peterson 1988; Somerville 1958; and Wells 1986).

For the area enclosed by a circle (see Fig. 21.1), we will employ the tools of integration, which are capable of giving even our first result. The area is found by summing up the height of the circle (off the x-axis) from $x = 0$ to r and then multiplying by *4* for the entire circle. We use trigonometric substitution in this integration.

$$A(r) = 4\int_0^r \sqrt{\left(r^2 - x^2\right)} \tag{21.4}$$

$$\sqrt{\left(r^2 - x^2\right)} = r\sin\theta; \quad x = r\cos\theta; \quad dx = -r\sin\theta\, d\theta \tag{21.5}$$

$$A(r) = 4r^2\int_0^{\pi/2} \sin^2\theta\, d\theta = 4r^2\int_0^{\pi/2} \frac{1-\cos(2\theta)}{2}\, d\theta = 2r^2\left[\theta \quad \frac{1}{2}\sin(2\theta)\right]_0^{\pi/2} = \pi r^2 \tag{21.6}$$

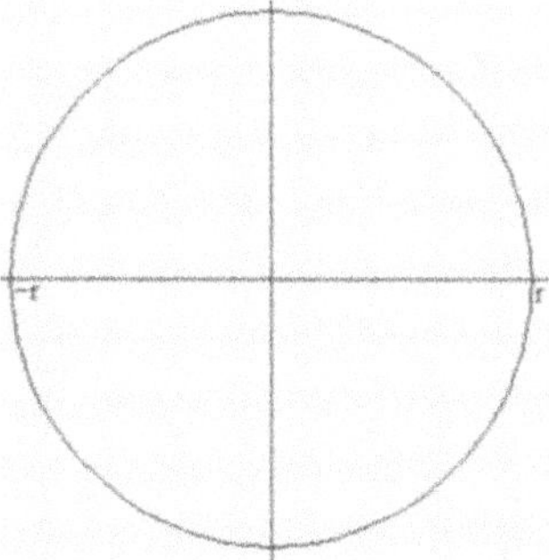

Fig. 21.1 A circle

To find the circumference we merely take the derivative of the area with respect to r, for the area could also have been found by integrating over circles of radii 0 to r, and differentiating reverses the process, giving the circumference of the outermost circle:

$$C(r) = \frac{dA}{dr} = 2\pi r \tag{21.7}$$

The volume enclosed by a sphere, now, is found by integrating the areas of its composite circles:

$$V^3(r) = 2\int_0^r \pi\left(\sqrt{\left(r^2 - x^2\right)}\right)^2 dx = 2\pi\left(r^2 x - \frac{1}{3}x^3\right)\Big|_0^r = \frac{4\pi}{3}r^3 \tag{21.8}$$

And the surface area is the derivative of the volume:

$$A^3(r) = \frac{dV}{dr} = 4\pi r^2 \tag{21.9}$$

Continuing this process of integrating over the content of (n–1)-spheres would give us the hyper-volume V^n of any n-sphere:

$$V^n(r) = 2\int_0^r V^{n-1}\left(\sqrt{\left(r^2 - x^2\right)}\right)^{n-1} dx = 2V^{n-1}(1)\int_0^r \left(\sqrt{\left(r^2 - x^2\right)}\right)^{n-1} dx \tag{21.10}$$

For general hyper-spheres, the challenge arises in developing a general formula for V^n. Differentiating the hyper-volume V^n of any n-sphere, the surface hyper-volume A^n) is given as a function of r in Table 21.1. We note that the rate $\frac{hypersphere - volume}{hpercube - volume}$ is rapidly decreasing with the dimension (for a sphere of radius r and a cube of ridge 2r).

Table 21.1. Hyper-sphere volume/area

n	V^n (r)	A^n (r)
1	2r	
2	πr^2	$2\pi r$
3	$4\pi/3\ r^3$	$4\pi r^2$
4	$\pi^2/2\ r^4$	$2\pi^2 r^3$
5	$8\pi^2/15\ r^5$	$8\pi^2/3\ r^4$
6	$\pi^3/6\ r^6$	$\pi^3 r^5$
7	$16\pi^3/105\ r^7$	$16\pi^3/15\ r^6$
8	$\pi^4/24\ r^8$	$\pi^4/3\ r^7$
9	$32\pi^4/945\ r^9$	$32\pi^4/105\ r^8$
10	$\pi^5/120\ r^{10}$	$\pi^5/12\ r^9$

21.3.3.2 Promising search areas

We first describe the non-adaptive part of the method, i.e. how to define promising search-areas by using hyper-spheres. Let the hyper-sphere $H = \{x\}$ be the search space, the set of the possible positions, and let D be its number of dimensions. Let ϕ be the objective function on H, S the target and ε the maximum admissible error (or minimum wanted accuracy). Let $f = |\phi - S|$ be the error function. We do not examine here the important problem of the stopping criterion for the iterations, and we assume we have at least one. When S is known, this stopping criterion is simply $f(x) < \varepsilon$ but usually it is just a given maximum number of function evaluations or a maximum processing time.

Let us define some components of a particle P:

$P.x(t)$ is the particle position at time t (i.e. a D-vector),

$P.p(t)$ is the best position found so far by the particle,

$P.g(t)$ is the best position found so far by its informers.

When there is no ambiguity, we will simply write x, p, and g.

All particles are generated in H in a purely random manner. Once generated, each particle moves according to the following equation

$$x(t+1) = w(x,\ p,\ g) \tag{21.11}$$

where w is a vector function of three positions.

Unlike *classical* PSOs, TRIBES has no explicit velocity, although, of course, $w - x(t)$ could be seen as such a velocity, to support intuition. The main difference is in the way w is defined to hopefully generate a good sample of positions in H. We now explain this inexplicit velocity concept, more clearly as in Clerc (2003).

Let H_p be the hyper-sphere centered on p and of radius $\phi = \|p - g\|$. Let H_g be the hyper-sphere centered on g and of radius ρ. We choose a random point p' in H_p by using a uniform distribution (it is not that easy, see the appendix). We also choose a random point g' in H_g. Then w is defined as a weighted combination of p' and g', according to their respective *performances*:

$$\begin{cases} w_p = 1/f(p) \\ w_g = 1/f(g) \\ w = \left(w_p \bullet p' + w_g \bullet gp\right)/(w_p + w_g) \end{cases} \tag{21.12}$$

For all possible values of p' and g', w describes a domain in the search space that can be seen as a promising area. Fig. 21.2 shows such a promising area for a two dimensional problem. We note that although this domain is also a hypersphere, the distribution-probability for w in it is not uniform (see Fig. 21.3).

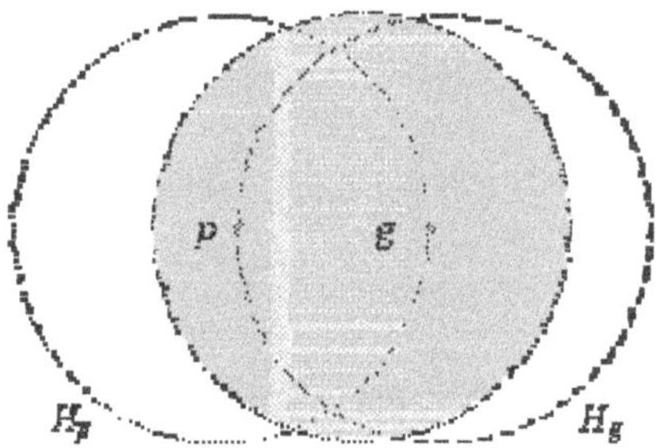

Fig. 21.2 Typical promising search-area for a 2D-problem (in Grey).

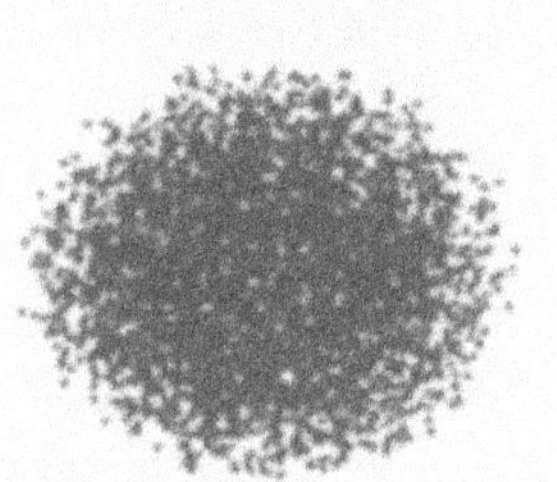

Fig. 21.3 Probability distribution of probability for a promising search-area (Clerc, 2003).

As can be seen, Fig. 21.3 looks like a normal one, or at least a *softball*. The multi-dimensional Gaussian distribution is a bit more complicated, and results do not seem to be better (Clerc, 2003). We notice that with the Gaussian variant, the non-divergence can be proved only if the standard deviation is smaller than the non-divergence can be proved only if the standard deviation is smaller than power(1.392,1/D)*norm(p-g) (probability 50% to be nearer of g than of p). In other words, it rapidly tends towards norm(p-g) when dimension *D* increases. We note that all previous PSO versions define a "moving" term such as $rand(0...\phi_1)(p-x)+rand(0...\phi_2)(g-x)$. The promising area is then an intersection of hyper-parallelepipeds and this introduces two biases: it is bigger than really necessary (and the higher the dimension, the more important is this bias), and it depends on the co-ordinates.

21.3.4 Adaptations

Adaptation rules are defined when a tribe or a particle is generated or removed, and when a particle becomes an informer for another. Using hyper-spheres instead of hyper-parallelepipeds is an improvement that any PSO can take advantage of. However, to be completely parameter free, a PSO also needs some rules to automatically define the number of particles and the relationships between them. Here, it means we have to define when a tribe or a particle is generated or removed, and when an informer is added to or removed from an i-group. The following components of adaptation, each of which, is described in the following sub-sections are as follows (Clerc 2003):

- Good particle and bad particle;
- Best particle and worst particle;
- Classification of good tribe and bad tribe;
- Rule for adding a tribe; and
- Rule for removing a tribe.

21.3.4.1 Good particle and bad particle

A particle improves its performance, and is said to be *good*, when the new position is strictly better than the best it has found so far

$$f(x(t+1)) < f(p) \tag{21.13}$$

Of course, as soon as it happens, we replace p by $x(t+1)$. A particle that is not good is said to be *bad*.

21.3.4.2 Best particle and worst particle

In a given set of particles E (tribe or i-group), a particle P_{best} is the *best* if its performance is better than all the others are

$$f(P_{best}.p) \leq f(P.p),\ \forall P \in E \tag{21.14}$$

If there are several candidates, one is chosen at random. A particle P_{worst} is the *worst* if it is bad and its performance is worse than all the others particles

$$\begin{cases} f(P_{worst}.x(t+1)) \geq f(P_{worst}.p), \ \forall P \in E \\ f(P.x) \leq f(P_{worst}.s)E \end{cases} \quad (21.15)$$

If there are several candidates, one is chosen at random. Note that if all particles have improved their performance, there is no worst one at all, according to this definition.

21.3.4.3 Classification of good tribe and bad tribe

A tribe is said to be bad if its best particle is bad (none of its particles is good). A tribe is said to be good if at least half of its particles are good. Let T be a tribe, $|T|$ its number of particles, and T^+ the subset of its good particles. We have then

$$\begin{cases} T \ is \ bad \Leftrightarrow |T^+| = 0 \\ T \ is \ good \Leftrightarrow |T^+| \geq |T/2| \end{cases} \quad (21.16)$$

We note that a tribe can perfectly be neither bad nor good. In fact (21.16) is just a simplification of a fuzzy rule saying that the quality of T is an increasing function of $|T^+|$. Also, a good mono particle tribe has no worst particle.

21.3.4.4 Rule for adding a tribe

When at least one tribe is bad (at time step t), a new empty tribe T' is generated. Then, each bad tribe "generates" a particle in T', whose position is a random one in the search space, with a probability equal to

$$\frac{|T|}{\sum_{tribes} |T_k|} \quad (21.17)$$

We note that the denominator is simply the swarm size, and thus this rule does not respect the "think locally, act locally" principle (Clerc 2002). So it is not very satisfying, and we believe an improvement is certainly possible. The *i*-group of the new particle contains T' (see the i-group definition), but also the best particle of the generating tribe, and vice versa in this "symmetric" version (i.e. the generated particle is added to the i-group of this best particle).

21.3.4.5 Rule for removing a tribe

When a tribe is good, its worst particle is removed, and the best one "plays its role". It means the *i*-group of the worst particle is merged into the one of the best particle, and, in all *i*-groups. If this worst particle is present, it is replaced by the best (if it is not already here).

21.3.5 Adaptive scheme

From time to time social adaptation is performed:

- for each tribe, the count of how many particles have improved their best performance (n) is made, such that
  if $n = 0$, the tribe is said *bad*
  if ($n \geq$ size(tribe)/2) then the tribe is said *good*
- if there is at least one bad tribe, a new tribe is generated
- for each bad tribe:
  - the best particle is found and "generates" a new particle in the new tribe, for the moment purely at random,
  - this new particle is added to the i-group of this best particle
- for each particle of the new tribe, the i-group contains
  i) the particle itself
  ii) the best particle of the new tribe,
  iii) the particle that has generated it (a case of symmetry; we note that with the *i*-group concept, one could build non symmetrical relationships)
- for each "good" tribe
  - the worst particle is removed (if it is bad)
  - in all *i*-groups it is replaced by the best one of the tribe (general case) or by the best particle of the i-group of the removed particle (mono-particle tribe case)
  - its *i*-group (except of course this removed particle itself) is merged to the one
  that replaces it.

21.3.6 Transformer

The position of a particle obtained from the TRIBES formulation is a rational one; it is not in an integer form. For combinatorial optimization problems the position of a particle must be converted into an integer form. Therefore, we do some transformation to obtain sequence that is of integer form, which enables us to find the solution from an objective function. The solutions obtained from the transformer are as good as those of PSO.

In TRIBES, the position of a particle, is a rational one defined as

$$x.u[t] \in \Re^{D} \tag{21.18}$$

The steps involved in the transformer are as follows:

Step 1: arrange $x.u[t]$ in an array $temp.u[t]$, in a non-descending order

Step 2: for each value of x in $x.u[t]$, find its position i in $temp.u[t]$ and store in $p.u[t]$

Step 3: the rational position $x.u[t]$ now has transformed sequence of $p.u[t]$

21.3.7 Local search

As a search heuristic, TRIBES does not have a local search; hence the result for combinatorial optimization problems is not satisfactory. After accessing the *transformer* to obtain discrete positions, the result of the objective function value obtained is not satisfactory. This is the motivation for including a *local search* engine to give a better result. The steps involved in the local search engine are as follows:

Step 1: generate two random permutations, pr and ps

Step 2: for each pair of permutation pr_i and ps_j, calculate the change in objective function due to a 'move'

Step 3: If the change in objective function due to a 'move' is negative, swap the positions of pr_i and ps_j in the TRIBES sequence, which was submitted to the local search engine.

21.3.8 The transformer-local search scheme

The transformer-local search strategies are carefully included in some segments of the TRIBES procedure in order to be effective. Three areas were identified, although further study will improve results. In order to transform particle positions of TRIBES normally in the form of rational numbers into a *sequence*, which is made up of discrete numbers, the transformer-local search strategies are inserted in the following positions

- after initialization;
- when swarm moves;
- during adaptation, when there is no enough improvement; and
- before every new iteration after adaptation.

21.3.9 Illustrating Tribes

At the very beginning the initial tribe is launched with three particles. The tribe definition, particle designation and the informer group are given as follows:

T0 = {0,1,2}

i-group table	
Particle	i-group
0	{0, 1, 2}
1	{0, 1, 2}
2	{0, 1, 2}

In adaptation *1*, T0 is "bad", best particle *1*. Therefore, a new tribe T1 is generated. Particle *3* is generated by particle *1* in the new tribe T1. The latest information becomes

T0 = {0,1,2}
T1 = {3}

i-group table	
Particle	i-group
0	{0, 1, 2}
1	{0, 1, 2, 3}
3	{1, 3}

In adaptation *2*, T0 is good, best particle 0, worst *1*, T1 is bad, best particle *3*. T2 is generated. Accordingly, particle *1* is removed (and replaced by 0 for the i-groups) and particle *4* is generated by particle *0* into T2. The latest information becomes

T0 = {0,2}
T1 = {3}
T2 = {4}

i-group table	
particle	i-group
0	{0, 2, 3}
2	{0, 2}
3	{0, 3, 4}
4	{3, 4}

In adaptation *3*, T0 is bad (best particle 2, worst particle 0), T1 is good, and T2 is bad. T3 is generated. Particle *0* is removed (replaced by 2 for i-groups); particle *6* is generated by particle *2* in T3; and particle *8* is generated by particle *4* in T3. The latest information becomes

T0 = {2,5}
T1 = {3}
T2 = {4}
T3 = {6,8}

i-group table	
particle	i-group
2	{2,5,6}
3	{3,7}
4	{4,8}
5	{2,5}
6	{2,6,8}
7	{3,7}
8	{4,8}

This process continues until adaptation *n*, and as we can see, the size of the tribe is dynamic.

21.4 The TRIBES Algorithm

In this sub-section, we bring together all the ingredients discussed above and formally present the TRIBES algorithm as follows.

Step 1. Initialize:

Set *two_pi: = 2Cos(ϕ)*
Set *best_result.size: = 0* {still no "best result" at all}
Set zzz: = 0 {to be returned by the main program}
Set *nb_eval_max: = 0* {maximum number of evaluations}
Set *nb_eval_min: = -3200* {minimum number of evaluations}
Set *begin: = 0* {just a counter}
Read data description

$$\text{Set } init\_size = \begin{cases} 3 \ for\ complete\ adaptive\ method \\ adapt\ \ for\ non-adaptive\ method \end{cases}$$

{initial size}

Set *param N: = init_size* {record initial size}
Set *cycle_size: = param. N* {cycle size}
Set *times: = 0* { }
Set *duration_tot:= 0* { }
Set *failure_tot:= 0* {number of failures}

Set *infinite:= 999999999* {very large number}
Set *min_eps:= infinite* {minimum admissible error}
Set *max_eps:= 0* { maximum admissible error }
Set *mean_eps:= 0* { mean admissible error }
Set *chance: = 0* {equal to 1 if a solution is found}
Set *eval_f_tot: = 0* {number of objective function evaluations}
Set *label: = 0* {label for the first particle: mainly for later visualization}
Set *almostzero:= 0.0000001* {to avoid overflow by dividing by too small a value}
Set *eval_f_max:= 1/almostzero* {high number of evaluations }
Set *eval_f:= 0* {number of evaluation counter}
Set *nb_no_progress:= 0* {implies failure if this number > parpet}
Set *Iter:= Iter + 1* {number of iterations}
Set *n_add:= 0* {number of swarms added}
Set *n_remove:= 0* {number of swarms removed}
Set *mean_swarm_size: = 0* {mean size of swarm}

Step 2. Swarm initialization:
Set *best_result.f:= infinite* {999999999}
Set *TR_size:= 1* {number of tribes}
Set *TR_tr[0].size:= param.N* {Tribe size}
For i: = 1 to *param.N* do {Generate particles & keep the best result found so far}
Tribe T0 is initialized with *n* distinct particles
If the position of particle i < the best position so far then the best position so far then *best_result:=* position of particle
end For
Transform rational position to discrete position with discrete sequence
Access local search engine to refine the solution
Memorize the best solution so far
If the best position so far < *min_esp* then *min_esp:=* best position
Initialize the i-group of the first particle
Set all other particles to have the same i-group as the first particle
Initialize the *rank_label* table
Set *cycle:= 0* {number of iterations since last adaptive update}

Step 3. Swarm moves
Set *Ite:= Iter + 1* {increment number of iterations}
For *i: = 1* to *TR.size* do {for each tribe}
For *j: = 1* to *TR.tr[i].size* do {for each particle}
Find the best particle *p* that gives useful information in all the tribes {“informer”}
Transform rational position to discrete position with discrete sequence
Access local search engine to refine the solution
Memorize the best solution so far
Move each particle in each tribe

```
        If the position of particle is better than the best so far then the best
        so far becomes this particle {updating the best result}
        If best_result.f < min_eps then min_eps:= best_result.f
                                                        {keep the min. eps}
        If best_result.f > max_eps then max_eps:= best_result.f
                                                        {keep the max. eps}
        If (chance = 1) goto  end_by_chance  {a solution has been found}
        If  failure condition is met then
        {
         failure_tot := failure_tot + 1
          goto  the_end  {arbitrary stop condition}
         }
       end For j
      end For i
Step 4. Adaptation:
      Set cycle_size:= init_size+ n_add_n_remove
      Set mean_swarm_size:= mean_swarm_size + cycle_size
          If (adapt = 0) goto  end_adapt  {end adaptation}
      Set cycle:= cycle + 1
          If (2*cycle < cycle_size) goto  end_adapt  {end adaptation}
      Set Trsize0:= TR.size {memorize the current number of tribes}
      For i: = 1 to TR.size0 do {check each tribe}
       Find the best in the current tribe
       Count how many particles in he current tribe that have improved
       their position
          If improve = 0  {if not enough improvement}
          then {
        Transform rational position to discrete position with
        discrete   sequence
              Access local search engine to refine the solution
              Memorize the best solution so far
              Calculate probability of generation
              If it is the first bad tribe, generate a new empty one
              Generate a particle in the new tribe
              Begin to build its i-group with the label of the best particle that
              has  generated he new particle and add it to the i-group of
              the new particle
              Increase the (new) tribe size
              Update the i-group of the "generating" best particle
               }
              If improve ≥ TR.tr[i].size/2   then {if enough improvement}
               {
              Find the worst particle in the current tribe
                 If the worst particle is good goto  exit_good {exit condition }
              Remove the worst particle from the tribe
              Replace the worst particle by the best particle in all i-groups
```

```
            Merge the worst particle's i-group into the i-groups of the
            best particle
             }
      End For i tribes
      Complete the i-groups of each particle of the new tribe (if there is one)
      with all particles of the tribe
      Remove empty tribes
     Set cycle:= 0 {initialize number of cycles}
      Transform rational position to discrete position with discrete sequence
      Access local search engine to refine the solution
       Memorize the best solution so far
      end_adapt: {end of adaptation}
      Goto Step 3
end_by_chance:
the_end:
      Output solution
      Stop
```

21.5 Experimental results

21.5.1 Parameter setting

TRIBES, does not need any parameters at all, except, of course, the ones defining the problem. These parameters are *function, dimension,* $x_{\min}$, x_{man}, *granularity* (normally required for the hyper-parallepipedic search space), *target*, and *admissible error*. The parameter, *all_different* is required if all components of the solution have to be different, as in most combinatorial problems. For a solution, where on some dimensions, is an *integer* point, or where it is even a rational one, the granularity information has to be given. For example, if for a given dimension, the solution is an integer then the user can set granularity to *1* for this dimension; the process may be faster. A positive granularity *g* on a given dimension is required if component values for this dimension are not continuous. More precisely, for any pair of values (x, x'), you must have abs(x-x') = k*d, where *k* is an integer. As you point out, for an integer dimension, g=1 In fact, for combinatorial problems, we must use *dimension = MM* (the size of the problem), *granularity =1, all_different =1*. Additionally, there is a parameter *adapt*, which describes the nature of adaptation used in the search. If *adapt = n*, it means a swarm of constant size *n* (and the neighborhood of each particle is the whole swarm) is being used. For classical test functions, and with *n* equal about *20*, this TRIBES version is often better than the other adaptive PSO versions that do not use hyper-spheres. If *adapt = 0*, it means you want to use complete adaptive method. This condition is not always better than the previous one, but the user does not even have to "guess" what the right swarm size is.

21.5.2 Comparison with other heuristics

In Table 21.2 we compare the results given by TRIBES to the ones given by PSO Type-1, as defined and used in Clerc and Kennedy (2002) and Kennedy and Eberhart (2001), together with other competitive search heuristics such as genetic algorithm (Goldberg 1989) and differential evolution (Storn and Price1995). For all our experiments, TRIBES is launched with full adaptation. With adaptation, i.e. when the user has no parameter at all to define, the mean swarm size tends to get somewhat large. In our examples the swarm size increased as much as *10*, less than *20*, which probably would have been closer to the optimum (Clerc 2002). When the size of swarm increases more than 20, then it means either particles are too easily generated or too rarely removed, or both. In such a case, it would be interesting to find better rules. In our case, the swarm size seems reasonable. The results show that results of TRIBES compared to its competitors are reasonable, although further work is possible to fine-tune the solutions.

Table 21.2: Comparison between DE, GA, PSO and TRIBES

Problem	DE	GA	PSO	TRIBES
8 x 15	139	143	139	134
10 x 25	208	205	201	219
15 x 25	258	248	246	268
20 x 50	475	468	464	490
25 x 75	715	673	710	726

21.6 Conclusion

In the work reported in this chapter, we have shown that it is possible to define a completely parameter free particle swarm optimizer, at least as good as a carefully tuned non-adaptive one. The research builds on the parameter free optimizer, which Clerc (2002) proposed for rational problems as a starting point, although some pseudo-discrete optimization problems could be solved using the version (fifty-fifty problem, etc.). The results show that results of TRIBES compared to its competitors are reasonable, although further work is possible to fine-tune the solutions. Further work includes finding better local search engines, which performs better than the currently implemented version. Extension of the applications of TRIBES to some other combinatorial optimization problems such as TSP and QAP is in progress; the results so far shows that opting for a better local search engine will give promising results.

Acknowledgement

Maurice Clerc is the initiator of TRIBES, and the work reported here extends his native version, which currently does not solve explicit discrete optimization problems such as flow shop, TSP and QAP problems.

References

Angeline PJ (1998) Using selection to improve particle swarm optimization, presented at *IEEE International Conference on Evolutionary Computation*, Anchorage, Alaska, May 4-9.

Carlisle A, Dozier G (1998) Adapting particle swarm optimization to dynamics environments, presented at *International Conference on Artificial Intelligence*, Monte Carlo Resort, Las Vegas, Nevada, USA.

Clerc M (1999) The swarm and the queen: towards a deterministic and adaptive particle swarm optimization, presented at *Congress on Evolutionary Computation*, Washington DC.

Clerc M, Kennedy J (2002) The Particle Swarm-Explosion, Stability, and Convergence in a Multidimensional Complex Space, *IEEE Transactions on Evolutionary Computation*, (6), 58-73.

Clerc M (2002) Think locally, act locally: a framework for adaptive particle swarm optimizers, *IEEE Journal of Evolutionary Computation*, (submitted).

Clerc M (2003) L'optimisation par essaim particulaire, *5ème congrès de la Société, Française de Recherche, Opérationnelle et d'Aide à la Décision*, 26, 27 et 28 février 2003, Avignon, Université d'Avignon et des Pays de Vaucluse, 74 rue Louis Pasteur 84000 AVIGNON.

Conway J H, Sloane N JA (1993) *Sphere Packings, Lattices, and Groups, 2nd ed.* New York: Springer-Verlag, p. 9.

Goldberg DE (1989) *Genetic Algorithms in Search, Optimization, and Machine Learning*, Reading, MA: Addison-Wesley

Hu X, Eberhart RC (2002) Adaptive particle swarm optimization: detection and response to dynamic systems, presented at *Congress on Evolutionary Computation*, Hawaii.

Kennedy J, Eberhart RC (1999) The particle swarm: social adaptation in information processing systems, in *New ideas in Optimization*, D. Corne, Dorigo, M., and Glover, F., Ed. London: McGraw-Hill.

Kennedy J, Eberhart RC, and Shi Y (2001) *Swarm Intelligence: Morgan Kaufmanns Academic Press.*

Peterson I (1988) *The Mathematical Tourist: Snapshots of Modern Mathematics.* New York: W. H. Freeman, p. 96-101.

Shi Y, Eberhart RC (2001) Fuzzy adaptive particle swarm optimization, presented at *Congress on Evolutionary Computation*, Seoul, Korea.

Somerville DMY (1958) *An Introduction to the Geometry of n Dimensions.* New York: Dover, p. 136.

Storn R, Price K (1995) Differential evolution - a simple and efficient adaptive scheme for global optimization over continuous spaces, *Technical Report TR-95-012, ICSI, March 1999*
(Available via ftp from ftp.icsi.berkeley.edu/pub/techreports/1995/tr-95-012.ps.Z).

Wells D (1986) *The Penguin Dictionary of Curious and Interesting Numbers.* Middlesex, England: Penguin Books.

22 Optimizing CNC Drilling Machine Operations: Traveling Salesman Problem-Differential Evolution Approach

Godfrey C Onwubolu

This chapter describes the application of differential evolution algorithm to the problem of minimizing the operating path of automated or computer numerically controlled drilling operations. The operating path is first defined as a traveling salesman problem. Then, the relatively new heuristic, differential evolution algorithm is applied to the traveling salesman problem. In a batch production of a large number of items to be drilled such as in printed circuit boards, the travel time of the drilling device is a significant portion of the overall manufacturing process. Hence, the differential evolution algorithm-traveling salesman problem heuristic, can play a significant role in reducing production costs.

22.1 Introduction

The traveling salesman problem (TSP) is well known in optimization. A traveling salesman has a number of N cities to visit. The sequence in which the salesperson visits different cities is called a tour. A tour should be such that every city on the list is visited once and only once, except that salesperson returns to the city from which he/she starts. The goal is to find a tour that minimizes the total distance the salesperson travel, among all the tours that satisfy this criterion. Much of the work on the TSP is not motivated by direct applications, but rather by the fact that the TSP provides an ideal platform for the study of general methods that can be applied to a wide range of discrete optimization problems. This is not to say, however, that the TSP does not find applications in many fields. Indeed, the numerous direct applications of the TSP bring life to the research area and help to direct future work. The TSP naturally arises as a sub-problem in many transportation and logistics applications, for example the problem of arranging school bus routes to pick up the children in a school district. More recent applications involve the scheduling of service calls at cable firms, the delivery of meals to homebound persons, the scheduling of stacker cranes in warehouses, the routing of trucks for parcel post pickup, and a host of others.

Although transportation applications are the most natural setting for the TSP, the simplicity of the model has led to many interesting applications in other areas. A classic example is the scheduling of a machine to drill holes in a circuit board or other object. In this case the holes to be drilled are the cities, and the cost of travel is the time it takes to move the drill head from one hole to the next. The technology for drilling varies from one industry to another, but whenever the travel time of the drilling device is a significant portion of the overall manufacturing process then the TSP can play a role in reducing costs. In a recent work by one of the authors of this article, TSP was applied to the scheduling of a computer-controlled machine to drill holes in a circuit board; this machine was designed and fabricated in-house.

Similar application is in complete and partial changeover times in numerically controlled (NC) tool machine problem. Suppose we are given N jobs to be performed on one machine. Total processing time, is fixed by unit processing time and batch size. Total set-up time depends on the sequence in which jobs will be run. For instance, in group-technology, parts in the same family may require little or no changeover. Parts in different families may require complete breakdown and rebuilding of machine tooling. Changing sizes in a packaging line may be relatively easy, but changing package contents may require extensive cleaning and switching of machinery. Common to both these examples is the fact that individual changeover times depend only on the current and next product. For bottleneck facilities, maximising the production of saleable product is the prime objective. This situation corresponds to the travelling salesman problem (TSP). In the TSP, a salesman must visit each city in his territory and then return home. In our case, the worker must perform each job and then return to the starting condition. The problem can be visualised on a graph. Each city (job) becomes a node. Arc lengths correspond to the distance between the attached cities (job changeover times). The salesman wants to find the shortest tour of the graph. A tour is a complete cycle. Starting at a home city, each city must be visited exactly one time before returning home. Each leg of the tour travels on an arc between two cities. If arc lengths differ depending on the direction of the arc, the TSP is said to be asymmetric. Although TSP is one of the most studied problems in combinatorial optimisation, it is NP-hard (Lawler *et al.* 1985; Reinelt 1994). For a broader collection of papers the reader is referred to the excellent book (Lawler *et al.* 1985) and to the more recent bibliography (Junger *et al.* 1995).

The effectiveness and efficiency of operating a manufacturing cell is affected by the problem of machine tooling and production activities sequencing within the cell (Askin and Standridge 1993). There are several ways productivity may be maximised: (i) by sequencing batches to minimise tooling changeovers; (ii) by sequencing machining of jobs (a batch of one or more similar parts) to minimise idle time of material handling devices; (iii) by optimising the layout of the work-cell. This chapter addresses the first problem, which is that of sequencing batches to minimise tooling changeovers. For bottleneck work-centres, maximising the production of goods that meet customer requirements is the prime objective of a manufacturing enterprise. Therefore, any strategy that can be adopted to minimise the production time will have much impact on achieving the firm's objectives.

22.2 Travelling Salesman Problem (TSP)

In the TSP, a salesman must visit each city in his designated area and then return home. In our case, the worker (tool) must perform each job and then return to the starting condition. The problem can be visualised on a graph. Each city (job) becomes a node. Arc lengths correspond to the distance between the attached cities (job changeover times). The salesman wants to find the shortest tour of the graph. A tour is a complete cycle. Starting at a home city, each city must be visited exactly one time before returning home. Each leg of the tour travels on an arc between two cities. The length of the tour is the sum of the lengths of the arcs selected. Figure 22.1 illustrates a five-city TSP. Trip lengths are shown on the arcs in Figure 22.1*a*, the distance from city *i* to *j* is denoted by c_{ij}. We have assumed in the figure that all paths (arcs) are bi-directional. If arc lengths differ depending on the direction of the arc the TSP-formulation is said to be asymmetric, otherwise it is symmetric. A possible tour is shown in Figure 22.1*b*. The cost of this tour is $c_{12} + c_{24} + c_{43} + c_{35} + c_{51}$.

Several mathematical formulations exist for the TSP. One approach is to let x_{ij} be *1* if city *j* is visited immediately after *i*, and be *0* if otherwise. A formal statement of TSP is given as follows:

$$\text{minimise} \sum_{i=1}^{N} \sum_{j=1}^{N} c_{ij} x_{ij} \tag{22.1}$$

$$\text{subject to} \sum_{j=1}^{N} x_{ij} = 1; \quad \forall i \tag{22.2}$$

$$\sum_{i=1}^{N} x_{ij} = 1; \quad \forall j \tag{22.3}$$

$$\textit{No subtours} \tag{22.4}$$

$$x_{ij} = 0 \text{ or } 1$$

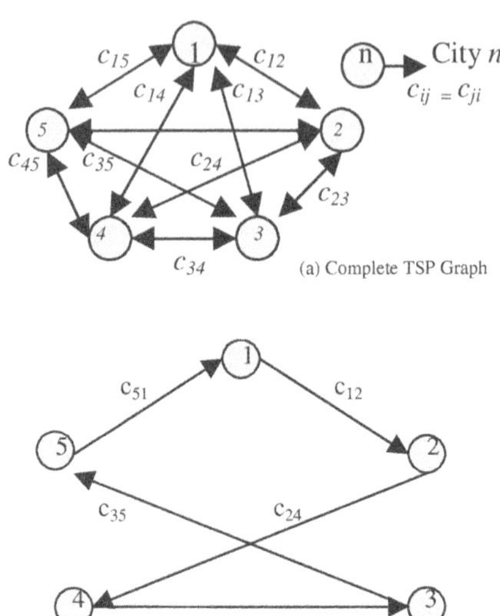

Fig. 22.1 TSP illustrated on a graph

No *subtours* mean that there is no need to return to a city prior to visiting all other cities. The objective function accumulates time as we go from city i to j. Constraint 22.2 ensures that we leave each city. Constraint 22.3 ensures that we visit (enter) each city. A subtour occurs when we return to a city prior to visiting all other cities. Restriction 22.4 enables the TSP-formulation, differ from a linear assignment programming (LAP) formulation. Unfortunately, the non-subtour constraint significantly complicates model solution. One reasonable construction procedure for solving TSP is the *closest insertion algorithm*. This is now discussed.

22.3 TSP using Closest Insertion Algorithm

The closest insertion algorithm starts by selecting any city. We then proceed through $N - 1$ stages, adding a new city to the sequence at each stage. Thus a partial sequence is always maintained, and the sequence grows by one city each stage. At each stage we select the city from those currently unassigned that is closest to any city in the partial sequence. We add the city to the location that causes the smallest increase in the tour length. The closest insertion algorithm can be shown to produce a solution with a cost no worse than twice the optimum when the cost matrix is symmetric and satisfies the triangle inequality. In fact, the closest insertion algorithm may be a useful seed-solution for combinatorial search methods when large problems are solved. Symmetric implies $c_{ij} = c_{ji}$ where c_{ij} is the cost to go from city i directly to city j. Unfortunately, symmetry need not exist in our changeover problem. Normally, the triangular inequality ($c_{ij} \leq c_{ik} + c_{kj}$) will be satisfied, but this alone does not suffice to ensure the construction of a good solution. We may also try repeated application of the algorithm choosing a different starting city each time and then choose the best sequence found. Of course, this increases our workload by a factor of N. Alternatively, a different starting city may be chosen randomly for a specific number of times, less than the total number of cities. This option is preferred for large problem instances.

We now state the algorithm formally. Let S_a be the set of available (unassigned) cities at any stage. S_p will be the partial sequence in existence at any stage and is denoted $S_p = \{s_1, s_2, ..., s_n\}$, implying that city s_2 immediately follows s_1. For each unassigned city j, we use $r(j)$ to keep track of the city in the partial sequence that is closest to j. We store $r(j)$ only to avoid repeating calculations at each stage. Last, bracketed subscripts $[i]$ refer to the ith city in the current partial sequence. The steps involved are:

STEP 0. Initialize. $N = 1$. $S_p = \{1\}$. $S_a = \{2, ..., N\}$. For $j = 2,...,N$. $r(j) = 1$.

STEP 1. Select new city. Find $j^* = \arg\min_{j \in S_a} \left\{ c_{j,c(j)}, \ or \ c_{c(j),j} \right\}$.

Set $n = n + 1$.

STEP 2. Insert j^*, update $r(j)$. $S_a = S_a - j^*$. Find city $i^* \in S_p$ such that $i^* = \text{argmin}_{[i] \in Sp} \{c_{[i]j^*} + c_{j^*,[i+1]} - c_{[i],[i+1]}\}$. Update $S_p = \{s_1,..,i^*, j^*, i^* + 1,...,s_n\}$. For all $j \in S_a$, if $\min\{c_{j,j^*}, c_{j^*j}\} < c_{j,r(j)}$ then $r(j) = j^*$. If $n < N$, go to 2.

As can be seen, the closest insertion algorithm is a constructive method. In order to understand the steps involved, let us consider an example related to change overtimes for a flexible manufacturing cell (FMC).

Example 22.1

Table 22.1 shows the changeover times for a flexible manufacturing-cell. A machine is finishing producing batch T1 and other batches are yet to be completed. Use the closest insertion heuristic to find a job sequence, treating the problem as a TSP.

Table 22.1. Changeover times

Changeover times (hrs)					
From/To	T1	T2	T3	T4	T5
T1	-	8	14	10	12
T2	4	-	15	11	13
T3	12	17	-	1	3
T4	12	17	5	-	3
T5	13	18	15	2	-

Solution

Step 0: $S_p = \{1\}$, $S_a = \{2, 3, 4, 5\}$, $\underset{j \in S_a}{r(j) = 1}$; j = 2,…,5. This is equivalent to choosing the first city from the partial-list and eliminating this city form the available list.

Step 1: Select the new city: find $j^* = \arg\min_{j \in S_a} \{c_{j,r[j]}\}$ and set $n = n + 1$

$\min_{j \in S_a} \{c_{12}, c_{13}, c_{14}, c_{15}, c_{21}, c_{31}, c_{4,1}, c_{5,1}\} = \min \{8, 14, 10, 12, 0, 0, 0, 0\} = 8; j^* = 2$

But ignore c_{21}, c_{31}, $c_{4,1}$, and $c_{5,1}$ because city 1 is already considered in S_a.

Step 2: Insert city 2, and update *r(j)* for the remaining jobs 3, 4, and 5
$S_p = \{1, 2\}$; $S_a = \{3, 4, 5\}$ $c_{12} + c_{21} - c_{11} = 8 + 4 - 0 = 12$

Step 1: Select new city
$\min \{c_{23}, c_{24}, c_{25}, c_{32}, c_{42}, c_{52}\} = \{15, 11, 13, 17, 17, 18\} = 11$;

$j^* = 4$; n = 3. So we have job 4 after job 1 or 2

Step 2: Insert job 4
There are the following possibilities from {1, 2}: {1, 2, 4} or {1, 4, 2}

For {1, 2, 4}, $c_{12} + c_{24} - c_{14} = 8 + 11 - 10 = 9$

For {1, 4, 2}, $c_{14} + c_{42} - c_{12} = 10 + 17 - 8 = 19$

The minimum occurs for inserting job 2 after job 4. Update *r(j)* for remaining jobs 3, 5 i.e., r(3) = r(5) = 4

Step 1: Select new job

$\min_{j \in S_a} \{c_{34}, c_{5,4}, c_{4,3}, c_{4,5}\} = \min\{1, 2, 5, 3\}$; $j^* = 3$

min = C_{34}, but 4 is already considered. Hence, $j^* = 3$.

Step 2: Insert job 3

There are the following possibilities from {1, 2, and 4}:

{1, 2, 4, 3}	:	$c_{43} + c_{31} - c_{41} = 5 + 12 - 12 = 5$
{1, 2, 3, 4}	:	$c_{23} + c_{34} - c_{24} = 15 + 1 - 11 = 5$
{1, 3, 2, 4}	:	$c_{13} + c_{32} - c_{12} = 14 + 17 - 8 = 25$

Choosing {1, 2, 4, and 3} breaks the tie. Updating *r*[5] = 3

Step 1: Select new job.

Since job 5 remains, $j^* = 5$

Step 2: Insert job 5

There are the following possibilities from {1, 2, 4, and 3}

{1, 2, 4, 3, 5}	:	$c_{35} + c_{51} - c_{31} = 3 + 13 - 12 = 4$
{1, 2, 4, 5, 3}	:	$c_{45} + c_{53} - c_{43} = 3 + 15 - 5 = 13$
{1, 2, 5, 4, 3}	:	$c_{25} + c_{54} - c_{24} = 13 + 2 - 11 = 4$
{1, 5, 2, 4, 3}	:	$c_{15} + c_{52} - c_{12} = 12 + 18 - 8 = 22$

Choosing {1, 2, 4, 3, and 5} breaks the tie as the final sequence. The cost = $c_{12} + c_{24} + c_{43} + c_{35} = 8 + 11 + 5 + 3 = 27$.

The TSP construction is shown in Figure 22.2. The meaning of this solution is that batch *1* is first produced, followed by batch *2*, then batch *4*, then batch *3*, and finally batch *5*.

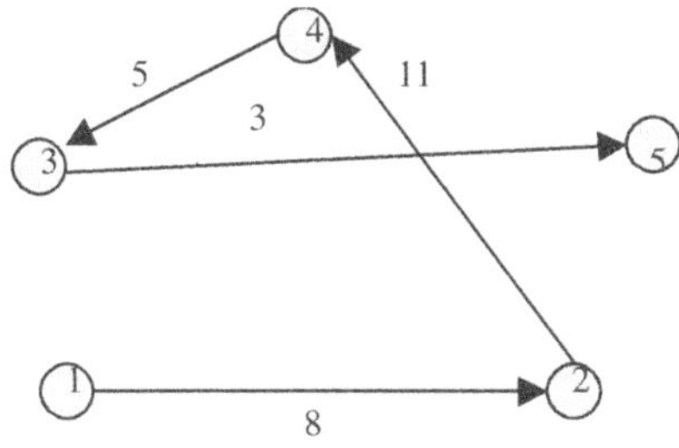

Fig. 22.2 TSP solution for Example 22.1

22.4 TSP using Differential Evolution

In recent years, several search procedures have emerged as having great potential to address practical optimisation problems. These emerging optimisation techniques include simulated annealing (Kirkpatrick *et al.* 1983), genetic algorithms (Goldberg, 1989), tabu search (Glover, 1989, 1990), and ant colony optimisation (Dorigo, 1992). Consequently, over the past few years, several researchers have demonstrated the applicability of these methods, to combinatorial optimisation problems such as the travelling salesman problem. The travelling salesman problem has been solved using simulated annealing (Lin *et al.* 1993), genetic algorithms (Whitley *et al.* 1989), ant colony optimisation (Dorigo, 1993; Dorigo and Gambardella 1997), evolutionary programming (Fogel, 1993), and elastic net method (Durbin and Willshaw, 1987).

More recently, a novel optimization method based on differential evolution (exploration) algorithm (Storn and Price, 1995) has been developed, which originally focused on solving non-linear programming problems containing continuous variables. Since Storn and Price (1995) invented the differential evolution (exploration) algorithm, the challenge has been to employ the algorithm to different areas of problems other than those areas that the inventors originally focussed on. Differential evolution (exploration) algorithm was invented by Storn and Price (1995) initially to solve possibly nonlinear and on differentiable continuous space functions. However, recently there has been scarce research to apply the new optimization technique to discrete problems such as design of gear train, pressure vessels and springs (Lampinen and Zelinka, 1999) and flow-shop scheduling (Onwubolu, 2001).

This paper presents a new approach based on differential evolution (exploration) algorithm (Storn and Price, 1995) for solving the travelling salesman problem. To date, the authors are not aware of any published work on the application of differential evolution (exploration) algorithm for solving the travelling salesman problem. TSP is a discrete optimisation problem, which can be readily solved using the formulation presented by Onwubolu (2001).

22.4.1 Differential Evolution Method

Whether in industry or in research, users generally demand that a practical optimization technique should fulfill three requirements:

(1) the method should find the true global minimum, regardless of the initial system parameter values;

(2) convergence should be fast; and

(3) the program should have a minimum of control parameters so that it will be easy to use.

Storn and Price (1995) invented the differential evolution (exploration) algorithm in a search for a technique that would meet the above criteria. DE is a method, which is not only astonishingly simple, but also performs extremely well on a wide variety of test problems. It is inherently parallel because it is a population based approach and hence lends itself to computation via a network of computers or processors. The basic strategy employs the difference of two randomly selected parameter vectors as the source of random variations for a third parameter vector. In the following subsection, we follow a more graphical approach of Storn and Price (1995) for presenting the new optimization method which, we have employed in TSP. The parameters used are $\Im$ = cost or the value of the objective function, D = problem dimension, NP = population size, P = population of X-vectors, G = generation number, G_{max} = maximum generation number, X = vector composed of D parameters, V = trial vector composed of D parameters, CR = crossover factor. Others are F = scaling factor ($0 < F \leq 1.2$), (U) = upper bound, (L) = lower bound, **u**, and **v** = trial vectors, $x_{best}^{(G)}$ = vector with minimum cost in generation G, $x_i^{(G)}$ = i^{th} vector in generation G, $b_i^{(G)}$ = i^{th} buffer vector in generation G, $x_{r1}^{(G)}, x_{r2}^{(G)}$ = randomly selected vector, L = random integer ($0 < L < D - 1$). In the formulation, N = number of cities. Some integers used are i, j.

Differential Evolution (DE) is a novel parallel direct search method, which utilizes NP parameter vectors

$$x_i^{(G)},\ i = 0, 1, 2, \ldots, NP\text{-}1. \tag{22.5}$$

as a population for each generation, G. The population size, NP does not change during the minimization process. The initial population is generated randomly assuming a uniform probability distribution for all random decisions if there is no initial intelligent information for the system. The crucial idea behind DE is a new scheme for generating trial parameter vectors. DE generates new parameter vectors by adding the weighted difference vector between two population members to a third member. If the resulting vector yields a lower objective function value than a predetermined population member, the newly generated vector replaces the vector with which it was compared. The comparison vector can, but need not be part of the generation process mentioned above. In addition the best parameter vector $x_{best}^{(G)}$, is evaluated for every generation G in order to keep track of the progress that is made during the minimization process. Extracting distance and direction information from the population to generate random deviations results in an adaptive scheme with excellent convergence properties (Storn and Price 1995).

Descriptions for the earlier two most promising variants of DE (later known as DE2 and DE3) are presented in order to clarify how the search technique works, then a complete list of the variants to date are given thereafter.

22.4.1.1 Scheme DE2

Initialization

As with all evolutionary optimization algorithms, DE works with a population of solutions, not with a single solution for the optimization problem. Population *P* of generation *G* contains *NP* solution vectors called individuals of the population and each vector represents potential solution for the optimization problem:

$$P^{(G)} = X_i^{(G)} \quad i = 1,...,NP;\; G = 1,...,G_{\max} \tag{22.6}$$

Additionally, each vector contains *D* parameters:

$$X_i^{(G)} = x_{j,i}^{(G)} \quad i = 1,...,NP;\; j = 1,...,D \tag{22.7}$$

In order to establish a starting point for optimum seeking, the population must be initialized. Often there is no more knowledge available about the location of a global optimum than the boundaries of the problem variables. In this case, a natural way to initialize the population $P^{(0)}$ (initial population) is to seed it with random values within the given boundary constraints:

$$P^{(0)} = x_{j,i}^{(0)} = x_j^{(L)} + rand_j[0,1] \bullet \left(x_j^{(U)} - x_j^{(L)}\right) \;\forall i \in [1, NP];\; \forall j \in [1, D] \tag{22.8}$$

where $rand_j[0,1]$ represents a uniformly distributed random value that ranges from zero to one. The lower and upper boundary constraints are, $x^{(L)}$ and $x^{(U)}$, respectively:

$$x_j^{(L)} \le x_j \le x_j^{(U)} \;\; \forall j \in [1, D] \tag{22.9}$$

For this scheme and other schemes, three operators are crucial: mutation, crossover and selection. These are now briefly discussed.

Mutation

The first variant of DE works as follows: for each vector $x_i^{(G)}$, i = 0, 1, 2, ... , *NP*-1, a trial vector v is generated according to

$$v_{j,i}^{(G+1)} = x_{j,r1}^{(G)} + F \bullet \left(x_{j,r2}^{(G)} - x_{j,r3}^{(G)}\right) \tag{22.10}$$

where $i \in [1, NP];\; j \in [1, D]$, *F* > *0*, and the integers $r1,\; r2,\; r3 \in [1, NP]$ are generated randomly selected, except: $r1 \ne r2 \ne r3 \ne i$.

Three randomly chosen indexes, *r1*, *r2*, and *r3* refer to three randomly chosen vectors of population. They are mutually different from each other and also different from the running index *i*. New random values for *r1*, *r2*, and *r3* are assigned for each value of index *i* (for each vector). A new value for the random number $rand[0,1]$ is assigned for each value of index *j* (for each vector parameter). *F* is a real and constant factor, which controls the amplification of the differential variation. A two dimensional example that illustrates the different vectors which

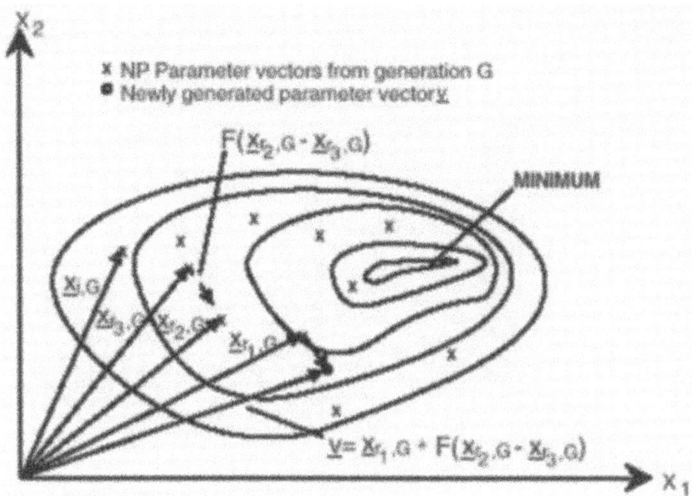

play a part in DE2 are shown in Figure 22.3.

Fig. 22.3. Contour lines and the process for generating v in scheme DE1.

Crossover

In order to increase the diversity of the parameter vectors, the vector

$$u = (u_1, u_2, ..., u_D)^T \tag{22.11}$$

with

$$u_j^{(G)} = \begin{cases} v_j^{(G)} \; for \; j = \langle n \rangle_D, \langle n+1 \rangle_D, ..., \langle n+L-1 \rangle_D \\ \left(x_i^{(G)}\right)_j \; otherwise \end{cases} \tag{22.12}$$

is formed where the acute brackets $\langle \; \rangle_D$ denote the modulo function with modulus *D*. This means that a certain sequence of the vector elements of *u* are identical to the elements of v, the other elements of u acquire the original values of $x_i^{(G)}$. Choosing a subgroup of parameters for mutation is similar to a process known as crossover in genetic algorithm. This idea is illustrated in Figure 22.4 for D = 7, n = 2 and L = 3. The starting index n in (22.12) is a randomly chosen integer from the interval [0, D-1]. The integer L is drawn from the interval [0, D-1] with the probability $\Pr(L = v) = (CR)^v$ CR $\in [0,1]$ is the crossover probability and constitutes a control variable for the DE2-scheme. The random decisions for both n and L are made anew for each trial vector v.

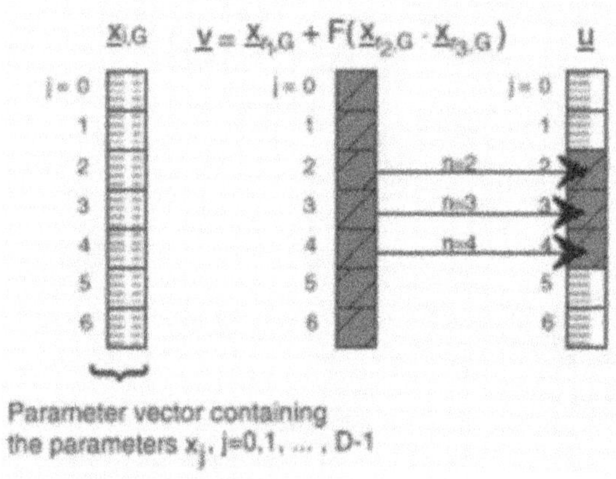

Fig. 22.4. Crossover process for D = 7, n = 2 and L = 3.

Selection

In order to decide whether the new vector u shall become a population member of generation $G+1$, it will be compared to $x_i^{(G)}$. If vector u yields a smaller objective function value than $x_i^{(G)}$, $x_i^{(G+1)}$ is set to u, otherwise the old value $x_i^{(G)}$ is retained.

22.4.1.2 Scheme DE3

Basically, scheme DE3 works the same way as DE2 but generates the vector v according to

$$v = x_i^{(G)} + \lambda \bullet \left(x_{best}^{(G)} - x_i^{(G)}\right) + F \bullet \left(x_{r2}^{(G)} - x_{r3}^{(G)}\right) \quad (22.13)$$

introducing an additional control variable λ. The idea behind λ is to provide a means to enhance the greediness of the scheme by incorporating the current best vector $x_{best}^{(G)}$. This feature can be useful for non-critical objective functions. Figure 22.5 illustrates the vector-generation process defined by (22.13). The construction of u from v and $x_i^{(G)}$ as well as the decision process are identical to DE2.

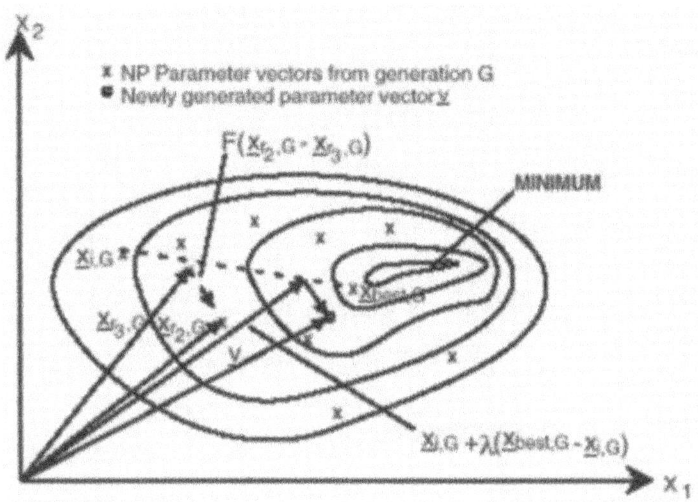

Fig. 22.5 Contour lines and the process for generating v in scheme DE3.

22.4.1.3 DE Strategies

Price and Storn (2001) have suggested ten different working strategies of DE and some guidelines in applying these strategies for any given problem (see Table 22.2). Different strategies can be adopted in the DE algorithm depending upon the type of problem for which it is applied. The strategies can vary based on the vector to be perturbed, number of difference vectors considered for perturbation, and finally the type of crossover used.

The general convention used above is as follows: DE/x/y/z. DE stands for differential evolution algorithm, x represents a string denoting the vector to be perturbed, y is the number of difference vectors considered for perturbation of x, and z is the type of crossover being used. Other notations are exp: exponential; bin: binomial). Thus, the working algorithm outline by Price and Storn (1997) is the seventh strategy of DE, that is, DE/rand/1/bin. Hence the perturbation can be either in the best vector of the previous generation or in any randomly chosen vector. Similarly for perturbation, either single or two vector differences can be used. For perturbation with a single vector difference, out of the three distinct randomly chosen vectors, the weighted vector differential of any two vectors is added to the third one. Similarly for perturbation with two vector differences, five distinct vectors other than the target vector are chosen randomly from the current population. Out of these, the weighted vector difference of each pair of any four vectors is added to the fifth one for perturbation.

Table 22.2 Ten different working strategies

Strategy	Formulation
1: DE/best/1/exp:	$v = x_{best}^{(G)} + F \bullet \left(x_{r2}^{(G)} - x_{r3}^{(G)}\right)$
2: DE/rand/1/exp:	$v = x_{r1}^{(G)} + F \bullet \left(x_{r2}^{(G)} - x_{r3}^{(G)}\right)$
3: DE/rand-to-best/1/exp:	$v = x_{i}^{(G)} + \lambda \bullet \left(x_{best}^{(G)} - x_{i}^{(G)}\right) + F \bullet \left(x_{r1}^{(G)} - x_{r2}^{(G)}\right)$
4: DE/best/2/exp:	$v = x_{best}^{(G)} + F \bullet \left(x_{r1}^{(G)} + x_{r2}^{(G)} - x_{r3}^{(G)} - x_{r4}^{(G)}\right)$
5: DE/rand/2/exp:	$v = x_{r5}^{(G)} + F \bullet \left(x_{r1}^{(G)} + x_{r2}^{(G)} - x_{r3}^{(G)} - x_{r4}^{(G)}\right)$
6: DE/best/1/bin:	$v = x_{best}^{(G)} + F \bullet \left(x_{r2}^{(G)} - x_{r3}^{(G)}\right)$
7: DE/rand/1/bin:	$v = x_{r1}^{(G)} + F \bullet \left(x_{r2}^{(G)} - x_{r3}^{(G)}\right)$
8: DE/rand-to-best/1/bin:	$v = x_{i}^{(G)} + \lambda \bullet \left(x_{best}^{(G)} - x_{i}^{(G)}\right) + F \bullet \left(x_{r1}^{(G)} - x_{r2}^{(G)}\right)$
9: DE/best/2/bin:	$v = x_{best}^{(G)} + F \bullet \left(x_{r1}^{(G)} + x_{r2}^{(G)} - x_{r3}^{(G)} - x_{r4}^{(G)}\right)$
10: E/rand/2/bin:	$v = x_{r5}^{(G)} + F \bullet \left(x_{r1}^{(G)} + x_{r2}^{(G)} - x_{r3}^{(G)} - x_{r4}^{(G)}\right)$

In exponential crossover, the crossover is performed on the *D* (the dimension or number of variables to be optimized) variables in one loop until it is within the *CR* bound. For discrete optimization problems, the first time a randomly picked number between *0* and *1* goes beyond the *CR* value, no crossover is performed and the remaining *D* variables are left intact. In binomial crossover, the crossover is performed on each the *D* variables whenever a randomly picked number between *0* and *1* is within the *CR* value. Hence, the exponential and binomial crossovers yield similar results.

22.4.2 Differential Evolution Method for TSP

The problem formulation is already discussed in Section 22.1. Solving TSP requires discrete variables. To achieve this, we enhanced the capability of DE by developing two strategies known as forward and backward transformation techniques respectively.

22.4.2.1 Forward Transformation and Backward Transformation Technique

The forward transformation method transforms integer variables into continuous variables for the internal representation of vector values since in its canonical form, the DE algorithm is only capable of handling continuous variables. We also present a backward transformation method for transforming a population of continuous variables obtained after mutation back into integer variables for evaluating the objective function. Onwubolu (2001) proposed both forward and backward transformations for dealing with the DE floating-point features when solving discrete, combinatorial optimization problems. Both forward and backward transformations are utilized in implementing the DE algorithm used in the present study for the travelling salesman problem. For more information on the application of the forward and backward transformations in DE, see Davendra (2001).

Forward Transformation

In integer variable optimization a set of integer number is normally generated randomly as an initial solution. Let this set of integer number be represented as:

$$z_i^{'} \in z^{'} \tag{22.14}$$

The equivalent continuous variable for $z_i^{'}$ is given as

$$1x10^2 < 5x10^2 \le 10^3 x10^2 \tag{22.15}$$

Then $z_i = -1 + \dfrac{z_i^{'} * f * 5}{10^3 - 1}$ (22.16)

Applying a scaling factor, *f* = *100* gives

$$z_i = -1 + \frac{z_i^{'} * f * 5}{10^3 - 1} = -1 + \frac{z_i^{'} * 500}{10^3 - 1} \tag{22.17}$$

Equation (22.17) is used to transform any integer variable into an equivalent continuous variable, which is then used for the DE internal representation of the population of vectors. Without this transformation, it is not possible to make useful moves towards the global optimum in the solution space using the mutation mechanism of DE, which works better on continuous variables.

For example in a five-city problem, suppose the sequence of visits given as {2, 4, 3, 1, 5 and back to 2}. This sequence is not directly used in DE internal representation. Rather, applying equation (22.16), the sequence is transformed into a continuous form. The floating-point equivalence of the first entry of the given sequence, $z_i^{'} = 2$, is $z_i = -1 + \frac{2*500}{10^3 - 1} = 0.001001$. Other values are similarly obtained and the sequence is therefore represented internally in the DE scheme as {0.001001, 1.002, 0.501502, -0.499499, 1.5025}.

In the technique we present, no rounding-off or truncation is required since such a truncation method often gives less than optimal results because no attempt is made during optimization to evaluate only realizable systems (Price and Storn, 1999).

Backward Transformation

Integer variables are used to evaluate the objective function. The DE self-referential population mutation scheme is quite unique. After the mutation of each vector, the trial vector is evaluated for its objective function in order to decide whether or not to retain it. This means that the objective function values of the current vectors in the population need to be also evaluated. These vector variables are continuous (from the forward transformation scheme) and have to be transformed into their integer number equivalence. It is not enough to round off these values for the class of problems we are solving. The backward transformation technique is used for converting floating point numbers to their integer number equivalence. The scheme is given as follows:

$$z_i^{'} = \frac{(1+z_i)*\left(10^3 - 1\right)}{5*f} = \frac{(1+z_i)*\left(10^3 - 1\right)}{500} \tag{22.18}$$

In this present form the backward transformation function is not able to properly discriminate between variables. Some modifications are required as follows:

$$\alpha = \text{int}(z_i^{'} + 0.5) \tag{22.19}$$

$$\beta = \alpha - z_i^{'} \tag{22.20}$$

$$z_i^{*} = \begin{cases} (\alpha - 1)\ if\ \ \beta > 0.5 \\ \alpha \quad\ \ if\ \ \beta < 0.5 \end{cases} \tag{22.21}$$

Equation (22.21) gives z_i^{*}, which is the transformed value used for computing the objective function.

As an example, we consider a set of trial vector, such as this one $z_i = \{-0.33, -0.17, 0.67, 0.84, 1.5\}$ obtained after mutation. The integer values corresponding to the trial vector values are obtained using equation (22.21) as follows:

$$z_1^{'} = (1-0.33)*(10^3-1)/500 = 1.33866$$
$$z_2^{'} = (1+0.67)*(10^3-1)/500 = 3.3367$$
$$z_3^{'} = (1-0.17)*(10^3-1)/500 = 1.65834$$
$$z_4^{'} = (1+1.50)*(10^3-1)/500 = 4.9950$$
$$z_5^{'} = (1+0.84)*(10^3-1)/500 = 3.6763$$
$$\alpha_1 = \text{int}(1.333866+0.5) = 2$$
$$\beta_1 = 2-1.33866 = 0.66134 > 0.5$$
$$z_1^{*} = 2-1 = 1$$
$$\alpha_2 = \text{int}(3.3367+0.5) = 4$$
$$\beta_2 = 4-3.3367 = 0.6633 > 0.5$$
$$z_2^{*} = 4-1 = 3$$
$$\alpha_3 = \text{int}(1.65834+0.5) = 2$$
$$\beta_3 = 2-1.65834 = 0.34166 < 0.5$$
$$z_3^{*} = 2$$
$$\alpha_4 = \text{int}(4.995+0.5) = 5$$
$$\beta_4 = 5-4.995 = 0.005 < 0.5$$
$$z_4^{*} = 5$$
$$\alpha_5 = \text{int}(3.673+0.5) = 4$$
$$\beta_5 = 4-3.673 = 0.3237 < 0.5$$
$$z_5^{*} = 4$$

The set of integer values is given as $z_i^{*} = \{1, 3, 2, 5, 4\}$. This set is used to obtain the objective function values.

22.4.3 Parameter Setting

The values of NP, CR, λ and F parameters are fixed empirically following certain heuristics (Price and Storn, 1997). F and λ are usually equal and are between 0.5 and 1.0. CR usually should be 0.3, 0.6, 0.8 or 1.0 to start with. NP should be of the order of 10*D and should be increased in case of mis-convergence. If NP is increased then usually F has to be decreased. In the present study, some trials were carried out and the best parameter values obtained from combining these parameters during experimentation are *NP =150, F = 0.3*, and *CR = 0.6*. Although the value of NP = 150 is adequate for small-size problems, a more general approach used for experimentation of NP being equal to the ten times the number of jobs. This problem-size dependent value of NP caters for small, medium and large problem sizes. Tests also show that the best strategy is strategy 7.

22.4.4 An Example

A simple 9x9-problem data set is shown in Figure 22.6 for depicting the convergence of DE. Both DE and ACS find the optimum value of 135. Figure 22.7 shows the corresponding graph of the objective function value as a function of the generation number. The maximum number of iterations was kept as 100; however, in all the runs during experimentation the algorithm converged within 10 generations. It is seen from Figure 22.7 that DE converges very well, reaching the optimum solutions before 10 generations.

$$\begin{bmatrix} - & 10 & 20 & 20 & 28 & 32 & 40 & 50 & 60 \\ 10 & - & 10 & 30 & 20 & 20 & 30 & 40 & 50 \\ 20 & 10 & - & 42 & 32 & 28 & 20 & 50 & 40 \\ 20 & 30 & 42 & - & 8 & 12 & 20 & 30 & 42 \\ 28 & 20 & 32 & 8 & - & 5 & 20 & 22 & 32 \\ 32 & 20 & 28 & 12 & 5 & - & 20 & 22 & 28 \\ 40 & 30 & 20 & 20 & 20 & 20 & - & 30 & 20 \\ 50 & 40 & 50 & 30 & 22 & 22 & 30 & - & 20 \\ 60 & 50 & 40 & 42 & 32 & 28 & 20 & 20 & - \end{bmatrix}$$

Fig. 22.6 The 9x9-problem data

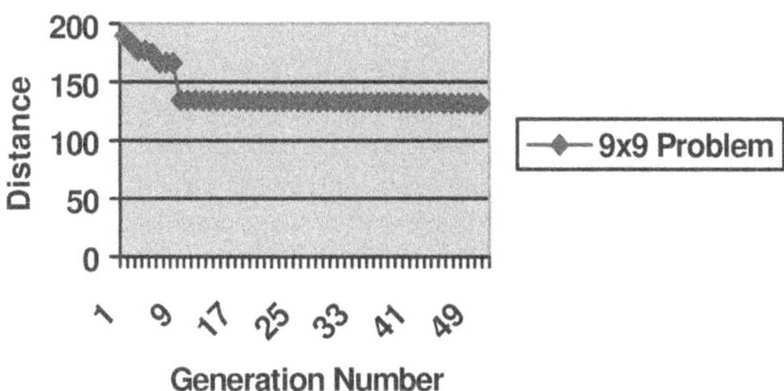

Fig. 22.7 Convergence of DE

22.4.5 Experimentation

In order to compare the DE method with other global minimizing strategies, we looked for an approach that has been accepted as one of the best in the literature, where the source code is readily available, and which is capable of coping with variants of TSP problems. Accordingly we chose ant colony system (ACS) by Dorigo and Gambardella (1997). Table 22.3 shows the results of five different types of 50x50 problems reported in Dorigo and Gambardella (1997). It can be seen that DE performs well compared to ant colony system (ACS), simulated annealing (SA), elastic net (EN), and self-organizing map (SOM).

Table 22.3 DE compared with other heuristics: random, symmetric instances

nxn	DE⁺	ACS	SA	EN	SOM
City set 1	2.237	5.88	5.88	5.98	6.06
City set 2	2.243	6.05	6.01	6.03	6.25
City set 3	2.318	5.58	5.56	5.70	5.83
City set 4	2.132	5.74	5.81	5.86	5.87
City set 5	2.392	6.18	6.33	6.49	6.70

$^{+}$Averages are over 5 trials

22.5 TSP/Differential Evolution Application in CNC Drilling of PCB

This section of the chapter describes the application of TSP/Differential Evolution to the optimization of drilling of holes in an automated drilling machine, which the author and others produced in-house (see Onwubolu *et al.* 2002). The application considered here is that of drilling holes on printed circuit board (PCBs). The PC based drilling system developed in-house at the Department of Engineering, University of the South Pacific, Fiji, was funded by the University Research Grant. The machine operates on the movement of the work-piece placed on a rigid and stable work-holder in *x* and *y* directions. This combination of process along with drilling are controlled by a set of program codes linked to actuators (stepper motors) via electronic interfacing circuits (interfacing card). The basic framework of the machine is shown in Figure 22.8. The remaining part of this chapter describes the production of PCBs, the differential evolution/TSP optimization techniques, and an example of how the optimization technique is applicable to the automated drilling machine drilling operations for PCBs, which has the potentials of cost reduction in manufacturing enterprises.

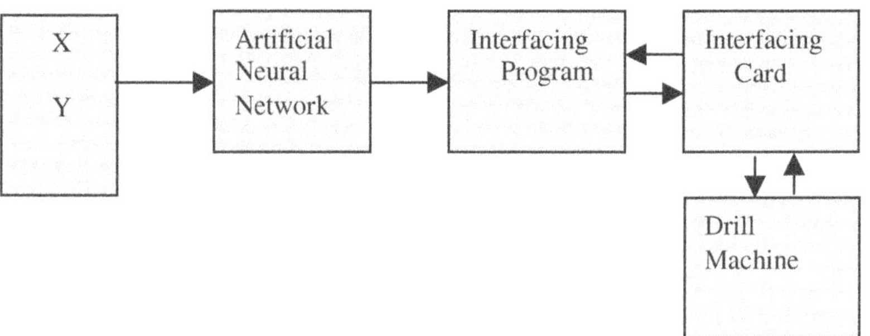

Fig. 22.8 PC based drilling machine framework

A PCB is a laminated flat panel of insulated material (usually polymer composites reinforced with glass fabrics or paper) designed to provide electrical interconnections between electronic components attached to it. The conducting paths are made of copper and are known as *tracks*. Other copper areas for attaching and electrically connecting components are also available and are called *lands*. Thin conducting paths on the surface of the board or in alternate layers sandwiched between layers of insulating material are used for interconnecting electronic components.

There are basically three principal types of PCB. They are (i) single-sided board, in which copper foil is only on one side of the insulating substrate; (ii) double-sided board, in which copper foil is only on both sides of the insulating substrate; and (iii) multi-layer board, in which layers of copper foil and insulating substrates alternate. The insulating materials in PCB are usually polymer composites reinforced with glass fabrics or paper. Polymers commonly used include phenolic, and polyamide. In epoxy PCBs, E-glass is the usual fiber used in glass-reinforcing fabrics. In phenolic PCBs, paper is a common reinforcing layer used.

22.5.1 PCB Manufacturing

The manufacturing processes involved in PCB are (1) starting boards, (2) board preparation, (3) circuit pattern imaging and etching, (4) hole drilling, (5) plating, and (6) testing. These are now discussed.

22.5.1.1 Starting Boards

Production of starting boards consists of pressing multiple sheets of woven glass fiber that have been impregnated with partially cured thermosetting polymer such as epoxy. The number of sheets determines thickness of the final board used in the starting sandwich. Copper foil is placed on one, both sides and alternate layer of the epoxy-glass laminated stack, depending on whether single-sided board, double-sided board, or multi-layer boards are to be produced. The copper foil used to clad the starting boards is produced by a continuous electro-forming process, in which a rotating smooth metal drum is partially submerged in an electrolytic bath containing copper ions. In this arrangement the drum is the cathode so that the copper plates on its surface. Thin copper foil is peeled from the board, which consists of a glass-fabric-reinforced epoxy panel, clad with copper over its surface area on one or more sides is now ready for the circuit fabrication.

22.5.1.2 Board Preparation

The starting panel may have to be sheared to size to fit fabricator's equipment. Drilling or punching are used for making tooling holes for positioning the PCB during subsequent processing. Tabs, slots, and similar features may be made on the PCB to facilitate handling subsequent processing. Usually, three tooling holes per PCB are sufficient. In this preparation stage, bar coding is used to identify PCBs. The board surface is then cleaned to remove small particles of dirt, which could lower the performance of the PCB.

22.5.1.3 Circuit Pattern Imaging and Etching

Photolithography is a commonly used method by which the circuit pattern is transferred to the copper surface of the board. CAD packages such as Circuit Maker® facilitate the design of circuit patterns. Photolithography is a method in which a light-sensitive resist material coated on the board surface is exposed through a mask to determine where etching of the copper will occur to create the tracks and lands of the circuit.

Most circuit fabricators use *negative photoresists*, which may be in the form of a liquid or dry film. Dry film resists consists of three layers. A polyester support sheet and a removable plastic cover sheet sandwich a film of photosensitive polymer. During use, the cover sheet is peeled off, and the resist film is placed on the copper surface to which it readily adheres. The resist should be uniformly pressed to the surface using a hot roll. Contact printing is used to expose the resist, which is beneath the mask. The resist is then developed, which involves removal of the unexposed regions of the negative resist from the surface. After resist development, certain areas of the copper surface remain covered by resist while other areas are now uncovered. The covered areas correspond to the circuit tracks and land, while the uncovered areas correspond to the open region between circuit tracks and land.

Etching is the process that removes the copper cladding in the uncovered regions from the surface of the board, leaving the covered regions to form the interconnections for an electrical circuit. Etching is carried out in an etching chamber by immersing the board in a solution of etchant. Several etchants in use include ammonium hydroxide (NH_4OH), ammonium persulfate ($(NH_4)_2S2O_4$), and ferric chloride ($FeCl_3$). The controlling parameters, which must be controlled, are etchant concentration, duration of etching, and temperature to avoid over-etching or under-etching, which is certainly undesirable in IC fabrication. Thereafter, the board is rinsed to remove any solution and the remaining resist is chemically stripped from the surface of the board.

22.5.1.4 Hole Drilling

Different types of holes are drilled on the PCB once the tracks have been formed. Insertion holes are drilled for inserting component leads in through-hole PCBs. Via holes are drilled to act as conducting paths from the outer boards to the sandwich boards; these holes are copper-plated to enhance conductivity. Fastening holes are drilled to fasten certain components such as connectors and heat sinks to the PCB.

Some operational issues that must be addressed in PCB hole drilling are (1) quality, (2) drill size, (3) unique work material, (4) high accuracy in hole-location, and (5) drill speed.

The quality requirement in drilling hole includes producing cleaner holes smooth hole wall, and absence of burrs on the holes. Thin sheets of material are often placed on both sides of the stack of boards to reduce burr formation on the boards themselves.

One of the main challenges in PCB hole drilling is the small hole-size involved, since such drill bits lack strength. Drill bits of less than 1.27mm or 0.05inch is common, requiring a tight tolerance of about ±0.01mm. The small hole-size combined with the stacking of several boards result in a high depth-to-diameter ratio, which increases the problem of chip extraction from the hole.

The unique work material, which is a composite of epoxy-glass, is coated with copper foil. The drill bit must pass through the metal (copper) foil and then also cu through the epoxy-glass composite. In normal drilling practice, different drills would be used for the different materials, but in PCB drilling, a single drill has to be specified.

High accuracy in hole-location is required in PCB hole drilling because the holes determine where the electronic components have to be placed. Electronic components must be accurately placed to make the PCB circuitry perform well. In order to automate PCB hole drilling, a computer numerically controlled (CNC) drill press is used. This automated equipment has a moveable worktable, which moves in x and y directions using stepper motors. The worktable, which carries the PCB, brings a particular point to be drilled directly underneath a drill, which is supported on a gantry structure. The CNC drill press receives its programming instructions from the design database; this has the advantage that the holes located are accurate according to the geometry of the PCB.

Cutting speed is one of the cutting parameters that have to be carefully chosen for a cutting tool. The very small drill-sizes used for hole drilling require that high speeds be chosen. This creates some problems when the speed is exceptionally high because special motors and spindle bearing would be needed.

22.5.1.5 Plating

Plating is needed on the hole-surfaces of PCBs in order to facilitate conduction in the conductive path via-holes. Electroplating and electroless plating are the two plating processes commonly used. Electroplating requires that the coated surface is conductive, while electroless plating does not require a conductive surface.

The walls of the via-holes and insertion holes after drilling consist of epoxy-glass insulation material, which hinders conduction. Therefore, electroless plating must be used to provide conductive path. Thereafter, the copper coating thickness is increased by means of electroplating process.

22.5.1.6 Testing

Before a fabricated PCB is dispatched from the production lien for use in electronic assembly, it must be tested to ensure that it conforms to the design specifications. Visual inspection and continuity testing are the two commonly used testing procedures. The PCB is visually checked for any defect at various stages of production. For continuity testing, contact probes are brought simultaneously into contact with track and land areas on the PCB surface. This procedure can inform the fabricator whether or not electrical connections exist between contact points.

22.5.1.7 Finishing Operations

After testing the PCB for proper functionality, two finishing operations are necessary: (i) application of a thin solder layer on the track and land surfaces, and (ii) application of a coating of solder resist to all areas of the PCB surface except the land. Applying a thin solder layer on the track and land protects the copper from oxidation and contamination. The practice used is to bring the copper track and landing in contact with rotating roller, which is partially submerged in molten solder; alternatively, electroplating process is used. Application of a coating of solder resist to all areas of the PCB surface except the land ensures that solder adheres only to land areas in subsequent soldering process.

The remaining of this chapter describes the application of TSP/DE to the hole-drilling operation of PCB since this is an important step in the production of PCBs.

22.5.2 Automated Drilling Location and Hit Sequencing using DE

Consider an automated drilling machine that drills holes on a flat sheet. The turret has to be loaded with all the tools required to hit the holes. There is no removing or adding of tools. The machine bed carries the flat plate and moves from home, locating each scheduled hit on the flat plate under the machine turret. Then the turret rotates and aligns the proper tool under the hit. This process continues until all hits are completed and the bed returns to the home position.

There are two problems to be solved here. One is to load tools to the turrets and the other is to locate or sequence hits. The objective is to minimize the cycle time such that the appropriate tools are loaded and the best hits-sequence is obtained. The problem can therefore be divided into two: (i) solve a TSP for the inter-hit sequencing; (ii) solve a quadratic assignment problem (QAP) for the tool loading. Walas and Askin (1984) developed a mathematical formulation to this problem and iterated between the TSP and QAP. Once the hit sequence is known, the sequence of tools to be used is then fixed since each hit requires specific tool. On the other hand, if we know the tool assignment on the turret, we need to know the inter-hit sequence. Connecting each hit in the best sequence is definitely a TSP, where we consider the machine bed home as the home for the TSP, and each hit, a city. Inter-hit travel times and the rotation of the turret are the costs involved and we take the maximum between them, i.e. inter-hit cost = max (inter-hit travel time, turret rotation travel time). The cost to place tool k in position i and tool l in position j is the time it takes the turret to rotate from i to j multiplied by the number of times the turret switches from tool k to l.

The inter-hit travel times are easy to estimate from the geometry of the plate to be punched and the tools required per punch. The inter-hits times are first estimated and then adjusted according to the turret rotation times. This information constitutes the data for solving the TSP. Once the hit sequence is obtained from the TSP, the tools are placed, by solving the QAP. Let us illustrate the TSP-QAP solution procedure by considering an example.

Example 22.3

A numerically controlled (NC) machine is to punch holes on a flat metal sheet and the hits are shown in Figure 22.9. The inter-hit times are shown in Table 22.4. There are four tools {a, b, c, and d} and the hits are {1, 2, 3, 4, 5, 6, and 7}. The machine turret can hold five tools and rotates in clockwise or anti-clockwise direction. When the turret rotates from one tool position to an adjacent position, it takes *60* time units. It takes *75* time units and *90* time units to two locations and three locations respectively. The machine bed home is marked *0*. Assign tools to the turret and sequence the hits.

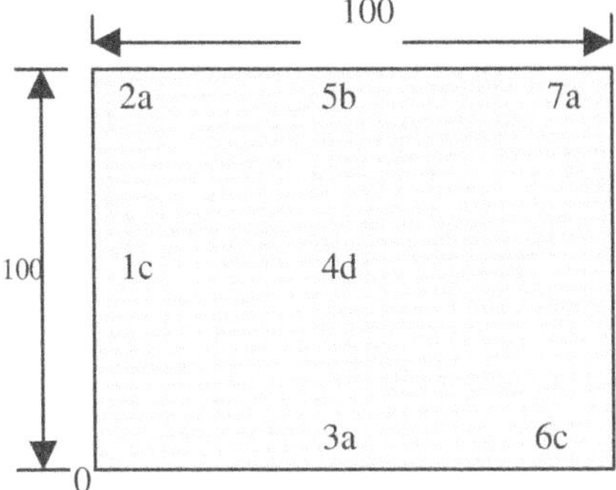

Fig. 22.9 Flat metal sheet to be punched

Table 22.4 Inter-hit travel times

Inter-hit travel times								
Hit								
Hit	0	1	2	3	4	5	6	7
0	-	50	100	50	100	150	100	200
1	50	-	50	100	50	100	150	150
2	100	50	-	150	100	50	200	100
3	50	100	150	-	50	100	50	150
4	100	50	100	50	-	50	100	100
5	150	100	50	100	50	-	150	50
6	100	150	200	50	100	150	-	100
7	200	150	100	150	100	50	100	-

Solution

From the given inter-hit times, modified inter-hit times have to be calculated using the condition: inter-hit cost = max (inter-hit travel time, turret rotation travel time). For example, for inter-hit between locations *1* and *2*, the inter-hit travel time is *50* time units. Now, the tool for hit *1* is *c* while the tool for hit *2* is *a*. This means there is change in tools because that the turret will rotate. The cost of rotation is *60* time units, which exceeds the *50* inter-hit time unit. This means that the modified inter-hit time between locations *1* and *2* is *60* time units. From the home to any hit is not affected. The modified inter-hit times are shown in Table 22.5. This information is used for TSP. One TSP solution for Table 22.4 is {0, 1, 2, 3, 4, 5, 6, and 7}, with a cost of *830*. We used the DE heuristic to obtain tool sequence of c→ d → b → a → c → a, and the cost is *410*. As can be seen a better solution is obtain by the latter. Let us explain how we obtained the tool sequence. Solving the TSP using DE, the sequence obtaineed is {2, 5, 6, 8, 7, 4, 1, and 3} or {1, 4, 5, 7, 6, 3, 0, and 2}. What we do is to refer to Figure 22.9 and get the labels that corrspond to this sequence as {c, d, b, a, c, a, a }. Hence the optimum sequence is c-d-b-a-c.

Table 22.5 Modified inter-hit travel times (considering turrent movements)

Modified inter-hit travel times								
Hit								
Hit	0	1	2	3	4	5	6	7
0	-	50	100	50	100	150	100	200
1	50	-	60	100	60	100	150	150
2	100	60	-	150	100	60	200	100
3	50	100	150	-	60	100	60	150
4	100	60	100	60	-	60	100	100
5	150	100	60	100	60	-	150	60
6	100	150	200	60	100	150	-	100
7	200	150	100	150	100	60	100	-

22.6 Summary

The Differential Evolution method (DE) for minimizing discrete space functions has been introduced and shown to compete with other emerging optimization approaches. Since DE is inherently parallel, a further significant speedup can be obtained if the algorithm is executed on a parallel machine or a network of computers. This is especially true for real world problems where computing the objective function requires a significant amount of time. Despite these already promising results, DE is still in its infancy and can most probably be improved. Further research might include a mathematical convergence proof like the one that exists for simulated annealing. A theoretically sound analysis to determine why DE converges so well would also be of great interest. Whether or not an annealed version of DE, or the combination of DE with other optimization approaches is of practical use, is still unanswered. Finally, it is important for practical applications to gain more knowledge on how to choose the control variables for DE.

References

Askin RG, Standridge CR (1993) *Modeling and Analysis of Manufacturing Systems*, John Wiley & Sons: Toronto

Davendra D (2001) Differential Evolution Algorithm for flow shop scheduling, Unpublished Report, The University of the South Pacific

Dorigo M (1992) Optimisation, Learning and Natural Algorithms (Ottimizzazione, aprendimento automatico, et algoritmi basati su metafora naturale), *PhD. Dissertation, Dipartimento Elettronica e Informazione, Politecnico di Milano*, Italy

Dorigo M, Gambardella L (1997) Ant colony system: a cooperative learning approach otthe traveling salesman problem, *IEEE Transactions on Evolutionary Computation*, 1, 53-66

Durbin R, Willshaw D (1987) An analogue approach to the traveling salesman problem using an elastic net method, *Nature*, 326, 689-691

Fogel DB (1993) Applying evolutionary programming to selected traveling salesman problem, *Cybernetic System: International Journal*, 24, 27-36

Glover F (1989) Tabu search-Part, *ORSA Journal of Computing* 1/3, 190-206.

Glover F (1990) Tabu search-Part II, *ORSA Journal of Computing* 2/1, 4-32.

Goldberg DE (1989) *Genetic Algorithms in Search, Optimization, and Machine Learning*, Reading, MA: Addison-Wesley

Jünger M, Reinelt G, Rinaldi G (1995) The traveling salesman problem, in: *Handbooks in Operations Research and Management Science, Volume 7*, Ball, M. O., Magnanti, T., Monma, C. L. and Nemhauser, G. (eds.), Elsevier Science B. V., 225-330.

Kirkpatrick S, Gelatt CD, Vechhi MP (1983) Optimization by simulated annealing, *Science*, 220 (4568), 671-680.

Lampinen J, Zelinka I (1999) Mechanical engineering design optimization by differential evolution, *New Ideas in Optimization*, (Eds.) Corne, D., Dorigo, M., and Glover, McGraw Hill, International (UK), 127-146.

Lawler EL, Lenstra JK, Rinnoy-Kan AGH, Shmoys DG (Eds.), (1985), *The Traveling Salesman Problem: A Guided Tour of combinatorial Optimization*, Chichester, U.K: Wiley

Lin F-T, Kao C-Y, Hsu C-C (1993) Applying the genetic approach to simulated annealing in solving some NP-hard problems, *IEEE Transaction Systems, Man, Cybernetics*, 23, 1752-1767

Onwubolu GC (2001) Optimization using differential evolution, *The University of the South Pacific Institute of Applied Science Technical Report, TR-2001/05.*

Onwubolu GC, Aborhey S, Singh R, Reddy H, Prasad M, Kumar S, Singh S (2002) Development of a PC-based computer numerical control drilling machine, to appear in *Journal of Engineering Manufacture, Short Communications in Manufacture & Design.*

Price K, Storn R (2001) Differential evolution homepage (Web site of Price and Storm) as at 2001. http://www.ICSI.Berkeley.edu/~storn/code.html

Reinelt G (1994) *The Traveling Salesman Problem: computational solutions for TSP applications*, Berlin: Springer-Verlag

Storn R, Price K (1995) Differential evolution - a simple and efficient adaptive scheme for global optimization over continuous spaces, *Technical Report TR-95-012, ICSI, March 1999 (Available via ftp from ftp.icsi.berkeley.edu/pub/techreports/1995/tr-95-012.ps.Z)*

Whitely D, Starkweather T, Fuquay D (1989) Scheduling problems and traveling salesman: the genetic edge recombination operator, in Proceedings of the 3rd International Conference on Genetic Algorithms, Palo Alto, CA: Morgan Kaufmanns, 133-140

23 Particle Swarm Optimization for the Assignment of Facilities to Locations

Godfrey C Onwubolu and Anuraganand Sharma

This chapter describes a new heuristic approach, for minimizing discrete space functions. The new heuristic, particle swarm optimization is applied to the quadratic assignment problem. It is observed from experimentation that the particle swarm optimization approach delivers competitive solutions when compared to ant system, ant system with non-deterministic hill climbing, simulated annealing, tabu search, genetic algorithm, evolutionary strategy, and sampling & clustering for the quadratic assignment problem. By comparing results from the particle swarm optimization and the results of these other best-known heuristics, it will be demonstrated that the particle swarm optimization method converges as much as best-known heuristics for the QAP. The new method requires few control variables, is versatile, is robust, easy to implement and easy to use.

23.1 Introduction

In a standard industrial engineering problem in location theory, we are given a set of n locations and n facilities, and the goal is to assign each facility to a location. To measure the cost of each possible assignment, there are n! of them. We multiply the prescribed flow between each pair of facilities by the distance between their assigned locations, and sum over all the pairs. Our aim is to find the assignment that minimizes this cost, and this problem is precisely a quadratic assignment problem. In practice, as the problems get larger, it becomes much, much more difficult to find the optimal solution. As n grows large it becomes impossible to enumerate all the possible assignments, even by fast computers. Consequently, heuristics are used which do not need to find optimal solutions, but are good enough to find solutions very close to the optimum within reasonable computation time.

The quadratic assignment problem (QAP) is an important problem in theory and practice. Many practical problems like backboard wiring (Steinberg, 1961), campus and hospital layout (Dickey and Hopkin, 1972; Elshafei, 1977), scheduling (Geoffrion and Graves, 1976) and many others (Eiselt & Larporte, 1991; Larporte and Mercure, 1988) can be formulated as QAPs.

The QAP is a *NP* -hard optimization problem (Sahni and Gonzale, 1976); even finding a solution within a factor of $1 + \varepsilon$ of the optimal one remains *NP* -hard. It is considered as one of the hardest optimization problems as general instances of size $n \geq 20$ cannot be solved to optimality (Dorigo *et al.* 1996). Therefore, in order to practically solve the QAP one has to apply heuristic algorithms which find very high quality solutions in short computation time. Several such heuristic algorithms have been proposed, which include algorithms like iterative improvement, ant algorithms (Dorigo *et al.* 1996), evolution strategies (Nissen, 1994), genetic algorithms (Fleurent and Ferland, 1994; Merz and Freisleben, 1997), and greedy randomized adaptive search procedure (Li *et al.* 1994). Others include scatter search (Cung *et al.* 1997), simulated annealing (Burkard and Rendl, 1984; Connolly, 1990), and tabu search (Battiti and Tecchiolli, 1994; Skorin-Kapov, 1990; Taillard, 1991).

More recently, a novel optimization method based on particle swarm optimization (Kennedy and Eberhart, 1995) has been developed. Since Kennedy and Eberhart (1995) invented the particle swarm optimization, the challenge has been to employ the algorithm to different areas of problems other than those areas that the inventors originally focused on.

In this chapter we give an overview of the existing applications of the PSO adaptive search heuristic to the QAP and present computational results which confirm that PSO algorithms are among the best performing algorithms for this problem. The chapter is structured as follows. We first outline how the PSO adaptive search heuristic can be applied to solve QAPs. To exemplify the computational results obtained with PSO algorithm, we compare its performance and other algorithms proposed for the QAP.

23.2 Problem Formulation

More formally, given n facilities and n locations, two $n \times n$ matrices $A = \lfloor a_{ij} \rfloor$ and $B = \lfloor b_{rs} \rfloor$, where a_{ij} is the distance between locations i and j and b_{rs} is the flow between facilities r and s, the QAP can be stated as follows:

$$\min_{p \in S(n)} \sum_{i=1}^{n} \sum_{j=1}^{n} b_{ij} a_{p(i)p(j)} \tag{23.1}$$

where $S(n)$ is the set of all permutations (corresponding to the assignments) of the set of

$\{1,...,n\}$, and $p(i)$ gives the location of facility i in the current solution $p\in S(n)$. Here $b_{ij}a_{p(i)p(j)}$ describes the cost contribution of simultaneously assigning facility i to location $p(i)$ and facility j to location $p(j)$. In the following we denote with $\Im_p$ the objective function value of permutation p.

The term *quadratic* stems from the formulation of the QAP as an integer optimization problem with a quadratic objective function. Let x_{ij} be a binary variable, which takes value *1* if facility i is assigned to location j and *0* otherwise. Then the problem can be formulated

as:

$$\Im_p = \min_{p\in S(n)} \sum_{i=1}^{n}\sum_{j=1}^{n}\sum_{k=1}^{n}\sum_{l=1}^{n} a_{ij}b_{kl}x_{ik}x_{jl} \tag{23.2}$$

$$\text{s.t.} \quad \sum_{i=1}^{n} x_{ij} = 1 \tag{23.3}$$

$$\sum_{j=1}^{n} x_{ij} = 1 \tag{23.4}$$

$$x\in \{0,1\}$$

23.3 Application of the PSO to the QAP

Whether in industry or in research, users generally demand that a practical optimization technique should fulfill three requirements:

(1) the method should find the true global minimum, regardless of the initial system parameter values;
(2) convergence should be fast; and
(3) the program should have a minimum of control parameters so that it will be easy to use.

Kennedy and Eberhart (1995) invented the particle swarm optimization algorithm in a search for a technique that would meet the above criteria. PSO is a method, which is not only astonishingly simple, but also performs extremely well on a wide variety of test problems. It is inherently parallel because it is a population based approach and hence lends itself to computation via a network of computers or processors. In the following subsection, we follow a more graphical approach for presenting the new optimization method which, we have employed in TSP.

Particle swarm optimization (PSO) is a novel parallel direct search method, which utilizes *NP* parameter vectors

$$x_i^{(G)}, \mathrm{i} = 0, 1, 2, \ldots, NP\text{-}1. \tag{23.5}$$

as a population for each generation, *G*. The initial population is generated randomly assuming a uniform probability distribution for all random decisions if there is no initial intelligent information for the system.

As with all evolutionary optimization algorithms, PSO works with a population of solutions, not with a single solution for the optimization problem. Population *P* of generation *G* contains *NP* solution vectors called individuals of the population and each vector represents potential solution for the optimization problem:

$$P^{(G)} = X_i^{(G)} = x_{j,i}^{(G)} \quad i = 1,\ldots, NP;\ G = 1,\ldots, G_{\max} \tag{23.6}$$

Each particle also maintains a memory of its previous best position, represented as

$$p_i = (p_{i1}, p_{i2},\ldots, p_{iD}) \tag{23.7}$$

A particle in a swarm is moving hence it must have a velocity, which can be represented as (Carlisle and Dozier, 1998)

$$v_i = (v_{i1}, v_{i2},\ldots, v_{iD}) \tag{23.8}$$

At each iteration the velocity of each particle is adjusted so that it can move towards the best position in the neighborhood and the global best position that any particle present in the swarm attains.

A simple theorem of vector addition and constant multiplication is used to define the new position and new velocity. We are considering a particle *i* whose current position at time *t* is represented as $x_i(t)$, previous best position $p_i(t)$ [cognitive], neighborhood's best position $p_g(t)$ [social] and the particle *i* is moving with velocity $v_i(t)$.

The next position after the iteration (i.e. at time (t+1)) can be given by

$$x_i(t+1) = x_i(t) + v_i(t) \tag{23.9}$$

But velocity can also be represented as

$$v_i(t) = x_i(t+1) - x_i(t) \text{ or}$$

$$v_i(t) = p_i - x_i(t); \qquad where\ \ x_i(t+1) = p_i \tag{23.10}$$

Hence, equation 23.7 can be modified to

$$x_i(t+1) = x_i(t) + (p_i - x_i(t)) \tag{23.11}$$

By observing the equation 23.11, it can be deduced that the particle would return immediately to the previous best and search would be ended. A new vector is introduced to the velocity of change and a vector of difference is weighted by a random number c whose upper limit is a constant parameter of the system, usually set to value of 2.0 (Kennedy and Eberhart, 1997). The new change of velocity vector is given as follows:

$$v_i(t+1) = v_i(t) + c \bullet (p_i - x_i(t)) \tag{23.12}$$

The change of vector introduces a tendency to return towards the previous best position (Kennedy and Eberhart, 1997).

But this procedure is not much effective too. Some new constraints have to be introduced into the equation 12. Thus far the operations have no social component, but rather individual search focusing on region of the problem space and hence do not perform very well. Social influence makes the particle swarm work (Kennedy, 1999). These new equations will involve social component. Social/cognitive coefficients are usually randomly chosen, at each time step, in given intervals. Of course, a number of researcher have done considerable amount of work to study and to generalize this method (Eberhart and Kennedy 1995; Kennedy and Eberhart 1997; Kennedy 1997; Angeline 1998; Kennedy and Spears 1998; Shi and Eberhart 1998). In particular, we use a *no-hope/re-hope* technique as defined in (Clerc 1999), and the convergence criterion proved in Clerc and Kennedy (2002).

At each time step, the behavior of a given particle is a compromise between three possible choices:

- To follow its own way
- To go towards its best previous position
- To go towards the best neighbor's best previous position, or towards the best neighbor (variant)

The social/cognitive influence that makes the PSO algorithm to work is now presented:

$$v_i(t+1) = v_i(t) + c_1 \bullet \left(p_i - x_i(t)\right) + c_2 \bullet \left(p_g - x_i(t)\right) \tag{23.13}$$

$$x_i(t+1) = x_i(t) + v_i(t+1) \tag{23.14}$$

where c_1 = cognitive influence factor, c_2 = social influence factor, p_i is the particle's best so far, and p_g is the best in the neighborhood.

We have used Figure 23.1 to get the velocity vector $v_i(t+1)$, and by observing the Figure 23. It can be stated that particle i has moved towards the neighborhood's best position and its own best position.

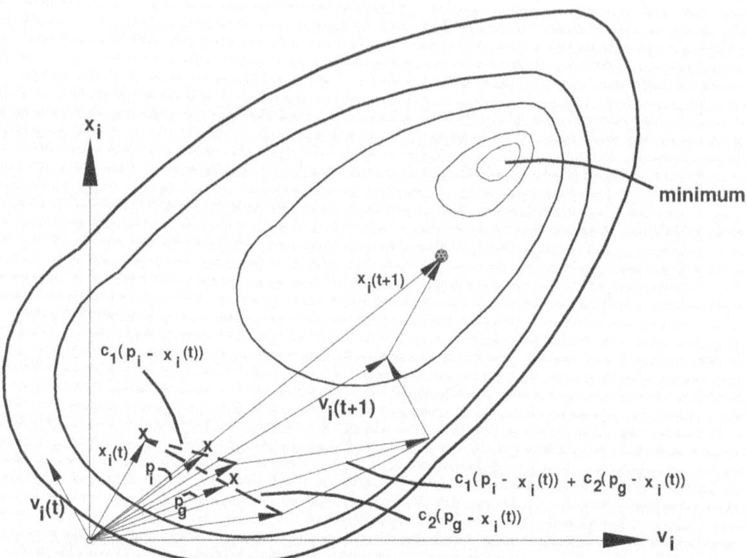

Fig. 23.1. Contour lines and the process for generating new x in PSO scheme

We note that from Figure 23.1 the operators' modification has been observed until the finest form is obtained. Few points are noted during the modification: (1) the new position is converging towards the best positions region; (2) the velocity could explode towards infinity. Hence new position on that situation is undefined.

23.3.1 Explosion Control

Constriction parameters have a great impact on the operations. Varying these parameters has the effect of varying the strength of the pull "towards" the two best positions, which could be verified by observing Figure 23.1. If accidentally the coefficients c_1 and c_2 exceeds the value *4.0*, both the velocities and positions explode towards infinity. Thus almost all implementation of the PSO limit each of the two coefficients c_1 and c_2 to 2.0 (Kennedy and Eberhart, 1997). To control the explosion of the system a new constriction coefficient is used, which is called "inertia weight". A large inertial weight facilitates global exploration while a small inertial weight tends to facilitate local exploration to fine-tune the current search area (Kennedy, 1997). Hence, equation 23.13 can be modified to have a place for new constriction constant, inertial weight (χ) (Clerc and Kennedy, 2002).

$$v_i(t+1) = \chi \bullet \{v_i(t) + c_1 \bullet (p_i - x_i(t)) + c_2 \bullet (p_g - x_i(t))\} \quad (23.15)$$

where inertial weight (χ), is given as

$$\chi = \frac{k}{abs\left[\frac{1-\frac{\beta}{2}-\sqrt{|\beta^2-4\beta|}}{2}\right]}$$

$k \in (0,1]$ and $\beta = c_1 + c_2$ such that $\beta > 4$.

Equation 23.15 is generic operator for all types of objective functions. This is to tune up these coefficients for any particular kind of problem domain.

23.3.2 Particle Swarm Optimization Operators

The detailed descriptions of the four basic operators involved in the PSO algorithm (Clerc 2002) are discussed; we note that the time is discrete: time step = 1:

23.3.2.1 Position minus Position: subtraction

$$(position, position) \xrightarrow{\text{minus}} velocity$$

Particles move from one place to another in search for a better position. To move a particle it must have some velocity to move in specified direction (towards better position).

Let say particle p has position x and particle p' has position x'. To move the particle p from position x to x' velocity v is given.

$$v = x' \Theta x \tag{23.16}$$

with Θ defined as follows.

Suppose position x and x' are represented in the form of array of dimension-4:

$$x = \begin{bmatrix} 0 \\ 1 \\ 2 \\ 3 \end{bmatrix}, \qquad x' = \begin{bmatrix} 0 \\ 2 \\ 3 \\ 1 \end{bmatrix}$$

then the velocity v will be:

$$v = \begin{bmatrix} 0 \\ 2 \\ 3 \\ 1 \end{bmatrix} \Theta \begin{bmatrix} 0 \\ 1 \\ 2 \\ 3 \end{bmatrix} = \begin{bmatrix} 1 & \leftrightarrow & 2 \\ 1 & \leftrightarrow & 3 \end{bmatrix}.$$

meaning: on x, exchange 1 and 2, *then* exchange 1 and 3. It is easy to see that the final result is indeed x'. More generally, knowing x and x', a simple and small algorithm builds a correct sequence of transpositions (not necessarily the shortest one).

The mathematical idea behind this procedure to calculate velocity of the particle is clear. We just memorize what kind of transformation the velocity will eventually do when applied to a position (see below "Position plus velocity"). Another important operator is coefficient times velocity. Detail of this operator is described next.

23.3.2.2 Coefficient times velocity:

external multiplication $(real\_number, velocity) \xrightarrow{\text{times}} velocity$

This operator is stochastic, and defined only for a coefficient between *0* and *1*. For a coefficient greater than *1*, say *coeff* = $k + c$, with k is integer part and c is decimal part, then simply k times *velocity plus velocity* and one times *coefficient times velocity* is used (Clerc, 2002).

Suppose a velocity v and coefficient c are given, then c x v can be computed for a coefficient c given as

$$\begin{cases} c \in [0,1],\ c' = random(0,1) \\ c' \le c \Rightarrow (i \leftrightarrow j) \rightarrow (i \leftrightarrow i) \\ c' > c \Rightarrow (i \leftrightarrow j) \rightarrow (i \leftrightarrow j) \end{cases} \qquad (23.17)$$

An example is given below to have a better idea about how this operator works. For a given v then *0.5xv* (c < *1* = *0.5*) is given as

$$0.5 \otimes \overset{v}{\begin{bmatrix} 0 & \leftrightarrow & 1 \\ 2 & \leftrightarrow & 3 \\ 3 & \leftrightarrow & 0 \\ 4 & \leftrightarrow & 4 \\ 0 & \leftrightarrow & 2 \\ 1 & \leftrightarrow & 0 \end{bmatrix}} = \begin{bmatrix} 0 & \leftrightarrow & 1 \\ 2 & \leftrightarrow & 2 \\ 3 & \leftrightarrow & 3 \\ 4 & \leftrightarrow & 4 \\ 0 & \leftrightarrow & 2 \\ 1 & \leftrightarrow & 1 \end{bmatrix} = \overset{v_{new}}{\begin{bmatrix} 0 & \leftrightarrow & 1 \\ 0 & \leftrightarrow & 2 \end{bmatrix}}$$

This operator is mainly used to change the velocity. The size of velocity is also changed when this operator is imposed. This operator is helpful when velocity of a particle needs to be changed. Either move away or move closer to another particle. From the previous sections the velocity of the particle was deduced. Now particle has a velocity to move into new position. One other important operator is *velocity plus velocity*, which is now discussed.

23.3.2.3 Velocity plus velocity: addition

$$(velocity, velocity) \xrightarrow{\text{plus}} velocity$$

Suppose a particle *p*, has two velocity components v_1 and v_2, then the new velocity v_{new} will be:

$$v_1 \oplus v_2 = v_{new} \tag{23.18}$$

The technique of addition is as follows. The sequence of transpositions describing v_2 is simply "added" to the one describing v_1. An example of this operator is now described.

$$\underset{v_1}{\begin{bmatrix} 0 & \leftrightarrow & 1 \\ 2 & \leftrightarrow & 3 \end{bmatrix}} \oplus \underset{v_2}{\begin{bmatrix} 0 & \leftrightarrow & 2 \\ 3 & \leftrightarrow & 4 \\ 3 & \leftrightarrow & 1 \end{bmatrix}} = \begin{bmatrix} 0 & \leftrightarrow & 1 \\ 2 & \leftrightarrow & 3 \\ 0 & \leftrightarrow & 2 \\ 3 & \leftrightarrow & 4 \\ 3 & \leftrightarrow & 1 \end{bmatrix} = \underset{v_{new}}{\begin{bmatrix} 0 & \leftrightarrow & 3 \\ 1 & \leftrightarrow & 2 \\ 1 & \leftrightarrow & 4 \end{bmatrix}}$$

Note that as this operator is not commutative, we usually do not have $v_1 \oplus v_2 = v_2 \oplus v_1$.

23.3.2.4 Position plus velocity:

$$\text{move } (position, velocity) \xrightarrow{\text{move (plus)}} position$$

Particles can be moved according to their current velocity. Suppose particle *p* has position *x* and velocity *v* then the new position becomes

$$x_{new} = x + v \tag{23.19}$$

This function is used to obtain the global best particle of the swarm. Let say particle p has position x, velocity v and new position x_{new}. Suppose position x has some arbitrary values of independent variables (a, b, c, d, e, f) and velocity v has values (b, d) in first component and values (a, f) in second component, with then the x_{new} is shown as

$$\overset{x}{\begin{bmatrix} a \\ b \\ c \\ d \\ e \\ f \end{bmatrix}} \oplus \overset{v}{\begin{bmatrix} b & \leftrightarrow & d \\ a & \leftrightarrow & f \end{bmatrix}} = \overset{x_{new}}{\begin{bmatrix} f \\ d \\ c \\ b \\ e \\ a \end{bmatrix}}$$

The technique of transformation is that each component of the velocity (that is a transposition) is successively applied, first to x, then to the position obtained.

23.3.3 Particle Swarm Optimization Neighborhood

The particle swarm algorithm is an adaptive algorithm based on a social-psychological metaphor. Each particle is influenced by a success of their topological neighbors (Kennedy and Spear, 1998). This external function provides a particle its neighbor of a given type. Also, there are many ways to define a "neighborhood" (Kennedy 1999), but we can distinguish three classes:

23.3.3.1 Social Neighborhood

The *social* neighborhood, just takes *relationships* into account. In practice, for each particle, its neighborhood is defined as a list of particles at the very beginning, and does not change. Note that, when the process converges, a social neighborhood becomes a physical one. A social neighborhood is a type when a particle chooses k nearest particles according to its location. Mathematically, social neighborhood is defined for a particle to simply get $k/2$ particles on each side of 1-D array.

23.3.3.2 Physical Neighborhood

The *physical* neighborhood, takes distances into account. In practice, distances are recomputed at each time step, which is quite costly, but some clustering techniques need this information In this type of neighborhood, a particle “chooses” k

best particles from the entire swarm; the distance between a particle and k particles in the globe is normally calculated. Normally the social neighborhood becomes a physical neighborhood during the process of algorithm.

23.3.3.3 Queens

Instead of using the best neighbor/neighbor's previous best of each particle, we can use an extra-particle, which "summarizes" the neighborhood. This method is a combination of the (unique) queen method defined in (Clerc 1999) and of the multi-clustering method described in (Kennedy 2000). For each neighborhood, we iteratively build a gravity center and take it as best neighbor. This method needs some mathematical computations. The coefficient is obtained as follows:

$$(c_i) = \sum_j \frac{(f_o + 1)}{(f_o + 1) + (f_j + 1)} \quad (23.20)$$

Since PSO is an adaptive algorithm based on social-psychological metaphor. The population of individuals adapt by returning stochastically towards previous successful regions in the search space. Move towards is the most important external function in PSO, because it explains the movement of the particles (Kennedy and Spears, 1998).

23.3.4 Particle Swarm Optimization Improvement Strategies

As in other generic optimization techniques such as genetic algorithm, differential evolution, ant colony optimization, etc., PSO has some strategies to avoid being stuck in local maximum or minimum. For PSO, in particular, the *No-Hope/Re-Hope process* is used in order to improve search performances; in other words, for the search to be adaptive.

23.3.4.1 No-Hope tests

For any objective function of any type, the decision has to be made regarding the optimum value. The PSO algorithm has a task to decide whether the optimum value is achievable or not, and if achievable then how good is it.

If the PSO algorithm is unable to output the desired or expected value, then the swarm is in *no hope* state. There are some conditions for the swarm to be in this status. Firstly, if individual particles are not moving then there is no question of movement of swarm, hence no better position is expected. Secondly, a situation may arise in which no effective movement is occurring i.e. either the swarm has reduced very much or the movement is extremely slow, hence there is no need to move the swarm. Finally when the PSO algorithm is producing the *same best* value for a number of times greater than the *threshold* value, then the better solution than the *same best* value is unexpected and the swarm is in *no hope* state. When the PSO algorithm gets into a *no-hope* state, the only way out is to either accept the result of the current situation or *re-hope* for better result. The following criteria are useful for the *no-hope* tests (Clerc 1999).

Criterion 0

If a particle has to move *toward* another one, which is at distance 1, either it does not move at all or it goes exactly to the same position as this other one, depending on the social/confidence coefficients. It may be possible that all moves computed according to equation 16 are null. In this case there is absolutely no hope to improve the current best solution.

Criterion 1

The No-Hope test defined in (Clerc 1999) is *swarm too small*. In this test, the swarm diameter is computed at each time step, which is costly. But in a discrete case, as soon as the distance between two particles tends to become *too small*, the two particles become identical, usually first by positions and then by velocities. Hence, at each time step, a *reduced* swarm is computed, in which all particles are different, which is not very expensive, and the No-Hope test becomes *swarm too reduced*, say by half the original size.

Criterion 2

Another criterion has been added to the *no-hope* test criterion is the *swarm too slow*. This criterion compares the velocities of all particles to a threshold, either individually or globally. In one version of the algorithm, this threshold is in fact modified at each time step, according to the best result obtained so far and to the statistical distribution of arc values.

Criterion 3

Another very simple criterion that is defined is the *no improvement for too many times*. In practice, it appears that criteria 1 and 2 are sufficient.

23.3.4.2 Re-Hope Process

PSO has a well-defined procedure to move out from the *no-hope* state; this is called *re-hope*. As soon as there is *no hope*, the swarm is re-expanded. The idea behind this method is to check if there is still hope to reach the better solution. If there is no hope, then swarm is moved (Clerc, 1999). The particles are moving slowly and continuously to get better position. The movement of particles is categorized in four: (i) lazy descent method, (ii) energetic descent method, (iii) local iterative leveling, and (iv) adaptive re-hope method. There are a number of re-hope strategies defined for PSO. The re-hope strategies in Clerc (1999) and He *et al.* (1999) inspire the first two methods described here.

23.3.4.2.1 *Lazy Descent Method (LDM)*

Each particle goes back to its previous best position and, from there, moves randomly and slowly (i.e. size of velocity is *1*) and stop as soon as it finds a better position or when a maximal number of moves (problem size) is reached. If the current swarm is smaller than the initial one, it is completed by a new set of particles randomly chosen.

23.3.4.2.2 *Energetic Descent Method (EDM)*

Each particle goes back to its previous best position and, from there moves slowly (i.e. velocity size is *1*) as long as it finds a better position in at most maximum number of move (problem size). If the current swarm is smaller than the initial one, it is completed by a new set of particles randomly chosen. The only drawback of this method is it is more expensive than LDM.

23.3.4.2.3 *Local Iterative Leveling (LIL)*

This method is more expensive and more powerful. This method is used when EDM fails to find a better position. For each immediate physical neighbor y (at distance *1*) of the particle p, a temporary objective function value $f(y)$is computed by using the following algorithm:

find all neighbors at a distance 1;

find the best distance, i.e. y_{min};

assign y to the temporary objective function value $f(y)=\frac{f(y_{\min})+f(x)}{2}$;

move p towards its best neighbor ;

Usually this algorithm's big O order ranges from polynomial to exponential hence this procedure is only used when the swarm is in *no-hope* state (Clerc, 1999).

23.3.4.2.4 *Adaptive Re-Hope Method (ARM)*

The three above methods can be automatically used in an adaptive way, according to how long (number of time steps) the best solution has not been improved. An example of adaptive re-hope strategy is shown in Table 23.1.

Table 23.1: Adaptive Re-Hope conditions

Same best[+]	Re-Hope type
0	No Re-Hope
1	Lazy Descent Method (type = 0)
2	Energetic Descent Method (type = 1)

⩾3	Local Iterative Leveling (type = 2)

+Number of time steps without improvement

23.3.4.2.5 Parallel and Sequential Versions

The algorithm can run either in (simulated) parallel mode or in sequential mode. In the parallel mode, at each time step, new positions are computed for all particles and then the swarm is globally moved. In sequential mode, each particle is moved at a time on a cycling way. So, in particular, the best neighbor used at time t+1 may be not anymore the same as the best neighbor at time t, even if the iteration is not complete. We note that equation 14 implicitly supposes a parallel mode, but in practice there is no clear difference in performances, and the sequential method is a bit less expensive.

23.4 Experimentation

In this section we report on the experimental results obtained with PSO on some QAP instances of QAPLIB (accessible via the WWW at address http://www.opt.math.tu-graz.ac.at/qaplib/inst.html) and compare them with some known heuristics which have a reputation for being very powerful in solving the QAP. The heuristics chosen for comparison are ant system (AS), ant system with non-deterministic hill climbing (ASN), simulated annealing (SA), tabu search (TS), genetic algorithm (GA), evolutionary strategy (ES), and sampling & clustering (SC). In (Taillard, 1995) it has been argued that the type of problem instance has a strong influence on the performance of the different algorithmic approaches proposed for solving the QAP. We present the computational results of PSO and compare them to the above mentioned algorithms.

23.4.1 Parameter settings

The original PSO by Kennedy and Eberhert (1995) requires extensive experimentation for parameter setting. However, the version by Clerc (1999) which, we have adopted in the work reported in this chapter requires minimal experimentation for parameter setting. Only three parameters need to be set. These are: the number of iterations, convergence-case and move-type. From experimentation, we have set number of iterations = *50*, convergence-case = *4* and move-type = *1*.

23.4.2 Computational results

The test problems used are those known as Elshafei (1977), Krarup (1978), and Nugent (1968). These problems are classified as *real-life* instances from practical applications of the QAP (Taillard, 1995). For example, Elshafei (1977), and Krarup (1978) instances define the layout problem for a hospital. The real-life instances have in common that the flow matrices have (in contrast to the previously mentioned randomly generated instances) many zero entries and the remaining entries are clearly not uniformly distributed.

As can be seen in Table 23.2, the performance of the particle swarm optimization heuristic competes well with other heuristics except for the Nugent 30 and Krarup 30 problems.

Table 23.2: Comparison of the PSO with other heuristic approaches⁺

	Nugent (15)	Nugent (20)	Nugent (30)	Elshafei (19)	Krarup (30)
Best known	1150	2570	6124	17212548	88900
Particle swarm opt. (PSO)	1162	2592	6204	20000000	93500
Ant System (AS)	1150	2598	6232	18122850	92490
ASN	1150	2570	6128	17212548	88900
Simulated Annealing (SA)	1150	2570	6128	17937024	89800
Tabu Search (TS)	1150	2570	6124	17212548	90090
Genetic Algorithm (GA)	1160	2688	6784	17640548	108830
Evolution Strategy (ES)	1168	2654	6308	19600212	97880
Sampling & Cluster (SC)	1150	2570	6154	17212548	88900

⁺Results are averaged over five runs

23.5 Conclusion

This chapter presents a particle swarm optimization for the quadratic assignment problem. It is observed from experimentation that the particle swarm optimization approach delivers competitive solutions when compared to ant system (AS), ant system with non deterministic hill climbing (ASN), simulated annealing (SA), tabu search (TS), genetic algorithm (GA), evolutionary strategy (ES), and sampling & clustering (SC). The successful implementation of the particle swarm optimization approach is demonstrated by addressing a problem data set. The data set problem instances, which are classified as *real-life* instances from practical applications of the QAP, were taken from the QAPLIB, which contains standard problems and their optimum solutions for benchmarking of heuristics. Specifically, the Elshafei (1997), Kraup (1978), and Nugent (1968) problems were used for comparing the PSO with other proven heuristics listed above.

In summary, the computational results show that algorithms based on the PSO adaptive search heuristic are currently among the best available algorithms for real-life, structured QAP instances. It will be interesting to continue to explore other areas of applications of PSO for discrete optimization problems.

References

Angeline PJ (1998) Using selection to improve particle swarm optimization, *IEEE International Conference on Evolutionary Computation, Anchorage, Alaska, May, 4-9.*

Battiti R, Tecchiolli G (1994) The reactive tabu search. *ORSA Journal on Computing*, 6(2), 126-140.

Burkard RE, Rendl F (1984) A thermodynamically motivated simulation procedure for combinatorial optimization problems, *European Journal of Operational Research*, 17, 169-174.

Carlisle A, Dozier G (1998) Adapting Particle Swarm Optimization to Dynamics Environments," presented at *International Conference on Artificial Intelligence, Monte Carlo Resort, Las Vegas, Nevada, USA.*

Clerc M (1999) The swarm and the queen: towards a deterministic and adaptive particle swarm optimization, *Congress on Evolutionary Computation, Washington D. C., IEEE Service Center, Piscataway, NJ*, 1951-1957.

Clerc M, Kennedy J (2002) The particle swarm: explosion, stability, and convergence in a multidimensional complex space, *IEEE Transactions on Evolutionary Computation*, **6**, 58-73.

Connolly DT (1990) An improved annealing scheme for the QAP, *European Journal of Operational Research*, 46, 93-100.

Cung V-D, Mautor T, Michelon P, Tavares A (1997) A scatter search based approach for the quadratic assignment problem. In Baeck, T., Michalewicz, Z., and Yao, X., editors, *Proceedings of ICEC'97*, 165-170, IEEE Press.

Dickey JW, Hopkins JW (1972) Campus building arrangement using TOPAZ. *Transportation Science*, 6, 59-68.

Dorigo M, Maniezzo V, Colorni A (1996) The Ant System: optimization by a colony of cooperating agents, *IEEE Transactions on Systems, Man, and Cybernetics, Part B*, 26(1), 29-41.

Eberhart RC, Kennedy J (1995) A new optimizer using particle swarm theory, *Proceedings, Sixth International Symposium on Micro Machine and human science, Nagoya, Japan, IEEE Service Center, Piscataway, NJ*, 39-43.

Eberhart RC, Shi Y (1998) Evolving artificial neural networks, presented at *International Conference on Neural Networks and Brain, Beijing*, P.R.C., PL5-PL13.

Eberhart RC, Shi Y (2000) Comparing inertia weights and constriction factors in particle swarm optimization, presented at *International Congress on Evolutionary Computation, San Diego, California*, 84-88.

Eiselt HA, Laport G (1991), A combinatorial optimization problem arising in dartboard design, *Journal of the Operational Research Society*, 42, 113-118.

Elshafei AN (1977) Hospital layout as a quadratic assignment problem, *Operations Research Quarterly*, 28, 167-179.

Fleurent C Ferland JA (1994) Genetic hybrids for the quadratic assignment problem. In Pardalos, P. M., and Wolkowicz, H., editors, *Quadratic assignment and related problems, DIMACS Series on Discrete Mathematics and Theoretical Computer Science*, 16, American Mathematical Society 173-187.

Geoffrion AM, Graves GW (1976) Scheduling parallel production lines with changeover costs: practical applications of a quadratic assignment/LP approach, *Operations Research*, 24, 595-610.

He Z, Wei, C (1999) A new population-based incremental learning method for the traveling salesman problem, *Congress on Evolutionary Computation, Washington D.C., IEEE.*

Johnson DS, McGeoch LA (1997) The travelling salesman problem: a case study in local optimization. In Aarts, E. H. L., and Lenstra, J.K., editors, *Local Search in Combinatorial Optimization*, John Wiley & Sons 215-310.

Kennedy J (1997) The particle swarm: social adaptation of knowledge, *IEEE International Conference on Evolutionary Computation, Indianapolis, Indiana, IEEE Service Center, Piscataway, NJ.*

Kennedy J (1999) Small worlds and mega-minds: effects of neighborhood topology on particle swarm performance, *Congress on Evolutionary Computation, Washington D.C, IEEE.*

Kennedy J (2000) Stereotyping: Improving Particle Swarm Performance With Cluster Analysis," presented at Congress on Evolutionary Computation.

Kennedy J, Eberhart RC (1995) Particle swarm optimization, *IEEE Proceedings of the International Conference on Neural Networks IV (Perth, Australia), IEEE Service Center, Piscataway, NJ,* 1942-1948.

Kennedy J, Eberhart RC (1997) A discrete binary version of the particle swarm algorithm, *International Conference on Systems, Man, and Cybernetics.*

Kennedy J (1999) Small worlds and mega-minds: effects of neighborhood topology on particle swarm performance, *Proceedings of the Congress on Evolutionary Computation, Washington D.C, IEEE Service Center, Piscataway, NJ,* 1931-1938.

Krarup J, Pruzan PM (1978) Computer-aided layout design, *Mathematical Programming Study*, 9, 75-94.

Laporte G, Mercure H (1988) Balancing hydraulic turbine runners: a quadratic assignment problem, *European Journal of Operational Research*, 35, 378-381.

Li Y, Pardalos PM, Resende MGC (1994) A greedy randomized adaptive search procedure for the quadratic assignment problem. In Pardalos, P. M., and Wolkowicz, H., editors, *Quadratic assignment and related problems, DIMACS Series on Discrete Mathematics and Theoretical Computer Science*, 16, 237-261, American Mathematical Society, 1994.

Merz P, Freisleben B (1997) A genetic local search approach to the quadratic assignment problem. In BÄack, T., editor, *Proceedings of the Seventh International Conference on Genetic Algorithms (ICGA'97)*, Morgan Kaufmann 465-472.

Nissen V (1994) Solving the Quadratic assignment problem with clues from nature, *IEEE Transactions on Neural Networks*, 5(1), 66-72.

Nugent CE, Vollman TE, Ruml J (1968) An experimental comparison of techniques for the assignment of facilities to locations, *Operations Research*, 16, 150-173.

Sahni S, Gonzalez T (1976) P-complete approximation problems, *Journal of the ACM*, 23, 555-565.

Shi Y, Eberhart RC (1998) A modified particle swarm optimizer, *IEEE International Conference on Evolutionary Computation, Anchorage, Alaska, May 4-9.*

Skorin-Kapov J (1990) Tabu Search applied to the quadratic assignment problem, *ORSA Journal on Computing*, 2, 33-45.

Steinberg L (1961) The backboard wiring problem: a placement algorithm, *SIAM Review*, 3, 37-50.

Taillard ED (1991) Robust taboo search for the quadratic assignment problem, *Parallel Computing*, 17, 443-455.

Taillard ED (1995) Comparison of iterative searches for the quadratic assignment problem, *Location Science*, 3, 87-105.

Tate DM, Smith AE (1995) A genetic approach to the quadratic assignment problem, *Computers & Operations Research*, 22(1), 73-83.

24 Differential Evolution for the Flow Shop Scheduling Problem

Godfrey C Onwubolu

The classical problem of scheduling n jobs on m machines in a flow shop is to minimize the throughput time of all the jobs under the assumption that all jobs are processed on all machines at the same sequence. This scheduling problem leads to the permutation situation in which there are n-factorial possible sequences to be considered which, for large number of jobs leads to combinatorial explosion. The flow shop scheduling-problem is among the combinatorial optimization problems because for large number of jobs, the search for the best job-sequence can be very demanding in terms of computational time. A summary of different methods for scheduling flow shop problem is found in Dudek *et al* (1992). Optimization algorithm such as branch-and-bound, has been employed by some researchers (see, for instance, Ignall and Linus (1965), and Hariri (1981)). Due to the difficulties associated with the computational requirements of optimization algorithms for large-sized problems, many researchers have opted for heuristic methods, which though do not guarantee optimal solutions, do produce satisfactory and reliable solutions with a reasonably small amount of computational efforts.

24.1 Introduction

In general, when discussing non-linear programming, the variables of the object function are usually assumed to be continuous. However, in practical real-life engineering applications it is common to have the problem variables under consideration being discrete or integer values. Real-life, practical engineering optimization problems are commonly integer or discrete because the available values are limited to a set of commercially available standard sizes. For example, the number of automated guided vehicles, the number of unit loads, the number of storage units in a warehouse operation are integer variables, while the size of a pallet, the size of billet for machining operation, etc., are often limited to a set of commercially available standard sizes. Another class of interesting optimization problem is finding the best order or sequence in which jobs have to be machined. None of these engineering problems has a continuous objective function rather each of these engineering problems has either an integer objective function or discrete objective function. In this paper we deal with the scheduling of jobs in a flow-shop manufacturing system.

In manufacturing, production scheduling comprises of the important tasks of dispatching jobs and allocating resources. Production scheduling is the key link in carrying out the production plan and process. The larger the production scale or wider the product variety, the more important the action of scheduling becomes. Although the objects to be produced are different in manufacturing industries, the basic attributes of production schedule are the same. The main goal is to seek an optional program for allocating resources under a production goal and related constraint conditions. There are several different production goals, e.g., shortest production time, minimum production cost, JIT mode, which all will seriously effect the production management and organization.

The constraint conditions, e.g., operation time, product priority, equipment implementation, production rate, production batch, due date, etc., are dependent upon the enterprise features including objects, environment, technology and process. The production scheduling includes the pre-process schedule for initiating the production process as static requirement, as well as the post-process schedule to feed new products on the production line for maintaining the continuous running of the production process as a dynamic requirement. The former is usually called the static schedule and the latter the dynamic schedule. Operations research, combinatorial mathematics, especially the branch and bound method and heuristic inference method, have all made a great contribution to production scheduling, it is still problematic to meet the modern production demands of increasing variety, expanding production scale, responding promptly, reforming production mode, etc. There is no unified model to deal with the problems with success.

The flow-shop scheduling-problem is a production planning-problem in which n jobs have to be processed in the same sequence on m machines. The assumptions are that there are no machine breakdowns and that all jobs are pre-emptive. This is commonly the case in many manufacturing systems where jobs are transferred from machine to machine by some kind of automated material handling systems.

For large problem instances, typical of practical manufacturing settings, most researchers have focused on developing heuristic procedures that yield near optimal-solutions within a reasonable computation time. Most of these heuristic procedures focus on the development of permutation schedules and use makespan as a performance measure. Some of the well-known scheduling heuristics, which have been reported in the literature, include Palmer (1965), Campbell, Dudek, and Smith (1970), Gupta (1971), Dannenbring (1977), Hundal and Rajgopal (1988) and Ho and Chang (1991). Cheng and Gupta (1989) and Baker and Scudder (1990) presented a comprehensive survey of research work done in flow shop scheduling.

In recent years, a growing body of literature suggests the use of heuristic search procedures for combinatorial optimisation problems. Several search procedures that have been identified as having great potential to address practical optimisation problems include simulated annealing (Kirkpatrick *et al.* 1983), genetic algorithms (Goldberg, 1989), tabu search (Glover, 1989, 1990), and ant colony optimisation (Dorigo, 1992). Consequently, over the past few years, several researchers have demonstrated the applicability of these methods, to combinatorial optimisation problems such as the flow-shop scheduling (see for example, Widmer and Hertz (1989), Ogbu and Smith (1990), Taillard (1990), Chen *et al.* (1995) and Onwubolu (2000)). More recently, a novel optimization method based on differential evolution (exploration) algorithm (Storn and Price, 1995) has been developed, which originally focused on solving non-linear programming problems containing continuous variables. Since Storn and Price (1995) invented the differential evolution (exploration) algorithm, the challenge has been to employ the algorithm to different areas of problems other than those areas that the inventors originally focussed on. Although application of DE to combinatorial optimization problems encountered in engineering is scarce, researchers have used DE to design complex digital filters (Storn, 1999), and to design mechanical elements such as gear train, pressure vessels and springs (Lampinen and Zelinka, 1999). Onwubolu (2001) applied the differential evolution approach to solving the flow shop minimum makespan problem, and the total tardiness flow shop scheduling problem. This approach is now discussed.

This chapter presents a new approach based on differential evolution algorithm for solving the problem of scheduling n jobs on m machines when all jobs are available for processing and the objective is to minimise the makespan. Other objective functions considered in the present work include mean flowtime and total tardiness.

24.2 Problem Formulation for the flow shop schedules

Machine idleness and/or job waiting are the two situations that constantly arise in a flow shop schedule (see Figure 24.1). Machine idleness occurs when the second machine waits for a unit of time (machine slack) during which the second job completes processing on the first machine. Job-waiting can be illustrated by the third job, which after completion of processing on the first machine, waits for unit time on the second machine until the second job completes processing on the second machine.

Let $t(i, m)$ represent the processing time of the i-th job on the m-th machine, $S_M(i,m)$ represent the machine slack on the last machine between the $(i-1)th$ and *the i-th* jobs, $S_J(n,m)$ represent the job slack experienced by the n-th job on the m-th machine, then the total makespan $T_S(n,m)$ can be formulated in two possible ways:

$$T_S(n,m) = \sum_{k=1}^{m-1} t(1,k) + \sum_{i=1}^{n} t(i,m) + \sum_{i=2}^{n} S_M(i,m) \tag{24.1}$$

$$T_S(n,m) = \sum_{k=1}^{m-1} t(1,k) + \sum_{i=1}^{n} t(i,m) + \sum_{i=2}^{n} S_M(i,m) \tag{24.2}$$

The standard three-field notation (Lawler *et al.* 1995) used is that for representing a scheduling problem as α|β|F(C), where α describes the machine environment, β describes the deviations from standard scheduling assumptions, and *F(C)* describes the objective *C* being optimised. In the work reported in this chapter, we are solving the $n/m/F \,||\, F(C_{\max})$ problem. Other problems solved include $F(C) = F(\sum C_i)$ and $F(C) = F(\sum T_j)$. Here $\alpha = n/m/F$ describes the multiple-machines flowshop problem, β = *null*, and $F(C) = F(C_{\max}, \sum C_i, and \sum T_j)$ for makespan, mean flowtime, and total tardiness respectively. The following notation applies: m = number of machines, n = number of jobs, $\sum C_i$ = flowtime, $C_{\max}$ = makespan, and $\sum T_j$ = total tardiness.

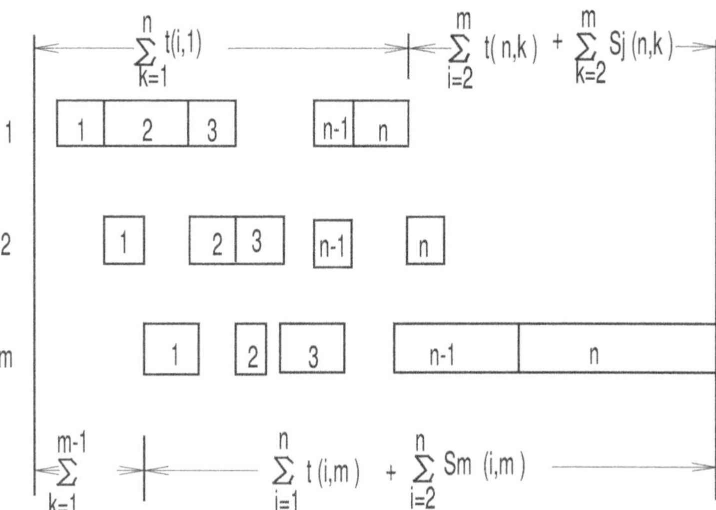

Fig. 24.1: Gantt chart for sequence {1,2........, n} in flow shop

24.3 Differential Evolution

The differential evolution (exploration) [DE] algorithm introduced by Storn and Price (1995) is a novel parallel direct search method, which utilizes *NP* parameter vectors as a population for each generation *G*. DE can be categorized into a class of *floating-point encoded, evolutionary optimization algorithms*. Currently, there are several variants of DE. The particular variant used throughout this investigation is the *DE/rand/1/bin* scheme. This scheme will be discussed here only briefly, since more detailed descriptions are provided (Price and Storn 1995). Since the DE algorithm was originally designed to work with continuous variables, the optimization of continuous problems is discussed first. Handling discrete variables is explained later. In this sections the parameters used are $\Im$ = cost or the value of the objective function, D = problem dimension, NP = population size, P = population of X-vectors, G = generation number, G_{max} = maximum generation number, X = vector composed of D parameters, V = trial vector composed of D parameters, and CR = crossover factor. Other parameters are F = scaling factor ($0 < F \leq 1.2$), (U) =upper bound, (L) = lower bound, **u,** and **v** = trial vectors, $x_{best}^{(G)}$ = vector with minimum cost in generation G. Finally, we have $x_i^{(G)}$ = i^th^ vector in generation G, $b_i^{(G)}$ = i^th^ buffer vector in generation G, $x_{r1}^{(G)}, x_{r2}^{(G)}$ = randomly selected vector, and L = random integer ($0 < L < D - 1$), and i, j = integers.

Generally, the function to be optimized, $\Im$, is of the form:

$$\Im(X): R^D \to R \tag{24.3}$$

The optimization target is to minimize the value of this objective function $\Im(X)$,

$$\min(\Im(X)) \tag{24.4}$$

by optimizing the values of its parameters:

$$X = \{x_1, x_2, \ldots, x_D\}, X \in R^D \tag{24.5}$$

where X denotes a vector composed of D objective function parameters. Usually, the parameters of the objective function are also subject to lower and upper boundary constraints, $x^{(L)}$ and $x^{(U)}$, respectively:

$$x_j^{(L)} \le x_j \le x_j^{(U)} \quad \forall j \in [1, D] \tag{24.6}$$

As with all evolutionary optimization algorithms, DE works with a population of solutions, not with a single solution for the optimization problem. Population *P* of generation *G* contains *NP* solution vectors called individuals of the population and each vector represents potential solution for the optimization problem:

$$P^{(G)} = X_i^{(G)} \quad i = 1,\ldots, NP;\; G = 1,\ldots, G_{\max} \tag{24.7}$$

Additionally, each vector contains *D* parameters:

$$X_i^{(G)} = x_{j,i}^{(G)} \quad i = 1,\ldots, NP;\; j = 1,\ldots, D \tag{24.8}$$

In order to establish a starting point for optimum seeking, the population must be initialized. Often there is no more knowledge available about the location of a global optimum than the boundaries of the problem variables. In this case, a natural way to initialize the population $P^{(0)}$ (initial population) is to seed it with random values within the given boundary constraints:

$$P^{(0)} = x_{j,i}^{(0)} = x_j^{(L)} + rand_j[0,1] \bullet \left(x_j^{(U)} - x_j^{(L)}\right) \quad \forall i \in [1, NP];\; \forall j \in [1, D] \tag{24.9}$$

where $rand_j[0,1]$ represents a uniformly distributed random value that ranges from zero to one.

The self-referential population recombination scheme of DE is different from the other evolutionary algorithms. From the first generation onward, the population of the subsequent generation $P^{(G+1)}$ is obtained on the basis of the current population $P^{(G)}$. First a temporary or trial population of candidate vectors for the subsequent generation, $P'^{(G+1)} = V^{(G+1)} = v_{j,i}^{(G+1)}$, is generated as follows:

$$v_{j,i}^{(G+1)} = \begin{cases} x_{j,r3}^{(G)} + F \bullet \left(x_{j,r1}^{(G)} - x_{j,r2}^{(G)}\right) if\ rand_j[0,1] < CR \vee j = k \\ x_{i,j}^{(G)}\ if\ otherwise \end{cases} \tag{24.10}$$

where $i \in [1, NP];\; j \in [1, D]$, F = greediness scaling factor (0 < F < 1.2)

$r1,\ r2,\ r3 \in [1, NP]$, randomly selected, except: $r1 \ne r2 \ne r3 \ne i$

$$k = (\text{int}(rand_i[0,1] \bullet D) + 1)$$

$$CR \in [0,1],\ F \in (0,1]$$

Three randomly chosen indexes, *r1*, *r2*, and *r3* refer to three randomly chosen vectors of population. They are mutually different from each other and also different from the running index *i*. New random values for *r1*, *r2*, and *r3* are assigned for each value of index *i* (for each vector). A new value for the random number $rand[0,1]$ is assigned for each value of index *j* (for each vector parameter).

The index *k* refers to a randomly chosen vector parameter and it is used to ensure that at least one vector parameter of each individual trial vector $V^{(G+1)}$ differs from its counterpart in the previous generation $X^{(G)}$. A new random integer value is assigned to *k* for each value of the index *i* (prior to construction of each trial vector). *F* and *CR* are DE control parameters. Both values remain constant during the search process. Both values as well as the third control parameter, *NP* (population size), remains constant during the search process. *F* is a real-valued factor in range [0.0, 1.0] that controls the amplification of differential variations. *CR* is a real-valued crossover factor in the range [0.0, 1.0] that controls the probability that a trial vector will be selected form the randomly chosen, mutated vector, $V_{j,i}^{(G+1)}$ instead of from the current vector, $x_{j,i}^{(G)}$. Generally, both *F* and *CR* affect the convergence rate and robustness of the search process. Their optimal values are dependent both on objective function characteristics and on the population size, *NP*. Usually, suitable values for *F*, *CR* and *NP* can be found by experimentation after a few tests using different values. Practical advice on how to select control parameters *NP* , *F* and *CR* can be found in Storn and Price (1995, 1996, 1997).

The selection scheme of DE also differs from the other evolutionary algorithms. On the basis of the current population $P^{(G)}$ and the temporary population $P'^{(G+1)}$, the population of the next generation $P^{(G+1)}$ is created as follows:

$$X_i^{(G+1)} = \begin{cases} V_i^{(G+1)} \ \textit{if} \ \Im\left(V_i^{(G+1)}\right) \leq \Im\left(X_i^{(G)}\right) \\ X_i^{(G)} \ \textit{if otherwise} \end{cases} \tag{24.11}$$

Thus, each individual of the temporary or trial population is compared with its counterpart in the current population. The one with the lower value of cost function $\Im(X)$ to be minimized will propagate the population of the next generation. As a result, all the individuals of the next generation are as good or better than their counterparts in the current generation. The interesting point concerning the

DE selection scheme is that a trial vector is only compared to one individual vector, not to all the individual vectors in the current population.

24.3.1 Constraint Handling

24.3.1.1 Boundary constraints

It is important to notice that the recombination operation of DE is able to extend the search outside of the initialized range of the search space (Equations 23.9 and 23.10). It is also worthwhile to notice that sometimes this is a beneficial property in problems with no boundary constraints because it is possible to find the optimum that is located outside of the initialized range. However, in boundary constrained problems, it is essential to ensure that parameter values lie inside their allowed ranges after recombination. A simple way to guarantee this is to replace parameter values that violate boundary constraints with random values generated within the feasible range:

$$u_{j,i}^{(G+1)} = \begin{cases} x_j^{(L)} + rand_j[0,1] \bullet \left(x_j^{(U)} - x_j^{(L)}\right) \ if \ u_{j,i}^{(G+1)} < x_j^{(L)} \vee u_{j,i}^{(G+1)} > x_j^{(U)} \\ u_{i,j}^{(G+1)} \ if \ otherwise \end{cases} \tag{24.12}$$

where $i \in [1, NP]; \ j \in [1, D]$.

This is the method that was used for this work. Another simple but less efficient method is to reproduce the boundary constraint violating values according to Equation 9 as many times as is necessary to satisfy the boundary constraints. Yet another simple method that allows bounds to be approached asymptotically while minimizing the amount of disruption that results from resetting out of bound values (Storn and Price, 1999) is:

$$u_{j,i}^{(G+1)} = \begin{cases} (x_{j,i}^{(G)} + x_j^{(L)})/2 \ if \ u_{j,i}^{(G+1)} < x_j^{(L)} \\ (x_{j,i}^{(G)} + x_j^{(U)})/2 \ if \ u_{j,i}^{(G+1)} > x_j^{(U)} \\ u_{j,i}^{(G+1)} \ if \ otherwise \end{cases} \tag{24.13}$$

24.3.1.2 Constraint functions

Storn (1999), and Lampinen and Zelinka (1999) elaborate soft-constraint (penalty) handling approaches for constraint functions. The constraint function introduces a distance measure from the feasible region, but is not used to reject unrealizable solutions, as it is in the case of hard-constraints. One possible soft-constraint approach is to formulate the cost-function as follows:

$$\Im(X) = (\Im(X) + a) \bullet \prod_{i=1}^{m} c_i^{b_i} \tag{24.14}$$

$$\text{where } c_i = \begin{cases} 1.0 + s_i \bullet g_i(X) \; if \; g_i(X) > 0 \\ 1 \qquad\qquad otherwise \end{cases}$$

$$s_i \geq 1$$
$$b_i \geq 1$$
$$\min(\Im(X)) + a > 0$$

The constant, a, is used to ensure that only non-negative values will be assigned to $\Im$. When the value of a is set high enough, it does not otherwise affect the search process. The constant, s, is used for appropriate scaling of the constraint function value. The exponent, b, modifies the shape of the optimization surface. Generally, higher values of s and b are used when the range of the constraint function, $g(X)$, is expected to be low. Often setting $s = 1$ and $b=1$ works satisfactorily and only if one of the constraint functions, $g_i(X)$, remains violated after the optimization run, it will be necessary to use higher values for s_i or/and b_i.

In many real-world engineering optimization problems, the number of constraint functions is relatively high and the constraints are often non-trivial. It is possible that the feasible solutions are only a small subset of the search space. Feasible solutions may also be divided into separated islands around the search space. Furthermore, the user may easily define totally conflicting constraints so that no feasible solutions exist at all. For example, if two or more constraints conflict, so that no feasible solution exists, DE is still able to find the nearest feasible solution. In the case of non-trivial constraints, the user is often able to judge which of the constraints are conflicting on the basis of the nearest feasible solution. It is then possible to reformulate the objective function in order to address these issues or reconsider the problem setting itself to resolve the conflict.

Another benefit of the soft-constraint approach is that the search space remains continuous. Multiple hard constraints often split the search space into many separated islands of feasible solutions. This discontinuity introduces stalling points for some genetic searches and also raises the possibility of new, locally optimal areas near the island borders. For these reasons a soft-constraint approach is considered essential. It should be mentioned that many traditional optimization methods are only able to handle hard-constraints. For evolutionary optimization, the soft-constraint approach was found to be a natural approach.

24.3.2 Integer and Discrete Optimization by Differential Evolution Algorithm

Several approaches have been used to deal with discrete variable optimization. Most of them round off the variable to the nearest available value before evaluating each *trial vector*. To keep the population robust, successful trial vectors must enter the population with all of the precision with which they were generated (Price and Storn, 1997).

24.3.2.1 Conventional Technique

In its canonical form, the differential evolution algorithm is only capable of handling continuous variables. Extending it for optimization of integer variables, however, is rather easy. Lampinen and Zelinka (1999) discuss how to modify DE for mixed variable optimization. They suggest that only a couple of simple modifications are required. First, integer values should be used to evaluate the objective function, even though DE itself may still works internally with continuous floating-point values. Thus,

$$\Im(y_i),\ i \in [1, D] \tag{24.15}$$

where $$y_i = \begin{cases} x_i & for\ continuous\ variables \\ INT(x_i) & for\ integer\ variables \end{cases}$$

$$x_i \in X$$

INT() is a function for converting a real value to an integer value by truncation. Truncation is performed here only for purposes of cost-function value evaluation. Truncated values are not elsewhere assigned. Thus, DE works with a population of continuous variables regardless of the corresponding object variable type. This is essential for maintaining the diversity of the population and the robustness of the algorithm. Second, in case of integer variable, instead of Equation 23.9, the population should be initialized as follows:

$$P^{(0)} = x_{i,j}^{(0)} = x_j^{(L)} + rand_j[0,1] \bullet \left(x_j^{(U)} - x_j^{(L)}\right)\ \forall i \in [1, NP];\ \forall j \in [1, D] \tag{24.16}$$

Additionally, instead of Equation 11, the boundary constraint handling integer variables should be performed as follows:

$$u_{i,j}^{(G+1)} = \begin{cases} x_j^{(L)} + rand_j[0,1] \bullet \left(x_j^{(U)} - x_j^{(L)} + 1\right) & if\ INT(u_{i,j}^{(G+1)}) < x_j^{(L)} \vee INT(u_{i,j}^{(G+1)}) > x_j^{(U)} \\ u_{i,j}^{(G+1)} & if\ otherwise \end{cases} \quad (24.17)$$

where $i \in [1, NP]$; $j \in [1, D]$.

Discrete values can also be handled in a straightforward manner. Suppose that the subset of discrete variables, $X^{(d)}$, contains l elements that can be assigned to variable x:

$$X^{(d)} = x_i^{(d)}, \ \ i \in [1, l] \quad (24.18)$$

where $x_i^{(d)} < x_{i+1}^{(d)}$

Instead of the discrete value x_i itself, we may assign its index, i, to x. Now the discrete variable can be handled as an integer variable that is boundary constrained to range 1,…,l. To evaluate the objective function, the discrete value, x_i , is used instead of its index i. In other words, instead of optimizing the value of the discrete variable directly, we optimize the value of its index i. Only during evaluation is the indicated discrete value used. Once the discrete problem has been converted into an integer one, the previously described methods for handling integer variables can be applied (Equations 24.15–24.17).

24.3.2.2 Forward Transformation and Backward Transformation Technique

In this paper, we present a forward transformation method for transforming integer variables into continuous variables for the internal representation of vector values since in its canonical form, the DE algorithm is only capable of handling continuous variables. We also present a backward transformation method for transforming a population of continuous variables obtained after mutation back into integer variables for evaluating the objective function (Onwubolu, 2001). Both forward and backward transformations are utilized in implementing the DE algorithm used in the present study for the flowshop scheduling problems.

Forward Transformation

In integer variable optimization a set of integer number is normally generated randomly as an initial solution. Let this set of integer number be represented as:

$$z_i^{'} \in z^{'} \tag{24.19}$$

The equivalent continuous variable for $z_i^{'}$ is given as

$$1x10^2 < 5x10^2 \leq 10^3 x10^2 \tag{24.20}$$

then $$z_i = -1 + \frac{z_i^{'} * f * 5}{10^3 - 1} \tag{24.21}$$

Applying a scaling factor, $f = 100$ gives

$$z_i = -1 + \frac{z_i^{'} * f * 5}{10^3 - 1} = -1 + \frac{z_i^{'} * 500}{10^3 - 1} \tag{24.22}$$

Equation (24.2) is used to transform any integer variable into an equivalent continuous variable, which is then used for the DE internal representation of the population of vectors. Without this transformation, it is not possible to make useful moves towards the global optimum in the solution space using the mutation mechanism of DE, which works better on continuous variables.

For example in a five-job problem, suppose the sequence is given as {2, 4, 3, 1, and 5}. This sequence is not directly used in DE internal representation. Rather, applying equation (24.22), the sequence is transformed into a continuous form. The floating-point equivalence of the first entry of the given sequence, $z_i^{'} = 2$, is $z_i = -1 + \frac{2 * 500}{10^3 - 1} = 0.001001$. Other values are similarly obtained and the sequence is therefore represented internally in the DE scheme as {0.001001, 1.002, 0.501502, -0.499499, 1.5025}.

In the technique we present, no rounding-off or truncation is required since such a truncation method often gives less than optimal results because no attempt is made during optimization to evaluate only realizable systems (Price 1999).

Backward Transformation

Integer variables are used to evaluate the objective function. The DE self-referential population mutation scheme is quite unique. After the mutation of each vector, the trial vector is evaluated for its objective function in order to decide whether or not to retain it. This means that the objective function values of the current vectors in the population need to be also evaluated. These vector variables are continuous (from the forward transformation scheme) and have to be transformed into their integer number equivalence. It is not enough to round off these values for the class of problems we are solving. The backward transformation technique is used for converting floating point numbers to their integer number equivalence. The scheme is given as follows:

$$z_i^{'} = \frac{(1+z_i)*\left(10^3-1\right)}{5*f} = \frac{(1+z_i)*\left(10^3-1\right)}{500} \tag{24.23}$$

In this present form the backward transformation function is not able to properly discriminate between variables. Some modifications are required as follows:

$$\alpha = \text{int}(z_i^{'} + 0.5) \tag{24.24}$$

$$\beta = \alpha - z_i^{'} \tag{24.25}$$

$$z_i^{*} = \begin{cases} (\alpha - 1)\ if\ \beta > 0.5 \\ \alpha \qquad if\ \beta < 0.5 \end{cases} \tag{24.26}$$

In these expressions, $z_i^{'}$ = intermediate integer number equivalence of a floating point (not rounded off), z_i^{*} = integer number equivalence of a floating point, z_i = floating point equivalence of an integer number, and α, and β = parameters. Equation (24.26) gives z_i^{*}, which is the transformed value used for computing the objective function.

As an example, we consider a set of trial vector, $z_i = \{-0.33, -0.17, 0.67, 0.84, 1.5\}$ obtained after mutation. The integer values corresponding to the trial vector values are obtained using equation (24.26) as follows:

$$z_1^{'} = (1-0.33)*(10^3-1)/500 = 1.33866$$

$$z_2^{'} = (1+0.67)*(10^3-1)/500 = 3.3367$$

$$z_3^{'} = (1-0.17)*(10^3-1)/500 = 1.65834$$

$$z_4^{'} = (1+1.50)*(10^3-1)/500 = 4.9950$$

$$z_5' = (1+0.84)*(10^3-1)/500 = 3.6763$$

$$\alpha_1 = \text{int}(1.333866+0.5) = 2$$

$$\beta_1 = 2-1.33866 = 0.66134 > 0.5$$

$$z_1^* = 2-1 = 1$$

$$\alpha_2 = \text{int}(3.3367+0.5) = 4$$

$$\beta_2 = 4-3.3367 = 0.6633 > 0.5$$

$$z_2^* = 4-1 = 3$$

$$\alpha_3 = \text{int}(1.65834+0.5) = 2$$

$$\beta_3 = 2-1.65834 = 0.34166 < 0.5$$

$$z_3^* = 2$$

$$\alpha_4 = \text{int}(4.995+0.5) = 5$$

$$\beta_4 = 5-4.995 = 0.005 < 0.5$$

$$z_4^* = 5$$

$$\alpha_5 = \text{int}(3.673+0.5) = 4$$

$$\beta_5 = 4-3.673 = 0.3237 < 0.5$$

$$z_5^* = 4$$

The set of integer values is given as $z_i^* = \{1, 3, 2, 5, 4\}$. This set is used to obtain the objective function values.

24.3.2.3 DE Strategies

Price and Storn (2001) have suggested ten different working strategies of DE and some guidelines in applying these strategies for any given problem. Different strategies can be adopted in the DE algorithm depending upon the type of problem for which it is applied. The strategies can vary based on the vector to be perturbed, number of difference vectors considered for perturbation, and finally the type of crossover used. The following are the ten different working strategies proposed by Price and Storn (2001):

(1) DE/best/1/exp
(2) DE/rand/1/exp
(3) DE/rand-to-best/1/exp
(4) DE/best/2/exp
(5) DE/rand/2/exp
(6) DE/best/1/bin
(7) DE/rand/1/bin
(8) DE/rand-to-best/1/bin
(9) DE/best/2/bin
(10) DE/rand/2/bin

Strategy 1: DE/best/1/exp: $v = x_{best}^{(G)} + F \bullet \left(x_{r2}^{(G)} - x_{r3}^{(G)}\right)$

Strategy 2: DE/rand/1/exp: $v = x_{r1}^{(G)} + F \bullet \left(x_{r2}^{(G)} - x_{r3}^{(G)}\right)$

Strategy 3: DE/rand-to-best/1/exp:
$v = x_{i}^{(G)} + \lambda \bullet \left(x_{best}^{(G)} - x_{i}^{(G)}\right) + F \bullet \left(x_{r1}^{(G)} - x_{r2}^{(G)}\right)$

Strategy 4: DE/best/2/exp: $v = x_{best}^{(G)} + F \bullet \left(x_{r1}^{(G)} + x_{r2}^{(G)} - x_{r3}^{(G)} - x_{r4}^{(G)}\right)$

Strategy 5: DE/rand/2/exp: $v = x_{r5}^{(G)} + F \bullet \left(x_{r1}^{(G)} + x_{r2}^{(G)} - x_{r3}^{(G)} - x_{r4}^{(G)}\right)$

Strategy 6: DE/best/1/bin: $v = x_{best}^{(G)} + F \bullet \left(x_{r2}^{(G)} - x_{r3}^{(G)}\right)$

Strategy 7: DE/rand/1/bin: $v = x_{r1}^{(G)} + F \bullet \left(x_{r2}^{(G)} - x_{r3}^{(G)}\right)$

Strategy 8: DE/rand-to-best/1/bin:
$v = x_{i}^{(G)} + F \bullet \left(x_{best}^{(G)} - x_{i}^{(G)}\right) + F \bullet \left(x_{r1}^{(G)} - x_{r2}^{(G)}\right)$

Strategy 9: DE/best/2/bin: $v = x_{best}^{(G)} + F \bullet \left(x_{r1}^{(G)} + x_{r2}^{(G)} - x_{r3}^{(G)} - x_{r4}^{(G)}\right)$

Strategy 10: DE/rand/2/bin: $v = x_{r5}^{(G)} + F \bullet \left(x_{r1}^{(G)} + x_{r2}^{(G)} - x_{r3}^{(G)} - x_{r4}^{(G)}\right)$

The general convention used above is as follows: DE/x/y/z. DE stands for differential evolution algorithm, x represents a string denoting the vector to be perturbed, y is the number of difference vectors considered for perturbation of x, and z is the type of crossover being used (exp: exponential; bin: binomial). Thus, the working algorithm outline by Price and Storn (1997) is the seventh strategy of DE, that is, DE/rand/1/bin. Hence the perturbation can be either in the best vector of the previous generation or in any randomly chosen vector. Similarly for perturbation, either single or two vector differences can be used. For perturbation with a single vector difference, out of the three distinct randomly chosen vectors, the weighted vector differential of any two vectors is added to the third one. Similarly for perturbation with two vector differences, five distinct vectors other than the target vector are chosen randomly from the current population. Out of these, the weighted vector difference of each pair of any four vectors is added to the fifth one for perturbation.

In exponential crossover, the crossover is performed on the D (the dimension or number of variables to be optimized) variables in one loop until it is within the CR bound. For discrete optimization problems, the first time a randomly picked number between 0 and 1 goes beyond the CR value, no crossover is performed and the remaining D variables are left intact. In binomial crossover, the crossover is performed on each the D variables whenever a randomly picked number between 0 and 1 is within the CR value. Hence, the exponential and binomial crossovers yield similar results.

The various ingredients of DE discussed so far are put together into a pseudo-code.

Pseudo-code of the DE Algorithm

The pseudo-code of the DE algorithm used in the present study is shown below:

- Initialize the values of D, NP, CR, F, λ and maximum number of generations G_{max}.
- Initialize all the vectors of the population randomly between a given lower bound (L), and upper bound (U)
  for i =1 to NP
  for j = 1 to D
  $(b_{i,0})_j$ = (L) + rand[0, 1]•((U) – (L)) /* for i for j ..end */
- Forward transformation: convert buffer discrete variables to floating point variables
  $$\left(x_{i,0}\right)_j \leftarrow \left(b_{i,0}\right)_j$$
- Evaluate the cost of each buffer vector. Cost here is the value of the objective function to be minimized.
  for i = 1 to NP
  $$\Im_i = \left(n/m/F\left\|F\left(\sum C_i, C_{\max}\right)\right)\right|_{x_{i,0}}$$
- Find out the vector with lowest cost i.e., the best vector so far
  $\Im_{\min} = \Im_1$ and $best$ = 1
  for i = 2 to NP
  if ($\Im_i < \Im_{\min}$)

then $\Im_{\min} = \Im_i$ and *best* = i

- While the current generation is less than the maximum number of generations perform recombination, mutation, recombination and evaluation of the objective function.
  while $(G < G_{max})$ *do*
  {
  for i =1 to NP
  {
  - Select two distinct vectors randomly from the population other than the vector $x_i^{(G)}$
    do r_1 = rand[0, 1]•NP while(r_1 = i)
    do r_2 = rand[0, 1]•NP while((r_2 = i) OR ($r_2 = r_1$))
  - Perform D binomial trails, change at least one parameter of the trial vector $u_i^{(G)}$ and perform mutation.
    j = rand[0, 1]•D
    for n = 1 to D
    {
    if ((rand[0, 1] < CR) OR (n = (D-1))) then

$$\left(u_i^{(G)}\right)_j = \left(x_i^{(G)}\right)_j + \lambda \bullet \left(\left(x_{best}^{(G)}\right)_j - \left(x_i^{(G)}\right)_j\right) + F \bullet \left(\left(x_{r1}^{(G)}\right)_j - \left(x_{r2}^{(G)}\right)_j\right)$$

else $\left(u_i^{(G)}\right)_j = \left(x_i^{(G)}\right)_j$

$$j = \langle n+1 \rangle_D$$

} /* n = 1 to D ends */

- Backward transformation: convert floating point trial vector variables $\left(u^{(G)}\right)_j$ to discrete vector variables

- Evaluate the cost of the trial vector.

$$\Im_{trial} = \left(n/m/F\middle\|F\left(\sum C_i, C_{\max}\right)\right)\Big|_{u_i^{(G)}}$$

- If the cost of the trial vector is less than the parent vector then select the trial vector to the next generation.
  if $(\Im_{trial} \leq \Im_i)$
  {
  $\Im_i = \Im_{trial}$
  if $(\Im_{trial} < \Im_{\min})$
  $\Im_{\min} = \Im_{trial}$ and *best* = i
  } /* if ends */

} /* for i = 1 to NP ends */

- Copy the new vectors $u_i^{(G)}$ to $x_i^{(G)}$ and increment G
  $G = G + 1$
- Check for convergence and stop if converged
  } /* while G ... ends */
- Print the results.

24.4 Illustrative Example

In order to understand the details outlined in this paper regarding the operations of DE for flowshop scheduling, a 5-machine, 5-job problem is given as an example. The processing times and due dates are given in Table 24.1.

Table 24.1: Processing times and due dates for a 5x5 problem

	Job				
Machine	1	2	3	4	5
1	5	7	4	3	6
2	6	5	7	6	7
3	7	8	3	8	5
4	8	6	5	5	8
5	4	4	8	7	3
Due date	30	30	46	50	55

Since the number of jobs is 5, then the number of parameters, (D), in the object vector is also 5. The population size, *NP*, is selected to be 150. Strategy 7, which is defined as *DE/rand/1/bin*, is selected along with the values of $F = 0.3$ and $CR = 0.9$. These values are used based on experimentation of different combinations, which is discussed in detail in Section 5. The value for NP also satisfies the criterion for $NP > 2D$ as stated by Price (1999).

The initial population is randomly generated. Since the random number generator only generates floating-point numbers between [0,1], equation 8 is used to transform the values into discrete numbers within the lower bound 1 and upper bound 5. The discrete values are passed to the array, but all the values are checked against the values already in the array. If a repeated value is detected, that value is rejected. The completed initial population will look somewhat as in the Figure 24.2. Vector-1 has a sequence of {3, 5, 2, 1, and 4}, while Vector-150 has a sequence of {3, 1, 4, 5, and 2}.

	i →						*NP*			
1	*2*	*3*	*4*	*5*	-	-	*147*	*148*	*149*	*150*
3	5	4	4	2	-	-	1	5	2	3
5	2	1	2	5	-	-	3	3	4	1
2	4	5	3	4	-	-	5	4	3	4
1	3	3	5	1	-	-	4	2	5	5
4	1	2	1	3	-	-	2	1	1	2

Fig. 24.2. Discrete population array.

At this point the objective function is calculated. The completion time matrix shown below has a makespan value of 58, which is the last entry in the matrix:

$$[C] = \begin{bmatrix} 4 & 10 & 17 & 22 & 25 \\ 11 & 18 & 23 & 29 & 35 \\ 14 & 23 & 31 & 38 & 46 \\ 19 & 31 & 37 & 46 & 51 \\ 27 & 34 & 41 & 50 & 58 \end{bmatrix}$$

Similar objective function values are calculated for all the vectors in the in the initial population. These objective function values are shown in Figure 24.3.

i →..NP										
1	2	3	4	5	-	-	147	148	149	150
58	58	57	56	62	-	-	57	62	58	55

Fig. 24.3. Objective function array.

At this point the lowest *makespan* for the initial population is obtained, for example 55, for Vector-150. Another population is also obtained for the floating-point numbers from the original array. The mutation scheme for DE only works on floating point numbers so a separate array for the equivalent floating point numbers is obtained.

24.4.1 Mutation Scheme

For the mutation scheme the discrete numbers of the object array are passed to the function for conversion. The first array is {3, 5, 2, 1, and 4}. The conversion is done using equation 21 as previously discussed. The transformed values for the discrete values {3, 5, 2, 1, and 4} are {0.5015, 1.5035, 0.001, –0.4995, 1.002}.

Thus the floating-point array is shown in Figure 24.4. The *mutation scheme* acts upon the population at this point.

i →..				NP	
1	2	3	-	149	150
0.5015	1.5035	1.002	-	0.001	0.5015
1.5035	0.001	-0.499	-	1.002	-0.499
0.001	1.002	1.5035	-	0.5015	1.002
-0.499	0.5015	0.5015	-	1.5035	1.5035
1.002	-0.499	0.001	-	-0.499	0.001

Fig. 24.4. Initial floating-point population.

The one hundred and forty eighth vector of object parameters $i = 148$, is used to illustrate the *mutation* scheme. The self referential mutation scheme is used to mutate the object vector with respect to the whole population. A random number *rnd* is generated between the bounds [0,1], for example 0.467 and checked against the value for *CR*, 0.9. If the random generated number *rnd* is less than *CR*, as in the case for this array, the value in the object array can then be acted upon by the *mutation scheme*.

The value n of the object array is chosen at random between 1 and 5. The value chosen is 3. Thus the first vector-member to be mutated is vector-member 3 in the one hundred and forty eighth vector as shown in Figure 24.5.

$i = 148$	D
1.5035	1
0.5015	2
1.002 ←	3
0.001	$= n$
-0.499	4
	5

Fig. 24.5 Vector for mutation operation.

Thus the first value to be mutated is 1.002. For Strategy 7 three random numbers *r1, r2, and r3* are generated. These numbers are within the population range [0, 150]. The number generated are $r1 = 3$, $r2 = 5$ and $r3 = 149$. The *mutation scheme* is outlined in Figure 24.6. First the value pointed to by *r1* and n is obtained as shown in Figure 24.6.

i = 3		i = 5		i = 149	
1.002		0.001		0.001	
-0.499		1.5035		1.002	
1.5035	←	1.002	←	0.5015	←
0.5015		-0.499		1.5035	
0.001		0.5015		-0.499	

Fig. 24.6 Mutation for (*r1*,*n*), (*r2*,*n*) and (*r3*,*n*)

Since *r1* is 3 and $i = 3$ the value is 1.5035. Secondly the value pointed to by *r2* and *n* is obtained is also shown. Since *r2* is 3 and *i* is 5, the value is 1.002. Finally the value pointed to by *r3* and *n* is obtained as also shown. Since *r3* is 3 and *i* is 149, the value is 0.5015. The weighted-mutation now takes place through the equation for Strategy 7, which is reproduced below.

Mutated Value = (*r1,n*) + *F* x ((*r2,n*) - (*r3,n*))

The value obtained for the first mutation is:

1.5035 + 0.3*(1.002 - 0.5015) = 1.65365

Now the value of *n* is incremented by 1, i.e. $n = 4$. The whole process of obtaining the values of (*r1,n*), (*r2,n*) and (*r3,n*) are then repeated. This is done till *n* has looped through all the values *D* meaning that $n = 1, 2, 3, 4, 5$.

The trial mutated values for the rest of objects are given as follows:

n = 4: 0.5015 + 0.3*(-0.499 - 1.5035) = -0.09925

n = 5: 0.001 + 0.3*(0.5015 - (-0.499)) = 0.30115

n = 1: 1.002 + 0.3*(0.001 - 0.001) = 1.002

n = 2: -0.499 + 0.3*(1.5035 - 1.002) = -0.34855

The trial mutated array obtained after the mutation is shown in Figure 24.7.

1.00200
-0.34855
1.65365
-0.09925
0.30115

Fig. 24.7 Trial mutated array for i = 148.

The mutated trial array is then transformed back into discrete values. This is achieved through the formulation of Onwubolu (2001) as outlined in Section 3.2 using equations 22 to 25. For example,

$$z_1' = (1 + 1.002) * (10^3 - 1) / 500 = 3.9999$$

$$\alpha_1 = \text{int}(3.9999 + 0.5) = 4$$

$$\beta_1 = 4 - 3.9999 = 0.0001 < 0.5$$

$$z_1^* = 4$$

Other conversions are similarly made, leading to the trial vector sequence {4, 1, 5, 2, and 3}.

24.4.2 Selection

Selection is the mechanism for determining whether the trial vector replaces the current vector or not according to the criterion already outlined in Section 3, equation 10. The objective function of the trial-vector {4, 1, 5, 2, and 3}is now determined as follows.

$$[C] = \begin{bmatrix} 3 & 8 & 14 & 21 & 25 \\ 9 & 15 & 22 & 27 & 34 \\ 17 & 24 & 29 & 37 & 40 \\ 22 & 32 & 40 & 46 & 51 \\ 29 & 36 & 41 & 43 & 59 \end{bmatrix}$$

The makespan value for this trial vector is 59. Since this value is less than the objective function value, 62, of the current vector-148 being examined, the trial vector replaces vector-148. This results in the objective function array of Figure 2 being updated as shown in Figure 24.8. The mutated value is in bold in position 148 of the updated array.

i→							*NP*
1	*2*	*3*	-	-	*148*	*149*	*150*
58	58	57	-	-	**59**	58	55

Fig. 24.8 Updated objective function array.

24.5 Experimentation

The differential evolution algorithm was written in C++ and runs on a PC with a Pentium, 65MB, 733MHz processor. The flowshop scheduling-problem was solved using flowtime, makespan, and tardiness as objective functions. The set of problem types, which were used for experimentation include randomly generated data set used in a previous study.

24.5.1 Parameter Setting

The values of NP, CR, λ and F are fixed empirically following certain heuristics (Price and Storn, 1997). F and λ are usually equal and are between 0.5 and 1.0. CR usually should be 0.3, 0.7, 0.9 or 1.0 to start with. NP should be of the order of 10*D and should be increased in case of mis-convergence. If NP is increased then usually F has to be decreased. In the present study, eighteen trials were carried out consisting of two levels of NP (50 and 150), three levels of F (0.3, 0.5, and 0.9) and three levels of CR (0.8, 0.9, and 1.0). The best values obtained from combining these parameters during experimentation are *NP =150, F = 0.3*, and *CR = 0.9.* Although the value of NP = 150 is adequate for small-size problems, a more general approach used for experimentation of NP being equal to the ten times the number of jobs. This problem-size dependent value of NP caters for small, medium and large problem sizes. Table 24.2 shows the comparison of the 10 DE-strategies using the 10x25-problem data set. The table shows the best strategy is strategy 7, while strategies 5 and 6 are found to compete with each another and closely follow strategy 7. Figure 24.9 shows the graph of the objective function value as a function of the generation number for the 10x25-problem data set. The maximum number of iterations was kept as 50; however, in all the runs during experimentation the algorithm converged within 30 generations. Tables 3, 4, and 5 show the comparison between GA developed in a previous study for flow-shop scheduling (Onwubolu and Mutingi, 1999) and the DE developed in the present study discussed in this paper. Table 24.3 shows the results for makespan as the objective function. Table 24.4 shows the results for total tardiness as the objective function, while Table 24.5 shows the results for mean flowtime as the objective function.

As can be seen, the differential evolution algorithm performs better than genetic algorithm for small-sized problems, and competes appreciably with genetic algorithm for medium to large-sized problems.

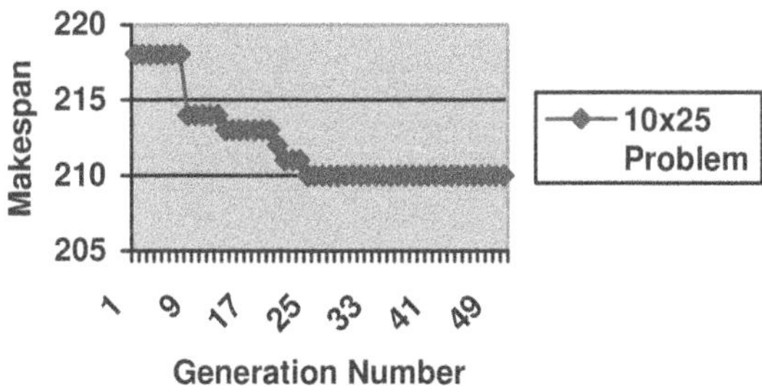

Fig. 24.9 Convergence of DE

Table 24.2 Comparison of 10 DE-strategies using the *10x25* problem data set

	Strategy									
	1	2	3	4	5	6	7^{+}	8	9	10
Make-span	211.8	209.2	212.2	212.4	208.6	210.6	207.8	212.4	210.0	207.2
Total tardiness	3001.8	3034.6	3021.4	3089.2	3008.0	2987.8	2936.4	3034.2	2982.8	2990.6
Mean flowtime	105.75	105.11	105.52	107.71	104.68	103.03	103.17	105.32	104.70	104.16

$^{+}$Strategy 7 is the best

Table 24.3 Makespan

mxn	Generated problems	GA	DE	$(\text{Solution})_{GA/DE}$
4x4	5	44.00	39.00	-
5x10	5	79.00	79.00	-
8x15	5	143.00	138.00	-
10x25	5	205.00	202.00	-
15x25	5	248.00	253.00	98.02
20x50	5	468.00	470.00	99.57
25x75	5	673.00	715.40	94.07
30x100	5	861.00	900.40	95.62

Table 24.4 Total tardiness

mxn	Generated problems	GA	DE	$(\text{Solution})_{GA/DE}$
4x4	5	54.00	52.60	-
5x10	5	285.00	307.00	92.83
8x15	5	1072.00	1146.00	93.54
10x25	5	2869.00	2957.00	97.02
15x25	5	3726.00	3839.40	97.06
20x50	5	13683.00	14673.6	93.25
25x75	5	30225.00	33335.6	90.67
30x100	5	51877.00	55735.6	93.07

Table 24.5 Mean flowtime

mxn	Generated problems	GA	DE	$(\text{Solution})_{GA/DE}$
4x4	5	21.38	22.11	-
5x10	5	35.30	36.34	97.14
8x15	5	63.09	66.41	95.00
10x25	5	98.74	103.89	95.04
15x25	5	113.85	122.59	93.03
20x50	5	216.00	234.32	92.18
25x75	5	317.00	354.77	89.35
30x100	5	399.13	435.49	91.56

24.6 Summary

This chapter presents a differential evolution algorithm for flow shop scheduling problem in which makespan, mean flowtime, and total tardiness are the performance measures. It is observed from experimentation that the differential evolution approach delivers competitive makespan, mean flowtime, and total tardiness when compared to genetic algorithm. The successful implementation of the differential evolution approach is demonstrated by addressing a problem data set. The data set problem instances were randomly generated for processing times of jobs on machines. From experimentation, the differential evolution algorithm is found to perform better than genetic algorithm for small-sized problems, and competes appreciably with genetic algorithm for medium to large-sized problems.

In this chapter, it is shown that the differential evolution algorithm was successfully applied to the makespan, mean flowtime, and total tardiness for flow shop scheduling problems. The described method is relatively simple, easy to implement and easy to use. It is, however, capable of optimizing all integer, discrete and continuous variables and capable of handling non-linear objective functions with multiple non-trivial constraints. Further research need to be carried out by comparing solution quality of DE with other emerging optimization techniques such as tabu search and ant colony optimization techniques. It will be also useful to investigate the application of DE to other combinatorial optimization problems such as traveling salesman problem, etc., since up till date these problems have not dealt with.

References

Baker, K.R., and Scudder, G.D., 1990, Sequencing with earliness and tardiness penalties: a review, *Operations Research*, 38, 22-36.

Campbell, H.G., Dudek, R.A., and Smith, M. L., 1970, A heuristic algorithm for the n job, m - machine sequencing problem, *Management Science* 16/B, 630-637.

Chen, C., Vempati, V. S., and Aljaber, N., 1995, An application of genetic algorithms for the flow shop problems, European Journal of Operations Research, 80, 359-396.

Cheng, T.C.E., and Gupta, M.C., 1989, Survey of scheduling research involving due-date determination decisions, *European Journal of Operations Research* 38, 156-166.

Dannenbring, D.G., 1977, An evaluation of flow-shop sequencing heuristics, *Management Science* 23, 1174-1182

Dorigo, M., *1992*, Optimisation, Learning and Natural Algorithms (Ottimizzazione, aprendimento automatico, et algoritmi basati su metafora naturale), *PhD. Dissertation, Dipartimento Elettronica e Informazione, Politecnico di Milano*, Italy

Dudek, R.A., Panwalkar, S. S., and Smith, M.L., 1992, The lessons of flowshop scheduling research. *Operations Research*, 40 (1), 7-13

Glover, F., 1989, Tabu search-Part, *ORSA Journal of Computing* 1/3, 190-206.

Glover, F., 1990, Tabu search-Part II, *ORSA Journal of Computing* 2/1, 4-32.

Goldberg, D. E., 1989, *Genetic Algorithms in Search, Optimization, and Machine Learning*, Reading, MA: Addison-Wesley

Gupta, J.N.D., 1971, A functional heuristic algorithm for the flow-shop scheduling problem, *Operational Research Quarterly* 22, 39-47.

Hariri, A. M. A., 1981, Scheduling using branch and bound techniques, *Ph.D. thesis, University of Keele, Keele*, 1981

Ho, Y.C., and Chang, Y-L., 1991, A new heuristic method for the n job, m - machine flow-shop problem, *European Journal of Operational Research.* 52, 194-202

Hundal, T.S., and Rajagopal, J., 1988, An extension of Palmers' heuristic for the flow-shop scheduling problem, *International Journal of Production Research* 26, 1119-1124.

Kirkpatrick, S., Gelatt, C. D., and Vechhi, M. P., 1983, Optimization by simulated annealing, *Science*, 220 (4568), 671-680.

Ignall, E., and Linus, S., 1965, Application of the Branch and Bound technique to some flowshop scheduling problems. *Operations Research*, 13, 400-412

Lampinen, J. and Zelinka, I., 1999, Mechanical engineering design optimization by differential evolution, *New Ideas in Optimization,* (Eds.) Corne, D., Dorigo, M., and Glover, McGraw Hill, International (UK*)*, 127-146.

Lawler, E. L., Lensta, J. K., Rinnooy Kan, A. H. G., and Shmoys, D. B., 1995, Sequencing and scheduling: algorithms and complexity. *In Logistics of Production and Inventory*, Graves, S. C., Rinnooy Kan, A. H. G., and Zipkin, P. H., (eds.). (Amsterdam, The Netherlands: North Holland), 445-522.

Ogbu, F. A., and Smith, D. K., 1990, The application of the simulated annealing algorithm to the solution of the $n/m/C_{max}$ flowshop problem, Computers in Operations Research, 17 (3), 243-253.

Onwubolu, G. C., and Mutingi, M., 1999, Genetic algorithm for minimizing tardiness in flow-shop scheduling, *Production Planning & Control*, 10 (15), 462-471.

Onwubolu, G. C., 2000, Ants can schedule, *Industrial Engineering Research Conference*, Cleveland Ohio: USA, May 22-24, 2000, In CDROM.

Onwubolu, G. C., 2001, Optimization using differential evolution, *Institute of Applied Science Technical Report, TR-2001/05.*

Palmer, D.S., 1965, Sequencing jobs through a multi-stage process in the minimum total time - A quick method of obtaining a near optimum, *Operational Research Quarterly* 16, 101-107

Price, K., 1999, An introduction to differential evolution. *New Ideas in Optimization*, (Eds.) Corne, D., Dorigo, M., and Glover, McGraw Hill, International (UK), 79-108.

Price, K., and Storn, R., 2001, Differential evolution homepage (Web site of Price and Storm) as at 2001. http://www.ICSI.Berkeley.edu/~storn/code.html

Storn, R., 1996, On the usage of differential evolution for function optimization, *NAFIPS*, 1996, Berkeley, 519 - 523.

Storn, R. and Price, K., 1995, Differential evolution - a simple and efficient adaptive scheme for global optimization over continuous spaces, *Technical Report TR-95-012, ICSI, March 1999*
(Available via ftp from ftp.icsi.berkeley.edu/pub/techreports/1995/tr-95-012.ps.Z).

Storn, R. and Price, K., 1997, Differential evolution – a simple evolution strategy for fast optimization. *Dr. Dobb's Journal*, April 1997, 18–24 and 78.

Storn, R., 1999, Designing digital filters with differential evolution, *New Ideas in Optimization*, (Eds.) Corne, D., Dorigo, M., and Glover, McGraw Hill, International (UK), 109-125.

Taillard, E., 1990, Some efficient heuristic methods for the flow shop sequencing problem, *European Journal of Operational Research*, 47, 65-74

Widmer, M., and Hertz, A., 1989, A new heuristic method for the flow shop sequencing problem, *European Journal of Operational Research*, 41, 186-193

25 Evaluation of Form Errors to Large Measurement Data Sets Using Scatter Search

Mu-Chen Chen and Kai-Ying Chen

25.1 Introduction

The automatic acquisition and interpretation of the information of product features is an important procedure in modern manufacturing systems. Roundness and sphericity are the basic geometric forms expected from part features. There is a requirement to develop an automatic inspection method that will cater the needs of assessing roundness and sphericity. To produce better quality parts, manufacturers can adequately control the production process and accurately evaluate the form errors. The assessment algorithms for form errors are typically applied to meet the intent of the ASME Y14.5M standard (1994). This standard identifies that a substitute feature (geometric form that best fits the measured points) must be established from the sampled data to minimize the maximum deviation between the substitute (reference) feature and the actual feature concerned.

For evaluating the roundness error from the actual measurement, a reference circle must be established from the measurement data to minimize the maximum deviation between the reference circle and the actual one. The roundness error is then defined as the maximum peak-to-valley distance from the reference circle. Sphericity is a type of form error, which is broadly used in industry for the geometrical measurement of precision balls. The sphericity is defined as the minimum shell width (minimum zone) between the concentric circumscribed and inscribed spheres enclosing all the measurement data. The roundness and sphericity assessments involve the optimization step to find the center coordinates and radii of the reference features. The evaluation of roundness and sphericity based on the minimum zone condition is a non-linear and non-convex problem that is difficult to solve mathematically. Therefore, the efficient and accurate assessment of roundness and sphericity is an important issue.

The data sampled from measuring instruments are treated by an evaluation algorithm to accurately estimate the form errors. There exists no precise mathematical definition of the constraints implied by tolerance specifications in the ASME

Y14.5M standard. Tolerance is not interpreted in agreement by different measuring systems, which leads to inconsistency between the requirements of tolerances and the results of software. Current evaluation algorithms are based on the least-squares solution, which minimizes the sum of the squared errors of the measured points from the reference feature. The least-squares solution is only capable of obtaining an approximate solution that does not guarantee to meet the minimum zone criterion. The least-squares solution can result in a possible overestimation of the form error. While evaluation algorithms successfully reject bad parts, they may also reject some good parts (Gass et al. 1998). An approximate method for calculating the least-squares center and radius is given in ANSI B89.3.1 (1972). Chen and Papadoupoulos (1996) compared the linear and non-linear least-squares fitting algorithm and concluded that there is no practical difference between the two algorithms. The method of least-squares is not necessarily optimal if the irregularity significantly departs from a Gaussian distribution (2000).

The least-squares circle (LSC) is the most widely used reference circle for the assessment of roundness error (Takiyama and Ono, 1989; Thomas and Chan, 1989; Chaudhuri and Kundu, 1993; Cooper 1993) due to its computational simplicity. However, the deviations obtained using LSC may be large. For example, the result of LSC can be larger than the actual roundness error by 20% (ANSI B89.3.1, 1972). For the roundness assessment, estimating the limacon circle in a 2D inspection plane is a linear programming problem (Chetwynd, 1979; Chetwynd and Phillipson, 1980; Carpinetti and Chetwynd, 1994). The major limitation of the limacon approximation is that the workpiece must be well centered for a roundness measuring instrument or the radial coordinate data representing the profile must be expressed relative to an origin that lies not far from the best fit center. The adequacy of the limacon circle is completely determined by the ratio of the center eccentricity to other radius.

For the sphericity assessment, Gass et al. (1998) developed a linear programming approach for determining reference spheres and minimum zone sphericity based on coordinate measuring machine (CMM) measurements of spherical parts. Kanada (1995) modeled the minimum zone sphere (MZS) as non-linear programs and applied simplex search to estimate the sphericity. Kanada compared the minimum zone method in which simplex search is applied to the iterative least-squares method. Observing from the results in Kanada, the minimum zone approach is comparably superior to the least-squares one. Fan and Lee (1999) evaluated the minimum zone sphericity based on a direct search method starting from the least-squares results. Using the 3D Voronoi diagram, Huang (1999) solved the minimum zone solution for sphericity problems. Chen and Liu (2000) evaluated sphericity errors by directly resolving the simultaneous linear algebraic equations.

Form fitting algorithms have become increasingly important in modern dimensional measurement systems. This is particularly true for coordinate measuring systems in which a large set of measurement data are sampled (Hopp 1993). The algorithms for form fitting which convert measured data to the reference geometry can be major source of error in a measurement system. Form fitting can be viewed as an optimization problem: find the parameters of reference geometry that minimize a particular fitting objective for a set of points.

Algorithms for estimating the reference circle and sphere involve constrained nonlinear constraints. The constrained optimization of geometric dimensioning and tolerancing (GD&T) attempts to preserve functionality from the design (Yeh and Ni 1996). The formulated constrained optimization models for reference features simulate functional gauging processes described in current standards. Engineers who are familiar with current GD&T standards can utilize the optimization models as easily as they use GD&T symbols.

Roundness and sphericity play an important role in industry, and their design standards can be found in ASME (1994). The evaluation of roundness and sphericity based on the minimum zone criterion is a non-linear and non-convex problem that is hard to handle in mathematics (Huang 1999). Assessment algorithms apply linear approximation approaches can give incorrect results to non-linear form-fitting problems (Phillips et al. 1993). By building the mathematical models for the circular and spherical form errors, Scatter Search (SS), which is a robust non-linear optimization method, can be introduced to assess roundness and sphericity. SS is comparably simple; it does not require differential information for the objective function, and it has a good convergence feature (Glover 1997, Glover 1999, Glover et al. 2000).

25.2 Mathematical Models for Roundness

25.2.1 Roundness

For evaluating the roundness error from the actual measurement, a reference circle must be established from the measurement data to minimize the maximum deviation between the reference circle and the actual one. The roundness error is then defined as the maximum peak-to-valley distance from the reference circle. Four reference circles are internationally accepted for roundness measurement (ASME Y14.5M 1994): (1) least-squares circles (LSC), (2) maximum inscribing circles (MIC), (3) minimum circumscribing circles (MCC), and (4) minimum zone circles (MZC).

The specification of geometric tolerances on the above basis is given in the standard (ASME Y14.5M 1994). However, it does not specify the methods by which the reference geometrical features are to be established. The LSC is the circle chosen so that the sum of the squares of the radial distance of all data points from the fitting circle is a minimum. Given a set of data points in two dimensions $\mathbf{M} = \{p_i(x_i, y_i),\ i = 1, 2, \cdots, N\}$ which represents the profile of a workpiece, it is possible to find explicitly the circle parameters by minimizing the least-squares errors between the given set of data points and the curve (Thomas and Chan 1989).

The normal deviation (e_i) between a data point (x_i, y_i) and the circle of radius R and center (x_c, y_c) is given by

$$e_i = \left[(x_i - x_c)^2 + (y_i - y_c)^2\right]^{1/2} - R \tag{25.1}$$

A close-form solution to the circle fitted by the MIC, MCC or MZC does not exist. However, for a well centered trace of the profile, the deviation can be approximated by a linear relation (Chetwynd 1979)

$$e_i' = r_i - (R + x_c \cos\theta_i + y_c \sin\theta_i) \tag{25.2}$$

where e_i' is the linear deviation and (r_i, θ_i) is the polar coordinates of the point (x_i, y_i). Eq. (25.2) is known as the limacon approximation. In the form of Eq. (25.2), the limacon is linear in its parameters (x_c, y_c, R).

Four roundness evaluation methods are suggested for roundness measurements. They are: least-squares circle (LSC) method, maximum inscribed circle (MIC) method, minimum circumscribed circle (MCC) method and minimum zone circle (MZC) method. The LSC method has been sufficiently described in the literature. The mathematical formalization of MIC, MCC and MZC methods can be restructured as follows (Chen et al. 1999, Chen 2000).

25.2.2 The maximum inscribed circle

The MIC method attempts to inscribe the given set of measured points, $p_i(x_i, y_i)$, $i = 1, 2, \cdots, N$, using a maximum circle, and then estimate its center (x_c, y_c) and radius R. Using the center point and radius of MIC, this approach can find the minimum circular zone, which all the data points are exterior to this circle. The mathematical model of MIC can be formulated as follows:

$$\begin{aligned} &\text{Maximize } R \\ &\text{Subject to: } \left[(x_i - x_c)^2 + (y_i - y_c)^2\right]^{1/2} - R \geq 0, \\ &\qquad i = 1, 2, \cdots, N; \\ &\qquad R_L \leq R \leq R_U;\ x_L \leq x_c \leq x_U;\ y_L \leq y_c \leq y_U \end{aligned} \tag{25.3}$$

where N is the number of measured points, R_L, x_L and y_L are the lower bounds of R, x_c and y_c; R_U, x_U and y_U are the upper bounds of R, x_c and y_c.

The roundness error for MIC, e_{MIC}, is then estimated by measuring the maximal peak-to-valley deviation between the measurement data and the reference circle. It can be expressed as

$$e_{MIC} = \max_{i=1,2,\ldots,N} \{e_i = [(x_i - x_c)^2 + (y_i - y_c)^2]^{1/2} - R\} \tag{25.4}$$

25.2.3 The minimum circumscribed circle

Unlike the MIC method, the MCC method attempts to circumscribe the given set of measured points, using a minimum circle. Using the center point and radius of MCC, this method can obtain the minimum circular zone, which encloses all the measured points. Mathematically, the model of MCC takes the following form

$$\text{Minimize } R \tag{25.5}$$
$$\text{Subject to: } \left[(x_i - x_c)^2 + (y_i - y_c)^2\right]^{1/2} - R \le 0,$$
$$i = 1,\ 2,\ \cdots,\ N;$$
$$R_L \le R \le R_U;\ x_L \le x_c \le x_U;\ y_L \le y_c \le y_U.$$

The roundness error for MCC, e_{MCC}, can be defined as

$$e_{MCC} = \max_{i=1,2,\ldots,N} \{e_i = R - [(x_i - x_c)^2 + (y_i - y_c)^2]^{1/2}\} \tag{25.6}$$

25.2.4 The minimum zone circle

The MZC method is proposed to comply with the ANSI tolerancing definition (ASME Y14.5M 1994) for form tolerances. For MZC method, a pair of concentric circles C_1 and C_2 with the minimum radius separation are determined, such that the measured points are contained between C_1 and C_2, where R_1 and R_2 are the radii of C_1 and C_2, respectively, as well as the center is positioned at a point (x_c, y_c). The mathematical model can be formulated as

$$\text{Minimizc } R_1 - R_2 \tag{25.7}$$
$$\text{Subject to: } \left[(x_i - x_c)^2 + (y_i - y_c)^2\right]^{/2} \le R_1,$$
$$i = 1,\ 2,\ \cdots,\ N;$$
$$\left[(x_i - x_c)^2 + (y_i - y_c)^2\right]^{1/2} \ge R_2,$$
$$i = 1,\ 2,\ \cdots,\ N;$$
$$R_1 - R_2 \ge 0;$$
$$R_L \le R_1,\ R_2 \le R_U;\ x_L \le x_c \le x_U;\ y_L \le y_c \le y_U.$$

The roundness error for MZC, e_{MZC}, can be simply expressed as

$$e_{MZC} = R_1 - R_2. \tag{25.8}$$

25.3 Mathematical Models for Sphericity

25.3.1 Sphericity

The sphericity error is determined by selecting a center point for the reference sphere, with the value being difference between the largest and smallest radii to the measured points (ASME Y14.5M 1994). To allow for presumed differences in function, four criteria for the substitute sphere fitting are internationally defined (ASME Y14.5M 1994). They are: (1) least-squares sphere (LSS); (2) maximum inscribing sphere (MIS); (3) minimum circumscribing sphere (MCS); and (4) minimum zone sphere (MZS).

A LSS is the sphere chosen so that the sum of the squares of the radial distance of all data points from the fitting sphere is a minimum. Given a set of data points in 3-D, $\mathbf{M}=\{p_i(x_i,y_i,z_i),\ i=1,\ 2,\ \cdots,\ N\}$ which represents the surface of a workpiece, it is possible to find the sphere parameters by minimizing the least-squares errors between the given set of data points and the surface. The normal deviation (e_i) between a data point (x_i,y_i,z_i) and the sphere of radius R and center (x_c,y_c,z_c) is given by

$$e_i=\left[(x_i-x_c)^2+(y_i-y_c)^2+(z_i-z_c)^2\right]^{1/2}-R \tag{25.9}$$

A shell is formed by the minimum inscribed sphere and the maximum circumscribed sphere having the least-squares center. Just about all CMMs compute the sphericity from LSS (Gass et al. 1998). However, the deviations obtained using LSS may be large.

For sphericity assessments, there exists a distinction between the specification of tolerances in design and the assessment methods in metrology community. By using the normal deviation defined in Eq. (25.9), the fitting methods for estimating substitute sphere of MIS, MCS and MZS involve nonlinear constraints. Provided that numerous measured points on the part surfaces are obtained, the fitting optimization models for MIS, MCS and MZS are non-linear programs with a huge set of constraints. These mathematical models hence have a high degree of computational complexity. To evaluate the sphericity error, a substitute sphere has to be established from the collected data. The mathematical models of MIS, MCS and MZS for sphericity assessments are formulated as follows (Chen 2002).

25.3.2 Maximum inscribed sphere

The MIS attempts to find a sphere which is the largest sphere being inscribed within the set of measured points, $\mathbf{M}=\{p_i(x_i,y_i,z_i),\ i=1,\ 2,\ \cdots,\ N\}$. The center coordinates (x_c,y_c,z_c) and radius R are then can be estimated. Using the center coordinates and radius of MIS, this approach can find the minimum shell width, which all the measured points are exterior to this sphere. Common practice seeks to maximize the reference radius R while maintaining the condition that the profile is disclose, that is, subject to the geometrical constraints. The constrained non-linear program of MIS takes the form as

$$\begin{aligned}
&\text{Maximize } R \\
&\text{Subject to: } \left[(x_i - x_c)^2 + (y_i - y_c)^2 + (z_i - z_c)^2\right]^{1/2} \geq R, \\
&\quad i = 1, 2, \cdots, N; \\
&\quad R_L \leq R \leq R_U; \\
&\quad x_L \leq x_c \leq x_U;\ y_L \leq y_c \leq y_U;\ z_L \leq z_c \leq z_U
\end{aligned} \tag{25.10}$$

where R_L, x_L, y_L and z_L are the lower bounds of R, x_c, y_c and z_c; R_U, x_U, y_U and z_U are the upper bounds.

The optimization algorithm resolves the MIS model to find the reference sphere. The sphericity error for MIS, e_{MIS}, is then estimated by measuring the maximal peak-to-valley deviation between the measurement data and the reference sphere. It can be expressed as

$$e_{MIS} = \max_{i=1,2,\ldots,N} \{e_i = [(x_i - x_c)^2 + (y_i - y_c)^2 + (z_i - z_c)^2]^{1/2} - R\}. \tag{25.11}$$

25.3.3 Minimum circumscribed sphere

The MCS attempts to find a sphere, which circumscribes the given set of measured points, using a smallest sphere. Using the center coordinates and radius of MCS, we can obtain the minimum shell width, which encloses all the measured points. Mathematically, the MCS model tries to minimize the reference radius R while maintaining the condition that the profile is enclosed. The mathematical programming model of MCS can be formulated as

$$\begin{aligned}
&\text{Minimize } R \\
&\text{Subject to: } \left[(x_i - x_c)^2 + (y_i - y_c)^2 + (z_i - z_c)^2\right]^{1/2} \leq R, \\
&\quad i = 1, 2, \cdots, N; \\
&\quad R_L \leq R \leq R_U; \\
&\quad x_L \leq x_c \leq x_U;\ y_L \leq y_c \leq y_U;\ z_L \leq z_c \leq z_U.
\end{aligned} \tag{25.12}$$

Similarly, the sphericity error for MCS, e_{MCS}, can be defined as

$$e_{MCS} = \max_{i=1,2,\ldots,N} \{e_i = R - [(x_i - x_c)^2 + (y_i - y_c)^2 + (z_i - z_c)^2]^{1/2}\} \tag{25.13}$$

25.3.4 Minimum zone sphere

The MZS method is proposed to comply with the ANSI tolerancing definition (ASME Y14.5M 1994) for form tolerances. The MZS defines a pair of concentric spheres, a circumscribed sphere S_1 and an inscribed sphere S_2 that just contains the measured points with a minimum shell width. The mathematical problem of finding the minimum zone can be formulated as a constrained non-linear program. This is a rather complicated non-linear problem, and exact algorithms for solving it are not readily available (Gass et al. 1998). The center of this pair of concentric spheres is positioned at a point (x_c, y_c, z_c). The mathematical model can be formulated as

$$\begin{aligned}
&\text{Minimize } R_1 - R_2 \qquad (25.14)\\
&\text{Subject to: } \left[(x_i - x_c)^2 + (y_i - y_c)^2 + (z_i - z_c)^2\right]^{1/2} \le R_1,\\
&\qquad i = 1,\ 2,\ \cdots,\ N;\\
&\qquad \left[(x_i - x_c)^2 + (y_i - y_c)^2 + (z_i - z_c)^2\right]^{1/2} \ge R_2,\\
&\qquad i = 1,\ 2,\ \cdots,\ N;\\
&\qquad R_1 - R_2 \ge 0;\ R_L \le R_1,\ \ R_2 \le R_U;\\
&\qquad x_L \le x_c \le x_U;\ y_L \le y_c \le y_U;\ z_L \le z_c \le z_U
\end{aligned}$$

where R_1 and R_2 are the radii of S_1 and S_2, respectively.

The sphericity error for MZS, e_{MZS}, is the minimum shell zone. It can be simply expressed as

$$e_{MZS} = R_1 - R_2 . \qquad (25.15)$$

After the form error has been obtained, one can check the conformance of measured features and design specifications, and then make the acceptance/ rejection decision for the inspected part.

25.4 Scatter Search

25.4.1 Overview of scatter search

In this section, a general description of scatter search (SS) is presented. For a more detailed description of the SS methodology, readers are referred to references (Glover 1997, Glover 1999, Glover et al. 2000, Laguna and Marti 2003). SS provides a way of considerably improving the performance of simple heuristic procedures. The search strategies proposed by SS result in iterative-procedures with the ability to escape local optimal points. Metaheuristics such as Simulated Annealing and Tabu Search typically maintain only one solution by applying mechanisms to change this solution iteratively. On the other hand, Genetic Algorithms (GAs) and SS are search approaches designed to operate on a set of solutions that is maintained from iteration to iteration.

SS operates with a population of solutions, rather than with a single solution at a time, and applies the solution combination procedures to generate new ones by exploiting the combinations of solutions (Glover 1997). One of the main distinguished features of SS is its close connection with the Tabu Search, and it therefore can benefit by incorporating special forms of adaptive memory (Glover and Laguna 1997). SS operates on a population of solutions, namely *reference set*, by combining these solutions to generate new ones. A new solution is created from the linear combination of two or more other solutions such that the reference set evolves. The process of SS is designed to (1) capture information not contained separately in the original vectors, (2) take advantage of auxiliary heuristic solution methods (to evaluate the combinations produced and to actively generate new vectors), and (3) make dedicated use of strategy instead of randomization to carry out component steps (Glover et al. 2000).

The SS is designed to maintain a set of reference points. New reference points are added by applying the linear combination method. The SS approach may be sketched in basic outline as follows (Glover 1997):

1. An initial set of solutions is generated to guarantee a significant level of diversity, and sequentially heuristic processes designed for the optimization problem are applied to improve these solutions. Particularly, a solution may be added to the reference set if the diversity of the set improves even when the objective value of the solution is inferior to other solutions competing for admission into the reference set.
2. New solutions are generated such that they consist of structured combinations of subsets of the current reference solutions. The structured combinations include creating solutions both inside and outside the convex regions spanned by the reference solutions, and modifying these solutions to yield acceptable solutions.
3. The designated heuristic processes are applied to further improve the newly generated solutions. These heuristic processes must be able to operate on infeasible solutions and may or may not yield feasible solutions.
4. A set of the "best" improved solutions is added into the reference set. The search procedure is repeated until the reference set does not change. The notion of "best" is not limited to a measure given exclusively by the evaluation of the objective function. Particularly, a solution may be added to the reference set if the diversity of the set improves.

25.4.2 Scatter search template

The mechanisms within SS are not restricted to a single standardized design. The operations that take place within each of the methods in the framework of SS have been illustrated above. The following notations are used in SS.

P	= the set of solutions generated by the *Diversification Generation Method*;
RefSet	= the set of solutions in the reference set;
$RefSet_1$	= the subset of the reference set that contains the best solutions as measured by the objective function value;
$RefSet_2$	= the subset of the reference set that contains the diverse solutions;
Psize	= the size of the set of diverse solutions generated by the *Diversification Generation Method*;
b	= the size of the reference set;
b_1	= the size of the high quality subset;
b_2	= the size of the diverse subset;
MaxIter	= maximum number of iterations.

The template including five methods for implementing SS is described as follows. Specific processes for carrying out these methods are described in Glover (1997) and Laguna and Marti (2003).

1. *Diversification Generation Method*: It generates a collection of diverse trial solutions by using an arbitrary trial solution (or seed solution) as an input. The diversification generator is used at the beginning of the search to generate a set of solutions P, the cardinality of the set *PSize*, generally set at max (100, 5b) diverse solutions in which b is the size of the reference set.
2. *Improvement Method*: It transforms a trial solution into one or more improved trial solutions. This method must be able to handle starting solutions that are either feasible or infeasible. The Nelder and Mead's simplex search (Nelder and Mead 1965) is recommended as the *Improvement Method* by Glover (Glover et al. 2000).
3. *Reference Set Update Method*: This method is used to build and maintain a *reference set* consisting of the b "best" solutions found in which the value of b is typically small (e.g., no more than 20). The reference set is organized to provide efficient accessing by other parts of the method. The reference set *RefSet* consists of both high quality solutions ($RefSet_1$ with size b_1) and diverse solutions ($RefSet_2$ with size b_2). The reference solutions are used to generate new solutions. The construction of initial reference set begins with the selection of the best b_1 solutions from P. For each improved solution in P – *RefSet*, the minimum of the Euclidean distances, $d(x, y)$, to the solutions in *RefSet* is computed. Then, the solution with the maximum of these minimum distances,

$d_{\min}(x) = \min_{y \in RefSet} \{d(x, y)\}$, is selected. The reference set is dynamically updated during the application of the *Solution Combination Method.* A newly created solution may gain membership to the reference set according to their quality (the new solution has a better objective function value) or their diversity (the new solution has a worse $d_{\min}(x)$).

4. *Subset Generation Method*: This method operates on the reference set to produce a subset of solutions as a basis for creating solutions with *Solution Combination Method.* Typically, this method generates the following types of subsets: all 2-element subsets, 3-element subsets, 4-element subsets and the subsets consisting of the best *i* elements, for *i* = 5 to *b*.
5. *Solution Combination Method:* This method transforms a given subset of solutions produced by the *Subset Generation Method* into one or more combined solution vectors. In general, the *Solution Combination Method* is problem-specific due to the various solution representations. For the non-linear optimization problem, Glover and Laguna (1997) suggest the linear combination approach. Three types of combinations are utilized herein, assuming that the reference solutions are x_1 and x_2:

   C1: $x = x_1 - d$;

   C2: $x = x_1 + d$;

   C3: $x = x_2 + d$;

   where $d = r\dfrac{x_2 - x_1}{2}$ and *r* is a random number in the range (0, 1). The following rules are applied to create solutions with these three types of linear combinations:

   - If both x_1 and x_2 are elements of *RefSet*$_1$, then generate 4 solutions by applying C1 and C3 once and C2 twice.
   - If only one of x_1 and x_2 is a member of *RefSet*$_1$, then generate 3 solutions by applying C1, C2 and C3 once.
   - If neither x_1 nor x_2 is a member of *RefSet*$_1$, then generate 2 solutions by applying C2 once and randomly choosing between applying C1 or C3.

SS is designed to maintain a set of reference points. New reference points are added by applying the linear combination method. The solution procedure of SS continues in a loop that consists of applying the *Solution Combination Method* followed by the *Improvement Method* and the *Reference Update Method.* The procedure terminates when the reference set does not change and all the subsets have already been subjected to the *Solution Combination Method.* At this point, the *Diversification Generation Method* is used to construct a new *RefSet*$_2$ and the search restart.

25.4.3 The scatter search procedure

By implementing the above SS template, the overall view of the procedure is now can be introduced as follows (Glover et al. 2000):

Step 1. (Initial solution generation) Create one or more initial solutions in a random manner to initiate the search procedure.

Step 2. (Diversification generation) Apply the *Diversification Generation Method* to create diverse trial solutions from the initial solutions.

Step 3. (Solution improvement and reference set update) For each trial solution created in Step 2, generate one or more enhanced trial solutions by the *Improvement Method.* During successive applications of this step, maintain and update a reference set consisting of the *b* best solutions by the *Reference Set Update Method.*

Step 4. (Repeat) Execute Steps 2 and 3 until generating some designated total number of enhanced trial solutions as a source of candidates for the reference set.

Step 5. (Subset generation) Generate subsets of the reference set as a basis for creating combined solutions.

Step 6. (Solution combination) For each subset *X* generated in Step 5, generate a set *C*(*X*) that consists of one or more combined solutions by the *Solution Combination Method.* Treat each member of *C*(*X*) as a trial solution for the following step.

Step 7. (Solution improvement and reference set update) For each solution generated in Step 6, generate one or more enhanced trial solutions by the *Improvement Method*, while continuing to maintain and update the reference set.

Step 8. (Repeat) Executes Steps 5-7 in repeated sequence, until the specified maximum number of iterations *MaxIter* is reached or the reference set does not change.

According to the above procedure steps, SS contrasts with GAs by providing unifying principles for joining solutions based on generalized path constructions in Euclidean space and by utilizing strategic designs in which GAs resort to randomization. Additionally, SS provides intensification and diversification mechanisms that exploit adaptive memory, drawing on foundations that link SS to Tabu search. There exist significant differences between classical GAs implementations and SS. While classical GAs heavily rely on randomization and a somewhat limiting operation to create new solutions (e.g., one-point crossover on binary strings), SS employs strategic choices and memory along with (convex and non-convex) linear combinations of solutions to create new solutions. SS explicitly encourage the use of additional heuristics to process selected reference points in search for improved solutions. This is advantageous in settings where heuristics that exploit the problem structure can either be developed or are already available. SS has been proved highly successful for solving a diverse array of complex optimization problems (Glover 1999, Glover et al. 2000).

The algorithm-specific parameters of SS are set with regarding to the suggestion by Glover (1999) and Glover et al. (2000), and they are as follows: *PSize* (the size of the reference set) = 100, b_1 (the size of the reference set) = 10, b_2 (the size of the diverse subset) = 10, and *MaxIter* (maximum number of iterations) = 20. The SS-based evaluation approach for roundness and sphericity is implemented with the C codes presented in Laguna and Marti (2003). All experiments of the SS-based evaluation approach are run on an IBM compatible PC with a Pentium III 800 MHz processor.

25.5 Computational Experience

25.5.1 Roundness measurement

The proposed SS-based algorithm is tested by using a real part for the roundness measurement (Chen 2000). The real data of an oil seal (refer to Fig. 25.1) are obtained by a machine vision system. The machine vision techniques for inspection are gaining recognition as the trend in industry. The development of a machine vision system for inspection has received considerable attention (Chin 1988, Harding, 1996). The advantages of using machine vision include a decrease in the time required for measurement as well as the greater accuracy of measurement and better flexibility than the conventional methods. Furthermore, machine vision systems can provide a non-contact measurement process of 100% inspection for measuring a wide class of objects in small-batch and mass production.

The machine vision system adopted herein for the roundness inspection consists of one TOSHIBA CCD camera, one FF1 DSP frame grabber, one digitizer and one lighting mechanism. The contour of tested object is used to test the proposed SS-based algorithm for MIC, MCC and MZC under the conditions of large data sets. The material of the oil seal is rubber. The size of the inspected oil seal is small for investigating the viability of the machine vision-based system and the SS-based evaluation algorithms.

The computational results of the real object are summarized in Table 25.1. Due to the measurement data of the machine vision are given in pixel (integer value), the computational results of real parts are also presented in pixel. However, the results (roundness error and radius) can be easily transformed into units in mm. The results shown in Table 25.1 indicate that SS compares favorably with GAs (Chen 2000). The CPU time is between 8-11 seconds for the real object with large boundary points (say 413 sampled points). The required computational time indicates the propriety of the proposed SS-based approach for MIC and MCC and MZC.

25.5.2 Sphericity measurement

The proposed SS-based evaluation approach for sphericity measurement is tested using a CMM measurement data set taken from Huang (1999). This data set is illustrated in Table 25.2. Since manufacturers have realized the utilities and economic advantages of geometric dimensioning and tolerancing (GD&T), CMMs are widely used in manufacturing to verify and control the dimensional accuracy of the manufactured parts. Through an assortment of mechanisms, CMMs collect the measurement data sets, and the form errors are then estimated. The precision of a CMM in measurement depends not only on its design and construction, but also the evaluation algorithms used to deal with the sampled data.

Fig. 25.1. The oil seal.

Table 25.1. The results for roundness measurements.

		SS	GA [a]
MIC	R	63.92	63.87
	(x_c, y_c)	(266.91, 184.48)	(266.87, 184.03)
	e	2.464	2.796
	T	8.4	11.0
MCC	R	65.93	66.04
	(x_c, y_c)	(266.11, 184.38)	(266.16, 184.33)
	e	2.812	2.872
	T	8.3	11.4
MZC	R_1	66.36	65.98
	R_2	63.91	63.27
	(x_c, y_c)	(266.91, 184.46)	(266.28, 184.44)
	e	2.454	2.711
	T	10.8	14.4

[a] The CPU times (T) of GA are based on an IBM compatible PC with a Pentium II processor.

CMMs have emerged to be important inspection tools owing to the recent advancements in numerical control and precision machining. However, CMMs are slowed down by three difficult problems (Walker 1988):

1. Develop suitable measurement techniques to obtain the data points, which accurately represent the parts being inspected.
2. Correctly and unambiguously interpret the definition of tolerances given in ASME Y14.5M standard, and formulate the problems precisely and systematically as optimization problems.
3. Develop evaluation algorithms which are consistent with ASME Y14.5M standard, highly efficient, robust, and easy to use.

The results are summarized in Table 25.3. As shown in this table, the SS-based approach performs equal or comparably better than GAs reported in Huang (1999) and Chen (2002). The CPU times of SS are between 2.6-3.4 seconds for these examples. Based on the experimental results described above, the proposed SS-based optimization algorithm is effective for sphericity measurements in CMMs. Additionally, the evaluation algorithms developed by previous studies (e.g., Chen and Liu 2000, Fan and Lee 1999, Huang 1999) are only suited to MZS, and not to MIS and MCS. However, the SS-based evaluation approach is a general method for MIS, MCS and MZS.

25.6 Summary

With the advent of low-cost computational hardware, the modern measuring systems such as machine visions and coordinate measuring machines (CMMs) have emerged as the financially feasible devices in automated inspection. The measurement data set must be analyzed and condensed to yield the desired parameters to fit the geometry. The coordinate measuring systems can potentially generate a large set of boundary coordinates. As the functionality of products becomes more complicated and their tolerance becomes more rigid, speed and accuracy of measurement methods grow to be an essential basis in manufacturing industries. The measurement errors depend on many factors including systematic and pseudo-random machine error, thermal error, form error, surface finish, evaluation algorithm correctness, sampling method and sampling density. The effect of these errors on the measurements can be minimized by cautiously modeling the process leading to errors (Yang et al. 1999). Form fitting algorithms have become increasingly important in modern dimensional measurement systems, which generate large measurement data sets. SS has been demonstrated highly successful for solving a diverse array of complex optimization problems. It is a population-based approach that shares features with the evolutionary methods. A form error evaluation approach based on SS is a viable tool for roundness and sphericity. The results reported in the experimental study show the effectiveness of SS in evaluating form errors.

Table 25.2. Data set for sphericity assessment.

No.	X	Y	Z	No.	X	Y	Z
1	8.42055	3.46726	1.19323	51	4.80000	6.40000	6.00000
2	-2.12100	-8.61568	2.36505	52	-1.36655	-2.55289	8.57942
3	0.86853	5.11876	8.40479	53	1.89727	-4.47838	-8.04041
4	3.17803	1.82449	9.29814	54	-7.94820	4.58889	-1.09601
5	-4.75718	8.23602	-0.01232	55	-2.73783	1.44044	8.67116
6	-0.00288	-0.00001	-9.22020	56	7.26417	2.57809	-5.19624
7	8.02534	2.43064	-3.34796	57	6.73556	3.02284	-6.53863
8	-0.77363	-1.36294	9.73536	58	-3.23473	7.59841	-4.09855
9	-0.42047	-0.05775	9.89921	59	-7.91122	-3.08604	-3.61944
10	-2.48279	-1.09659	9.20123	60	-0.94722	1.19895	8.94372
11	-1.32236	-0.33044	8.94394	61	4.53577	-5.59322	-5.70760
12	-8.65863	-2.71179	2.06558	62	-4.80000	6.40000	6.00000
13	6.94724	4.54788	4.56109	63	-6.29337	4.51725	5.89396
14	1.23285	-4.10410	8.21792	64	-1.84698	-0.03353	9.58379
15	-4.98020	-8.66566	-0.00066	65	2.31123	-2.20250	8.46912
16	9.55234	0.02726	-0.01951	66	6.79260	5.21214	-2.80259
17	0.01829	0.00006	9.63208	67	1.47478	5.43919	7.04760
18	2.18897	8.46413	3.92633	68	0.16864	-5.92638	7.31883
19	-3.91959	8.01503	-1.25551	69	8.77527	2.10675	1.40999
20	2.35189	-6.47936	-5.90072	70	0.00000	-9.00000	0.00000
21	-4.93565	-6.79583	3.45841	71	-3.76405	-3.05133	-7.83327
22	6.91250	2.39367	5.84698	72	0.49092	9.03545	0.68276
23	7.54898	3.28552	4.06809	73	0.75328	-6.19756	6.88526
24	3.21574	8.02367	4.36928	74	-6.04510	7.58341	-0.86041
25	5.43679	-5.10013	5.28983	75	0.32479	9.39478	-2.96664
26	2.93158	2.45466	-8.24089	76	8.91620	1.36278	-0.15665
27	-3.06343	4.81048	-7.94254	77	-8.40008	1.03735	3.90549
28	7.24483	4.85365	-4.62014	78	-1.19949	-9.61640	-2.36150
29	-3.57018	1.37046	8.96134	79	0.00000	0.00000	-10.0000
30	-0.00414	0.11571	9.21327	80	3.05448	3.63890	7.67445
31	-5.50164	7.07481	3.52308	81	0.23151	-0.47424	9.12475
32	3.59470	3.35799	-8.69633	82	-5.67083	-7.29763	-3.60701
33	5.77727	7.19318	3.81492	83	-5.54879	2.48093	6.93936
34	1.09155	5.21130	-7.36348	84	-1.67709	-6.77174	5.92368
35	1.05627	9.86747	-0.86997	85	-4.66332	1.19213	8.29668
36	1.69428	6.09696	7.46172	86	1.82616	8.61360	3.44446
37	0.33632	-6.18998	-6.86310	87	-0.98304	-3.28498	-9.05523
38	-1.36144	-7.06321	5.74638	88	-6.80116	5.56668	3.16158
39	1.21537	-3.61139	9.02052	89	2.66568	-3.94312	7.86528
40	2.67713	-4.21685	-8.13177	90	-1.78909	2.91725	-8.51074
41	-8.50578	-2.75549	-2.48880	91	1.94884	2.94550	-9.05638
42	4.36823	-3.35307	7.41081	92	-0.67512	0.22825	9.94209
43	-3.40167	-6.65031	-5.24305	93	6.10308	5.99957	-3.17171
44	0.77619	1.66759	9.25988	94	-5.38958	3.65589	6.78410
45	6.40027	-6.67380	-3.49152	95	0.00000	9.00000	0.00000
46	-4.47850	-7.55166	-2.62989	96	2.06263	-8.17541	3.89862
47	-0.01253	0.03167	9.51814	97	-0.04665	-0.34500	9.67874
48	-4.24576	-8.30408	0.54593	98	7.15160	6.27399	2.33940
49	9.21459	0.78511	-0.27609	99	-5.54207	6.67783	2.70456
50	4.23950	-8.49940	1.74151	100	3.74134	-7.56337	3.36648

Table 25.3. The results for sphericity measurements.

		SS [a]	GA [a]	Huang's method
MIS	R	9.0000	9.0000	
	(x_c, y_c, z_c)	(-0.0146, 0.0000, -0.0035)	(0.0009, 0.0000, 0.0011)	
	e	1.0090	1.0003	
	T	2.6	1.3	
MCS	R	9.9800	10.0000	
	(x_c, y_c, z_c)	(-0.0092, 0.0069, -0.0010)	(0.0000, 0.0000, 0.0000)	
	e	0.9868	1.0004	
	T	2.7	1.5	
MZS	R_1	8.9938	10.0000	
	R_2	9.9800	9.0000	
	(x_c, y_c, z_c)	(-0.0046, -0.0002, -0.0086)	(0.0000, 0.0000, 0.0000)	
	e	0.9862	1.0000	1.0000
	T	3.4	2.0	

[a] The CPU times (T) of both SS and GA are based on an IBM compatible PC with a Pentium III processor.

References

ANSI B89.3.1 (1972) Measurement of out of Roundness. American Society of Mechanical Engineers, New York

ASME Y14.5M (1994) Dimensioning and Tolerancing. American Society of Mechanical Engineers, New York

Carpinetti LCR, Chetwynd DG (1994) A new strategy for inspecting roundness features. Precision Engineering: 16 283-289

Chaudhuri BB, Kundu P (1993) Optimum circular fit to weighted data in multidimensional space. Pattern Recognition Letters: 14 1-6

Chen M-C (2000) Roundness inspection strategies for machine visions using nonlinear programs and genetic algorithms. International Journal of Production Research: 38 2967-2988

Chen M-C 2002 Analysis of spherical form errors to coordinate measuring machine data. JSME International Journal Series C: 45 647-656

Chen C-K, Liu C-H (2000) A study on analyzing the problem of the spherical form error. Precision Engineering: 24 119-126

Chen KW, Papadoupoulos AS (1996) Comparison of linear least-squares and nonlinear least-squares spheres. Microelectronics Reliability: 36 37-46

Chen M-C, Tsai D-M, Tseng H-Y (1999) A stochastic optimization approach for roundness measurements. Pattern Recognition Letters: 20 207-219

Chetwynd DG (1979) Roundness measurement using limacons. Precision Engineering: 1 137-141

Chetwynd DG, Phillipson PH (1980) An investigation of reference criteria used in roundness measurement. Journal of Physics E: Scientific Instruments: 13 530-538

Chin RT (1988) Survey of automated visual inspection: 1981 to 1987. Computer Vision, Graphics and Image Processing: 41 346-381

Cooper LD (1993) Circle fitting by linear and nonlinear least squares. Journal of Optimization Theory and Applications: 76 381-388.

Fan K-C, Lee J-C (1999) Analysis of minimum zone sphericity error using minimum potential energy theory. Precision Engineering: 23 65-72

Gass SI, Witzgall C, Harary HH (1998) Fitting circles and spheres to coordinate measuring machine data. The International Journal of Flexible Manufacturing Systems:10 5-25

Glover F (1997) A template for scatter search and path relinking. In: Hao JK, Lutton E, Ronald E, Schoenauer M, Snyers D (eds.) Lecture Notes in Computer Science 1363, Springer-Verlag, Berlin, pp 13-54

Glover F (1999) Scatter search and path relinking. In: Corne D, Dorigo M, Glover F (eds.) New Ideas in Optimization, Wiley, New York

Glover F, Laguna M, Marti R (2000) Fundamentals of scatter search and path relinking. Control and Cybernetics: 39 653-684

Glover F, Laguna M (1997) Tabu Search, New York, Kluwer Academic Publishers

Harding K (1996) Machine vision based gaging of manufactured parts. Vision, MVA/SME quarterly: 12 1-6

Hopp TH (1993) Computational metrology. Manufacturing Review: 6 295-304

Huang J (1999) An exact minimum zone solution for sphericity evaluation. Computer-Aided Design: 31 845-853

Kanada T (1995) Evaluation of spherical form errors-computation of sphericity by means of minimum zone method and some examinations with using simulated data. Precision Engineering,: 17 281-289

Laguna M, Marti R (2003) Scatter Search, Kluwer Academic Publishers, Boston.

Nelder JA, Mead R (1965) A Simplex method for function minimization. Computer Journal: 7 308-313

Phillips SD, Borchardt B, Gaskey G (1993) Measurement uncertainty considerations for coordinate measuring machines. NISTIR 5170, NIST, Gaithersburg, MD

Takiyama R, Ono N (1989) A least square error estimation of the center and radii of concentric arcs. Pattern Recognition Letters: 10 237-242

Thomas SM, Chan YT (1989) A simple approach for estimation of circular arc center and its radius. Computer Vision, Graphics and Image Processing: 45 362-370

Walker R (1988) GIDEP Alert No. X1-A-88-01. Technical Report, Government-Industry Data Exchange Program, August 22

Yang CC, Marefat MM, Ciarallo FW (1999) Modeling errors for dimensional inspection using active vision. Robotics and Computer-Integrated Manufacturing: 15 23-37

26 Mechanical engineering problem optimization by SOMA

Ivan Zelinka and Jouni Lampinen

26.1 Mechanical engineering problem optimization by SOMA

To discover the effectiveness of the techniques just proposed in Chapter 7, three numerical examples were optimized using SOMA (Table 26.1). These non-linear, engineering design optimization problems with discrete, integer and continuous variables were first investigated by Eric Sandgren [1] and subsequently by many other researchers [2], [3], [4], [5], [6], [7], [8], [9], [10], [11] and [12] who applied a variety of optimization techniques (Table 26.2). These problems represent optimization situations involving discrete, integer and continuous variables that are similar to those encountered in everyday mechanical engineering design tasks. Because the problems are clearly defined and relatively easy to understand, they form a suitable basis for comparing alternative optimization methods

Table 26.1. Test problems

Summary of Test Problems					
Example	Description	Number of variables			
		Total	Discrete	Integer	Continuous
1	Gear Train	4	0	4	0
2	Pressure Vessel	4	2	0	2
3	Coil Spring	3	1	1	1

Table 26.2. Alternative methods used to solve the test problems.

Compared Methods		
Reported by	Solution technique	Reference
Sandgren	Branch & Bound using Sequential Quadratic Programming	[1]
Fu, Fenton & Gleghorn	Integer-Discrete-Continuous Non-Linear Programming	[2]
Loh & Papalambros	Sequential Linearization Algorithm	[3], [4]
Zhang & Wang	Simulated Annealing	[5]
Chen & Tsao	Genetic Algorithm	[6]
Li & Chow	Non-Linear Mixed-Discrete Programming	[7]
Wu & Chow	Meta-Genetic Algorithm	[8]
Lin, Zhang & Wang	Modified Genetic Algorithm	[9]
Thierauf & Cai	Two-level Parallel Evolution Strategy	[10]
Cao & Wu	Evolutionary Programming	[11]
Lampinen & Zelinka	Differential Evolution	[12]
Zelinka &Lampinen	SOMA	This article

26.1.1 Designing a gear train

In the first example the problem is to optimize the gear ratio for the compound gear train arrangement shown in Fig. 26.1. The gear ratio for a reduction gear train is defined as the ratio of the angular velocity of the output shaft to that of the input shaft. In order to produce the desired overall gear ratio, the compound gear train is constructed out of two pairs of gearwheels, *d-a* and *b-f*. The overall gear ratio, i_{tot}, between the input and output shafts can be expressed as:

$$i_{tot} = \frac{\omega_o}{\omega_i} = \frac{z_d z_b}{z_a z_f} \tag{26.1}$$

Variables ω_o and ω_i are the angular velocities of the output and input shafts, respectively, and z denotes the number of teeth on each gearwheel.

The optimization problem is to find the number of teeth for gearwheels *d, a, b* and *f* in order to produce a gear ratio, i_{tot}, as close as possible to the target ratio: i_{trg} = 1/6.931 (= 0.1443). For each gear, the minimum number of teeth is 12 and the maximum is 60.

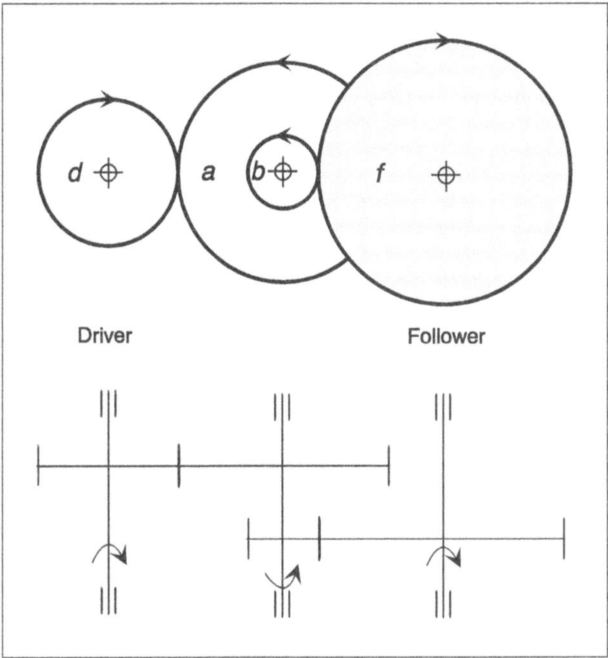

Fig. 26.1. Compound gear train for Example 1.

The problem is formulated as follows:

$$\begin{aligned}
&\text{Find} \\
&X = (x_1, x_2, x_3, x_4) = (z_d, z_b, z_a, z_f), \quad x \in \{12, 13, \ldots, 60\} \\
&\text{to minimize} \\
&f(X) = (i_{trg} - i_{tot})^2 = \left(\frac{1}{6.931} - \frac{x_1 x_2}{x_3 x_4}\right)^2 \\
&\text{subject to} \\
&12 \le x_i \le 60 \ , \qquad i = 1,2,3,4.
\end{aligned} \tag{26.2}$$

Thus, the goal is to find optimum values for four integer variables that will minimize the squared difference between the desired gear ratio, i_{trg}, and the current gear ratio, i_{tot}. For this problem, each variable is subject only to upper and lower boundary constraints.

The gear train problem was solved using the SOMA (see Table 26.4), i.e. by AllToOne and All ToAll strategies. The integer techniques described in Chapter 7 were invoked to handle boundary constraints. Because no constraint functions were involved, the objective (cost) function was simply defined to be the squared error, i.e., *f(X)*:

$$f_{cost}(X) = f(X) \tag{26.3}$$

Table 26.3 lists the various gear train solutions and compares DE's result with those reported in [1], [2], [4], [5], [9], [8] and [11]. In Table 26.4 are the results of the SOMA algorithm for mutual comparison.

Table 26.3. Optimal solutions for the gear train problem.

Item	Optimum solution				Type
	Sandgren [1]	Fu et al. [2]	Loh and Papalambros [4]	Zhang and Wang [5]	
x_1 (z_d)	18	14	19	30	integer
x_2 (z_b)	22	29	16	15	integer
x_3 (z_a)	45	47	42	52	integer
x_4 (z_f)	60	59	50	60	integer
f(x)	5.7×10^{-6}	4.5×10^{-6}	0.233×10^{-6}	2.36×10^{-9}	
Gear Ratio	0.146666	0.146411	0.144762	0.144231	
Item	Optimum solution				Type
	Lin et al. [9]	Wu and Chow [8]	Cao & Wu [11]	Lampinen & Zelinka [12] *	
x_1 (z_d)	19	19	30	16	integer
x_2 (z_b)	16	16	15	19	integer
x_3 (z_a)	49	43	52	43	integer
x_4 (z_f)	43	49	60	49	integer
f(x)	$\underline{2.7\times10^{-12}}$	$\underline{2.7\times10^{-12}}$	2.36×10^{-9}	2.7×10^{-12}	
Gear Ratio	0.144281	0.144281	0.144231	0.144281	

* Also alternative solutions with an equal target function value were obtained from run to run.

Table 26.4. Optimal solutions for the gear train problem by SOMA.

Item	Optimum solution by SOMA*				Type
	AllToOne PathLength=3, Step=0.3, PopSize =100, PRT=0.61, Migrations=20, MinDiv = negative Average cost value = 2.05×10^{-10}		AllToAll PathLength=3, Step=0.3, PopSize=100, PRT=0.61, Migrations=5, MinDiv = negative Average cost value = 2.7×10^{-12}		
	the worst case	the best case	the worst case	the best case	
x_1 (z_d)	24	19	16	19	integer
x_2 (z_b)	13	16	19	16	integer
x_3 (z_a)	47	43	49	43	integer
x_4 (z_f)	46	49	43	49	integer
f(x)	$9.9\text{x}10^{-10}$	2.7×10^{-12}	2.7×10^{-12}	2.7×10^{-12}	
Gear Ratio	0.14311	0.144281	0.144231	0.144281	

* Also alternative solutions with an equal target function value were obtained from run to run as in the case of DE, see [12].

Solutions obtained by SOMA in Table 26.6 were obtained after 100 simulations for each SOMA version. From Table 26.4 it is visible that the solution found by SOMA was equally as good as the best solution in the literature or in the DE solutions. In addition, it should be noted that SOMA as well as DE provided different results from run to run with the same objective function value (Table 26.5).

Table 26.5. Alternative solutions for the gear train problem found by SOMA.

Alternative solutions for gear train problem by SOMA					
Version	Solution	z_d	z_b	z_a	z_f
AllToOne	1	24	13	47	46
	2	13	24	46	47
AllToAll	1	19	16	43	49
	2	16	19	43	49

Trough close inspection Eq. (26.2), it is obvious that there are four global optima. Because SOMA as well as any evolutionary algorithm in general can work with a population of solutions rather than just a single solution, it is capable of finding multiple global optima for this problem. By using a sufficiently large population, it is possible to obtain all four alternative solutions in a single run. Despite that, only one solution was extracted from the population of the last generation because the other three solutions can be found in a trivial way based on one solution and the symmetry of Eq. (26.2). In practice, however, there exist optimization tasks with multiple global optima that cannot be detected so simply. One may wonder why finding a single, globally optimal solution is not always sufficient when multiple global optima exist. One reason is that the sensitivity of the objective function to small changes in its variables may be different at the alternative optimal points. In practice, it is often important to select the most robust solution, i.e., the global optima with the least sensitivity to noise. For example, if a machine design is subject to optimization, it is possible that the optimized design variables cannot effectively be manufactured. Alternatively it may be that the design parameters change during the lifetime of the machine due, for example, to normal wear of its components. In such cases, robust global optima are to be preferred over those that exhibit a high sensitivity to design implementation errors.

Results for SOMA were repeated 100 times (see Fig. 26.2) and are based on parameters of the SOMA algorithm (Table 26.4). All runs yielded the reported value of *f(X)*. As mentioned, different solutions were obtained from run to run because of the existence of multiple global optima (Table 26.5).

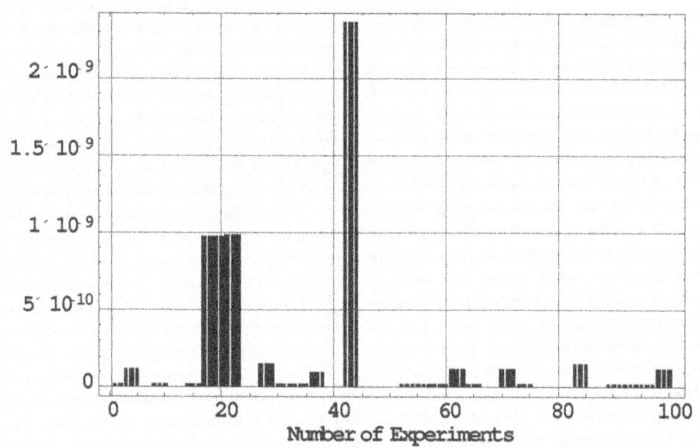

Fig. 26.3. Typical result of repeated optimization by SOMA (AllToOne)

26.1.2 Designing a pressure vessel

The second test problem is to design a compressed air storage tank with a working pressure of 3,000 psi and a minimum volume of 750 ft^3. As Fig. 26.3 shows, the cylindrical pressure vessel is capped at both ends by hemispherical heads. Using rolled steel plate, the shell is to be made in two halves that are joined by two longitudinal welds to form a cylinder. Each head is forged and then welded to the shell.

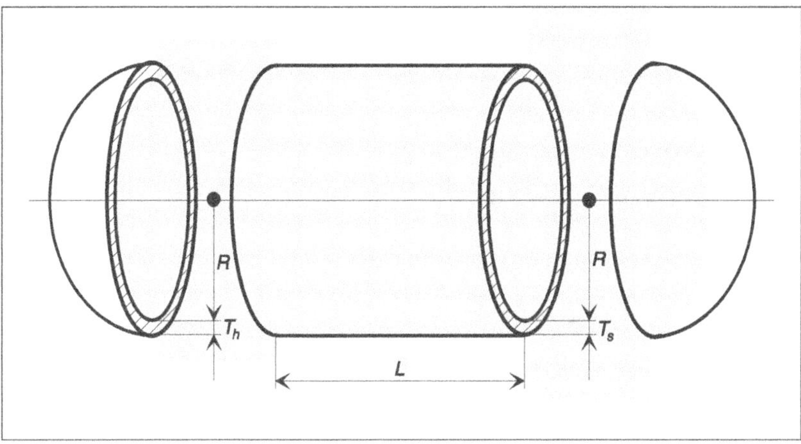

Fig. 26.3. Pressure vessel.

The objective is to minimize the manufacturing cost of the pressure vessel. The cost is a combination of the material cost, welding cost and forming cost. Refer to [1] for more details.

The design variables are shown in Fig 26.3. Variables L and R are continuous while T_s and T_h are discrete. The thickness of the shell, T_s, and the head, T_h, are both required to be standard sizes. For this example, steel plates were available in multiples of 0.0625 inch.

The problem can be formulated as follows:

Find

$$X = (x_1, x_2, x_3, x_4) = (T_s, T_h, R, L) \tag{26.4}$$

to minimize

$$f(X) = 0.6224x_1x_3x_4 + 1.7781x_2x_3^2 + 3.1611x_1^2x_4 + 19.84x_1^2x_3$$

subject to

$$g_1(X) = 0.0193x_3 - x_1 \le 0$$

$$g_2(X) = 0.00954x_3 - x_2 \le 0$$

$$g_3(X) = 750.0 \times 1728.0 - \pi x_3^2 x_4 - \frac{4}{3}\pi x_3^3 \le 0$$

$$g_4(X) = x_4 - 240.0 \le 0$$

$$g_5(X) = 1.1 - x_1 \le 0$$

$$g_6(X) = 0.6 - x_2 \le 0$$

The objective function, $f(X)$, represents the total manufacturing cost of the pressure vessel as a function of the design variables. The constraints, $g_1, \ldots, g_6$, quantify the restrictions to which the pressure vessel design must adhere. These limits arise from a variety of sources. For example, the minimal wall thickness of the shell, T_s (g_1), and heads, T_h (g_2), with respect to the shell radius are limited by the pressure vessel design code. The volume of the vessel must be at least the specified 750 ft^3 (g_3). Available rolling equipment limits the length of the shell, L, to no more than 20 feet (g_4). According to the pressure vessel design code, the thickness of the shell, T_s, is not to be less than 1.1 inches (g_5) and the thickness of the head, T_h, is not to be less than 0.6 inches (g_6).

The SOMA control variables used to solve the pressure vessel design problem are described in Table 26.8 (Case A), Table 26.10 (Case B) and Table 26.12 (Case C). The problem statements do not define the boundaries for the design variables, but the constraints, g_4, g_5 and g_6 are pure boundary constraints, so they were handled as lower boundary constraints for x_1 and x_2, and as an upper boundary constraint for x_4, respectively. The lower boundaries for x_3 and x_4 can be set to zero, since common sense demands that they must be non-negative values. The upper boundaries for x_1, x_2 and $x3$, however, must still be specified in order to define the search space. Consequently, these bounds were arbitrarily set high enough to make it highly probably that the global optimum lies inside of the defined search space. Since the possibility existed that the global optimum was outside of the initially defined search space, these estimated bounds were used only for initializing the population. SOMA was then allowed to extend the search beyond these boundaries. The possibility of using this kind of "loose" boundary constraint for variables are one of the advantages of modern evolutionary algorithms such as is for example DE. In practical engineering design work, it is not unusual for one or more boundaries to be unknown so that the distance to the optimum cannot be reliably estimated. The boundary constraints used for each variable are shown in Table 5. The other constraints, g_1, g_2 and g_3 were handled as constraint functions.

Table 26.6. Boundary constraints used for the pressure vessel example.

Boundary constraints for pressure vessel example		
Lower limitation	Constraint	Upper limitation
Constraint g_5	$1.1 \le x_1 \le 12.5$	Roughly guessed *
Constraint g_6	$0.6 \le x_2 \le 12.5$	Roughly guessed *
non-negative value of x_3	$0.0 \le x_3 \le 240.0$	Roughly guessed *
non-negative value of x_4	$0.0 \le x_4 \le 240.0$	Constraint g_4

* The value of this boundary is not given among the problem statements. Thus, the value is estimated roughly and used only for initialization of population. SOMA was allowed to extend the search beyond this limit.

The cost function for optimization was formulated as follows:

$$f_{\text{cost}}(X) = f(X) \cdot \prod_{i=1}^{3} c_i^2 \tag{26.5}$$

where

$$c_i = \begin{cases} 1.0 + s_i \cdot g_i(X) & \text{if} \quad g_i(X) > 0 \\ 1 & \text{otherwise} \end{cases}$$

$$s_1 = 1.0 \cdot 10^{10}, \, s_2 = s_3 = 1.0$$

Notice that it is not necessary to evaluate the constraint functions, g_4, g_5 and g_6, because they were handled as boundary constraints, and thus there was no reason to generate a candidate vector that violated any of them.

In researching this problem, at least three different formulations were found in the literature. To enable a more comprehensive comparison with the other methods, all three cases were solved using SOMA. Case A is an exception to the problem statements above, since all of the variables are treated as continuous. Table 26.8 contains results for SOMA (Case A) and can be compared with Table 26.7. Case B is formulated according to the original problem statements and results are in Table 26.10. Case C, reported in Table 26.12, also differs from the original problem statements. For some unknown reason, [7], [10] and [11] have used a slightly reformulated constraint-function, g_5:

$$g_5(X) = 1.0 - x_1 \le 0 \tag{26.6}$$

This modification extends the region of feasible solutions and also makes it possible to obtain a significantly lower objective function value. All of the solutions of [7], [10] and [11] lie in the extended region of the search space. Because of that, the results of [7], [10] and [11] cannot be fairly compared with the results obtained using Sandgren's original problem statements [1]. Thus, Cases A, B and C are not comparable because they represent differently formulated problems.

As Table 26.8, Table 26.10 and Table 26.12, show, SOMA, usually similar to DE, found a better solution for each pressure vessel design problem than the best solution found in the literature. Computations were based on parameters as mentioned in these tables and were repeated 100 times. In addition, all solutions were within the feasible design domain

Table 26.7. Optimal solutions for the pressure vessel problem. Case A: all variables are treated as continuous.

Item	Optimum solution, Case A			Type
	Sandgren [1]	Fu et al. [2]	Lampinen & Zelinka [12]	
x_1 (T_s) [inch]	1.1	1.100001	1.100000	continuous
x_2 (T_h) [inch]	0.6	0.600016	0.600000	continuous
x_3 (R) [inch]	47.7008	48.35145	56.99482	continuous
x_4 (L) [inch]	117.701	111.9893	51.00125	continuous
f(X) [$]	7867.0	7790.588	7019.031	

Table 26.8. Optimal solutions for the pressure vessel problem by SOMA. Case A: all variables are treated as continuous.

Item	Optimum solution by SOMA – Case A				Type
	AllToOne PathLength=3, Step=0.11, PopSize=20, PRT=0.1, Migrations=100, MinDiv = negative Average cost value = 7057.77		AllToAll PathLength=3, Step=0.11, PopSize=20, PRT=0.9, Migrations=20, MinDiv = negative Average cost value = 7030.4		
	the worst case	the best case	the worst case	the best case	
x_1 (T_s)	1.10019	1.10015	1.1791	1.1	cont.
x_2 (T_h)	0.600052	0.600001	0.6	0.6	cont.
x_3 (R)	54.5505	57.0024	61.0933	56.9948	cont.
x_4 (L)	65.923	50.9591	29.069	51.0012	cont.
f(x)	7199.75	7019.32	7098.15	7019.03	

Table 26.9. Optimal solutions for the pressure vessel problem. Case B: solved according to Sandgren's original problem statements [1].

Item	Optimum solution, Case B *					Type
	Sand-gren [1]	Fu et al. [2]	Loh and Pa-palambros [4]	Wu and Chow [8]	Lampinen & Zelinka [12]	
X_1 (T_s)	1.125	1.125	1.125	1.125	1.125	discrete
X_2 (T_h)	0.625	0.625	0.625	0.625	0.625	discrete
X_3 (R)	48.97	48.38070	58.2901 **	58.1978	58.29016	continuous
X_4 (L)	106.72	111.7449	43.693	44.2930	43.69266	continuous
f(x)	7982.5	8048.619	7197.734 **	7207.497	7197.729	

* In [5] and [9] it is reported that the value of f(X) = 7197.7 was reached. Because neither a more accurate result, nor details of the result were provided, they are not included in this comparison.

** No values for constraint functions were reported in [4]. Also, the optimum solution was too inaccurately reported to reconstruct the results properly. The reported value, $x_3 = 58.290$, causes a violation of constraint g_3. Because of that, 58.2901 was used here for reconstructing the constraint functions and target function values. In [4], a value of f(X) = 7197.7 was originally reported.

Table 26.10. Optimal solutions for the pressure vessel problem. Case B: solved according to Sandgren's original problem statements [1].

Item	Optimum solution by SOMA – Case B				Type
	AllToOne PathLength=3, Step=0.11, PopSize=20, PRT=0.5, Migrations=20, MinDiv = negative Average cost value = 7256.71		AllToAll PathLength=3, Step=0.11, PopSize =20, PRT=0.5, Migrations=20, MinDiv = negative Average cost value = 7197.73		
	the worst case	the best case	the worst case	the best case	
x_1 (T_s)	1.125	1.125	1.125	1.125	discr.
x_2 (T_h)	0.625	0.625	0.625	0.625	discr.
x_3 (R)	55.8592	55.8592	58.2902	58.2902	cont.
x_4 (L)	57.7315	57.7315	43.6927	43.6927	cont.
f(x)	7359.2	7197.73	7197.73	7197.73	

Table 26.11. Optimal solutions for the pressure vessel problem. Case C: different formulation of constraint-function, g_5, with respect to Sandgren's original problem statements [1].

Item	Optimum solution, Case C				Type
	Li and Chou [7]	Cao & Wu [11]	Thierauf and Cai [10]	Lampinen & Zelinka [12]	
x_1 (T_s)	1.000	1.000	1.000	1.000	discrete
x_2 (T_h)	0.625	0.625	0.625	0.625	discrete
x_3 (R)	51.250	51.1958	51.812	51.81347	continuous
x_4 (L)	90.991	90.7821	84.591	84.57853	Continuous
f(x)	7127.3	7108.6160	7006.9	7006.358	

Table 26.12. Optimal solutions for the pressure vessel problem. Case C: different formulation of constraint-function, g_5, with respect to Sandgren's original problem statements [1].

Item	Optimum solution by SOMA – Case C				Type
	AllToOne PathLength=3, Step=0.11, PopSize =20, PRT=0.5, Migrations=20, MinDiv = negative Average cost value = 7050.45		AllToAll PathLength=3, Step=0.11, PopSize =20, PRT=0.5, Migrations=20, MinDiv = negative Average cost value = 7006.36		
	the worst case	the best case	the worst case	the best case	
x_1 (T_s)	1.000	1.000	1.000	1.000	discr.
x_2 (T_h)	0.625	0.625	0.625	0.625	discr.
x_3 (R)	51.8128	51.8128	51.8135	51.8135	cont.
x_4 (L)	84.5839	84.5839	84.5785	84.5785	cont.
f(x)	7106.53	7006.42	7006.36	7006.36	

26.1.3 Designing a coil compression spring

The third example involves the design of a coil compression spring (Fig. 26.4). The spring is to be a helical compression spring to which a strictly axial and constant load will be applied.

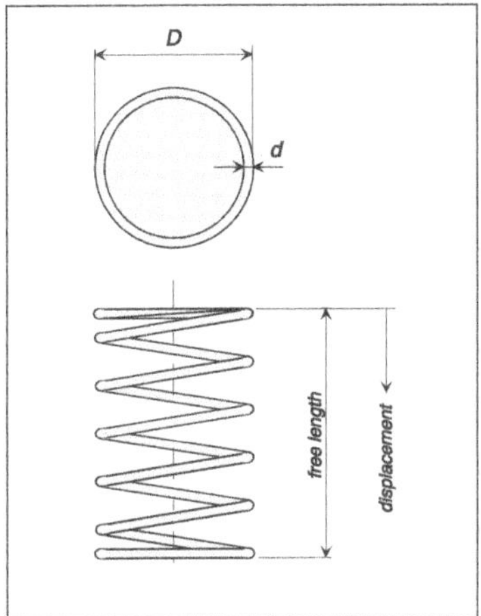

Fig. 26.4. Coil spring for Example 3.

The objective is to minimize the volume of spring steel wire needed to manufacture the spring. The design variables are the number of spring coils, N, the outside diameter of the spring, D, and the spring wire diameter, d. This example contains integer, discrete and continuous variables. The number of spring coils, N, is an integer variable and the outside diameter, D, is a continuous variable. Additionally, the spring wire diameter, d, is only available in the standard (discrete) sizes shown in Table 26.13.

Table 26.13. Allowable spring steel wire diameters for the coil spring design problem.

Allowable wire diameters [inch]					
0.009	0.0095	0.0104	0.0118	0.0128	0.0132
0.014	0.015	0.0162	0.0173	0.018	0.020
0.023	0.025	0.028	0.032	0.035	0.041
0.047	0.054	0.063	0.072	0.080	0.092
0.105	0.120	0.135	0.148	0.162	0.177
0.192	0.207	0.225	0.244	0.263	0.283
0.307	0.331	0.362	0.394	0.4375	0.500

The problem is formulated as follows:

$$
\begin{aligned}
&\text{Find} \\
&X = (x_1, x_2, x_3) = (N, D, d) \\
&\text{to minimize} \\
&f(X) = \frac{\pi^2 x_2 x_3^2 (x_1 + 2)}{4} \\
&\text{subject to} \\
&g_1(X) = \frac{8C_f F_{max} x_2}{\pi x_3^3} - S \le 0 \\
&g_2(X) = l_f - l_{max} \le 0 \\
&g_3(X) = d_{min} - x_3 \le 0 \\
&g_4(X) = x_2 - D_{max} \le 0 \\
&g_5(X) = 3.0 - \frac{x_2}{x_3} \le 0 \\
&g_6(X) = \sigma_p - \sigma_{pm} \le 0 \\
&g_7(X) = \sigma_p + \frac{F_{max} - F_p}{K} + 1.05(x_1 + 2)x_3 - l_f \le 0 \\
&g_8(X) = \sigma_w - \frac{F_{max} - F_p}{K} \le 0 \\
&\text{where} \\
&C_f = \frac{4(x_2/x_3) - 1}{4(x_2/x_3) - 4} + \frac{0.615 x_3}{x_2} \le 0 \\
&K = \frac{G x_3^4}{8 x_1 x_2^3} \\
&\sigma_p = \frac{F_p}{K} \\
&l_f = \frac{F_{max}}{K} + 1.05(x_1 + 2)x_3
\end{aligned}
\tag{26.7}
$$

The objective function, *f(X)*, computes the volume of spring steel wire as a function of the design variables. The design constraints are specified as follows:

a) The maximum working load is: F_{max} = 1000.0 lb.
b) The allowable maximum shear stress is: S = 189000.0 psi (g_1).
c) The maximum free length is: l_{max} = 14.0 inch (g_2).
d) The minimum wire diameter is: d_{min} = 0.2 inch (g_3).
e) The maximum outside diameter of the spring is: D_{max} = 3.0 inch (g_4).
f) The pre-load compression force is: F_p = 300.0 lb.
g) The allowable maximum deflection under pre-load is: σ_{pm} = 6.0 inch (g_6).
h) The deflection from pre-load position to maximum load position is: σ_w = 1.25 inch (g_8).
i) The combined deflections must be consistent with the length, i.e., the spring coils should not touch each other under the maximum load at which the maximum spring deflection occurs (g_7).
j) The shear modulus of the material is: $G = 11.5\times10^6$.
k) The spring is guided, so the buckling constraint is bypassed.
l) The outside diameter of the spring, D, should be at least three times greater than the wire diameter, d, to avoid lightly wound coils (g_5).

A more detailed explanation about the coil spring design procedure can be found in [1], [8] and in [13]. The SOMA control variable settings that solved the coil spring problem are described in Table 26.16 (Case A) and Table 26.18 (Case B). Although the problem statements do not define the boundaries for design variables, the constraints, g_3 and g_4 are pure boundary constraints and were treated as a lower boundary constraint for x_3 and as an upper boundary constraint for x_2, respectively. Furthermore, $g5$ can also be handled as a boundary constraint. In order to define the search space, the other boundary constraints were chosen based on the problem statements and the simple geometric space limitations elaborated in Table 26.14. The remaining constraints were handled as soft-constraint functions.

Table 26.14. Boundary constraints used for the coil spring example.

Boundary constraints for coil spring example		
Lower limitation	Constraint	Upper limitation
At least one spring coil is required to form a spring.	$1 \le x_1 \le \frac{l_{\max}}{d_{\min}}$	Upper and lower surfaces of unloaded spring coils touch each other.
Constraints g_3 and g_5 together	$3d_{\min} \le x_2 \le D_{\max}$	constraint g_4
Constraint g_3	$d_{\min} \le x_3 \le \frac{D_{\max}}{3}$	Constraints g_4 and g_5 together

The cost function for optimization was formulated as follows:

$$f_{cost}(X) = f(X) \cdot \prod_{i=1}^{2} c_i^3 \cdot \prod_{i=6}^{8} c_i^3 \qquad (26.8)$$

where

$$c_i = \begin{cases} 1.0 + s_i \cdot g_i(X) & \text{if} \quad g_i(X) > 0 \\ 1 & \text{otherwise} \end{cases}$$

$$s_1 = s_2 = s_6 = 1.0 \,,\ s_7 = s_8 = 1.0 \cdot 10^{10}$$

Constraint functions: g_3, g_4 and g_5 did not have to be evaluated because they were handled as boundary constraints and SOMA was not allowed to generate a candidate vector that violated any of them.

Table 26.15. Optimal solutions for the coil spring problem. Case A: continuous solution.

Item	Optimum solution, Case A		Type
	Sandgren [1]	Lampinen & Zelinka [12]	
x_1 (N)	9.1918	7.117621	Continuous
x_2 (D)	1.2052	1.372407	Continuous
x_3 (d)	0.2814	0.2909652	Continuous
f(x)	2.6353	2.61388	

Table 26.16. Optimal solutions for the coil spring problem by SOMA. Case A: continuous solution.

Item	Optimum solution by SOMA – Case A				Type
	AllToOne PathLength=3, Step=0.14, PopSize =40, PRT=0.3, Migrations=30, MinDiv = negative Average cost value = 2.65614		AllToAll PathLength=3, Step=1.5, PopSize =40, PRT=0.1, Migrations=30, MinDiv = negative Average cost value = 2.61389		
	the worst case	the best case	the worst case	the best case	
x_1 (N)	12.6744	7.11045	7.08773	7.11854	cont.
x_2 (D)	1.02875	1.37328	1.37538	1.37232	cont.
x_3 (d)	0.270741	0.291027	0.291131	0.29096	cont.
f(X)	2.73034	2.6146	2.61393	2.61388	

Table 26.17. Optimal solutions for the coil spring problem. Case B: discrete solution.

Item	Optimum solution, Case B				Type
	Sandgren [1]	Chen and Tsao [6]	Wu and Chow [8]	Lampinen & Zelinka [12]	
X_1 (N)	10	9	9	9	integer
X_2 (D)	1.180701	1.2287	1.227411	1.2230410	continuous
x_3 (d)	0.283	0.283	0.283	0.283	discrete
f(X)	2.7995	2.6709	2.6681	2.65856	

Table 26.18. Optimal solutions for the coil spring problem by SOMA. Case B: discrete solution.

Item	Optimum solution by SOMA – Case B				Type
	AllToOne PathLength=3, Step=0.14, PopSize =40, PRT=0.5, Migrations=30, MinDiv = negative Average cost value = 2.71043		AllToAll PathLength=3, Step=1.5, PopSize =40, PRT=0.5, Migrations=30, MinDiv = negative Average cost value = 2.66346		
	the worst case	the best case	the worst case	the best case	
x_1 (N)	3	9	9	9	integer
x_2 (D)	2.17372	1.22304	1.22305	1.22304	cont.
x_3 (d)	0.331	0.283	0.283	0.283	discr.
f(X)	2.93811	2.65855	2.67599	2.65859	

Table 26.15 and Table 26.17 compare SOMA's solution with results obtained by other researchers. Table 26.16 and Table 26.18 shows results from the SOMA algorithm. Two different versions of the coil spring problem were solved. Case A (Table 26.15 and Table 26.16) reports the solutions generated when all variables are assumed to be continuous. Case B (Table 26.17 and Table 26.18) reports the mixed-variable solution according to the problem statements above. In both cases, SOMA generally obtained a better solution than the best solution found in the literature (except and comparable mostly with DE). In order to demonstrate the robustness of the SOMA algorithm, 100 independent optimization trials were performed for both Case A and Case B. All trials yielded the reported value of *f(X)* or better. All of the solutions were also within the feasible region of the design space.

26.2 Conclusion

This chapter describes the use of SOMA in mechanical engineering problems: it is a new algorithm for global optimization. The basic principles of this algorithm were introduced in Chapter 7, including versions of the algorithm, an overview of selected applications, as well as testing for robustness and the handling of various constraints. The methods described for handling constraints are relatively simple, easy to implement and easy to use. SOMA is capable of optimizing integer, discrete and continuous variables and capable of handling non-linear objective functions with multiple non-trivial constraints as well as differential evolution or the other evolutionary algorithms which use techniques described in Chapter 7.

A soft-constraint (penalty) approach is applied for the handling of constraint functions. Some optimization methods require a feasible initial solution as a starting point for a search. Preferably, this solution should be rather close to a global optimum to ensure convergence to it instead of to a local optimum. If non-trivial constraints are imposed, it may be difficult or impossible to provide a feasible initial solution.

The efficiency, effectiveness and robustness of many methods are often highly dependent on the quality of the starting point. The combination of the SOMA algorithm and the soft-constraint approach does not require any initial solution, but yet it can take advantage of a high quality initial solution if one is available.

For example, this initial solution can be used for initialization of the population in order to establish an initial population that is biased towards a feasible region of the search space. If there are no feasible solutions in the search space, as is the case for totally conflicting constraints, SOMA algorithms with the soft-constraint approach are still able to find the nearest feasible solution. This is often important in practical engineering optimization applications, because many non-trivial constraints are involved.

After using SOMA to optimize the well-known test functions (Chapter 7), it was then used on three optimization problems from the field of mechanical engineering. This set of problems is useful for testing because **11** other algorithms (Table 26.2) from the EA domain were already used there. Hence, it offers quite a rich source of information for comparing new algorithms. All these problems comprised variables from the continual, integer and discrete domain for the cost function arguments. All tests have shown that results from SOMA are fully comparable with other algorithms especially with DE, which was proven here to be the best of all methods used. For all problems, 100 simulations were carried out for two versions of SOMA (AllToOne and AllToAll). All these results are described discussed in this chapter.

The SOMA algorithm is undoubtedly one of the most promising and novel methods for non-linear optimization that can be applied generally, and they work with minimum assumptions with respect to the objective function.

The algorithm requires only the cost value returned from the objective function for guidance of its seeking for the optimum. No derivatives or other auxiliary information are needed. Including the algorithm's extensions discussed in this article, the SOMA algorithm can be applied to a wide range of optimization problems, which practitioners in the field of modern optimization would like to solve.

All results described here can be very easy checked and repeated by means of C source code and Mathematica software - all is accessible at [14]. All examples are predefined there and thus it is easy to use them.

Acknowledgements

This work was supported by the grant No. MSM 26500014 of the Ministry of Education of the Czech republic and by grants of Grant Agency of Czech Republic GACR 102/03/0070 and GACR 102/02/0204.

References

[1] Sandgren, E. (1990). *Nonlinear integer and discrete programming in mechanical design optimization.* Transactions of the ASME, Journal of Mechanical Design, 112(2):223–229, June 1990. ISSN 0738-0666

[2] Fu, J.-F., Fenton, R. G. and Cleghorn, W. L. (1991). *A mixed integer-discrete-continuous programming method and its application to engineering design optimization.* Engineering Optimization, 17(4):263–280, 1991. ISSN 0305-2154

[3] Loh, Han Tong and Papalambros, P. Y. (1991). *A sequential linearization approach for solving mixed-discrete nonlinear design optimization problems.* Transactions of the ASME, Journal of Mechanical Design, 113(3):325–334, September 1991.

[4] Loh, Han Tong and Papalambros, P. Y. (1991). *Computational implementation and tests of a sequential linearization algorithm for mixed-discrete nonlinear design optimization.* Transactions of the ASME, Journal of Mechanical Design, 113(3):335–345, September 1991.

[5] Zhang, Chun and Wang, Hsu-Pin (1993). *Mixed-discrete nonlinear optimization with simulated annealing.* Engineering Optimization, 21(4):277–291, 1993. ISSN 0305-215X

[6] Chen, J. L. and Tsao, Y. C. (1993). *Optimal design of machine elements using genetic algorithms.* Journal of the Chinese Society of Mechanical Engineers, 14(2):193–199, 1993.

[7] Li, H.-L. and Chou, C.-T. (1994). *A global approach for nonlinear mixed discrete programming in design optimization.* Engineering Optimization, 22():109–122, 1994.

[8] Wu, S.-J. and Chow, P.-T. (1995). *Genetic algorithms for nonlinear mixed discrete-integer optimization problems via meta-genetic parameter optimization.* Engineering Optimization, 24(2):137–159, 1995. ISSN 0305-215X

[9] Lin, Shui-Shun, Zhang, Chun and Wang, Hsu-Pin (1993). *On mixed-discrete nonlinear optimization problems: A comparative study.* Engineering Optimization, 23(4):287–300, 1995. ISSN 0305-215X

[10] Thierauf, G. and Cai, J. (1997). *Evolution strategies – parallelisation and application in engineering optimization.* In B.H.V. Topping (ed.) (1997). *Parallel and distributed processing for computational mechanics.* Saxe-Coburg Publications, Edinburgh (Scotland). ISBN 1-874672-03-2

[11] Cao, Y. J. and Wu, Q. H. (1997). *Mechanical design optimization by mixed-variable evolutionary programming.* Proceedings of the *1997 IEEE Confer-*

ence on Evolutionary Computation, IEEE Press, pp. 443–446.

[12] Lampinen Jouni, ZELINKA, Ivan. New Ideas in Optimization - Mechanical Engineering Design Optimization by Differential Evolution. Volume 1. London : McGraw-Hill, 1999. 20 p. ISBN 007-709506-5.

[13] Siddall, James N. (1982). *Optimal engineering design: principles and applications.* Mechanical engineering series / 14. Marcel Dekker Inc. ISBN 0-8247-1633-7

[14] Articles about SOMA algorithm (source codes, graphics animated gallery, bibliography, see http://www.ft.utb.cz/people/zelinka/soma

27 Scheduling and Production & Control: MA

Pablo Moscato, Alexandre Mendes and Carlos Cotta

27.1 Introduction

This chapter addresses three main problems of the production planning area: Single Machine Scheduling (SMS), Parallel Machine Scheduling (PMS) and Flowshop Scheduling (FS). Many situations in production environments found in industries around the world can be modelled as one or a set of them (Baker 1974). It is easy to find dozens of problem variants in the literature due to different production constraints and objective functions. Due to the overwhelming number of combinations, we focused our discussion on only one of each problem class. They are:

- SMS with sequence-dependent setup times and with the objective of minimizing of the total tardiness.
- PMS with sequence-dependent setup times and minimization of the makespan.
- FS with families of jobs, sequence-dependent setup times and minimization of the makespan.

The three problems are NP-hard. That means finding that under the current conjecture that the computational complexity classes P and NP are not equal, it is widely conjectured that the optimal solution of an arbitrary instance of these problems cannot be found in polynomial time (Lawler et al. 1993). They also share sequence-dependent setup times, which are common in industrial environments and dramatically increase the complexity of the problem. These times come from the fact that the machinery usually requires some kind of maintenance (cleaning, alignment, tooling change, etc.) before starting the production of a given job, or sets of jobs. The objective functions treated are the *total tardiness* and the *makespan*.

The total tardiness obliges the scheduling to respect production due dates, defined by the clients that make the orders. They are utilized when the industry wants to deliver the customers' orders on time, or with the least delay. Moreover, no penalty is applied if products are completed before their due dates. In other words, no after-production inventory costs are considered.

The makespan measures the total time of production, marked by the time span between the production start and the time when the last job leaves the shop floor. When no setup times are considered, the makespan is constant, but with setup times, the minimization of this value becomes a challenge. Other important considerations of the problems addressed are as follows:

- A set on independent, single-operation, non-preemptive jobs is available with *ready-time*[1] zero.
- The machines are continuously available and have no limitation of resources.
- Tooling availability is not a production-limitation factor.
- No machine breakdowns or other contingencies are considered.
- All parameters defining the production environment – processing times, setup times and due dates – are previously known.

27.2 The single machine scheduling problem

The Single Machine Scheduling is one of the first studied class of problems in the scheduling area (Graham et al. 1979; Graves 1981). There are many different types of SMS problems (generally due to different input data and objective functions). One of the simplest to state, but not easy to solve at all, is the problem of sequencing n jobs, given its processing times and due dates (distinct for each job), and with the objective function being to minimize the total tardiness. Tardiness is a *regular* performance measure, which means that an optimal schedule cannot have idle times between jobs. Moreover, the tardiness increases only if at least one of the completion times[2] in the schedule also increases. Under these circumstances, a valid solution of the problem can be defined as a simple permutation of the jobs. Using the permutation space representation, the solution space has $n!$ configurations, all of them representing valid solutions of the problem.

The SMS problem addressed can be extended by the inclusion of precedence constraints for the jobs, ready-times, resources limitations, etc. From an objective function's point-of-view, we may want to minimize the *makespan*, the *total tardiness*, the *mean tardiness*, the number of *tardy jobs*, or even a combination of these objectives, which would characterize a *multi-criteria*[3] problem.

It is easy to see the large variety of problems that we may face in practice (Gen and Cheng 1997). Even looking at very complex industrial manufacturing sys-

---

[1] Ready-time is the time that the job becomes ready to start being processed.

[2] Completion time is the time when a given job finishes its processing.

[3] In multi-criteria problems, two or more performance criteria, sometimes confronting ones, shall be optimised. These problems are harder to solve, mainly because of the difficulty to compare the solutions' quality. In such cases, Pareto's optimality criteria should be employed.

tems, it is not hard to find situations in which a simple SMS should be solved (Ow and Morton 1989). In multiple-operation processes, a single part usually goes through several machines, where each machine performs a specific job at the part. Depending on the processes involved, there might be bottlenecks in this sequence, which can be modelled as SMS problems. Intricate jobs, performed by dedicated machines fit in that category. These machines are usually very expensive and probably there will be only one of them in the production environment, being used continuously. This situation represents a bottleneck and the schedule determination for this machine will affect the production as a whole. Next, it is shown the description of the SMS problem addressed in this chapter:

Input: Let n be the number of jobs to be processed in one machine. Let $P = \{p_0, p_2, ..., p_n\}$ be the list of processing times, $D = \{d_1, d_2, ..., d_n\}$ be the list of due dates for each job and $S_0 = \{s_{01}, s_{02}, ..., s_{0n}\}$ be the list of initial setup times. Let $\{S_{i,j}\}$ be a matrix of setup times, where $s_{i,j}$ is the time required to set up job j after the machine has just finished processing job i.

Task: Find a permutation that minimizes the total tardiness of the schedule. Suppose that the jobs are scheduled in the order $\{\pi(1), \pi(2), ..., \pi(n)\}$. Then the total tardiness can be calculated using the Eq. 27.1.

$$Total_{tardiness} = \sum_{k=1}^{n} \max\left[0, c_{\pi(k)} - d_{\pi(k)}\right] \tag{27.1}$$

Where $c_{\pi(k)}$ represents the *completion time* of job $\pi(k)$. The $c_{\pi(k)}$ values can be calculated using the Eq. 27.2.

$$c_{\pi(k)} = s_{0\pi(1)} + p_{\pi(1)} + \sum_{i=2}^{k}\left[s_{\pi(i-1),\pi(i)} + p_{\pi(i)}\right] \tag{27.2}$$

It is well known that the problem of sequencing jobs in one machine without setup times is already NP-hard (Du and Leung 1990). Despite its application to real world settings, the SMS problem addressed in this chapter has received little attention in the scheduling literature. In (Ragatz 1993) a *branch-and-bound* (B&B) method is proposed, but only small instances could be solved to optimality. The papers of (Raman et al. 1989) and (Lee et al. 1997) use dispatch rules based on the calculation of a priority index to build an approximate schedule, which is then improved by the application of a local search procedure. The ATCS heuristic presented in (Lee et al. 1997) had an impressive performance for this problem, considering its simplicity. Three papers using metaheuristics have been proposed so far. In (Rubin and Ragatz 1995) a new crossover operator was developed and a *genetic algorithm* (GA) was applied to a set of test problems. The results obtained by the GA approach were compared with the ones from a B&B and with a *multiple start* (MS) algorithm and they concluded that MS outperformed the B&B and the GA in many instances, considering running time and quality of solutions as performance measures. Of course, the instances in which MS outperformed B&B were the ones where the exact method did not find an optimal solution before a limit on the number of nodes was reached. The B&B was truncated in such cases, returning sub-optimal schedules. Given these results, (Tan and Narasimhan 1997) chose the MS technique as a baseline benchmark for conducting comparisons with the *simulated annealing* (SA) approach they proposed. Their conclusion was that SA outperformed MS in all but three instances, with percentage improvements not greater than 6%. More recently, (França et al. 2001) proposed a new *memetic algorithm* (MA) that outperformed the previous approaches. The results presented in this chapter are an improvement over previously reported results. The BOX crossover, together with the multi-population approach (both equal to those described in Chap. 18) produced a big performance leap.

27.2.1 The test instances

The test set used for the experimental analysis consists of instances that vary from 17 to 100 jobs. This set was constructed from *Asymmetric Travelling Salesman Problem* (ATSP) instances which optimal solutions are available in the literature. All ATSP instances used for this purpose in this work, as well as their optimal solutions, are available at the TSPLIB website[4].

It is clear that the ATSP and the SMS problems share similarities and they have been highlighted for decades. The most important relates the SMS's *setup times* with the ATSP's *distance matrix*. This correspondence suggests a transformation procedure to create an "interesting set" of SMS instances starting from an arbitrary ATSP instance with known optimal solution. We propose a three-step procedure to generate such SMS instances. Each SMS instance is composed of a matrix of setup times and three lists of n integers. In the transformation, the setup times matrix will be equal or be a multiple of the ATSP's distance matrix. Then, a simple procedure generates the three lists required to complete the SMS instance. Suppose that the permutation $\{\pi(1), \pi(2), ..., \pi(n)\}$ represents the sequence of cities in the optimal tour for the ATSP instance.

Step 1 - Generation of the S_0 list: Let $[\pi(k), \pi(k+1)]$ be the pair of adjacent nodes of the ATSP optimal tour with the largest distance between cities (named *dist*). Let $s_{0,\pi(k+1)}$ be equal to the cost of this arc, $dist_{\pi(k),\pi(k+1)}$. Let all other s_{0j}, $j \neq \pi(k+1)$ be greater than $s_{0,\pi(k+1)}$. Moreover, the initial job of the optimal SMS solution will be $\pi(k+1)$.
Step 2 - Generation of the P list. Initially, re-enumerate the nodes so that $\pi(1) \Rightarrow \pi(k+1)$; $\pi(2) \Rightarrow \pi(k+2)$; ...; $\pi(n) \Rightarrow \pi(k)$. Then, construct the list $\{p_{\pi(1)}, p_{\pi(2)}, ..., p_{\pi(n)}\}$ in such a way that $p_{\pi(1)} < p_{\pi(2)} < ... < p_{\pi(n)}$.
Step 3 - Generation of the D list: Construct the due dates in such a way that $d_{\pi(1)} < d_{\pi(2)} < ... < d_{\pi(n)}$ and $d_{\pi(i)} \in [c_{\pi(i)} - p_{\pi(i)}, c_{\pi(i)}]$ $\forall$ $\pi(i)$: $i = 1, ..., n$, where $c_{\pi(i)}$ is the completion time of job $\pi(i)$ following the ATSP optimal sequence.

It is worth emphasizing that we are not stating that the transformation procedure described above preserves the ATSP optimal tour permutation as the SMS

[4] http://www.crpc.rice.edu/softlib/tsplib/

optimal sequence. Consequently, it is an open problem to find the optimal tardiness value of the resulting SMS instances. Nevertheless, in all tests we have executed (using the TSPLIB instances as the basis of the transformation) we have never reached a better tardiness than the one obtained with the jobs in the ATSP optimal permutation. As a consequence, we will assume that such values can be considered at least as very high-quality upper bounds.

We will now describe the intuition behind the suggested transformation. The direct use of the distance matrix as the setup times between jobs is pretty clear. Suppose that all processing times are zero in the SMS problem. In this case, the SMS problem is basically a generalization of the ATSP on which we have the extra constraint of due dates to visit each city.

The idea of starting the sequence with the setup corresponding to the largest distance belonging to the optimal ATSP tour aims to avoid that a change in the initial job produces a reduction in the sum of the completion times, what could make the optimal sequence be lost. Another important point is the requirement that $dist_{\pi(k+1),\pi(k)} < dist_{\pi(k),\pi(k+1)}$. However, these rules are not enough. Let us show an example where the transformation does not preserve optimality (see Fig. 27.1).

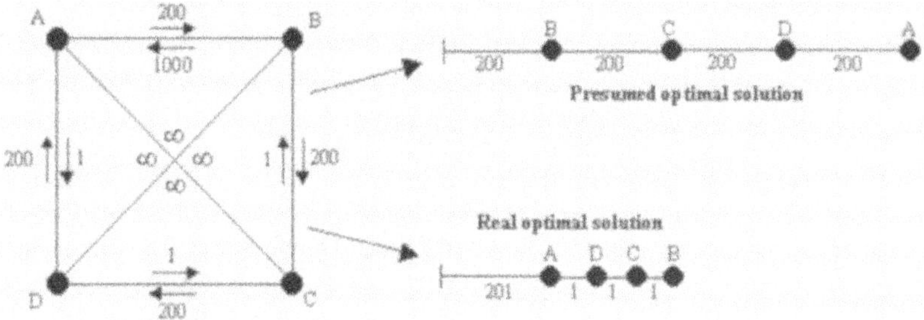

Fig. 27.1. An example where the optimal solution of the ATSP instance is not the optimal permutation of the SMS instance.

In Fig. 27.1, on the left side there is an ATSP instance composed of four cities. As the internal arcs are valued "infinite", the only two tours that avoid using one of these edges are [B C D A] – with a total length of 800 – and [A D C B] – with a total lenght 1003. Therefore, consider the optimal tour [B C D A]. For the sake of simplicity, ignore the processing times and due dates of the corresponding SMS problem as we are interested only in the effect caused by the setup times. On the upper-right portion of the Fig. 27.1, the presumed optimal sequence is shown, with a makespan of 800 – equal to the optimal ATSP tour length. Consider the situation where the initial arc has a very high value in the opposite direction. In this case, according to the transformation procedure, it could be replaced by a value greater than $s_{0,\pi(k+1)}$ but smaller than the original. In the example, it was substituted by 201, what is much smaller that 1000. The consequence is that the real optimum makespan becomes lower than the presumed one and the transformation fails to preserve optimality. If in Step 1, the restriction $dist_{\pi(k+1),\pi(k)} < dist_{\pi(k),\pi(k+1)}$ is not observed, the transformation might not be valid. Thus, a necessary condition is that the initial arc must always have a larger value when it is considered in the direction of the presumed optimal solution than when it is considered in the opposite direction.

The steps 2 and 3 have the same function. When we order the processing times and the due dates following an increasing sequence, we avoid that any change in a job position leads to a reduction in total tardiness. The way that the P and D lists are generated produces instances that are solvable to optimality by the *Shortest Processing Time* (SPT) and by the *Earliest Due Date* (EDD) heuristics. Methods that use such procedures, or their variants, will solve the instances instantly, consequently producing biased results. Nevertheless, in order to evaluate the memetic algorithm, which does not employ any such heuristics in its operators, the instances are absolutely adequate.

27.2.2 The memetic algorithm approach

The MA utilized in the SMS problem is very similar to that presented for the Gate Matrix Layout problem (see Chap. 18). Therefore, we refer the reader to that chapter and we will focus only on a few minor differences between them.

The representation chosen is quite intuitive, with a solution being represented as a chromosome with the alleles assuming different integer values in the $[1, n]$ interval. In other words, a solution is a permutation of n integers.

Two crossover operators were tested: BOX and OX. In the first one, after choosing two parents, several fragments of the chromosome from one of them are randomly selected and copied into the offspring. In a second phase, the offspring's empty positions are sequentially filled with the alleles present in the chromosome of the other parent, from left to right. The procedure tends to perpetuate the relative order of the jobs and has a better overall performance over the original OX crossover (Goldberg 1989), which selects a single block of one of the parents to be copied into the offspring. In this implementation, the parent that copies its block(s) into the offspring is the leader of the cluster, and the supporter will complete the empty positions (see Fig. 27.2).

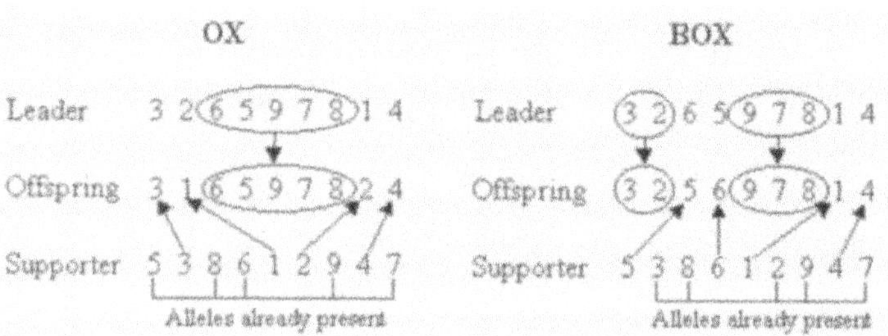

Fig. 27.2. Crossover operators for the SMS problem – OX and BOX

The number of individuals created in every generation is very high – two times the number of individuals in the population. This value is consequence of the acceptance policy for the offspring, which is very restrictive. Therefore in order to balance the large number of discarded individuals, we must create many of them every generation.

In this implementation, we employed the *all-pairs* and the *insertion* neighbourhoods, just like in the VLSI problem. Due to the complexity of each individual evaluation, a neighbourhood reduction based on the setup times' values had to be developed. It was observed that most good schedules have a common characteristic: the setup times between jobs in these solutions are very small. This is reasonable, since schedules with small setup times between jobs will be less lengthy, generating fewer delays. Next, we show two diagrams illustrating how the schedule is modified if jobs i and j swap their positions (see Fig. 27.3) and if job i is inserted immediately before job j (see Fig. 27.4).

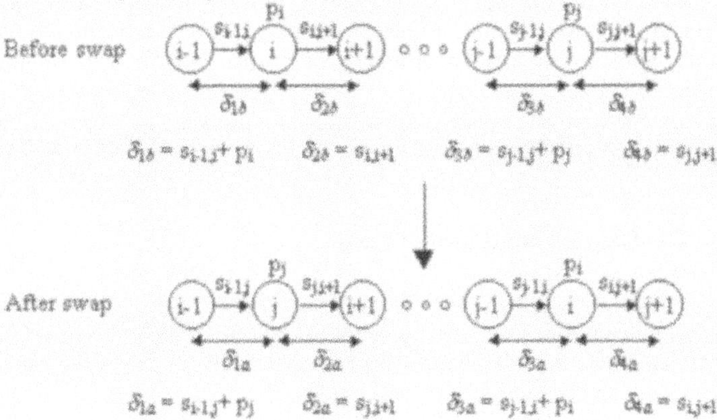

Fig. 27.3. Diagram of a swap move in an all-pairs neighbourhood and its implications on the schedule

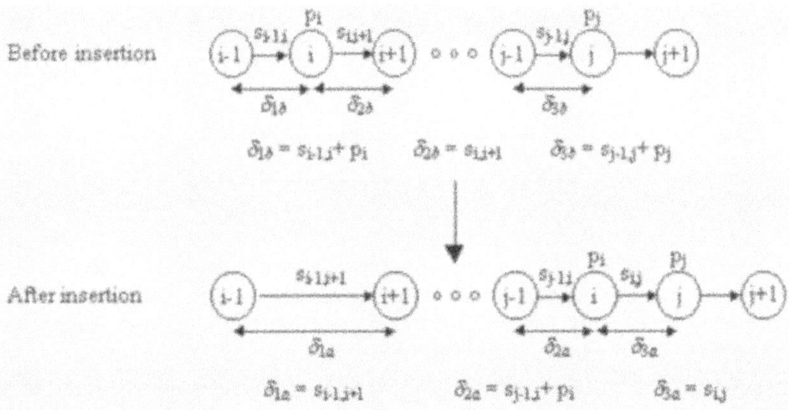

Fig. 27.4. Diagram of a move in an insertion neighborhood and its implications on the schedule

Based on the δ variations, it is possible to create rules to determine if a given move should be evaluated or not. Among the several possibilities of combinations, the best results came, for the all-pairs neighbourhood, if we evaluate the swap move only if '*at least one δ value improves in both modified regions*'. For the insertion neighbourhood, the best policy is to test an insertion move only '*if the sum of the δ variables in at least one modified region improves*'. In mathematical form, these rules can be written as logical clauses:

- $[(\delta_{1a} < \delta_{1b}) \vee (\delta_{2a} < \delta_{2b})] \wedge [(\delta_{3a} < \delta_{3b}) \vee (\delta_{4a} < \delta_{4b})]$, for the all-pairs neighbourhood.
- $(\delta_{1a} < \delta_{1b} + \delta_{2b}) \vee (\delta_{2a} + \delta_{3a} < \delta_{3b})$, for the insertion neighbourhood.

Since each individual evaluation requires a considerable computational effort[5] reductions like these save a lot of time. In fact, although the reduction policies are very strict, reducing the neighbourhoods to less than 10% of their original size, they maintain the search focused on promising moves. More information on the several reduction policies tested, as well as performance comparisons, can be found in (França et al. 2001).

27.2.3 The SMS computational results

As said before, the SMS instances use the ATSP distance matrices. The only flexible parameters are the processing times and the due dates. By combining these two components of the input, we classified the instances in four groups. The parameters are generated according to the rules:

Processing Times:

LOW:	$p_k \in [0, \frac{1}{4}.\max(s_{ij})]$	$\forall\ k: k = 1, \ldots, n$
HIGH:	$p_k \in [0, 2.\max(s_{ij})]$	$\forall\ k: k = 1, \ldots, n$

Due dates:

HARD:	$d_k = c_k$	$\forall\ k: k = 1, \ldots, n$
SOFT:	$d_k \in [c_k - p_k, c_k]$	$\forall\ k: k = 1, \ldots, n$

The *LOW* rule makes the setup times a more critical aspect in the problem, emphasizing its "ATSP-like character". In these instances, the processing times are small, compared to the setup times. The *HIGH* policy, on the other hand, makes the processing times be more relevant in the schedule determination. Their values can be up to two times larger than the maximum setup time.

The *HARD* policy generates instances in which the total tardiness of the schedule that corresponds to the optimal ATSP solution equals to zero. This value in

[5] Each individual evaluation is $O(n)$ and the complete neighbourhoods are $O(n^2)$.

guaranteed because the due dates are placed over the jobs' completion times, when the production follows the optimal ATSP tour. Nevertheless, it is a quite difficult problem, since only one – or perhaps a few – of the $n!$ solutions has a zero tardiness. The *SOFT* instances, on the other hand, have greater-than-zero optimal total tardiness and are better suited for making relative comparisons, where percentage deviations from optimal value are numerically necessary. Next, in Table 27.1, we show some characteristics of the original ATSP instances such as number of cities and distance between cities – minimum and maximum $dist_{ij}$.

Table 27.1. Number of cities (jobs), and distance-between-cities interval (setup times) of the ATSP instances

Instance name	n	Minimum $dist_{ij}$	Maximum $dist_{ij}$
br17	17	0	74
ftv33	34	7	332
ftv55	56	6	324
ftv70	71	5	348
kro124p	100	81	4,545

The instances tried to cover a wide range of number of jobs. Moreover, the resulting SMS setup times also cover a wide range. The Tables 27.2 and 27.3 show all the results obtained for the set of instances created. The name of the instance is in the first column and is divided in two parts. The initial part refers to the name of the original ATSP instance. Then, the two capital letters indicate if the instance is LOW or HIGH, and HARD or SOFT. The second column indicates the presumed optimal tardiness. Next to it, we have the average total tardiness found using the OX and the BOX recombination operators, considering 10 runs for each instance. Written in bold letter, subscript, there is the number of times that the optimal solution was found, out of 10 trials. Finally, in the columns labeled 'CPU time' it is shown the average CPU time.

27.2.3.1 Single population tests

The Table 27.2 shows the results for the single population memetic algorithm, using the OX and BOX crossovers. The maximum CPU time was fixed at four minutes, but several times the "presumed optimal" (i.e. the permutation corresponding to the optimum ATSP tour) was reached before this limit. The equipment utilized is a 366 MHz Pentium II Celeron with 128 MB RAM.

The results in Table 27.2 show that the memetic algorithm successfully solved most of the instances. The OX crossover had a worse overall performance, compared to the BOX, failing in many of the 100-job instances. Concerning this criterion, the finding of 167 and 180 optimal solutions – out of 200 possible – for the OX and BOX crossovers, respectively, can be considered a very reasonable rate of success. Based on the results, we can also conclude that the 100-job instances are in the threshold of the search capability of the algorithms, because the number of optimal solutions decreased considerably. For the *kro124pLS*, for example, the "presumed optimal" was found only once in twenty trials. Probably, for instances with more than 100 jobs, the method might fail to find any optimal solutions and more features will have to be included in the algorithm in order to increase its search power. Another important point is the CPU time required by each configuration. In general, the BOX needs less CPU time than the OX to find the optimal solutions. This characteristic becomes more evident in the *ftv70* and *kro124p* sets of instances.

Table 27.2. Results for the single population memetic algorithm

Instance name	Presumed optimal tardiness	OX Average Tardiness	BOX average tardiness	OX CPU time (sec.)	BOX CPU time (sec.)
br17LH	0	0_{10}	0_{10}	0.1	0.1
br17LS	52	52_{10}	52_{10}	0.1	0.1
br17HH	0	0_{10}	0_{10}	0.1	0.1
br17HS	547	547_{10}	547_{10}	0.1	0.1
ftv33LH	0	0_{10}	0_{10}	2.1	2.1
ftv33LS	664	664_{10}	664_{10}	2.8	2.6
ftv33HH	0	0_{10}	0_{10}	1.3	1.3
ftv33HS	5,324	$5{,}324_{10}$	$5{,}324_{10}$	1.5	1.4
ftv55LH	0	0_{10}	0_{10}	17.4	13.8
ftv55LS	1,170	$1{,}170_{10}$	$1{,}170_{10}$	50.9	19.8
ftv55HH	0	0_{10}	0_{10}	6.9	8.0
ftv55HS	8,515	$8{,}515_{10}$	$8{,}515_{10}$	12.9	14.5
ftv70LH	0	918.6_7	0_{10}	148.5	46.0
ftv70LS	1,506	$1{,}557.4_8$	$1{,}506_{10}$	165.0	70.1
ftv70HH	0	0_{10}	0_{10}	41.8	23.6
ftv70HS	12,368	$12{,}368_{10}$	$12{,}368_{10}$	40.3	35.3
kro124pLH	0	$23{,}872.0_2$	$22{,}717.1_3$	229.5	205.5
kro124pLS	26,111	$74{,}636.3_0$	$68{,}618.8_1$	240.0	224.3
kro124pHH	0	$4{,}710.6_7$	0_{10}	157.3	107.1
kro124pHS	223,890	$238{,}114.2_3$	$231{,}997.5_6$	220.4	185.5
* Total optimal solutions found		*167	*180		

27.2.3.2 Multiple population tests

The multiple population memetic algorithm aims to validate the use of multiple populations for the SMS problem. Using the results presented in the Chap. 18 – Gate Matrix Layout Problem – as a guideline, the number of populations was fixed at four and the migration policy used was the so-called *1-Migrate*. The maximum CPU time remained at four minutes and the results are presented in Table 27.3.

The multiple population approach had a better performance than the single population one. The number of presumed optimal solutions found was greater – 170 and 189 against 167 and 180 in the single population version. With the use of the BOX crossover, we attained very strong results even for the 100-job instances, what makes us believe that the method could deal with larger problems. In this test, the relation between CPU time's requirements was the same. The BOX is much more efficient than the OX, reaching the optimal solutions much faster. Still regarding the CPU time, sometimes the multiple population approach is slower than the single population, especially for the smaller instances. This was already expected since the algorithm had to evolve four populations instead of only one. The problem must have a sufficient size – and complexity – in order to take advantage of the *genetic drift* effect (see Chap. 18) and be noticeable from computer experiments.

Table 27.3. Results for the multiple population memetic algorithm

Instance name	Presumed optimal tardiness	OX average tardiness	BOX Average Tardiness	OX CPU time (sec.)	BOX CPU time (sec.)
br17LH	0	0_{10}	0_{10}	0.1	0.1
br17LS	52	52_{10}	52_{10}	0.1	0.1
br17HH	0	0_{10}	0_{10}	0.1	0.1
br17HS	547	547_{10}	547_{10}	0.1	0.1
ftv33LH	0	0_{10}	0_{10}	1.7	2.1
ftv33LS	664	664_{10}	664_{10}	3.4	2.6
ftv33HH	0	0_{10}	0_{10}	1.2	1.6
ftv33HS	5,324	$5{,}324_{10}$	$5{,}324_{10}$	1.6	2.0
ftv55LH	0	0_{10}	0_{10}	43.8	13.6
ftv55LS	1,170	$1{,}170_{10}$	$1{,}170_{10}$	40.1	22.5
ftv55HH	0	0_{10}	0_{10}	8.4	8.1
ftv55HS	8,515	$8{,}515_{10}$	$8{,}515_{10}$	12.8	8.7
ftv70LH	0	0_{10}	0_{10}	122.7	48.3
ftv70LS	1,506	$1{,}684.7_{8}$	$1{,}506_{10}$	172.9	53.6
ftv70HH	0	0_{10}	0_{10}	39.2	18.0
ftv70HS	12,368	$12{,}368_{10}$	$12{,}368_{10}$	50.8	35.1
kro124pLH	0	$48{,}536.5_{1}$	$26{,}175.9_{6}$	231.7	206.6
kro124pLS	26,111	$72{,}553.7_{0}$	$53{,}807.4_{4}$	240.0	227.7
kro124pHH	0	$3{,}337.5_{8}$	0_{10}	159.3	118.9
kro124pHS	223,890	$240{,}213.1_{3}$	$233{,}074.3_{9}$	228.1	141.3
* Total optimal solutions found		*170	*189		

27.3 The parallel machine scheduling problem

The second part of this chapter addresses the PMS problem. It consists of scheduling a given set of n jobs to m identical parallel machines with the objective of minimizing the *makespan*. As in the SMS problem, there are sequence-dependent setup times between jobs. The PMS can be considered a generalization of the SMS since it takes into account several machines, instead of only one. In fact, the solution found to deal with bottleneck situations like those described in Sect. 27.2 is usually the addition of more machines, transforming the SMS problem into a PMS one. Of course, sometimes this cannot be done, generally due to financial limitations. The resulting problem can be faced as two problems. The first is how to assign the jobs to the several machines and the second is how to schedule the jobs in each machine. The PMS with setup times is frequently found in real world settings where the setups are not negligible, especially in the paper, chemical and textile industries.

Previous PMS-related works are dated back to 1984. In that year, Dearing and Henderson (1984) developed an integer linear programming model for loom assignment in a textile weaving operation. They reported results found through the rounding of the solutions obtained by the linear relaxation of the integer model. In 1987, Sumichrast and Baker (1987) proposed a heuristic method based on the solution of a series of 0-1 integer subproblems, improving the results obtained in (Dearing and Henderson 1984). These two articles deal with a slightly different problem because they assume that a job can be split among several machines. The case being addressed in this chapter is combinatorially more complex. To the authors' knowledge, very few papers have reported computational results for this problem. The work of Frederickson et al. (1978) present approximate algorithms derived from an equivalence between the PMS and the *Travelling Salesman Problem* (TSP). França et al. (1996) used a *tabu search*-based (TS) heuristic in connection with a powerful neighborhood scheme that employs the concept of local and global neighbors. Later, Mendes et al. (2002) proposed a MA approach for the problem, comparing the results with the ones found in (França et al. 1996).

The PMS problem is a difficult combinatorial problem proved to be NP-hard in a strong sense because it is equivalent to the TSP when the number of processors equals one (Baker 1974). Next, it is shown the description of the PMS problem.

Input: Let n be the number of jobs to be processed in m identical machines. Let $P = \{p_1, p_2, ..., p_n\}$ be the list of the jobs' processing times, and $S_0 = \{s_{01}, s_{02}, ..., s_{0n}\}$ be the list of initial setup times. Let $\{S_{i,j}\}$ be a matrix of setup times, where $s_{i,j}$ is the time required to set up job j after the machine has just finished processing job i.

Task: Assign the jobs to the machines and find the permutation in each machine that minimizes the production makespan. In order to calculate the makespan, initially consider that $\pi(k,l)$ represents the k-th job of machine l. The total production time in machine l, represented by Γ_l, can be calculated by the Eq. 27.3.

$$\Gamma_l = s_{0\pi(1,l)} + p_{\pi(1,l)} + \sum_{i=2}^{n_l} \left[s_{\pi(i-1,l),\pi(i,l)} + p_{\pi(i,l)} \right] \tag{27.3}$$

Where n_l is the number of jobs processed by machine l. Thus, the makespan is the maximum production time among all machines, being represented by the Eq. 27.4.

$$Makespan = \max_i \left[\Gamma_i \right] \tag{27.4}$$

27.3.1 The test instances

The instances used in the computational experiments were randomly generated. The number of jobs was fixed at 20, 40, 60 and 80, and the number of machines at 2, 4, 6 and 8. Processing times were generated following a discrete uniform distribution DU(1, 100). Setup times were divided into two categories: small setup times – with values in the interval [1, 10] – and large setup times – with values in the interval [1, 100]. The setup times were also generated according to two possibilities: structured and non-structured. The structured setup times follow the triangular inequality, that is, $s_{ij} \leq s_{ik} + s_{kj}$, $\forall\ i, j, k;\ k \neq i, j$. The non-structured setup times do not follow the triangular inequality. We considered 10 replications for each combination of number of jobs and machines.

27.3.2 The memetic algorithm approach

The MA utilized for the PMS problem is very similar to the GMLP (see Chap. 18, this book) and the SMS. The differences are concentrated on the crossover type – the BOX was replaced by the OX – and on the use of a single population, instead of several ones. Now, let us begin by addressing the individual representation for this problem.

The representation chosen for the PMS is a chromosome with its alleles assuming different integer values in the [1, n] interval. In order to include information about the m machines, m-1 *cut-points* (represented by a '*' symbol) were introduced to define the jobs assignment to each machine. For instance, a chromosome [1 4 6 * 3 7 2 10 * 8 9 5] is a possible solution for a problem with 10 jobs and 3 machines. The cut-points are in positions 4 and 9. Thus, machine 1 executes operations [1 4 6], in this order; machine 2 executes operations [3 7 2 10] and machine 3 performs operations [8 9 5]. As the machines are identical, no distinction must be made between the cut-points.

As said before, the crossover operator implemented is the Order Crossover (OX). Because of the cut-points' role and their effect in the schedule, the OX in this case behaves similarly to the BOX. The reason is that any cut-point position change influences the job/machine assignment of the entire sequence, resembling the kind of perturbation that would be caused by the BOX. The difference between the OX implemented and the originally introduced by (Goldberg 1989), is that the offspring is filled from the beginning of the sequence and not after the piece copied from the first parent. The Fig. 27.5 shows a diagram of how the crossover works.

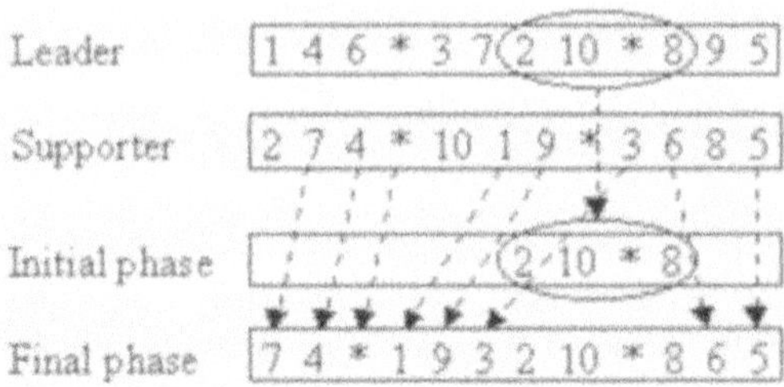

Fig. 27.5. Diagram of the OX crossover used for the PMS problem

In the Fig. 27.5, the initial fragment inherited from the *leader* parent consists of the alleles [2 10 * 8], and they are copied to the same positions in the offspring. The offspring's empty positions were then filled according to the order that the alleles appear in the chromosome of the *supporter* parent. Repeated alleles are skipped as well as cut-points, if there are already m-1 cut-points present in the offspring.

The local search utilized is based on the *all-pairs* swap and the *insertion* neighbourhoods and is applied to every new individual. No reduction scheme was employed for this problem, since the largest instance consisted of 80 jobs and 8 machines, which is still a computationally tractable size.

27.3.3 The PMS computational results

The previous best results in the literature on the PMS problem are presented in (França et al. 1996) and (Mendes et al. 2002). In (França et al. 1996) it was also introduced a set of instances for which optimal solutions or high quality lower bounds were calculated. In Table 27.4, we compare the MA against these values and also against upper bounds obtained by an intensively executed *tabu search* (TS) algorithm, named *long tabu*. The long tabu usually required several hours of execution, but at the end, high quality *upper bounds* were found.

The percentage deviation from the lower bound/optimal solution is the performance measure for problems with 2 and 4 machines. The exception is the 4-machine problem with structured instances and $s_{ij} \in [1, 10]$. For this problem set and also for all structured problems with 6 and 8 machines, the deviation is related to the long tabu upper bounds.

A variant of the long tabu method is also presented in (França et al. 1996). It is a faster, less complex TS named *fast tabu*. Although less powerful, the fast tabu method is very efficient in finding good schedules for the PMS problem. Since the computational time required by this algorithm was much lower compared to the long tabu, we concluded it would be a fair opponent for the MA. Thus, the CPU times utilized by the MA are the same of the fast tabu, which vary from a few seconds up to five minutes, depending on the instance size.

Table 27.4. Results for the PMS instances

	Non-structured problems				Structured problems			
	$s_{ij} \in [1, 10]$		$s_{ij} \in [1, 100]$		$s_{ij} \in [1, 10]$		$s_{ij} \in [1, 100]$	
n / m	TS	MA	TS	MA	TS	MA	TS	MA
20 / 2	1.6	1.8	2.1	3.4	0.5	0.5	0.6	0.3
40 / 2	3.8	3.0	4.6	4.4	0.9	0.6	0.7	0.3
60 / 2	5.1	3.9	5.2	5.2	0.9	0.5	0.5	0.3
80 / 2	5.4	4.3	5.1	4.7	1.1	0.5	0.5	0.2
20 / 4	5.3	4.4	6.0	6.6	0.3	0.0	1.9	1.7
40 / 4	6.5	8.4	6.8	8.9	0.2	-0.1	1.6	1.3
60 / 4	7.0	9.1	7.2	9.9	0.3	-0.1	1.1	0.9
80 / 4	6.7	8.6	6.9	9.5	0.3	-0.1	0.8	0.8
20 / 6	0.8	1.3	1.6	1.1	0.7	-0.2	0.4	-0.8
40 / 6	1.0	4.1	0.8	3.8	0.2	0.3	0.2	0.1
60 / 6	0.4	4.5	0.7	5.0	0.2	0.2	0.2	0.1
80 / 6	0.7	4.2	0.3	4.5	0.2	0.1	0.3	0.3
20 / 8	1.4	1.7	1.0	0.6	1.0	-0.2	1.4	-0.7
40 / 8	1.0	6.6	1.1	4.1	0.2	0.2	0.7	0.2
60 / 8	0.7	5.3	0.8	5.2	0.1	0.3	0.3	0.4
80 / 8	0.4	4.8	0.9	5.5	0.1	0.2	0.3	0.5
Average	3.0	4.7	3.3	5.1	0.5	0.2	0.7	0.4

The MA was programmed in Java JDK 2.0 and run using a 366 MHz Pentium II Celeron. The TS were programmed in C and executed in a Sun Sparc 10 Workstation. The Java JDK 2.0 and the C compiler are very similar in terms of speed, with some advantage for the C compiler. On the other hand, the Sun Sparc 10 is a little slower than the Pentium II Celeron. For this reason we believe the speeds of both systems are somewhat equivalent, although this conclusion might not be accurate.

The Table 27.4 shows a comparison between the fast tabu and the MA approaches for the PMS problem. Looking at the averages row, it is clear that the methods had very different behaviours considering the instances structures. The non-structured instances were easier for the TS than for the MA. The opposite occurred for the non-structured ones. The deviations are very small in terms of percentage points, except for a few configurations where some figures are close to 10%. The negative figures in some of the MA columns mean that the makespans found were better than the upper bounds provided by the long tabu strategy.

There is a clear degradation in the MA as the number of machines increases, while the TS maintains an average performance, independently of the problem size. There is a strong probability that this was due to the local search employed in the MA. The simple *all-pairs* and *insertion* neighborhoods could be better suited for this problem if they utilized information about the cut-points. In fact, both local search and crossover operators are dealing with the PMS the same way they would deal with a SMS problem; all alleles are being treated equally. There is no distinction between jobs and cut-points from the genetic operators' point-of-view. More intelligent operators should use information of the problem's structure and the cut-points presence in the chromosome. For instance, the separation of the *job-*

to-machine assignment and the *intra-machine* scheduling in the local search seems to be a reasonable starting point. A neighborhood reduction similar to that employed in the SMS should also be a good choice. Nevertheless, despite the relative lack of problem-driven properties, the general-use crossover and local search operators performed quite well against the well-tailored fast tabu. That increases our belief that the MA's refined structure is playing an important role in the algorithm's performance. Furthermore, as there is plenty of space for improvement in the operators, we believe that future contributions in this issue will probably make the MA surpass the TS in most instances types.

27.4 The flowshop scheduling problem

The last part of this chapter addresses a *flowshop* problem (FS), which main characteristic is to group the jobs in families. This is a quite common real-manufacturing characteristic since manufacturers want to take advantage of *group technology* (GT) environment (Schaller et al. 2000). In the GT scheduling problem, a part family is composed of several parts (in this work represented by the jobs) that have similar requirements in terms of tooling, setup costs and operations sequences. Usually, the families are assigned to a manufacturing cell based on operation sequences so that materials flow and scheduling are simplified. This process may result in a situation where each family is processed by a certain set of machines, and all jobs (parts) are processed following the same technological order.

In this production environment, the manufacturing cells resemble the traditional flowshops except for the existence of multiple part families. Since the jobs in the same family share similar tooling and setup requirements, usually a negligible or minor setup is needed to change from one part to another and thus can be included in the processing times of the jobs. However, a major sequence-dependent setup is needed to change the processing environment between two part families.

The FS with families of jobs is a difficult combinatorial problem proved to be NP-hard. When the number of jobs in each family equals one, the problem becomes a traditional FS problem with sequence-dependent setup times. This problem is proved to be NP-hard when the number of machines is greater than one (Gupta and Darrow 1986). Next, it is shown the description of the FS problem being addressed in this section.

Input: Let n be the number of jobs and m be the number of machines. All the jobs are processed following the same technological order, creating the flowshop structure. Let f be the number of families. Consider also a setup time to change the production from one family to another. Let $\{S_{i,j}^{l}\}$ represent these setup times, where $s_{i,j}^{l}$ is the setup time of family j after family i was processed, in machine l. Finally, let $\{P_{i,j}\}$ be a matrix of processing times, where $p_{i,j}$ is the processing time of job i in machine j.

Task: Find the permutation of the families, and of the jobs within these families, which minimizes the production *makespan*. Calculating the makespan for this problem is not an easy task. Let us initially suppose that the families are scheduled in the order $\{\pi(1), \pi(2), ..., \pi(f)\}$and the order of the jobs within family f is given by the sequence $\{\sigma_f(1), \sigma_f(2), ..., \sigma_f(n_f)\}$, where n_f is the number of jobs in family f. Moreover, let $tt^m_{\sigma_f(i)}$ be the total processing time within family f, until job $\sigma_f(i)$, in machine m; i.e., the time span from when the machine finished its setup and is ready to process the first job of family f until the job $\sigma_f(i)$ is finished. This value can be calculated as:

$$tt^m_{\sigma_f(i)} = it^m_{\sigma_f(i)} + \sum_{z=1}^{i} p_{\sigma_f(z),m} \tag{27.5}$$

Where $it^m_{\sigma_f(i)}$ is the *idle time*[6] within family f accumulated until job $\sigma_f(i)$. In the first machine, no idle times are allowed, and the production flows without any interruption. That simplifies the equation, making it become the sum of the job's processing times within family f. But in the other machines, idle times might occur, depending on the schedule. The completion time of the i-th job of the f-th family in machine m, represented by $c^m_{\sigma_{\pi(f)}(i)}$ can thus be calculated as:

$$c^m_{\sigma_{\pi(f)}(i)} = \underbrace{\sum_{z=1}^{f-1}\left[tt^m_{\sigma_{\pi(z)}(n_{\pi(z)})} + s^m_{\pi(z),\pi(z+1)}\right]}_{\text{total time before family } \pi(f)} + \underbrace{tt^m_{\sigma_{\pi(f)}(i)}}_{\text{total time within family } \pi(f)} \tag{27.6}$$

The first part of the Eq. 27.6 calculates the total processing time before the f-th family, taking into account all the setup times, processing times and idle times before it. The second part calculates the total processing time within family $\pi(f)$ (processing times + idle times) until job $\sigma_{\pi(f)}(i)$. Now let us explicit the idle times calculation. Idle times occur always when a machine finishes processing a certain job, or completes the family setup, and the next job is still being processed in the previous machine. That creates a gap in the schedule, forcing the machine to wait until the next job becomes available. The idle time within family $\pi(f)$, accumulated until job $\sigma_{\pi(f)}(i)$ can be calculated as:

$$it^m_{\sigma_{\pi(f)}(i)} = \underbrace{\max\left(0, c^{m-1}_{\sigma_{\pi(f)}(1)} - c^m_{\sigma_{\pi(f-1)}(n_{\pi(f-1)})} + s^m_{\pi(f-1),\pi(f)}\right)}_{\text{idle time just before the first job of the } f\text{-th family}} + \tag{27.7}$$

$$+ \underbrace{\sum_{z=1}^{i} \max\left(0, +c^{m-1}_{\sigma_{\pi(f)}(z)} - c^m_{\sigma_{\pi(f)}(z-1)}\right)}_{\text{idle times within jobs and before job } \sigma_{\pi(f)}(i)}$$

The first part of the Eq. 27.7 calculates the idle time before the *first job* of the f-th family. For doing so, it uses information about the completion time of the previous family's *last job*, plus the setup time between the f-th family and its prede-

[6] Idle times are periods when the machine is not operating because it is waiting for the next job to become available.

cessor. The second part adds up the idle times between every two consecutive jobs of the f-th family, until job $\sigma_{\pi(f)}(i)$. The completion times of the previous job in the present machine (m) and of the present job in the previous machine (m-1) are utilized. The makespan is then calculated iteratively, job by job, machine by machine, being represented by the last job's completion time in the last machine.

In view of the NP-hard nature of the general FS problem, most researchers have focused on developing heuristic procedures that provide good permutation schedules (in which the order of job processing is the same on all machines) within a reasonable amount of computational time. However, there is no guarantee that a permutation schedule will be optimal when the shop contains several machines. In fact, it is very likely that the optimal schedule will have different job-permutations in each machine. However, considering different job-permutations increases the resulting computational complexity so much that the problem becomes intractable even for very small instances. The usual approach is then to assume the same job-permutation for all machines, reducing the problem's complexity.

Recent reviews (Allahverdi et al. 1999; Cheng et al. 2000) showed that most prior research on manufacturing cell scheduling has assumed sequence-independent setup times. For the flowshop manufacturing cell scheduling problem involving sequence-dependent setup times, (Hitomi et al. 1977) described a simulation model and showed that the scheduling rules considering explicitly sequence-dependent setups outperformed rules which did not explicitly do so. Realizing this, (Schaller et al. 2000) developed and tested several heuristic algorithms for minimizing the makespan in a FS with sequence-dependent family setup times.

27.4.1 The test instances

The instances utilized in this paper are the same presented in (Schaller et al. 2000) and are divided into three classes. In each class, processing times are random integers following a discrete uniform distribution DU(1, 10). As in the previously presented scheduling problems, the hardness depends on the balance between average processing times and the average setup times. Due to this processing times/setup times relation, three different classes of problems were used in the computational experiments. The setup times follow discrete uniform distributions in the ranges:

- Small setup times (SSU): DU(1, 20)
- Medium setup times (MSU): DU(1, 50)
- Large setup times (LSU): DU(1, 100)

According to these definitions, in the SSU-class instances, the ratio defined by average family setup time/average job processing time is approximately 2:1; in the MSU class, the ratio is 5:1, and in the LSU the ratio jumps to 10:1.

Moreover, problems were generated with the number of families varying between 3 and 10, and the number of jobs per family between 1 and 10. The number of machines varied between 3 and 10. For each combination of problem parameters, there are 30 problem instances. As an example of the notation, consider the

LSU108 set of problems: it consists of 30 instances with setup times in the [1, 100] interval, 10 families of jobs, 10 jobs per family at maximum and 8 machines. Considering all configurations tested, we obtain a total of 900 problem instances.

27.4.2 The memetic algorithm approach

The MA utilized for the FS problem follows the same structure of the PMS one. In order to describe the peculiarities of the algorithm for this specific problem, we will begin with the individual representation.

The FS scheduling problem has a structure that allows its division into two parts: the schedule of the families and the schedule of the jobs within the families. The representation adopted takes this division into account and is illustrated in Fig. 27.6. It consists of an arbitrary solution for a problem with twelve jobs and four families. Family 1 consists of jobs [1, 2, 3]; family 2 consists of jobs [4, 5, 6, 7]; family 3 of jobs [8, 9] and family 4 consists of jobs [10, 11, 12]. The processing sequence of the families is [1, 4, 2, 3]. Moreover, the jobs in family 1 are processed in the sequence [2, 1, 3]; family 2 in the order [7, 4, 6, 5], family 3 in the sequence [8, 9] and finally family 4 in the order [11, 12, 10].

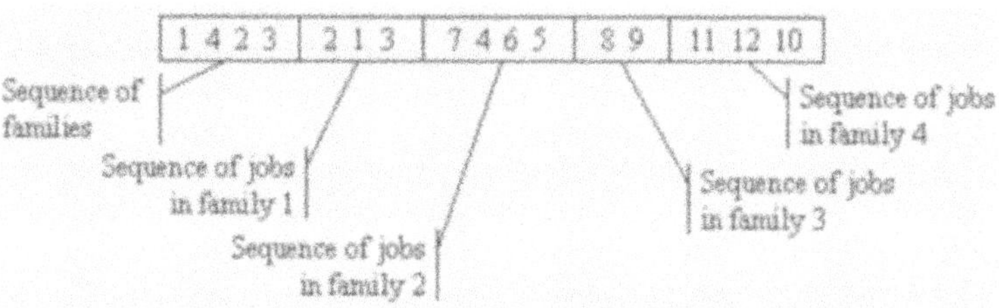

Fig. 27.6. Diagram of the individual representation utilized in the FS problem

The representation divides the solution into f+1 independent parts making the local search, crossover, mutation and other chromosome-level operators be executed separately within each part, without affecting the rest of the solution – which is reminiscent to divide-and-conquer strategies. The computational effort required by the local search is especially reduced due to this chromosome division.

The crossover utilized was the OX, already described in previous sections. The difference is that the OX is applied within each part of the chromosome, separately (see Fig. 27.7).

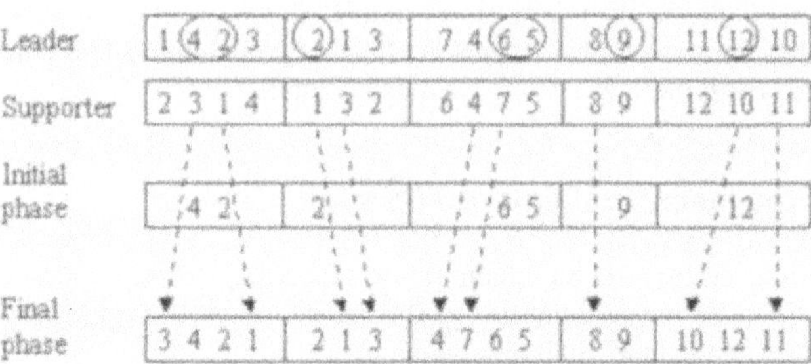

Fig. 27.7. Diagram of the OX crossover utilized in the FS problem

In Fig. 27.7, the OX crossover starts by selecting parts of the *leader*'s genetic material to be copied into the offspring. In the example, these parts are circled. This information is copied into the chromosome labeled as 'Initial phase'. Note that only one piece is copied from each part of the chromosome. In the *final phase*, the *supporter* parent completes the offspring with its genetic information, with each part being filled from left to right, following the sequence of non-repeated alleles.

The local search scheme adopted is the same utilized in the PMS problem: *all-pairs* plus *insertion* neighborhoods, without neighborhood reduction. In this problem, the separation of the chromosome into parts has reduced the local search computational effort and no neighborhood reductions were necessary. Just to illustrate the effort-reduction effect when applying local search in an individual with five families and 10 jobs per family, the algorithm had in fact to apply the local search in six 10-allele sized individuals. A much easier task than applying a single local search in an individual with 60 alleles.

27.4.3 The flowshop computational results

The previous best results for the FS problem available in the literature were present in (Schaller et al. 2000). In that work, high quality lower bounds are provided, and the best heuristic for the problem is named CMD. It employs a dispatch rule together with a local search-based method.

The MA was programmed in Java JDK 2.0 and run using a Pentium II 266 MHz. Since the CMD is a constructive heuristic with a local search phase, it runs very quickly. The CPU times are below three seconds. For the MA, after a few considerations, we concluded that a 30-second limit was sufficient, given the instances' sizes. Therefore, the algorithm stops only if the lower bound is reached - i.e., the optimal solution is found – or after 30 seconds of CPU time. The Tables 27.5, 27.6 and 27.7 show the results for the MA and the CMD algorithm considering the LSU, MSU and SSU-type instances, respectively.

Table 27.5. Results for the LSU-type flowshop instances

	Memetic algorithm				CMD heuristic			
Instance	Min.	Ave.	Max.	CPU	Min.	Ave.	Max.	CPU
LSU33	0.00_{27}	0.07	1.12	3.1	0.00_{21}	0.91	8.41	0.04
LSU34	0.00_{20}	0.32	2.43	10.1	0.00_{12}	1.08	16.39	0.06
LSU44	0.00_{20}	0.20	1.09	10.1	0.00_{8}	1.95	10.27	0.12
LSU55	0.00_{18}	0.28	1.86	12.1	0.00_{4}	2.49	9.57	0.21
LSU56	0.00_{9}	0.51	2.42	21.1	0.00_{4}	3.37	17.07	0.26
LSU65	0.00_{15}	0.31	2.42	15.1	0.00_{3}	3.29	10.02	0.30
LSU66	0.00_{15}	0.19	1.36	15.3	0.00_{3}	3.03	10.47	0.44
LSU88	0.00_{6}	0.58	1.86	24.7	0.28_{0}	6.25	18.17	0.95
LSU108	0.00_{1}	0.47	1.19	29.8	0.12_{0}	6.22	11.25	1.82
LSU1010	0.00_{1}	0.77	2.27	29.5	0.53_{0}	6.30	11.42	2.37
Average	0.00_{132}	0.37	1.80	17.1	0.09_{55}	3.49	12.30	0.66

The results in Table 27.5 reveal an impressive performance for the MA. The figures represent percentage deviations from the lower bounds. In the 'Min'-labeled columns, the subscript figure is the number of times that the algorithm reached the lower bound, finding the optimal solution. In the 'Average' row, the subscript marks the sum of all optimal solutions found. The MA surpassed the CMD performance in all instance configurations. The large number of optimal solutions found is an indicative of the high quality of the lower bounds presented in (Schaller et al. 2000). We must emphasize that each instance configuration set was composed of 30 different instances. The total number of instances tested in each table is 300. Analyzing the MA against the CMD, the MA found 132 (44% of the entire set) optimal instances, more than twice the number found by the CMD. All the averages were better, except the CPU time, as expected. These results do not cause surprise, since the MA employs much more problem-driven features than the CMD algorithm.

An interesting characteristic is present in the results and is worth to be emphasized. The average percentage deviation from the lower bounds remains at low levels for all instances' sizes. Usually, the search methods lose performance for the larger instances, in an indication that they are gradually getting beyond the algorithm's search capabilities. This has occurred with the CMD algorithm, which began with an average deviation of 0.91% and ended with a 6.30% deviation. However, this was not observed for the MA, leading to the conclusion that probably the algorithm is still very far from its limit and would be able to deal with instances larger than the ones utilized. Although there is a slight increase in the MA's averages as the instances become larger, this is probably due to the declining lower bounds' quality.

Table 27.6. Results for the MSU-type flowshop instances

	Memetic algorithm				CMD heuristic			
Instance	Min.	Ave.	Max.	CPU	Min.	Ave.	Max.	CPU
MSU33	0.00_{23}	0.37	3.37	7.1	0.00_{21}	0.92	11.46	0.04
MSU34	0.00_{17}	0.56	2.29	13.1	0.00_{11}	2.00	16.28	0.05
MSU44	0.00_{11}	0.50	2.32	19.1	0.00_{4}	1.96	11.11	0.13
MSU55	0.00_{15}	0.45	2.09	15.1	0.00_{4}	3.10	8.48	0.19
MSU56	0.00_{6}	0.87	3.12	24.1	0.00_{1}	3.58	13.13	0.27
MSU65	0.00_{14}	0.36	1.22	16.2	0.00_{3}	3.68	8.88	0.30
MSU66	0.00_{8}	0.50	1.63	22.7	0.00_{1}	4.59	15.77	0.40
MSU88	0.00_{3}	0.99	2.98	27.3	0.63_{0}	5.68	12.68	0.97
MSU108	0.00_{1}	0.86	1.80	30.1	2.89_{0}	6.11	10.83	1.84
MSU1010	0.15_{0}	1.15	2.53	30.8	2.29_{0}	5.73	9.92	2.37
Average	0.01_{98}	0.66	2.33	20.6	0.58_{45}	3.73	11.85	0.65

The Table 27.6 shows a similar performance, with the MA surpassing the CMD in all criteria but the CPU time. The MA has reached fewer optimal solutions (98 in total, almost 33% of the total), but that is a number still two times larger than the one obtained by the CMD algorithm. The averages are a little worse than the ones for the LSU-type instances, but we believe this is due to the decreasing quality of the lower bounds. Both instance sets (LSU and MSU) have similar characteristics, such as number of machines and families. Moreover, the MA does not employ any procedure, like local search reduction policies, which could be affected by a change in the setup time's interval. Therefore, no reduction in the number of optimal solutions found should be observed at all. The lower bounds might in fact work better with larger setup times. This vision is reinforced by a reduction in the number of optimal solutions found by the CMD algorithm, too.

Table 27.7. Results for the SSU-type flowshop instances

	Memetic algorithm				CMD heuristic			
Instance	Min.	Ave.	Max.	CPU	Min.	Ave.	Max.	CPU
SSU33	0.00_{23}	0.31	2.47	7.1	0.00_{18}	0.67	4.13	0.04
SSU34	0.00_{15}	0.83	2.94	15.1	0.00_{8}	1.85	8.46	0.05
SSU44	0.00_{15}	0.57	2.90	15.0	0.00_{8}	1.94	8.33	0.12
SSU55	0.00_{4}	0.92	2.34	26.0	0.00_{1}	3.15	6.61	0.20
SSU56	0.00_{3}	1.56	3.08	27.4	0.00_{1}	4.02	8.93	0.25
SSU65	0.00_{6}	0.99	3.44	24.0	0.00_{2}	3.00	6.90	0.32
SSU66	0.00_{1}	1.28	2.72	29.1	1.24_{0}	4.06	9.16	0.44
SSU88	0.28_{0}	1.85	3.31	30.3	3.20_{0}	5.62	8.59	0.91
SSU108	0.72_{0}	1.77	2.90	30.7	2.58_{0}	5.63	8.96	1.82
SSU1010	0.59_{0}	2.33	3.65	30.6	4.07_{0}	6.86	9.11	2.21
Average	0.16_{67}	1.24	2.98	23.5	1.11_{38}	3.68	7.92	0.64

In Table 27.7, one can see that the MA maintained its behaviour, obtaining good general average performance followed by a decrease in the number of optimal solutions found. The number dropped to 22% of all instances. The loss of performance with the increasing instances' size is now clear, with the MA being unable to reach the lower bounds for instances with eight or more families of jobs.

Given previous experiences with local searches and how they behave with different sizes of solution spaces, it is likely that the number of optimal solutions found is much larger than the ones reported. Finding the optimal permutation in 10-allele sequences is an easy task when all-pairs and insertion neighbourhoods are employed.

27.5 Discussion

This chapter presented three job scheduling problems: Single Machine Scheduling (SMS), Parallel Machine Scheduling (PMS) and Flowshop Scheduling (FS). In order to deal with these problems a MA was employed. The algorithm was very similar to that introduced in Chap. 18 for the Gate Matrix Layout problem. The MA had an impressive performance for the problems, what reinforces the belief that the algorithm's main features are general enough to deal with a much broader variety of problems.

In the SMS section, a procedure to transform Asymmetrical Travelling Salesman Problem's (ATSP) instances into SMS ones was described. By using it, one can create a SMS instance with known high-quality solutions if an ATSP instance with a known optimal tour is utilized as the starting point. Although this procedure is not yet proved to create optimal SMS instances, during the tests using TSPLIB instances, no counter-example was ever found. If the procedure is proved to be correct – or under which circumstances it is correct – it would solve a problem quite common is SMS tests: The lack of large optimal instances to test algorithms. In the literature, there are ATSP instances with thousands of cities solved to optimality. Such instances could be used to create SMS ones, also with thousands of jobs and known optimal solution. The procedure has some drawbacks, like a limitation on how the processing times and the due dates must be generated and the fact that the resulting instances are EDD/SPT-optimally solvable. Nevertheless, such instances can be very useful to test the performance of general-use metaheuristics.

The MA was able to find optimal solutions for instances with up to 100 jobs, with a high rate of success and CPU times no longer than four minutes, in average. The instances with 71 jobs or less had an impressive 100% rate of success. The 100-job instances also had a high rate of success, but apparently, they mark the beginning of the MA's search power exhaustion. For larger instances, new features will probably have to be added to the MA in order to sustain the performance, or more CPU time will have to be given.

In the PMS problem, the MA is compared to a well-tailored *tabu search* (TS), named *fast tabu*. The TS results were the best previously available in the literature (França et al. 1996) for the problem. Instances with up to 80 jobs and 8 machines were tested, as well as four setup times configurations. The MA performance was comparable to the TS especially for the so-called *structured* instances, which setup times follow the triangular inequality, and for problems with fewer machines, independently of the number of jobs. In this problem, the lack of a neighborhood-reduction policy has weighted against the MA. Although the performance was not disappointing, it could have been better with a reduction policy being applied. This is the logical next step for this problem. Both MA and TS were evaluated against lower bounds and optimal solutions for the smaller instances. For the larger instances, upper bounds obtained through a high-performance TS, named *long tabu*, were the benchmark performance measure. However, it is important to emphasize that the long tabu algorithm requires long CPU times, usually hours, against the few seconds or minutes required by the MA and the fast tabu.

The last problem addressed in the chapter was the flowshop scheduling (FS). In addition to the ordinary FS problem, this one also considers that the jobs are grouped in families, with jobs in the same family requiring similar tooling and machinery. That makes the setup times between jobs within the same family be very small, so that they can be added to the job's processing time. However, major setup times are required to change the production from one family to another. The most remarkable feature of the MA for this problem was the representation utilized. It reduced the search space for the problem so much that even large-size instances, with 10 families of jobs and 10 machines were still below the MA's search capability.

Although its importance for the industry, this FS-variant has received little attention in the past, and the previous best method available in the literature was a constructive rule followed by a local search procedure, named CMD (Schaller et al. 2000). In this same paper, several lower bounds were calculated and utilized as benchmarks. The MA presented in this chapter is probably the first metaheuristic approach for it. The performance of the MA has dramatically surpassed the CMD algorithm, as was expected. But most impressive was the number of optimal solutions found, a situation resultant from the high quality of the lower bounds and the superior performance of the MA.

The MA utilized in all problems presented in this chapter is available in the NP-Opt Framework (Mendes et al. 2001). As said in the Chap. 18, the NP-Opt is an object-oriented, Java-based software, which is updated and improved continuously by a team of collaborators. For more information, please refer to the NP-Opt homepage[7], where the latest version is always available for download, as well as the software guide and the instances utilized in this chapter.

[7] http://www.densis.fee.unicamp.br/~smendes/NP-Opt.html.

References

Allahverdi A, Gupta JND, Aldowaisan T (1999) A survey of scheduling research involving setup considerations. OMEGA – International Journal of Management Science 27:219–239

Baker KR (1974) Introduction to Sequencing and Scheduling. John Wiley & Sons, New York

Cheng TCE, Gupta JND, Wang G (2000) A review of flowshop scheduling research with setup times. Production and Operations Management 9:283–302

Dearing PM, Henderson RA (1984) Assigning looms in a textile weaving operation with changeover limitations. Production and Inventory Management 25:23–31

Du J, Leung JYT (1990) Minimizing total tardiness on one machine is NP-hard. Mathematics of Operations Research 15:483–495

França PM, Gendreau M, Laport G, Muller F (1996) A tabu search heuristic for the multiprocessor scheduling problem with sequence dependent setup times. International Journal of Production Economics 43:79–89

França PM, Mendes AS, Moscato P (2001) A Memetic Algorithm for the total tardiness Single Machine Scheduling Problem. European Journal of Operational Research 132:224–242

Frederickson G, Hecht MS, Kim CE (1978) Approximation algorithm for some routing problems. SIAM Journal on Computing 7:178–193

Gen M, Cheng R (1997) Genetic Algorithms and Engineering Design. John Wiley & Sons, New York

Goldberg DE (1989) Genetic Algorithms in Search, Optimization, and Machine Learning. Addison-Wesley

Graham RL, Lawler EL, Lenstra JK, Rinooy Kan AHG (1979) Optimization and approximation in deterministic sequencing and scheduling: A survey. Annals of Discrete Mathematics 5:287–326

Graves SC (1981) A review of production scheduling. Operations Research 29:646–675

Gupta JND, Darrow WP (1986) The two-machine sequence dependent flowshop scheduling problem. European Journal of Operational Research 24:439–446

Hitomi K, Nakamura N, Yoshida T, Okuda K (1977) An Experimental Investigation of Group Production Scheduling. Proceedings of the 4th International Conference on Production Research, pp 608–617

Lawler EL, Lenstra JK, Rinooy Kan AHG, Shmoys DB (1993) Sequencing and Scheduling: Algorithms and Complexity. In: Handbooks in Operations Research and Management Science Vol. 4. North-Holland, pp 445–522

Lee YH, Bhaskaran K, Pinedo M (1997) A heuristic to minimize the total weighted tardiness with sequence-dependent setups. IIE Transactions 29:45–52

Mendes AS, França PM, Moscato P (2001) NP-Opt: An Optimization Framework for NP Problems. Proceedings of the POM2001 – International Conference of the Production and Operations Management Society, pp 82–89

Mendes AS, Muller F, França PM, Moscato P (2002) Comparing Meta-Heuristic Approaches for Parallel Machine Scheduling Problems. Production Planning and Control 13:143–154

Ow PS, Morton TE (1989) The single machine early/tardy problem. Management Science 35:177–191

Ragatz GL (1993) A branch-and-bound method for minimum tardiness sequencing on a single processor with sequence dependent setup times. Proceedings of the 24th Annual Meeting of the Decision Sciences Institute, pp 1375–1377

Raman N, Rachamadugu RV, Talbot FB (1989) Real time scheduling of an automated manufacturing center. European Journal of Operational Research 40:222–242

Rubin PA, Ragatz GL (1995) Scheduling in a sequence dependent setup environment with genetic search. Computers & Operations Research 22:85–99

Schaller JE, Gupta JND, Vakharia AJ (2000) Scheduling a Flowline Manufacturing Cell with Sequence Dependent Family Setup Times. European Journal of Operational Research 125:324–339

Sumichrast R, Baker JR (1987) Scheduling parallel processors: an integer linear programming based heuristic for minimizing setup time. International Journal of Production Research 25:761–771

Tan KC, Narasimhan R (1997) Minimizing tardiness on a single processor with sequence-dependent setup times: a simulated annealing approach. OMEGA – International Journal of Management Science 25:619–634

28 Determination of Optimal Machining Conditions Using Scatter Search

Mu-Chen Chen and Kai-Ying Chen

28.1 Introduction

The advent of modern computer technology and a new generation of manufacturing equipment, particularly computer numerical control (CNC) machine, have brought enormous changes to the manufacturing industry. The process planning of CNC machining runs through all necessary steps and information to manufacture a part. Process plans are established through rational determination of machines, tools, cutting fluids, machining conditions (parameters), etc., for each operation of a workpiece. In process planning of CNC machining, selecting reasonable machining parameters is necessary to satisfy requirements involving machining economics, machining quality and machining safety. The machining parameters in turning operations consist of cutting speed, feed, depth of cut and number of passes. These machining parameters significantly affect the cost, productivity and quality of machined parts.

Traditionally, machining parameters have been established using data delivered by the tool manufacturer or cutting data handbooks. These handbooks include tables with recommended machining parameters for a wide variety of workpiece-tool materials. This approach of selecting machining conditions has several drawbacks (Boogert et al. 1996):

1. The storage space of all the data will grow very fast owing to a combinational explosion.
2. It is difficult to ensure the consistency of the data.
3. The recommended values cover a wide range; the user needs to interpolate between the upper and lower values and only cutting experts know the mutual influences of the various machining conditions.
4. The handbook approach is not suitable for the automatic generation of optimal machining conditions because the optimization criteria and machining constraints cannot be included in data table.

To overcome the aforementioned drawbacks, considerable research has been done to establish an adequate description of the machining characteristics by mathematical models. The machining models can reduce the storage requirements and facilitate the automatic generation of machining conditions using optimization techniques. Equations are used to predict the machining process characteristics, such as cutting force, required power, surface finish, tool life and chip-tool interface temperature. Generation of optimal machining conditions is generally based on optimizing certain economic criteria, which are subjected to a set of machining constraints. Using CNC machines, cutting processes can be repeated easily and the improvements in modern shop floor control systems allow data acquisition during processing from which machining models can be adopted.

Various solution approaches have been used to optimize turning operations. However, there exists few generalized solution method for all machining optimization problems (Chang et al. 1991). The existing methods differ in their reliability, efficiency and sensitivity to initial solution. Furthermore, they are only useful for a specific problem or are inclined to obtain a local optimal solution. Simulated annealing (SA) (Metropolis et al. 1953, Kirkpatrick et al. 1983) and genetic algorithm (GA) (Holland 1975, Goldberg 1989) are two types of heuristic search techniques, which are promising for investigation of complex optimization problems. Both these techniques are modeled on processes in nature (thermodynamics and natural evolution, respectively). They both can solve optimization problems possessing quite arbitrary degrees of nonlinearity, discontinuity and stochasticity. Some general approaches based on SA and GA were successfully implemented in machining condition optimization (Chen and Tsai 1996, Khan et al. 1996, Chen and Su 1998, Chen and Tseng 1998, Su and Chen 1999).

More recently, Scatter Search (SS)(Glover 1997, Glover 1999, Glover et al. 2000, Laguna and Marti 2003) increasingly gets focus in the area of evolutionary approach. SS originates from strategies for combining decision rules and surrogate constraints. The advantages of this approach for solving a diverse array of complex optimization problems (e.g., vehicle routing, arc routing quadratic assignment, job shop scheduling and form error assessment) have been demonstrated in the recent research (Glover 1999, Glover et al. 2000). The potential of SS for machining condition optimization is investigated herein.

28.2 Fundamentals of CNC Turning

Modern computer technology has a significant influence on engineering design and manufacturing in terms of reducing design and manufacturing cost, inventory, lead time as well as increasing productivity and product quality. The CNC turning machines continue to be the mainstay of rotational-part metal-removal machine tools. Before the topic of machining condition optimization for turning operations is discussed, it is worthwhile to discuss the basic concepts of turning. Readers are referred to Lin (1994) for the further details.

28.2.1 CNC turning machine axes

There are a variety of CNC turning machines, ranging from a simple two-axis lathe to a multi-axis machining center. Most CNC controls for turning machines provide for control of two basic axes and, with the expansion option, up to four axes (Lin 1994). In this chapter, only two-axis CNC turning machine is discussed.

A basic CNC turning machine uses only two axes: Z and X. The Z-axis is parallel to the spindle axis and represents the carriage travel in the longitudinal direction. The X-axis is perpendicular to the Z-axis and represents the cross slide travel. The direction designations for the Z and X axes are given as follows:

- +Z: Tool carriage moves away from the spindle head;
- -Z: Tool carriage moves toward the spindle head;
- +X: Cross slide moves away from the spindle axis;
- -X: Cross slide moves toward the spindle axis.

28.2.2 CNC turning operations

Various types of machining operations can be performed on a CNC turning machine. These operations generally are of seven types (Lin 1994): turning (straight, taper, circular turning), facing, grooving, drilling, boring, parting off and threading. The most common type of turning operations is external turning, which removes material by rotating the workpiece against a single-point cutter. Machined parts on the CNC turning machines typically have continuous external profiles, which require linear and circular interpolations. This chapter focuses on the optimization of machining conditions for turned parts requiring external turning of linear interpolation.

Turning using linear interpolation simply means machining in straight lines. These lines may be horizontal, vertical or at an angle to Z direction. When turning, each cut under linear interpolation is often followed by a rapid move, back to the start point, ready to take a subsequent cut. Before returning to the start point, it is practical to retract the tool slightly, from the surface of the workpiece, to avoid scouring the work surface. Only straight turning is considered herein. Straight turning involves the cutting of a workpiece in the longitudinal (Z) direction to produce a constant stock diameter.

28.2.3 CNC turning conditions

The functions included in process planning are: raw material preparation, process selection, process sequencing, machining parameter selection, tool path planning, machine selection, tool selection and fixture selection (Lin 1994). The process planner then uses part features to specify machining processes and their operation sequence, types of cutting tools, cutter path, work holding devices or fixtures to be used, and the machining parameters (number of cuts, cutting speed, feed and depth of cut).

The two most important considerations in any CNC operations are accuracy and efficiency (Lin 1994). Accuracy refers to the dimensional precision of the workpiece, which is determined by two factors: program accuracy and machine accuracy. Machine accuracy is a complex measure that relies on the quality of the servo control system, measuring devices, machine drive system as well as the suitable selection of machining parameters. Efficiency refers to the production rate or production cost, that is, the time or money it takes to produce a workpiece. The determination of machining conditions (parameters) dictates the efficiency of the machining process.

Machine tool manufacturers normally supply machining data, including cutting speed, feed, and depth of cut, for their machine tools. This should be the starting point for determining cutting speed and feed. A trial-and-error approach is then followed to determine the machining conditions for a particular operation. This approach of selecting machining conditions has several drawbacks as discussed in the previous section. To overcome the aforementioned drawbacks, many researchers try to establish machining models and develop optimization algorithms, which can reduce the storage requirements and facilitate the automatic generation of machining conditions using optimization techniques.

28.2.3.1 Cutting speed

Cutting speed in machining operations refers to the speed at which the cutting edge of the tool passes over the surface of the workpiece (Lin 1994). It is also called the surface speed, and defined as the maximum linear speed between the tool and workpiece. The cutting speed for turning can be determined as a function of the workpiece and rotation speed. The following formula is generally used to determine cutting speed for turning operations:

$$V = \frac{\pi DN}{1000} \tag{28.1}$$

where V = surface cutting speed, m/min (MPM);
D = diameter of rotating part, mm; and
N = rotation speed of the spindle, revolutions/min (RPM).

28.2.3.2 Feed and feed rate

Feed can be defined as the relative lateral movement between the tool and workpiece during a machining operation (Lin 1994). It corresponds to the thickness of the chip produced by the operation. In turning operations, feed is defined as the advancement of the cutter per revolution of workpiece. The typical unit is MMPR (mm per revolution). Feed rate is defined as the speed of feed; the unit is MMPM (mm per minute). In turning operations, the following formula can be used to convert feed to feed rate:

$$f_m = N \times f \tag{28.2}$$

where f_m = feed rate, feed per minute; and
f = feed, feed per revolution.

28.2.3.3 Depth of cut

Depth of cut is the distance the cutter penetrates into the workpiece and is measured in the direction perpendicular to the cutter motion direction (Lin 1994). The depth of cut for roughing cuts is generally larger than for finishing cuts.

28.3 Literature Review

28.3.1 Machining optimization for turning operations

Numerous researchers have formulated mathematical models and used optimization techniques to select the optimal turning parameters. In the area of machining optimization, the following three basic criteria are used for the selection of machining parameters.

1. *The minimum production cost criterion*: This refers to producing a piece at minimum cost.
2. *The minimum production time or the maximum production rate criterion*: This refers to producing a part at the fastest rate.
3. *The maximum profit rate criterion*: This refers to maximization of profit generated per unit time.

The earlier studies (Okushima and Hitomi 1964, Wu and Ermer 1966, Ermer 1971, Boothroyd and Rusek 1976) were limited to the single-pass turning operations without consideration of any constraints. It is now evident that various machining constraints should be satisfied in actual manufacturing applications. The machining constraints usually consider the CNC machine specifications, CNC machine dynamics, cutting tool dynamics and machined part design specifications. Multi-pass operations are used if the total depth of cut to be removed in turning exceeds the maximum allowable depth of cut. The maximum depth of cut for an individual pass is usually restricted by either chatter or the physical dimensions of a particular tool. In addition, the multi-pass operations are preferred to the single-pass machining for economics reasons (Ermer and Kromodihardjo 1981, Tsai 1986, Kee 1994). However, the additional machining parameter, number of passes, makes the solution procedure more difficult than that for a single-pass problem. The more recent studies have been concentrated on determining the optimal machining conditions for multi-pass operations. In the machining models of multi-pass operations (Hati and Rao 1976, Lambert and Walvekar 1978, Yellowley 1978, Ermer and Kromodihardjo 1981, Gopalkrishnan and Al-Khayal 1991), the number of passes and depth of cut for each pass are fixed *a priori* and are kept outside the scope of optimization. Recently, the studies involving machining condition optimization have been directed towards the simultaneous determination of optimal values of cutting speed, feed, depth of cut for an individual pass and number of passes (Agapiou 1992a, Agapiou 1992b, Gupta et al. 1994, Gupta et al. 1995, Kee 1994, Kee 1995, Kee 1996, Mesquita et al. 1995, Narang and Fischer 1993, Shin and Joo 1992, Tan and Creese 1995, Yeo, 1995, Chen and Tsai 1996, Chen and Su 1998, Chen and Tseng 1998, Su and Chen 1999).

28.3.2 Review of machining optimization techniques

There are a host of methods and algorithms in the literature, which can be used to resolve machining optimization models. The solution methodologies used in the previous machining optimization studies have been briefly reviewed in Chen and Tseng (1998). Table 28-1 summaries these methods and algorithms.

Okushima and Hitomi (1964), Armarego and Russell (1966), Wu and Ermer (1966), Yellowley (1978) and Philipson and Ravindran (1979) have developed optimization models for turning conditions and applied the calculus differential based approaches to obtain the optimal conditions. These approaches can obtain the truly optimal solutions. However, they are very restrictive for solving small machining optimization problems.

The geometric programming (GP) (Ermer 1971, Petropoulos 1971, Philipson and Ravindran 1979, Somlo and Nagy 1979, Narang and Fischer 1993) has been frequently utilized to solve machining optimization problems. By introduction of suitable variables and modification of the objective function and constraint set, the machining model can be converted into posynomial form which becomes amenable to solution by a hybrid method combining geometric and linear programming (GP-LP) (Ermer and Kromodihardjo 1981, Gupta et al.1994). GP and GP-LP are the special cases of the general nonlinear programs in which the objective function and constraints are posynomials. The utility of GP and GP-LP for machining optimization problems has been demonstrated by applying these two methods to relatively simple machining models. Somlo and Nagy (1979) stated that GP is more powerful than other methods in optimizing the machining models in which the optimal conditions are restricted by one or two constraints. As the number of constraints increased, other optimization methods should be employed.

Iwata et al. (1974) proposed a probabilistic approach for selecting the optimal cutting conditions and applied sequential unconstrained minimization technique (SUMT) to solve the equivalent deterministic problem. Also, Hati and Rao (1976) utilized SUMT to solve a probabilistic model and a deterministic model formulated for a multi-pass turning operation. However, the number of passes and depth of cut for each pass are prefixed and are not selected by the optimization algorithm.

Iwata et al. (1977) have presented a multi-pass turning model as well as applied an algorithm based on dynamic programming (DP) and stochastic programming to simultaneously determine the optimal values of cutting speed, feed and depth of cut for an individual pass along with the optimal number of passes. They have considered the probabilistic nature of the objective function and constraints and have applied stochastic programming to solve the single pass problem.

Lambert and Walvekar (1978) have also developed a machining model for the multi-pass turning operation under constraints of cutting force, power and surface finish. A two-stage solution procedure combining DP and GP has been utilized to optimize this turning model under minimum production cost. However, they have considered two-pass problems only. DP coupled with GP has also been used to optimize the multi-pass turning conditions by Unklesbay and Creighton (1978). More recently, Agapiou (1992) has used a DP model similar to that of Iwata et al. (1977) for optimizing machining conditions. They have presented a deterministic model with a combined objective function (weighted sum of production cost and time) and used simplex search to determine the optimal machining parameters and objective function.

In addition, Shin and Joo (1992) have developed a machining model for multi-pass turning operations. They have suggested DP coupled with Fibonacci search to resolve this turning problem. DP has been used to the selection of depth of cut for individual passes. The final finish pass is fixed on the basis of the minimal allowable depth of cut and the remaining depth of cut is divided into a number of rough passes of equal sizes such that minimum production cost is obtained. It has been noticed that the above multi-pass turning models are not simple DP problems since the optimal numbers of passes (number of stages in a multi-stage decision system) are not predetermined and they are to be determined during optimization.

Groover (1975) has introduced the Monte-Carlo simulation technique for determining the optimal turning conditions. This technique is mainly based on a mathematical model, which is described by some assumed probability distribution. The actual machining process is replaced by its mathematical model (machining model). The variables (machining parameters) in the mathematical model are then sampled by means of the random number generator.

Some researchers have utilized the direct search techniques as general nonlinear programming methods to select the optimal machining parameters. As mentioned previously, Agapiou (1992) has applied dynamic programming (DP) coupled with simplex search to optimize the machining problems under a combined objective function. Also, Mesquita et al. (1995) have presented an interactive system, MECCANO2, for computer-aided selection of the optimum cutting conditions in multi-pass rough turning operations. Combinations of three criteria including minimum production cost, minimum production time and minimum number of passes are considered simultaneously. Their priority is given by users. Mesquita et al. have used a combination of the direct search methods, the random search and Hooke-Jeeves pattern search. The solution procedure is performed three times from different starting points chosen with respect to feasibility and criterion values out of randomly generated points. Kee (1994, 1995, 1996) has put his efforts into the development of optimization analysis and strategies for multi-pass rough turning on lathes with practical constraints. The multi-pass solution has been solved by using a combination of mathematical analysis of the theoretical economic trends approach and numerical search techniques. Nevertheless, numerical search technique is only used in the final stage of the proposed solution procedure to optimize the integer number of passes and the depth of cut distribution.

Table 28.1. List of methods and algorithms for machining optimization.

Methods and algorithms	References
Calculus differential approach	Okushima and Hitomi (1964), Armarego and Russell (1966), Wu and Ermer (1966), Yellowley (1978), Philipson and Ravindran (1979)
Lagrangian multiplier method	Bhattacharyya et al. (1970)
Geometric programming	Ermer (1971), Petropoulos (1971), Philipson and Ravindran (1979), Somlo and Nagy (1979), Tsai (1986), Narang and Fischer (1993), Gopalkrishnan and Al-Khayal (1991), Choi and Bricker (1996)
GP + linear programming	Ermer and Kromodihardjo (1981), Gupta et al. (1994), Prasad et al. (1997)
Linear goal programming	Sundaram (1978), Philipson and Ravindran (1979)
SUMT	Iwata et al. (1974), Hati and Rao (1976)
Sequential quadratic programming	Yeo (1995)
Linear approximation method	Tan and Creese (1995)
Integer programming	Gupta et al. (1995)
Random search + pattern search	Mesquita et al. (1995)
DP + Simplex search	Agapiou (1992a, 1992b)
Numerical search	Kee (1994), Kee (1995), Kee (1996)
DP + stochastic programming	Iwata et al. (1977)
DP + Fibonacci search	Shin and Joo (1992)
DP + geometric programming	Lambert and Walvekar (1978), Unklesbay and Creighton (1978)
Fuzzy nonlinear programming	Chen et al. (1995)
Monte-Carlo simulation approach	Groover (1975)
Simulated annealing	Chen and Tsai (1996), Khan et al. (1996), Chen and Su (1998), Su and Chen (1999)
Genetic algorithm	Khan et al. (1996), Chen and Tseng (1998)

Yeo (1995) has proposed a multi-pass optimization strategy for CNC lathe operations. He has applied sequential quadratic programming (SQP) to solve this machining optimization problem. The optimization methodology is simplified and it attempts to give near-optimal solutions. A machining parameter determination model has been developed using multi-pass turning operation as a general type by Tan and Creese (1995). An approximation optimization approach, linear approximation method, has been defined for the machining model. The premise conditions for this solution technique are: (1) the initial point must be feasible, and (2) variables are continuous in the defined boundary region around the starting point. In the paper of Chen et al. (1995), a fuzzy expert system called Smart Assistant to Machinist, has been introduced for cutter selection and cutting condition design in turning operations. To accommodate the fuzzy information in their developed model, the fuzzy nonlinear programming method has been adopted. In this system, Chen et al. used Hill climbing method to solve the fuzzy nonlinear program. Recently, SA and GA have been successfully implemented in machining condition optimization (Chen and Tsai 1996, Su and Chen 1997, Chen and Su 1998, Su and Chen 1999).

The existing approaches can be classified into two categories: exact and approximate methods. The exact approaches such as calculus differential method and GP are only useful for a specific problem. The approximate approaches such as SUMT and direct search method mostly obtain the local optimal solutions. Recently, metaheuristics such as SA and GA have attracted attention due to their broad success of applications on complex optimization problems. The potentials of applying SA and GA to machining optimization problems have been investigated in the recent literature. More recently, Scatter Search (SS) increasingly gets focus in the area of metaheuristics. Owing to the high complexity in machining condition optimization, SS is adopted as the solution approach herein.

28.4 Notations in Machining Model

C_0 constant of the tool-life equation

C_I machine idle cost (\$/piece)

C_M cutting cost by actual time in cut (\$/piece)

C_R tool replacement cost (\$/piece)

C_T tool cost (\$/piece)

d_r, d_s depths of cut for each pass of rough and finish machining (mm)

d_{rL}, d_{rU} lower and upper bounds of depth of rough cut (mm)

d_{sL}, d_{sU} lower and upper bounds of depth of finish cut (mm)

d_t total depth of metal to be removed (mm)

D diameter of workpiece (mm)

f feed (mm/rev)

f_r, f_s	feeds in rough and finish machining (mm/rev)
f_{rL}, f_{rU}	lower and upper bounds of feed in rough machining (mm/rev)
f_{sL}, f_{sU}	lower and upper bounds of feed in finish machining (mm/rev)
F_r, F_s	cutting forces during rough and finish machining (kgf)
F_U	maximum allowable cutting force (kgf)
h_1, h_2	constants pertaining to tool travel and approach/depart time (min)
k_f	constant pertaining to cutting force
k_o	direct labor cost + overhead ($/min)
k_t	cutting edge cost ($/edge)
L	length of the workpiece (mm)
n	number of rough passes
N	spindle speed (RPM)
N_L, N_U	lower and upper bounds of n
P_r, P_s	cutting power during roughing and finishing (kW)
P_U	maximum allowable cutting power (kW)
R_a	maximum allowable surface roughness (μ m)
R_n	nose radius of cutting tool (mm)
t	tool life (min)
t_c	constant term of machine idling time (min)
t_e	tool exchange time (min)
t_p	tool life (min) considering roughing and finishing
t_r, t_s	tool lives (min) for roughing and finishing
t_v	variable term of machine idling time (min)
T_I	machine idling time (min)
T_L, T_U	lower and upper bounds of tool life
T_M	cutting time by actual machining (min)
T_{Mr}, T_{Ms}	cutting time by actual machining (min) for roughing and finishing
T_R	tool replacement time (min)
UC	unit production cost except material cost ($/piece)
V	cutting speed (m/min)
V_r, V_s	cutting speeds in rough and finish machining (m/min)
V_{rL}, V_{rU}	lower and upper bounds of cutting speed in rough machining (m/min)

V_{sL}, V_{sU}	lower and upper bounds of cutting speed in finish machining (m/min)
w	weight of tool-life equation
X	vector of machining parameters
α, β, γ	constants of tool-life equation
η	power efficiency
μ, υ	constants of cutting force equation

28.5 The Multi-Pass Turning Model

The objective of this machining model is to determine the optimal machining parameters including cutting speed, feed, depth of cut and number of rough cuts in order to minimize the unit production cost. In addition, such combination of machining parameters should not violate the imposed cutting constraints. A multi-pass turning model taken from Shin and Joo (1992) and Chen and Tsai (1996) is adopted as the test problem for the SS-based machining optimization approach.

28.5.1 The cost function

The unit production cost, *UC*, for multi-pass turning operations can be divided into four basic cost elements (Shin and Joo 1992): (1) cutting cost by actual time in cut, C_M ; (2) machine idle cost due to loading and unloading operations and idling tool motion, C_I ; (3) tool replacement cost, C_R ; and (4) tool cost, C_T . These four cost elements are defined as follows.

28.5.1.1 Cutting Cost

The turning process is divided into multi-pass roughing and finishing. The cutting cost, C_M , can be expressed as

$$C_M = k_o T_M \tag{28.3}$$

where T_M is the actual cutting time, which is given by

$$T_M = T_{Mr} + t_{Ms} \tag{28.4}$$

$$T_{Mr} = \frac{\pi DL}{1000V_r f_r}(n) = \frac{\pi DL}{1000V_r f_r}\left(\frac{d_t - d_s}{d_r}\right) \tag{28.5}$$

$$T_{Ms} = \frac{\pi DL}{1000V_s f_s}. \tag{28.6}$$

Hence,

$$C_M = k_o \left[\frac{\pi DL}{1000 V_r f_r} \left(\frac{d_t - d_s}{d_r} \right) + \frac{\pi DL}{1000 V_s f_s} \right]. \tag{28.7}$$

28.5.1.2 Machine idling cost

The machine idle time is divided into a constant term due to loading and unloading operations and a variable term due to idle tool motion (Shin and Joo 1992). The machine idle time T_I is given by

$$T_I = t_c + t_v. \tag{28.8}$$

where the variable term, idle tool motion time t_v, can be represented in terms of the number of passes and the length of workpiece, thus

$$t_v = (h_1 L + h_2)(n+1) = (h_1 L + h_2)\left(\frac{d_t - d_s}{d_r} + 1 \right). \tag{28.9}$$

Consequently, the machine idling cost, C_I, can be expressed as

$$C_I = k_O \left[t_c + (h_1 L + h_2)\left(\frac{d_t - d_s}{d_r} + 1 \right) \right]. \tag{28.10}$$

28.5.1.3 Tool replacement cost

The Taylor tool-life equation is given by (Armarego and Brown 1969)

$$V\ t^{\alpha} f^{\beta} d^{\gamma} = C. \tag{28.11}$$

The above equation can be rewritten for t

$$t = \frac{C^{1/\alpha}}{V^{1/\alpha} f^{\beta/\alpha} d^{\gamma/\alpha}} = \frac{C_o}{V^p f^q d^r}. \tag{28.12}$$

It is assumed that the same tool is used for the entire machining process of both roughing and finishing. The wear rate of tools usually differs between roughing and finishing because the machining condition is different. The tool life in such a situation can be expressed as

$$t_p = w t_r + (1-w) t_s \tag{28.13}$$

where $$t_r = \frac{C_o}{V_r^p f_r^q d_r^r}, \quad t_s = \frac{C_o}{V_s^p f_s^q d_s^r}. \tag{28.14}$$

The tool replacement time can be written in terms of tool life (t_p), time required to exchange a tool (t_e) and cutting time (T_M). It is given by

$$T_R = t_e \frac{T_M}{t_p}. \tag{28.15}$$

The tool replacement cost, C_R, is given by

$$C_R = k_o T_R. \tag{28.16}$$

By substituting Eqs. (28-4), (28-5), (28-6) and (28-15) into Eq. (28-16), the tool replacement cost can be expressed as

$$C_R = k_o \frac{t_e}{t_p}\left[\frac{\pi DL}{1000 V_r f_r}\left(\frac{d_t - d_s}{d_r}\right) + \frac{\pi DL}{1000 V_s f_s}\right]. \tag{28.17}$$

28.5.1.4 Tool cost

The tool cost, C_T, can be defined by

$$C_T = k_t \frac{T_M}{t_p}. \tag{28.18}$$

By using Eqs. (28-4), (28-5), (28-6) and (28-18), the tool cost can be expressed as

$$C_T = \frac{k_t}{t_p}\left[\frac{\pi DL}{1000 V_r f_r}\left(\frac{d_t - d_s}{d_r}\right) + \frac{\pi DL}{1000 V_s f_s}\right]. \tag{28.19}$$

28.5.1.5 Unit production cost

Based on the above discussion, the unit production cost is defined as the sum of cutting cost C_M, machine idling cost C_I, tool replacement cost C_R, and tool cost C_T. The unit production cost takes the form as

$$\begin{aligned} UC &= C_M + C_I + C_R + C_T \\ &= k_o\left[\frac{\pi DL}{1000 V_r f_r}\left(\frac{d_t - d_s}{d_r}\right) + \frac{\pi DL}{1000 V_s f_s}\right] + \\ &\quad k_o\left[t_c + (h_1 L + h_2)\left(\frac{d_t - d_s}{d_r} + 1\right)\right] + \\ &\quad k_o \frac{t_e}{t_p}\left[\frac{\pi DL}{1000 V_r f_r}\left(\frac{d_t - d_s}{d_r}\right) + \frac{\pi DL}{1000 V_s f_s}\right] + \\ &\quad \frac{k_t}{t_p}\left[\frac{\pi DL}{1000 V_r f_r}\left(\frac{d_t - d_s}{d_r}\right) + \frac{\pi DL}{1000 V_s f_s}\right]. \end{aligned} \tag{28.20}$$

28.5.2 Turning condition constraints

For the optimization of the unit production cost, practical constraints, which present the states of machining process need to be considered. The constraints imposed during the roughing and finishing operations include (1) parameter bounds; (2) tool-life constraint; (3) operating constraints consisting of surface finish constraint (only for finish machining); cutting force constraint and power constraint. They are discussed as follows (Shin and Joo 1992, Chen and Tsai 1996).

28.5.2.1 Rough machining

Parameter bounds

Bounds on cutting speed: The available range of cutting speed is expressed in terms of lower and upper bounds as

$$V_{rL} \leq V_r \leq V_{rU}. \tag{28.21}$$

The lower bound is provided to avoid the formation of built-up edge, whereas the upper bound is setup for the safety of the operator.

Bounds on feed: The feed is restricted as

$$f_{rL} \leq f_r \leq f_{rU}. \tag{28.22}$$

Bounds on depth of cut: The depth of cut is restricted as

$$d_{rL} \leq d_r \leq d_{rU}. \tag{28.23}$$

The bounds on feed and depth of cut are related to the type of cutting tool and the material of workpiece.

Tool-life constraint

Considering production economics and the quality of the machined part, the tool life should be within an acceptable range. The constraint on the tool life is taken as

$$T_L \leq t_r \leq T_U. \tag{28.24}$$

Operating Constraints

Cutting force constraint: It is necessary to put a constraint on the cutting force to limit the deflection of the workpiece or the cutting tool, which would result in dimensional errors, and to reduce the power required for the cutting process. The expression of cutting force constraint takes the form as

$$F_r = k_f f_r^{\mu} d_r^{\upsilon} \leq F_U. \tag{28.25}$$

Power constraint: The power required during the cutting operation should not exceed the available power of the machine tool. The power is given by

$$P_r = \frac{F_r V_r}{6120\eta}. \tag{28.26}$$

Hence, the power constraint becomes

$$P_r = \frac{k_f f_r^{\mu} d_r^{\upsilon} V_r}{6120\eta} \le P_U \tag{28.27}$$

28.5.2.2 Finish machining

All the constraints other than the surface finish constraint are similar for rough and finish machining.

Parameter bounds

Cutting speed: $$V_{sL} \le V_s \le V_{sU} \tag{28.28}$$

Feed: $$f_{sL} \le f_s \le f_{sU} \tag{28.29}$$

Depth of cut: $$d_{sL} \le d_s \le d_{sU} \tag{28.30}$$

Tool-life constraint: $$T_L \le t_s \le T_U \tag{28.31}$$

Operating Constraints

Cutting force constraint: $$F_s = k_f f_s^{\mu} d_s^{\upsilon} \le F_U \tag{28.32}$$

Power constraint: $$P_s = \frac{k_f f_s^{\mu} d_s^{\upsilon} V_s}{6120\eta} \le P_U \tag{28.33}$$

Surface finish constraint: The surface finish dominates the quality of the machined part, and is generally influenced by various factors such as speed, feed, depth of cut, tool geometry and material of tool. Furthermore, some undesirable machining conditions such as excessive tool wear, built-up edge or chatter, deteriorate the surface finish. Only feed and nose radius, however, are considered since they have the most influential effect on surface finish (Narang and Fischer 1993). This constraint takes the form as

$$\frac{f_s^2}{8R_n} \le R_a . \tag{28.34}$$

In addition, an equality constraint needs to be manipulated here. The depth of finish cut (d_s) should be equal to the total depth of cut (d_t) minus the total depth of the rough cut (nd_r). Therefore, this equality equation can be defined as

$$d_s = d_t - n d_r . \tag{28.35}$$

Before formulating the final optimization model for this cutting problem, we first eliminate the equality constraint by some mathematical manipulations. To a certain extent, the optimal number of rough cuts, *n*, should not be determined by the optimization algorithm as the valid range of *n* is small. The optimal value of *n* can be obtained in an exhaustive manner with a given value of depth of finish cut. From Eqs. (28-23) and (28-35) the number of rough cuts can be expressed as

$$n = \frac{d_t - d_s}{d_r} \tag{28.36}$$

and restricted to

$$N_L \leq n \leq N_U \tag{28.37}$$

where $N_L = \dfrac{d_t - d_s}{d_{rU}}$, $N_U = \dfrac{d_t - d_s}{d_{rL}}$. (28.38)

Eq. (28-36) can be rewritten for depth of one rough cut, d_r, as follow

$$d_r = \frac{d_t - d_s}{n} \tag{28.39}$$

once the depth of finish cut d_s is specified, the depth of rough cut d_r can be obtained from Eq. (28-39). Therefore, one variables d_r and the equality constraint, Eq. (28-35), can be eliminated in the optimization algorithm.

Based upon the previous discussions, an optimization model for multi-pass turning operations can be formulated. The multi-pass turning model is a constrained nonlinear programming problem with multiple variables (machining parameters). The complexity of machining optimization problem presents difficulties for some solution techniques. SS is one of the optimization techniques recently developed in the area of metaheuristics. SS provides a way of considerably improving the performance of simple heuristic procedures. The search strategies proposed by SS result in iterative-procedures with the ability to escape local optimal points. The proposed SS-based optimization algorithm is described in Chapter 24. In the next section, the effectiveness of SS for the machining condition optimization is investigated.

28.6 Computational Experience

An example for the multi-pass turning model taken from Shin and Joo (1992) and Chen and Tsai (1996) is addressed to evaluate the viability of SS in machining condition optimization. The data of this example is presented in Table 28-2. The initial solutions (starting points) for SS are picked in a random way. The experimental study takes 500 runs of this example for comparing to the results of simulated annealing (Chen and Tsai 1996). The output statistics are computed every 100 runs. The SS-based method for machining optimization is implemented with the C codes presented in Laguna and Marti (2003). The experimentation has been run on the PC with a Pentium III 800 MHz processor. The user-specified parameters of SS are presented in Table 28-3. The computational results are summarized in Table 28-4.

Table 28.2. Data of test example.

D = 50 mm	L = 300 mm	d_t = 6 mm
V_{rU} = 500 m/min	V_{rL} = 50 m/min	f_{rU} = 0.9 mm/rev
f_{rL} = 0.1 mm/rev	d_{rU} = 3.0 mm	d_{rL} =1.0 mm
V_{sU} = 500 m/min	V_{sL} = 50 m/min	f_{sU} = 0.9 mm/rev
f_{sL} = 0.1 mm/rev	d_{sU} = 3.0 mm	d_{sL} = 1.0 mm
p = 5	q = 1.75	r = 0.75
k_f = 108	μ = 0.75	υ = 0.95
η = 0.85	R = 1.2 mm	k_o = 0.5 $/min
k_q = 132	h_1 = 7×10^{-4}	h_2 = 0.3
k_t = 2.5 $/edge	t_c = 0.75 min/piece	t_e = 1.5 min/edge
C_0 = 6×10^{11}	T_U = 45 min	T_L = 25 min
F_U = 5.0 kgf	R_a = 10 μ	P_U = 200 kW

Based on the above results, SS performs competitively to simulated annealing in machining condition optimization problems. The computational results validate the advantage of SS in terms of solution quality and computational requirement. Similar to simulated annealing and genetic algorithms, SS is a generalized optimization methodology for machining problems since it has no restrictive assumptions about the objective function, parameter set and constraint set.

Table 28.3. User specified parameters for SS[a].

$Psize$	= the size of the set of diverse solutions generated by the Diversification Generation Method (100);
b	= the size of the reference set (20);
b_1	= the size of the high quality subset (10);
b_2	= the size of the diverse subset (10);
$MaxIter$	= maximum number of iterations (3).

[a]The definitions for these parameters can be found in Chapter 24.

Table 28.4. Computational results of test example.

	SS		SA	
Runs	Avg. final function values ($/piece)	Standard deviation	Avg. final function values ($/piece)	Standard deviation
1 ~ 100	1.9730	0.02611	2.2973	0.00080
101 ~ 200	1.9652	0.02718	2.2975	0.00068
201 ~ 300	1.9610	0.02515	2.2974	0.00070
301 ~ 400	1.9646	0.02678	2.2975	0.00065
401 ~ 500	1.9659	0.02690	2.2974	0.00076
CPU time[a]	14.3 seconds/run		19.3 seconds/run	

[a]SS and SA were run on a Pentium III based PC and an IBM PC 486/DX2, respectively.

28.7 Conclusions

Improving performance of multi-pass turning operations can lead to considerable savings. In the area of machining condition optimization, determination of optimal machining parameters is a substantial problem. The machining parameters in multi-pass turning operations consist of cutting speed, feed, depth of cut and number of passes. In process planning of CNC machining, selecting reasonable machining parameters is necessary to satisfy requirements involving machining economics, machining quality and machining safety. The machining models can reduce the storage requirement and facilitate the automatic generation of machining conditions using optimization techniques.

The machining condition optimization problem for multi-pass turning operations is a constrained nonlinear program with a high degree of computational complexity. Owing to the high complexity, a generalized optimization algorithm based on Scatter Search (SS) is proposed to resolve the machining optimization model. From the computational results of the test example, it is demonstrated that the SS-based optimization method can adequately solve the complex machining optimization problem.

The features of applying SS to machining condition optimization can be characterized as follows:

1. Machining optimization problems have been studied for quite a long time due to their high degrees of complexity. It is more probable to have good algorithms either efficient in computational time or excellent in optimality for these problems. Due to the essence of high complexity, machining optimization problems can have only excellent approaches in optimality, or quick but approximate approaches with unsatisfactory local solutions. SS is completely generalized and problem-independent since it has no restrictive assumptions about the objective function, parameter set and constraint set. Therefore, SS can be easily modified to optimize the turning operations under various economic criteria and numerous practical constraints. In addition, no special information regarding the solution surface, e.g., gradient and local curvature, must be identified.
2. The obtained machining conditions can be easily applied in multi-pass turning operations. SS can obtain a near-optimal solution within a reasonable execution time on PC. Potentially, it can be extended as an on-line adjustment system of machining parameters based on signals from sensors. In addition, the machining model and the SS-based optimization method can be integrated into a CAD/CAM system for determining the optimal machining parameters, thereby reducing the manufacturing cost in metal machining.

References

Agapiou JS (1992a) The optimization of machining operations based on a combined criterion, Part 1: The use of combined objectives in single-pass operations. Transactions of the ASME, Journal of Engineering for Industry: 114 500-506

Agapiou JS (1992b) The optimization of machining operations based on a combined criterion, Part 2: Multi-pass operations. Transactions of the ASME, Journal of Engineering for Industry: 114 507-513

Armarego EJA, Brown RH (1969) The Machining of Metal, Englewood Cliffs, New Jersey, Prentice-Hall.

Bhattacharyya A, Faria-Gonzalez R, Ham I (1970) Regression analysis for predicting surface finish and its applications in the determination of optimum machining conditions. Transactions of the ASME, Journal of Engineering for Industry: 92 711-714

Boogert RM, Kals HJJ, van Houten FJAM (1996) Tool paths and cutting technology in computer-aided process planning. International Journal of Advanced Manufacturing Technology: 11 186-197

Boothroyd G, Rusek P (1976) Maximum rate of profit criteria in machining. Transactions of the ASME, Journal of Engineering for Industry: 98 217-220

Chang TC, Wysk RA, Wang HP (1991) Computer-aided Manufacturing, Englewood Cliffs, New Jersey, Prentice-Hall.

Chen Y, Hui, A, Du R (1995) A fuzzy expert system for the design of machining operations. International Journal of Machine Tool Design and Manufacturing: 35 1605-1621

Chen M-C, Su, C-T (1998) Optimization of machining conditions for turning cylindrical stocks into continuous finished profiles. International Journal of Production Research: 36 2115-2130

Chen M-C, Tsai D-M (1996) A simulated annealing approach for optimization of multipass turning operations. International Journal of Production Research: 34 2803-2825

Chen M-C, Tseng H-Y (1998) Machining parameters selection for stock removal turning in process plans using a float encoding genetic algorithm. Journal of the Chinese Institute of Engineers: 21 493-506

Choi JC, Bricker DL (1996) Effectiveness of a geometric programming algorithm for optimization of machining economics models. Computers and Operations Research: 23 957-961

Ermer DS (1971) Optimization of the constrained machining economics problem by geometric programming. Transactions of the ASME, Journal of Engineering for Industry: 93 1067-1072

Ermer DS, Kromodihardjo S (1981) Optimization of multipass turning with constraints. Transactions of the ASME, Journal of Engineering for Industry: 103 462-468

Glover F (1997) A template for scatter search and path relinking. In: Hao JK, Lutton E, Ronald E, Schoenauer M, Snyers D (eds.) Lecture Notes in Computer Science 1363, Springer-Verlag, Berlin, pp 13-54

Glover F (1999) Scatter search and path relinking. In: Corne D, Dorigo M, Glover F (eds.) New Ideas in Optimization, Wiley, New York

Glover F, Laguna M, Marti R (2000) Fundamentals of scatter search and path relinking. Control and Cybernetics: 39 653-684

Goldberg DE (1989) Genetic Algorithms in Search, optimization, and Machine Learning. Addison-Wesley, Reading, MA

Groover MP (1975) Monte Carlo simulation of the machining economics problem. Transactions of the ASME, Journal of Engineering for Industry: 97 931-938

Gopalkrishnan B, Al-Khayal F (1991) Machining parameter selection for turning with constraints: an analytical approach based on geometric programming. International Journal of Production Research: 29 1897-1908

Gupta R, Batra JL, Lal GK (1994) Profit rate maximization in multipass turning with constraints: a geometric programming approach. International Journal of Production Research: 32 1557-1569

Gupta R, Batra JL, Lal GK (1995) Determination of optimal subdivision of depth of cut in multipass turning with constraints. International Journal of Production Research: 33 2555-2565

Hati SK, Rao SS (1976) Determination of optimum machining conditions-deterministic and probabilistic approaches. Transactions of the ASME, Journal of Engineering for Industry: 98 354-359

Holland JH (1975) Adaptation in Natural and Artificial Systems. University of Michigan press, Ann Arbor

Iwata K, Murotsu Y, Iwatsubo T, Fuju S (1972) A probabilistic approach to the determination of the optimum cutting conditions. Transactions of the ASME, Journal of Engineering for Industry: 94 1099-1107

Iwata K, Murotsu Y, Oba F (1977) Optimization of cutting conditions for multipass operations considering probabilistic nature in machining process. Transactions of the ASME, Journal of Engineering for Industry: 99 210-217

Kee PK (1994) Development of computer-aided machining optimization for multi-pass rough turning operations. International Journal of Production Economics: 37 215-227

Kee PK (1995) Alternative optimization strategies and CAM software for multi-pass rough turning operations. International Journal of Advanced Manufacturing Technology: 10 287-298

Kee PK (1996) Development of constrained optimization analyses and strategies for multi-pass rough turning operations. International Journal of Machine Tools and Manufacture: 36 115-127

Khan Z, Prasad B, Singh T (1997) Machining condition optimization by genetic algorithms and simulated annealing. Computers and Operations Research: 24 647-657

Kirkpatrick S, Gelatt Jr CD, Vecchi MP (1983) Optimization by simulated annealing. Science: 220 671-680

Lin SCJ (1994) Computer Numerical Control From Programming to Networking, Delmar Publisher, New York

Laguna M, Marti R (2003) Scatter Search, Kluwer Academic Publishers, Boston

Metropolis N, Rosenbluth A, Rosenbluth M, Teller A, Teller E (1953) Equation of state calculations by fast computing machines. Journal of Chemical Physics: 21 1087-1092

Mesquita R, Krasteva E, Doytchinov S (1995) Computer-aided selection of optimum parameters in multipass turning. International Journal of Advanced Manufacturing Technology: 10 19-26

Narang RV, Fischer GW (1993) Development of a framework to automate process planning functions and to determine machining parameters. International Journal of Production Research: 31 1921-1942

Okushima K, Hitomi K (1964) A study of economic machining: an analysis of maximum profit cutting speed. International Journal of Production Research: 3 73-78

Prasad AVSRK, Rao PN, Rao RK (1997) Optimal selection of process parameters for turning operations in a CAPP system. International Journal of Production Research: 35 1495-1522

Petropoulos PG (1973) Optimal selection of machine rate variables by geometric programming. International Journal of Production Research: 11 305-314

Philipson RH, Ravindran A (1979) Application of mathematical programming to metal cutting. Mathematical Programming Study: 11 116-134

Shin YC, Joo YS (1992) Optimization of machining conditions with practical constraints. International Journal of Production Research: 30 2907-2919

Somlo J, Nagy J (1977) A new approach to cutting data optimization. Advances in Computer-aided Manufacture, North-Holland

Su C-T, Chen M-C (1999) Computer-aided optimization of multi-pass turning operations for continuous forms on CNC lathes. IIE Transactions: 31 583-596

Sundaram AM (1978) An application of goal programming technique in metal cutting. International Journal of Production Research: 16 375-382

Tan FP, Creese RC (1995) A generalized multi-pass machining model for machining parameter selection in turning. International Journal of Production Research: 33 1467-1487

Tsai P (1986) An optimization algorithm and economic analysis for constrained machining model. Ph.D. thesis, West Virginia University.

Unklesbay K, Creighton DL (1978) The optimization of multi-pass machining processes. Engineering Optimization: 3 229-238

van Laarhoven PJM, Aarts EHL (1987) Simulated annealing: Theory and Applications, Reidel, Dordrecht, Holland

Walvekar AG, Lambert BK (1970) An application of geometric programming to machine variable selection. International Journal of Production Research: 8 122-133

Wu SM, Ermer DS (1966) Maximum profit as the criterion in the determination of the optimum cutting conditions. Transactions of the ASME, Journal of Engineering for Industry: 88 435-442

Yellowley I (1978) A fundamental examination of the economics of the two pass turning operation. International Journal of Production Research: 18 617-626

Yellowley I, Gunn EA (1989) The optimal subdivision of cut in multi-pass machining operations. International Journal of Production Research: 27 1573-1588

Yeo SH (1995) A multipass optimization strategy for CNC lathe operations. International Journal of Production Economics: 40 209-218

29 Extended Frontiers in Optimization Techniques

Sergiy Butenko and Panos M Pardalos

29.1 Recent Progress in Optimization Techniques

Optimization has been expanding in all directions at an astonishing rate during the last few decades. New algorithmic and theoretical techniques have been developed, the diffusion into other disciplines has proceeded at a rapid pace, and our knowledge of all aspects of the field has grown even more profound (Floudas and Pardalos 2002; Pardalos and Resende 2002). At the same time, one of the most striking trends in optimization is the constantly increasing emphasis on the interdisciplinary nature of the field. Optimization today is a basic research tool in all areas of engineering, medicine and the sciences. The decision making tools based on optimization procedures are successfully applied in a wide range of practical problems arising in virtually any sphere of human activities, including biomedicine, energy management, aerospace research, telecommunications and finance. In this chapter we will briefly discuss the current developments and emerging challenges in optimization techniques and their applications.

The problems of finding the ``best'' and the ``worst'' have always been of a great interest. For example, given n sites, what is fastest way to visit all of them consecutively? In manufacturing, how should one cut plates of a material so that the waste is minimized? Some of the first optimization problems were solved in ancient Greece and are regarded among the most significant discoveries of that time. In the first century A.D., the Alexandrian mathematician Heron solved the problem of finding the shortest path between two points by way of the mirror. This result, also known as the Heron's theorem of the light ray, can be viewed as the origin of the theory of geometrical optics. The problem of finding extreme values gained a special importance in the seventeenth century when it served as one of motivations in the invention of differential calculus. The soon after developed calculus of variations and the theory of stationary values lie in the foundation of the modern mathematical theory of optimization.

The invention of the digital computer served as a powerful spur to the field of numerical optimization. During the World War II optimization algorithms were used to solve military logistics and operations problems. The military applications motivated the development of linear programming (LP), which studies optimization problems with linear objective function and constraints. In 1947 George Dantzig invented the simplex method for solving linear programs arising in U.S. Air Force operations. Linear programming has become one of the most popular and well studied optimization topics ever since. Although Dantzig is widely regarded as the father of linear programming, as early as 1939 the Soviet scientist Leonid Kantorovich emphasized the importance of certain classes of linear programs for applications in use of complex resources management, equipment work distribution, rational material cutting, the optimal use of sowing area, and transportation. He also proposed the method of resolving multipliers to solve these problems. Unfortunately, Kantorovich's work remained unnoticed until the linear programming methodology became a soundly developed and widely used discipline. But eventually the Kantorovich's contributions to the area of applied optimization were recognized by the Nobel Prize in Economics in 1975.

Despite of a fine performance of the simplex method on a wide variety of practical instances, it has an exponential worst-case time complexity and therefore is unacceptably slow in some large-scale cases. The question of existence of a theoretically efficient algorithm for LP remained open until 1979, when Leonid Khachian published his polynomial-time ellipsoid algorithm for linear programming. This theoretical breakthrough was followed by the interior point algorithm of Narendra Karmarkar published in 1984. Not only this algorithm has a polynomial-time theoretical complexity, it is also extremely efficient practically, allowing for solving larger instances of linear programs. Nowadays, various versions of interior point methods are an integral part of the state-of-the-art optimization solvers.

In nonlinear optimization, one deals with optimizing a nonlinear function over a feasible domain described by a set of, in general, nonlinear functions. The pioneering works on the gradient projection method by J. B. Rosen (Rosen 1960, 1961) generated a great deal of research enthusiasm in the area of for nonlinear programming, resulting in a number of new techniques for solving large-scale problems. This research resulted in several powerful nonlinear optimization software packages, including MINOS (Murtagh and Saunders 1983) and Lancelot (Conn et al. 1992).

In many practically important situations in linear, as well as nonlinear programming, all or a fraction of variables are restricted to be integer, yielding integer or mixed integer programming problems. These problems are in general computationally intractable, and it is unlikely that a universal polynomial-time algorithm will be developed for integer programming. Linear and integer programming can be considered as special cases of a broad optimization area called combinatorial optimization. In fact, most of combinatorial optimization problems can be formulated as integer programs. The most powerful integer programming solvers used by modern optimization packages such as CPLEX (ILOG 2001) and Xpress (Dash Optimization 2001) usually combine branch-and-bound algorithms with cutting plane methods, efficient preprocessing schemes, including fast heuristics, and sophisticated decomposition techniques.

It is fascinating to observe how naturally nonlinear and combinatorial optimization are bridged with each other to yield new, better optimization techniques. Combining the techniques for solving combinatorial problems with nonlinear optimization approaches is especially promising since it provides an alternative point of view and leads to new characterizations of the considered problems. These ideas also give a fresh insight into the complexity issues and frequently guide to discovery of remarkable connections between problems of seemingly different nature. For example, the ellipsoid and interior point methods for linear programming mentioned above are based on nonlinear optimization techniques. Let us also mention that an integrality constraint of the form x belongs to $\{0,1\}$ is equivalent to the nonconvex quadratic constraint x^2-$x = 0$. This straightforward fact suggests that it is the presence of nonconvexity, not integrality that makes an optimization problem difficult (Horst et al. 2000).

Nonlinear programming techniques, in particular interior point methods, played a key role in the recent foundation and development of semidefinite programming. The remarkable result by Goemans and Williamson (Goemans and Williamson 1995) served as a major step forward in development of approximation algorithms and proved a special importance of semidefinite programming for combinatorial optimization.

The extensive field of global optimization copes with problems having multiple locally optimal solutions, which arise in various important applications. Since convexity of the objective function or the feasible region is difficult to verify for many problems, it makes sense to assume that these problems are multi-extreme and thus are of interest to the field of global optimization. Clearly, such problems are extremely difficult to solve, which explains the fact that only in the last few decades have solution techniques for global optimization problems been developed (Horst and Pardalos 1995). Recent advances in global optimization include efficient solution methods for nonconvex quadratic programming, general concave minimization, network optimization, Lipshchitz and DC (difference of convex) programming, multi-level and multi-objective optimization problems. More detail on these and other developments in global optimization can be found in (Horst and Pardalos 1995; Horst et al. 2000; Pardalos and Romeijn 2002).

In many optimization problems arising in supply chain management, resource allocation, inventory control, energy management, finance, and other applications, the input data, such as demand or cost, are stochastic. In addition to the difficulties encountered in deterministic optimization problems, the stochastic problems introduce the additional challenge of dealing with uncertainties. To handle such problems, one needs to utilize probabilistic methods alongside with optimization techniques. This led to the development of a new area called stochastic programming (Prekopa 1995), whose objective is to provide tools helping to design and control stochastic systems with the goal of optimizing their performance.

Due to a large size of most practical optimization problems, especially of the stochastic ones, the so-called decomposition methods were introduced. The decomposition techniques (Lasdon 1970) are used to subdivide a large-scale problem into subproblems of lower dimension which are easier to solve than the original problem. The optimal solution of the large problem is then found using the optimal solution of the subproblems. These techniques are usually applicable if the problem at hand has some special structural properties. Say, the Dantzig-Wolfe decomposition method (Dantzig and Wolfe 1960) applies to linear programs with block diagonal or block angular constraint matrices. Another popular technique used to solve large-scale linear programs of special structure is Benders decomposition (Benders 1962). One of the advantages of the decomposition approaches is that they can be easily parallelized and implemented in distributed computing environments.

The advances in parallel computing, including hardware, software, and algorithms, increase the limits of the sizes of problems that can be solved (Migdalas et al. 1997). In many cases, a parallel version of an algorithm allows for a reduction of the running time by several orders of magnitude compared to the sequential version. Recently, distributed computing environments were used to solve several extremely hard instances of some combinatorial optimization problems, for instance a 13,509-city instance of the traveling salesmen problem (Applegate et al. 1998) and an instance of the quadratic assignment problem of dimension 30 (Anstreicher et al. 2002). The increasing importance of parallel processing in optimization is reflected in the fact that modern commercial optimization software packages tend to incorporate parallelized versions of certain algorithms.

29.2 Heuristic Approaches

As a result of ongoing enhancement of the optimization methodology and of improvement of available computational facilities, the scale of the problems solvable to optimality is continuously rising. However, many large-scale optimization problems encountered in practice cannot be solved using traditional optimization techniques. A variety of new computational approaches, called heuristics, have been proposed for finding good sub-optimal solutions to difficult optimization problems. Etymologically, the word ''heuristic'' comes from the Greek *heuriskein* (to find). Recall the famous ''Eureka, Eureka!'' (I have found it! I have found it!) by Archimedes (287-212 B.C.).

A heuristic in optimization is any method that finds an ``acceptable'' feasible solution. Many classical heuristics are based on local search procedures, which iteratively move to a better solution (if such solution exists) in a neighborhood of the current solution. A procedure of this type usually terminates when the first local optimum is obtained. Randomization and restarting approaches used to overcome poor quality local solutions are often ineffective. More general strategies known as metaheuristics usually combine some heuristic approaches and direct them towards solutions of better quality than those found by local search heuristics. Heuristics and metaheuristics play a key role in the solution of large difficult applied optimization problems.

Sometimes in search for efficient heuristics people turn to nature, which seems to always find the best solutions. In the recent decades, new types of optimization algorithms have been developed and successfully tested, which essentially attempt to imitate certain natural processes. The natural phenomena observed in annealing processes, nervous systems and natural evolution were adopted by optimizers and led to design of the simulated annealing (Kirkpatrick et al. 1983), neural networks (Hopfield 1982) and evolutionary computation (Holland 1975) methods in the area of optimization. The ant colony optimization method presented in Chapter Five of this book is based on the behavior of natural ant colonies. Other popular metaheuristics include greedy randomized adaptive search procedures or GRASP (Feo and Resende 1995) and tabu search (Glover and Laguna 1997). Some of these and other heuristics and their applications in engineering were discussed in detail in previous chapters of this book. See also (Glover and Kochenberger 2003; Ribeiro and Hansen 2002).

29.2.1 Parallel Metaheuristics

Although metaheuristic approaches, in general, do not guarantee optimality, very often they offer the only practically feasible approach to solve large scale problems. However, with rapid growth of the scale of problems arising in science, engineering and industry, even metaheuristics may require computing times exceeding the limits of what is considered acceptable. *Parallel metaheuristics* are designed to deal with this and other problems encountered in heuristic approaches. In the remainder of this section we will give a brief review of developments and challenges in the area of parallel metaheuristics. More detailed surveys and further references can be found in (Correa et al. 2002; Glover and Kochenberger 2003; Migdalas et al. 1997).

The core idea of the parallel computing is to break up the assignment among several processors in order to accelerate the computation. In the case with metaheuristics, not only parallel computing helps to speed up finding good quality solutions, it also enhances the *robustness* of the approaches. The availability of multiple processors provides more opportunities for diversification of the search strategies thus widening the coverage of the solution space and yielding more reliable solutions. Improved robustness constitutes one of the most essential contributions of the parallel computing to metaheuristics.

The performance of a parallel algorithm depends on several key factors, such as the architectures of parallel machines used, the employed parallel programming environments, and the types of parallelism and models implemented.

The different architectures used in parallel computing include *shared-memory* machines and *distributed-memory* machines. The number of processors in modern shared memory multiprocessor machines (SMP) ranges from two to several hundred. One of the most popular currently used alternatives in parallel computing is represented by a *cluster of computers*, which is essentially a group of PCs or workstations connected through a network. The main advantage of the cluster systems is their good cost/performance ratios comparing to other parallel machines (Buyya 1999).

The basic choices of parallel computing environments consist of parallel programming languages, communication libraries, and programming with lightweight processes. Nowadays, there are many parallel computing tools available, such as Parallel Virtual Machine (PVM) (Geist et al. 1994), Message Passing Interface (MPI) (Gropp et al. 1998) and Linda (Carriero et al. 1994). In implementation of parallel metaheuristics, a proper programming tool can be selected depending on characteristics of a specific problem to be solved.

The two main types of parallelism are data parallelism and functional parallelism. In *data parallelism*, the same sequence of commands is carried out on subsets of the data, whereas in *functional parallelism* the program consists of cooperative tasks, which use different codes and can run asynchronously.

Apart from the different programming environments, there are two basic parallel programming models, namely centralized and distributed. The parallel implementations of metaheuristics are often based on hybrid models, in which a centralized model in shared memory multiprocessor machines is run under a distributed model used in the machines cluster.

Description of successful parallel implementations strategies of many various mataheuristics can be found in the literature. The metaheuristics that have been efficiently implemented in parallel environments include tabu search, GRASP, genetic algorithms, simulated annealing, ant colonies, and others. Recent surveys and references can be found in (Glover and Kochenberger 2003; Ribeiro and Hansen 2002).

29.3 Emerging Application Areas of Optimization

The fast pace of technological progress, the invention of Internet and wireless communications have changed the way people communicate and do business. This age of technology and electronic commerce creates new types of optimization problems. The enormous amounts of data generated in government, military, astronomy, finance, telecommunications, medicine, and other important applications pose new problems which require special interdisciplinary efforts and novel sophisticated techniques for their solution.

The problems brought by massive data sets include data storage, compression and visualization, information retrieval, nearest neighbor search, clustering, and pattern recognition among many others. These and other problems arising in massive data sets create enormous opportunities and challenges for representatives of many fields of science and engineering, including optimization. Some of these challenges are beginning to be addressed (Abello et al. 2002). For example, optimization algorithms on massive graphs have been recently applied to analyze the data modeled by these graphs (Boginski et al. 2003). There are still many questions to be formulated and answered in this extremely broad and significant area.

In many cases the data sets are too large to fit entirely inside the fast computer's internal memory, and a slower external memory (for example disks) needs to be used. The input/output communication (I/O) between these memories can result in an algorithm's slow performance. *External memory* (EM) algorithms and data structures are designed with aim to reduce the I/O cost by exploiting the locality. The first EM graph algorithm was developed by Ullman and Yannakakis in 1991 and dealt with the problem of transitive closure. Many other researchers contributed to the progress in this area ever since. Recently, external memory algorithms have been successfully applied for solving various problems, including finding connected components in graphs, topological sorting, and shortest paths. For more detail on external memory algorithms and data structures see (Abello and Vitter 1999; Vitter 2001).

Another area that has a huge potential for application of optimization techniques is biomedicine. In the last few years there have been successful attempts to employ optimization procedures in biomedical problems. For example, quadratic integer programming has been applied to find the optimal positioning of electrodes in the human brain used for detection and prediction of epileptic seizures (Iasemidis et al. 2001); network flow algorithms have been utilized in order to maximize efficiency and minimize risk in radiation therapy treatment; optimization has been used to improve the efficiency of cancer detection and treatment (Lee and Sofer 2003; Mangasarian et al. 1995). More references on applications of optimization techniques in biomedicine can be found in (Du et al. 2000, Pardalos and Principe 2002; Pardalos et al. 1996).

29.4 Concluding Remarks

The area of optimization is one in which recent developments have effected great changes in many other disciplines. This trend it seems will continue for the next several years. Taking into account the increasing demand for solving large scale optimization problems we are going to witness the development of powerful heuristics and their implementations in parallel computing environments in the near future. Although we singled out massive data sets and biomedicine as examples of extending applied optimization frontiers, many other important areas of human activities provide exciting opportunities and challenges for optimization techniques. In conclusion, we want to stress the greater than ever significance of optimization in solving interdisciplinary problems.

References

J. Abello, P.M. Pardalos, and M.G.C. Resende (editors). *Handbook of Massive Data Sets*. Kluwer Academic Publishers, 2002.

J. Abello and J. S. Vitter (editors). *External Memory Algorithms*. Vol. 50 of DIMACS Series in Discrete Mathematics and Theoretical Computer Science. American Mathematical Society, 1999.

K.Anstreicher, N. Brixius, J.-P. Goux, and J. Linderoth. Solving large quadratic assignment problems on computational girds. *Mathematical Programming, Series B* 91 563-588, 2002.

D. Applegate, R. Bixby, V. Chv´atal, and W. Cook. On the solution of traveling salesman problems. *Doc. Math. J. DMV*, Extra ICM III:645–656, 1998.

J.F. Benders. Partitioning procedures for solving mixed variables programming problems. *Numerische Mathematik*, 4, 1962.

V. Boginski, S. Butenko, and P.M. Pardalos. Modeling and optimization in massive graphs. In P.M. Pardalos and H. Wolkowicz (editors), *Novel Approaches to Hard Discrete Optimization*. American Mathematical Society, 2003.

R. Buyya (editor). *High performance cluster computing: Architectures and systems* (in 2 volumes). Prentice Hall, 1999.

N. Carriero, D. Gelernter, and T. Mattson. The Linda alternative to message-passing systems. *Parallel Computing*, 20: 458-633, 1994.

A.R. Conn, N.I.M. Gould, and Ph.L. Toint. *LANCELOT: a Fortran packagefor large-scale nonlinear optimization (Release A)*. Springer-Verlag, 1992.

R. Correa, I. Dutra, M. Fiallos, and F. Gomez (editors). *Models for Parallel and Distributed Computation: Theory, Algorithmic Techniques and Applications*. Kluwer Academic Publishers, 2002.

D.B. Dantzig and P. Wolfe. The decomposition principle for linear programs. *Operations Research*, 8, 1960.

Dash Optimization. Xpress, 2001. http://www.dashoptimization.com.

D.-Z. Du, P.M. Pardalos, and J. Wang (editors). *Discrete Mathematical Problems with Medical Applications*, DIMACS Series Vol. 55. American Mathematical Society, 2000.

T.A. Feo and M.G.C. Resende. Greedy randomized adaptive search procedures. *Journal of Global Optimization*, 6, 1995.

C. Floudas and P.M. Pardalos (editors). *Encyclopedia of Optimization (in 6 volumes)*. Kluwer Academic Publishers, 2002.

A. Geist, A. Beguelin, J. Dongarra, W. Jiang, R. Manchek, and V. Sunderman. *PVM: Parallel Virtual Machine – A User's Guide and Tutorial for Networked Parallel Computing*. MIT Press, 1994. Also available at http://www.netlib.org/pvm3/book/pvm-book.html.

F. Glover and G.A. Kochenberger. *Handbook of Metaheuristics*. Kluwer Academic Publishers, 2003.

F. Glover and M. Laguna. *Tabu Search*. Kluwer Academic Publishers, 1997.

M. X. Goemans and D. P. Williamson, Improved approximation algorithms for maximum cut and satisfiability problems using semidefinite programming, *Journal of ACM*, 42: 1115-1145, 1995.

W. Gropp, S. Huss-Lederman, A. Lumsdane, E. Lusk, B. Nitzberg, W. Saphir, and M. Snir. *MPI: the Complete Reference*. MIT Press, 1998. Also available at http://www.netlib.org/utk/papers/mpi-book/mpi-book.html.

J.H. Holland. *Adaption in Natural and Artificial Systems*. University of Michigan Press, 1975.

J.J. Hopfield. Neural networks and physical systems with emergent collective computational abilities. *Proceedings of the National Academy of Sciences*, 79: 2554-2558, 1982.

R. Horst and P.M. Pardalos. *Handbook of Global Optimization*. Kluwer Academic Publishers, 1995.

R. Horst, P.M. Pardalos, and N.V. Thoai. *Introduction to Global Optimization*. Kluwer Academic Publishers, 2nd edition, 2000.

L.D. Iasemidis, P.M. Pardalos, D.-S. Shiau, and J.C. Sackellares. Quadratic binary programming and dynamic system approach to determine the predictability of epileptic seizures. *Journal of Combinatorial Optimization*, 5: 9-26, 2001.

ILOG. ILOG CPLEX, 2001. http://www.ilog.com/products/cplex.

N. Karmarkar. A new polynomial-time algorithm for linear programming. *Combinatorica*, 4: 373-395, 1984.

L.G. Khachian. A polynomial algorithm in linear programming. *Soviet Mathematics Doklady*, 20:1093–1096, 1979.

S. Kirkpatrick, C.D. Gellat Jr., and M.P. Vecchi. Optimization by simulated annealing. *Science*, 220:671–680, 1983.

L.S. Lasdon. *Optimization Theory for Large Systems*. MacMillan, 1970.

E. Lee and A. Sofer (editors). *Optimization in Medicine*. Volume 19 of *Annals of Operations Research*. Kluwer Academic Publishers, 2003.

O.L. Mangasarian, W.N. Street, and W.H. Wolberg. Breast cancer diagnosis and prognosis via linear programming. *Operations Research*, 43: 570-577, 1995.

A. Migdalas, P.M. Pardalos, and S. Storøy (editors). *Parallel Computing in Optimization*. Kluwer Academic Publishers, 1997.

B.A. Murtagh and M.A. Saunders. *MINOS 5.0 User's Guide*. Technical Report 83-20R, Systems Optimization Laboratory, Stanford University, Stanford, CA, 1983.

P.M. Pardalos and J. Principe (editors). *Biocomputing*. Kluwer Academic Publishers, 2002.

P.M. Pardalos and M.G.C. Resende (editors). *Handbook of Applied Optimization*. Oxford University Press, 2002.

P.M. Pardalos and E. Romeijn. *Handbook of Global Optimization. Volume 2: Heuristic Approaches*. Kluwer Academic Publishers, 2002.

P.M. Pardalos, D. Shalloway, and G. Xue. *Global Minimization of Nonconvex Energy Functions: Molecular Conformation and Protein Folding*, DIMACS Series Vol. 23. American Mathematical Society, 1996.

A. Prekopa. *Stochastic Programming*. Kluwer Academic Publishers, 1995.

C. Ribeiro and P. Hansen (editors). *Essays and Surveys in Metaheuristics*. Kluwer Academic Publishers, 2002.

J.B. Rosen. The gradient projection method for nonlinear programming. Part I. Linear constraints. *J. Soc. Ind. Appl. Math.*, 8, 1960.

J.B. Rosen. The gradient projection method for nonlinear programming. Part II. Non-linear constraints. *J. Soc. Ind. Appl. Math.*, 9, 1961.

J. D. Ullman and M. Yannakakis. The input/output complexity of transitive closure. *Annals of Mathematics and Artificial Intelligence*, 3: 331-360, 1991.

J. S. Vitter. External memory algorithms and data structures: Dealing with MASSIVE DATA. *ACM Computing Surveys*, 33:209-271, 2001.

GPSR Compliance
The European Union's (EU) General Product Safety Regulation (GPSR) is a set of rules that requires consumer products to be safe and our obligations to ensure this.

If you have any concerns about our products, you can contact us on

ProductSafety@springernature.com

In case Publisher is established outside the EU, the EU authorized representative is:

Springer Nature Customer Service Center GmbH
Europaplatz 3
69115 Heidelberg, Germany

www.ingramcontent.com/pod-product-compliance
Ingram Content Group UK Ltd.
Pitfield, Milton Keynes, MK11 3LW, UK
UKHW021903190726
13853UKWH00003B/1392

* 9 7 8 3 6 4 2 5 3 5 6 2 8 *